CISA REVIEW MANUAL 26TH EDITION

ISACA is pleased to offer the 26th edition of the *CISA® Review Manual.* The purpose of this manual is to provide CISA candidates with updated technical information and references to assist in preparation and study for the Certified Information Systems Auditor exam.

The content in the manual has been substantially updated. Most of the changes made were to recognize and map to the new task and knowledge statements that resulted from the new CISA job practice analysis. Further details regarding the new job practice can be found in the section titled NEW–CISA Job Practice and can be viewed at *www.isaca.org/cisajobpractice* and in the ISACA Exam Candidate Information Guide at *www.isaca.org/examguide.* **The exam is based on the task and knowledge statements in the job practice.** The development of the task and knowledge statements involved thousands of CISAs and other industry professionals worldwide who served as committee members, focus group participants, subject matter experts and survey respondents.

The *CISA® Review Manual* is updated to keep pace with rapid changes in the IS audit, control and security professions. As with previous manuals, the 26th edition is the result of contributions from many qualified authorities who have generously volunteered their time and expertise. We respect and appreciate their contributions and hope their efforts provide extensive educational value to CISA manual readers.

Your comments and suggestions regarding this manual are welcomed. After taking the exam, please take a moment to complete the online questionnaire *(www.isaca.org/studyaidsevaluation).* Your observations will be invaluable for the preparation of the next edition of the *CISA® Review Manual.*

The sample questions contained in this manual are designed to depict the type of questions typically found on the CISA exam and to provide further clarity to the content presented in this manual. The CISA exam is a practice-based exam. Simply reading the reference material in this manual will not properly prepare candidates for the exam. The sample questions are included for guidance only. Scoring results do not indicate future individual exam success.

Certification has resulted in a positive impact on many careers, and the CISA designation is respected and acknowledged by organizations around the world. We wish you success with the CISA exam. Your commitment to pursue the leading certification in IS audit, assurance, control and security is exemplary.

ACKNOWLEDGMENTS

The 26th edition of the *CISA® Review Manual* is the result of the collective efforts of many volunteers. ISACA members from throughout the global IS audit, control and security professions participated, generously offering their talent and expertise. This international team exhibited a spirit and selflessness that has become the hallmark of contributors to this manual. Their participation and insight are truly appreciated.

Special thanks go to Ian J. Cooke, CISA, CGEIT, CRISC, CFE, COBIT Foundation, CPTS, ITIL-F, Six Sigma Green Belt, An Post, Ireland, and Jeffrey L. Roth, CISA, CGEIT, CISSP-ISSEP, QSA, USA, who worked on the 26th edition of the *CISA® Review Manual.*

Expert Reviewers
Rajeev Andharia, CISA, CISSP, COBIT 5 Assessor & Implementation, ITIL Expert, PMP, Business Technology Partner Pte Ltd., Singapore
Sunil Bakshi, CISA, CISM, CGEIT, CRISC, National Institute of Bank Management, India
Ishwar Chandra, CISA, FCA, I C & Associates, India
James T. Enstrom, CISA, CRISC, CIA, Chicago Board Options Exchange, USA
Mohamed Gohar, CISA, CISM, Egypt
Florin-Mihai Iliescu, CISA, Info-Logica Silverline SRL, Romania
Binoy Koonammavu, CISA, CISM, CRISC, CISSP, ValueMentor Infosec Pvt. Ltd, India
Shruti Shrikant Kulkarni, CISA, CRISC, CISSP, CPISI, CCSK, ITIL V3 Expert, Monitise Group Ltd, United Kingdom
S. Krishna Kumar, CISA, CISM, CGEIT, India
Juan Carlos López, CISA, CGEIT, CRISC, COBIT Implementation, COBIT Certified Assessor, ITIL, PMP, Exacta Consulting, Ecuador
Balakrishnan Natarajan, CISA, CISM, Pivotal Software, Inc., USA
S. Peter Nota, CISA, CISM, CISSP, MBCS, PCI-ISA, Premier Farnell plc, United Kingdom
Derek J. Oliver, Ph.D., CISA, CISM, CRISC, DBA, Ravenswood Consultants Ltd., United Kingdom
Opeyemi Onifade, CISA, CISM, CGEIT, Afenoid Enterprise Limited, Nigeria
Manuel (Manolo) Palao, CISA, CISM, CGEIT, Accredited COBIT 5 Trainer, COBIT 5 Certified Assessor, P&T: S, SLU | Innovation & Technology Trends Institute (iTTi), Spain
Robert D. Prince, CISA, CISSP, USA
Beth Pumo, CISA, CISM, Kaiser Permanente, USA
Sree Krishna Rao, CISA, Ernst & Young LLP, United Kingdom
Markus Schiemer, CISA, CGEIT, CRISC, Microsoft Österreich GmbH, Austria
Hilary Shreter, CISA, PMP, USA
Katalin Szenes, Ph.D., CISA, CISM, CGEIT, CISSP, Obuda University, Hungary
Hui Zhu, CISA, CISM, CGEIT, BlueImpact Ltd., Canada
Tichaona Zororo, CISA, CISM, CRISC, CGEIT, Certified COBIT 5 Assessor, CIA, CRMA, EGIT | Enterprise Governance of IT (Pty) Ltd, South Africa

ISACA has begun planning the next edition of the *CISA® Review Manual.* Volunteer participation drives the success of the manual. If you are interested in becoming a member of the select group of professionals involved in this global project, we want to hear from you. Please email us at *studymaterials@isaca.org.*

NEW—CISA JOB PRACTICE

BEGINNING IN 2016, THE CISA EXAM WILL TEST THE NEW CISA JOB PRACTICE.

An international job practice analysis is conducted at least every five years or sooner to maintain the validity of the CISA certification program. A new job practice forms the basis of the CISA exam beginning in June 2016.

The primary focus of the job practice is the current tasks performed and the knowledge used by CISAs. By gathering evidence of the current work practice of CISAs, ISACA is able to ensure that the CISA program continues to meet the high standards for the certification of professionals throughout the world.

The findings of the CISA job practice analysis are carefully considered and directly influence the development of new test specifications to ensure that the CISA exam reflects the most current best practices.

The new 2016 job practice reflects the areas of study to be tested and is compared below to the previous job practice. The complete CISA job practice can be found at *www.isaca.org/cisajobpractice.*

Previous CISA Job Practice	New 2016 CISA Job Practice
Domain 1: The Process of Auditing Information Systems (14%)	Domain 1: The Process of Auditing Information Systems (21%)
Domain 2: Governance and Management of IT (14%)	Domain 2: Governance and Management of IT (16%)
Domain 3: Information Systems Acquisition, Development and Implementation (19%)	Domain 3: Information Systems Acquisition, Development and Implementation (18%)
Domain 4: Information Systems Operations, Maintenance and Support (23%)	Domain 4: Information Systems Operations, Maintenance and Service Management (20%)
Domain 5: Protection of Information Assets (30%)	Domain 5: Protection of Information Assets (25%)

Page intentionally left blank

Table of Contents

CISA
Certified Information Systems Auditor
An ISACA Certification

CISA
Certified Information
Systems Auditor
An ISACA Certification

CISA
Certified Information Systems Auditor
An ISACA Certification

CISA
Certified Information Systems Auditor
An ISACA Certification

Chapter 5:

Protection of Information Assets ... 317

CISA
Certified Information Systems Auditor
An ISACA Certification

About This Manual

OVERVIEW

The *CISA® Review Manual 26th Edition* is intended to assist candidates in preparing for the CISA exam. **The manual is one source of preparation for the exam and should not be thought of as the only source or be viewed as a comprehensive collection of all the information and experience that is required to pass the exam.** No single publication offers such coverage and detail.

As candidates read through the manual and encounter a topic that is new to them or one in which they feel their knowledge and experience are limited, additional references should be sought. The exam is a combination of questions testing **candidates'** technical and practical knowledge, and their ability to apply the knowledge (based on experience) in given situations.

The *CISA® Review Manual 26th Edition* provides coverage of the knowledge and activities related to the various functions associated with the content areas as detailed in the CISA job practice and described in the *ISACA Exam Candidate Information Guide (www.isaca.org/examguide)*:

Domain 1	The Process of Auditing Information Systems	21 percent
Domain 2	Governance and Management of IT	16 percent
Domain 3	Information Systems Acquisition, Development and Implementation	18 percent
Domain 4	Information Systems Operations, Maintenance and Service Management	20 percent
Domain 5	Protection of Information Assets	25 percent

Note: Each chapter defines the tasks that CISA candidates are expected to know how to do and includes a series of knowledge statements required to perform those tasks. These constitute the current practices for the IS auditor. The detailed CISA job practice can be viewed at *www.isaca.org/cisajobpractice.* This exam is based on these task and knowledge statements.

The manual has been developed and organized to assist candidates in their study. CISA candidates should evaluate their strengths, based on knowledge and experience, in each of these areas.

FORMAT OF THIS MANUAL

Each of the five chapters of the *CISA® Review Manual 26th Edition* is divided into two sections for focused study.

Section one of each chapter includes:
- A definition of the domain
- Objectives for the domain as a practice area
- A listing of the task and knowledge statements for the domain
- A map of the relationship of each task to the knowledge statements for the domain
- A reference guide for the knowledge statements for the domain, including the relevant concepts and explanations
- References to specific content in section two for each knowledge statement
- Self-assessment questions and answers with explanations
- Suggested resources for further study

Section two of each chapter includes:
- Reference material and content that supports the knowledge statements
- Definitions of terms most commonly found on the exam

Material included is pertinent for CISA candidates' knowledge and/or understanding when preparing for the CISA certification exam.

The structure of the content includes numbering to identify the chapter where a topic is located and the headings of the subsequent levels of topics addressed in the chapter (i.e., 2.8.3 Risk Analysis Methods, is a subtopic of Risk Management in chapter 2). Relevant content in a subtopic is bolded for specific attention.

Understanding the material in this manual is one measurement of a candidate's knowledge, strengths and weaknesses, and an indication of areas where additional or focused study is needed. However, written material is not a substitute for experience. **CISA exam questions will test the candidate's practical application of this knowledge.** Case studies at the end of each chapter present situations within the profession and in specific areas of study. The scenarios involve topics addressed in the chapters and include practice questions which assist in understanding how a question could be presented on the CISA exam. The self-assessment questions in the first section of each chapter also serve this purpose and should not be used independently as a source of knowledge. Self-assessment questions should not be considered a measurement of one's ability to answer questions correctly on the CISA exam for that area. The questions are intended to familiarize candidates with question structure and general content, and may or may not be similar to questions that will appear on the actual exam. The reference material included in the first section of each chapter lists publications used in the creation of this manual.

At the end of the publication the candidate will find a glossary. The glossary includes both terms that are discussed in the text and terms that apply to the different areas but may not have been specifically discussed. The glossary can be another tool to identify areas in which candidates may need to seek additional references.

Although every effort is made to address the majority of information that candidates are expected to know, not all examination questions are necessarily covered in the manual, and candidates will need to rely on professional experience to provide the best answer.

Throughout the manual, the word "association" refers to ISACA. Also, please note that the manual has been written using standard American English.

Note: The *CISA® Review Manual 26th Edition* is a living document. As technology advances, the manual will be updated to reflect such advances. Further updates to this document before the date of the exam may be viewed at *www.isaca.org/studyaidupdates.*

EVALUATION OF THIS MANUAL

ISACA continuously monitors the swift and profound professional, technological and environmental advances affecting the IS audit, assurance, control and security professions. Recognizing these rapid advances, the *CISA® Review Manual* is updated periodically.

To assist ISACA in keeping abreast of these advances, please take a moment to evaluate the *CISA® Review Manual 26th Edition*. Such feedback is valuable to fully serve the profession and future CISA exam registrants.

To complete the evaluation on the web site, please go to *www.isaca.org/studyaidsevaluation.*

Thank you for your support and assistance.

ABOUT THE CISA REVIEW QUESTIONS, ANSWERS AND EXPLANATIONS MANUAL

Candidates may also wish to enhance their study and preparation for the exam by using the *CISA® Review Questions, Answers & Explanations Manual 11th Edition* or the CISA® Review Questions, Answers & Explanations Database – 12 month subscription.

The *CISA® Review Questions, Answers & Explanations Manual 11th Edition* consists of 1,000 multiple-choice study questions, answers and explanations arranged in the areas of the current CISA job practice. Many of these items appeared in previous editions of the *CISA® Review Questions, Answers & Explanations Manual*, but have been rewritten to correspond with current practice and/or be more representative of actual CISA exam items.

Another study aid that is available is the CISA® Review Questions, Answers & Explanations Database – 12 Month Subscription. It consists of the 1,000 questions, answers and explanations included in the *CISA® Review Questions, Answers & Explanations Manual 11th Edition*. With this product, CISA candidates can identify strengths and weaknesses by taking random sample exams of varying lengths and breaking the results down by domain. Sample exams also can be chosen by domain, allowing for concentrated study, one domain at a time, and other sorting features such as the omission of previous correctly answered questions are available.

Questions in these products are representative of the types of questions that have appeared on the exam and include an explanation of the correct and incorrect answers. Questions are sorted by the CISA domains and as a sample test. These products are ideal for use in conjunction with the *CISA® Review Manual 26th Edition*. These manuals can be used as study sources throughout the study process or as part of a final review to determine where a candidate may need additional study. Again, it should be noted that these questions and suggested answers are provided as examples; they are not actual questions from the exam and may differ in content from those that actually appear on the exam.

Note: When using the CISA review materials to prepare for the exam, please note that they cover a broad spectrum of information systems audit, control and security issues. **Do not assume that reading these manuals and answering review questions will fully prepare you for the exam.** Since actual exam questions often relate to practical experiences, candidates should refer to their own experiences and other reference sources, and draw on the experiences of colleagues and others who have earned the CISA designation.

CISA ONLINE REVIEW COURSE

The CISA Online Review Course is a web-based, self-paced study tool. There are no hard copy materials (books, study manuals, etc.) provided with the course. While it is significantly different in terms of how the information is delivered, the course is based on content from the *CISA® Review Manual 26th Edition* and from additional content provided by subject matter experts. The course includes practice questions as well as interactive activities and exercises and an online glossary to reinforce content comprehension.

To better evaluate whether this is an appropriate study tool for you, please view the course demonstration at *http://demo.certification-partners.com/demo/v3/index.htm*. To register for the course, please go to *www.isaca.org/elearningcampus.*

Chapter 1:

The Process of Auditing Information Systems

Section One: Overview

Section Two: Content

Section One: Overview

DEFINITION

This chapter is on the process of auditing information systems (IS) and encompasses the entire practice of IS auditing, including procedures and a thorough methodology that allows an IS auditor to perform an audit on any given IT area in a professional manner.

OBJECTIVES

The objective of this domain is to ensure that the CISA candidate has the knowledge necessary to provide audit services in accordance with IS audit standards to assist the organization with protecting and controlling information systems.

This area represents 21 percent of the CISA exam (approximately 32 questions).

TASK AND KNOWLEDGE STATEMENTS

TASKS

There are five tasks within the domain covering the process of auditing information systems:

T1.1 Execute a risk-based IS audit strategy in compliance with IS audit standards to ensure that key risk areas are audited.

T1.2 Plan specific audits to determine whether information systems are protected, controlled and provide value to the organization.

T1.3 Conduct audits in accordance with IS audit standards to achieve planned audit objectives.

T1.4 Communicate audit results and make recommendations to key stakeholders through meetings and audit reports to promote change when necessary.

T1.5 Conduct audit follow-ups to determine whether appropriate actions have been taken by management in a timely manner.

KNOWLEDGE STATEMENTS

The CISA candidate must have a good understanding of each of the topics or areas delineated by the knowledge statements. These statements are the basis for the exam.

There are 11 knowledge statements within the domain covering the process of auditing information systems:

K1.1 Knowledge of ISACA IS Audit and Assurance Standards, Guidelines, and Tools and Techniques, Code of Professional Ethics and other applicable standards

K1.2 Knowledge of risk assessment concepts and tools and techniques in planning, examination, reporting and follow-up

K1.3 Knowledge of fundamental business processes (e.g., purchasing, payroll, accounts payable, accounts receivable) and the role of IS in these processes

K1.4 Knowledge of control principles related to controls in information systems

K1.5 Knowledge of risk-based audit planning and audit project management techniques, including follow-up

K1.6 Knowledge of applicable laws and regulations which affect the scope, evidence collection and preservation and frequency of audits

K1.7 Knowledge of evidence collection techniques (e.g., observation, inquiry, inspection, interview, data analysis, forensic investigation techniques, computer-assisted audit techniques [CAATs]) used to gather, protect and preserve audit evidence

K1.8 Knowledge of different sampling methodologies and other substantive/data analytical procedures

K1.9 Knowledge of reporting and communication techniques (e.g., facilitation, negotiation, conflict resolution, audit report structure, issue writing, management summary, result verification)

K1.10 Knowledge of audit quality assurance (QA) systems and frameworks

K1.11 Knowledge of various types of audits (e.g., internal, external, financial) and methods for assessing and placing reliance on the work of other auditors or control entities

Relationship of Task to Knowledge Statements

The task statements are what the CISA candidate is expected to know how to perform. The knowledge statements delineate each of the areas in which the CISA candidate must have a good understanding in order to perform the tasks. The task and knowledge statements are mapped in **figure 1.1** insofar as it is possible to do so. Note that although there is often overlap, each task statement will generally map to several knowledge statements.

Figure 1.1—Task and Knowledge Statements Mapping

Task	Knowledge Statements
T1.1 Execute a risk-based IS audit strategy in compliance with IS audit standards to ensure that key risk areas are audited.	K1.1 Knowledge of ISACA IS Audit and Assurance Standards, Guidelines and Tools and Techniques, Code of Professional Ethics and other applicable standards K1.2 Knowledge of risk assessment concepts and tools and techniques in planning, examination, reporting and follow-up K1.3 Knowledge of fundamental business processes (e.g., purchasing, payroll, accounts payable, accounts receivable) and the role of IS in these processes K1.5 Knowledge of risk-based audit planning and audit project management techniques, including follow-up K1.6 Knowledge of applicable laws and regulations which affect the scope, evidence collection and preservation and frequency of audits K1.10 Knowledge of audit quality assurance systems and frameworks K1.11 Knowledge of various types of audits (e.g., internal, external, financial) and methods for assessing and placing reliance on the work of other auditors or control entities
T1.2 Plan specific audits to determine whether information systems are protected, controlled and provide value to the organization.	K1.1 Knowledge of ISACA IS Audit and Assurance Standards, Guidelines and Tools and Techniques, Code of Professional Ethics and other applicable standards K1.2 Knowledge of risk assessment concepts and tools and techniques in planning, examination, reporting and follow-up K1.3 Knowledge of fundamental business processes (e.g., purchasing, payroll, accounts payable, accounts receivable) and the role of IS in these processes K1.4 Knowledge of control principles related to controls in information systems K1.5 Knowledge of risk-based audit planning and audit project management techniques, including follow-up K1.6 Knowledge of applicable laws and regulations which affect the scope, evidence collection and preservation and frequency of audits K1.10 Knowledge of audit quality assurance (QA) systems and frameworks K1.11 Knowledge of various types of audits (e.g., internal, external, financial) and methods for assessing and placing reliance on the work of other auditors or control entities
T1.3 Conduct audits in accordance with IS audit standards to achieve planned audit objectives.	K1.1 Knowledge of ISACA IS Audit and Assurance Standards, Guidelines and Tools and Techniques, Code of Professional Ethics and other applicable standards K1.2 Knowledge of risk assessment concepts and tools and techniques in planning, examination, reporting and follow-up K1.3 Knowledge of fundamental business processes (e.g., purchasing, payroll, accounts payable, accounts receivable) and the role of IS in these processes K1.4 Knowledge of control principles related to controls in information systems K1.5 Knowledge of risk-based audit planning and audit project management techniques, including follow-up K1.6 Knowledge of applicable laws and regulations which affect the scope, evidence collection and preservation and frequency of audits K1.7 Knowledge of evidence collection techniques (e.g., observation, inquiry, inspection, interview, data analysis, forensic investigation techniques, computer-assisted audit techniques [CAATs]) used to gather, protect and preserve audit evidence K1.8 Knowledge of different sampling methodologies and other substantive/data analytical procedures K1.9 Knowledge of reporting and communication techniques (e.g., facilitation, negotiation, conflict resolution, audit report structure, issue writing, management summary, result verification) K1.10 Knowledge of audit quality assurance (QA) systems and frameworks K1.11 Knowledge of various types of audits (e.g., internal, external, financial) and methods for assessing and placing reliance on the work of other auditors or control entities.
T1.4 Communicate audit results and make recommendations to key stakeholders through meetings and audit reports to promote change when necessary.	K1.1 Knowledge of ISACA IS Audit and Assurance Standards, Guidelines and Tools and Techniques, Code of Professional Ethics and other applicable standards K1.2 Knowledge of risk assessment concepts and tools and techniques in planning, examination, reporting and follow-up K1.3 Knowledge of fundamental business processes (e.g., purchasing, payroll, accounts payable, accounts receivable) and the role of IS in these processes K1.6 Knowledge of applicable laws and regulations which affect the scope, evidence collection and preservation and frequency of audits K1.9 Knowledge of reporting and communication techniques (e.g., facilitation, negotiation, conflict resolution, audit report structure, issue writing, management summary, result verification) K1.10 Knowledge of audit quality assurance (QA) systems and frameworks K1.11 Knowledge of various types of audits (e.g., internal, external, financial) and methods for assessing and placing reliance on the work of other auditors or control entities

Figure 1.1—Task and Knowledge Statements Mapping *(cont.)*		
T1.5 Conduct audit follow-ups to determine whether appropriate actions have been taken by management in a timely manner.	K1.1	Knowledge of ISACA IS Audit and Assurance Standards, Guidelines and Tools and Techniques, Code of Professional Ethics and other applicable standards
	K1.2	Knowledge of risk assessment concepts and tools and techniques in planning, examination, reporting and follow-up
	K1.3	Knowledge of fundamental business processes (e.g., purchasing, payroll, accounts payable, accounts receivable) and the role of IS in these processes
	K1.4	Knowledge of control principles related to controls in information systems
	K1.5	Knowledge of risk-based audit planning and audit project management techniques, including follow-up
	K1.6	Knowledge of applicable laws and regulations which affect the scope, evidence collection and preservation and frequency of audits
	K1.7	Knowledge of evidence collection techniques (e.g., observation, inquiry, inspection, interview, data analysis, forensic investigation techniques, computer-assisted audit techniques [CAATs]) used to gather, protect and preserve audit evidence
	K1.8	Knowledge of different sampling methodologies and other substantive/data analytical procedures
	K1.9	Knowledge of reporting and communication techniques (e.g., facilitation, negotiation, conflict resolution, audit report structure, issue writing, management summary, result verification)
	K1.10	Knowledge of audit quality assurance (QA) systems and frameworks
	K1.11	Knowledge of various types of audits (e.g., internal, external, financial) and methods for assessing and placing reliance on the work of other auditors or control entities

Knowledge Statement Reference Guide

Each knowledge statement is explained in terms of underlying concepts and relevance of the knowledge statement to the IS auditor. It is essential that the exam candidate understand the concepts. The knowledge statements are what the IS auditor must know in order to accomplish the tasks. Consequently, only the knowledge statements are detailed in this section.

The sections identified in K1.1 through K1.11 are described in greater detail in section two of this chapter.

K1.1 Knowledge of ISACA IS Audit and Assurance Standards, Guidelines and Tools and Techniques, Code of Professional Ethics and other applicable standards

Explanation	Key Concepts	Reference in Manual
The credibility of any IS audit activity is largely determined by its adherence to commonly accepted standards. ISACA IS Audit and Assurance Standards, Guidelines and Tools and Techniques, and the Code of Professional Ethics, are developed, circulated for discussion among audit professionals and issued by ISACA in order to provide a framework of minimum and essential references regarding how an IS auditor should perform work and act in a professional manner. IS auditors should comply with ISACA IS Audit and Assurance Standards and follow guidelines, as relevant. Failure to follow standards or justify departure from guidelines may result in a violation of the Code of Professional Ethics. Although the CISA candidate is expected to have knowledge of these standards and guidelines, the exam will test the candidate's understanding of the application of the information rather than asking "definitional" questions that simply test information recall.	Code of Professional Ethics	1.3.1 ISACA Code of Professional Ethics
	IS Audit and Assurance Standards, Guidelines and Tools and Techniques	1.3.2 ISACA IS Audit and Assurance Standards 1.3.3 ISACA IS Audit and Assurance Guidelines 1.3.4 ISACA IS Audit and Assurance Tools and Techniques 1.3.5 Relationship Among Standards, Guidelines and Tools and Techniques
	Understanding ITAF™	1.3.6 ITAF™

K1.2 Knowledge of risk assessment concepts and tools and techniques in planning, examination, reporting and follow-up

Explanation	Key Concepts	Reference in Manual
The overall audit plan of the organization should be based on business risk related to the use of IT, and the IS auditor is expected to be aware of the need to focus on this risk. In addition, an audit must focus on the most critical elements of the function under review. For this reason, the IS auditor should be aware of, and be able to put into practice, the risk analysis techniques needed to identify and prioritize business risk within the audit scope. This approach allows the IS auditor to create an audit plan that applies finite audit resources to where they are most needed. Although business risk is the most important driver of the audit program, the IS auditor must also take steps to minimize associated elements such as sampling risk, detection risk, materiality of findings, etc., because these may impact the adequacy of the review.	Impact of risk assessment on IS auditing	1.4.1 Risk Analysis 1.5.3 Audit Methodology 1.5.4 Risk-based Auditing 1.5.5 Audit Risk and Materiality 1.5.7 IS Audit Risk Assessment Techniques
	Understanding risk analysis concepts within an auditing context	1.4.1 Risk Analysis
	Applying risk analysis techniques during audit planning	1.5.4 Risk-based Auditing 1.5.5 Audit Risk and Materiality 1.5.6 Risk Assessment and Treatment 1.5.7 IS Audit Risk Assessment Techniques
	Communicating results and following up on corrective actions and recommendations	1.6 Communicating Audit Results 1.6.1 Audit Report Structure and Contents 1.6.2 Audit Documentation

K1.3 Knowledge of fundamental business processes (e.g., purchasing, payroll, accounts payable, accounts receivable) and the role of IS in these processes

Explanation	Key Concepts	Reference in Manual
To effectively identify the enterprise's key risk, the IS auditor must obtain an understanding of the organization and its environment, specifically obtaining an understanding of the: • External and internal factors affecting the entity • Entity's selection and application of policies and procedures • Entity's objectives and strategies • Measurement and review of the entity's performance As part of obtaining this understanding, the IS auditor must also obtain an understanding of some key components, such as the entity's: • Strategic management • Business model • Corporate governance processes • Transaction types engaged in and with whom they are transacted One must understand how those transactions flow through and are captured into the information systems.	Understanding risk analysis concepts within an auditing context	1.4.1 Risk Analysis
	Understanding control objectives	1.4.2 Internal Controls 1.4.3 IS Control Objectives 1.4.4 COBIT 5 1.4.5 General Controls 1.4.6 IS Specific Controls

K1.4 Knowledge of control principles related to controls in information systems

Explanation	Key Concepts	Reference in Manual
IS auditing involves the assessment of IS-related controls put in place to ensure the achievement of control objectives. Understanding control objectives and identifying the key controls that help achieve a properly controlled environment are essential for the effectiveness and efficiency of the IS audit process. Auditing is, therefore, a process of ensuring that control objectives are appropriately addressed by the associated controls. COBIT provides a comprehensive control framework that can help the IS auditor benchmark control objectives. The CISA candidate will find COBIT to be an excellent source of information when preparing for the CISA exam. The CISA candidate should remember that the CISA exam will not include questions that ask for COBIT definitions nor will the candidate be asked to quote any particular COBIT reference.	Proper audit planning techniques	1.2.3 Audit Planning
	Understanding control objectives	1.4.2 Internal Controls 1.4.3 IS Control Objectives 1.4.4 COBIT 5 1.4.5 General Controls 1.4.6 IS Specific Controls

K1.5 Knowledge of risk-based audit planning and audit project management techniques, including follow-up

Explanation	Key Concepts	Reference in Manual
To achieve audit objectives within a precise scope and budget, the audit should be adequately planned. The performance of an IS audit does not differ substantially from a project. Accordingly, audit planning requires a similar level of preplanning to ensure an appropriate and efficient use of audit resources. Auditors need to understand project planning and management techniques to properly manage the audit and avoid an inefficient utilization of resources. The CISA exam will not include questions that are written for a project manager who is not an IS auditor.	Application of audit planning techniques	1.2.2 IS Audit Resource Management 1.2.3 Audit Planning 1.2.4 Effect of Laws and Regulations on IS Audit Planning
	Impact of IS environment on IS auditing practices and techniques	1.5.1 Audit Objectives 1.5.3 Audit Methodology 1.5.8 Audit Programs 2.11 Auditing IT Governance Structure and Implementation 2.13 Auditing Business Continuity 3.14 Auditing Application Controls 3.15 Auditing Systems Development, Acquisition and Maintenance 4.7 Auditing Infrastructure and Operations 5.5 Auditing Information Security Management Framework 5.6 Auditing Network Infrastructure Security

K1.6 Knowledge of applicable laws and regulations which affect the scope, evidence collection and preservation, and frequency of audits

Explanation	Key Concepts	Reference in Manual
Laws and regulations of any kind—including international treaties; central, federal or local government; or industry-related laws and regulations—affect the way that organizations conduct business, and very often determine scope, frequency and type of audits, and how reporting requirements are substantially affected. In fraud investigations or legal proceedings, maintaining the integrity of evidence throughout the evidence life cycle may be referred to as the chain of custody when the evidence is classified as forensic evidence. The CISA candidate is expected to be aware of, rather than a participant in, such specific evidence collection.	Factors to consider in collection, protection and chain of custody of audit evidence in an IS audit	1.5.11 Evidence 1.6.2 Audit Documentation
	Special considerations in audit documentation for evidence	1.8.2 Continuous Auditing

K1.7 Knowledge of evidence collection techniques (e.g., observation, inquiry, inspection, interview, data analysis, forensic investigation techniques, computer-assisted audit techniques [CAATs]) used to gather, protect and preserve audit evidence

<table>
<tr><th>Explanation</th><th>Key Concepts</th><th>Reference in Manual</th></tr>
<tr><td rowspan="6">One essential audit concept is that audit findings must be supported by objective evidence. Therefore, it is essential to know the techniques used to gather and preserve evidence. Information is gathered from the auditees or from a variety of alternative sources, including reference manuals; accountants, banks, suppliers, vendors, etc.; and other related functional areas of the business. Information is gathered through inquiry, observation and interviews, and analysis of data using computer-assisted audit techniques (CAATs). Electronic media, including the use of automated audit software, may be used for preserving evidence that supports audit findings, but care should be taken to preserve any hard copy that may constitute part of the audit evidence. In all cases, it is important that retention policies for electronic evidence are sufficient to preserve evidence that supports audit findings. As an international organization, ISACA recognizes that the rules of evidence will differ according to local and national legislation, regulation and culture; however, concepts such as the importance of forensic evidence are universal.

Audit conclusions should be supported by reliable and relevant evidence. Evidence collected during the course of an audit follows a life cycle. This life cycle includes collection, analysis, and preservation and destruction of evidence. The source of evidence should be reliable and qualified (i.e., from an appropriate, original source rather than obtained as a comment or hearsay) and originate directly from a trusted source to help ensure objectivity. As an example, system configuration settings copied by a system administrator to a spreadsheet and then presented to an auditor would not be considered as reliable because they would have been subject to alteration. Audit evidence should include information regarding date of creation and original source. Because electronic evidence is more dynamic than hard copy documents, security measures should be used to preserve the integrity of evidence collected and provide assurance that the evidence has not been altered in any way.

Continuous auditing is a process by which the effectiveness and efficiency of controls is measured primarily by automated reporting processes that enable management to be aware of emerging risks or control weaknesses, without the need for a regular audit. The result is that information flow to management and implementation of corrective measures occur sooner. The IS auditor should be aware of the techniques involved in continuous auditing in order to facilitate the introduction of these techniques, as appropriate. The IS auditor must not rely solely on continuous auditing techniques when there is a high business risk and the continuous auditing technique deployed is not considered elaborate and exhaustive. This is the case when continuous auditing as a process has been put in place recently—for example, when the impact of control failure would be considerable. In such cases, regular formal audits must be scheduled to support and reinforce continuous auditing.</td><td>Application and relative value of computer-assisted audit techniques</td><td>1.5.15 Computer-assisted Audit Techniques</td></tr>
<tr><td>Techniques for obtaining evidence</td><td>1.5.11 Evidence
1.5.12 Interviewing and Observing Personnel in Performance of Their Duties</td></tr>
<tr><td>Computer-assisted audit techniques</td><td>1.5.15 Computer-assisted Audit Techniques</td></tr>
<tr><td>Factors to consider in collection, protection and chain of custody of audit evidence in an IS audit</td><td>1.5.11 Evidence
1.6.2 Audit Documentation</td></tr>
<tr><td>Special considerations in audit documentation for evidence</td><td>1.8.2 Continuous Auditing</td></tr>
<tr><td>Continuous auditing techniques</td><td>1.8.2 Continuous Auditing</td></tr>
</table>

K1.8 Knowledge of different sampling methodologies and other substantive/data analytical procedures

Explanation	Key Concepts	Reference in Manual
Compliance testing is evidence gathering for the purpose of testing an enterprise's compliance with control procedures. This differs from substantive testing in which evidence is gathered to evaluate the integrity of individual transactions, data or other information. There is a direct correlation between the level of internal controls and the amount of substantive testing required. If the results of testing controls (compliance tests) reveal the presence of adequate internal controls, then the IS auditor is justified in minimizing the substantive procedures. Conversely, if the control testing reveals weaknesses in controls that may raise doubts about the completeness, accuracy or validity of the accounts, substantive testing can alleviate those doubts. The efficiency and effectiveness of this testing can be enhanced through the use of sampling. Sampling is performed when time and cost considerations preclude a total verification of all transactions or events in a predefined population. The population consists of the entire group of items that need to be examined. The subset of population members used to perform testing is called the sample. Sampling is used to infer characteristics about the entire population, based on the characteristics of the sample. For some time, there has been a focus on the IS auditor's ability to verify the adequacy of internal controls through the use of sampling techniques. This has become necessary because many controls are transactional in nature, which can make it difficult to test the entire population. However, sampling is not always warranted because software may allow the testing of certain attributes across the entire population. Although a CISA candidate is not expected to become a sampling expert, it is important for the candidate to have a foundational understanding of the general principles of sampling and how to design a relevant and reliable sample.	Relative use of compliance and substantive testing	1.5.10 Compliance Versus Substantive Testing
	Basic approaches to sampling and their relation to testing approaches	1.5.13 Sampling

K1.9 Knowledge of reporting and communication techniques (e.g., facilitation, negotiation, conflict resolution, audit report structure, issue writing, management summary, result verification)

Explanation	Key Concepts	Reference in Manual
Effective and clear communication can significantly improve the quality of audits and maximize their results. Audit findings should be reported and communicated to stakeholders with appropriate buy-in from the auditees for the audit process to be successful. Auditors should also take into account the motivations and perspectives of recipients of the audit report so that their concerns may be properly addressed. Communication skills (both written and verbal) determine the effectiveness of the audit reporting process. Communication and negotiation skills are required throughout the audit activity. Successful resolution of audit findings with auditees is essential so that auditees will adopt the recommendations in the report and initiate prompt corrective action. This goal may require the use of techniques such as facilitation, negotiation and conflict resolution. IS auditors should also understand the concept of materiality (i.e., the relative importance of findings based on business impact).	Understanding reporting standards	1.3.2 ISACA IS Audit and Assurance Standards (1400 Reporting)
	Applying various communications techniques to the reporting of audit results	1.6 Communicating Audit Results 1.6.1 Audit Report Structure and Contents 1.6.2 Audit Documentation
	Applying communication techniques to facilitation roles in control self-assessments	1.7 Control Self-assessment 1.7.4 Auditor Role in CSA

K1.10 Knowledge of audit quality assurance (QA) systems and frameworks

Explanation	Key Concepts	Reference in Manual
IS auditing is a branch of the broader field of auditing. Auditing standards refer to minimum parameters that should be taken into account when performing an audit. However, there may be guidelines and additional audit procedures that an auditor may wish to add in order to develop an opinion on the proper functioning of controls. Most of the basic auditing practices and techniques are equally relevant in an IS audit. The IS auditor should understand the impact of the IS environment on traditional auditing practices and techniques to ensure that the basic objective of the audit exercise is achieved. The practices and techniques to be used in a specific IS audit should be determined during the audit planning stage and incorporated into an audit program. ISACA does not define, or require knowledge of, any specific audit methodology, but expects the IS auditor to be aware of the general principles involved in planning and conducting an effective audit program. Control self-assessment (CSA) is a process in which an IS auditor can act in the role of facilitator to the business process owners to help them define and assess appropriate controls. The process owners and the personnel who run the processes use their knowledge and understanding of the business function to evaluate the performance of controls against the established control objectives, while taking into account the risk appetite of the enterprise. Process owners are in an ideal position to define the appropriate controls because they have a greater knowledge of the process objectives. The IS auditor helps the process owners understand the need for controls, based on risk to the business processes. Results must be interpreted with a certain level of skepticism because process owners are not always objective when assessing their own activities.	Impact of IS environment on IS auditing practices and techniques	1.5.1 Audit Objectives 1.5.3 Audit Methodology 1.5.8 Audit Programs 2.11 Auditing IT Governance Structure and Implementation 2.13 Auditing Business Continuity 3.14 Auditing Application Controls 3.15 Auditing Systems Development, Acquisition and Maintenance 4.7 Auditing Infrastructure and Operations 5.5 Auditing Information Security Management Framework 5.6 Auditing Network Infrastructure Security
	Points of relevance while using services of other auditors and experts	1.5.14 Using the Services of Other Auditors and Experts
	Advantages and disadvantages of CSA	1.7 Control Self-assessment 1.7.1 Objectives of CSA 1.7.2 Benefits of CSA 1.7.3 Disadvantages of CSA
	The role of the auditor in CSA	1.7.4 Auditor Role in CSA
	Relevance of different technology drivers for CSA in the current business environment	1.7.5 Technology Drivers for CSA 1.7.6 Traditional Versus CSA Approach
	Relevance of different approaches of CSA in a given context	
	Applying communication techniques to facilitation roles in control self-assessments	1.7 Control Self-assessment 1.7.4 Auditor Role in CSA
	Audit quality evaluation	1.5.16 Evaluation of the Control Environment

K1.11 Knowledge of various types of audits (e.g., internal, external, financial) and methods for assessing and placing reliance on the work of other auditors or control entities

Explanation	Key Concepts	Reference in Manual
The IS auditor must be aware of the variety of audits that can be performed along with the goals and objectives of each of these engagements. Furthermore, there are a wide variety of entities (Payment Card Industry Data Security Standard [PCI DSS] third-party providers, government regulators, third-party assessing organizations/independent validation and verification, etc.) auditing the same organization, often concurrently. IS auditors must understand the purpose, scope and timing of these audits, so they can take these audits into consideration during their audit planning, testing and reporting processes. Recognizing that many recent, current and upcoming audits may provide adequate depth and coverage of area within the IS auditor's audit scope could enable the IS auditor to place reliance on other auditors' or control entities' work, if the work of others meets the standards of professional practice and testing needed to provide reasonable assurance that the IS controls are operating effectively, efficiently and are aligned with both current and planned organizational goals and objectives.	Understanding the proper techniques to plan assigned audits while maximizing IS Audit resource utilization and preventing duplication of audit activities	1.3.2 ISACA IS Audit and Assurance Standards (1201 Engagement Planning)
	Understanding the other type of audits that may be performed by other auditors and experts	1.5.2 Types of Audits
	Application of audit planning techniques	1.2.2 IS Audit Resource Management 1.2.3 Audit Planning 1.2.4 Effect of Laws and Regulations on IS Audit Planning
	Understanding risk analysis concepts within an auditing context	1.4.1 Risk Analysis
	Points of relevance while using services of other auditors and experts	1.3.2 ISACA IS Audit and Assurance Standards (1206 Using the Work of Other Experts) 1.5.14 Using the Services of Other Auditors and Experts

SUGGESTED RESOURCES FOR FURTHER STUDY

Cascarino, Richard E.; *Auditor's Guide to IT Auditing and Software Demo, 2nd Edition*, Wiley, USA, 2012

Davis, Chris; Mike Schiller; Kevin Wheeler; *IT Auditing: Using Controls to Protect Information Assets, 2nd Edition*, McGraw Hill, USA, 2011

ISACA, COBIT 5, USA, 2012, *www.isaca.org/cobit*

ISACA, *COBIT 5 for Assurance*, USA, 2013, *www.isaca.org/cobit*

ISACA, *IT Control Objectives for Sarbanes-Oxley: Using COBIT® 5 in the Design and Implementation of Internal Controls Over Financial Reporting*, USA, 2014, *www.isaca.org/sox*

ISACA, *ITAF™: A Professional Practices Framework for IS Audit/Assurance, 3rd Edition*, USA, 2014, *www.isaca.org/ITAF*

IT Governance Institute, *Control Objectives for BASEL II: The Importance of Governance and Risk Management for Compliance*, USA, 2007

Senft, Sandra; Frederick Gallegos; Aleksandra Davis; *Information Technology Control and Audit, 4th Edition*, CRC Press, USA, 2012

Note: Publications in bold are stocked in the ISACA Bookstore.

SELF-ASSESSMENT QUESTIONS

CISA self-assessment questions support the content in this manual and provide an understanding of the type and structure of questions that have typically appeared on the exam. Questions are written in a multiple-choice format and designed for one best answer. Each question has a stem (question) and four options (answer choices). The stem may be written in the form of a question or an incomplete statement. In some instances, a scenario or a description problem may also be included. These questions normally include a description of a situation and require the candidate to answer two or more questions based on the information provided. Many times a question will require the candidate to choose the **MOST** likely or **BEST** answer among the options provided.

In each case, the candidate must read the question carefully, eliminate known incorrect answers and then make the best choice possible. Knowing the format in which questions are asked, and how to study and gain knowledge of what will be tested, will help the candidate correctly answer the questions.

1-1 Which of the following outlines the overall authority to perform an IS audit?

A. The audit scope, with goals and objectives
B. A request from management to perform an audit
C. The approved audit charter
D. The approved audit schedule

1-2 In performing a risk-based audit, which risk assessment is completed **INITIALLY** by the IS auditor?

A. Detection risk assessment
B. Control risk assessment
C. Inherent risk assessment
D. Fraud risk assessment

1-3 While developing a risk-based audit program, on which of the following would the IS auditor **MOST** likely focus?

A. Business processes
B. Administrative controls
C. Operational controls
D. Business strategies

1-4 Which of the following types of audit risk assumes an absence of compensating controls in the area being reviewed?

A. Control risk
B. Detection risk
C. Inherent risk
D. Sampling risk

1-5 An IS auditor performing a review of an application's controls finds a weakness in system software that could materially impact the application. The IS auditor should:

A. disregard these control weaknesses because a system software review is beyond the scope of this review.
B. conduct a detailed system software review and report the control weaknesses.
C. include in the report a statement that the audit was limited to a review of the application's controls.
D. review the system software controls as relevant and recommend a detailed system software review.

1-6 Which of the following is the **MOST** important reason why an audit planning process should be reviewed at periodic intervals?

A. To plan for deployment of available audit resources
B. To consider changes to the risk environment
C. To provide inputs for documentation of the audit charter
D. To identify the applicable IS audit standards

1-7 Which of the following is **MOST** effective for implementing a control self-assessment (CSA) within business units?

A. Informal peer reviews
B. Facilitated workshops
C. Process flow narratives
D. Data flow diagrams

1-8 The **FIRST** step in planning an audit is to:

A. define audit deliverables.
B. finalize the audit scope and audit objectives.
C. gain an understanding of the business' objectives.
D. develop the audit approach or audit strategy.

1-9 The approach an IS auditor should use to plan IS audit coverage should be based on:

A. risk.
B. materiality.
C. professional skepticism.
D. sufficiency of audit evidence.

1-10 A company performs a daily backup of critical data and software files, and stores the backup tapes at an offsite location. The backup tapes are used to restore the files in case of a disruption. This is a:

A. preventive control.
B. management control.
C. corrective control.
D. detective control.

ANSWERS TO SELF-ASSESSMENT QUESTIONS

1-1 A. The audit scope is specific to one audit and does not grant authority to perform an audit.
B. A request from management to perform an audit is not sufficient because it relates to a specific audit.
C. The approved audit charter outlines the auditor's responsibility, authority and accountability.
D. The approved audit schedule does not grant authority to perform an audit.

1-2 A. Detection risk assessment is performed only after the inherent and control risk assessment have been performed to determine ability to detect either errors within a targeted processes.
B. Control risk assessment is performed after the inherent risk assessment has been completed and is to determine the level of risk that remains after controls for the targeted process are in place.
C. Inherent risk exists independently of an audit and can occur because of the nature of the business. To successfully conduct an audit, it is important to be aware of the related business processes. To perform the audit, the IS auditor needs to understand the business process, and by understanding the business process, the IS auditor better understands the inherent risk.
D. Fraud risk assessments are a subset of a control risk assessment in which the auditor determines if the control risk addresses the ability of internal and/or external parties to commit fraudulent transactions within the system.

1-3 **A. A risk-based audit approach focuses on the understanding of the nature of the business and being able to identify and categorize risk. Business risk impacts the long-term viability of a specific business. Thus, an IS auditor using a risk-based audit approach must be able to understand business processes.**
B. Administrative controls, while an important subset of controls, are not the primary focus needed to understand the business processes within scope of the audit.
C. Like administrative controls, operational controls are an important subset of controls; however, they do not address high-level overarching business processes under review.
D. Business strategies are the drivers for business processes; however, in this case, the IS auditor is focusing on the business processes that were put in place to enable the organization to meet the strategy.

1-4 A. Control risk is the risk that a material error exists that will not be prevented or detected in a timely manner by the system of internal controls.
B. Detection risk is the risk that a material misstatement with a management assertion will not be detected by the auditor's substantive tests. It consists of two components, sampling risk and nonsampling risk.
C. The risk level or exposure without taking into account the actions that management has taken or might take is inherent risk.
D. Sampling risk is the risk that incorrect assumptions are made about the characteristics of a population from which a sample is taken. Nonsampling risk is the detection risk not related to sampling; it can be due to a variety of reasons, including, but not limited to, human error.

1-5 A. The IS auditor is not expected to ignore control weaknesses just because they are outside the scope of a current review.
B. The conduct of a detailed systems software review may hamper the audit's schedule, and the IS auditor may not be technically competent to do such a review at this time.
C. If there are control weaknesses that have been discovered by the IS auditor, they should be disclosed. By issuing a disclaimer, this responsibility would be waived.
D. The appropriate option would be to review the systems software as relevant to the review and recommend a detailed systems software review for which additional resources may be recommended.

1-6 A. Planning for deployment of available audit resources is determined by the audit assignments planned, which is influenced by the planning process.
B. Short- and long-term issues that drive audit planning can be heavily impacted by changes to the risk environment, technologies and business processes of the enterprise.
C. The audit charter reflects the mandate of top management to the audit function and resides at a more abstract level.
D. Applicability of IS audit standards, guidelines and procedures is universal to any audit engagement and is not influenced by short- and long-term issues.

1-7 A. Informal peer reviews would not be as effective because they would not necessarily identify and assess all control issues.
B. Facilitated workshops work well within business units.
C. Process flow narratives would not be as effective because they would not necessarily identify and assess all control issues.
D. Data flow diagrams would not be as effective because they would not necessarily identify and assess all control issues.

1-8 A. Defining audit deliverables is dependent upon having a thorough understanding of the business' objectives and purpose.
B. Finalizing the audit scope and objectives is dependent upon having a thorough understanding of the business' objectives and purpose.
C. The first step in audit planning is to gain an understanding of the business's mission, objectives and purpose, which in turn identifies the relevant policies, standards, guidelines, procedures and organization structure.
D. Developing the audit approach or strategy is dependent upon having a thorough understanding of the business' objectives and purpose.

1-9 **A. ISACA IS Audit and Assurance Standard 1202, Planning, establishes standards and provides guidance on planning an audit. It requires a risk-based approach.**
B. Materiality is addressed within ISACA IS Audit and Assurance Standard 1204: "IS audit and assurance professionals shall consider potential weaknesses or absences of controls while planning an engagement, and whether such weaknesses or absences of controls could result in a significant deficiency or a material weakness."
C. Professional skepticism is addressed within ISACA IS Audit and Assurance Standard 1207.2: "IS audit and assurance professionals shall maintain an attitude of professional skepticism during the engagement."
D. Sufficiency of audit evidence is addressed within ISACA IS Audit and Assurance Standard 1205.2: "IS audit and assurance professionals shall evaluate the sufficiency of evidence obtained to support conclusions and achieve engagement objectives."

1-10 A. Preventive controls are those that avert problems before they arise. Backup tapes cannot be used to prevent damage to files and, therefore, cannot be classified as a preventive control.
B. Management controls modify processing systems to minimize a repeat occurrence of the problem. Backup tapes do not modify processing systems and, therefore, do not fit the definition of a management control.
C. A corrective control helps to correct or minimize the impact of a problem. Backup tapes can be used for restoring the files in case of damage of files, thereby reducing the impact of a disruption.
D. Detective controls help to detect and report problems as they occur. Backup tapes do not aid in detecting errors.

Section Two: Content

1.1 QUICK REFERENCE

Quick Reference Review

Chapter 1 outlines the framework for performing IS auditing, specifically including those mandatory requirements regarding the IS auditor's mission and activity as well as good practices to achieve an appropriate IS auditing outcome. CISA candidates should have a sound understanding of the following items, not only within the context of the present chapter, but also to correctly address questions in related subject areas. It is important to keep in mind that it is not enough to know these concepts from a definitional perspective. The CISA candidate must also be able to identify which elements may represent the greatest risk and which controls are most effective at mitigating this risk. Key topics in this chapter include:

- IS auditor roles and associated responsibilities, including expected audit outcomes and differences between IS auditing tasks within an assurance assignment and those within a consulting assignment
- The need for audit independence and level of authority within the internal audit environment as opposed to an external context
- Minimum audit planning requirements for an IS audit assignment, regardless of the specific or particular audit objective and scope
- Understanding the required level of compliance with ISACA IS Audit and Assurance Standards, as well as for ISACA IS Audit and Assurance Guidelines
- When planning audit work, the importance of clear identification of the audit approach related to controls defined as general versus auditing controls that are defined as application controls
- Scope, field work, application and execution of the concepts included in audit risk versus business risk
- The key role of requirements-compliant audit evidence when supporting the credibility of audit results and reporting
- The reliance on electronic audit work papers and evidence
- Purpose and planning opportunities of compliance testing versus substantive testing
- Audit responsibility and level of knowledge when considering legal requirements affecting IT within an audit scope
- The IS risk-oriented audit approach versus the complementary need for IS auditors to be acquainted with diverse IS standards and frameworks
- Understanding the difference between the objectives of implemented controls and control procedures
- Understanding evidence collection, sampling and evidence analysis techniques and its importance while conducting an IS audit
- Understanding reporting and communication methods

1.2 MANAGEMENT OF THE IS AUDIT FUNCTION

The audit function should be managed and led in a manner that ensures that the diverse tasks performed and achieved by the audit team will fulfill audit function objectives, while preserving audit independence and competence. Furthermore, managing the audit function should ensure value-added contributions to senior management regarding the efficient management of IT and achievement of business objectives.

Note: Information systems (IS) are defined as the combination of strategic, managerial and operational activities involved in gathering, processing, storing, distributing and using information and its related technologies. Information systems are distinct from information technology (IT) in that an information system has an IT component that interacts with the process components. IT is defined as the hardware, software, communication and other facilities used to input, store, process, transmit and output data in whatever form. Therefore, the terms "IS" and "IT" will be used according to these definitions throughout the manual.

1.2.1 ORGANIZATION OF THE IS AUDIT FUNCTION

IS audit is the formal examination, interview and/or testing of information systems to determine whether:

- Information systems are in compliance with applicable laws, regulations, contracts and/or industry guidelines
- IS data and information have appropriate levels of confidentiality, integrity and availability
- IS operations are being accomplished efficiently and effectiveness targets are being met

An organization can use both externally or internally provided IS audit services. The fundamental elements of IS audit are listed in section 1.3, ISACA IS Audit and Assurance Standards and Guidelines.

The role of the IS internal audit function should be established by an audit charter approved by board of directors and the audit committee (senior management if these entities do not exist). IS audit can be a part of internal audit, function as an independent group, or integrated within a financial and operational audit to provide IT-related control assurance to the financial or management auditors. Therefore, the audit charter may include IS audit as an audit support function. The charter should clearly state management's responsibility and objectives for, and delegation of authority to, the IS audit function. This document should outline the overall authority, scope and responsibilities of the audit function. The highest level of management and the audit committee, if one exists, should approve this charter. Once established, this charter should be changed only if the change can be and is thoroughly justified. ISACA IS Audit and Assurance Standards require that the responsibility, authority and accountability of the IS audit function are appropriately documented in an audit charter or engagement letter (1001 Audit Charter). An **audit charter** is an overarching document that covers the entire scope of audit activities in an entity while an **engagement letter** is more focused on a particular audit exercise that is sought to be initiated in an organization with a specific objective in mind.

If IS audit services are provided by an external firm, the scope and objectives of these services should be documented in a formal contract or statement of work between the contracting organization and the service provider.

In either case, the internal audit function should be independent and report to an audit committee, if one exists, or to the highest management level such as the board of directors.

1.2.2 IS AUDIT RESOURCE MANAGEMENT

IS technology is constantly changing. Therefore, it is important that IS auditors maintain their competency through updates of existing skills and obtain training directed toward new audit techniques and technological areas. ISACA IS Audit and Assurance Standards require that the IS auditor be technically competent (1006 Proficiency), having the skills and knowledge necessary to perform the auditor's work. Further, the IS auditor is to maintain technical competence through appropriate continuing professional education. Skills and knowledge should be taken into consideration when planning audits and assigning staff to specific audit assignments.

Preferably, a detailed staff training plan should be drawn for the year based on the organization's direction in terms of technology and related risk that needs to be addressed. This should be reviewed periodically to ensure that the training efforts and results are aligned to the direction that the audit organization is taking. Additionally, IS audit management should also provide the necessary IT resources to properly perform IS audits of a highly specialized nature (e.g., tools, methodology, work programs).

1.2.3 AUDIT PLANNING

Annual Planning

Audit planning includes both short- and long-term planning. Short-term planning takes into account audit issues that will be covered during the year, whereas long-term planning relates to audit plans that will take into account risk-related issues regarding changes in the organization's IT strategic direction that will affect the organization's IT environment.

All of the relevant processes that represent the blueprint of the entity's business should be included in the audit universe. The audit universe ideally lists all of the processes that may be considered for audit. Each of these processes may be subjected to a qualitative or quantitative risk assessment by evaluating the risk in respect to defined, relevant risk factors. The risk factors are those factors that influence the frequency and/or business impact of risk scenarios. For example, for an entity engaged in retail business, reputation can be a critical risk factor. The evaluation of risk should ideally be based on inputs from the business process owners. Evaluation of the risk factors should be based on objective criteria, although subjectivity cannot be completely avoided. For example, in respect to reputation factor, the criteria based on which inputs can be solicited from the business may be rated as:

- **High**—A process issue may result in damage to the reputation of the entity that will take more than six months to recover
- **Medium**—A process issue may result in damage to the reputation of the entity that will take less than six months but more than three months to recover
- **Low**—A process issue may result in damage to the reputation of the entity that will take less than three months to recover

In this example, the defined time frame represents the objective aspect of the criteria, and the subjective aspect of the criteria can be found in the business process owners' determination of the time frame—whether it is more than six months or less than three months. After the risk is evaluated for each relevant factor, an overall criterion may be defined to determine the overall risk of each of the processes.

The audit plan can then be constructed to include all of the processes that are rated "high," which would represent the ideal annual audit plan. However, in practice, when the resources required to execute the ideal plan are agreed on, often the available resources are not sufficient to execute the entire ideal plan. This analysis will help the audit function to demonstrate to top management the gap in resourcing and give top management a good idea of the amount of risk that management is accepting if it does not add to or augment the existing audit resources.

Analysis of short- and long-term issues should occur at least annually. This is necessary to take into account new control issues; changes in the risk environment, technologies and business processes; and enhanced evaluation techniques. The results of this analysis for planning future audit activities should be reviewed by senior audit management and approved by the audit committee, if available, or alternatively by the board of directors and communicated to relevant levels of management. The annual planning should be updated if any key aspects of the risk environment have changed (e.g., acquisitions, new regulatory issues, market conditions).

Individual Audit Assignments

In addition to overall annual planning, each individual audit assignment must be adequately planned. The IS auditor should understand that other considerations, such as the results of periodic risk assessments, changes in the application of technology, and evolving privacy issues and regulatory requirements, may impact the overall approach to the audit. The IS auditor should also take into consideration system implementation/upgrade deadlines, current and future technologies, requirements from business process owners, and IS resource limitations.

When planning an audit, the IS auditor must have an understanding of the overall environment under review. This should include a general understanding of the various business practices and functions relating to the audit subject, as well as the types of information systems and technology supporting the activity. For example, the IS auditor should be familiar with the regulatory environment in which the business operates.

To perform audit planning, the IS auditor should perform the steps indicated in **figure 1.2**.

Figure 1.2—Steps to Perform Audit Planning

- Gain an understanding of the business's mission, objectives, purpose and processes, which include information and processing requirements such as availability, integrity, security and business technology and information confidentiality.
- Understand changes in business environment of the auditee.
- Review prior work papers.
- Identify stated contents such as policies, standards and required guidelines, procedures and organization structure.
- Perform a risk analysis to help in designing the audit plan.
- Set the audit scope and audit objectives.
- Develop the audit approach or audit strategy.
- Assign personnel resources to the audit.
- Address engagement logistics.

ISACA IS Audit and Assurance Standards require the IS auditor to plan the IS audit work to address the audit objectives and comply with applicable professional auditing standards (1201 Engagement Planning). The IS auditor should develop an audit plan that takes into consideration the objectives of the auditee relevant to the audit area and its technology infrastructure. Where appropriate, the IS auditor should also consider the area under review and its relationship to the organization (strategically, financially and/or operationally) and obtain information on the strategic plan, including the IS strategic plan. The IS auditor should have an understanding of the auditee's information technology architecture and technological direction to design a plan appropriate for the present and, where appropriate, future technology of the auditee.

Steps an IS auditor could take to gain an understanding of the business include:

- Reading background material including industry publications, annual reports and independent financial analysis reports
- Reviewing prior audit reports or IT-related reports (from external or internal audits, or specific reviews such as regulatory reviews)
- Reviewing business and IT long-term strategic plans
- Interviewing key managers to understand business issues
- Identifying specific regulations applicable to IT
- Identifying IT functions or related activities that have been outsourced
- Touring key organization facilities

Another basic component of planning is the matching of available audit resources to the tasks as defined in the audit plan. The IS auditor who prepares the plan should consider the requirements of the audit project, staffing resources and other constraints. This matching exercise should consider the needs of individual audit projects as well as the overall needs of the audit department.

1.2.4 EFFECT OF LAWS AND REGULATIONS ON IS AUDIT PLANNING

Each organization, regardless of its size or the industry within which it operates, will need to comply with a number of governmental and external requirements related to computer system practices and controls and to the manner in which computers, programs and data are stored and used. Additionally, business regulations can impact the way data are processed, transmitted and stored (stock exchange, central banks, etc.). Special attention should be given to these issues in industries that are closely regulated. The banking industry worldwide has severe penalties for banks and their officers should a bank be unable to provide an adequate level of service due to security breaches. Inadequate security in a bank's online portal can result in loss of customer funds. In several countries, Internet service providers (ISPs) are subject to laws regarding confidentiality and service availability.

Because of a growing dependency on information systems and related technology, several countries are making efforts to add legal regulations concerning IS audit. The content of these legal regulations pertains to:

- Establishment of regulatory requirements
- Responsibilities assigned to corresponding entities
- Financial, operational and IT audit functions

Management personnel as well as audit management, at all levels, should be aware of the external requirements relevant to the goals and plans of the organization, and to the responsibilities and activities of the information services department/function/activity.

There are two major areas of concern: legal requirements (laws, regulatory and contractual agreements) placed on audit or IS audit, and legal requirements placed on the auditee and its systems, data management, reporting, etc. These areas impact the audit scope and audit objectives. The latter is important to internal and external auditors. Legal issues also impact the organizations' business operations in terms of compliance with ergonomic regulations, the US Health Insurance Portability and Accountability Act (HIPAA), Protection of Personal Data Directives and Electronic Commerce within the European Community, fraud prevention within banking organizations, etc.

An example of strong control practices is the US Sarbanes-Oxley Act of 2002, which requires evaluating an organization's internal controls. Sarbanes-Oxley provides for new corporate governance rules, regulations and standards for specified public companies including US Securities and Exchange Commission (SEC) registrants. The SEC has mandated the use of a recognized internal control framework. Sarbanes-Oxley requires organizations to select and implement a suitable internal control framework. Similarly, Japan enacted the Tokyo Stock Exchange Principles. In March 2004, the Listed Company Corporate Governance Committee, established by the Tokyo Stock Exchange in December 2002, published *Principles of Corporate Governance for Listed Companies*. The *Internal Control—Integrated Framework* from the Committee of Sponsoring Organizations of the Treadway Commission (COSO) has become the most commonly adopted framework by public companies seeking to comply. Because the US Sarbanes-Oxley Act has as its objective increasing the level of control of business processes and the information systems supporting them, IS auditors must consider the impact of Sarbanes-Oxley as part of audit planning.

A similar example of regulatory impact is the Basel Accords (Basel I, Basel II and Basel III). The Basel Accords regulate the minimum amount of capital for financial organizations based on the level of risk they face. The Basel Committee on Banking Supervision recommends conditions and capital requirements that

should be fulfilled to manage risk exposure. These conditions will ideally result in an improvement in:
- Credit risk
- Operational risk
- Market risk

The following are steps an IS auditor would perform to determine an organization's level of compliance with external requirements:
- Identify those government or other relevant external requirements dealing with:
  - Electronic data, personal data, copyrights, e-commerce, e-signatures, etc.
  - Computer system practices and controls
  - The manner in which computers, programs and data are stored
  - The organization or the activities of information technology services
  - IS audits
- Document applicable laws and regulations.
- Assess whether the management of the organization and the IT function have considered the relevant external requirements in making plans and in setting policies, standards and procedures, as well as business application features.
- Review internal IT department/function/activity documents that address adherence to laws applicable to the industry.
- Determine adherence to established procedures that address these requirements.
- Determine if there are procedures in place to ensure contracts or agreements with external IT services providers reflect any legal requirements related to responsibilities.

It is expected that the organization would have a legal compliance function on which the IS control practitioner could rely.

Note: A CISA candidate will not be asked about any specific laws or regulations but may be questioned about how one would audit for compliance with laws and regulations. The examination will only test knowledge of accepted global practices.

1.3 ISACA IS AUDIT AND ASSURANCE STANDARDS AND GUIDELINES

1.3.1 ISACA CODE OF PROFESSIONAL ETHICS

ISACA sets forth this Code of Professional Ethics to guide the professional and personal conduct of members of the association and/or its certification holders.

Members and ISACA certification holders shall:
1. Support the implementation of, and encourage compliance with, appropriate standards, procedures for the effective governance and management of enterprise information systems and technology, including: audit, control, security and risk management.
2. Perform their duties with objectivity, due diligence and professional care, in accordance with professional standards.
3. Serve in the interest of stakeholders in a lawful manner, while maintaining high standards of conduct and character, and not discrediting their profession or the Association.
4. Maintain the privacy and confidentiality of information obtained in the course of their activities unless disclosure is required by legal authority. Such information shall not be used for personal benefit or released to inappropriate parties.
5. Maintain competency in their respective fields and agree to undertake only those activities they can reasonably expect to complete with the necessary skills, knowledge and competence.
6. Inform appropriate parties of the results of work performed, including the disclosure of all significant facts known to them that, if not disclosed, may distort the reporting of the results.
7. Support the professional education of stakeholders in enhancing their understanding of the governance and management of enterprise information systems and technology, including: audit, control, security and risk management.

Failure to comply with this Code of Professional Ethics can result in an investigation into a member's or certification holder's conduct and, ultimately, in disciplinary measures.

Note: A CISA candidate is not expected to have memorized the ISACA IS Audit and Assurance Standards, Guidelines, and Tools and Techniques and the ISACA Code of Professional Ethics (*www.isaca.org/certification/code-of-professional-ethics*), word for word. Rather, the candidates will be tested on their understanding of the standard, guideline or code, its objectives and how it applies in a given situation.

1.3.2 ISACA IS AUDIT AND ASSURANCE STANDARDS

The specialized nature of IS auditing and the skills and knowledge necessary to perform such audits require globally applicable standards that pertain specifically to IS auditing. One of the most important functions of ISACA is providing information (common body of knowledge) to support knowledge requirements. (See standard 1006 Proficiency.)

One of ISACA's goals is to advance standards to meet this need. The development and dissemination of the ISACA IS Audit and Assurance Standards is a cornerstone of the association's professional contribution to the audit community. The IS auditor needs to be aware that there may be additional standards, or even legal requirements, placed on the auditor.

Standards contain statements of mandatory requirements for IS audit and assurance. They inform:
- IS audit and assurance professionals of the minimum level of acceptable performance required to meet the professional responsibilities set out in the ISACA Code of Professional Ethics
- Management and other interested parties of the profession's expectations concerning the work of practitioners
- Holders of the Certified Information Systems Auditor (CISA) designation of their requirements. Failure to comply with these standards may result in an investigation into the CISA holder's conduct by the ISACA Board of Directors or appropriate ISACA group and, ultimately, in disciplinary action.

The framework for the ISACA IS Audit and Assurance Standards provides for multiple levels of documents:
- Standards define mandatory requirements for IS audit and assurance and reporting.
- Guidelines provide guidance in applying IS Audit and Assurance Standards. The IS auditor should consider them in determining how to achieve implementation of the above standards, use

professional judgment in their application and be prepared to justify any departure from the standards.
- Tools and techniques provide examples of processes an IS auditor might follow in an audit engagement. The tools and techniques documents provide information on how to meet the standards when completing IS auditing work, but do not set requirements.

> **Note:** The complete text of the ISACA IS Audit and Assurance Standards, Guidelines, and Tools and Techniques is available at *www.isaca.org/standards.*

There are three categories of standards and guidelines—general, performance and reporting:
- **General**—The guiding principles under which the IS assurance profession operates. They apply to the conduct of all assignments, and deal with the IS audit and assurance professional's ethics, independence, objectivity and due care as well as knowledge, competency and skill.
- **Performance**—Deal with the conduct of the assignment, such as planning and supervision, scoping, risk and materiality, resource mobilization, supervision and assignment management, audit and assurance evidence, and the exercising of professional judgment and due care
- **Reporting**—Address the types of reports, means of communication and the information communicated

General

- **1001 Audit Charter**
  - 1001.1 The IS audit and assurance function shall document the audit function appropriately in an audit charter, indicating purpose, responsibility, authority and accountability.
  - 1001.2 The IS audit and assurance function shall have the audit charter agreed upon and approved at an appropriate level within the enterprise.
- **1002 Organisational Independence**
  - 1002.1 The IS audit and assurance function shall be independent of the area or activity being reviewed to permit objective completion of the audit and assurance engagement.
- **1003 Professional Independence**
  - 1003.1 IS audit and assurance professionals shall be independent and objective in both attitude and appearance in all matters related to audit and assurance engagements.
- **1004 Reasonable Expectation**
  - 1004.1 IS audit and assurance professionals shall have reasonable expectation that the engagement can be completed in accordance with the IS audit and assurance standards and, where required, other appropriate professional or industry standards or applicable regulations and result in a professional opinion or conclusion.
  - 1004.2 IS audit and assurance professionals shall have reasonable expectation that the scope of the engagement enables conclusion on the subject matter and addresses any restrictions.
  - 1004.3 IS audit and assurance professionals shall have reasonable expectation that management understands its obligations and responsibilities with respect to the provision of appropriate, relevant and timely information required to perform the engagement.
- **1005 Due Professional Care**
  - 1005.1 IS audit and assurance professionals shall exercise due professional care, including observance of applicable professional audit standards, in planning, performing and reporting on the results of engagements.
- **1006 Proficiency**
  - 1006.1 IS audit and assurance professionals, collectively with others assisting with the assignment, shall possess adequate skills and proficiency in conducting IS audit and assurance engagements and be professionally competent to perform the work required.
  - 1006.2 IS audit and assurance professionals, collectively with others assisting with the assignment, shall possess adequate knowledge of the subject matter.
  - 1006.3 IS audit and assurance professionals shall maintain professional competence through appropriate continuing professional education and training.
- **1007 Assertions**
  - 1007.1 IS audit and assurance professionals shall review the assertions against which the subject matter will be assessed to determine that such assertions are capable of being audited and that the assertions are sufficient, valid and relevant.
- **1008 Criteria**
  - 1008.1 IS audit and assurance professionals shall select criteria, against which the subject matter will be assessed, that are objective, complete, relevant, measureable, understandable, widely recognised, authoritative and understood by, or available to, all readers and users of the report.
  - 1008.2 IS audit and assurance professionals shall consider the source of the criteria and focus on those issued by relevant authoritative bodies before accepting lesser-known criteria.

Performance

- **1201 Engagement Planning**
  - 1201.1 IS audit and assurance professionals shall plan each IS audit and assurance engagement to address:
    - Objective(s), scope, timeline and deliverables
    - Compliance with applicable laws and professional auditing standards
    - Use of a risk-based approach, where appropriate
    - Engagement-specific issues
    - Documentation and reporting requirements
  - 1201.2 IS audit and assurance professionals shall develop and document an IS audit or assurance engagement project plan, describing the:
    - Engagement nature, objectives, timeline and resource requirements
    - Timing and extent of audit procedures to complete the engagement
- **1202 Risk Assessment in Planning**
  - 1202.1 The IS audit and assurance function shall use an appropriate risk assessment approach and supporting methodology to develop the overall IS audit plan and determine priorities for the effective allocation of IS audit resources.
  - 1202.2 IS audit and assurance professionals shall identify and assess risk relevant to the area under review, when planning individual engagements.
  - 1202.3 IS audit and assurance professionals shall consider subject matter risk, audit risk and related exposure to the enterprise.
- **1203 Performance and Supervision**
  - 1203.1 IS audit and assurance professionals shall conduct the work in accordance with the approved IS audit plan to cover identified risk and within the agreed-on schedule.
  - 1203.2 IS audit and assurance professionals shall provide supervision to IS audit staff whom they have supervisory responsibility for so as to accomplish audit objectives and meet applicable professional audit standards.

- 1203.3 IS audit and assurance professionals shall accept only tasks that are within their knowledge and skills or for which they have a reasonable expectation of either acquiring the skills during the engagement or achieving the task under supervision.
- 1203.4 IS audit and assurance professionals shall obtain sufficient and appropriate evidence to achieve the audit objectives. The audit findings and conclusions shall be supported by appropriate analysis and interpretation of this evidence.
- 1203.5 IS audit and assurance professionals shall document the audit process, describing the audit work and the audit evidence that supports findings and conclusions.
- 1203.6 IS audit and assurance professionals shall identify and conclude on findings.

• **1204 Materiality**
- 1204.1 IS audit and assurance professionals shall consider potential weaknesses or absences of controls while planning an engagement, and whether such weaknesses or absences of controls could result in a significant deficiency or a material weakness.
- 1204.2 IS audit and assurance professionals shall consider audit materiality and its relationship to audit risk while determining the nature, timing and extent of audit procedures.
- 1204.3 IS audit and assurance professionals shall consider the cumulative effect of minor control deficiencies or weaknesses and whether the absence of controls translates into a significant deficiency or a material weakness.
- 1204.4 IS audit and assurance professionals shall disclose the following in the report:
  · Absence of controls or ineffective controls
  · Significance of the control deficiency
  · Probability of these weaknesses resulting in a significant deficiency or material weakness

• **1205 Evidence**
- 1205.1 IS audit and assurance professionals shall obtain sufficient and appropriate evidence to draw reasonable conclusions on which to base the engagement results.
- 1205.2 IS audit and assurance professionals shall evaluate the sufficiency of evidence obtained to support conclusions and achieve engagement objectives.

• **1206 Using the Work of Other Experts**
- 1206.1 IS audit and assurance professionals shall consider using the work of other experts for the engagement, where appropriate.
- 1206.2 IS audit and assurance professionals shall assess and approve the adequacy of the other experts' professional qualifications, competencies, relevant experience, resources, independence and quality-control processes prior to the engagement.
- 1206.3 IS audit and assurance professionals shall assess, review and evaluate the work of other experts as part of the engagement, and document the conclusion on the extent of use and reliance on their work.
- 1206.4 IS audit and assurance professionals shall determine whether the work of other experts, who are not part of the engagement team, is adequate and complete to conclude on the current engagement objectives, and clearly document the conclusion.
- 1206.5 IS audit and assurance professionals shall determine whether the work of other experts will be relied upon and incorporated directly or referred to separately in the report.
- 1206.6 IS audit and assurance professionals shall apply additional test procedures to gain sufficient and appropriate evidence in circumstances where the work of other experts does not provide sufficient and appropriate evidence.
- 1206.7 IS audit and assurance professionals shall provide an appropriate audit opinion or conclusion and include any scope limitation where required evidence is not obtained through additional test procedures.

• **1207 Irregularity and Illegal Acts**
- 1207.1 IS audit and assurance professionals shall consider the risk of irregularities and illegal acts during the engagement.
- 1207.2 IS audit and assurance professionals shall maintain an attitude of professional scepticism during the engagement.
- 1207.3 IS audit and assurance professionals shall document and communicate any material irregularities or illegal act to the appropriate party in a timely manner.

Reporting

• **1401 Reporting**
- 1401.1 IS audit and assurance professionals shall provide a report to communicate the results upon completion of the engagement including:
  · Identification of the enterprise, the intended recipients and any restrictions on content and circulation
  · The scope, engagement objectives, period of coverage and the nature, timing and extent of the work performed
  · The findings, conclusions, and recommendations
  · Any qualifications or limitations in scope that the IS audit and assurance professional has with respect to the engagement
  · Signature, date and distribution according to the terms of the audit charter or engagement letter
- 1401.2 IS audit and assurance professionals shall ensure that findings in the audit report are supported by sufficient and appropriate evidence.

• **1402 Follow-up Activities**
- 1402.1 IS audit and assurance professionals shall monitor relevant information to conclude whether management has planned/taken appropriate, timely action to address reported audit findings and recommendations.

Note: The CISA exam does not test whether a candidate knows the specific number of an IS auditing standard. The CISA exam tests how standards are applied within the audit process.

1.3.3 ISACA IS AUDIT AND ASSURANCE GUIDELINES

The objective of the ISACA IS Audit and Assurance Guidelines is to provide guidance and additional information on how to comply with the ISACA IS Audit and Assurance Standards. The IS auditor and assurance professional should:

- Consider them in determining how to implement the above standards.
- Use professional judgment in applying them to specific audits.
- Be able to justify any departure from the standards.

> **Note:** The CISA candidate is not expected to know the specific number of an IS Audit and Assurance Guideline. The CISA exam tests how guidelines are applied within the audit process. The IS auditor should review the IS Audit and Assurance Guidelines thoroughly to identify the subject matter that is truly needed in the job. The IS Audit and Assurance Guidelines are living documents. The most current documents may be viewed at *www.isaca.org/guidelines*.

The following are the Purpose sections of the guidelines, section 1.1.

General

- **2001 Audit Charter**
  - 1.1.1 The purpose of this guideline is to assist IS audit and assurance professionals in preparing an audit charter. The audit charter defines the purpose, responsibility, authority and accountability of the IS audit and assurance function.
  - 1.1.2 IS audit and assurance professionals should consider this guideline when determining how to implement the standard, use professional judgement in its application, be prepared to justify any departure and seek additional guidance if considered necessary.
- **2002 Organisational Independence**
  - 1.1.1 The purpose of this guideline is to address the independence of the IS audit and assurance function in the enterprise. Three important aspects are considered:
    - The position of the IS audit and assurance function within the enterprise
    - The level to which the IS audit and assurance function reports to within the enterprise
    - The performance of non-audit services within the enterprise by IS audit and assurance management and IS audit and assurance professionals
  - 1.1.2 This guideline provides guidance on assessing organisational independence and details the relationship between organisational independence and the audit charter and audit plan.
  - 1.1.3 IS audit and assurance professionals should consider this guideline when determining how to implement the standard, use professional judgement in its application, be prepared to justify any departure and seek additional guidance if considered necessary.
- **2003 Professional Independence**
  - 1.1.1 The purpose of this guideline is to provide a framework that enables the IS audit and assurance professional to:
    - Establish when independence may be, or may appear to be, impaired
    - Consider potential alternative approaches to the audit process when independence is, or may appear to be, impaired
    - Reduce or eliminate the impact on independence of IS audit and assurance professionals performing non-audit roles, functions and services
    - Determine disclosure requirements when required independence may be, or may appear to be, impaired
  - 1.1.2 IS audit and assurance professionals should consider this guideline when determining how to implement the standard, use professional judgement in its application, be prepared to justify any departure and seek additional guidance if considered necessary.
- **2004 Reasonable Expectation**
  - 1.1.1 The purpose of this guideline is to assist the IS audit and assurance professionals in implementing the principle of reasonable expectation in the execution of audit engagements. The main features over which the professionals should have reasonable expectation are that:
    - The audit engagement can be completed in accordance with these standards, other applicable standards or regulations, and result in a professional opinion or conclusion.
    - The scope of the audit engagement permits an opinion or conclusion to be expressed on the subject matter.
    - Management will provide them with appropriate, relevant and timely information required to perform the audit engagement.
  - 1.1.2 This guideline further assists the IS audit and assurance professionals in addressing scope limitations and provides guidance on accepting a change in terms.
  - 1.1.3 IS audit and assurance professionals should consider this guideline when determining how to implement the standard, use professional judgement in its application, be prepared to justify any departure and seek additional guidance if considered necessary.
- **2005 Due Professional Care**
  - 1.1.1 The purpose of this guideline is to clarify the term 'due professional care' as it applies to performing an audit engagement with integrity and care in compliance with the ISACA Code of Professional Ethics.
  - 1.1.2 This guideline explains how IS audit and assurance professionals should apply due professional care in planning, performing and reporting on an audit engagement.
  - 1.1.3 IS audit and assurance professionals should consider this guideline when determining how to implement the standard, use professional judgement in its application, be prepared to justify any departure and seek additional guidance if considered necessary.
- **2006 Proficiency**
  - 1.1.1 This guideline provides guidance to the IS audit and assurance professionals to acquire the necessary skills and knowledge and maintain the professional competences while carrying out audit engagements.
  - 1.1.2 IS audit and assurance professionals should consider this guideline when determining how to implement the standard, use professional judgement in its application, be prepared to justify any departure and seek additional guidance if considered necessary.
- **2007 Assertions**
  - 1.1.1 The purpose of this guideline is to detail the different assertions, guide IS audit and assurance professionals in assuring that the criteria, against which the subject matter is to be assessed, supports the assertions, and provide guidance on formulating a conclusion and drafting a report on the assertions.
  - 1.1.2 IS audit and assurance professionals should consider this guideline when determining how to implement the standard, use professional judgement in its application, be prepared to justify any departure and seek additional guidance if considered necessary.
- **2008 Criteria**
  - 1.1.1 The purpose of this guideline is to assist IS audit and assurance professionals in selecting criteria, against which the subject matter will be assessed, that are suitable, acceptable and come from a relevant source.

- 1.1.2 IS audit and assurance professionals should consider this guideline when determining how to implement the standard, use professional judgement in its application, be prepared to justify any departure and seek additional guidance if considered necessary.

Performance

- **2201 Engagement Planning**
  - 1.1.1 This guideline provides guidance to the IS audit and assurance professionals. Adequate planning helps to ensure that appropriate attention is devoted to important areas of the audit, potential problems are identified and resolved on a timely basis, and the audit engagement is properly organised, managed and performed in an effective and efficient manner.
  - 1.1.2 IS audit and assurance professionals should consider this guideline when determining how to implement the standard, use professional judgement in its application, be prepared to justify any departure and seek additional guidance if considered necessary.
- **2202 Risk Assessment in Planning**
  - 1.1.1 The level of audit work required to meet the audit objective is a subjective decision made by IS audit and assurance professionals. The purpose of this guideline is to reduce the risk of reaching an incorrect conclusion based on the audit findings and to reduce the existence of errors occurring in the area being audited.
  - 1.1.2 The guideline provides guidance in applying a risk assessment approach to develop an:
    - IS audit plan that covers all annual audit engagements
    - Audit engagement project plan that focuses on one specific audit engagement
  - 1.1.3 The guideline provides the details of the different types of risk the IS audit and assurance professionals encounter.
  - 1.1.4 IS audit and assurance professionals should consider this guideline when determining how to implement the standard, use professional judgement in its application, be prepared to justify any departure and seek additional guidance if considered necessary.
- **2203 Performance and Supervision**
  - 1.1.1 This guideline provides guidance to IS audit and assurance professionals in performing the audit engagement and supervising IS audit team members. It covers:
    - Performing an audit engagement
    - Roles and responsibilities, required knowledge and skills for performing audit engagements
    - Key aspects of supervision
    - Gathering evidence
    - Documenting work performed
    - Formulating findings and conclusions
  - 1.1.2 IS audit and assurance professionals should consider this guideline when determining how to implement the standard, use professional judgement in its application, be prepared to justify any departure and seek additional guidance if considered necessary.
- **2204 Materiality**
  - 1.1.1 The purpose of this guideline is to clearly define the concept 'materiality' for the IS audit and assurance professionals and make a clear distinction with the materiality concept used by financial audit and assurance professionals.
  - 1.1.2 The guideline assists the IS audit and assurance professionals in assessing materiality of the subject matter and considering materiality in relationship to controls and reportable issues.
  - 1.1.3 IS audit and assurance professionals should consider this guideline when determining how to implement the standard, use professional judgement in its application, be prepared to justify any departure and seek additional guidance if considered necessary.
- **2205 Evidence**
  - 1.1.1 The purpose of this guideline is to provide guidance to IS audit and assurance professionals in obtaining sufficient and appropriate evidence, evaluating the received evidence and preparing appropriate audit documentation.
  - 1.1.2 IS audit and assurance professionals should consider this guideline when determining how to implement the standard, use professional judgement in its application, be prepared to justify any departure and seek additional guidance if considered necessary.
- **2206 Using the Work of Other Experts**
  - 1.1.1 This guideline provides guidance to IS audit and assurance professionals when considering the use of work of other experts. The guideline assists in assessing the adequacy of the experts, reviewing and evaluating the work of other experts, assessing the need for performing additional test procedures and expressing an opinion for the audit engagement, while taking into account the work performed by other experts.
  - 1.1.2 IS audit and assurance professionals should consider this guideline when determining how to implement the standard, use professional judgement in its application, be prepared to justify any departure and seek additional guidance if considered necessary.
- **2207 Irregularity and Illegal Acts**
  - 1.1.1 The purpose of this guideline is to provide IS audit and assurance professionals with guidance on how to deal with irregularities and illegal acts.
  - 1.1.2 The guideline details the responsibilities of both management and IS audit and assurance professionals with regards to irregularities and illegal acts. It furthermore provides guidance on how to deal with irregularities and illegal acts during the planning and performance of the audit work. Finally, the guideline suggests good practices for internal and external reporting on irregularities and illegal acts.
  - 1.1.3 IS audit and assurance professionals should consider this guideline when determining how to implement the standards, use professional judgement in its application, be prepared to justify any departure and seek additional guidance if considered necessary.
- **2208 Sampling**
  - 1.1.1 The purpose of this guideline is to provide guidance to IS audit and assurance professionals to design and select an audit sample and evaluate sample results. Appropriate sampling and evaluation will help in achieving the requirements of sufficient and appropriate evidence.
  - 1.1.2 IS audit and assurance professionals should consider this guideline when determining how to implement related standards, use professional judgement in its application, be prepared to justify any departure and seek additional guidance if considered necessary.

Reporting

- **2401 Reporting**
  - 1.1.1 This guideline provides guidance for IS audit and assurance professionals on the different types of IS audit engagements and related reports.
  - 1.1.2 The guideline details all aspects that should be included in an audit engagement report and provides IS audit and assurance professionals with considerations to make when drafting and finalising an audit engagement report.
  - 1.1.3 IS audit and assurance professionals should consider this guideline when determining how to implement the standard, use professional judgement in its application, be prepared to justify any departure and seek additional guidance if considered necessary.
- **2402 Follow-up Activities**
  - 1.1.1 The purpose of this guideline is to provide guidance to IS audit and assurance professionals in monitoring if management has taken appropriate and timely action on reported recommendations and audit findings.
  - 1.1.2 IS audit and assurance professionals should consider this guideline when determining how to implement the standard, use professional judgement in its application, be prepared to justify any departure and seek additional guidance if considered necessary.

Note: The CISA candidate should be familiar with IS Audit and Assurance Guideline 2001 Audit Charter. Also important is 2207 Irregularities and Illegal Acts in relation to the standard 1207 Irregularities and Illegal Acts for the purpose of reporting irregularities such as fraud. In addition, the IS auditor should be familiar with the IS Audit and Assurance Guideline 2003 Professional Independence and the related standard 1003 Professional Independence. Knowledge on 2402 Follow-up Activities, should be further identified by the IS auditor in the IS Audit and Assurance Guidelines.

1.3.4 ISACA IS AUDIT AND ASSURANCE TOOLS AND TECHNIQUES

Tools and techniques developed by ISACA provide examples of possible processes an IS auditor may follow in an audit engagement. In determining the appropriateness of any specific tool and technique, IS auditors should apply their own professional judgment to the specific circumstances. The tools and techniques documents provide information on how to meet the standards when performing IS auditing work, but do not set requirements.

Tools and techniques are currently categorized into:

- White papers, *www.isaca.org/whitepapers* (complimentary PDF files)
- Audit/Assurance programs, *www.isaca.org/auditprograms* (complimentary Microsoft® Word files for ISACA members)
- COBIT 5 family of products, *www.isaca.org/cobit*
- Technical and Risk Management Reference series, *www.isaca.org/Knowledge-Center/ITAF-IS-Assurance-Audit-/Pages/Reference-Series.aspx* (available in the ISACA Bookstore)
- *ISACA® Journal* IT Audit Basics column, *www.isaca.org/Knowledge-Center/ITAF-IS-Assurance-Audit-/IT-Audit-Basics/Pages/IT-Audit-Basics-Articles.aspx* (complimentary access)

It is not mandatory for the IS auditor to follow these tools and techniques; however, following these procedures will provide assurance that the standards are being followed by the auditor.

Note: The ISACA IS Audit and Assurance Tools and Techniques are living documents. The most current documents may be viewed at *www.isaca.org/standards.*

1.3.5 RELATIONSHIP AMONG STANDARDS, GUIDELINES, AND TOOLS AND TECHNIQUES

Standards defined by ISACA are to be followed by the IS auditor. Guidelines provide assistance on how the auditor can implement standards in various audit assignments. Tools and techniques are not intended to provide exhaustive guidance to the auditor when performing an audit. Tools and techniques provide examples of steps the auditor may follow in specific audit assignments to implement the standards; however, the IS auditor should use professional judgment when using guidelines and tools and techniques.

There may be situations in which the legal/regulatory requirements are more stringent than the requirements contained in ISACA IS Audit and Assurance Standards. In such cases, the IS auditor should ensure compliance with the more stringent legal/regulatory requirements.

For example, section 2.3.2 of Guideline 2002 supporting Standard 1002 Organisational Independence states: "Activities that are routine and administrative or involve matters that are insignificant generally are deemed not to be management responsibilities and, therefore, would not impair independence. Non-audit services that would also not impair independence or objectivity if adequate safeguards are implemented include providing routine advice on information technology risk and controls." However, in some countries, regulatory enactments strictly prohibit auditors from accepting audit assignments from banks from which they have availed credit facilities. In such cases, IS auditors should give precedence to the applicable regulatory requirement and not accept the assignment, even though accepting the assignment would be in compliance with the requirement of the Guideline 2002. As stated throughout the IS Audit and Assurance Guidelines, the IS audit and assurance professionals should consider all guidelines when determining how to implement related standards, use professional judgment in its application, be prepared to justify any departure and seek additional guidance if considered necessary.

1.3.6 ITAF™

ITAF is a comprehensive and good practice-setting reference model that:

- Establishes standards that address IS audit and assurance professional roles and responsibilities; knowledge and skills; and diligence, conduct and reporting requirements
- Defines terms and concepts specific to IS assurance
- Provides guidance and tools and techniques on the planning, design, conduct and reporting of IS audit and assurance assignments

ITAF is focused on ISACA material and provides a single source through which IS audit and assurance professionals can seek guidance, research policies and procedures, obtain audit and assurance programs, and develop effective reports. ITAF 3rd Edition (*www.isaca.org/ITAF*) incorporates guidelines effective 1 September 2014. As new guidance is developed and issued, it will be indexed within the framework.

1.4 IS CONTROLS

In order for information systems to fully realize the benefits and risk and resource optimization goals, risk that could prevent or inhibit obtaining these goals needs to be addressed. Organizations design, develop, implement and monitor information systems through policies, procedures, practices and organizational structures to address these types of risk. The internal control life cycle is dynamic in nature and designed to provide reasonable assurance that business goals and objectives will be achieved and undesired events will be prevented or detected and corrected.

1.4.1 RISK ANALYSIS

Risk analysis is part of audit planning and helps identify risk and vulnerabilities so the IS auditor can determine the controls needed to mitigate risk.

In evaluating IT-related business processes applied by an organization, understanding the relationship between risk and control is important for IS audit and control professionals. IS auditors must be able to identify and differentiate risk types and the controls used to mitigate the risk. They must have knowledge of common business risk, related technology risk and relevant controls. They must also be able to evaluate the risk assessment and management techniques used by business managers, and to make assessments of risk to help focus and plan audit work. In addition to an understanding of business risk and control, IS auditors must understand that risk exists within the audit process.

Risk is the combination of the probability of an event and its consequence (*International Organization for Standardization [ISO] 31000:2009: Risk management—Principles and guidelines/ISO Guide 73:2009: Risk management—Vocabulary*). Business risk may negatively impact the assets, processes or objectives of a specific business or organization. The IS auditor is often focused on high-risk issues associated with the confidentiality, integrity or availability of sensitive and critical information and the underlying information systems and processes that generate, store and manipulate such information. In reviewing these types of IT-related business risk, IS auditors will often assess the effectiveness of the risk management process an organization uses.

In analyzing the business risk arising from the use of IT, it is important for the IS auditor to have a clear understanding of:
- Industry and or internationally accepted risk management processes
- The purpose and nature of business, the environment in which the business operates and related business risk
- The dependence on technology to process and deliver business information
- The business risk of using IT and how it impacts the achievement of the business goals and objectives
- A good overview of the business processes and the impact of IT and related risk on the business process objectives

ISACA's *Risk IT Framework* is based on a set of guiding principles and features business processes and management guidelines that conform to these principles. It is dedicated to helping enterprises manage IT-related risk. The collective experience of a global team of practitioners and experts and existing and emerging practices and methodologies for effective IT risk management have been consulted in the development of the Risk IT framework.

There are many definitions of risk, reflecting that risk means different things to different people. Perhaps one of the most holistic definitions of risk applicable throughout the information security business world is derived from *NIST Special Publication 800-30 Revision 1 Guide for Conducting Risk Assessments*:

> *Adverse impact(s) that could occur…to organizational operations (including mission, functions, image, reputation), organizational assets, individuals, other organizations…due to the potential for unauthorized access, use, disclosure, disruption, modification, or destruction of information and/or information systems.*

This definition is used commonly by the IT industry because it puts risk into an organizational context by using the concepts of assets and loss of value—terms that are easily understood by business managers.

The risk assessment process is characterized as an iterative life cycle that begins with identifying business objectives, information assets, and the underlying systems or information resources that generate, store, use or manipulate the assets (hardware, software, databases, networks, facilities, people, etc.) critical to achieving these objectives. Because IT risk is dynamic, it is strategic for management to recognize the need for and establish a dynamic IT risk management process that supports the business risk management process. The greatest degree of risk management effort may then be directed toward those considered most sensitive or critical to the organization. After sensitive and/or critical information assets are identified, a risk assessment is performed to identify vulnerabilities and threats, and determine the probability of occurrence and the resulting impact and additional safeguards that would mitigate this impact to a level acceptable to management.

Next, during the risk mitigation phase, controls are identified for mitigating identified risk. These controls are risk-mitigating countermeasures that should prevent or reduce the likelihood of a risk event occurring, detect the occurrence of a risk event, minimize the impact, or transfer the risk to another organization.

The assessment of countermeasures should be performed through a cost-benefit analysis where controls to mitigate risk are selected

to reduce risk to a level acceptable to management. This analysis process may be based on any of the following:
- The cost of the control compared to the benefit of minimizing the risk
- Management's appetite for risk (i.e., the level of residual risk that management is prepared to accept)
- Preferred risk-reduction methods (e.g., terminate the risk, minimize probability of occurrence, minimize impact, transfer the risk via insurance)

The final phase relates to monitoring performance levels of the risk being managed when identifying any significant changes in the environment that would trigger a risk reassessment, warranting changes to its control environment. It encompasses three processes—risk assessment, risk mitigation and risk reevaluation—in determining whether risk is being mitigated to a level acceptable to management. It should be noted that, to be effective, risk assessment should be an ongoing process in an organization that endeavors to continually identify and evaluate risk as it arises and evolves. See **figure 1.3** for the summary of the risk management process.

Figure 1.3—Risk Management Process

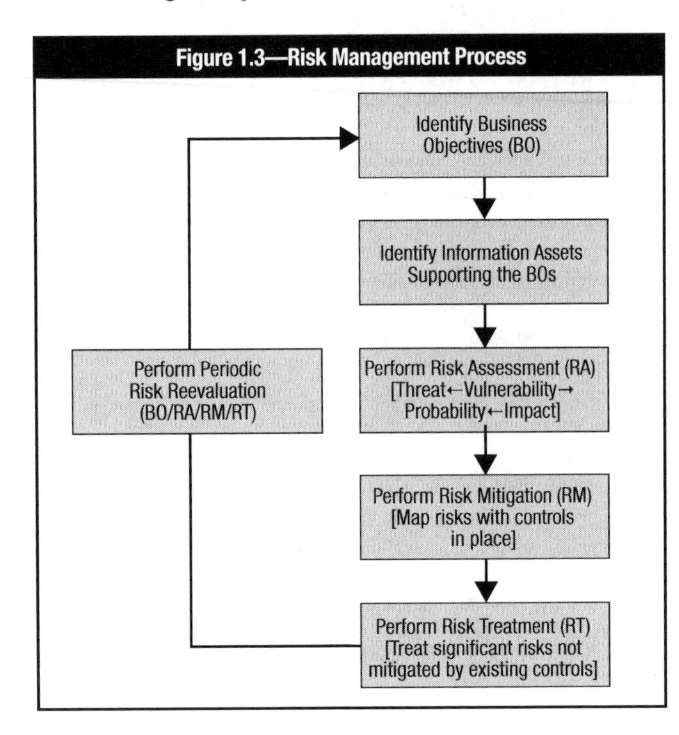

From the IS auditor's perspective, risk analysis serves more than one purpose:
- It assists the IS auditor in identifying risk and threats to an IT environment and IS system—risk and threats that would need to be addressed by management—and in identifying system-specific internal controls. Depending on the level of risk, this assists the IS auditor in selecting certain areas to examine.
- It helps the IS auditor in his/her evaluation of controls in audit planning.
- It assists the IS auditor in determining audit objectives.
- It supports risk-based audit decision making.

Figure 1.4 depicts the specific processes used by the IS auditor to realize the above listed objectives.

Figure 1.4—Risk Assessment Process

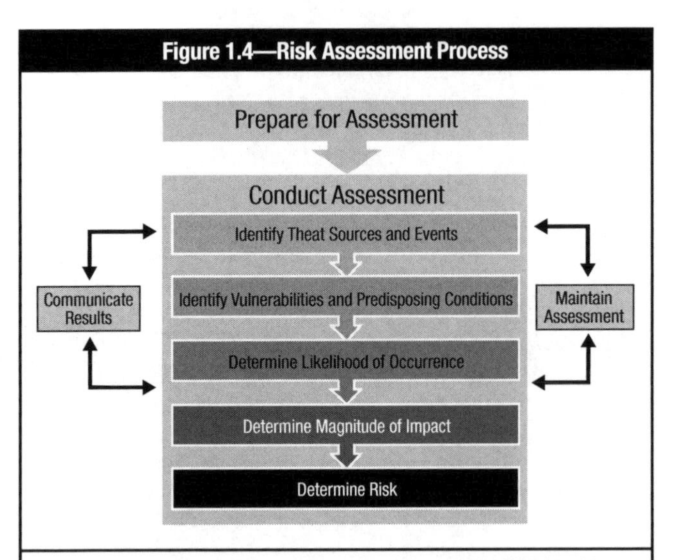

Source: National Institute of Standards and Technology (NIST), *NIST Special Publication 800-30, Revision 1: Information Security*, USA, 2012. Reprinted courtesy of the National Institute of Standards and Technology, U.S. Department of Commerce. Not copyrightable in the United States.

1.4.2 INTERNAL CONTROLS

Internal controls are normally composed of policies, procedures, practices and organizational structures that are implemented to reduce risk to the organization.

Internal controls are developed to provide reasonable assurance to management that the organization's business objectives will be achieved and risk events will be prevented, or detected and corrected. Internal control activities and supporting processes are either manual or driven by automated computer information resources. Internal controls operate at all levels within an organization to mitigate its exposures to risk that potentially could prevent it from achieving its business objectives. The board of directors and senior management are responsible for establishing the appropriate culture to facilitate an effective and efficient internal control system, and for continuously monitoring the effectiveness of the internal control system, although each individual within an organization must take part in this process.

There are two key aspects that controls should address: (1) what should be achieved and (2) what should be avoided. Internal controls address business/operational objectives and should also address undesired events through prevention, detection and correction.

Elements of controls that should be considered when evaluating control strength are classified as preventive, detective or corrective in nature.

Figure 1.5 displays control classifications, functions and usages.

Figure 1.5—Control Classifications

Class	Function	Examples
Preventive	• Detect problems before they arise. • Monitor both operation and inputs. • Attempt to predict potential problems before they occur and make adjustments. • Prevent an error, omission or malicious act from occurring. • Segregate duties (deterrent factor). • Control access to physical facilities. • Use well-designed documents (prevent errors).	• Employ only qualified personnel. • Establish suitable procedures for authorization of transactions. • Complete programmed edit checks. • Use access control software that allows only authorized personnel to access sensitive files. • Use encryption software to prevent unauthorized disclosure of data.
Detective	• Use controls that detect and report the occurrence of an error, omission or malicious act.	• Hash totals • Check points in production jobs • Echo controls in telecommunications • Error messages over tape labels • Duplicate checking of calculations • Periodic performance reporting with variances • Past-due account reports • Internal audit functions • Review of activity logs to detect unauthorized access attempts • Secure code reviews • Software quality assurance
Corrective	• Minimize the impact of a threat. • Remedy problems discovered by detective controls. • Identify the cause of a problem. • Correct errors arising from a problem. • Modify the processing system(s) to minimize future occurrences of the problem.	• Contingency/continuity of operations planning • Disaster recovery planning • Incident response planning • Backup procedures • Rerun procedures • System break/fix service level agreements

Note: A CISA candidate should know the differences between preventive, detective and corrective controls.

Control objectives are statements of the desired result or purpose to be achieved by implementing control activities (procedures). For example, control objectives may relate to the following concepts:
- Effectiveness
- Efficiency
- Confidentiality
- Integrity
- Availability
- Compliance
- Reliability

Control objectives apply to all controls, whether they are manual, automated or a combination (e.g., review of system logs). Control objectives in an IS environment do not differ from those in a manual environment; however, the way these controls are implemented may be different. Thus, control objectives need to be addressed relevant to specific IS-related processes.

1.4.3 IS CONTROL OBJECTIVES

IS control objectives provide a complete set of high-level requirements to be considered by management for effective control of each IT process. IS control objectives are:
- Statements of the desired result or purpose to be achieved by implementing controls around information systems processes
- Comprised of policies, procedures, practices and organizational structures
- Designed to provide reasonable assurance that business objectives will be achieved and undesired events will be prevented or detected and corrected

Enterprise management needs to make choices relative to these control objectives by:
- Selecting those that are applicable
- Deciding on those that will be implemented
- Choosing how to implement them (frequency, span, automation, etc.)
- Accepting the risk of not implementing those that may apply

Specific IS control objectives may include:
- Safeguarding assets: information on automated systems is secure from improper access and current
- Ensuring system development life cycle (SDLC) processes are established, in place and operating effectively to provide reasonable assurance that business, financial and/or industrial software systems and applications are developed in a repeatable and reliable manner to assure business objectives are met. (See chapter 3 Information Systems Acquisition, Development and Implementation, for more information.)
- Ensuring integrity of general operating system (OS) environments, including network management and operations
- Ensuring integrity of sensitive and critical application system environments, including accounting/financial and management information (information objectives) and customer data, through:
  - Authorization of the input. Each transaction is authorized and entered only once.
  - Validation of the input. Each input is validated and will not cause negative impact to the processing of transactions.
  - Accuracy and completeness of processing of transactions
  - All transactions are recorded accurately and entered into the system for the proper period.
  - Reliability of overall information processing activities

– Accuracy, completeness and security of the output
– Database confidentiality, integrity and availability
- Ensuring appropriate identification and authentication of users of IS resources (end users as well as infrastructure support)
- Ensuring the efficiency and effectiveness of operations (operational objectives)
- Complying with the users' requirements, organizational policies and procedures, and applicable laws and regulations (compliance objectives)
- Ensuring availability of IT services by developing efficient business continuity plans (BCPs) and disaster recovery plans (DRPs)
- Enhancing protection of data and systems by developing an incident response plan
- Ensuring integrity and reliability of systems by implementing effective change management procedures
- Ensuring that outsourced IS processes and services have clearly defined service level agreements (SLAs) and contract terms and conditions to ensure the organization's assets are properly protected and meet business goals and objectives

1.4.4 COBIT 5

COBIT 5, developed by ISACA, provides a comprehensive framework that assists enterprises in achieving their objectives for the governance and management of enterprise IT (GEIT). Simply stated, it helps enterprises create optimal value from IT by maintaining a balance between realizing benefits and optimizing risk levels and resource use. COBIT 5 enables IT to be governed and managed in a holistic manner for the entire enterprise, taking in the full end-to-end business and IT functional areas of responsibility, considering the IT-related interests of internal and external stakeholders. COBIT 5 is generic and useful for enterprises of all sizes, whether commercial, not-for-profit or in the public sector.

Figure 1.6—COBIT 5 Principles

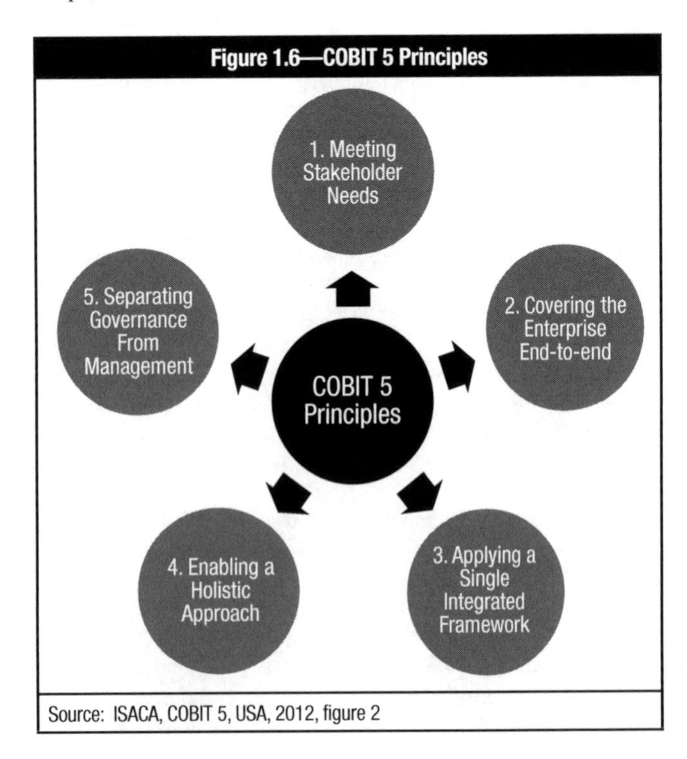

Source: ISACA, COBIT 5, USA, 2012, figure 2

COBIT 5 is based on five key principles for governance and management of enterprise IT (shown in **figure 1.6**):

- **Principle 1: Meeting Stakeholder Needs**—Enterprises exist to create value for their stakeholders, by maintaining a balance between the realization of benefits and the optimization of risk and use of resources. COBIT 5 provides all of the required processes and other enablers to support business value creation through the use of IT. Because every enterprise has different objectives, an enterprise can customize COBIT 5 to suit its own context through the goals cascade, translating high-level enterprise goals into manageable, specific, IT-related goals and mapping these to specific processes and practices.
- **Principle 2: Covering the Enterprise End-to-End**—COBIT 5 integrates governance of enterprise IT into enterprise governance:
  – It covers all functions and processes within the enterprise; COBIT 5 does not focus only on the "IT function," but treats information and related technologies as assets that need to be dealt with just like any other asset by everyone in the enterprise.
  – It considers all IT-related governance and management enablers to be enterprisewide and end-to-end (i.e., inclusive of everything and everyone—internal and external—that is relevant to governance and management of enterprise information and related IT).
- **Principle 3: Applying a Single, Integrated Framework**—There are many IT-related standards and good practices, each providing guidance on a subset of IT activities. COBIT 5 aligns with other relevant standards and frameworks at a high level, and thus can serve as the overarching framework for governance and management of enterprise IT.
- **Principle 4: Enabling a Holistic Approach**—Efficient and effective governance and management of enterprise IT requires a holistic approach, taking into account several interacting components. COBIT 5 defines a set of enablers to support the implementation of a comprehensive governance and management system for enterprise IT. Enablers are broadly defined as anything that can help to achieve the objectives of the enterprise. The COBIT 5 framework defines seven categories of enablers:
  – Principles, Policies and Frameworks
  – Processes
  – Organizational Structures
  – Culture, Ethics and Behavior
  – Information
  – Services, Infrastructure and Applications
  – People, Skills and Competencies
- **Principle 5: Separating Governance from Management**—The COBIT 5 framework makes a clear distinction between governance and management. These two disciplines encompass different types of activities, require different organizational structures and serve different purposes. COBIT 5's view on this key distinction between governance and management is:
  – Governance

Governance ensures that stakeholder needs, conditions and options are evaluated to determine balanced, agreed-on enterprise objectives to be achieved; setting direction through prioritization and decision making; and monitoring performance and compliance against agreed-on direction and objectives.

In most enterprises, overall governance is the responsibility of the board of directors under the leadership of the chairperson. Specific governance responsibilities may be delegated to special organizational structures at an appropriate level, particularly in larger, complex enterprises.

– Management

Management plans, builds, runs and monitors activities in alignment with the direction set by the governance body to achieve the enterprise objectives.

In most enterprises, management is the responsibility of the executive management under the leadership of the chief executive officer (CEO).

Together, these five principles enable the enterprise to build an effective governance and management framework that optimizes information and technology investment and use for the benefit of stakeholders.

Note: A CISA candidate will not be asked to specifically identify the COBIT process, the COBIT domains or the set of IT processes defined in each. However, candidates should know what frameworks are, what they do and why they are used by enterprises. Knowledge of the existence, structure and key principles of major standards and frameworks related to IT governance, assurance and security will also be advantageous. COBIT can be used as a supplemental study material in understanding control objectives and principles as detailed in this review material.

1.4.5 GENERAL CONTROLS

Controls include policies, procedures and practices (tasks and activities) established by management to provide reasonable assurance that specific objectives will be achieved.

General controls apply to all areas of the organization including IT infrastructure and support services. General controls include:

- Internal accounting controls that are primarily directed at accounting operations—controls that concern the safeguarding of assets and reliability of financial records
- Operational controls that concern day-to-day operations, functions and activities, and ensure that the operation is meeting the business objectives
- Administrative controls that concern operational efficiency in a functional area and adherence to management policies (administrative controls support the operational controls specifically concerned with these areas)
- Organizational security policies and procedures to ensure proper usage of assets
- Overall policies for the design and use of adequate documents and records (manual/automated) to help ensure proper recording of transactions—transactional audit trail
- Procedures and practices to ensure adequate safeguards over access to and use of assets and facilities
- Physical and logical security policies for all facilities, data centers and IT resources (e.g., servers and telecom infrastructure)

1.4.6 IS SPECIFIC CONTROLS

Each general control can be translated into an IS-specific control. A well-designed information system should have controls built in for all its sensitive or critical functions. For example, the general procedure to ensure that adequate safeguards over access to assets and facilities can be translated into an IS-related set of control procedures, covering access safeguards over computer programs, data and computer equipment. The IS auditor should understand the basic control objectives that exist for all functions. IS control procedures include:

- Strategy and direction of the IT function
- General organization and management of the IT function
- Access to IT resources, including data and programs
- Systems development methodologies and change control
- Operations procedures
- Systems programming and technical support functions
- Quality assurance (QA) procedures
- Physical access controls
- Business continuity (BCP)/disaster recovery planning (DRP)
- Networks and communications
- Database administration
- Protection and detective mechanisms against internal and external attacks

The IS auditor should understand concepts regarding IS controls and how to apply them in planning an audit.

Note: The IS controls listed in this section should be considered by the CISA candidate within the related job practice area (i.e., Protection of Information Assets).

1.5 PERFORMING AN IS AUDIT

Several steps are required to perform an audit. Adequate planning is a necessary first step in performing effective IS audits. To efficiently use IS audit resources, audit organizations must assess the overall risk for the general and application areas and related services being audited, and then develop an audit program that consists of objectives and audit procedures to satisfy the audit objectives. The audit process requires the IS auditor to gather evidence, evaluate the strengths and weaknesses of controls based on the evidence gathered through audit tests and prepare an audit report that presents those issues (areas of control weaknesses with recommendations for remediation) in an objective manner to management.

Audit management must ensure the availability of adequate audit resources and a schedule for performing the audits and, in the case of internal IS audit, for follow-up reviews on the status of corrective actions taken by management. The process of auditing includes defining the audit scope, formulating audit objectives, identifying audit criteria, performing audit procedures, reviewing and evaluating evidence, forming audit conclusions and opinions, and reporting to management after discussion with key process owners.

Project management techniques for managing and administering audit projects, whether automated or manual, include the following basic steps:

- **Plan the audit engagement**—Plan the audit considering project-specific risk.

- **Build the audit plan**—Chart out the necessary audit tasks across a time line, optimizing resource use. Make realistic estimates of the time requirements for each task with proper consideration given to the availability of the auditee.
- **Execute the plan**—Execute audit tasks against the plan.
- **Monitor project activity**—IS auditors report their actual progress against planned audit steps to ensure challenges are managed proactively and the scope is completed within time and budget.

1.5.1 AUDIT OBJECTIVES

Audit objectives refer to the specific goals that must be accomplished by the audit. In contrast, a control objective refers to how an internal control should function. An audit generally incorporates several audit objectives.

Audit objectives often focus on substantiating that internal controls exist to minimize business risk and that they function as expected. These audit objectives include assuring compliance with legal and regulatory requirements as well as the confidentiality, integrity, reliability and availability of information and IT resources. Audit management may give the IS auditor a general control objective to review and evaluate when performing an audit.

A key element in planning an IS audit is to translate basic and wide-ranging audit objectives into specific IS audit objectives. For example, in a financial/operational audit, a control objective could be to ensure that transactions are properly posted to the general ledger accounts. However, in the IS audit, the objective could be extended to ensure that editing features are in place to detect errors in the coding of transactions that may impact the account-posting activities.

The IS auditor must have an understanding of how general audit objectives can be translated into specific IS control objectives. Determining an audit's objectives is a critical step in planning an IS audit.

One of the basic purposes of any IS audit is to identify control objectives and the related controls that address the objective. For example, the IS auditor's initial review of an information system should identify key controls. The IS auditor should then decide whether to test these controls for compliance. The IS auditor should identify both key general and application controls after developing an understanding and documenting the business processes and the applications/functions that support these processes and general support systems. Based on that understanding, the IS auditor should identify the key control points.

Alternatively, an IS auditor may assist in assessing the integrity of financial reporting data, referred to as substantive testing, through computer-assisted audit techniques (CAATs).

1.5.2 TYPES OF AUDITS

The IS auditor should understand the various types of audits that can be performed, internally or externally, and the audit procedures associated with each:

- **Compliance audits**—Compliance audits include specific tests of controls to demonstrate adherence to specific regulatory or industry standards. These audits often overlap traditional audits but may focus on particular systems or data. Examples include Payment Card Industry Data Security Standard (PCI DSS) audits for companies that process credit card data and Health Insurance Portability and Accountability Act (HIPAA) audits for companies that handle health care data.
- **Financial audits**—The purpose of a financial audit is to assess the accuracy of financial reporting. A financial audit will often involve detailed, substantive testing, although increasingly, auditors are placing more emphasis on a risk- and control-based audit approach. This kind of audit relates to financial information integrity and reliability.
- **Operational audits**—An operational audit is designed to evaluate the internal control structure in a given process or area. IS audits of application controls or logical security systems are some examples of operational audits.
- **Integrated audits**—An integrated audit combines financial and operational audit steps. An integrated audit is also performed to assess the overall objectives within an organization, related to financial information and assets' safeguarding, efficiency and compliance. An integrated audit can be performed by external or internal auditors and would include compliance tests of internal controls and substantive audit steps.
- **Administrative audits**—These are oriented to assess issues related to the efficiency of operational productivity within an organization.
- **IS audits**—This process collects and evaluates evidence to determine whether the information systems and related resources adequately safeguard assets, maintain data and system integrity and availability, provide relevant and reliable information, achieve organizational goals effectively, consume resources efficiently and have, in effect, internal controls that provide reasonable assurance that business, operational and control objectives will be met and that undesired events will be prevented, or detected and corrected, in a timely manner.
- **Specialized audits**—Within the category of IS audits, a number of specialized reviews examine areas such as services performed by third parties. Because businesses are becoming increasingly reliant on third-party service providers, it is important that internal controls be evaluated in these environments. The Statement on Standards for Attestation Engagements 16 (SSAE 16), titled, "Reporting on Controls at a Service Organization," is a widely known auditing standard developed by the American Institute of Certified Public Accountants (AICPA). This standard replaced the previous standard, Statement on Auditing Standards 70 (SAS 70), titled "Reports on the Processing of Transactions by Service Organizations." This standard defines the professional standards used by a service auditor to assess the internal controls of a service organization. This type of audit has become increasingly relevant due to the current trend of outsourcing of financial and business processes to third-party service providers, which, in some cases, may operate in different jurisdictions or even different countries. It should be noted that a Type 2 SSAE 16 review is a more thorough variation of a regular SSAE 16 review, which is often required in connection with regulatory reviews. Many other countries have their own equivalent of this standard. An SSAE 16–type audit is important because it represents that a service organization has been through an in-depth audit of their control activities, which generally include controls over information technology and related

processes. SSAE 16–type reviews provide guidance to enable an independent auditor (service auditor) to issue an opinion on a service organization's description of controls through a service auditor's report, which then can be relied on by the IS auditor of the entity that utilizes the services of the service organization.

- **Forensic audits**—Forensic auditing has been defined as auditing specialized in discovering, disclosing and following up on fraud and crimes. The primary purpose of such a review is the development of evidence for review by law enforcement and judicial authorities. Forensic professionals have been called on to participate in investigations related to corporate fraud and cybercrime. In cases where computer resources may have been misused, further investigation is necessary to gather evidence for possible criminal activity that can then be reported to appropriate authorities. A computer forensic investigation includes the analysis of electronic devices such as computers, smartphones, disks, switches, routers, hubs and other electronic equipment. An IS auditor possessing the necessary skills can assist the information security manager in performing forensic investigations and conduct the audit of the systems to ensure compliance with the evidence collection procedures for forensic investigation. Electronic evidence is vulnerable to changes; therefore, it is necessary to handle electronic evidence with utmost care and controls should ensure that no manipulation can occur. Chain of custody for electronic evidence should be established to meet legal requirements.

  Improperly handled computer evidence is subject to being ruled inadmissible by judicial authorities. The most important consideration for a forensic auditor is to make a bit-stream image of the target drive and examine that image without altering date stamps or other information attributable to the examined files. Further, forensic audit tools and techniques such as data mapping for security and privacy risk assessment, and the search for intellectual property for data protection, are also being used for prevention, compliance and assurance.

1.5.3 AUDIT METHODOLOGY

An audit methodology is a set of documented audit procedures designed to achieve planned audit objectives. Its components are a statement of scope, audit objectives and audit programs.

The audit methodology should be set up and approved by audit management to achieve consistency in the audit approach. This methodology should be formalized and communicated to all audit staff.

Figure 1.7 lists the phases of a typical audit. An early and critical product of the audit process should be an audit program that is the guide for performing and documenting all of the audit steps and the extent and types of evidential matter reviewed.

Although an audit program does not necessarily follow a specific set of steps, the IS auditor typically would follow, as a minimum course of action, sequential program steps to gain an understanding of the entity under audit, evaluate the control structure and test the controls.

Figure 1.7—Audit Phases

Audit Phase	Description
Audit subject	• Identify the area to be audited.
Audit objective	• Identify the purpose of the audit. For example, an objective might be to determine whether program source code changes occur in a well-defined and controlled environment.
Audit scope	• Identify the specific systems, function or unit of the organization to be included in the review. For example, in the previous program changes example, the scope statement might limit the review to a single application system or to a limited period of time.
Preaudit planning	• Identify technical skills and resources needed. • Identify the sources of information for test or review such as functional flow charts, policies, standards, procedures and prior audit work papers. • Identify locations or facilities to be audited. • Develop a communication plan at the beginning of each engagement that describes who to communicate to, when, how often and for what purpose(s).
Audit procedures and steps for data gathering	• Identify and select the audit approach to verify and test the controls. • Identify a list of individuals to interview. • Identify and obtain departmental policies, standards and guidelines for review. • Develop audit tools and methodology to test and verify control.
Procedures for evaluating the test or review results	• Identify methods (including tools) to perform the evaluation. • Identify criteria for evaluating the test (similar to a test script for the auditor to use in conducting the evaluation). • Identify means and resources to confirm the evaluation was accurate (and repeatable, if applicable).
Procedures for communication with management	• Determine frequency of communication. • Prepare documentation for final report.
Audit report preparation	• Disclose follow-up review procedures. • Disclose procedures to evaluate/test operational efficiency and effectiveness. • Disclose procedures to test controls. • Review and evaluate the soundness of documents, policies and procedures.

Each audit department should design and approve an audit methodology as well as the minimum steps to be observed in any audit assignment.

All audit plans, programs, activities, tests, findings and incidents should be properly documented in work papers.

The format and media of work papers can vary depending on specific needs of the department. IS auditors should particularly consider how to maintain the integrity and protection of audit test evidence in order to preserve their value as substantiation in support of audit results.

Work papers can be considered the bridge or interface between the audit objectives and the final report. Work papers should provide a seamless transition—with traceability and support for the work performed—from objectives to report and from report to objectives. In this context, the audit report can be viewed as a particular work paper.

1.5.4 RISK-BASED AUDITING

Effective risk-based auditing is driven by two processes:

1. The risk assessment that drives the audit schedule (see section 1.5.6 Risk Assessment and Treatment)
2. The risk assessment that minimizes the audit risk during the execution of an audit (see section 1.5.5 Audit Risk and Materiality)

A risk-based audit approach is usually adapted to develop and improve the continuous audit process. This approach is used to assess risk and to assist an IS auditor in making the decision to perform either compliance testing or substantive testing. It is important to stress that the risk-based audit approach efficiently assists the auditor in determining the nature and extent of testing.

Within this concept, inherent risk, control risk or detection risk should not be of major concern, despite some weaknesses. In a risk-based audit approach, IS auditors are not just relying on risk; they also are relying on internal and operational controls as well as knowledge of the company or the business. This type of risk assessment decision can help relate the cost-benefit analysis of the control to the known risk, allowing practical choices.

Business risk includes concerns about the probable effects of an uncertain event on achieving established business objectives. The nature of business risk may be financial, regulatory or operational and may also include risk derived from specific technology. For example, an airline company is subject to extensive safety regulations and economic changes, both of which impact the continuing operations of the company. In this context, the availability of IT service and its reliability are critical.

By understanding the nature of the business, IS auditors can identify and categorize the types of risk that will better determine the risk model or approach in conducting the audit. The risk model assessment can be as simple as creating weights for the types of risk associated with the business and identifying the risk in an equation. On the other hand, risk assessment can be a scheme where risk has been given elaborate weights based on the nature of the business or the significance of the risk. A simplistic overview of a risk-based audit approach can be seen in **figure 1.8**.

1.5.5 AUDIT RISK AND MATERIALITY

Audit risk can be defined as the risk that information may contain a material error that may go undetected during the course of the audit. The IS auditor should also take into account, if applicable, other factors relevant to the organization: customer data, privacy, availability of provided services as well as corporate and public image as in the case of public organizations or foundations.

Figure 1.8—Risk-based Audit Approach

Gather Information and Plan

- Knowledge of business and industry
- Prior year's audit results
- Recent financial information
- Regulatory statutes
- Inherent risk assessments

Obtain Understanding of Internal Control

- Control environment
- Control procedures
- Detection risk assessment
- Control risk assessment
- Equate total risk

Perform Compliance Tests

- Identify key controls to be tested.
- Perform tests on reliability, risk prevention and adherence to organization policies and procedures.

Perform Substantive Tests

- Analytical procedures
- Detailed tests of account balances
- Other substantive audit procedures

Conclude the Audit

- Create recommendations.
- Write audit report.

Audit risk is influenced by:
- **Inherent risk**—As it relates to audit risk, it is the risk level or exposure of the process/entity to be audited without taking into account the controls that management has implemented. Inherent risk exists independent of an audit and can occur because of the nature of the business.
- **Control risk**—The risk that a material error exists that would not be prevented or detected on a timely basis by the system of internal controls. For example, the control risk associated with manual reviews of computer logs can be high because activities requiring investigation are often easily missed due to the volume of logged information. The control risk associated with computerized data validation procedures is ordinarily low if the processes are consistently applied.
- **Detection risk**—The risk that material errors or misstatements that have occurred will not be detected by the IS auditor.
- **Overall audit risk**—The probability that information or financial reports may contain material errors and that the auditor may not detect an error that has occurred. An objective in formulating the audit approach is to limit the audit risk in the area under scrutiny so the overall audit risk is at a sufficiently low level at the completion of the examination.

Note: Audit risk should not be confused with statistical sampling risk, which is the risk that incorrect assumptions are made about the characteristics of a population from which a sample is selected.

Specifically, this means that an internal control weakness or set of combined internal control weaknesses leaves the organization highly susceptible to the occurrence of a threat (e.g., financial loss, business interruption, loss of customer trust, economic sanction, etc.). The IS auditor should be concerned with assessing the materiality of the items in question through a risk-based audit approach to evaluating internal controls.

The IS auditor should have a good understanding of audit risk when planning an audit. An audit sample may not detect every potential error in a population. However, by using proper statistical sampling procedures or a strong quality control process, the probability of detection risk can be reduced to an acceptable level.

Similarly, when evaluating internal controls, the IS auditor should realize that a given system may not detect a minor error. However, that specific error, combined with others, could become material to the overall system.

The concept of materiality requires sound judgment from the IS auditor. The IS auditor may detect a small error that could be considered significant at an operational level, but may not be viewed as significant to upper management. Materiality considerations combined with an understanding of audit risk are essential concepts for planning the areas to be audited and the specific test to be performed in a given audit.

1.5.6 RISK ASSESSMENT AND TREATMENT

Assessing Risk

To develop a more complete understanding of audit risk, the IS auditor should also understand how the organization being audited approaches risk assessment and treatment.

Risk assessments should identify, quantify and prioritize risk against criteria for risk acceptance and objectives relevant to the organization. The results should guide and determine the appropriate management action, priorities for managing information security risk and priorities for implementing controls selected to protect against risk.

Risk assessments should also be performed periodically to address changes in the environment, security requirements and in the risk situation (e.g., in the assets, threats, vulnerabilities, impacts) and when significant changes occur. These risk assessments should be undertaken in a methodical manner capable of producing comparable and reproducible results.

The scope of a risk assessment can be either the entire organization, parts of the organization, an individual information system, specific system components or services where this is practicable, realistic and helpful.

Treating Risk

Before considering the treatment of risk, the organization should decide the criteria for determining whether risk can be managed within the risk appetite. Risk may be accepted if, for example, it is assessed that the risk is low or that the cost of treatment is not cost-effective for the organization. Such decisions should be recorded.

Risk identified in the risk assessment needs to be treated. Possible risk response options include:
- **Risk mitigation**—Applying appropriate controls to reduce the risk
- **Risk acceptance**—Knowingly and objectively not taking action, providing the risk clearly satisfies the organization's policy and criteria for risk acceptance
- **Risk avoidance**—Avoiding risk by not allowing actions that would cause the risk to occur
- **Risk transfer/sharing**—Transferring the associated risk to other parties (e.g., insurers or suppliers)

For risk where the risk treatment decision has been to apply appropriate controls, controls should be selected to ensure that risk is reduced to an acceptable level, taking into account:
- Requirements and constraints of national and international legislation and regulations
- Organizational objectives
- Operational requirements and constraints
- Cost-effectiveness (the need to balance the investment in implementation and operation of controls against the harm likely to result from security failures)

Controls can be selected from professional or industry standards, or new controls can be designed to meet the specific needs of the organization. It is necessary to recognize that some controls may not be applicable to every information system or environment and might not be practical for all organizations.

Information security controls should be considered at the systems and project requirements specification and design stage. Failure to do so can result in additional costs and less effective solutions and, in a worst case scenario, the inability to achieve adequate security.

No set of controls can achieve complete security. Additional management action should be implemented to monitor, evaluate and improve the efficiency and effectiveness of security controls to support the organization's aims.

1.5.7 IS AUDIT RISK ASSESSMENT TECHNIQUES

When determining which functional areas should be audited, the IS auditor could face a large variety of audit subjects. Each of these subjects may represent different types of risk. The IS auditor should evaluate these various risk candidates to determine the high-risk areas that should be audited.

There are many risk assessment methodologies, computerized and noncomputerized, from which the IS auditor may choose. These range from simple classifications based on the IS auditor's judgment of high, medium and low to complex scientific calculations that provide a numeric risk rating.

One such risk assessment approach is a scoring system that is useful in prioritizing audits based on an evaluation of risk factors. The system considers variables such as technical complexity, level of control procedures in place and level of financial loss. These variables may or may not be weighted. The risk values are then compared to each other, and audits are scheduled accordingly. Another form of risk assessment is judgmental, where an independent decision is made based on business knowledge, executive management directives, historical perspectives, business goals and environmental factors. A combination of techniques may be used as well. Risk assessment methods may change and develop over time to best serve the needs of the organization. The IS auditor should consider the level of complexity and detail appropriate for the organization being audited.

Using risk assessment to determine areas to be audited:
- Enables management to effectively allocate limited audit resources
- Ensures that relevant information has been obtained from all levels of management, including boards of directors, IS auditors and functional area management. Generally, this information assists management in effectively discharging its responsibilities and ensures that the audit activities are directed to high-risk areas, which will add value for management.
- Establishes a basis for effectively managing the audit department
- Provides a summary of how the individual audit subject is related to the overall organization as well as to the business plans

1.5.8 AUDIT PROGRAMS

An audit program is a step-by-step set of audit procedures and instructions that should be performed to complete an audit. Audit programs for financial, operational, integrated, administrative and IS audits are based on the scope and objective of the particular assignment. IS auditors often evaluate IT functions and systems from different perspectives such as security (confidentiality, integrity and availability), quality (effectiveness, efficiency), fiduciary (compliance, reliability), service and capacity. The audit work program is the audit strategy and plan—it identifies scope, audit objectives and audit procedures to obtain sufficient, relevant and reliable evidence to draw and support audit conclusions and opinions.

General audit procedures are the basic steps in the performance of an audit and usually include:
- Obtaining and recording an understanding of the audit area/subject
- A risk assessment and general audit plan and schedule
- Detailed audit planning that would include the necessary audit steps and a breakdown of the work planned across an anticipated time line
- Preliminary review of the audit area/subject
- Evaluating the audit area/subject
- Verifying and evaluating the appropriateness of controls designed to meet control objectives
- Compliance testing (tests of the implementation of controls and their consistent application)
- Substantive testing (confirming the accuracy of information)
- Reporting (communicating results)
- Follow-up in cases where there is an internal audit function

The IS auditor must understand the procedures for testing and evaluating IS controls. These procedures could include:
- The use of generalized audit software to survey the contents of data files (including system logs)
- The use of specialized software to assess the contents of OS database and application parameter files (or detect deficiencies in system parameter settings)
- Flow-charting techniques for documenting automated applications and business processes
- The use of audit logs/reports available in operation/application systems
- Documentation review
- Inquiry and observation
- Walk-throughs
- Reperformance of controls

The IS auditor should have a sufficient understanding of these procedures to allow for the planning of appropriate audit tests.

Note: For audit program examples, visit *www.isaca.org/auditprograms*

1.5.9 FRAUD DETECTION

The use of information technology for business has immensely benefited enterprises in terms of significantly increased quality of delivery of information. However, the widespread use of information technology and the Internet leads to risk that enables the perpetration of errors and fraud.

Management is primarily responsible for establishing, implementing and maintaining a framework and design of IT controls to meet the control objectives. A well-designed internal control system provides good opportunities for deterrence and/or timely detection of fraud. Internal controls may fail where such controls are circumvented by exploiting vulnerabilities or through management perpetrated weakness in controls or collusion among people.

Legislation and regulations relating to corporate governance cast significant responsibilities on management, auditors and the audit committee regarding detection and disclosure of any fraud, whether material or not.

IS auditors should observe and exercise due professional care (1005 Due Professional Care) in all aspects of their work. IS auditors entrusted with assurance functions should ensure reasonable care while performing their work and be alert to the possible opportunities that allow fraud to materialize.

The presence of internal controls does not altogether eliminate fraud. IS auditors should be aware of the possibility and means of perpetrating fraud, especially by exploiting the vulnerabilities and overriding controls in the IT-enabled environment. IS auditors should have knowledge of fraud and fraud indicators, and be alert to the possibility of fraud and errors while performing an audit.

During the course of regular assurance work, the IS auditor may come across instances or indicators of fraud. After careful evaluation, the IS auditor may communicate the need for a detailed investigation to appropriate authorities. In the case of the auditor identifying a major fraud or if the risk associated with the detection is high, audit management should also consider communicating in a timely manner to the audit committee.

Regarding fraud prevention, the IS auditor should be aware of potential legal requirements concerning the implementation of specific fraud detection procedures and reporting fraud to appropriate authorities.

1.5.10 COMPLIANCE VERSUS SUBSTANTIVE TESTING

Compliance testing is evidence gathering for the purpose of testing an organization's compliance with control procedures. This differs from substantive testing in which evidence is gathered to evaluate the integrity of individual transactions, data or other information.

A compliance test determines whether controls are being applied in a manner that complies with management policies and procedures. For example, if the IS auditor is concerned about whether production program library controls are working properly, the IS auditor might select a sample of programs to determine whether the source and object versions are the same. The broad objective of any compliance test is to provide IS auditors with reasonable assurance that the particular control on which the IS auditor plans to rely is operating as the IS auditor perceived in the preliminary evaluation.

It is important that the IS auditor understands the specific objective of a compliance test and of the control being tested. Compliance tests can be used to test the existence and effectiveness of a defined process, which may include a trail of documentary and/or automated evidence—for example, to provide assurance that only authorized modifications are made to production programs.

A substantive test substantiates the integrity of actual processing. It provides evidence of the validity and integrity of the balances in the financial statements and the transactions that support these balances. IS auditors could use substantive tests to test for monetary errors directly affecting financial statement balances or other relevant data of the organization. Additionally, an IS auditor might develop a substantive test to determine whether the tape library inventory records are stated correctly. To perform this test, the IS auditor might take a thorough inventory or might use a statistical sample, which will allow the IS auditor to develop a conclusion regarding the accuracy of the entire inventory.

There is a direct correlation between the level of internal controls and the amount of substantive testing required. If the results of testing controls (compliance tests) reveal the presence of adequate internal controls, then the IS auditor is justified in minimizing the substantive procedures. Conversely, if the control testing reveals weaknesses in controls that may raise doubts about the completeness, accuracy or validity of the accounts, substantive testing can alleviate those doubts.

Examples of compliance testing of controls where sampling could be considered include user access rights, program change control procedures, documentation procedures, program documentation, follow-up of exceptions, review of logs, software license audits, etc.

Examples of substantive tests where sampling could be considered include performance of a complex calculation (e.g., interest) on a sample of accounts or a sample of transactions to vouch for supporting documentation, etc.

The IS auditor could also decide during the preliminary assessment of the controls to include some substantive testing if the results of this preliminary evaluation indicate that implemented controls are not reliable or do not exist.

Figure 1.9 shows the relationship between compliance and substantive tests and describes the two categories of substantive tests.

> **Note:** The IS auditor should be knowledgeable on when to perform compliance tests or substantive tests.

1.5.11 EVIDENCE

Evidence is any information used by the IS auditor to determine whether the entity or data being audited follows the established criteria or objectives and supports audit conclusions. It is a requirement that the auditor's conclusions be based on sufficient, relevant and competent evidence. When planning the IS audit, the IS auditor should take into account the type of audit evidence to be gathered, its use as audit evidence to meet audit objectives and its varying levels of reliability.

Audit evidence may include:
- The IS auditor's observations (presented to management)
- Notes taken from interviews
- Results of independent confirmations obtained by the IS auditor from different stakeholders
- Material extracted from correspondence and internal documentation or contracts with external partners
- The results of audit test procedures

While all evidence will assist the IS auditor in developing audit conclusions, some types of evidence are more reliable than others. The rules of evidence and sufficiency as well as the competency of evidence must be taken into account as required by audit standards.

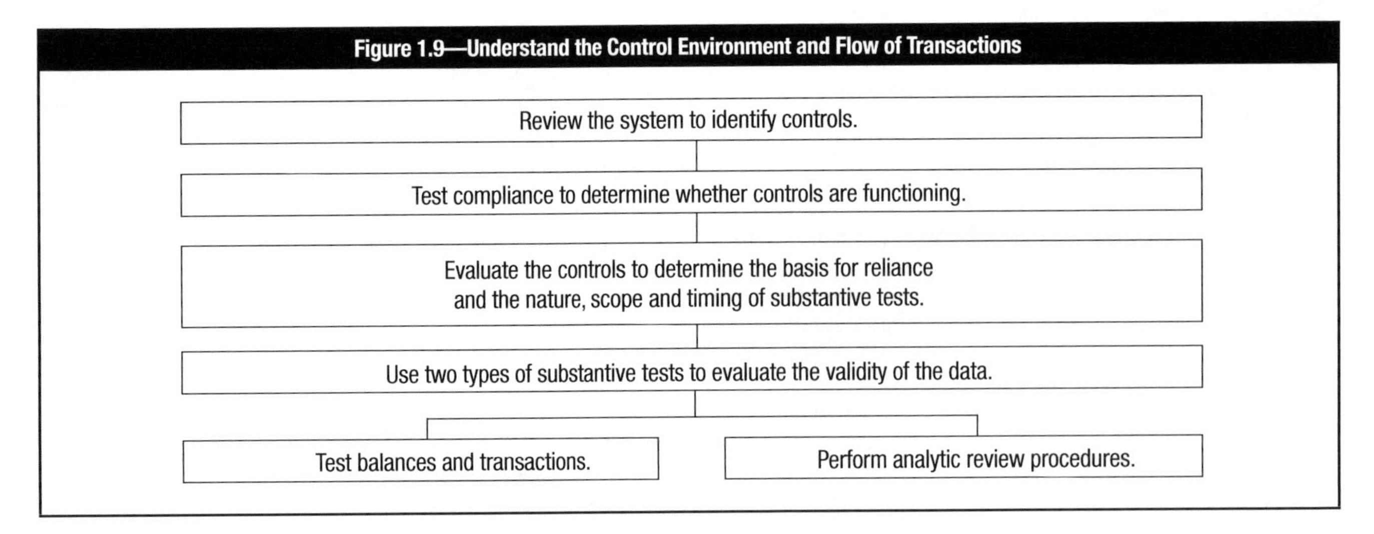
Figure 1.9—Understand the Control Environment and Flow of Transactions
Review the system to identify controls.
Test compliance to determine whether controls are functioning.
Evaluate the controls to determine the basis for reliance and the nature, scope and timing of substantive tests.
Use two types of substantive tests to evaluate the validity of the data.
Test balances and transactions.
Perform analytic review procedures.

Determinants for evaluating the reliability of audit evidence include:

- **Independence of the provider of the evidence**—Evidence obtained from outside sources is more reliable than from within the organization. This is why confirmation letters are used for verification of accounts receivable balances. Additionally, signed contracts or agreements with external parties could be considered reliable if the original documents are made available for review.
- **Qualifications of the individual providing the information/evidence**—Whether the providers of the information/evidence are inside or outside of the organization, the IS auditor should always consider the qualifications and functional responsibilities of the persons providing the information. This can also be true of the IS auditor. If an IS auditor does not have a good understanding of the technical area under review, the information gathered from testing that area may not be reliable, especially if the IS auditor does not fully understand the test.
- **Objectivity of the evidence**—Objective evidence is more reliable than evidence that requires considerable judgment or interpretation. An IS auditor's review of media inventory is direct, objective evidence. An IS auditor's analysis of the efficiency of an application, based on discussions with certain personnel, may not be objective audit evidence.
- **Timing of the evidence**—The IS auditor should consider the time during which information exists or is available in determining the nature, timing and extent of compliance testing and, if applicable, substantive testing. For example, audit evidence processed by dynamic systems, such as spreadsheets, may not be retrievable after a specified period of time if changes to the files are not controlled or the files are not backed up.

The IS auditor gathers a variety of evidence during the audit. Some evidence may be relevant to the objectives of the audit, while other evidence may be considered peripheral. The IS auditor should focus on the overall objectives of the review and not the nature of the evidence gathered.

The quality and quantity of evidence must be assessed by the IS auditor. These two characteristics are referred to by the International Federation of Accountants (IFAC) as competent (quality) and sufficient (quantity). Evidence is competent when it is both valid and relevant. Audit judgment is used to determine when sufficiency is achieved in the same manner that is used to determine the competency of evidence.

An understanding of the rules of evidence is important for IS auditors because they may encounter a variety of evidence types.

Gathering of evidence is a key step in the audit process. The IS auditor should be aware of the various forms of audit evidence and how evidence can be gathered and reviewed. The IS auditor should understand ISACA IS Audit and Assurance Standard 1205 Evidence and should obtain evidence of a nature and sufficiency to support audit findings.

Note: A CISA candidate, given an audit scenario, should be able to determine which type of evidence gathering technique would be best.

The following are techniques for gathering evidence:

- **Reviewing IS organization structures**—An organizational structure that provides an adequate separation or segregation of duties is a key general control in an IS environment. The IS auditor should understand general organizational controls and be able to evaluate these controls in the organization under audit. Where there is a strong emphasis on cooperative distributed processing or on end-user computing, IT functions may be organized somewhat differently than the classic IS organization, which consists of separate systems and operations functions. The IS auditor should be able to review these organizational structures and assess the level of control they provide.
- **Reviewing IS policies and procedures**—An IS auditor should review whether appropriate policies and procedures are in place, determine whether personnel understand the implemented policies and procedures, and ensure that policies and procedures are being followed. The IS auditor should verify that management assumes full responsibility for formulating, developing, documenting, promulgating and controlling policies covering general aims and directives. Periodic reviews of policies and procedures for appropriateness should be carried out.
- **Reviewing IS standards**—The IS auditor should first understand the existing standards in place within the organization.

- **Reviewing IS documentation**—A first step in reviewing the documentation for an information system is to understand the existing documentation in place within the organization. This documentation could be a hard copy or stored electronically. If the latter is the case, controls to preserve the document integrity should be evaluated by the IS auditor. The IS auditor should look for a minimum level of IS documentation. Documentation may include:
  - Systems development initiating documents (e.g., feasibility study)
  - Documentation provided by external application suppliers
  - SLAs with external IT providers
  - Functional requirements and design specifications
  - Tests plans and reports
  - Program and operations documents
  - Program change logs and histories
  - User manuals
  - Operations manuals
  - Security-related documents (e.g., security plans, risk assessments)
  - BCPs
  - QA reports
  - Reports on security metrics
- **Interviewing appropriate personnel**—Interviewing techniques are an important skill for the IS auditor. Interviews should be organized in advance with objectives clearly communicated, follow a fixed outline and be documented by interview notes. An interview form or checklist prepared by an IS auditor is a good approach. The IS auditor should always remember that the purpose of such an interview is to gather audit evidence. Procedures to gather audit evidence include inquiry, observation, inspection, confirmation, performance and monitoring. Personnel interviews are discovery in nature and should never be accusatory; the interviewer should help people feel comfortable, encouraging them to share information, ideas, concerns and knowledge. The IS auditor should verify the accuracy of the notes with the interviewee whether or not these notes would be necessary to support conclusions.
- **Observing processes and employee performance**—The observation of processes is a key audit technique for many types of review. The IS auditor should be unobtrusive while making observations and should document everything in sufficient detail to be able to present it, if required, as audit evidence at a later date. In some situations, the release of the audit report may not be timely enough to use this observation as evidence. This may necessitate the issuance of an interim report to management of the area being audited. The IS auditor may also wish to consider whether documentary evidence would be useful as evidence (e.g., photograph of a server room with doors fully opened).
- **Reperformance**—The reperformance process is a key audit technique that generally provides better evidence than the other techniques and is therefore used when a combination of inquiry, observation and examination of evidence does not provide sufficient assurance that a control is operating effectively.
- **Walk-throughs**—The walk-through is an audit technique to confirm the understanding of controls.

All of these techniques for gathering evidence are part of an audit, but an audit is not considered only review work. An audit includes examination, which incorporates by necessity the testing of controls and audit evidence, and therefore, includes the results of audit tests.

IS auditors should recognize that with systems development techniques such as computer-aided software engineering (CASE) or prototyping, traditional systems documentation will not be required or will be in an automated form rather than on paper. However, the IS auditor should look for documentation standards and practices within the IS organization.

The IS auditor should be able to review documentation for a given system and determine whether it follows the organization's documentation standards. In addition, the IS auditor should understand the current approaches to developing systems such as object orientation, CASE tools or prototyping, and how the documentation is constructed. The IS auditor should recognize other components of IS documentation such as database specifications, file layouts or self-documented program listings.

1.5.12 INTERVIEWING AND OBSERVING PERSONNEL IN PERFORMANCE OF THEIR DUTIES

Observing personnel in the performance of their duties assists an IS auditor in identifying:

- **Actual functions**—Observation could be an adequate test to ensure that the individual who is assigned and authorized to perform a particular function is the person who is actually doing the job. It allows the IS auditor an opportunity to witness how policies and procedures are understood and practiced. Depending on the specific situation, the results of this type of test should be compared with the respective logical access rights.
- **Actual processes/procedures**—Performing a walk-through of the process/procedure allows the IS auditor to gain evidence of compliance and observe deviations, if any. This type of observation could prove to be useful for physical controls.
- **Security awareness**—Security awareness should be observed to verify an individual's understanding and practice of good preventive and detective security measures to safeguard the company's assets and data. This type of information could be complemented with an examination of previous and planned security training.
- **Reporting relationships**—Reporting relationships should be observed to ensure that assigned responsibilities and adequate segregation of duties are being practiced. Often, the results of this type of test should be compared with the respective logical access rights.
- **Observation drawbacks**—The observer may interfere with the observed environment. Personnel, upon noticing that they are being observed, may change their usual behavior. Interviewing information processing personnel and management should provide adequate assurance that the staff has the required technical skills to perform the job. This is an important factor that contributes to an effective and efficient operation.

1.5.13 SAMPLING

Sampling is used when time and cost considerations preclude a total verification of all transactions or events in a predefined population. The population consists of the entire group of items that need to be examined. The subset of population members used to perform testing is called a sample. Sampling is used to infer characteristics about a population based on the characteristics of a sample.

> **Note:** Increasing regulation of organizations has led to a major focus on the IS auditor's ability to verify the adequacy of internal controls through the use of sampling techniques. This has become necessary because many controls are transactional in nature, which can make it difficult to test the entire population. Although a candidate is not expected to become a sampling expert, it is important for the candidate to have a foundational understanding of the general principles of sampling and how to design a sample that is reliable.

The two general approaches to audit sampling are statistical and nonstatistical:

- **Statistical sampling**—An objective method of determining the sample size and selection criteria
  - Statistical sampling uses the mathematical laws of probability to: (1) calculate the sampling size, (2) select the sample items, and (3) evaluate the sample results and make the inference.
  - With statistical sampling, the IS auditor quantitatively decides how closely the sample should represent the population (assessing sample precision) and the number of times in 100 that the sample should represent the population (the reliability or confidence level). This assessment will be represented as a percentage. The results of a valid statistical sample are mathematically quantifiable.
- **Nonstatistical sampling (often referred to as judgmental sampling)**—Uses auditor judgment to determine the method of sampling, the number of items that will be examined from a population (sample size) and which items to select (sample selection)
  - These decisions are based on subjective judgment as to which items/transactions are the most material and most risky.

When using either statistical or nonstatistical sampling methods, the IS auditor should design and select an audit sample; perform audit procedures; and evaluate sample results to obtain sufficient, reliable, relevant and useful audit evidence. These methods of sampling require the IS auditor to use judgment when defining the population characteristics and, thus, are subject to the risk that the IS auditor will draw the wrong conclusion from the sample (sampling risk). However, statistical sampling permits the IS auditor to quantify the probability of error (confidence coefficient). To be a statistical sample, each item in the population should have an equal opportunity or probability of being selected. Within these two general approaches to audit sampling, there are two primary methods of sampling used by IS auditors—attribute sampling and variable sampling. Attribute sampling, generally applied in compliance testing situations, deals with the presence or absence of the attribute and provides conclusions that are expressed in rates of incidence. Variable sampling, generally applied in substantive testing situations, deals with population characteristics that vary, such as monetary values and weights (or any other measurement), and provides conclusions related to deviations from the norm.

Attribute sampling refers to three different but related types of proportional sampling:

- **Attribute sampling (also referred to as fixed sample-size attribute sampling or frequency-estimating sampling)**—A sampling model that is used to estimate the rate (percent) of occurrence of a specific quality (attribute) in a population. Attribute sampling answers the question of "how many?" An example of an attribute that might be tested is approval signatures on computer access request forms.
- **Stop-or-go sampling**—A sampling model that helps prevent excessive sampling of an attribute by allowing an audit test to be stopped at the earliest possible moment. Stop-or-go sampling is used when the IS auditor believes that relatively few errors will be found in a population.
- **Discovery sampling**—A sampling model that can be used when the expected occurrence rate is extremely low. Discovery sampling is most often used when the objective of the audit is to seek out (discover) fraud, circumvention of regulations or other irregularities.

Variable sampling—also known as dollar estimation or mean estimation sampling—is a technique used to estimate the monetary value or some other unit of measure (such as weight) of a population from a sample portion. An example of variable sampling is a review of an organization's balance sheet for material transactions and an application review of the program that produced the balance sheet.

Variable sampling refers to a number of different types of quantitative sampling models:

- **Stratified mean per unit**—A statistical model in which the population is divided into groups and samples are drawn from the various groups; used to produce a smaller overall sample size relative to unstratified mean per unit
- **Unstratified mean per unit**—A statistical model in which a sample mean is calculated and projected as an estimated total
- **Difference estimation**—A statistical model used to estimate the total difference between audited values and book (unaudited) values based on differences obtained from sample observations

To perform attribute or variable sampling, the following statistical sampling terms need to be understood:

- **Confidence coefficient (also referred to as confidence level or reliability factor)**—A percentage expression (90 percent, 95 percent, 99 percent, etc.) of the probability that the characteristics of the sample are a true representation of the population. Generally, a 95 percent confidence coefficient is considered a high degree of comfort. If the IS auditor knows internal controls are strong, the confidence coefficient may be lowered. The greater the confidence coefficient, the larger the sample size.
- **Level of risk**—Equal to one minus the confidence coefficient. For example, if the confidence coefficient is 95 percent, the level of risk is five percent (100 percent minus 95 percent).
- **Precision**—Set by the IS auditor, it represents the acceptable range difference between the sample and the actual population.

For attribute sampling, this figure is stated as a percentage. For variable sampling, this figure is stated as a monetary amount or a number. The higher the precision amount, the smaller the sample size and the greater the risk of fairly large total error amounts going undetected. The smaller the precision amount, the greater the sample size. A very low precision level may lead to an unnecessarily large sample size.
- **Expected error rate**—An estimate stated as a percent of the errors that may exist. The greater the expected error rate, the greater the sample size. This figure is applied to attribute sampling formulas but not to variable sampling formulas.
- **Sample mean**—The sum of all sample values, divided by the size of the sample. The sample mean measures the average value of the sample.
- **Sample standard deviation**—Computes the variance of the sample values from the mean of the sample. Sample standard deviation measures the spread or dispersion of the sample values.
- **Tolerable error rate**—Describes the maximum misstatement or number of errors that can exist without an account being materially misstated. Tolerable rate is used for the planned upper limit of the precision range for compliance testing. The term is expressed as a percentage. Precision range and precision have the same meaning when used in substantive testing.
- **Population standard deviation**—A mathematical concept that measures the relationship to the normal distribution. The greater the standard deviation, the larger the sample size. This figure is applied to variable sampling formulas but not to attribute sampling formulas.

Key steps in the construction and selection of a sample for an audit test are seen in **figure 1.10**.

It is important to know that tools exist to analyze all of the data, not just those available through CAATs.

> **Note:** The IS auditor should be familiar with the different types of sampling techniques and when it is appropriate to use each of them.

1.5.14 USING THE SERVICES OF OTHER AUDITORS AND EXPERTS

Due to the scarcity of IS auditors and the need for IT security specialists and other subject matter experts to conduct audits of highly specialized areas, the audit department or auditors entrusted with providing assurance may require the services of other auditors or experts. Outsourcing of IS assurance and security services is increasingly becoming a common practice. External experts could include experts in specific technologies such as networking, automated teller machines, wireless, systems integration and digital forensics or subject matter experts such as specialists in a particular industry or area of specialization such as banking, securities trading, insurance, legal experts, etc.

When a part or all of IS audit services are proposed to be outsourced to another audit or external service provider, the following should be considered with regard to using the services of other auditors and experts:
- Restrictions on outsourcing of audit/security services provided by laws and regulations
- Audit charter or contractual stipulations
- Impact on overall and specific IS audit objectives
- Impact on IS audit risk and professional liability
- Independence and objectivity of other auditors and experts
- Professional competence, qualifications and experience
- Scope of work proposed to be outsourced and approach
- Supervisory and audit management controls
- Method and modalities of communication of results of audit work
- Compliance with legal and regulatory stipulations
- Compliance with applicable professional standards

Based on the nature of assignment, the following may also require special consideration:
- Testimonials/references and background checks
- Access to systems, premises and records
- Confidentiality restrictions to protect customer-related information
- Use of CAATs and other tools to be used by the external audit service provider
- Standards and methodologies for performance of work and documentation
- Nondisclosure agreements

The IS auditor or entity outsourcing the auditing services should monitor the relationship to ensure the objectivity and independence throughout the duration of the arrangement.

It is important to understand that although a part of or the whole of the audit work may be delegated to an external service provider, the related professional liability is not necessarily delegated. Therefore, it is the responsibility of the IS auditor or entity employing the services of external service providers to:
- Clearly communicate the audit objectives, scope and methodology through a formal engagement letter.
- Put in place a monitoring process for regular review of the work of the external service provider with regard to planning, supervision, review and documentation. For example, the work papers of other IS auditors or experts should be reviewed

Figure 1.10—Steps in the Selection of a Sample for an Audit Test

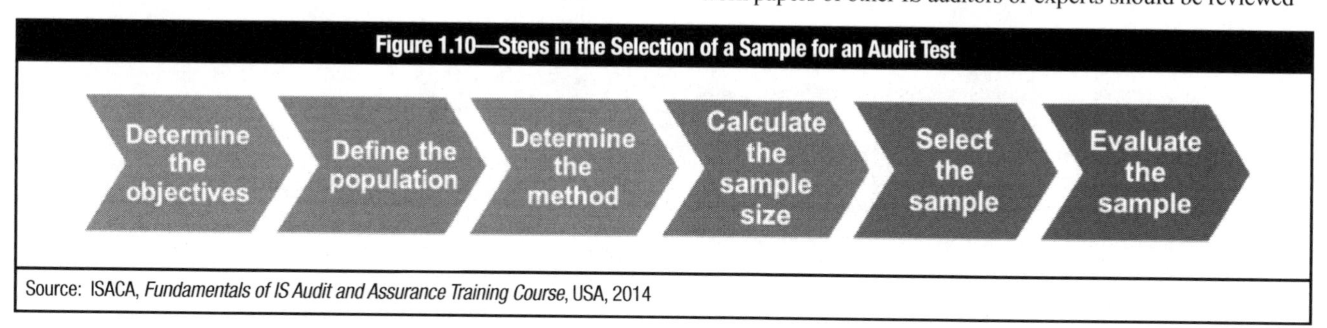

Source: ISACA, *Fundamentals of IS Audit and Assurance Training Course*, USA, 2014

to confirm the work was appropriately planned, supervised, documented and reviewed and to consider the appropriateness and sufficiency of the audit evidence provided. Another example is the reports of other IS auditors or experts should be reviewed to confirm the scope specified in the audit charter, terms of reference or letter of engagement has been met, that any significant assumptions used by other IS auditors or experts have been identified, and the findings and conclusions reported have been agreed on by management.
- Assess the usefulness and appropriateness of reports of such external providers, and assess the impact of significant findings on the overall audit objectives.

Note: The IS auditor should be familiar with ISACA Audit and Assurance Standard 1203 Performance and Supervision and the IS Audit and Assurance Guideline 2206 Using the Work of Other Experts focusing on the rights of access to the work of other experts.

1.5.15 COMPUTER-ASSISTED AUDIT TECHNIQUES

During the course of an audit, the IS auditor is to obtain sufficient, relevant and useful evidence to effectively achieve the audit objectives. The audit findings and conclusions should be supported by appropriate analysis and interpretation of the evidence. Current information processing environments pose a significant challenge to the IS auditor to collect sufficient, relevant and useful evidence because the evidence may only exist in electronic form.

CAATs are important tools for the IS auditor in gathering information from these environments. When systems have different hardware and software environments, data structures, record formats or processing functions, it is almost impossible for the auditors to collect certain evidence without a software tool to collect and analyze the records.

CAATs also enable IS auditors to gather information independently. CAATs provide a means to gain access and analyze data for a predetermined audit objective and to report the audit findings with emphasis on the reliability of the records produced and maintained in the system. The reliability of the source of the information used provides reassurance on findings generated.

CAATs include many types of tools and techniques such as generalized audit software (GAS), utility software, debugging and scanning software, test data, application software tracing and mapping, and expert systems.

GAS refers to standard software that has the capability to directly read and access data from various database platforms, flat-file systems and ASCII formats. GAS provides IS auditors an independent means to gain access to data for analysis and the ability to use high-level, problem-solving software to invoke functions to be performed on data files. Features include mathematical computations, stratification, statistical analysis, sequence checking, duplicate checking and recomputations. The following functions are commonly supported by GAS:
- **File access**—Enables the reading of different record formats and file structures
- **File reorganization**—Enables indexing, sorting, merging and linking with another file
- **Data selection**—Enables global filtration conditions and selection criteria
- **Statistical functions**—Enables sampling, stratification and frequency analysis
- **Arithmetical functions**—Enables arithmetic operators and functions

The effective and efficient use of software requires an understanding of its capabilities and limitations.

Utility software is a subset of software—such as report generators of the database management system—that provides evidence to auditors about system control effectiveness. Test data involve the auditors using a sample set of data to assess whether logic errors exist in a program and whether the program meets its objectives. The review of an application system will provide information about internal controls built in the system. The audit-expert system will give direction and valuable information to all levels of auditors while carrying out the audit because the query-based system is built on the knowledge base of the senior auditors or managers.

These tools and techniques can be used in performing various audit procedures:
- Tests of the details of transactions and balances
- Analytical review procedures
- Compliance tests of IS general controls
- Compliance tests of IS application controls
- Network and OS vulnerability assessments
- Penetration testing
- Application security testing and source code security scans

The IS auditor should have a thorough understanding of CAATs and know where and when to apply them. Please see ISACA Audit and Assurance Standard 2207 Irregularity and Illegal Acts, sections 2.2.4 and 2.5.3. Professionals should review the results of engagement procedures to determine whether there are indications that irregularities or illegal acts may have occurred. Using CAATs could aid significantly in the effective and efficient detection of irregularities or illegal acts.

An IS auditor should weigh the costs and benefits of CAATs before going through the effort, time and expense of purchasing or developing them. Issues to consider include:
- Ease of use, both for existing and future audit staff
- Training requirements
- Complexity of coding and maintenance
- Flexibility of uses
- Installation requirements
- Processing efficiencies (especially with a PC CAAT)
- Effort required to bring the source data into the CAATs for analysis
- Ensuring the integrity of imported data by safeguarding their authenticity
- Recording the time stamp of data downloaded at critical processing points to sustain the credibility of the review
- Obtaining permission to install the software on the auditee servers
- Reliability of the software
- Confidentiality of the data being processed

When developing CAATs, the following are examples of documentation to be retained:
- Online reports detailing high-risk issues for review
- Commented program listings
- Flowcharts
- Sample reports
- Record and file layouts
- Field definitions
- Operating instructions
- Description of applicable source documents

CAATs documentation should be referenced to the audit program and clearly identify the audit procedures and objectives being served. When requesting access to production data for use with CAATs, the IS auditor should request read-only access. Any data manipulation by the IS auditor should be applied to copies of production files in a controlled environment to ensure that production data are not exposed to unauthorized updating. Most of the CAATs provide for downloading production data from production systems to a stand-alone platform and then conducting analysis from the standalone platform, thereby insulating the production systems from any adverse impact.

CAATs as a Continuous Online Audit Approach

An increasingly important advantage of CAATs is the ability to improve audit efficiency through continuous online auditing techniques. To this end, IS auditors must develop audit techniques that are appropriate for use with advanced computerized systems. In addition, they must be involved in the creation of advanced systems at the early stages of development and implementation, and must make greater use of automated tools that are suitable for their organization's automated environment. This takes the form of the continuous audit approach. (For more detailed information on continuous online auditing, see chapter 3 Information Systems Acquisition, Development and Implementation.)

1.5.16 EVALUATION OF THE CONTROL ENVIORNMENT

The IS auditor will review evidence gathered during the audit to determine if the operations reviewed are well controlled and effective. This is also an area that requires the IS auditor's judgment and experience. The IS auditor should assess the strengths and weaknesses of the controls evaluated and then determine if they are effective in meeting the control objectives established as part of the audit planning process.

A control matrix is often utilized in assessing the proper level of controls. Known types of errors that can occur in the area under review are placed on the top axis and known controls to detect or correct errors are placed on the side axis. Then, using a ranking method, the matrix is filled with the appropriate measurements. When completed, the matrix will illustrate areas where controls are weak or lacking.

In some instances, one strong control may compensate for a weak control in another area. For example, if the IS auditor finds weaknesses in a system's transaction error report, the IS auditor may find that a detailed manual balancing process over all transactions compensates for the weaknesses in the error report. The IS auditor should be aware of compensating controls in areas where controls have been identified as weak.

While a compensating control situation occurs when one stronger control supports a weaker one, overlapping controls are two strong controls. For example, if a data center employs a card key system to control physical access and a guard inside the door requires employees to show their card key or badge, an overlapping control exists. Either control might be adequate to restrict access, but the two complement each other.

Normally, a control objective will not be achieved by considering one control adequate. Rather, the IS auditor will perform a variety of testing procedures and evaluate how these relate to one another. Generally a group of controls, when aggregated together, may act as compensating controls, and thereby minimize the risk. An IS auditor should always review for compensating controls prior to reporting a control weakness.

The IS auditor may not find each control procedure to be in place but should evaluate the comprehensiveness of controls by considering the strengths and weaknesses of control procedures.

Judging the Materiality of Findings

The concept of materiality is a key issue when deciding which findings to bring forward in an audit report. Key to determining the materiality of audit findings is the assessment of what would be significant to different levels of management. Assessment requires judging the potential effect of the finding if corrective action is not taken. A weakness in computer security physical access controls at a remote distributed computer site may be significant to management at the site but will not necessarily be material to upper management at headquarters. However, there may be other matters at the remote site that would be material to upper management.

The IS auditor must use judgment when deciding which findings to present to various levels of management. For example, the IS auditor may find that the transmittal form for delivering tapes to the offsite storage location is not properly initialed or authorization evidenced by management as required by procedures. If the IS auditor finds that management otherwise pays attention to this process and that there have been no problems in this area, the IS auditor may decide that the failure to initial transmittal documents is not material enough to bring to the attention of upper management. The IS auditor might decide to discuss this only with local operations management. However, there may be other control problems that will cause the IS auditor to conclude that this is a material error because it may lead to a larger control problem in other areas. The IS auditor should always judge which findings are material to various levels of management and report them accordingly.

1.6 COMMUNICATING AUDIT RESULTS

The exit interview, conducted at the end of the audit, provides the IS auditor with the opportunity to discuss findings and recommendations with management. During the exit interview, the IS auditor should:
- Ensure that the facts presented in the report are correct
- Ensure that the recommendations are realistic and cost-effective and, if not, seek alternatives through negotiation with auditee management
- Recommend implementation dates for agreed-on recommendations

The IS auditor will frequently be asked to present the results of audit work to various levels of management. The IS auditor should have a thorough understanding of the presentation techniques necessary to communicate these results.

Presentation techniques could include the following:
- **Executive summary**—An easy-to-read, concise report that presents findings to management in an understandable manner. Findings and recommendations should be communicated from a business perspective. Detailed attachments can be more technical in nature because operations management will require the detail to correct the reported situations.
- **Visual presentation**—May include slides or computer graphics

IS auditors should be aware that ultimately they are responsible to senior management and the audit committee of the board of directors. IS auditors should feel free to communicate issues or concerns to such management. An attempt to deny access by levels lower than senior management would limit the independence of the audit function.

Before communicating the results of an audit to senior management, the IS auditor should discuss the findings with the management staff of the audited entity. The goal of such a discussion would be to gain agreement on the findings and develop a course of corrective action. In cases where there is disagreement, the IS auditor should elaborate on the significance of the findings, risk and effects of not correcting the control weakness. Sometimes the auditee's management may request assistance from the IS auditor in implementing the recommended control enhancements. The IS auditor should communicate the difference between the IS auditor's role and that of a consultant and give careful consideration to how assisting the auditee may adversely affect the IS auditor's independence.

After agreement has been reached with the auditee, IS audit management should brief senior management of the audited organization. A summary of audit activities will be presented periodically to the audit committee. Audit committees typically are composed of individuals who do not work directly for the organization, and thus, provide the auditors with an independent route to report sensitive findings.

1.6.1 AUDIT REPORT STRUCTURE AND CONTENTS

Audit reports are the end product of the IS audit work. They are used by the IS auditor to report findings and recommendations to management. The exact format of an audit report will vary by organization; however, the skilled IS auditor should understand the basic components of an audit report and how it communicates audit findings to management.

> **Note:** The CISA candidate should become familiar with the ISACA IS Audit and Assurance Standards 1401 Reporting and 1402 Follow-up Activities.

There is no specific format for an IS audit report; the organization's audit policies and procedures will dictate the general format. Audit reports will usually have the following structure and content:

- An introduction to the report, including a statement of audit objectives, limitations to the audit and scope, the period of audit coverage, and a general statement on the nature and extent of audit procedures conducted and processes examined during the audit, followed by a statement on the IS audit methodology and guidelines
- Audit findings included in separate sections and often grouped in sections by materiality and/or intended recipient
- The IS auditor's overall conclusion and opinion on the adequacy of controls and procedures examined during the audit, and the actual potential risk identified as a consequence of detected deficiencies
- The IS auditor's reservations or qualifications with respect to the audit
  - This may state that the controls or procedures examined were found to be adequate or inadequate. The balance of the audit report should support that conclusion, and the overall evidence gathered during the audit should provide an even greater level of support for the audit conclusions.
- Detailed audit findings and recommendations
  - The IS auditor decides whether to include specific findings in an audit report. This should be based on the materiality of the findings and the intended recipient of the audit report. An audit report directed to the audit committee of the board of directors, for example, may not include findings that are important only to local management but have little control significance to the overall organization. The decision of what to include in various levels of audit reports depends on the guidance provided by upper management.
- A variety of findings, some of which may be quite material while others are minor in nature
  - The auditor may choose to present minor findings to management in an alternate format such as by memorandum.

The IS auditor should make the final decision about what to include or exclude from the audit report. Generally, the IS auditor should be concerned with providing a balanced report, describing not only negative issues in terms of findings but positive constructive comments regarding improving processes and controls or effective controls already in place. Overall, the IS auditor should exercise independence in the reporting process.

Auditee management evaluates the findings, stating corrective actions to be taken and timing for implementing these anticipated corrective actions.

Management may not be able to implement all audit recommendations immediately. For example, the IS auditor may recommend changes to an information system that is also undergoing other changes or enhancements. The IS auditor should not necessarily expect that the other changes will be suspended until the IS auditor's recommendations are implemented. Rather, all may be implemented at once.

The IS auditor should discuss the recommendations and any planned implementation dates while in the process of releasing the audit report. The IS auditor must realize that various constraints—such as staff limitations, budgets or other projects—may limit immediate implementation. Management should develop a firm program for corrective actions. It is important to obtain a commitment from the auditee/management on the date by which the action plan will be implemented (the solution can be something which takes a long time for implementation) and the manner in which it will be performed because the corrective action may bring risk that may be avoided if identified while discussing and finalizing the audit report. If appropriate, the IS auditor may want to report to upper management on the progress of implementing recommendations.

ISACA IS Audit and Assurance Standard 1401 Reporting and the ISACA IS Audit and Assurance Guideline 2401 Reporting state that the report should include all significant audit findings. When a finding requires explanation, the IS auditor should describe the finding, its cause and risk. When appropriate, the IS auditor should provide the explanation in a separate document and make reference to it in the report. For example, this approach may be appropriate for highly confidential matters. The IS auditor should also identify the organizational, professional and governmental criteria applied, such as COBIT. The report should be issued in a timely manner to encourage prompt corrective action. When appropriate, the IS auditor should promptly communicate significant findings to the appropriate persons prior to the issuance of the report. Prior communication of significant findings should not alter the intent or content of the report.

Note: The CISA candidate should review the detail from the ISACA IS Audit and Assurance Guideline 2401 Reporting.

1.6.2 AUDIT DOCUMENTATION

Audit documentation should include, at a minimum, a record of the following:
- Planning and preparation of the audit scope and objectives
- Description and/or walk-throughs on the scoped audit area
- Audit program
- Audit steps performed and audit evidence gathered
- Use of services of other auditors and experts
- Audit findings, conclusions and recommendations
- Audit documentation relation with document identification and dates

It is also recommended that documentation include:
- A copy of the report issued as a result of the audit work
- Evidence of audit supervisory review

Documents should include audit information that is required by laws and regulations, contractual stipulations and professional standards. Audit documentation is the necessary evidence supporting the conclusions reached and should be clear, complete, easily retrievable and sufficiently comprehensible. Audit documentation is generally the property of the auditing entity and should be accessible only to authorized personnel under specific or general permission. Where access to audit documentation is requested by external parties, the auditor should obtain appropriate prior approval of senior management and legal counsel.

The IS auditor/IS audit department should also develop policies regarding custody, retention requirements and release of audit documentation.

Note: The CISA candidate should be familiar with the detailed content of the ISACA IS Audit and Assurance Guideline 2203 Performance and Supervision.

The documentation format and media are optional, but due diligence and good practices require that work papers are dated, initialed, page-numbered, relevant, complete, clear, self-contained and properly labeled, filed and kept in custody. Work papers may be automated. IS auditors should particularly consider how to maintain integrity and protection of audit test evidence to preserve their proof value in support of audit results.

Audit documentation or work papers can be considered the bridge or interface between the audit objectives and the final report. They should provide a seamless transition—with traceability and accountability—from objectives to report and from report to objectives. The audit report, in this context, can be viewed as a set of particular work papers.

Audit documentation should support the findings and conclusions/opinion. Time of evidence can be crucial to supporting audit findings and conclusions. The IS auditor should take enough care to ensure that the evidence gathered and documented will be able to support audit findings and conclusions. An IS auditor should be able to prepare adequate working papers, narratives, questionnaires and understandable system flowcharts.

IS auditors are a scarce and expensive resource. Any technology capable of increasing the audit productivity is welcome. Automating work papers affects productivity directly and indirectly (granting access to other auditors, reusing documents or parts of them in recurring audits, etc.).

The quest for integrating work papers in the auditor's e-environment has resulted in all major audit and project management packages, CAATs and expert systems offering a complete array of automated documentation and import-export features.

ISACA IS Audit and Assurance Standards and Guidelines set forth many specifications about work papers, including the need to document the audit plan, program and evidence (2205 Evidence); how to use those of other auditors (2206 Using the Work of Other Experts); or the use of sampling (2208 Sampling).

1.6.3 CLOSING FINDINGS

IS auditors should realize that auditing is an ongoing process. The IS auditor is not effective if audits are performed and reports issued, but no follow-up is conducted to determine whether management has taken appropriate corrective actions. IS auditors should have a follow-up program to determine if agreed-on corrective actions have been implemented. Although IS auditors who work for external audit firms may not necessarily follow this process, they may achieve these tasks if agreed to by the audited entity.

The timing of the follow-up will depend on the criticality of the findings and would be subject to the IS auditor's judgment. The results of the follow-up should be communicated to appropriate levels of management.

The level of the IS auditor's follow-up review will depend on several factors. In some instances, the IS auditor may merely need to inquire as to the current status. In other instances, the IS auditor who works in an internal audit function may have to perform certain audit steps to determine whether the corrective actions agreed on by management have been implemented.

1.7 CONTROL SELF-ASSESSMENT

Control self-assessment (CSA) is an assessment of controls made by the staff and management of the unit or units involved. It is a management technique that assures stakeholders, customers and other parties that the internal control system of the organization is reliable. It also ensures that employees are aware of the risk to the business and they conduct periodic, proactive reviews of controls. It is a methodology used to review key business objectives, risk involved in achieving the business objectives and internal controls designed to manage business risk in a formal, documented and collaborative process.

In practice, CSA is a series of tools on a continuum of sophistication ranging from simple questionnaires to facilitated workshops, designed to gather information about the organization by asking those with a day-to-day working knowledge of an area as well as their managers. The basic tools used during a CSA project are the same whether the project is technical, financial or operational. These tools include management meetings, client workshops, worksheets, rating sheets and the CSA project approach. Like the continuum of tools used to gather information, there are diverse approaches to the levels below management that are queried; some organizations even include outsiders (such as clients or trading partners) when making CSA assessments.

The CSA program can be implemented by various methods. For small business units within organizations, it can be implemented by facilitated workshops where functional management and control professionals such as auditors can come together and deliberate how best to evolve a control structure for the business unit.

In a workshop, the role of a facilitator is to support the decision-making process. The facilitator creates a supportive environment to help participants explore their own experiences and those of others, identify control strengths and weaknesses and share their knowledge, ideas and concerns. If appropriate, a facilitator may also offer his/her own expertise in addition to facilitating the exchange of ideas and experience. A facilitator does not have to be an expert in a certain process or subject matter; however, the facilitator should have basic skills such as:

- Active listening skills and the ability to ask good questions, including questions that probe the topics and move the discussions forward
- Good verbal communication skills, including the ability to pose questions in a nonthreatening manner and the ability to summarize material
- The ability to manage the dynamics of the group, including managing various personalities so that a few members do not dominate the discussions and managing processes so that goals are met
- The ability to resolve conflicts
- The ability to manage time and keep the proceedings on schedule

In organizations with offices located at geographically dispersed locations, it may not be practical to organize facilitated workshops. In this case, a hybrid approach is needed. A questionnaire based on the control structure can be used. Operational managers can periodically complete the questionnaire, which can be analyzed and evaluated for effectiveness of the controls. However, a hybrid approach will be effective only if the analysis and readjustment of the questionnaire is performed using a life cycle approach, as shown in **figure 1.11**.

Figure 1.11—Hybrid Approach for a CSA

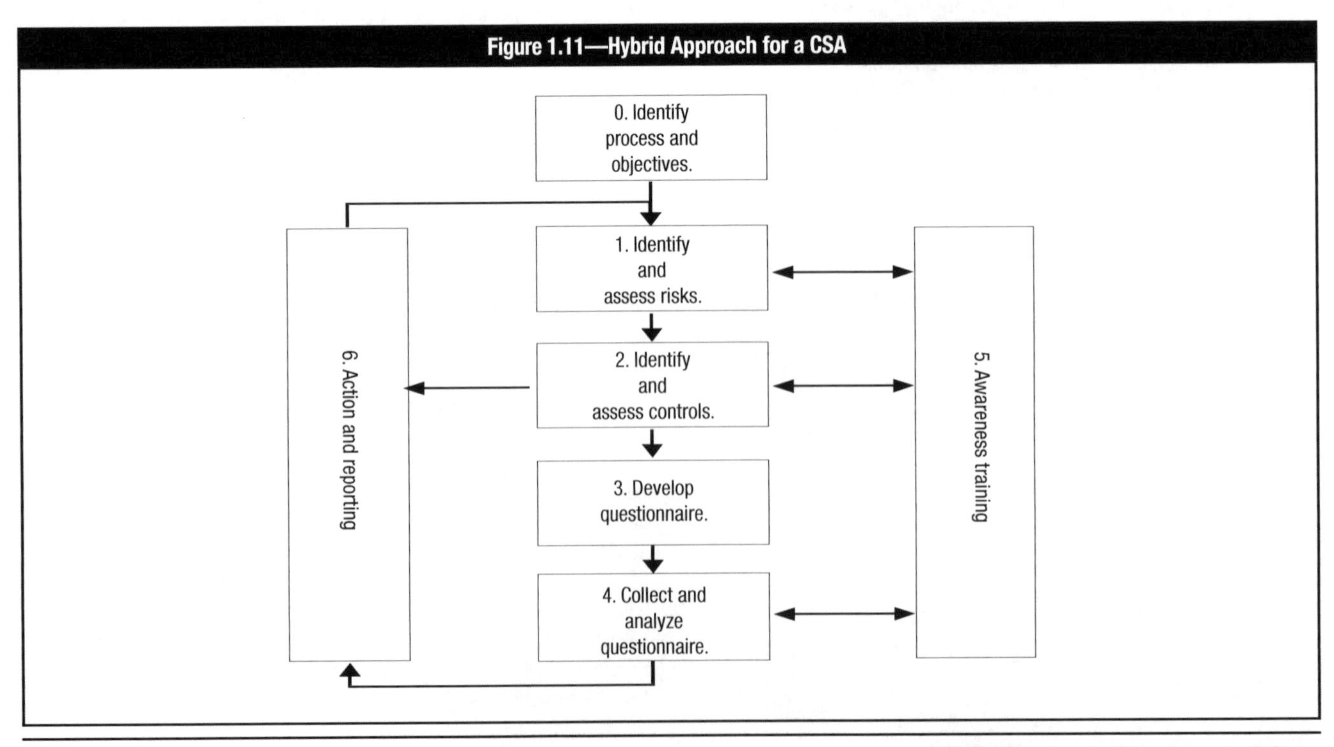

1.7.1 OBJECTIVES OF CSA

There are several objectives associated with adopting a CSA program. The primary objective is to leverage the internal audit function by shifting some of the control monitoring responsibilities to the functional areas. It is not intended to replace audit's responsibilities but to enhance them. Auditees, such as line managers, are responsible for controls in their environment; the managers also should be responsible for monitoring the controls. CSA programs also must educate management about control design and monitoring, particularly concentration on areas of high risk. These programs are not just policies requiring clients to comply with control standards. Instead, they offer a variety of support ranging from written suggestions outlining acceptable control environments to in-depth workshops. When workshops are included in the program, an additional objective—the empowerment of workers to assess or even design the control environment—may be included in the program.

When employing a CSA program, measures of success for each phase (planning, implementation and monitoring) should be developed to determine the value derived from CSA and its future use. One critical success factor (CSF) is to conduct a meeting with the business unit representatives (including appropriate and relevant staff and management) to identify the business unit's primary objective—to determine the reliability of the internal control system. In addition, actions that increase the likelihood of achieving the primary objective should be identified.

A generic set of goals and metrics for each process, which can be used in designing and monitoring the CSA program, has been provided in COBIT.

COBIT is a governance and control framework that provides guidance in the development of the control assessment method. One could develop a CSA method by identifying the tasks and processes that are relevant to the business environment and then defining the controls for relevant activities. A CSA questionnaire can be developed using the statements in the relevant control objectives of the identified IT processes. Various components of the COBIT framework—such as input-output matrix, RACI chart, goals, metrics and maturity model—can be converted into the form of a CSA questionnaire to assess each of the areas as required.

1.7.2 BENEFITS OF CSA

Some of the benefits of a CSA include the following:
- Early detection of risk
- More effective and improved internal controls
- Creation of cohesive teams through employee involvement
- Developing a sense of ownership of the controls in the employees and process owners and reducing their resistance to control improvement initiatives
- Increased employee awareness of organizational objectives, and knowledge of risk and internal controls
- Increased communication between operational and top management
- Highly motivated employees
- Improved audit rating process
- Reduction in control cost
- Assurance provided to stakeholders and customers
- Necessary assurance given to top management about the adequacy of internal controls as required by the various regulatory agencies and laws such as the US Sarbanes-Oxley Act

1.7.3 DISADVANTAGES OF CSA

CSA contains several disadvantages, including:
- It could be mistaken as an audit function replacement
- It may be regarded as an additional workload (e.g., one more report to be submitted to management)
- Failure to act on improvement suggestions could damage employee morale
- Lack of motivation may limit effectiveness in the detection of weak controls

1.7.4 AUDITOR ROLE IN CSA

The auditor's role in CSAs should be considered enhanced when audit departments establish a CSA program. When these programs are established, auditors become internal control professionals and assessment facilitators. Their value in this role is evident when management takes responsibility and ownership for internal control systems under their authority through process improvements in their control structures, including an active monitoring component.

For an auditor to be effective in this facilitative and innovative role, the auditor must understand the business process being assessed. This can be attained via traditional audit tools such as a preliminary survey or walk-through. Also, the auditors must remember that they are the facilitators and the management client is the participant in the CSA process. For example, during a CSA workshop, instead of the auditor performing detailed audit procedures, the auditor will lead and guide the auditees in assessing their environment by providing insight about the objectives of controls based on risk assessment. The managers, with a focus on improving the productivity of the process, might suggest replacement of preventive controls. In this case, the auditor is better positioned to explain the risk associated with such changes.

1.7.5 TECHNOLOGY DRIVERS FOR CSA

The development of techniques for empowerment, information gathering and decision making is a necessary part of a CSA program implementation. Some of the technology drivers include the combination of hardware and software to support CSA selection, and the use of an electronic meeting system and computer-supported decision aids to facilitate group decision making. Group decision making is an essential component of a workshop-based CSA where employee empowerment is a goal. In case of a questionnaire approach, the same principle applies for the analysis and readjustment of the questionnaire.

1.7.6 TRADITIONAL VERSUS CSA APPROACH

The traditional approach can be summarized as any approach in which the primary responsibility for analyzing and reporting on internal control and risk is assigned to auditors, and to a lesser extent, controller departments and outside consultants. This approach has created and reinforced the notion that auditors and consultants, not management and work teams, are responsible for

assessing and reporting on internal control. The CSA approach, on the other hand, emphasizes management and accountability over developing and monitoring internal controls of an organization's sensitive and critical business processes.

A summary of attributes or focus that distinguishes each from the other is described in **figure 1.12**.

Figure 1.12—Traditional and CSA Attributes

Traditional	CSA
Assigns duties/supervises staff	Empowered/accountable employees
Policy/rule-driven	Continuous improvement/learning curve
Limited employee participation	Extensive employee participation and training
Narrow stakeholder focus	Broad stakeholder focus
Auditors and other specialists	Staff at all levels, in all functions, are the primary control analysts.

1.8 THE EVOLVING IS AUDIT PROCESS

The IS audit process must continually change to keep pace with innovations in technology. Topics to address these evolving changes include areas such as integrated auditing and continuous auditing.

1.8.1 INTEGRATED AUDITING

Dependence of business processes on information technology has necessitated that traditional financial and operational auditors develop an understanding of IT control structures and IS auditors develop an understanding of the business control structures. Integrated auditing can be defined as the process whereby appropriate audit disciplines are combined to assess key internal controls over an operation, process or entity.

The integrated approach focuses on risk. A risk assessment aims to understand and identify risk arising from the entity and its environment, including relevant internal controls. At this stage, the role of IT audit is typically to understand and identify risk under topical areas such as information management, IT infrastructure, IT governance and IT operations. Other audit specialists will seek to understand the organizational environment, business risk and business controls. A key element of the integrated approach is discussion of the risk arising among the whole audit team, with consideration of impact and likelihood.

Detailed audit work then focuses on the relevant controls in place to manage this risk. IT systems frequently provide a first line of preventive and detective controls, and the integrated audit approach depends on a sound assessment of their efficiency and effectiveness.

The integrated audit process typically involves:
- Identification of risk faced by the organization for the area being audited
- Identification of relevant key controls
- Review and understanding of the design of key controls
- Testing that key controls are supported by the IT system
- Testing that management controls operate effectively
- A combined report or opinion on control risk, design and weaknesses

The integrated audit demands a focus on business risk and a drive for creative control solutions. It is a team effort of auditors with different skill sets. Using this approach permits a single audit of an entity with one comprehensive report. An additional benefit is that this approach assists in staff development and retention by providing greater variety and the ability to see how all of the elements (functional and IT) mesh together to form the complete picture. See **figure 1.13** for an integrated auditing approach.

Figure 1.13—An Integrated Audit

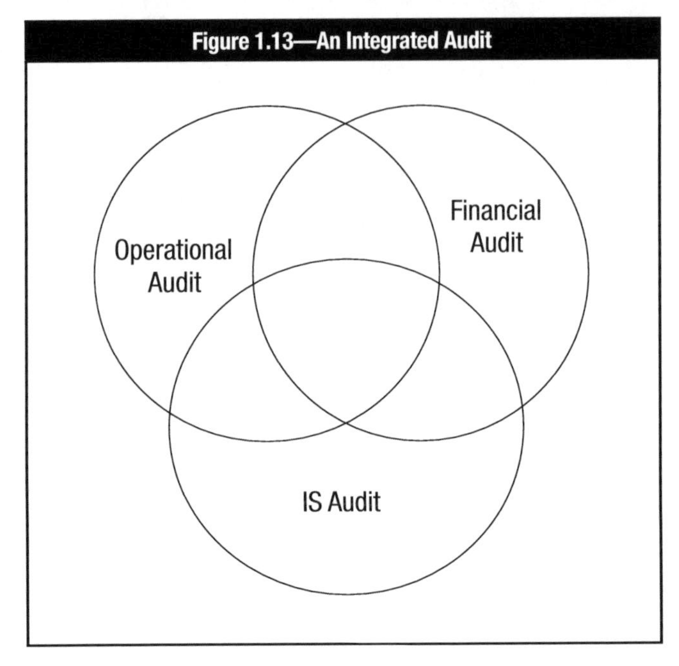

The integrated audit concept has also radically changed the manner in which audits are looked on by the different stakeholders. Employees or process owners better understand the objectives of an audit because they are able to see the linkage between controls and audit procedures. Top management better understands the linkage between increased control effectiveness and corresponding improvements in the allocation and utilization of IT resources. Shareholders are able to better understand the linkage between the push for a greater degree of corporate governance and its impact on the generation of financial statements that can be relied on. All these developments have led to greater impetus for the growing popularity of integrated audits.

Note: This topic on integrated auditing, though important, is not specifically tested in the CISA exam.

1.8.2 CONTINUOUS AUDITING

The focus on increased effectiveness and efficiency of assurance, internal auditing and control has spurred the development of new studies and examination of new ideas concerning continuous auditing as opposed to more traditional periodic auditing reviews. Several research studies and documents addressing the subject carry different definitions of continuous auditing. All studies, however, recognize that a distinctive character of continuous auditing is the short time lapse between the facts to be audited, the collection of evidence and audit reporting.

Traditional financial reports and the traditional audit style sometimes prove to be insufficient because they lack the essential element in the current business environment—updated information. Therefore, continuous auditing appears to be gaining more and more followers.

Some of the drivers of continuous auditing are a better monitoring of financial issues within a company, ensuring that real-time transactions also benefit from real-time monitoring; prevention of financial fraud and audit scandals such as Enron and Tesco Plc; and the use of software to determine that financial controls are proper. Continuous auditing involves a large amount of work because the company practicing continuous auditing will not provide one report at the end of a quarter, but will provide financial reports on a more frequent basis. Audit functions in organizations that use ERP platforms are increasingly using automated governance, risk and compliance (GRC) tools, which flag transactions that meet predefined criteria on a real-time basis. These tools are set up at the database level and pull data that meet the predefined criteria. Such data may include purchase invoices that have the same or similar address as that of an employee. The advantage of using these tools is that voluminous data are analyzed at a high speed to highlight relevant patterns of data that may be of interest to the auditors.

Continuous auditing is not a recent development. Traditional application systems may contain embedded audit modules. These would allow an auditor to trap predefined types of events or to directly inspect abnormal or suspect conditions and transactions.

Most current commercial applications could be customized with such features. However, cost and other considerations and the technical skills that would be required to establish and operate these tools tend to limit the usage of embedded audit modules to specific fields and applications.

To properly understand the implications and requirements of continuous auditing, a clear distinction has to be made between continuous auditing and continuous monitoring:

- **Continuous monitoring**—This is provided by IS management tools and typically based on automated procedures to meet fiduciary responsibilities. For instance, real-time antivirus or intrusion detection systems (IDSs) may operate in a continuous monitoring fashion.
- **Continuous auditing**— According to the *Global Technology Audit Guide 3: Continuous Auditing: Implications for Assurance Monitoring and Risk Assessment*, continuous auditing is "a method to automatically perform control and risk assessments on a more frequent basis. Continuous auditing changes the audit paradigm from periodic reviews of a sample of transactions to ongoing audit testing of 100 percent of transactions. It becomes an integral part of modern auditing at many levels." Continuous IS (and non-IS) auditing is typically completed using automated audit procedures.

Continuous auditing should be independent of continuous control or monitoring activities. When both continuous monitoring and auditing take place, continuous assurance can be established. In practice, continuous auditing is the precursor to management adopting continuous monitoring as a process on a day-to-day basis. Often, the audit function will hand over the techniques used in continuous auditing to the business, which will then run the continuous monitoring. This collaboration has led to increased appreciation among process owners of the value that the audit function brings to the organization, leading to greater confidence and trust between the business and auditors. Nevertheless, the lack of independence and objectivity inherent in continuous monitoring should not be overlooked, and continuous monitoring should never be considered as a substitute for the audit function.

Continuous auditing efforts often incorporate new IT developments, increased processing capabilities of current hardware, software, standards and artificial intelligence (AI) tools. Continuous auditing attempts to facilitate the collection and analysis of data at the moment of the transaction. Data must be gathered from different applications working within different environments, transactions must be screened, the transaction environment has to be analyzed to detect trends and exceptions, and atypical patterns (i.e., a transaction with significantly higher or lower value than typical for a given business partner) must be exposed. If all of this must happen in real time, perhaps even before final sign-off of a transaction, it is mandatory to adopt and combine various top-level IT techniques. The IT environment is a natural enabler for the application of continuous auditing because of the intrinsic automated nature of its underlying processes.

Continuous auditing aims to provide a more secure platform to avoid fraud and a real-time process aimed at ensuring a high-level of financial control.

Prerequisites/preconditions for continuous auditing to succeed include:

- A high degree of automation
- An automated and highly reliable process in producing information about subject matter soon after or during the occurrence of events underlying the subject matter
- Alarm triggers to report timely control failures
- Implementation of highly automated audit tools that require the IS auditor to be involved in setting up the parameters
- Quickly informing IS auditors of the results of automated procedures, particularly when the process has identified anomalies or errors
- The quick and timely issuance of automated audit reports
- Technically proficient IS auditors
- Availability of reliable sources of evidence
- Adherence to materiality guidelines
- A change of mind-set required for IS auditors to embrace continuous reporting
- Evaluation of cost factors

Simpler continuous auditing and monitoring tools are already built into many ERP packages and most OS and network security packages. These environments, if appropriately configured and populated with rules, parameters and formulas, can output exception lists on request while operating against actual data. Therefore, they represent an instance of continuous auditing. The difficult but significant added value to using these features is that they postulate a definition of what would be a "dangerous" or exception condition. For instance, whether a set of granted IS access permissions is to be deemed risk-free will depend on

having well-defined rules of segregation of duties. On the other hand, it may be much harder to decide if a given sequence of steps, taken to modify and maintain a database record, points to a potential risk.

IT techniques that are used to operate in a continuous auditing environment must work at all data levels—single input, transaction and databases—and include:
- Transaction logging
- Query tools
- Statistics and data analysis (CAAT)
- Database management system (DBMS)
- Data warehouses, data marts, data mining
- Intelligent agents
- Embedded audit modules (EAM)
- Neural network technology
- Standards such as Extensible Business Reporting Language (XBRL)

Intelligent software agents may be used to automate the evaluation processes and allow for flexibility and dynamic analysis capabilities. The configuration and application of intelligent agents (sometimes referred to as bots), allows for continuous monitoring of systems settings and the delivery of alert messages when certain thresholds are exceeded or when certain conditions are met.

Full continuous auditing processes have to be carefully built into applications and work in layers. The auditing tools must operate in parallel to normal processing—capturing real-time data, extracting standardized profiles or descriptors and passing the result to the auditing layers.

Continuous auditing has an intrinsic edge over point-in-time or periodic auditing because it captures internal control problems as they occur, preventing negative effects. Implementation can also reduce possible or intrinsic audit inefficiencies such as delays, planning time, inefficiencies of the audit process, overhead due to work segmentation, multiple quality or supervisory reviews, or discussions concerning the validity of findings.

Full top management support, dedication and extensive experience and technical knowledge are all necessary to accomplish continuous auditing, while minimizing the impact on the underlying audited business processes. The auditing layers and settings may also need continual adjustment and updating. Besides difficulty and cost, continuous auditing has an inherent disadvantage in that internal control experts and auditors might be resistant to trust an automated tool in lieu of their personal judgment and evaluation. Also, mechanisms have to be put in place to eliminate false negatives and false positives in the reports generated by such audits so that the report generated continues to inspire stakeholders' confidence in the accuracy of the report.

The implementation of continuous auditing involves many factors; however, the task is not impossible. There is an increasing desire to provide auditing over information in a real-time environment (or as close to real time as possible).

1.9 CASE STUDIES

The following case studies are included as a learning tool to reinforce the concepts introduced in this chapter.

1.9.1 CASE STUDY A

The IS auditor has been asked to perform preliminary work that will assess the readiness of the organization for a review to measure compliance with new regulatory requirements. These requirements are designed to ensure that management is taking an active role in setting up and maintaining a well-controlled environment, and accordingly, will assess management's review and testing of the general IT control environment. Areas to be assessed include logical and physical security, change management, production control and network management, IT governance, and end-user computing. The IS auditor has been given six months to perform this preliminary work, so sufficient time should be available. It should be noted that in previous years, repeated problems have been identified in the areas of logical security and change management, so these areas will most likely require some degree of remediation. Logical security deficiencies noted included the sharing of administrator accounts and failure to enforce adequate controls over passwords. Change management deficiencies included improper segregation of incompatible duties and failure to document all changes. Additionally, the process for deploying OS updates to servers was found to be only partially effective. In anticipation of the work to be performed by the IS auditor, the chief information officer (CIO) requested direct reports to develop narratives and process flows describing the major activities for which IT is responsible. These were completed, approved by the various process owners and the CIO, and then forwarded to the IS auditor for examination.

CASE STUDY A QUESTIONS	
A1.	What should the IS auditor do **FIRST**? A. Perform an IT risk assessment. B. Perform a survey audit of logical access controls. C. Revise the audit plan to focus on risk-based auditing. D. Begin testing controls that the IS auditor feels are most critical.
A2.	When testing program change management, how should the sample be selected? A. Change management documents should be selected at random and examined for appropriateness. B. Changes to production code should be sampled and traced to appropriate authorizing documentation. C. Change management documents should be selected based on system criticality and examined for appropriateness. D. Changes to production code should be sampled and traced back to system-produced logs indicating the date and time of the change.
See answers and explanations to the case study questions at the end of the chapter (page 65).	

1.9.2 CASE STUDY B

An IS auditor is planning to review the security of a financial application for a large company with several locations worldwide. The application system is made up of a web interface, a business logic layer and a database layer. The application is accessed locally through a LAN and remotely through the Internet via a virtual private network (VPN) connection.

CASE STUDY B QUESTIONS	
B1.	The **MOST** appropriate type of CAATs tool the auditor should use to test security configuration settings for the entire application system is: A. generalized audit software (GAS). B. test data. C. utility software. D. expert system.
B2.	Given that the application is accessed through the Internet, how should the auditor determine whether to perform a detailed review of the firewall rules and VPN configuration settings? A. Documented risk analysis B. Availability of technical expertise C. Approach used in previous audit D. IS auditing guidelines and best practices
B3.	During the review, if the auditor detects that the transaction authorization control objective cannot be met due to a lack of clearly defined roles and privileges in the application, the auditor should **FIRST**: A. review the authorization on a sample of transactions. B. immediately report this finding to upper management. C. request that auditee management review the appropriateness of access rights for all users. D. use GAS to check the integrity of the database.
See answers and explanations to the case study questions at the end of the chapter (page 66).	

1.9.3 CASE STUDY C

An IS auditor has been appointed to carry out IS audits in an entity for a period of two years. After accepting the appointment the IS auditor noted that:

- The entity has an audit charter that detailed, among other things, the scope and responsibilities of the IS audit function and specifies the audit committee as the overseeing body for audit activity.
- The entity is planning a major increase in IT investment, mainly on account of implementation of a new ERP application, integrating business processes across units dispersed geographically. The ERP implementation is expected to become operational within the next 90 days. The servers supporting the business applications are hosted offsite by a third-party service provider.
- The entity has a new incumbent as chief information security officer (CISO) who reports to the chief financial officer (CFO).
- The entity is subject to regulatory compliance requirements that require its management to certify the effectiveness of the internal control system as it relates to financial reporting. The entity has been recording consistent growth over the last two years at double the industry average. However, the entity has seen increased employee turnover as well.

CASE STUDY C QUESTIONS	
C1.	The **FIRST** priority of the IS auditor in year one should be to study the: A. previous IS audit reports and plan the audit schedule. B. audit charter and plan the audit schedule. C. impact of the new incumbent as CISO. D. impact of the implementation of a new ERP on the IT environment and plan the audit schedule.
C2.	How should the IS auditor evaluate backup and batch processing within computer operations? A. Plan and carry out an independent review of computer operations. B. Rely on the service auditor's report of the service provider. C. Study the contract between the entity and the service provider. D. Compare the service delivery report to the service level agreement.
See answers and explanations to the case study questions at the end of the chapter (page 66).	

1.10 ANSWERS TO CASE STUDY QUESTIONS

ANSWERS TO CASE STUDY A QUESTIONS

A1. **A** An IT risk assessment should be performed first to ascertain which areas present the greatest risk and what controls mitigate that risk. Although narratives and process flows have been created, the organization has not yet assessed which controls are critical. All other choices would be undertaken after performing the IT risk assessment.

A2. **B** When testing a control, it is advisable to trace from the item being controlled to the relevant control documentation. When a sample is chosen from a set of control documents, there is no way to ensure that every change was accompanied by appropriate control documentation. Accordingly, changes to production code provide the most appropriate basis for selecting a sample. These sampled changes should then be traced to appropriate authorizing documentation. In contrast, selecting from the population of change management documents will not reveal any changes that bypassed the normal approval and documentation process. Similarly, comparing production code changes to system-produced logs will not provide evidence of proper approval of changes prior to their being migrated to production.

ANSWERS TO CASE STUDY B QUESTIONS

B1. **C** When testing the security of the entire application system—including OSs, database and application security—the auditor will most likely use a utility software that assists in reviewing the configuration settings. In contrast, the auditor might use GAS to perform a substantive testing of data and configuration files of the application. Test data are normally used to check the integrity of the data and expert systems are used to inquire on specific topics.

B2. **A** In order to decide if the audit scope should include specific infrastructure components (in this case, the firewall rules and VPN configuration settings), the auditor should perform and document a risk analysis in order to determine which sections present the greatest risk and include these sections in the audit scope. The risk analysis may consider factors such as previous revisions to the system, related security incidents within the company or other companies of the same sectors, resources available to do the review and others. Availability of technical expertise and the approach used in previous audits may be taken into consideration; however, these should be of secondary importance. IS auditing guidelines and best practices provide a guide to the auditor on how to comply with IS audit standards, but by themselves they would not be sufficient to make this decision.

B3. **A** The auditor should first review the authorization on a sample of transactions in order to determine and be able to report the impact and materiality of this issue. Whether the auditor would immediately report the issue or wait until the end of the audit to report this finding will depend on the impact and materiality of the issue, which would require reviewing a sample of transactions. The use of GAS to check the integrity of the database would not help the auditor assess the impact of this issue.

ANSWERS TO CASE STUDY C QUESTIONS

C1. **D** In terms of priority, because the implementation of the new ERP will have far reaching consequences on the way IS controls are configured in the system, the IS auditor should study the impact of implementation of the ERP and plan the audit schedule accordingly. Preferably, the IS auditor should discuss the audit plan with the external auditor and the internal audit division of the entity to make the audit more effective and useful for the entity.

C2. **D** The service delivery report that captures the actual performance of the service provider against the contractually agreed-on levels provides the best and most objective basis for evaluation of the computer operations. The service auditor's report is likely to be more useful from a controls evaluation perspective for the external auditor of the entity.

Chapter 2:

Governance and Management of IT

Section One: Overview

Section Two: Content

Section One: Overview

DEFINITION

Governance and management of IT is an integral part of enterprise governance and consists of the leadership and organizational structures and processes that ensure that the enterprise's IT sustains and extends the enterprise's strategy and objectives (adapted from IT Governance Institute, *Board Briefing on IT Governance, 2nd Edition*, USA, 2003). Knowledge of IT governance is fundamental to the work of the IS auditor, and it forms the foundation for the development of sound control practices and mechanisms for management oversight and review.

OBJECTIVES

The objective of this domain is to ensure that the CISA candidate understands and can provide assurance that the necessary leadership and organizational structures and processes are in place to achieve the objectives and to support the enterprise's strategy.

This domain represents 16 percent of the CISA examination (approximately 24 questions).

TASK AND KNOWLEDGE STATEMENTS

TASKS

There are 10 tasks within the IT governance domain:

T2.1 Evaluate the IT strategy, including the IT direction, and the processes for the strategy's development, approval, implementation and maintenance for alignment with the organization's strategies and objectives.

T2.2 Evaluate the effectiveness of the IT governance structure to determine whether IT decisions, directions and performance support the organization's strategies and objectives.

T2.3 Evaluate IT organizational structure and human resources (personnel) management to determine whether they support the organization's strategies and objectives.

T2.4 Evaluate the organization's IT policies, standards and procedures and the processes for their development, approval, release/publishing, implementation and maintenance to determine whether they support the IT strategy and comply with regulatory and legal requirements.

T2.5 Evaluate IT resource management, including investment, prioritization, allocation and use for alignment with the organization's strategies and objectives.

T2.6 Evaluate IT portfolio management, including investment, prioritization and allocation, for alignment with the organization's strategies and objectives.

T2.7 Evaluate risk management practices to determine whether the organization's IT-related risks are identified, assessed, monitored, reported and managed.

T2.8 Evaluate IT management and monitoring of controls (e.g., continuous monitoring, quality assurance [QA]) for compliance with the organization's policies, standards and procedures.

T2.9 Evaluate monitoring and reporting of IT key performance indicators (KPIs) to determine whether management receives sufficient and timely information.

T2.10 Evaluate the organization's business continuity plan (BCP), including the alignment of the IT disaster recovery plan (DRP) with the BCP, to determine the organization's ability to continue essential business operations during the period of an IT disruption.

KNOWLEDGE STATEMENTS

The CISA candidate must have a good understanding of each of the topics or areas delineated by the knowledge statements. These statements are the basis for the exam.

There are 17 knowledge statements within the domain covering the governance and management of IT:

K2.1 Knowledge of the purpose of IT strategy, policies, standards and procedures for an organization and the essential elements of each

K2.2 Knowledge of IT governance, management, security and control frameworks and related standards, guidelines and practices

K2.3 Knowledge of organizational structure, roles, and responsibilities related to IT, including segregation of duties (SoD)

K2.4 Knowledge of relevant laws, regulations and industry standards affecting the organization

K2.5 Knowledge of the organization's technology direction and IT architecture and their implications for setting long-term strategic directions

K2.6 Knowledge of the processes for the development, implementation and maintenance of IT strategy, policies, standards and procedures

K2.7 Knowledge of the use of capability and maturity models

K2.8 Knowledge of process optimization techniques

K2.9 Knowledge of IT resource investment and allocation practices, including prioritization criteria (e.g., portfolio management, value management, personnel management)

K2.10 Knowledge of IT supplier selection, contract management, relationship management and performance monitoring processes including third party outsourcing relationships

K2.11 Knowledge of enterprise risk management (ERM)

K2.12 Knowledge of practices for monitoring and reporting of controls performance (e.g., continuous monitoring, quality assurance [QA])

K2.13 Knowledge of quality management and quality assurance (QA) systems

K2.14 Knowledge of practices for monitoring and reporting of IT performance (e.g., balanced scorecards [BSCs], key performance indicators [KPIs])

K2.15 Knowledge of business impact analysis (BIA)

K2.16 Knowledge of the standards and procedures for the development, maintenance and testing of the business continuity plan (BCP)

K2.17 Knowledge of procedures used to invoke and execute the business continuity plan and return to normal operations

Relationship of Task to Knowledge Statements

The task statements are what the CISA candidate is expected to know how to do. The knowledge statements delineate each of the areas in which the CISA candidate must have a good understanding in order to perform the tasks. The task and knowledge statements are mapped in **figure 2.1** insofar as it is possible to do so. Note that although there is often overlap, each task statement will generally map to several knowledge statements.

Figure 2.1—Task and Knowledge Statements Mapping

Task Statement	Knowledge Statements
T2.1 Evaluate the IT strategy, including the IT direction, and the processes for the strategy's development, approval, implementation and maintenance for alignment with the organization's strategies and objectives.	K2.1 Knowledge of the purpose of IT strategy, policies, standards and procedures for an organization and the essential elements of each K2.4 Knowledge of relevant laws, regulations and industry standards affecting the organization K2.5 Knowledge of the organization's technology direction and IT architecture and their implications for setting long-term strategic directions K2.6 Knowledge of the processes for the development, implementation and maintenance of IT strategy, policies, standards and procedures K2.9 Knowledge of IT resource investment and allocation practices, including prioritization criteria (e.g., portfolio management, value management, personnel management) K2.11 Knowledge of enterprise risk management (ERM)
T2.2 Evaluate the effectiveness of the IT governance structure to determine whether IT decisions, directions and performance support the organization's strategies and objectives.	K2.2 Knowledge of IT governance, management, security and control frameworks, and related standards, guidelines and practices K2.4 Knowledge of relevant laws, regulations and industry standards affecting the organization K2.5 Knowledge of the organization's technology direction and IT architecture and their implications for setting long-term strategic directions
T2.3 Evaluate IT organizational structure and human resources (personnel) management to determine whether they support the organization's strategies and objectives.	K2.3 Knowledge of organizational structure, roles and responsibilities related to IT, including segregation of duties (SoD) K2.4 Knowledge of relevant laws, regulations and industry standards affecting the organization K2.5 Knowledge of the organization's technology direction and IT architecture and their implications for setting long-term strategic directions K2.9 Knowledge of IT resource investment and allocation practices, including prioritization criteria (e.g., portfolio management, value management, personnel management)
T2.4 Evaluate the organization's IT policies, standards and procedures, and the processes for their development, approval, release/publishing, implementation and maintenance to determine whether they support the IT strategy and comply with regulatory and legal requirements.	K2.1 Knowledge of the purpose of IT strategy, policies, standards and procedures for an organization and the essential elements of each K2.2 Knowledge of IT governance, management, security and control frameworks, and related standards, guidelines and practices K2.3 Knowledge of organizational structure, roles and responsibilities related to IT, including segregation of duties (SoD) K2.4 Knowledge of relevant laws, regulations and industry standards affecting the organization K2.5 Knowledge of the organization's technology direction and IT architecture and their implications for setting long-term strategic directions K2.6 Knowledge of the processes for the development, implementation and maintenance of IT strategy, policies, standards and procedures K2.7 Knowledge of the use of capability and maturity models K2.8 Knowledge of process optimization techniques
T2.5 Evaluate IT resource management, including investment, prioritization, allocation and use for alignment with the organization's strategies and objectives.	K2.3 Knowledge of organizational structure, roles and responsibilities related to IT, including segregation of duties (SoD) K2.4 Knowledge of relevant laws, regulations and industry standards affecting the organization K2.5 Knowledge of the organization's technology direction and IT architecture and their implications for setting long-term strategic directions K2.6 Knowledge of the processes for the development, implementation and maintenance of IT strategy, policies, standards and procedures K2.7 Knowledge of the use of capability and maturity models K2.8 Knowledge of process optimization techniques K2.9 Knowledge of IT resource investment and allocation practices, including prioritization criteria (e.g., portfolio management, value management, personnel management) K2.10 Knowledge of IT supplier selection, contract management, relationship management and performance monitoring processes including third party outsourcing relationships K2.12 Knowledge of practices for monitoring and reporting of controls performance (e.g., continuous monitoring, quality assurance [QA]) K2.14 Knowledge of practices for monitoring and reporting of IT performance (e.g., balanced scorecards [BSCs], key performance indicators [KPIs])

Figure 2.1—Task and Knowledge Statements Mapping *(cont.)*

Task Statement	Knowledge Statements
T2.6 Evaluate IT portfolio management, including investment, prioritization and allocation, for alignment with the organization's strategies and objectives.	K2.4 Knowledge of relevant laws, regulations and industry standards affecting the organization K2.6 Knowledge of the processes for the development, implementation and maintenance of IT strategy, policies, standards and procedures K2.7 Knowledge of the use of capability and maturity models K2.8 Knowledge of process optimization techniques K2.9 Knowledge of IT resource investment and allocation practices, including prioritization criteria (e.g., portfolio management, value management, personnel management) K2.10 Knowledge of IT supplier selection, contract management, relationship management and performance monitoring processes including third party outsourcing relationships K2.12 Knowledge of practices for monitoring and reporting of controls performance (e.g., continuous monitoring, quality assurance [QA]) K2.14 Knowledge of practices for monitoring and reporting of IT performance (e.g., balanced scorecards [BSCs], key performance indicators [KPIs])
T2.7 Evaluate risk management practices to determine whether the organization's IT-related risks are identified, assessed, monitored, reported and managed.	K2.4 Knowledge of relevant laws, regulations and industry standards affecting the organization K2.5 Knowledge of the organization's technology direction and IT architecture and their implications for setting long-term strategic directions K2.6 Knowledge of the processes for the development, implementation and maintenance of IT strategy, policies, standards and procedures K2.7 Knowledge of the use of capability and maturity models K2.8 Knowledge of process optimization techniques K2.11 Knowledge of enterprise risk management (ERM) K2.12 Knowledge of practices for monitoring and reporting of controls performance (e.g., continuous monitoring, quality assurance [QA]) K2.14 Knowledge of practices for monitoring and reporting of IT performance (e.g., balanced scorecards [BSCs], key performance indicators [KPIs]) K2.15 Knowledge of business impact analysis (BIA)
T2.8 Evaluate IT management and monitoring of controls (e.g., continuous monitoring, quality assurance) for compliance with the organization's policies, standards and procedures.	K2.2 Knowledge of IT governance, management, security and control frameworks, and related standards, guidelines, and practices K2.4 Knowledge of relevant laws, regulations and industry standards affecting the organization K2.5 Knowledge of the organization's technology direction and IT architecture and their implications for setting long-term strategic directions K2.6 Knowledge of the processes for the development, implementation and maintenance of IT strategy, policies, standards and procedures K2.7 Knowledge of the use of capability and maturity models K2.8 Knowledge of process optimization techniques K2.12 Knowledge of practices for monitoring and reporting of controls performance (e.g., continuous monitoring, quality assurance [QA]) K2.13 Knowledge of quality management and quality assurance systems K2.14 Knowledge of practices for monitoring and reporting of IT performance (e.g., balanced scorecards [BSCs], key performance indicators [KPIs])

Figure 2.1—Task and Knowledge Statements Mapping *(cont.)*

Task Statement	Knowledge Statements
T2.9 Evaluate monitoring and reporting of IT key performance indicators to determine whether management receives sufficient and timely information.	K2.2 Knowledge of IT governance, management, security and control frameworks and related standards, guidelines and practices K2.4 Knowledge of relevant laws, regulations and industry standards affecting the organization K2.5 Knowledge of the organization's technology direction and IT architecture and their implications for setting long-term strategic directions K2.6 Knowledge of the processes for the development, implementation and maintenance of IT strategy, policies, standards and procedures K2.7 Knowledge of the use of capability and maturity models K2.8 Knowledge of process optimization techniques K2.10 Knowledge of IT supplier selection, contract management, relationship management and performance monitoring processes including third party outsourcing relationships K2.11 Knowledge of enterprise risk management (ERM) K2.12 Knowledge of practices for monitoring and reporting of controls performance (e.g., continuous monitoring, quality assurance [QA]) K2.13 Knowledge of quality management and quality assurance systems K2.14 Knowledge of practices for monitoring and reporting of IT performance (e.g., balanced scorecards [BSCs], key performance indicators [KPIs]) K2.15 Knowledge of business impact analysis (BIA)
T2.10 Evaluate the organization's business continuity plan (BCP), including the alignment of the IT disaster recovery plan (DRP) with the BCP, to determine the organization's ability to continue essential business operations during the period of an IT disruption.	K2.3 Knowledge of organizational structure, roles and responsibilities related to IT, including segregation of duties (SoD) K2.4 Knowledge of relevant laws, regulations and industry standards affecting the organization K2.5 Knowledge of the organization's technology direction and IT architecture and their implications for setting long-term strategic directions K2.6 Knowledge of the processes for the development, implementation and maintenance of IT strategy, policies, standards and procedures K2.7 Knowledge of the use of capability and maturity models K2.8 Knowledge of process optimization techniques K2.11 Knowledge of enterprise risk management (ERM) K2.15 Knowledge of business impact analysis (BIA) K2.16 Knowledge of the standards and procedures for the development, maintenance, and testing of the business continuity plan (BCP) K2.17 Knowledge of procedures used to invoke and execute the business continuity plan and return to normal operations

Knowledge Statement Reference Guide

Each knowledge statement is explained in terms of underlying concepts and relevance of the knowledge statement to the IS auditor. It is essential that the exam candidate understand the concepts. The knowledge statements are what the IS auditor must know in order to accomplish the tasks. Consequently, only the knowledge statements are detailed in this section.

The sections identified in K2.1 through K2.17 are described in greater detail in section two of this chapter.

K2.1 Knowledge of the purpose of IT strategy, policies, standards and procedures for an organization and the essential elements of each

Explanation	Key Concepts	Reference in Manual
In order to be effective, IT governance efforts require a formal framework. Specifically, organizations depend on the IT governance framework (COBIT®, ISO 38000, etc.) to provide reasonable assurance that IT solutions automate processes to achieve business goals and objectives. Furthermore, it enables the organization to focus IT deployment in a manner consistent with business/organization strategy and objectives. Organizations should define IT strategies, policies, standards and operating procedures in line with organizational goals and objectives. The framework addresses the key elements within the IT governance model that enable the effective management and monitoring of an IT organization. This end-state is only possible when the organization's strategies, policies, standards and procedures are documented and adopted and implemented across the organization. The strategies, policies, standards and procedures also should contain specific management practices used to govern IT activities that support business needs at all levels.	Management provides strategic direction on the basis of which IT decisions and performance is taken across the enterprise	2.3 Governance of Enterprise IT 2.3.1 Good Practices for Governance of Enterprise IT
	Nature and purpose of IT strategies and how the governance and related frameworks enable an organization to meet goals and objectives	2.4 Information Systems Strategy

K2.2 Knowledge of IT governance, management, security and control frameworks, and related standards, guidelines and practices

Explanation	Key Concepts	Reference in Manual
In order to provide assurance to stakeholders that IT services are aligned with the business vision, mission and objectives, top management should implement an IT governance framework. IT governance frameworks include: • Strategic alignment of IT objectives with business objectives • Value delivery from IT • Risk management • Resource management • Performance management The IT governance framework enables stakeholders to be assured that the IT strategy, together with its interpretation into activities, is wholly aligned to the business. This includes the effective role of business executive management in the creation, maintenance and implementation of the IT governance and strategy through board- and executive-level committees. The committees, made up of "business/organization senior leaders," will examine and approve the IT strategy—together with its associated standards, procedures and guidelines—against the business strategy, goals and objectives to ensure that: • Technology will enable the achievement of those business/organization objectives through the timely implementation and adequate performance of the necessary facilities. • IT costs will be minimized in the provision of those facilities to obtain the best value from IT resources. • Roles and responsibilities, within both IT and business functions, are clearly defined. At all times, the governance framework will consider business risk associated with IT to ensure that risk is adequately and appropriately addressed.	Understanding IT governance frameworks	2.3.1 Good Practices for Governance of Enterprise IT 2.3.4 Information Security Governance
	Understanding roles and responsibilities as they relate to IT governance	2.10 IT Organizational Structure and Responsibilities 2.10.1 IT Roles and Responsibilities
	Good practices and how they are aligned with IT governance	2.7.1 Policies
	Current sourcing practices and their impact on IT governance	2.9.2 Sourcing Practices
	Impact of IT governance requirements on contractual commitments	2.11.2 Reviewing Contractual Commitments
	Purpose of control frameworks and how control frameworks are used in performance and resource management in an IT organization	2.3 Governance of Enterprise IT 2.9.7 Performance Optimization

K2.2 Knowledge of IT governance, management, security and control frameworks, and related standards, guidelines and practices (cont.)

Explanation	Key Concepts	Reference in Manual
Various standards, based on generally accepted good practices, are followed by organizations. These standards are generic in nature and should be adopted by enterprises, based on their specific needs. International IT standards and guidelines provide a wealth of benchmarking information for IT governance and facilitate a uniform approach to IT governance practices on a global basis. Knowledge of international IT standards and guidelines provides a ready reference to the IS auditor in evaluating IT governance initiatives and the current posture of organizations. In order to mitigate risk, organizations identify controls that they regard as being critical to the good management of the enterprise. Each control objective is derived from the risk it is addressing. Knowledge of various control frameworks helps in identifying appropriate control objectives required for the organization. Control frameworks such as COBIT, International Organization for Standardization (ISO) publications and other recognized and relevant standards are used to guide management in establishing IT practices; to monitor, measure and improve the performance of those practices; and to offer specific good practices that can be suited to particular business needs. These frameworks support IT governance processes within an organization and are important repositories of IT governance practices. Knowledge of different control frameworks assists the IS auditor in benchmarking controls identified by the organization. Knowledge and understanding of these control frameworks and their relevance to IT governance are essential to drive efficiencies and effectiveness in IT governance efforts. When implemented, control frameworks allow an IT organization to monitor and measure performance against IT strategies, policies and practices by outlining specific controls, procedures and best practices that can be used in IT governance. Frameworks provide the structure needed to implement key performance management, compliance management and IT resource management policies. Because these frameworks are considered to be generally accepted, they are also used to measure the performance of key IT service providers, vendors and outsourcing partners. The CISA exam will test the IS auditor's understanding of the frameworks and how the frameworks may be used to ensure the security, integrity and availability of information and processing.		

K2.3 Knowledge of organizational structure, roles and responsibilities related to IT, including segregation of duties (SoD)

Explanation	Key Concepts	Reference in Manual
Enterprises must clearly define organizational structure to enable resources to be deployed in a manner that will achieve the appropriate value and service delivery, security, risk management, and quality of information required by the organization. Defining organizational structure requires the outlining and documenting of the responsibilities of major organizational/business functions to ensure both proper segregation of duties (SoD) and to identify who in the organization uses and manages various information and related resources. The IS auditor should have a clear understanding of the organizational structure and the roles and responsibilities of personnel at all levels within the IT management structure and other areas of the organization in which responsibility for IT facilities or functions may exist (e.g., system and data owners) so that the requirements of each responsible person are transparent.	Understanding the relative roles of each level of organizational structure in IT governance	2.3.2 IT Governing Committees 2.3.4 Information Security Governance 2.9.3 Organizational Change Management 2.10 IT Organizational Structure and Responsibilities 2.10.1 IT Roles and Responsibilities

K2.4 Knowledge of relevant laws, regulations and industry standards affecting the organization

Explanation	Key Concepts	Reference in Manual
The complex nature of IT and global connectivity has introduced various types of risk within the organization's information life cycle—from receipt, processing, storage, transmission/distribution through destruction. In order to protect stakeholder interests, various legal and regulatory requirements have been enacted. The major compliance requirements that are considered globally recognized include protection of privacy and confidentiality of personal data, intellectual property rights and reliability of financial information. In addition, there are some compliance requirements that are industry specific. All of these drivers demand the development and implementation of well maintained, timely, relevant and actionable, organizational business policies, procedures and processes. Legislative and regulatory requirements pertaining to the access and use of IT resources, systems and data should be reviewed to assess whether the IT organization is protecting IT assets and effectively managing associated risk. For the CISA exam, the IS auditor must be aware of these globally recognized concepts; however, knowledge of specific legislation and regulations will not be tested.	Impact of legislative requirements on organizations standards, policies, procedures and processes	2.7.1 Policies 2.8.2 Risk Management Process 2.9.2 Sourcing Practices 2.9.6 Information Security Management 2.10.2 Segregation of Duties Within IT 2.10.3 Segregation of Duties Controls 2.11 Auditing IT Governance Structure and Implementation 2.11.1 Reviewing Documentation 2.11.2 Reviewing Contractual Commitments

K2.5 Knowledge of the organization's technology direction and IT architecture and their implications for setting long-term strategic directions

Explanation	Key Concepts	Reference in Manual
Effective IT strategic planning involves a consideration of the enterprise's requirements for new and revised IT systems and the IT organization's capacity to deliver new functionality through well-governed projects. Determining requirements for new and revised IT systems will involve a systematic consideration of the enterprise's strategic intentions, how these translate into specific objectives and business initiatives and what IT capabilities will be needed to support these objectives and initiatives. In assessing IT capabilities, the existing system's portfolio should be reviewed in terms of functional fit, cost and risk. The strategic IT plan should balance the cost of maintenance of existing systems against the cost of new initiatives or systems to support the business strategies. The IS auditor should be aware that a key input to determining the long-term strategic direction of an IT organization is the review, analysis and assessment of its IT architecture. The review, analysis and assessment may take the form of a road map and may illustrate current and future states. Review of the enterprise's IT architecture and its usage can help to determine whether management is following its IT strategy and whether that strategy needs to be adapted to changing business needs.	Relevance of different elements of enterprise architecture and their impact on IT governance	2.3.5 Enterprise Architecture 2.4.1 Strategic Planning
	Alignment of policies with enterprise architecture and their relation to IT governance	2.7.1 Policies

K2.6 Knowledge of the processes for the development, implementation and maintenance of IT strategy, policies, standards and procedures

Explanation	Key Concepts	Reference in Manual
Senior management should define a process for developing IT strategies that achieve business objectives. These IT strategies must be based wholly on defined business objectives with a clear understanding of the relevant laws, regulations and industry standards that the organization must comply with across all locations within the enterprise. The successful integration of both sound IT strategy and compliance processes enable organization to achieve business objectives. Key to this success is the quality of governance processes related to the development and implementation of IT strategic and tactical policies, standards and procedures. The IT strategy must be subjected to periodic review to ensure that the strategy continues to address both emerging and developing business needs and regulatory and industry risk. Specifically, good IT governance requires that all the dynamic industry and regulatory influences to be identified, considered as their impacts, approved by business executive management and subsequently monitored. These practices form part of the IT governance program and should be understood by the IS auditor.	Factors that contribute to the development and implementation of an IT strategy	2.4.1 Strategic Planning 2.8.2 Risk Management Process
	Factors that contribute to effective information security governance and management	2.3.4 Information Security Governance 2.9.6 Information Security Management

K2.7 Knowledge of the use of capability and maturity models

Explanation	Key Concepts	Reference in Manual
The effectiveness and efficiency of IT governance efforts in the organization are dependent on the quality management strategies and policies that are embedded in the IT governance framework. The integration of defined processes and corresponding process management techniques across the organization's enterprise is related to the effectiveness and efficiency of the IS organization. Quality management strategies and policies outline how the IT strategies, policies, procedures and standards are maintained, used and improved over time as the organization changes. The IS auditor needs to understand how the development, implementation and integration of capability and maturity modeling quality tools, techniques and processes (TTPs) will facilitate and foster the quality of enterprise IT policies and procedures. These TTPs can be based on a variety of standard frameworks. The use of quality standards within an IS organization enhances the ability of the IT organization to realize greater value and mission success.	Understanding management techniques to continuously improve IT performance	2.5 Maturity and Process Improvement Models
	Knowledge of quality standards	2.9.5 Quality Management 2.9.7 Performance Optimization

K2.8 Knowledge of process optimization techniques

Explanation	Key Concepts	Reference in Manual
Maturity and process improvement models help enterprises evaluate the current state of their internal controls environment in comparison to the desired state and help identify activities for moving toward the desired state. A variety of improvement and optimization methodologies are available that complement simple, internally developed approaches. These include: • Continuous improvement methodologies, such as the Plan-Do-Check-Act cycle and specifically as implemented during agile development/project management • Comprehensive best practices, such as ITIL® • Frameworks, such as COBIT and Val IT™ • The Zachman Framework™ This evaluation is important to the IS auditor because the results illustrate to executive management the effectiveness, compliance and relevance of its IT procedures, tools and processes in support of alignment with business needs. This evaluation can be further used to review management practices within IT to determine compliance with organizational IT strategies and policies.	Current practices in measuring the maturity state of the organization	2.3.5 Enterprise Architecture 2.5 Maturity and Process Improvement Models
	Impact of sourcing practices on the current maturity state and desired maturity state	2.9.2 Sourcing Practices
	Role of quality management in bridging the gap between current state and desired state	2.9.5 Quality Management 2.9.7 Performance Optimization

K2.9 Knowledge of IT resource investment and allocation practices, including prioritization criteria (e.g., portfolio management, value management, personnel management)

<table>
<tr><th>Explanation</th><th>Key Concepts</th><th colspan="2">Reference in Manual</th></tr>
<tr><td rowspan="5">Organizations deploy IT resources to ensure that service delivery and value meet established goals and objectives. Furthermore, they evaluate service delivery and value in relation to investment in IT. Knowledge of IT resource investment and allocation practices is essential to justify the investment in IT governance to stakeholders. Methods for allocating resources to IT investments allow a predictable and consistent approach to authorizing funds to IT initiatives that demonstrate tangible benefits to the organization. Specific practices to evaluate IT initiatives, such as cost-benefit analysis and planned and forecasted resource consumption, are executed to ensure that management is funding projects and initiatives that meet the needs of the organization. The costs and benefits should be reviewed on a periodic basis throughout the execution of those initiatives.

The increased automation of business processes has created challenges in optimal management of human resources (HR) and in addressing the control gaps that are created when job roles are combined through automation of tasks. Performance evaluations, compensation plans and succession planning are important. The IS auditor must understand the need for good management of HR in relation to IT, most notably the need to remove unnecessary risk by verifying the qualifications, history and references of applicants; verifying the necessary skill sets required for the achievement of IT objectives, including training requirements; and recognizing the potential need for employee termination to be immediate rather than allowing a "notice period."

Process optimization techniques help enterprises prioritize investment initiatives, eliminate unnecessary activities and either re-purpose and/or reallocate the underutilized resources to maintain alignment with organization's enterprise goals and objectives. Portfolio management coupled with value and personnel management enable agile response to environmental factors affecting the enterprise. Process optimization requires evaluating the current state of the environment in comparison to an optimum design and then identifying activities that can be eliminated in order to migrate to the desired state.

This includes ongoing embedded process analysis that can detect and allow management to correct variance and anomalous processes in a timely manner. The IS auditor needs to understand the criticality of these IT resource management allocation processes in order to evaluate how the governance processes are fully integrated within the enterprise's IT architecture.</td><td>Awareness of current practices in IT investment and resource allocation</td><td>2.3.5</td><td>Enterprise Architecture</td></tr>
<tr><td>Role of financial management practices in IT portfolio management</td><td>2.6

2.9.4</td><td>IT Investment and Allocation Practices
Financial Management Practices</td></tr>
<tr><td>Role of HR processes and policies on IT governance</td><td>2.9.1
2.9.3</td><td>Human Resource Management
Organizational Change Management</td></tr>
<tr><td>HR management and its responsiveness to the changing needs of the IT organization</td><td>4.8.5</td><td>Organization and Assignment of Responsibilities</td></tr>
<tr><td>Current practices in process optimization of IT resources</td><td>2.9.7</td><td>Performance Optimization</td></tr>
</table>

K2.10 Knowledge of IT supplier selection, contract management, relationship management and performance monitoring processes including third party outsourcing relationships

Explanation	Key Concepts	Reference in Manual
Critical to an organization's enterprise IT operations are its IT supplier, business partnering and relationship management processes. Outsourcing IT (and related solutions such as process management and infrastructure management) can help reduce costs and/or complement an enterprise's own expertise. Organizations deploy IT resources to ensure service delivery and value, and they evaluate service delivery and value in relation to investment in IT. Knowledge of IT supplier selection, contract management, relationship management and performance monitoring processes, including third-party outsourcing relationships, plays a critical role in the overall management of the enterprise IT portfolio. With the increasing trend of outsourcing IT infrastructure to third-party service providers, specific practices to evaluate these IT initiatives have been developed (e.g., cost-benefit analysis). It is essential that the IS auditor understand the latest approaches in contract strategies, processes and management practices, and how outsourcing may introduce additional risk. Thus, it is essential for the IS auditor to understand the soundest approaches in contract strategies, processes and management practices, such as what critical concepts must be included in an outsourcing contract and business case requirements.	Awareness of current practices in IT investment and resource allocation	2.6 IT Investment and Allocation Practices
	Role of financial management practices in IT portfolio management	2.9.4 Financial Management Practices
	Impact of sourcing practices on IT governance	2.9.2 Sourcing Practices
	Relationship between vendor management and IT governance of the outsourcing entity	2.10.1 IT Roles and Responsibilities
	Contractual terms and their impact on driving IT governance of the outsourcing entity	2.11.2 Reviewing Contractual Commitments

K2.11 Knowledge of enterprise risk management (ERM)

Explanation	Key Concepts	Reference in Manual
Oversight of the enterprise's IT-related business risk is essential to achieve effective governance. In turn, knowledge of risk management methodologies and tools is essential to assessing and mitigating the organization's IT-related business risk. Enterprises may follow different risk management models to manage risk. The IS auditor should be aware of concepts related to risk management, such as risk identification, assessment evaluation, risk response, risk monitoring, risk governance, etc. The IS auditor needs to be aware of risk response techniques such as avoid, mitigate, share/transfer and accept. The IS auditor also should be aware that the controls are identified, designed and implemented based on regulatory, contractual and organizational mission impact. Also within the evaluation of a mitigating controls implementation is a feasibility analysis looking at both organizational risk appetite and cost-benefit analysis where the risk appetite is not exceeded and the benefits derived from the risk mitigation do not exceed the cost of the control.	Risk management process and applying various risk analysis methods	2.8 Risk Management 2.8.1 Developing a Risk Management Program 2.8.2 Risk Management Process 2.8.3 Risk Analysis Methods

K2.12 Knowledge of practices for monitoring and reporting of controls performance (e.g., continuous monitoring, quality assurance)

<table>
<tr><th>Explanation</th><th>Key Concepts</th><th colspan="2">Reference in Manual</th></tr>
<tr><td rowspan="3">Enterprises are governed by generally accepted good or best practices, ensured by the establishment of controls. Good practices guide organizations in determining how to use resources. Results are measured and reported, providing input to the cyclical revision and maintenance of controls. In order to evaluate, maintain and enhance system of control, the enterprise IT organization must establish both quality metrics (key performance indicators [KPIs]) and monitoring processes to enable agile response to changes within the enterprise and/or industry.

The IS auditor needs to understand the key business/organizational goals and objectives and both the risk management processes and risk environment the enterprise IT operates. This knowledge and understanding better enables the IS auditor to evaluate the effectiveness and degree of fidelity the organization's controls are performing along with the relevance and accuracy of monitoring and reporting on these controls.</td><td>Accepted good practices for control performance monitoring and reporting</td><td>2.3.1</td><td>Good Practices for Governance of Enterprise IT</td></tr>
<tr><td>Components of the IT balanced scorecard and its relevance for IT governance</td><td rowspan="2">2.3.3</td><td rowspan="2">IT Balanced Scorecard</td></tr>
<tr><td>Use of KPIs in driving performance optimization for effective IT governance</td></tr>
</table>

K2.13 Knowledge of quality management and quality assurance systems

<table>
<tr><th>Explanation</th><th>Key Concepts</th><th colspan="2">Reference in Manual</th></tr>
<tr><td rowspan="2">The integrity and reliability of enterprise IT processes are directly attributed to the quality assurance (QA) processes in place and integrated within the enterprise. The QA program and respective policies, procedures and processes are encompassed within a planned and systematic pattern of all actions necessary to provide adequate confidence that an item or product conforms to established technical requirements. QA helps the IT department to ensure that personnel are following prescribed quality processes. For example, QA will set up procedures (e.g., ISO 9001–compliant) to facilitate widespread use of quality management/assurance.

The degree and level of quality within the enterprise IT operations can be measured and analyzed. This information can be used to correct existing deviations from desired performance and to predict and prevent future deficiencies.

The IS auditor needs to understand the QA concepts, structures, and roles and responsibilities within the organization.</td><td>Structures, roles and responsibilities of the QA function with the enterprise</td><td>2.10.1</td><td>IT Roles and Responsibilities</td></tr>
<tr><td>Use of key performance indicators (KPIs) in driving performance optimization for effective IT governance</td><td>2.9.7</td><td>Performance Optimization</td></tr>
</table>

K2.14 Knowledge of practices for monitoring and reporting of IT performance (e.g., balanced scorecards [BSCs], key performance indicators [KPIs])

Explanation	Key Concepts	Reference in Manual
Corporate IT governance provides the structure through which the objectives of the company are set and the means of attaining those objectives and monitoring performance are determined. Progress of the organization along the path of IT governance must be measured and monitored through effective tools such as balanced scorecards (BSCs) and key performance indicators (KPIs). BSCs and KPIs translate the expectations from IT governance into terms that the process owner understands. The results provide insight into the capabilities of the IT organization to meet its objectives and can be used to determine whether changes are required to the IT strategy over the long term as the IT organization strives to meet the needs of the enterprise. The IS auditor must both understand how KPIs and BSCs are integrated within the organization's governance processes and if the data being provided actually map to meaningful measures of enterprise IT performance.	Concepts related to establishing, monitoring and reporting processes needed by the governance team to evaluate performance and provide direction to senior management	2.3.1 Good Practices for Governance of Enterprise IT 2.3.3 IT Balanced Scorecard 2.3.4 Information Security Governance 2.9.7 Performance Optimization

K2.15 Knowledge of business impact analysis (BIA)

Explanation	Key Concepts	Reference in Manual
An IS auditor must be able to determine whether a business impact analysis (BIA) and business continuity plan (BCP) are suitably aligned. To be effective and efficient, the BCP should be based on a well-documented BIA. A BIA drives the focus of the BCP efforts of an organization and helps in balancing costs to be incurred with the corresponding benefits to the organization. A good understanding of the BIA concept is essential for the IS auditor in order to audit the effectiveness and efficiency of a BCP.	Understanding the BIA as a key driver of the BCP/disaster recovery process	2.12.6 Business Impact Analysis

K2.16 Knowledge of the standards and procedures for the development, maintenance and testing of the business continuity plan

Explanation	Key Concepts	Reference in Manual
An IS auditor should be well-versed in the practices and techniques followed for development and maintenance of business continuity plans (BCPs)/disaster recovery plans (DRPs), including the need to coordinate recovery plans across the organization. Plans should be tailored to fit the individual needs of organizations because differences in industry, size and scope of an organization, and even geographic location, can affect the contents of the plans. The size and nature of the selected recovery facility for technology will materially depend on the overall risk associated with disruption. In essence, the faster the required recovery, as determined by the recovery time objective (RTO), the greater the potential cost. Once established, recovery plans must be kept up to date with changes in the organization and associated risk.	Understanding the life cycle of BCP/DRP development and maintenance	2.12.1 IT Business Continuity Planning 2.12.3 Business Continuity Planning Process 2.12.4 Business Continuity Policy 2.12.5 Business Continuity Planning Incident Management 2.12.7 Development of Business Continuity Plans 2.12.8 Other Issues in Plan Development 2.12.9 Components of a Business Continuity Plan 2.12.10 Plan Testing 2.12.11 Summary of Business Continuity
An IS auditor should know the testing approaches and methods for BCP/DRP to evaluate the effectiveness of the plans. To ensure that the BCP/DRP will work in the event of a disaster, it is important to periodically test the BCP/DRP, also ensuring that the testing effort is efficient. The role of the IS auditor is to observe tests, ensure that all "lessons learned" are properly recorded and reflected in a revised plan, and review write-ups that document previous tests. Key items to look for include the degree to which the test leverages resources or extensive preplanning meetings that would not be available during an actual disaster. The objective of a test should be to identify gaps that can be improved on, rather than to have a flawless test. Another important aspect of BCP/DRP testing is to provide training for management and staff who may be involved in the recovery process.	Understanding the types of BCP tests, factors to consider when choosing the appropriate test scope, methods for observing recovery tests and analyzing test results	2.12.10 Plan Testing 2.13 Auditing Business Continuity

K2.17 Knowledge of procedures used to invoke and execute the business continuity plan and return to normal operations

Explanation	Key Concepts	Reference in Manual
The IS auditor needs to not only evaluate the content of the organization's business continuity plan (BCP) but also determine if the methodology, processes and procedures are in place to realistically initiate the business continuity and resumption of normal operations after the event causing disruption of business. Specifically, the IS auditor should verify that initiating triggers are based on the service level thresholds identified in the business impact analysis (BIA). Furthermore, the procedures being evoked must be validated to reasonably assure these processes accurately reflect the actions required to both compensate for immediate business disruption and promptly resume normal business operations. For example, an organization's critical patient billing portal transaction processing capacity drops to 10 claims an hour (well below the 100,000 claims per minute service level established for the clearinghouse). The organization's incident response team suspects a denial-of-service attack. The IS auditor needs to evaluate the procedures initiated by the organization to assure prompt recovery and resumption of portal operation. Specific areas to be addressed include, but are not limited to: • Incident triage (what will trigger the response plan to be initiated) • Notification and escalation processes • Implementation of compensating controls if applicable to enable the portal to meet service level agreements as close as possible to normal operating parameters until the incident is resolved • All resources (hardware/software, communication links, personnel within the correct organizational structure and properly level of authority to carry out the assigned responsibilities)	Understanding how the BIA defines the triggers to initiate the various actions within the business continuity plan (BCP)/disaster recovery (DRP) process	2.12.6 Business Impact Analysis
	Ability to evaluate the procedures and processes within the BCP provide reasonable assurance in enabling prompt resumption of normal operations.	2.12.7 Development of Business Continuity Plans 2.12.8 Other Issues in Plan Development 2.12.9 Components of a Business Continuity Plan 2.12.10 Plan Testing

SUGGESTED RESOURCES FOR FURTHER STUDY

Burtles, Jim; *Principles and Practice of Business Continuity: Tools and Techniques*, Rothstein Associates Inc., USA, 2007

Graham, Julia; David Kaye; *A Risk Management Approach to Business Continuity*, Rothstein Associates Inc., USA, 2006

Hiles, Andrew; *The Definitive Handbook of Business Continuity Management, 3rd Edition*, John Wiley & Sons Inc., USA, 2011

ISACA, COBIT 5, USA, 2012, *www.isaca.org/cobit*

International Organization for Standardization (ISO), *ISO/IEC 38500:2015: Information technology — Governance of IT for the organization*, Switzerland, 2015

IT Governance Institute, *Board Briefing on IT Governance, 2nd Edition*, USA, 2003, *www.isaca.org*

Ramos, Michael J.; *How to Comply With Sarbanes-Oxley Section 404, 3rd Edition*, John Wiley & Sons Inc., USA, 2008

Raval, Vasant; Ashok Fichadia; *Risks, Controls, and Security: Concepts and Applications*, John Wiley & Sons, USA, 2007, Chapter 6: System Availability and Business Continuity

Sherwood, John; Andrew Clark; David Lynas; *Enterprise Security Architecture: A Business-Driven Approach*, UK, 2008

Tarantino, Anthony; *Manager's Guide to Compliance: Sarbanes-Oxley, COSO, ERM, COBIT, IFRS, BASEL II, OMB's A-123, ASX 10, OECD Principles, Turnbull Guidance, Best Practices, and Case Studies*, John Wiley & Sons Inc., USA, 2006

***Note:* Publications in bold are stocked in the ISACA Bookstore.**

SELF-ASSESSMENT QUESTIONS

CISA self-assessment questions support the content in this manual and provide an understanding of the type and structure of questions that have typically appeared on the exam. Questions are written in a multiple-choice format and designed for one best answer. Each question has a stem (question) and four options (answer choices). The stem may be written in the form of a question or an incomplete statement. In some instances, a scenario or a description problem may also be included. These questions normally include a description of a situation and require the candidate to answer two or more questions based on the information provided. Many times a question will require the candidate to choose the **MOST** likely or **BEST** answer among the options provided.

In each case, the candidate must read the question carefully, eliminate known incorrect answers and then make the best choice possible. Knowing the format in which questions are asked, and how to study and gain knowledge of what will be tested, will help the candidate correctly answer the questions.

2-1 In order for management to effectively monitor the compliance of processes and applications, which of the following would be the **MOST** ideal?

A. A central document repository
B. A knowledge management system
C. A dashboard
D. Benchmarking

2-2 Which of the following would be included in an IS strategic plan?

A. Specifications for planned hardware purchases
B. Analysis of future business objectives
C. Target dates for development projects
D. Annual budgetary targets for the IT department

2-3 Which of the following **BEST** describes an IT department's strategic planning process?

A. The IT department will have either short- or long-range plans depending on the organization's broader plans and objectives.
B. The IT department's strategic plan must be time- and project-oriented but not so detailed as to address and help determine priorities to meet business needs.
C. Long-range planning for the IT department should recognize organizational goals, technological advances and regulatory requirements.
D. Short-range planning for the IT department does not need to be integrated into the short-range plans of the organization since technological advances will drive the IT department plans much quicker than organizational plans.

2-4 The **MOST** important responsibility of a data security officer in an organization is:

A. recommending and monitoring data security policies.
B. promoting security awareness within the organization.
C. establishing procedures for IT security policies.
D. administering physical and logical access controls.

2-5 What is considered the **MOST** critical element for the successful implementation of an information security program?

A. An effective enterprise risk management (ERM) framework
B. Senior management commitment
C. An adequate budgeting process
D. Meticulous program planning

2-6 An IS auditor should ensure that IT governance performance measures:

A. evaluate the activities of IT oversight committees.
B. provide strategic IT drivers.
C. adhere to regulatory reporting standards and definitions.
D. evaluate the IT department.

2-7 Which of the following tasks may be performed by the same person in a well-controlled information processing computer center?

A. Security administration and change management
B. Computer operations and system development
C. System development and change management
D. System development and system maintenance

2-8 Which of the following is the **MOST** critical control over database administration (DBA)?

A. Approval of DBA activities
B. Segregation of duties (SoD) in regard to access right granting/revoking
C. Review of access logs and activities
D. Review of the use of database tools

2-9 When a complete segregation of duties (SoD) cannot be achieved in an online system environment, which of the following functions should be separated from the others?

A. Origination
B. Authorization
C. Recording
D. Correction

2-10 In a small organization where segregation of duties (SoD) is not practical, an employee performs the function of computer operator and application programmer. Which of the following controls should the IS auditor recommend?

A. Automated logging of changes to development libraries
B. Additional staff to provide SoD
C. Procedures that verify that only approved program changes are implemented
D. Access controls to prevent the operator from making program modifications

ANSWERS TO SELF-ASSESSMENT QUESTIONS

2-1 A. A central document repository provides a great deal of data but not necessarily the specific information that would be useful for monitoring and compliance.
B. A knowledge management system provides valuable information but is generally not used by management for compliance purposes.
C. A dashboard provides a set of information to illustrate compliance of the processes, applications and configurable elements and keeps the enterprise on course.
D. Benchmarking provides information to help management adapt the organization, in a timely manner, according to trends and environment.

2-2 A. Specifications for planned hardware purchases are not strategic items.
B. IS strategic plans must address the needs of the business and meet future business objectives. Hardware purchases may be outlined, but not specified, and neither budget targets nor development projects are relevant choices.
C. Target dates for development projects are not strategic items.
D. Annual budgetary targets for the IT department are not strategic items.

2-3 A. Typically, the IT department will have short- or long-range plans that are consistent and integrated with the organization's plans.
B. These plans must be time- and project-oriented and address the organization's broader plans toward attaining its goals.
C. Long-range planning for the IT department should recognize organizational goals, technological advances and regulatory requirements.
D. Short-range planning for the IT department should be integrated into the short-range plans of the organization to better enable the IT department to be agile and responsive to needed technological advances that align with organizational goals and objectives.

2-4 **A. A data security officer's prime responsibility is recommending and monitoring data security policies.**
B. Promoting security awareness within the organization is one of the responsibilities of a data security officer, but it is not as important as recommending and monitoring data security policies.
C. The IT department, not the data security officer, is responsible for establishing procedures for IT security policies recommended by the data security officer.
D. The IT department, not the data security officer, is responsible for the administration of physical and logical access controls.

2-5 A. An effective enterprise risk management (ERM) framework is not a key success factor for an information security program.
B. Commitment from senior management provides the basis to achieve success in implementing an information security program.
C. Although an effective information security budgeting process will contribute to success, senior management commitment is the key element.
D. Program planning is important, but will not be sufficient without senior management commitment.

2-6 **A. Evaluating the activities of boards and committees providing oversight is an important aspect of governance and should be measured.**
B. Providing strategic IT drivers is irrelevant to the evaluation of IT governance performance measures.
C. Adhering to regulatory reporting standards and definitions is irrelevant to the evaluation of IT governance performance measures.
D. Evaluating the IT department is irrelevant to the evaluation of IT governance performance measures.

2-7 A. The roles of security administration and change management are incompatible functions. The level of security administration access rights could allow changes to go undetected.
B. Computer operations and system development is the incorrect choice because this would make it possible for an operator to run a program that he/she had amended.
C. The combination of system development and change control would allow program modifications to bypass change control approvals.
D. It is common for system development and maintenance to be undertaken by the same person. In both, the programmer requires access to the source code in the development environment but should not be allowed access in the production environment.

2-8 A. Approval of database administration (DBA) activities does not prevent the combination of conflicting functions. Review of access logs and activities is a detective control.
B. Segregation of duties (SoD) will prevent combination of conflicting functions. This is a preventive control, and it is the most critical control over DBA.
C. If DBA activities are improperly approved, review of access logs and activities may not reduce the risk.
D. Reviewing the use of database tools does not reduce the risk because this is only a detective control and does not prevent combination of conflicting functions.

2-9 A. Origination in conjunction with recording and correction does not enable the transaction to be authorized for processing and committed within the system of record.
B. Authorization should be separated from all aspects of record keeping (origination, recording and correction). Such a separation enhances the ability to detect the recording of unauthorized transactions.
C. Recording in conjunction with origination and correction does not enable the transaction to be authorized for processing and committed within the system of record.
D. Correction in conjunction with origination and recording does not enable the transaction to be authorized for processing and committed within the system of record.

2-10 A. Logging changes to development libraries would not detect changes to production libraries.
B. In smaller organizations, it generally is not appropriate to recruit additional staff to achieve a strict segregation of duties (SoD). The IS auditor must look at alternatives.
C. The IS auditor should recommend processes that detect changes to production source and object code, such as code comparisons, so the changes can be reviewed by a third party on a regular basis. This would be a compensating control process.
D. Access controls to prevent the operator from making program modifications require a third party to do the changes, which may not be practical in a small organization.

Section Two: Content

2.1 QUICK REFERENCE

Quick Reference Review

Chapter 2 addresses the need for IT governance. An IS auditor must be able to understand and provide assurance that the organization has the structure, policies, accountability mechanisms and monitoring practices in place to achieve the requirements of corporate governance of IT. For an IS auditor, knowledge of IT governance forms the foundation for evaluating control practices and mechanisms for management oversight and review.

CISA candidates should have a sound understanding of the following items. It is important to keep in mind that it is not enough to know these concepts from a definitional perspective. Examples of key topics in this chapter include:

- An objective of corporate governance is to resolve the conflicting objectives of exploiting available opportunities to increase stakeholder value while keeping the organization's operations within the limits of regulatory requirements and social obligations. Applied to IT, governance helps ensure the alignment of IT and enterprise objectives. IT governance is concerned with two issues: that IT delivers value to the business and that IT risk is managed. The first is driven by strategic alignment of IT with the business. The second is driven by establishing risk governance and management as well as accountability into the enterprise. IT governance is the responsibility of the board of directors and executive management, and the key IT governance practices for executive management include an IT strategy committee, a risk management process and an IT balanced scorecard.
- Governance of enterprise IT is a governance view that ensures that information and related technology support and enable the enterprise strategy and the achievement of enterprise objectives; this also includes the functional governance of IT (i.e., ensuring that IT capabilities are provided efficiently and effectively). Effective IT strategic planning involves consideration of the organization's demand for IT and its IT supply capacity. The strategy is governed by a steering committee and the strategy is guided and controlled by policies and procedures, including the information security policy. Strategies, policies and procedures should be evaluated by the IS auditor to determine the importance placed on the planning process; the involvement of senior IT management in the overall business strategy; and policy compliance, relevance and applicability to third parties.
- An IT strategy/steering committee monitors IT value, risk and performance and provides information to the board to support decision making on IT strategies. The IS auditor must evaluate the effectiveness of IT governance structure to ensure adequate board control over the decisions, direction and performance of IT, so that it supports the organization's strategies and objectives.
- A key aspect of IT governance is the governance of information security. Information is one of an organization's most valuable assets and must be adequately protected regardless of how it is created, received, handled, processed, transported, stored or disposed. Information security includes all information processes, physical and electronic, regardless of whether they involve people, technology or relationships with trading partners, customers and third parties. It ensures that information security risk is appropriately managed and enterprise information resources are used responsibly.
- The governance of information security should be executed and supported with information security strategies, policies and organization structure. Information security governance must be the responsibility of the board of directors/senior management to approve policy and penalties for noncompliance, and the mandate of information security may be delegated to a chief information security officer (CISO).

Quick Reference Review ***(cont.)***

- IT governance encompasses minimizing IT risk to the organization. Risk management is the process of identifying vulnerabilities and threats to the information resources used by an organization in achieving business objectives and deciding what countermeasures (safeguards or controls), if any, to take in reducing risk to an acceptable level (i.e., residual risk), based on the value of the information resource to the organization. This process begins with understanding the organization's appetite for risk and then determining the risk exposure on its IT assets. From this identification, risk management strategies and responsibilities are defined. Depending on the type of risk and its significance to the business, risk can be avoided, mitigated, transferred or accepted. The result of a risk occurring is called an impact and can result in losses, such as financial, legal, reputational and efficiency.
- Risk is measured using a qualitative analysis (defining risk in terms of high/medium/low), semiquantitative analysis (defining risk according to a numeric scale) or quantitative analysis (applying several values to risk, including financial, and calculating the risk's probability and impact). After risk has been identified, existing or new controls are designed and measured for their strength and likelihood of effectiveness. Controls may be preventive, detective or corrective; manual or automated; and formal (i.e., documented) or *ad hoc*. Moreover, compensating controls may also be present. Residual risk can be used by management to determine which areas require more control and whether the benefits of such controls outweigh the costs. This entire process of IT risk management needs to be managed at multiple levels in the organization, including the operational, project and strategic levels, and should form part of the IT business management practice. Risk analysis and risk management plans should be periodically reviewed as the environment and organization changes.
- Key management processes that will shape the effectiveness of an IT department and outline controls on strategy and use of resources are human resource management, change management, financial practices, quality management, information security management and performance optimization practices. Management's control and governance of the IS environment can be evaluated based on the review of its organizational structure. Charts should provide a clear definition of the department's hierarchy and authorities, and the specific roles and responsibilities. The structure should define the role of each area in the IT department and indicate appropriate segregation of duties (SoD) within the IT department.
- The purpose of segregation of duties is to prevent fraud and error by splitting tasks and authority to accomplish a process among multiple employees or managers. Specifically, the duties that should be segregated are custody of the assets, authorization and recording of transactions. If combined roles are required, then compensating controls should be described and applied as appropriate for the organization. While assigning new roles or modifying existing ones, it is important ensure that incompatible roles are not assigned. Roles should be reviewed periodically to ensure against function creep.

Quick Reference Review *(cont.)*

This chapter also addresses the need for business continuity and disaster recovery within an organization. Most organizations have some degree of disaster recovery plans (DRPs) in place for the recovery of IT infrastructure, critical systems and associated data. However, many organizations have not taken the next step and developed plans for how key business units will function during a period of IT disruption. CISA candidates should be aware of the components of DRPs and business continuity plans (BCPs), the importance of aligning one with the other, and aligning DRPs and BCPs with the organization's risk appetite and tolerance.

In summary, for an IS auditor, all IT business management practices should be evaluated to determine management's governance over IT, including documentation regarding IT strategies, budgets, policies and procedures; control over information security as it relates to compartmentalization of access rights; as well as the structure of the IT department, because each of these elements illustrates how effective an organization is at ensuring that IT delivers value to the business and IT risk is managed.

2.2 CORPORATE GOVERNANCE

Ethical issues, decision making and overall practices within an organization must be fostered through corporate governance practices. Corporate governance has been defined as "the system by which business corporations are directed and controlled" (International Finance Corporation; Vietnam Ministry of Finance; Organisation for Economic Cooperation and Development; *International Corporate Governance Meeting: Why Corporate Governance Matters for Vietnam*, Hanoi, Vietnam, 6 December 2004). More specifically, corporate governance is a set of responsibilities and practices used by an organization's management to provide strategic direction, thereby ensuring that goals are achievable, risk is properly addressed and organizational resources are properly utilized. In its *Principles of Corporate Governance* (2004), the Organisation for Economic Cooperation and Development (OECD) states: "Corporate governance involves a set of relationships between a company's management, its board, its shareholders and other stakeholders. Corporate governance also provides the structure through which the objectives of the company are set, and the means of attaining those objectives and monitoring performance are determined. Good corporate governance should provide proper incentives for the board and management to pursue objectives that are in the interests of the company and its shareholders and should facilitate effective monitoring."

This framework is being increasingly utilized by government bodies of different countries in an effort to reduce the frequency and impact of inaccurate financial reporting and provide greater transparency and accountability. Many of these government regulations include a requirement that senior management sign off on the adequacy of internal controls and include an assessment of organizational internal controls in the organization's financial reports.

2.3 GOVERNANCE OF ENTERPRISE IT

Governance of enterprise IT (GEIT) implies a system in which all stakeholders, including the board, senior management, internal customers and departments such as finance, provide input into the decision-making process.

GEIT is the management system used by board of directors. In other words, GEIT is about the stewardship of IT resources on behalf of all stakeholders (internal and external stakeholders) who expect their interests to be met. The board of directors responsible for this stewardship will look to management to implement the necessary systems and IT controls.

GEIT is the responsibility of the board of directors and executive management.

The purpose of GEIT is to direct IT endeavors to ensure that IT performance meets the objectives of aligning IT with the enterprise's objectives and the realization of promised benefits. Additionally, IT should enable the enterprise by exploiting opportunities and maximizing benefits. IT resources should be used responsibly, and IT-related risk should be managed appropriately.

Implementing the GEIT framework addresses these two issues by implementing practices that provide feedback on value delivery and risk management. The broad processes are:

- IT resource management—Focuses on maintaining an updated inventory of all IT resources and addresses the risk management process
- Performance measurement—Focuses on ensuring that all IT resources perform as expected to deliver value to the business and also extends to identifying risk early on. This process is based on performance indicators that are optimized for value delivery and from which any deviation might lead to a materialization of risk.
- Compliance management—Focuses on implementing processes that address legal and regulatory policy and contractual compliance requirements

ISACA's COBIT 5 framework makes a clear distinction between governance and management. These two disciplines encompass different types of activities, require different organizational structures and serve different purposes. COBIT 5's view on this key distinction between governance and management is:

- **Governance**—Ensures that stakeholder needs, conditions and options are evaluated to determine balanced, agreed-on enterprise objectives to be achieved; setting direction through prioritization and decision making; and monitoring performance and compliance against agreed-on direction and objectives.
- **Management**—Plans, builds, runs and monitors activities in alignment with the direction set by the governance body to achieve the enterprise objectives.

GEIT, one of the domains of enterprise governance, comprises the body of issues addressed in considering how IT is applied within the enterprise.

Effective enterprise governance focuses individual and group expertise and experience on specific areas where they can be most effective. IT, long considered only an enabler of an organization's strategy, is now regarded as an integral part of that strategy. Chief executive officers (CEOs), chief operating officers (COOs), chief financial officers (CFOs), chief information officers (CIOs) and chief technology officers (CTOs) agree that strategic alignment between IT and enterprise objectives is a critical success factor. IT governance helps achieve this critical

success factor by economically, efficiently and effectively deploying secure, reliable information and applied technology. IT is so critical to the success of enterprises that it cannot be relegated to either IT management or IT specialists, but must receive the attention of both under the guidance and supervision of senior management and oversight by the board of directors. A key element of GEIT is the alignment of business and IT, leading to the achievement of business value.

Fundamentally, GEIT is concerned with two issues: that IT delivers value to the business and that IT risk is managed. The first is driven by strategic alignment of IT with the business. The second is driven by embedding accountability into the enterprise.

2.3.1 GOOD PRACTICES FOR GOVERNANCE OF ENTERPISE IT

GEIT integrates and institutionalizes good practices to ensure that the enterprise's IT supports the business objectives. GEIT enables the enterprise to take full advantage of its information, thereby maximizing benefits, capitalizing on opportunities and gaining competitive advantage. GEIT is a structure of relationships and processes used to direct and control the enterprise toward achievement of its goals by adding value while balancing risk versus return over IT and its processes.

The topics that executive management must address to govern IT within the enterprise are described in three focus areas: benefits realization, risk optimization and resource optimization (**figure 2.2**).

GEIT has become significant due to a number of factors:
- Business managers and boards demanding a better return from IT investments (i.e., that IT deliver what the business needs to enhance stakeholder value)
- Concern over the generally increasing level of IT expenditure
- The need to meet regulatory requirements for IT controls in areas such as privacy and financial reporting (e.g., the US Sarbanes-Oxley Act, Basel Accords) and in specific sectors such as finance, pharmaceuticals and health care
- The selection of service providers and the management of service outsourcing and acquisition (e.g., cloud computing)
- IT governance initiatives that include adoption of control frameworks and good practices to help monitor and improve critical IT activities to increase business value and reduce business risk
- The need to optimize costs by following, where possible, standardized rather than specially developed approaches
- The growing maturity and consequent acceptance of well-regarded frameworks
- The need for enterprises to assess how they are performing against generally accepted standards and their peers (benchmarking)

The processes to evaluate, direct and monitor (**figure 2.3**) are integrated end to end into the governance process and focus on evaluation, direction and monitoring of the following:
- Conformance and performance
- The system of internal controls
- Compliance with external requirements

Governance of Enterprise IT and Management Frameworks

Examples of GEIT frameworks include the following:
- **COBIT 5** was developed by ISACA to support GEIT by providing a framework to ensure that IT is aligned with the business, IT enables the business and maximizes benefits, IT resources are used responsibly, and IT risk is managed appropriately. COBIT provides tools to assess and measure the performance of IT processes within an organization. COBIT 5 includes five principles, five domains, 37 processes and 210 practices.
- **The International Organization for Standardization (ISO)/International Electrotechnical Commission (IEC) 27001 (ISO 27001)** series of standards is a set of best practices that provides guidance to organizations implementing and maintaining information security programs. ISO 27001 has become a well-known standard in the industry.

Figure 2.2—The Governance Objective: Value Creation

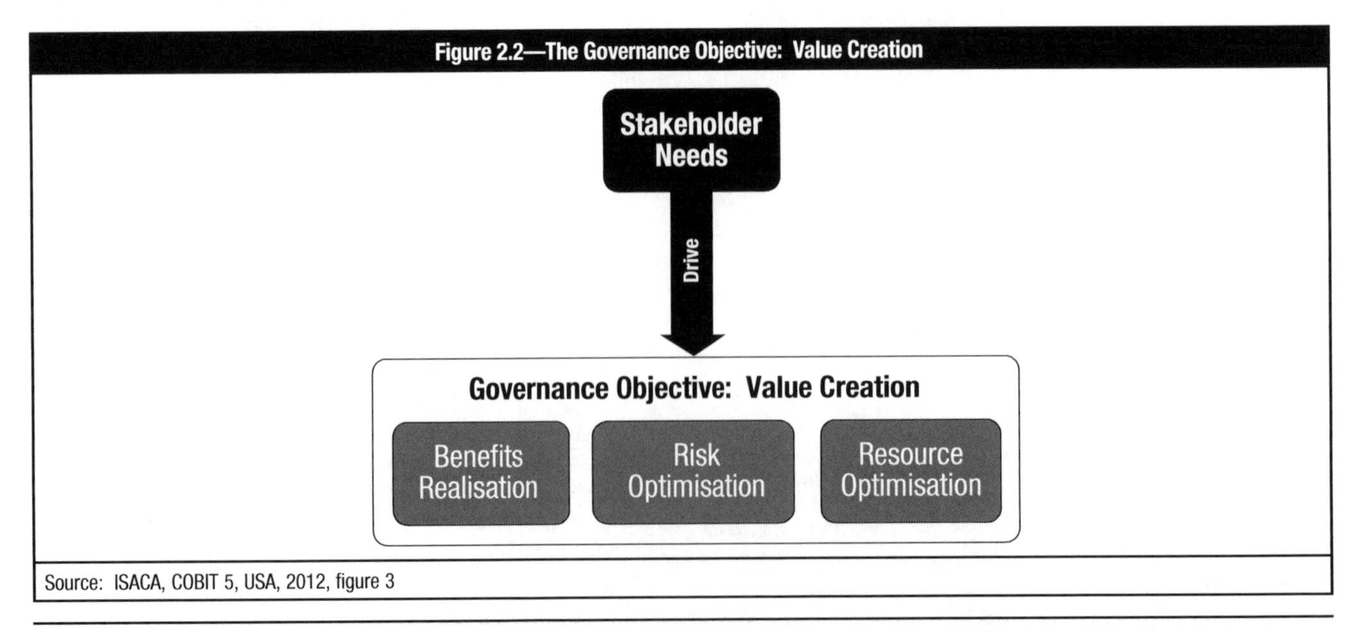

Source: ISACA, COBIT 5, USA, 2012, figure 3

Figure 2.3—COBIT 5 Governance and Management Areas

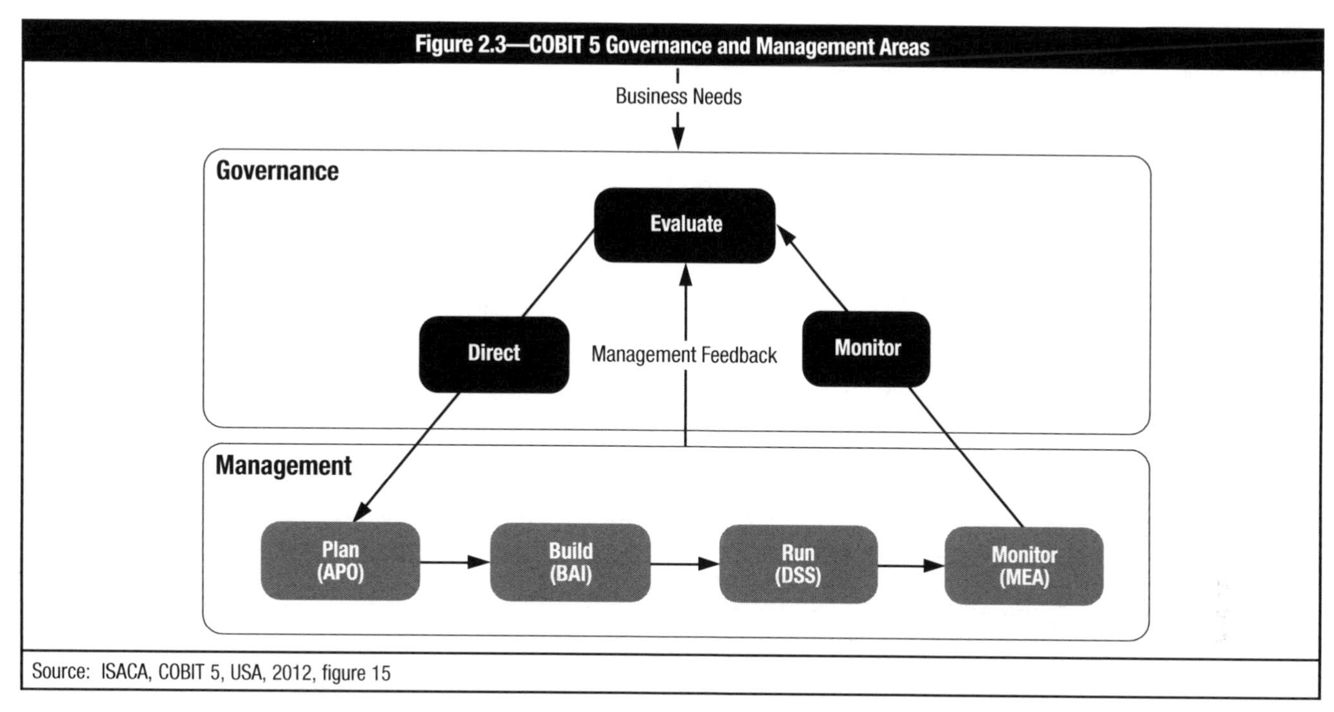

Source: ISACA, COBIT 5, USA, 2012, figure 15

- The **Information Technology Infrastructure Library (ITIL®)** was developed by the UK Office of Government Commerce (OGC), in partnership with the IT Service Management Forum, and is a detailed framework with hands-on information regarding how to achieve successful operational service management of IT and also includes business value delivery.
- The **IT Baseline Protection Catalogs, or IT-Grundschutz Catalogs**, previously known as the IT Baseline Protection Manual, are a collection of documents from the German Federal Office for Security in Information Technology (FSI). The documents are useful for detecting and combating security weak points in the IT environment.
- **The Information Security Management Maturity Model (ISM3)** is a process-based ISM maturity model for security.
- **ISO/IEC 38500:2008 Corporate governance of information technology** (very closely based on AS8015-2005) provides a framework for effective governance of IT. ISO/IEC 38500 assists those at the highest organizational level to understand and fulfill their legal, regulatory and ethical obligations in respect to their organizations' use of IT. ISO/IEC 38500 is applicable to organizations of all sizes, including public and private companies, government entities and not-for-profit organizations. This standard provides guiding principles for board of directors of organizations on the effective, efficient and acceptable use of IT within their organizations.
- **ISO/IEC 20000** is a specification for service management that is aligned with ITIL's service management framework. It is divided into two parts. ISO/IEC 20000-1:2011 consists of specific requirements for service management improvement, and ISO/IEC 20000-2:2012 provides guidance and examples for the application of ISO/IEC 20000-1:2011.

Audit Role in Governance of Enterprise IT

Enterprises are governed by generally accepted good practices, ensured by the establishment of controls. Good practices guide organizations in determining how to use resources. Results are measured and reported, providing input to the cyclical revision and maintenance of controls.

Similarly, IT is governed by good practices, which ensure that the organization's information and related technology support the enterprise's business objectives (i.e., strategic alignment), deliver value, use resources responsibly, manage risk appropriately and measure performance.

Audit plays a significant role in the successful implementation of GEIT within an organization. Audit is well positioned to provide leading practice recommendations to senior management to help improve the quality and effectiveness of the IT governance initiatives implemented.

As an entity that monitors compliance, audit helps ensure compliance with GEIT initiatives implemented within an organization. The continual monitoring, analysis and evaluation of metrics associated with GEIT initiatives require an independent and balanced view to ensure a qualitative assessment that subsequently facilitates the qualitative improvement of IT processes and associated GEIT initiatives.

Reporting on GEIT involves auditing at the highest level in the organization and may cross divisional, functional or departmental boundaries. The IS auditor should confirm that the terms of reference state the:

- Scope of the work, including a clear definition of the functional areas and issues to be covered
- Reporting line to be used, where GEIT issues are identified to the highest level of the organization
- IS auditor's right of access to information both within the organization and from third-party service providers

The organizational status and skill sets of the IS auditor should be considered for appropriateness with regard to the nature of the planned audit. Where this is found insufficient, the hiring of an independent third party to manage or perform the audit should be considered by an appropriate level of management.

In accordance with the defined role of the IS auditor, the following aspects related to GEIT need to be assessed:
- How enterprise governance and GEIT are aligned
- Alignment of the IT function with the organization's mission, vision, values, objectives and strategies
- Achievement of performance objectives (e.g., effectiveness and efficiency) established by the business and the IT function
- Legal, environmental, information quality, fiduciary, security and privacy requirements
- The control environment of the organization
- The inherent risk within the IS environment
- IT investment/expenditure

2.3.2 IT GOVERNING COMMITTEES

Traditionally, organizations have had executive-level steering committees to handle IT issues that are relevant organizationwide. There should be a clear understanding of both the IT strategy and steering levels. ISACA issued a document where a clear analysis is made (**figure 2.4**). The IS auditor should be aware that organizations may have other executive- and mid-management-led committees guiding IT operations, such as an IT executive committee, IT governance committee, IT investment committee and/or IT management committee.

Note: The Analysis of IT Steering Committee Responsibilities is information the CISA should know.

2.3.3 IT BALANCED SCORECARD

The IT balanced scorecard (BSC), **figure 2.5**, is a process management evaluation technique that can be applied to the GEIT process in assessing IT functions and processes. The technique goes beyond the traditional financial evaluation, supplementing it with measures concerning customer (user) satisfaction, internal (operational) processes and the ability to innovate. These additional measures drive the organization toward optimal use of IT, which is aligned with the organization's strategic goals, while keeping all evaluation-related perspectives in balance.

To apply the BSC to IT, a multi-layered structure (determined by each organization) is used in addressing four perspectives:
- **Mission**—for example:
  - Become the preferred supplier of information systems.
  - Deliver economic, effective and efficient IT applications and services.
  - Obtain a reasonable business contribution from IT investments.
  - Develop opportunities to answer future challenges.
- **Strategies**—for example:
  - Develop superior applications and operations.
  - Develop user partnerships and greater customer services.
  - Provide enhanced service levels and pricing structures.
  - Control IT expenses.
  - Provide business value to IT projects.
  - Provide new business capabilities.

Figure 2.4—Analysis of Steering Committee Responsibilities

Level	IT Strategy Committee	IT Steering Committee
Responsibility	• Provides insight and advice to the board on topics such as: – The relevance of developments in IT from a business perspective – The alignment of IT with the business direction – The achievement of strategic IT objectives – The availability of suitable IT resources, skills and infrastructure to meet the strategic objectives – Optimization of IT costs, including the role and value delivery of external IT sourcing – Risk, return and competitive aspects of IT investments – Progress on major IT projects – The contribution of IT to the business (i.e., delivering the promised business value) – Exposure to IT risk, including compliance risk – Containment of IT risk – Direction to management relative to IT strategy – Drivers and catalysts for the board's IT	• Decides the overall level of IT spending and how costs will be allocated • Aligns and approves the enterprise's IT architecture • Approves project plans and budgets, setting priorities and milestones • Acquires and assigns appropriate resources • Ensures that projects continuously meet business requirements, including reevaluation of the business case • Monitors project plans for delivery of expected value and desired outcomes, on time and within budget • Monitors resource and priority conflict between enterprise divisions and the IT function as well as between projects • Makes recommendations and requests for changes to strategic plans (priorities, funding, technology approaches, resources, etc.) • Communicates strategic goals to project teams • Is a major contributor to management's IT governance responsibilities and practices
Authority	• Advises the board and management on IT strategy • Is delegated by the board to provide input to the strategy and prepare its approval • Focuses on current and future strategic IT issues	• Assists the executive in the delivery of the IT strategy • Oversees day-to-day management of IT service delivery and IT projects • Focuses on implementation
Membership	• Board members and specialist non-board members	• Sponsoring executive • Business executive (key users) • Chief information officer (CIO) • Key advisors as required (IT, audit, legal, finance)

Figure 2.5—Generic IT Balanced Scorecard

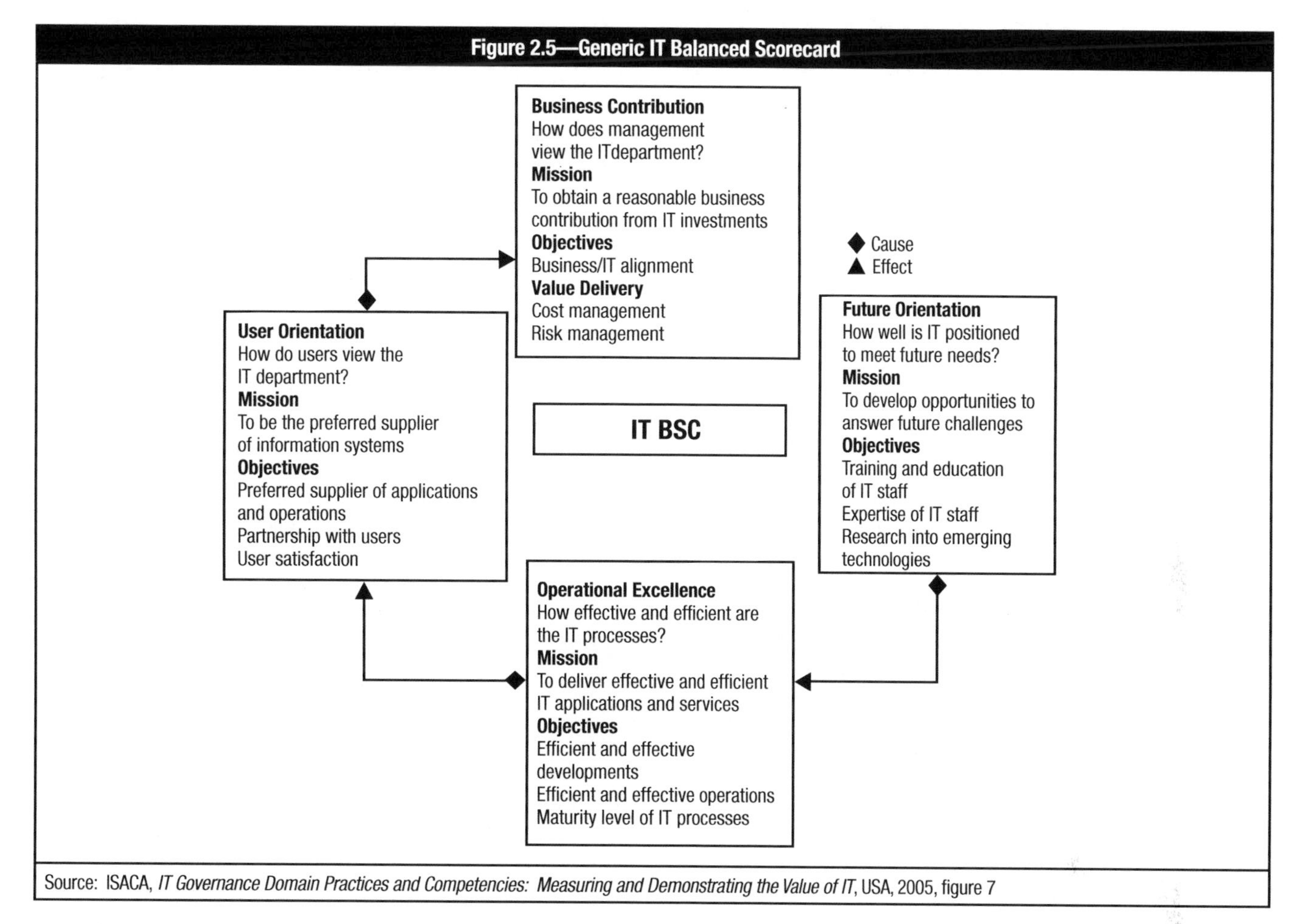

Source: ISACA, *IT Governance Domain Practices and Competencies: Measuring and Demonstrating the Value of IT*, USA, 2005, figure 7

- Train and educate IT staff and promote excellence.
- Provide support for research and development.

- **Measures**—for example:
  - Provide a balanced set of metrics (i.e., key performance indicators [KPIs]) to guide business-oriented IT decisions.
- **Sources**—for example:
  - End-user personnel (specific by function)
  - COO
  - Process owners

Use of an IT BSC is one of the most effective means to aid the IT strategy committee and management in achieving IT governance through proper IT and business alignment. The objectives are to establish a vehicle for management reporting to the board, foster consensus among key stakeholders about IT's strategic aims, demonstrate the effectiveness and added value of IT, and communicate IT's performance, risk and capabilities.

> **Note:** A CISA candidate should know the elements of the IT BSC.

2.3.4 INFORMATION SECURITY GOVERNANCE

Within IT governance processes, information security governance has risen to one of the highest levels of focused activity with specific value drivers: confidentiality, integrity and availability of information, continuity of services and protection of information assets. Security has become a significant governance issue as a result of global networking, rapid technological innovation and change, increased dependence on IT, increased sophistication of threat agents and exploits, and an extension of the enterprise beyond its traditional boundaries. Therefore, information security should become an important and integral part of IT governance. Negligence in this regard will diminish an organization's capacity to mitigate risk and take advantage of IT opportunities for business process improvement. With this said, board of directors and CEOs globally are realizing their combined accountability and responsibility for information security governance. This accountability and responsibility are what the information security governance shareholders expect as an integral element of enterprise IT. The CEO is accountable to the board of directors for information security governance and responsible for its discharge through the executive management and the organization and resources under his/her charge.

The members of senior management who approve security policies should come from various operations and staff functions within the enterprise in order to ensure that there is a fair representation of the enterprise as a whole. This is to minimize any potential leaning toward a specific business priority or technology overhead or security concerns. Typically, the board-level committee approving security policies may include directors, CEO, COO, CFO, chief risk officer (CRO), CIO, CTO, head of human resources (HR), chief of audit (auditor's independence will be a lesser concern in this context), chief compliance officer

(CCO) and legal. Policy approval should be, to the greatest extent possible, based on consensus.

Information is a key resource for all enterprises, and from the time that information is created or received to the moment that it is destroyed, technology plays a significant role. Information technology is increasingly advanced and has become pervasive in enterprises and in social, public and business environments. As a result, today, more than ever, enterprises and their executives strive to:
- Maintain high quality information to support business decisions
- Generate business value from IT-enabled investments (i.e., achieve strategic goals and realize business benefits through effective and innovative use of IT)
- Achieve operational excellence through the reliable and efficient application of technology
- Maintain IT-related risk at an acceptable level
- Optimize the cost of IT services and technology
- Comply with ever-increasing relevant laws, regulations, contractual agreements and policies

Until recently, protection efforts have focused on the information systems that collect, process and store information rather than the information itself. This approach has become too narrow to accomplish the overall security that is necessary. Information security takes the broader view that data, as well as the information and knowledge based on them, must be adequately protected regardless of where is the data are created, received, processed, transported or stored and disposed. This applies particularly to situations in which data are shared easily over the Internet through blogs, newsfeeds, peer-to-peer or social networks, or web sites. Thus, the reach of protection efforts should encompass not only the process that generates the information, but also the continued preservation of information generated as a result of the controlled processes.

Some of the major trends that global business is experiencing today include the outsourcing of in-house processes and increased use of cloud computing. Information security coverage extends beyond the geographic boundary of the enterprise's premises in onshoring and offshoring models being adopted by organizations. The promise of cloud computing is arguably revolutionizing the IT services world by transforming computing into a ubiquitous utility. These trends have also changed the way in which information security is managed.

Information security includes the security of technology and is typically driven from the CIO level. Information security related to privacy of information, and information security itself, addresses the universe of risk, benefits and processes involved with information and must be driven by executive management (e.g., CEO, CFO, CTO, CIO) and supported by the board of directors.

Information security governance is the responsibility of the board of directors and executive management and must be an integral and transparent part of enterprise governance. Information security governance consists of the leadership, organizational structures and processes that safeguard information.

The basic outcomes of effective security governance should include strategic alignment, risk management, compliance and value delivery. These outcomes are enabled through the development of:
- **Performance measurement**—Measure, monitor and report on information security processes to ensure that SMART (specific, measurable, attainable, realistic and timely) objectives are achieved. The following should be accomplished to achieve performance measurement:
  - A defined, agreed-on and meaningful set of metrics properly aligned with strategic objectives
  - A measurement process that will help identify shortcomings and provide feedback on progress made in resolving issues
  - Independent assurance provided by external assessments and audits
- **Resource management**—Utilize information security knowledge and infrastructure efficiently and effectively. To achieve resource management consider the following:
  - Ensure that knowledge is captured and available
  - Document security processes and practices
  - Develop security architecture(s) to define and utilize infrastructure resources efficiently
- **Process integration**—This focuses on the integration of an organization's management assurance processes for security. Security activities are at times fragmented and segmented in silos with different reporting structures. This makes it difficult, if not impossible, to seamlessly integrate them. Process integration serves to improve overall security and operational efficiencies.

Effective Information Security Governance

The strategic direction of the business will be defined by business goals and objectives. Information security must support business activities to be of value to the organization. Information security governance is a subset of corporate governance that provides strategic direction for security activities and ensures that objectives are achieved. It ensures that information security risk is appropriately managed and enterprise information resources are used responsibly. According to the National Institute of Standards and Technology (NIST) Special Publication 800-100, *Information Security Handbook: A Guide for Managers*:

> *Information security governance can be defined as the process of establishing and maintaining a framework and supporting management structure and processes to provide assurance that information security strategies are aligned with and support business objectives, are consistent with applicable laws and regulations through adherence to policies and internal controls, and provide assignment of responsibility, all in an effort to manage risk.*

To achieve effective information security governance, management must establish and maintain a framework to guide the development and management of a comprehensive information security program that supports business objectives.

The information security governance framework will generally consist of:
- A comprehensive security strategy intrinsically linked with business objectives
- Governing security policies that address each aspect of strategy, controls and regulation

- A complete set of standards for each policy to ensure that procedures and guidelines comply with policy
- An effective security organizational structure void of conflicts of interest
- Institutionalized monitoring processes to ensure compliance and provide feedback on effectiveness

This framework provides the basis for the development of a cost-effective information security program that supports the organization's business goals. The objective of the information security program is a set of activities that provide assurance that information assets are given a level of protection commensurate with their value or the risk their compromise poses to the organization.

Roles and Responsibilities of Senior Management and Boards of Directors

Information security governance requires strategic direction and impetus. It requires commitment, resources and assignment of responsibility for information security management as well as a means for the board to determine that its intent has been met.

BOARDS OF DIRECTORS/SENIOR MANAGEMENT

Effective information security governance can be accomplished only by involvement of the board of directors and/or senior management in approving policy, ensuring appropriate monitoring and reviewing metrics, reports and trend analysis.

Members of the board need to be aware of the organization's information assets and their criticality to ongoing business operations. This can be accomplished by periodically providing the board with the high-level results of comprehensive risk assessments and business impact analysis (BIA). It may also be accomplished by business dependency assessments of information resources. These activities should include approval by board members of the assessment of key assets to be protected, which helps ensure that protection levels and priorities are appropriate to a standard of due care.

The tone at the top must be conducive to effective security governance. It is unreasonable to expect lower-level personnel to abide by security measures if they are not exercised by senior management. Senior management endorsement of intrinsic security requirements provides the basis for ensuring that security expectations are met at all levels of the enterprise. Penalties for noncompliance must be defined, communicated and enforced from the board level down.

SENIOR MANAGEMENT

Implementing effective security governance and defining the strategic security objectives of an organization is a complex, arduous task. As with any other major initiative, it must have leadership and ongoing support from executive management to succeed. Developing an effective information security strategy requires integration with and cooperation of business process owners. A successful outcome is the alignment of information security activities in support of business objectives. The extent to which this is achieved will determine the cost-effectiveness of the information security program in achieving the desired objective of providing a predictable, defined level of assurance for business information and processes and an acceptable level of impact from adverse events.

INFORMATION SECURITY STANDARDS COMMITTEE

To some extent, security affects all aspects of an organization. To be effective, security must be pervasive throughout the enterprise. To ensure that all stakeholders impacted by security considerations are involved, many organizations use a steering committee comprised of senior representatives of affected groups. This facilitates achieving consensus on priorities and trade-offs. It also serves as an effective communications channel and provides an ongoing basis for ensuring the alignment of the security program with business objectives. It can also be instrumental in achieving modification of behavior toward a culture more conducive to good security.

The chief information security officer (CISO) will primarily drive the information security program to have realistic policies, standards, procedures and processes that are implementable and auditable and to achieve a balance of performance in relation to security. However, It is necessary to involve the affected groups in a deliberating committee, which may be called information security standards committee (ISSC). The ISSC includes members from C-level executive management and senior managers from IT, application owners, business process owners, operations, HR, audit and legal. The committee will deliberate on the suitability of recommended controls and good practices in the context of the organization, including secure configuration of operating systems (OSs) and databases. The auditor's presence is required to make the systems auditable by providing for suitable audit trails and logs. Legal is required to advise on liability and conflict with the law issues. This is not a prescriptive list of members to be included on the ISSC. Members of the committee may be modified to suit the context of the organizations, and other members may be co-opted as necessary to suit the control objectives in question.

CHIEF INFORMATION SECURITY OFFICER

All organizations have a CISO whether anyone holds the exact title. The responsibilities may be performed by the CIO, CTO, CFO or, in some cases, the CEO even when there is an information security office or director in place. The scope and breadth of information security is such that the authority required and the responsibility taken will inevitably make it a senior officer or top management responsibility. This could include a position such as a CRO or a CCO. Legal responsibility will, by default, extend up the command structure and ultimately reside with senior management and the board of directors. Failure to recognize this and implement appropriate governance structures can result in senior management being unaware of this responsibility and the attendant liability. It also usually results in a lack of effective alignment of business objectives and security activities. Increasingly, prudent management is elevating the position of information security officer to a senior management position (i.e., CISO), as organizations begin to understand their dependence on information and the growing threats to it.

Matrix of Outcomes and Responsibilities

The relationships between the outcomes of effective security governance and management responsibilities are shown in **figure 2.6**. This matrix is not meant to be comprehensive but merely to indicate some primary tasks and the management level responsible for those tasks. Depending on the nature of the

Figure 2.6—Relationships of Security Governance Outcomes to Management Responsibilities

Management Level	Strategic Alignment	Risk Management	Value Delivery	Performance Measurement	Resource Management	Process Assurance
Board of directors	Require demonstrable alignment.	• Establish risk tolerance. • Oversee a policy of risk management. • Ensure regulatory compliance.	Require reporting of security activity costs.	Require reporting of security effectiveness.	Oversee a policy of knowledge management and resource utilization.	Oversee a policy of assurance process integration.
Executive management	Institute processes to integrate security with business objectives.	• Ensure that roles and responsibilities include risk management in all activities. • Monitor regulatory compliance.	Require business case studies of security activities.	Require monitoring and metrics for security initiatives.	Ensure processes for knowledge capture and efficiency metrics.	Provide oversight of all assurance functions and plans for integration.
Steering committee	• Review and assist security strategy and integration efforts. • Ensure that business owners support integration.	Identify emerging risks, promote business unit security practices and identify compliance issues.	Review and advise on the adequacy of security initiatives to serve business functions.	Review and advise whether security initiatives meet business objectives.	Review processes for knowledge capture and dissemination.	• Identify critical business processes and assurance providers. • Direct assurance integration efforts.
CISO/ information security management	Develop the security strategy, oversee the security program and initiatives, and liaise with business process owners for ongoing alignment.	• Ensure that risk and business impact assessments are conducted. • Develop risk mitigation strategies. • Enforce policy and regulatory compliance.	Monitor utilization and effectiveness of security resources.	Develop and implement monitoring and metrics approaches, and direct and monitor security activities.	Develop methods for knowledge capture and dissemination, and develop metrics for effectiveness and efficiency.	• Liaise with other assurance providers. • Ensure that gaps and overlaps are identified and addressed.
Audit executives	Evaluate and report on degree of alignment.	Evaluate and report on corporate risk management practices and results.	Evaluate and report on efficiency.	Evaluate and report on degree of effectiveness of measures in place and metrics in use.	Evaluate and report on efficiency or resource management.	Evaluate and report on effectiveness of assurance processes performed by different areas of management.

Source: ISACA, *Information Security Governance: Guidance for Information Security Managers*, 2008. All rights reserved. Used by permission.

organization, the titles may vary, but the roles and responsibilities should exist even if different labels are used.

Note: While **figure 2.6** is not specifically tested in the CISA exam, the CISA candidate should be aware of this information.

2.3.5 ENTERPRISE ARCHITECTURE

An area of IT governance that is receiving increasing attention is enterprise architecture (EA). Essentially, EA involves documenting an organization's IT assets in a structured manner to facilitate understanding, management and planning for IT investments. An EA often involves both a current state and optimized future state representation (e.g., a road map).

The current focus on EA is a response to the increasing complexity of IT, the complexity of modern organizations, and an enhanced focus on aligning IT with business strategy and ensuring that IT investments deliver real returns.

The Framework for Enterprise Architecture, a groundbreaking work in the field of EA, was first published by John Zachman in the late 1980s. The Zachman framework continues to be a starting point for many contemporary EA projects. Zachman reasoned that constructing IT systems had considerable similarities to building construction. In both cases there is a range of participants who become involved at differing stages of the project. In building construction, one moves from the abstract to the physical using models and representations (such as blueprints, floor plans and wiring diagrams). Similarly with IT, different artifacts (such as diagrams, flowcharts, data/class models and code) are used to convey different aspects of an organization's systems at progressively greater levels of detail.

The basic Zachman framework is shown in **figure 2.7**.

Figure 2.7—Zachman Framework for Enterprise Architecture

	Data	Functional (Application)	Network (Technology)	People (Organization)	Process (Workflow)	Strategy
Scope						
Enterprise model						
Systems model						
Technology model						
Detailed representation						

The ultimate objective is to complete all cells of the matrix. At the outset of an EA project, most organizations will have difficulty providing details for every cell, particularly at the highest level.

In attempting to complete an EA, organizations can address the challenge either from a technology perspective or a business process perspective.

Technology-driven EA attempts to clarify the complex technology choices faced by modern organizations. The idea is to provide guidance on issues such as whether and when to use advanced technical environments (e.g., JavaEE or .NET) for application development, how to better connect intra- and interorganizational systems, how to "web enable" legacy and enterprise resource planning (ERP) applications (without extensive rewrite), whether to insource or outsource IT functions, and whether and when to use solutions such as virtualization and cloud computing.

Business process-driven EA attempts to understand an organization in terms of its core value-adding and supporting processes. The idea is that by understanding processes, their constituent parts and the technology that supports them, business improvement can be obtained as aspects are progressively redesigned and replaced. The genesis for this type of thinking can be traced back to the work of Harvard professor Michael Porter, and particularly his business value chain model. The effort to model business processes is being given extra impetus by a number of industrywide business models such as the telecommunications industry's enhanced Telecom Operations Map (eTOM) and the Supply Chain Operations Reference (SCOR) model. The contents from a business process model can be mapped to upper tiers of the Zachman framework. After the mapping is completed, an organization can consider the optimum mix of technologies needed to support its business processes.

For example, a US federal organization is required by law to develop an EA and set up an EA governance structure that ensures that the EA is referenced and maintained in the planning and budgeting activities of all systems. The Federal Enterprise Architecture (FEA) was developed to guide this process. The FEA is described as "a business and performance based framework to support cross-agency collaboration, transformation and government-wide improvement" *(www.whitehouse.gov/omb/e-gov/fea)*. The FEA has a hierarchy of five reference models:

- **Performance reference model**—A framework to measure the performance of major IT investments and their contribution to program performance
- **Business reference model**—A function-driven framework that describes the functions and subfunctions performed by the government, independent of the agencies that actually perform them
- **Service component reference model**—A functional framework that classifies the service components that support business and performance objectives
- **Technical reference model**—A framework that describes how technology supports the delivery, exchange and construction of service components
- **Data reference model**—A framework that describes the data and information that support program and business line operations

The documentation on corporate architecture and the FEA are primarily used for maintaining and describing technological coherence, continually describing and evaluating the technology that is being managed by the IT department.

Relevant aspects of IT governance regarding the management of an IT department are the processes for selection and/or the methodologies that are used to change strategic technologies. This relevant topic affects management decisions and is subject to great business risk.

2.4 INFORMATION SYSTEMS STRATEGY

Information systems are crucial in the support, sustainability and growth of enterprises. Previously, governing boards and senior management executives could minimize their involvement in the direction and development of IS strategy and direction, leaving most decisions to functional management. However, this approach is no longer acceptable or possible with increased or total dependency on IS for day-to-day operations and successful growth. Along with the near complete dependence on IS for functional and operational activities, organizations also face numerous internal and external threats ranging from IS resource abuse, cybercrime, fraud, errors and omissions. IS strategic processes are integral components within the organization governance structure to provide reasonable assurance that both existing and emerging business goals and objectives will be attained as a critical facilitator for enhancement of competitive advantage.

2.4.1 STRATEGIC PLANNING

Strategic planning from an IS standpoint relates to the long-term direction an enterprise wants to take in leveraging IT for improving its business processes.

Under the responsibility of top management, factors to consider include identifying cost-effective IT solutions in addressing problems and opportunities that confront the enterprise, and developing action plans for identifying and acquiring needed resources. In developing strategic plans, generally three to five years in duration, enterprises should ensure that the plans are fully aligned and consistent with the overall organizational goals and objectives. IT department management, along with the IT steering committee and the strategy committee (which provides valuable strategic input related to stakeholder value), play a key role in the development and implementation of the plans.

Effective IS strategic planning involves a consideration of the enterprise's requirements for new and revised IS systems and the IT organization's capacity to deliver new functionality through well-governed projects. Determining requirements for new and revised IS systems will involve a systematic consideration of the enterprise's strategic intentions, how these translate into specific objectives and business initiatives, and what IT capabilities will be needed to support these objectives and initiatives. In assessing IT capabilities, the existing system's portfolio should be reviewed in terms of functional fit, cost and risk. Assessing IT's capacity to deliver involves a review of the organization's technical IT infrastructure and key support processes (e.g., project management, software development and maintenance practices, security administration and help desk services) to determine whether expansion or improvement is necessary. It is important that the strategic planning process encompasses the delivery of new systems and technology and considers return on investment (ROI) on existing IT and the decommissioning of legacy systems. The strategic IT plan should balance the cost of maintenance of existing systems against the cost of new initiatives or systems to support the business strategies.

The IS auditor should pay full attention to the importance of IS strategic planning, taking management control practices into consideration. In addition, the IT governance objective requires that IT strategic plans be synchronized with the overall business strategy. An IS auditor must focus on the importance of a strategic planning process or planning framework. Particular attention should be paid to the need to assess how operational, tactical or business development plans from the business are taken into account in IT strategy formulation, contents of strategic plans, requirements for updating and communicating plans, and monitoring and evaluation requirements. The IS auditor should consider how the CIO or senior IT management are involved in the creation of the overall business strategy. A lack of involvement of IT in the creation of the business strategy indicates that there is a risk that the IT strategy and plans will not be aligned with the business strategy.

2.4.2 IT STEERING COMMITTEE

The enterprise's senior management should appoint a planning or steering committee to oversee the IT function and its activities. A high-level steering committee for information systems is an important factor in ensuring that the IT department is in harmony with the corporate mission and objectives. Although not a common practice, it is highly desirable that a member of the board of directors who understands the risk and issues is responsible for IT and is chair of this committee. The committee should include representatives from senior management, each line of business, corporate departments, such as HR and finance, and the IT department.

The committee's duties and responsibilities should be defined in a formal charter. Members of the committee should know IT department policies, procedures and practices. Each member should have the authority to make decisions within the group for his/her respective areas.

Such a committee typically serves as a general review board for major IS projects and should not become involved in routine operations. Primary functions performed by this committee include:

- Review the long- and short-range plans of the IT department to ensure that they are in accordance with the corporate objectives.
- Review and approve major acquisitions of hardware and software within the limits approved by the board of directors.
- Approve and monitor major projects and the status of IS plans and budgets, establish priorities, approve standards and procedures and monitor overall IS performance.
- Review and approve sourcing strategies for select or all IS activities, including insourcing or outsourcing, and the globalization or offshoring of functions.
- Review adequacy of resources and allocation of resources in terms of time, personnel and equipment.
- Make decisions regarding centralization versus decentralization and assignment of responsibility.
- Support development and implementation of an enterprisewide information security management program.
- Report to the board of directors on IS activities.

Note: Responsibilities will vary from enterprise to enterprise, and the responsibilities listed are the most common responsibilities of the IT steering committee. Each enterprise should have formally documented and approved terms of reference for its steering committee, and the IS auditor should familiarize him/herself with the IT steering committee documentation and understand the major responsibilities that are assigned to its members. Many enterprises may refer to this committee with a different name. The IS auditor needs to identify the group that performs the previously mentioned functions.

The IT steering committee should receive the appropriate management information from IT departments, user departments and the audit department to effectively coordinate and monitor the enterprise's IS resources. The committee should monitor performance and institute appropriate action to achieve desired results. The committee should meet regularly and report to senior management. Formal minutes of the IS steering committee meetings should be maintained to document the committee's activities and decisions.

2.5 MATURITY AND PROCESS IMPROVEMENT MODELS

Implementation of IT governance requires ongoing performance measurement of an organization's resources that contribute to the execution of processes that deliver IT services to the business. Maintaining consistent efficiency and effectiveness of processes requires implementing a process maturity framework. The framework can be based on various models such as Capability Maturity Model Integration (CMMI®), the Initiating, Diagnosing, Establishing, Acting and Learning (IDEAL) model, etc. This section presents several maturity process and improvement models that CISA candidates may find in an organization.

The **COBIT Process Assessment Model (PAM)**, using COBIT 5, has been developed to address the need to improve the rigor and reliability of IT process reviews. The model serves as a reference document for conducting capability assessments of an organization's current IT processes and defines the minimum set of requirements for conducting an assessment to ensure that the outputs are consistent, repeatable and representative of the processes assessed. It is aligned with ISO/IEC 15504-2 and uses process capability and process performance indicators to determine whether process attributes have been achieved.

The **IDEAL** model is a software process improvement (SPI) program model, developed by the Software Engineering Institute (SEI) at Carnegie Mellon University. It forms an infrastructure to guide enterprises in planning and implementing an effective software process improvement program and consists of five phases: initiating, diagnosing, establishing, acting and learning (IDEAL).

CMMI is a process improvement approach that provides enterprises with the essential elements of effective processes. It can be used to guide process improvement across a project, division or entire organization. CMMI helps integrate traditionally separate organizational functions, set process improvement goals and priorities, and provide guidance for quality processes and a point of reference for appraising current processes. The process capability model is based on the internationally recognized *ISO/IEC 15504 Information Technology—Process Assessment* standard. This model will achieve the same overall objectives of process assessment and process improvement support (i.e., it will provide a means to measure the performance of any of the COBIT 5 governance [EDM-based] processes or management [PBRM-based] processes and will allow areas for improvement to be identified).

2.6 IT INVESTMENT AND ALLOCATION PRACTICES

Each enterprise faces the challenge of using its limited resources, including people and money, to achieve its goals and objectives. When an organization invests its resources in a given effort, it incurs opportunity costs because it is unable to pursue other efforts that could bring value to the enterprise. An IS auditor should understand an organization's investment and allocation practices to determine whether the enterprise is positioned to achieve the greatest value from the investment of its resources.

Traditionally, when IT professionals and top managers discussed the ROI of an IT investment, they were thinking about financial benefits. Today, business leaders also consider the nonfinancial benefits of IT investments. Where feasible, nonfinancial benefits should be made visible and tangible by using algorithms that transform them into monetary units to understand their impact and improve their analysis.

Financial benefits include impacts on the organization's budget and finances (e.g., cost reductions or revenue increases).

Nonfinancial benefits include impacts on operations or mission performance and results (e.g., improved customer satisfaction, better information, shorter cycle time).

2.6.1 VALUE OF IT

Decision makers make IT project selection decisions based upon the perceived value of the investment. IT's value is determined by the relationship between what the organization will pay (costs) and what it will receive (benefits). The larger the benefit in relation to cost, the greater the value of the IT project.

IT portfolio management is distinct from IT financial management in that it has an explicitly directive, strategic goal in determining what the enterprise will invest or continue to invest in versus what the enterprise will divest.

In COBIT 5, process EDM02 *Ensure benefits delivery* optimizes the value contribution to the business from the business processes, IT services and IT assets resulting from investments made by IT at acceptable cost. Key governance practices in the process include:
1. Evaluate value optimization.
2. Direct value optimization.
3. Monitor value optimization.

2.6.2 IMPLEMENTING IT PORTFOLIO MANAGEMENT

Implementation methods include risk profile analysis; diversification of projects, infrastructure and technologies; continuous alignment with business goals; and continuous improvement.

There is no single best way to implement IT portfolio management; therefore, a variety of approaches can be applied.

2.6.3 IT PORTFOLIO MANAGEMENT VERSUS BALANCED SCORECARD

The biggest advantage of IT portfolio management is its agility in adjusting investments. While BSCs also emphasize the use of vision and strategy in any investment decision, the oversight and control of operations budgets is not the goal. IT portfolio management allows organizations to adjust investments based upon the built-in feedback mechanism.

2.7 POLICIES AND PROCEDURES

Policies and procedures reflect management guidance and direction over information systems, related resources and IT department processes.

2.7.1 POLICIES

Policies are high-level documents that represent the corporate philosophy of an organization. To be effective, policies must be clear and concise. Management must create a positive control environment by assuming responsibility for formulating, developing, documenting, promulgating and controlling policies covering general goals and directives. Management should take the steps necessary to ensure that employees affected by a specific policy receive a full explanation of the policy and understand its intent. In addition, policies may also apply to third parties and outsourcers, who will need to be bound to follow the policies through contracts or statements of work (SOW).

In addition to corporate policies that set the tone for the organization as a whole, individual divisions and departments should define lower-level policies. The lower-level policies should be consistent with the corporate-level policies. These would apply to the employees and operations of these units and would focus at the operational level.

Management should review all policies periodically. Ideally, these documents should specify a review date, which the IS auditor should check for currency. Policies need to be updated to reflect new technology, changes in environment (e.g., regulatory compliance requirements) and significant changes in business processes in exploiting information technology for efficiency and effectiveness in productivity or competitive gains. Policies formulated must support achievement of business objectives and implementation of IS controls. However, management must be responsive to the needs of the customers and change policies that may hinder customer satisfaction or the organization's ability to achieve business objectives. This consideration must take into account matters of confidentiality and information security, which may run counter to the convenience of the customer. The broad policies at a higher level and the detailed policies at a lower level need to be in alignment with the business objectives.

IS auditors should understand that policies are a part of the audit scope and test the policies for compliance. IS controls should flow from the enterprise's policies, and IS auditors should use policies as a benchmark for evaluating compliance. However, if policies exist that hinder the achievement of business objectives, these policies must be identified and reported for improvement. The IS auditor should also consider the extent to which the policies apply to third parties or outsourcers, the extent to which third parties or outsourcers comply with the policies, and whether the policies of the third parties or outsourcers are in conflict with the enterprise's policies.

Information Security Policy

An information security policy communicates a coherent security standard to users, management and technical staff. A security policy for information and related technology is a first step toward building the security infrastructure for technology-driven organizations. Policies will often set the stage in terms of what tools and procedures are needed for the organization. Information security policies must balance the level of control with the level of productivity. Also, the cost of a control should never exceed the expected benefit to be derived. In designing and implementing these policies, the organizational culture will play an important role. The information security policy must be approved by senior management, and should be documented and communicated, as appropriate, to all employees, service providers and business partners (i.e., suppliers). The information security policy should be used by IS auditors as a reference framework for performing various IS audit assignments. The adequacy and appropriateness of the security policy could also be an area of review for the IS auditor.

INFORMATION SECURITY POLICY DOCUMENT

The information security policy should state management's commitment and set out the organization's approach to managing information security. The ISO/IEC 27001 standard (or equivalent standards) as well as the 27002 guideline may be considered a benchmark for the content covered by the information security policy document.

The policy document should contain:
- A definition of information security, its overall objectives and scope, and the importance of security as an enabling mechanism for information sharing
- A statement of management intent, supporting the goals and principles of information security in line with the business strategy and objectives
- A framework for setting control objectives and controls, including the structure of risk assessment and risk management
- A brief explanation of the information security policies, principles, standards and compliance requirements of particular importance to the organization including:
  - Compliance with legislative, regulatory and contractual requirements
  - Information security education, training and awareness requirements
  - Business continuity management
  - Consequences of information security policy violations
- A definition of general and specific responsibilities for information security management, including reporting information security incidents
- References to documentation that may support the policy (e.g., more detailed security policies, standards, and procedures for specific information systems or security rules with which users should comply)

This information security policy should be communicated throughout the organization to users in a form that is accessible and understandable to the intended reader. The information security policy might be a part of a general policy document and may be suitable for distribution to third parties and outsourcers of the organization as long as care is taken not to disclose sensitive organizational information. All employees or third parties having access to information assets should be required to sign off on their understanding and willingness to comply with the information security policy at the time they are hired and on a regular basis thereafter (e.g., annually) to account for policy changes over time.

Depending upon the need and appropriateness, organizations may document information security policies as a set of policies. Generally, the following policy concerns are addressed:
- A **High-level Information Security Policy** should include statements on confidentiality, integrity and availability.

- A **Data Classification Policy** should describe the classifications, levels of control at each classification and responsibilities of all potential users including ownership.
- An **Acceptable Use Policy** is a comprehensive policy that includes information for all information resources (hardware, software, networks, Internet, etc.) and describes the organizational permissions for the usage of IT and information-related resources.
- An **End-user Computing Policy** describes the parameters and usage of desktop, mobile computing and other tools by users.
- **Access Control Policies** describe the method for defining and granting access to users to various IT resources.

ACCEPTABLE USE POLICY

Inappropriate use of IT resources by users exposes an enterprise to risk, including virus attacks, compromise of network systems and services, and legal issues. To address this, an organization defines a set of guidelines and/or rules that are put into effect to control how its information system resources will be used. These guidelines and/or rules are referred to as an acceptable use policy (AUP). It is common practice to require new members of an enterprise to sign an acknowledgment before receiving access to information systems.

The AUP should explain what the enterprise considers to be acceptable computer use, with the goal of protecting both the employee and the enterprise from the ramifications of illegal actions. For this reason, the AUP must be concise and clear, while at the same time covering the most important points such as defining who is considered to be a user and what the user is allowed to do with the IS systems. The AUP should refer users to the more comprehensive security policy, where relevant. The AUP should also clearly define which sanctions will be applied if the user fails to comply with the AUP, up to and including termination. Compliance with this policy should be measured by regular audits. This policy should state (in terms drafted by legal counsel) the right of the company to keep logs, backups and copies, to conduct manual or automated forensic analyses, and to take the evidence to court, while preserving privacy rights.

The most common form of an AUP is the Acceptable Internet Usage Policy, which prescribes the code of conduct that governs the behavior of a user while connected to the network/Internet. The code of conduct may include "netiquette"—a description of language that is considered appropriate to use while online. The code of conduct also should outline what is considered illegal or an excessive personal activity. Adherence to a code of conduct helps ensure that activities embarked on by a user will not expose the enterprise to information security risks.

REVIEW OF THE INFORMATION SECURITY POLICY

The information security policy should be reviewed at planned intervals (at least annually) or when significant changes to the enterprise, its business operations or inherent security-related risk occur to ensure its continuing suitability, adequacy and effectiveness. The information security policy should have an owner who has approved management responsibility for the development, review and evaluation of the policy. The review should include assessing opportunities for improvement to the organization's information security policy and approach to managing information security in response to changes to the organizational environment, business circumstances, legal conditions or technical environment.

The maintenance of the information security policy should take into account the results of these reviews. There should be defined management review procedures, including a schedule or period for the review.

The input to the management review should include:
- Feedback from interested parties
- Results of independent reviews
- Status of preventive, detective and corrective actions
- Results of previous management reviews
- Process performance and information security policy compliance
- Changes that could affect the organization's approach to managing information security, including changes to the organizational environment; business circumstances; resource availability; contractual, regulatory and legal conditions; or technical environment
- Usage of the consideration of outsourcers or offshore of IT or business functions
- Trends related to threats and vulnerabilities
- Reported information security incidents
- Recommendations provided by relevant authorities

The output from management review should include any decisions and actions related to:
- Improvement in the alignment of information security with business objectives
- Improvement of the organization's approach to managing information security and its processes
- Improvement of control objectives and controls
- Improvement in the allocation of resources and/or responsibilities

A record of management reviews should be maintained and management approval for the revised policy should be obtained.

> **Note:** This review is performed by management to address the changes in environmental factors.

While reviewing the policies, the IS auditor needs to assess the following:
- Basis on which the policy has been defined—generally, it is based on a risk management process
- Appropriateness of these policies
- Contents of policies
- Exceptions to the policies—clearly noting in which area the policies do not apply and why—(e.g., password policies may not be compatible with legacy applications)
- Policy approval process
- Policy implementation process
- Effectiveness of implementation of policies
- Awareness and training
- Periodic review and update process

2.7.2 PROCEDURES

Procedures are documented, defined steps for achieving policy objectives. They must be derived from the parent policy and must implement the spirit (intent) of the policy statement. Procedures must be written in a clear and concise manner so they may be easily and properly understood by those governed by them. Procedures document business and aligned IT processes (administrative and operational) and the embedded controls. Procedures are formulated by process owners as an effective translation of policies.

Generally, procedures are more dynamic than their respective parent policies. Procedures must reflect the regular changes in business and aligned IT focus and environment. Therefore, frequent reviews and updates of procedures are essential if they are to be relevant. IS auditors review procedures to identify/evaluate and, thereafter, test controls over business and aligned IT processes. The controls embedded in procedures are evaluated to ensure that they fulfill necessary control objectives while making the process as efficient and practical as possible. Where operational practices do not match documented procedures or where documented procedures do not exist, it is difficult (for management and auditors) to identify controls and ensure that they are in continuous operation.

One of the most critical aspects related to procedures is that they should be well known by the people they govern. A procedure that is not thoroughly known by the personnel who are to use it is, essentially, ineffective. Therefore, attention should be paid to deployment methods and automation of mechanisms to store, distribute and manage IT procedures.

Quite often procedures are embedded in information systems, which is an advisable practice to further integrate these practices within the enterprise.

2.8 RISK MANAGEMENT

Risk management is the process of identifying vulnerabilities and threats to the information resources used by an organization in achieving business objectives and deciding what countermeasures (safeguards or controls), if any, to take in reducing risk to an acceptable level (i.e., residual risk), based on the value of the information resource to the organization.

Effective risk management begins with a clear understanding of the organization's appetite for risk. This drives all risk management efforts and, in an IT context, impacts future investments in technology, the extent to which IT assets are protected and the level of assurance required. Risk management encompasses identifying, analyzing, evaluating, treating, monitoring and communicating the impact of risk on IT processes. Having defined risk appetite and identified risk exposure, strategies for managing risk can be set and responsibilities clarified. Depending on the type of risk and its significance to the business, management and the board may choose to:

- **Avoid**—Eliminate the risk by eliminating the cause (e.g., where feasible, choose not to implement certain activities or processes that would incur risk).
- **Mitigate**—Lessen the probability or impact of the risk by defining, implementing and monitoring appropriate controls.
- **Share/Transfer (deflect, or allocate)**—Share risk with partners or transfer via insurance coverage, contractual agreement or other means.
- **Accept**—Formally acknowledge the existence of the risk and monitor it.

Therefore, risk can be avoided, reduced, transferred or accepted. An organization can also choose to reject risk by ignoring it, which can be dangerous and should be considered a red flag by the IS auditor.

2.8.1 DEVELOPING A RISK MANAGEMENT PROGRAM

Steps to developing a risk management program include:

- **Establish the purpose of the risk management program**—The first step is to determine the organization's purpose for creating a risk management program. The program's purpose may be to reduce the cost of insurance or reduce the number of program-related injuries. By determining its intention before initiating risk management planning, the organization can define key performance indicators (KPIs) and evaluate the results to determine its effectiveness. Typically, senior management, with the board of directors, sets the tone and goals for the risk management program.
- **Assign responsibility for the risk management plan**—The second step is to designate an individual or team responsible for developing and implementing the organization's risk management program. While the team is primarily responsible for the risk management plan, a successful program requires the integration of risk management within all levels of the organization. Operations staff and board members should assist the risk management committee in identifying risk and developing suitable loss control and intervention strategies.

2.8.2 RISK MANAGEMENT PROCESS

To ensure that an enterprise manages its risk consistently and appropriately, an organization should identify and establish a repeatable process to manage its IT risk. COBIT 5 provides a risk management process, APO12 Manage risk. The key management practices include:

1. Collect data.
2. Analyze risk.
3. Maintain a risk profile.
4. Articulate risk.
5. Define a risk management action portfolio.
6. Respond to risk.

Step 1: Asset Identification

The first step in the process is the identification and collection of relevant data to enable effective IT-related risk identification, analysis and reporting. This will help to identify information resources or assets that need protection because they are vulnerable to threats. In this context, a threat is any circumstance or event with the potential to cause harm (such as destruction, disclosure, modification of data and/or denial of service) to an information resource. The purpose of the classification may be either to prioritize further investigation and identify appropriate protection (simple classification based on asset value) or to

enable a standard model of protection to be applied (classification in terms of criticality and sensitivity). Examples of typical assets associated with information and IT include:
- Information and data
- Hardware
- Software
- Documents
- Personnel

Other more traditional business assets for consideration are buildings, stock of goods (inventory), and cash and intangible assets such as goodwill or image/reputation.

Step 2: Evaluation of Threats and Vulnerabilities to Assets

The second step in the process is to assess threats and vulnerabilities associated with the information resource and the likelihood of their occurrence. Common classes of threats are:
- Errors
- Malicious damage/attack
- Fraud
- Theft
- Equipment/software failure

IT risk occurs because of threats (or predisposing conditions) that have the potential to exploit vulnerabilities associated with use of information resources. Vulnerabilities are characteristics of information resources that can be exploited by a threat to cause harm. Examples of vulnerabilities are:
- Lack of user knowledge
- Lack of security functionality
- Inadequate user awareness/education (e.g., poor choice of passwords)
- Untested technology
- Transmission of unprotected communications

In order for a vulnerability to be realized, there must be either a human or environmental threat to exploit the vulnerability. Typical human threat actors are:
- Novices (Kiddie scripters)
- Hacktivists
- Criminal
- Terrorists
- Nation states
- Riots and civil unrest

Typical environmental threats include the following:
- Floods
- Lightning
- Tornados
- Hurricanes
- Earthquakes

Step 3: Evaluation of the Impact

The result of a threat agent exploiting a vulnerability is called an impact. The impact can vary in magnitude, affected by severity and duration. In commercial organizations, threats usually result in a direct financial loss in the short term or an ultimate (indirect) financial loss in the long term. Examples of such losses include:
- Direct loss of money (cash or credit)
- Breach of legislation (e.g., unauthorized disclosure)
- Loss of reputation/goodwill
- Endangering of staff or customers
- Breach of confidence
- Loss of business opportunity
- Reduction in operational efficiency/performance
- Interruption of business activity

Step 4: Calculation of Risk

After the elements of risk have been established, they are combined to form an overall view of risk. A common method of combining the elements is to calculate for each threat: probability of occurrence × magnitude of impact. This will give a measure of overall risk.

The risk is proportional to the estimated likelihood of the threat and the value of the loss/damage.

Step 5: Evaluation of and Response to Risk

After risk has been identified, existing controls can be evaluated or new controls designed to reduce the vulnerabilities to an acceptable level. These controls are referred to as countermeasures or safeguards and include actions, devices, procedures or techniques (i.e., people, processes or products). The strength of a control can be measured in terms of its inherent or design strength and the likelihood of its effectiveness. Characteristics of controls that should be considered when evaluating control strength include whether the controls are preventive, detective or corrective, manual or automated, and formal (i.e., documented in procedure manuals and evidence of their operation is maintained) or *ad hoc*.

Residual risk, the remaining level of risk after controls have been applied, can be used by management to further reduce risk by identifying those areas in which more control is required. An acceptable level of risk target can be established by management (risk appetite). Risk in excess of this level should be reduced by the implementation of more stringent controls. Risk below this level should be evaluated to determine whether an excessive level of control is being applied and whether cost savings can be made by removing these excessive controls. Final acceptance of residual risk takes into account:
- Organizational policy
- Risk appetite
- Risk identification and measurement
- Uncertainty incorporated in the risk assessment approach
- Cost and effectiveness of implementation
- Cost of control versus benefit

It is important to realize that IT risk management needs to operate at multiple levels, including:
- **The operational level**—At the operational level, one is concerned with risk that could compromise the effectiveness and efficiency of IT systems and supporting infrastructure, the ability to bypass system controls, the possibility of loss or unavailability of key resources (e.g., systems, data, communications, personnel, premises), and failure to comply with laws and regulations.
- **The project level**—Risk management needs to focus on the ability to understand and manage project complexity and, if this is not done effectively, to handle the consequent risk that the project objectives will not be met.

- **The strategic level**—The risk focus shifts to considerations such as how well the IT capability is aligned with the business strategy, how it compares with that of competitors and the threats (as well as the opportunities) posed by technological change.

The identification, evaluation and management of IT risk at various levels will be the responsibility of different individuals and groups within the organization. However, these individuals and groups should not operate separately because risk at one level or in one area may also impact risk in another. A major system malfunction could impair an organization's ability to deliver customer service or deal with suppliers, and it could have strategic implications that require top management attention. Similarly, problems with a major project could have strategic implications. Also, as projects deliver new IT systems and infrastructure, the new operational risk environment needs to be considered.

In summary, the risk management process should achieve a cost-effective balance between the application of security controls as countermeasures and the significant threats. Some of the threats are related to security issues that can be extremely sensitive for some industries.

2.8.3 RISK ANALYSIS METHODS

This section discusses qualitative, semiquantitative and quantitative risk management methods, and the advantages and limitations of the latter.

Qualitative Analysis Methods

Qualitative risk analysis methods use word or descriptive rankings to describe the impacts or likelihood. They are the simplest and most frequently used methods. They are normally based on checklists and subjective risk ratings such as high, medium or low.

Such approaches lack the rigor that is customary for accounting and management.

Semiquantitative Analysis Methods

In semiquantitative analysis, the descriptive rankings are associated with a numeric scale. Such methods are frequently used when it is not possible to utilize a quantitative method or to reduce subjectivity in qualitative methods. For example, the qualitative measure of "high" may be given a quantitative weight of 5, "medium" may be given 3 and "low" may be given 1. The total weight for the subject area that is evaluated may be the aggregate of the weights so derived for the various factors being considered.

Quantitative Analysis Methods

Quantitative analysis methods use numeric values to describe the likelihood and impacts of risk, using data from several types of sources such as historic records, past experiences, industry practices and records, statistical theories, testing, and experiments.

Many quantitative risk analysis methods are currently used by military, nuclear, chemical, financial and other areas.

Quantitative risk analysis expresses risk in numeric (e.g., monetary) terms. A quantitative risk analysis is generally performed during a BIA. The main problem within this process is the valuation of information assets. Different individuals may assign different values to the same asset, depending on the relevance of information to the individuals. In the case of technology assets, it is not the cost of the asset that is considered but also the cost of replacement and the value of information processed by that asset.

2.9 INFORMATION TECHNOLOGY MANAGEMENT PRACTICES

IT management practices reflect the implementation of policies and procedures developed for various IT-related management activities. In most organizations, the IT department is a service (support) department. The traditional role of a service department is to help production (line) departments conduct their operations more effectively and efficiently. However, IT has become an integral part of every facet of the operations of an organization. Its importance continues to grow year after year, and there is little likelihood of a reversal of this trend. IS auditors must understand and appreciate the extent to which a well-managed IT department is crucial to achieving the organization's objectives.

Management activities to review the policy/procedure formulations and their effectiveness within the IT department includes practices such as HR (personnel) management, sourcing and IT change management.

2.9.1 HUMAN RESOURCE MANAGEMENT

HR management relates to organizational policies and procedures for recruiting, selecting, training and promoting staff, measuring staff performance, disciplining staff, succession planning, and staff retention. The effectiveness of these activities, as they relate to the IT function, impacts the quality of staff and the performance of IT duties.

Note: The IS auditor should be aware of HR management issues, but this information is not tested in the CISA exam due to its subjectivity and organizational specific subject matter.

Hiring

An organization's hiring practices are important to ensure that the most effective and efficient staff is chosen and that the company is in compliance with legal recruitment requirements. Some of the common controls include:

- Background checks (e.g., criminal, financial, professional, references, qualifications)
- Confidentiality agreements or nondisclosure agreements. Specific provision may be made in these agreements to abide by the security policies of the previous employer and not to exploit the knowledge of internal controls in that organization.
- Employee bonding to protect against losses due to theft, mistakes and neglect (Note: Employee bonding is not always an accepted practice all over the world; in some countries, it is not legal.)

- Conflict of interest agreements
- Codes of professional conduct/ethics
- Noncompete agreements
- Nondisclosure agreements

Control risk includes:
- Staff may not be suitable for the position they are recruited to fill.
- Reference checks may not be carried out.
- Temporary staff and third-party contractors may introduce uncontrolled risk.
- Lack of awareness of confidentiality requirements may lead to the compromise of the overall security environment.

Employee Handbook

Employee handbooks, distributed to all employees at time of hire, should explain items such as:
- Security policies and procedures
- Acceptable and unacceptable conduct
- Organizational values and ethics code
- Company expectations
- Employee benefits
- Vacation (holiday) policies
- Overtime rules
- Outside employment
- Performance evaluations
- Emergency procedures
- Disciplinary actions for:
  - Excessive absence
  - Breach of confidentiality and/or security
  - Noncompliance with policies

In general, there should be a published code of conduct for the organization that specifies the responsibilities of all employees.

Promotion Policies

Promotion policies should be fair and equitable and understood by employees. Policies should be based on objective criteria and consider an individual's performance, education, experience and level of responsibility.

The IS auditor should ensure that the IT organization has well-defined policies and procedures for promotion and is adhering to them.

Training

Training should be provided on a regular basis to all employees based on the areas where employee expertise is lacking. Training is particularly important for IT professionals, given the rapid rate of change of technology and products. It assures more effective and efficient use of IT resources and strengthens employee morale. Training must be provided when new hardware and/or software is being implemented. Training should also include relevant management, project management and technical training.

Cross-training means having more than one individual properly trained to perform a specific job or procedure. This practice has the advantage of decreasing dependence on one employee and can be part of succession planning. It also provides a backup for personnel in the event of absence for any reason and, thereby, provides for continuity of operations. However, in using this approach, it would be prudent to have first assessed the risk of any person knowing all parts of a system and what exposure this may cause.

Scheduling and Time Reporting

Proper scheduling provides for more efficient operation and use of computing resources. Time reporting allows management to monitor the scheduling process. Management can then determine whether staffing is adequate and whether the operation is running efficiently. It is important that the information being entered or recorded into such a system is accurate.

Time reporting can be an excellent source of information for IT governance purposes. One of the scarcest resources in IT is time, and its proper reporting will definitively help to better manage this finite resource. This input can be useful for cost allocation, invoicing, chargeback, key goal indicator (KGI) and KPI measurement, and activities analysis (e.g., how many hours the organization dedicates to application changes versus new developments).

Employee Performance Evaluations

Employee assessment/performance evaluations must be a standard and regular feature for all IT staff. The HR department should ensure that IT managers and IT employees set mutually agreed-on goals and expected results. Assessment can be set against these goals only if the process is objective and neutral.

Salary increments, performance bonuses and promotions should be based on performance. The same process can also allow the organization to gauge employee aspirations and satisfaction and identify problems.

Required Vacations

A required vacation (holiday) ensures that once a year, at a minimum, someone other than the regular employee will perform a job function. This reduces the opportunity to commit improper or illegal acts. During this time, it may be possible to discover fraudulent activity as long as there has been no collusion between employees to cover possible discrepancies.

Job rotation provides an additional control (to reduce the risk of fraudulent or malicious acts) because the same individual does not perform the same tasks all the time. This provides an opportunity for an individual other than the regularly assigned person to perform the job and notice possible irregularities. In addition, job rotation also guards against the risk of over dependence on key staff by spreading experience in procedures and controls as well as specific technologies. Without this, an enterprise could be vulnerable should a key employee be unavailable.

> **Note:** A CISA should be familiar with ways to mitigate internal fraud. Mandatory leave is such a control measure.

Termination Policies

Written termination policies should be established to provide clearly defined steps for employee separation. It is important that policies be structured to provide adequate protection for the

organization's computer assets and data. Termination practices should address voluntary and involuntary (e.g., immediate) terminations. For certain situations, such as involuntary terminations under adverse conditions, an organization should have clearly defined and documented procedures for escorting the terminated employee from the premises. In all cases, however, the following control procedures should be applied:
- **Return of all devices, access keys, ID cards and badges**—To prevent easy physical access
- **Deletion/revocation of assigned logon IDs and passwords**—To prohibit system access
- **Notification**—To appropriate staff and security personnel regarding the employee's status change to "terminated"
- **Arrangement of the final pay routines**—To remove the employee from active payroll files
- **Performance of a termination interview**—To gather insight on the employee's perception of management

Note: Changes in job role and responsibilities, such as a transfer to a different department, may necessitate revocation and reissuance of system and work area access rights similar to termination procedures.

2.9.2 SOURCING PRACTICES

Sourcing practices relate to the way in which the organization will obtain the IT functions required to support the business. Organizations can perform all the IT functions in-house (known as "insourcing") in a centralized fashion, or outsource all functions across the globe. The sourcing strategy should consider each IT function and determine which approach allows the IT function to meet the enterprise's goals.

Delivery of IT functions can include:
- **Insourced**—Fully performed by the organization's staff
- **Outsourced**—Fully performed by the vendor's staff
- **Hybrid**—Performed by a mix of the organization's and vendor's staffs; can include joint ventures/supplemental staff

IT functions can be performed across the globe, taking advantage of time zones and arbitraging labor rates, and can include:
- **Onsite**—Staff work onsite in the IT department.
- **Offsite**—Also known as nearshore, staff work at a remote location in the same geographic area.
- **Offshore**—Staff work at a remote location in a different geographic region.

The organization should evaluate its IT functions and determine the most appropriate method of delivering the IT functions, giving consideration to the following:
- Is this a core function for the organization?
- Does this function have specific knowledge, processes and staff critical to meeting its goals and objectives, and that cannot be replicated externally or in another location?
- Can this function be performed by another party or in another location for the same or lower price, with the same or higher quality, and without increasing risk?
- Does the organization have experience managing third parties or using remote/offshore locations to execute IS or business functions?
- Are there any contractual or regulatory restrictions preventing offshore locations or use of foreign nationals?

On completion of the sourcing strategy, the IT steering committee should review and approve the strategy. At this point, if the organization has chosen to use outsourcing, a rigorous process should be followed, including the following steps:
- Define the IT function to be outsourced.
- Describe the service levels required and minimum metrics to be met.
- Know the desired level of knowledge, skills and quality of the expected service provider desired.
- Know the current in-house cost information to compare with third-party bids.
- Conduct due diligence reviews of potential service providers.
- Confirm any architectural considerations to meeting contractual or regulatory requirements.

Using this information, the organization can perform a detailed analysis of the service provider bids and determine whether outsourcing will allow the organization to meet their goals in a cost-effective manner, with limited risk.

The same process should be considered when an organization chooses to globalize or take their IT functions offshore.

Outsourcing Practices and Strategies

Outsourcing practices relate to contractual agreements under which an organization hands over control of part or all of the functions of the IT department to an external party. Most IT departments use information resources from a wide array of vendors and, therefore, need a defined outsourcing process for effectively managing contractual agreements with these vendors.

The contractor provides the resources and expertise required to perform the agreed-on service. Outsourcing is becoming increasingly important in many organizations. The IS auditor must be aware of the various forms outsourcing can take and the associated risk.

The specific objectives for IT outsourcing vary from organization to organization. Typically, the goal is to achieve lasting, meaningful improvement in business processes and services through corporate restructuring to take advantage of a vendor's core competencies. As with the decision to downsize or rightsize, the decision to outsource services and products requires management to revisit the control framework on which it can rely.

Reasons for embarking on outsourcing include:
- A desire to focus on core activities
- Pressure on profit margins
- Increasing competition that demands cost savings
- Flexibility with respect to organization, structure and market size

An IS auditor should determine whether an enterprise considered the advantages, the disadvantages and business risk, and the risk reduction options depicted in **figure 2.8** as it developed its outsourcing practices and strategies.

Figure 2.8—Advantages, Disadvantages and Business Risk, and Risk Reduction Options Related to Outsourcing		
Possible Advantages	**Possible Disadvantages and Business Risk**	**Risk Reduction Options**
• Commercial outsourcing companies can achieve economies of scale through the deployment of reusable component software. • Outsourcing vendors are likely to be able to devote more time and to focus more effectively and efficiently on a given project than in-house staff. • Outsourcing vendors are likely to have more experience with a wider array of problems, issues and techniques than in-house staff. • The act of developing specifications and contractual agreements using outsourcing services is likely to result in better specifications than if developed only by in-house staff. • Because vendors are highly sensitive to time-consuming diversions and changes, feature creep or scope creep is substantially less likely with outsourcing vendors.	• Costs exceeding customer expectations • Loss of internal IT experience • Loss of control over IT • Vendor failure (ongoing concern) • Limited product access • Difficulty in reversing or changing outsourced arrangements • Deficient compliance with legal and regulatory requirements • Contract terms not being met • Lack of loyalty of contractor personnel toward the customer • Disgruntled customers/employees as a result of the outsource arrangement • Service costs not being competitive over the period of the entire contract • Obsolescence of vendor IT systems • Failure of either company to receive the anticipated benefits of the outsourcing arrangement • Reputational damage to either or both companies due to project failures • Lengthy, expensive litigation • Loss or leakage of information or processes	• Establishing measurable, partnership-enacted shared goals and rewards • Software escrow to ensure maintenance of the software • Using multiple suppliers or withholding a piece of business as an incentive • Performing periodic competitive reviews and benchmarking/benchtrending • Implementing short-term contracts • Forming a cross-functional contract management team • Including contractual provisions to consider as many contingencies as can reasonably be foreseen

In addition, an enterprise should consider the following provisions in its outsourcing contracts:

- Incorporating service quality expectations, including usage of ISO/IEC 15504 (Software Process Improvement and Capability dEtermination [SPICE]), CMMI, ITIL or ISO methodologies
- Ensuring adequate contractual consideration of access control/security administration, whether vendor- or owner-controlled
- Ensuring that violation reporting and follow-up are required by the contract
- Ensuring any requirements for owner notification and cooperation with any investigations
- Ensuring that change/version control and testing requirements are contractually required for the implementation and production phases
- Ensuring that the parties responsible and the requirements for network controls are adequately defined and any necessary delineation of these responsibilities established
- Stating specific, defined performance parameters that must be met; for example, minimum processing times for transactions or minimum hold times for contractors
- Incorporating capacity management criteria
- Providing contractual provisions for making changes to the contract
- Providing a clearly defined dispute escalation and resolution process
- Ensuring that the contract indemnifies the company from damages caused by the organization responsible for the outsourced services
- Requiring confidentiality agreements protecting both parties
- Incorporating clear, unambiguous "right to audit" provisions, providing the right to audit vendor operations (e.g., access to facilities, access to records, right to make copies, access to personnel, provision of computerized files) as they relate to the contracted services
- Ensuring that the contract adequately addresses business continuity and disaster recovery provisions, and appropriate testing
- Establishing that the confidentiality, integrity and availability (sometimes referred to as the CIA triad) of organization-owned data must be maintained, and clearly establishing the ownership of the data
- Requiring that the vendor comply with all relevant legal and regulatory requirements, including those enacted after contract initiation
- Establishing ownership of intellectual property developed by the vendor on behalf of the customer
- Establishing clear warranty and maintenance periods
- Providing software escrow provisions
- Protecting intellectual property rights
- Complying with legislation
- Establishing clear roles and responsibilities between the parties
- Requiring that the vendor follow the organization's policies, including their information security policy, unless the vendor's policies have been agreed to in advance by the organization
- Requiring the vendor to identify all subcontract relationships and requiring the organization's approval to change subcontractors

Outsourcing requires management to actively manage the relationship and the outsourced services. Because the outsourcing agreement is governed by the contract terms, the contract with the outsourced service provider should include a description of the means, methods, processes and structure accompanying the offer of IT services and products, and the control of quality. The formal or legal character of these agreements depends on the relationship between the parties and the demands placed by principals on those performing the engagement.

After the outsourcer has been selected, the IS auditor should regularly review the contract and service levels to ensure that they are appropriate. In addition, the IS auditor could review the outsourcer's documented procedures and results of their quality programs—which could include, for example, ISO/IEC 15504 (SPICE), CMMI, ITIL and ISO methodologies. These quality programs require regular audits to certify that the process and procedures meet the quality standard.

Outsourcing is not only a cost decision; it is a strategic decision that has significant control implications for management. Quality of service, guarantees of continuity of service, control procedures, competitive advantage and technical knowledge are issues that need to be part of the decision to outsource IT services. Choosing the right supplier is extremely important, particularly when outsourcing is a long-term strategy. The compatibility of suppliers in terms of culture and personnel is an important issue that should not be overlooked by management.

The decision to outsource a particular service currently within the organization demands proper attention to contract negotiations. A well-balanced contract and service level agreement (SLA) is of great importance for quality purposes and future cooperation between the concerned parties.

Above all, an SLA should serve as an instrument of control. If the outsourcing vendor is from another country, the organization should be aware of cross-border legislation.

SLAs stipulate and commit a vendor to a required level of service and support options. This includes providing for a guaranteed level of system performance regarding downtime or uptime and a specified level of customer support. Software or hardware requirements are also stipulated. SLAs also provide for penalty provisions and enforcement options for services not provided and may include incentives such as bonuses or gain-sharing for exceeding service levels.

SLAs are a contractual means of helping the IT department manage information resources that are under the control of a vendor.

Industry Standards/Benchmarking

Most outsourcing organizations must adhere to a well-defined set of standards that can be relied on by their clients. These industry standards provide a means of determining the level of performance provided by similar information processing facility environments. These standards can be obtained from vendor user groups, industry publications and professional associations. Examples include *ISO 9001:2008: Quality Management Systems—Requirements* and CMMI.

Globalization Practices and Strategies

Many organizations have chosen to globalize their IT functions in addition to outsourcing functions. The globalization of IT functions is performed for many of the same reasons cited for outsourcing; however, the organization may choose not to outsource the function. Globalizing IT functions requires management to actively oversee the remote or offshore locations.

Where the organization performs functions in-house, it may choose to move the IT functions offsite or offshore. The IS auditor can assist in this process by ensuring that IT management considers the following risk and audit concerns when defining the globalization strategy and completing the subsequent transition to remote offshore locations:

- **Legal, regulatory and tax issues**—Operating in a different country or region may introduce new risk about which the organization may have limited knowledge.
- **Continuity of operations**—Business continuity and disaster recovery may not be adequately provided for and tested.
- **Personnel**—Needed modifications to personnel policies may not be considered.
- **Telecommunication issues**—Network controls and access from remote or offshore locations may be subject to more frequent outages or a larger number of security exposures.
- **Cross-border and cross-cultural issues**—Managing people and processes across multiple time zones, languages and cultures may present unplanned challenges and problems. Cross-border data flow may also be subject to legislative requirements (e.g., that data must be encrypted during transmission).
- **Planned globalization and/or important expansion**

Cloud Computing

One issue surrounding the cloud and its related services is the lack of agreed-upon definitions. As with all emerging technologies, the lack of clarity and agreement often hinders the overall evaluation and adoption of that technology. Two groups that have offered a baseline of definitions are the National Institute of Standards and Technology (NIST) and the Cloud Security Alliance® (CSA). They both define cloud computing as a model for enabling convenient, on-demand network access to a shared pool of configurable computing resources (e.g., networks, servers, storage, applications and services) that can be rapidly provisioned and released with minimal management effort or service provider interaction. Another way to describe services offered in the cloud is to liken them to that of a utility. Just as enterprises pay for the electricity, gas and water they use, they now have the option of paying for IT services on a consumption basis.

The cloud model can be thought of as being composed of three service models (**figure 2.9**), four deployment models (**figure 2.10**) and five essential characteristics (**figure 2.11**). Overall risk and benefits will differ per model, and it is important to note that when considering different types of service and deployment models, enterprises should consider the risk that accompanies them.

Cloud storage may involve additional legal requirements of which the IS auditor should be aware. Some legislation, for example, requires data stored outside of the region to be subjected to additional security controls including strong encryption.

Outsourcing and Third-party Audit Reports

One method for the IS auditor to have assurance of the controls implemented by a service provider requires the provider to periodically submit a third-party audit report. These reports cover the range of issues related to confidentiality, integrity and availability of data. In some industries, third-party audits may fall under regulatory oversight and control, such as Statement

Figure 2.9—Cloud Computing Service Models		
Service Model	**Definition**	**To Be Considered**
Infrastructure as a Service (IaaS)	Capability to provision processing, storage, networks and other fundamental computing resources, offering the customer the ability to deploy and run arbitrary software, which can include operating systems and applications. IaaS puts these IT operations into the hands of a third party.	Options to minimize the impact if the cloud provider has a service interruption
Platform as a Service (PaaS)	Capability to deploy onto the cloud infrastructure customer-created or acquired applications created using programming languages and tools supported by the provider	• Availability • Confidentiality • Privacy and legal liability in the event of a security breach (as databases housing sensitive information will now be hosted offsite) • Data ownership • Concerns around e-discovery
Software as a Service (SaaS)	Capability to use the provider's applications running on cloud infrastructure. The applications are accessible from various client devices through a thin client interface such as a web browser (e.g., web-based email).	• Who owns the applications? • Where do the applications reside?
ISACA, *Cloud Computing: Business Benefits With Security, Governance and Assurance Perspectives*, USA, 2009, figure 1, page 5, *www.isaca.org/Knowledge-Center/Research/ResearchDeliverables/Pages/Cloud-Computing-Business-Benefits-With-Security-Governance-and-Assurance-Perspective.aspx*		

Figure 2.10—Cloud Computing Deployment Models		
Deployment Model	**Description of Cloud Infrastructure**	**To Be Considered**
Private cloud	• Operated solely for an organization • May be managed by the organization or a third party • May exist on-premise or off-premise	• Cloud services with minimum risk • May not provide the scalability and agility of public cloud services
Community cloud	• Shared by several organizations • Supports a specific community that has shared mission or interest. • May be managed by the organizations or a third party • May reside on-premise or off-premise	• Same as private cloud, plus: • Data may be stored with the data of competitors.
Public cloud	• Made available to the general public or a large industry group • Owned by an organization selling cloud services	• Same as community cloud, plus: • Data may be stored in unknown locations and may not be easily retrievable.
Hybrid cloud	A composition of two or more clouds (private, community or public) that remain unique entities but are bound together by standardized or proprietary technology that enables data and application portability (e.g., cloud bursting for load balancing between clouds)	• Aggregate risk of merging different deployment models • Classification and labeling of data will be beneficial to the security manager to ensure that data are assigned to the correct cloud type.
ISACA, *Cloud Computing: Business Benefits With Security, Governance and Assurance Perspectives*, USA, 2009, figure 2, page 5, *www.isaca.org/Knowledge-Center/Research/ResearchDeliverables/Pages/Cloud-Computing-Business-Benefits-With-Security-Governance-and-Assurance-Perspective.aspx*		

Figure 2.11—Cloud Computing Essential Characteristics	
Characteristic	Definition
On-demand self-service	The cloud provider should have the ability to automatically provision computing capabilities, such as server and network storage, as needed without requiring human interaction with each service's provider.
Broad network access	According to NIST, the cloud network should be accessible anywhere, by almost any device (e.g., smart phone, laptop, mobile devices).
Resource pooling	The provider's computing resources are pooled to serve multiple customers using a multitenant model, with different physical and virtual resources dynamically assigned and reassigned according to demand. There is a sense of location independence. The customer generally has no control or knowledge over the exact location of the provided resources. However, he/she may be able to specify location at a higher level of abstraction (e.g., country, region or data center). Examples of resources include storage, processing, memory, network bandwidth and virtual machines.
Rapid elasticity	Capabilities can be rapidly and elastically provisioned, in many cases automatically, to scale out quickly and rapidly released to scale in quickly. To the customer, the capabilities available for provisioning often appear to be unlimited and can be purchased in any quantity at any time.
Measured service	Cloud systems automatically control and optimize resource use by leveraging a metering capability (e.g., storage, processing, bandwidth and active user accounts). Resource usage can be monitored, controlled and reported, providing transparency for both the provider and customer of the utilized service.
ISACA, *Cloud Computing: Business Benefits With Security, Governance and Assurance Perspectives*, USA, 2009, figure 3, page 6, *www.isaca.org/Knowledge-Center/Research/ResearchDeliverables/Pages/Cloud-Computing-Business-Benefits-With-Security-Governance-and-Assurance-Perspective.aspx*	

on Standards for Attestation Engagements (SSAE) 16 and an audit guide by the American Institute of Certified Public Accountants (AICPA), which provides a framework for three Service Organization Control (SOC) reporting options (SOC 1, SOC 2 and SOC 3 reports). These reporting standards represent significant changes from the Statement on Auditing Standards (SAS) 70 report, as organizations increasingly became interested in risks beyond financial statement reporting (e.g., privacy). The International Auditing and Assurance Standards Board (IAASB) also issued new guidance in this regard—the International Standard on Assurance Engagements (ISAE) 3402, Assurance Reports on Controls at a Service Organization.

An IS auditor should be familiar with the following:
- Management assertions and how well these address the services being provided by the service provider
- SSAE 16 reports as follows:
  - SOC 1: Report on the service organization's system controls likely to be relevant to user entities' internal control over financial reporting
  - SOC 2: Report on the service organization's system controls relevant to security, availability, processing integrity, confidentiality or privacy, including the organization's compliance with its privacy practices
  - SOC 3: Similar to a SOC 2 report, but does not include the detailed understanding of the design of controls and the tests performed by the service auditor
- How to obtain the report, review it and present results to management for further action

Governance in Outsourcing

Outsourcing is the mechanism that allows organizations to transfer the delivery of services to third parties. Fundamental to outsourcing is accepting that, while service delivery is transferred, accountability remains firmly with the management of the client organization, which must ensure that the risk is properly managed and there is continued delivery of value from the service provider. Transparency and ownership of the decision-making process must reside within the purview of the client.

The decision to outsource is a strategic, not merely a procurement, decision. The organization that outsources is effectively reconfiguring its value chain by identifying those activities that are core to its business, retaining them and making noncore activities candidates for outsourcing. Understanding this in the light of governance is key, not only because well-governed organizations have been shown to increase shareholder value, but more importantly, because organizations are competing in an increasingly aggressive, global and dynamic market.

Establishing and retaining competitive and market advantage requires the organization to be able to respond effectively to competition and changing market conditions. Outsourcing can support this, but only if the organization understands which parts of its business truly create competitive advantage.

Governance of outsourcing is the set of responsibilities, roles, objectives, interfaces and controls required to anticipate change and manage the introduction, maintenance, performance, costs and control of third-party provided services. It is an active process that the client and service provider must adopt to provide a common, consistent and effective approach that identifies the necessary information, relationships, controls and exchanges among many stakeholders across both parties.

The decision to outsource and subsequently successfully manage that relationship demands effective governance. Most people who conduct outsourcing contracts include basic control and service execution provisions; however, one of the main objectives of the outsourcing governance process, as defined in the outsourcing contract, is to ensure continuity of service at the appropriate levels and profitability and added value to sustain the commercial viability of both parties. Experience has shown that many companies make assumptions about what is included in the outsource proposition. Whereas it is neither possible nor cost-effective to contractually define every detail and action, the governance process provides the mechanism to balance risk, service demand, service provision and cost.

The governance of outsourcing extends both parties' (i.e., client and supplier) responsibilities into:
- Ensuring contractual viability through continuous review, improvement and benefit gain to both parties
- Inclusion of an explicit governance schedule to the contract
- Management of the relationship to ensure that contractual obligations are met through SLAs and operating level agreements (OLAs)
- Identification and management of all stakeholders, their relationships and expectations
- Establishment of clear roles and responsibilities for decision making, issue escalation, dispute management, demand management and service delivery
- Allocation of resources, expenditure and service consumption in response to prioritized needs
- Continuous evaluation of performance, cost, user satisfaction and effectiveness
- Ongoing communication across all stakeholders

The increasing size of the technology solution space is driven by the pace of technological evolution. Acquiring, training and retaining qualified staff is becoming more expensive in an increasingly global, dynamic and mobile economy. Investing in costly technology implementation and training is seen as less of an organizational core activity than is the ability to work effectively across the value chain by integrating the outsourcing of services where appropriate.

Although the term "business alignment" is often used, what it encompasses is not always clear. In the widest sense, it involves making the services provided by the corporate IT function more closely reflect the requirements and desires of the business users. When organizations recognize what is core to their business, which services provide them differential advantage and then outsource the activities that support these services, business alignment can be achieved. If the degree to which this alignment is approached is to be understood, the implication is that SLAs and OLAs must be established, monitored and measured in terms of performance and user satisfaction. Business alignment should be driven by the service end user.

Governance should be preplanned and built into the contract as part of the service cost optimization. The defined governance processes should evolve as the needs and conditions of the outsourcing relationship adapt to changes to service demand and delivery and to technology innovation.

It is critical for the IS auditor to understand right-to-audit clauses and controls in outsourcing activities involving confidential information and sensitive processes. This understanding includes, but is not limited to:
- How auditing of the outsourced service provider is allowed to be conducted under the terms of the contract
- What visibility the IS auditor has into the internal controls being implemented by the outsourced service provider to provide reasonable assurance that confidentiality, integrity and availability and preventive, detective and corrective controls are in place and effective
- SLAs regarding problem management to include incident response are documented and communicated to all parties affected by these outsourcing agreements

Capacity and Growth Planning

Given the strategic importance of IT in companies and the constant change in technology, capacity and growth planning are essential. This activity must be reflective of the long- and short-range business plans and must be considered within the budgeting process. Changes in capacity should reflect changes in the underlying infrastructure and in the number of staff available to support the organization. A lack of appropriately qualified staff may delay projects that are critical to the organization or result in not meeting agreed-on service levels. This can lead some organizations to choose outsourcing as a solution for growth.

Third-party Service Delivery Management

Every organization using the services of third parties should have a service delivery management system in place to implement and maintain the appropriate level of information security and service delivery in line with third-party service delivery agreements.

The organization should check the implementation of agreements, monitor compliance with the agreements and manage changes to ensure that the services delivered meet all requirements agreed to with the third party.

SERVICE DELIVERY

Security controls, service definitions and delivery levels included in the third-party service delivery agreement should be implemented, operated and maintained by the third party.

Service delivery by a third party should include the agreed-on security arrangements, service definitions and aspects of service management. In case of outsourcing arrangements, the organization should plan the necessary transitions (of information, data center facilities and anything else that needs to be moved) and ensure that security is maintained throughout the transition period.

The organization should ensure that the third party maintains sufficient service capability together with workable plans designed to ensure that agreed-on service continuity levels are maintained following major service failures or disaster.

MONITORING AND REVIEW OF THIRD-PARTY SERVICES

The services, reports and records provided by the third party should be regularly monitored and reviewed, and audits should be carried out regularly. Monitoring and review of third-party

services should ensure that the information security terms and conditions of the agreements are being adhered to, and information security incidents and problems are managed properly. This should involve a service management relationship and process between the organization and the third party to:

- Monitor service performance levels to check adherence to the agreements
- Review service reports produced by the third party and arrange regular progress meetings as required by the agreements
- Provide information about information security incidents and review of this information by the third party and the organization as required by the agreements and any supporting guidelines and procedures
- Review third-party audit trails and records of security events, operational problems, failures, tracing of faults and disruptions related to the service delivered
- Resolve and manage any identified problems

CLOUD GOVERNANCE

The strategic direction of the business and of IT in general is the main focus when considering the use of cloud computing. As enterprises look to the cloud to provide IT services that traditionally have been managed internally, they will need to make some changes to help ensure that they continue to meet performance objectives, their technology provisioning and business are strategically aligned, and risk is managed. Ensuring that IT is aligned with the business, systems are secure and risk is managed is challenging in any environment and even more complex in a third-party relationship. Typical governance activities such as goal setting, policy and standard development, defining roles and responsibilities, and managing risk must include special considerations when dealing with cloud technology and its providers.

As with all organizational changes, it is expected that some adjustments will need to be made to the way business processes are handled. Business/IT processes such as data processing, development and information retrieval are examples of potential change areas. Additionally, processes detailing the way information is stored, archived and backed up will need revisiting.

The cloud presents many unique situations for businesses to address. One large governance issue is that business unit personnel, who were previously forced to go through IT for service, can now bypass IT and receive service directly from the cloud. Policies must be modified or developed to address the process of sourcing, managing and discontinuing the use of cloud services.

The responsibility for managing the relationship with a third party should be assigned to a designated individual or service management team. In addition, the organization should ensure that the third party assigns responsibilities for checking for compliance and enforcing the requirements of the agreements. Sufficient technical skills and resources should be made available to monitor whether requirements of the agreement, in particular the information security requirements, are being met. Appropriate action should be taken when deficiencies in the service delivery are observed.

The organization should maintain sufficient overall control and visibility into all security aspects for sensitive or critical information or information processing facilities accessed, processed or managed by a third party. The organization should ensure that they retain visibility in security activities such as change management, identification of vulnerabilities and information security incident reporting/response through a clearly defined reporting process, format and structure. When outsourcing, the organization needs to be aware that the ultimate responsibility for information processed by an outsourcing party remains with the organization.

MANAGING CHANGES TO THIRD-PARTY SERVICES

Changes to the provision of services, including maintaining and improving existing information security policies, procedures and controls, should be managed taking into account the criticality of business systems and processes involved and reassessing risk.

The process of managing changes to a third-party service needs to take into account:

- Changes made by the organization to implement:
  - Enhancements to the current services offered
  - Development of any new applications and systems
  - Modifications or updates of the organization's policies and procedures
  - New controls to resolve information security incidents and to improve security
  - Updates to policies, including the IT security policy
- Changes in third-party services to implement:
  - Changes and enhancements to networks
  - Use of new technologies
  - Adoption of new products or newer versions/releases
  - New development tools and environments
  - Changes to physical location of service facilities
  - Change of vendors or subcontractors

Service Improvement and User Satisfaction

SLAs set the baseline by which outsourcers perform the IT function. In addition, organizations can set service improvement expectations into the contracts with associated penalties and rewards. Examples of service improvements include:

- Reductions in the number of help desk calls
- Reductions in the number of system errors
- Improvements to system availability

Service improvements should be agreed on by users and IT with the goals of improving user satisfaction and attaining business objectives. User satisfaction should be monitored by interviewing and surveying users.

2.9.3 ORGANIZATIONAL CHANGE MANAGEMENT

Organizational change management involves use of a defined and documented process to identify and apply technology improvements at the infrastructure and application level that are beneficial to the organization and involve all levels of the organization impacted by the changes. This level of involvement and communication will ensure that the IT department fully understands the users' expectations and changes are not resisted or ignored by users after they are implemented.

The IT department is the focal point for such changes by leading or facilitating change in the organization. This includes staying abreast of technology changes that could lead to significant business process improvements and obtaining senior management commitment for the changes or projects that will be required at the user level.

After senior management support is obtained to move forward with the changes or projects, the IT department can begin working with each functional area and its management to obtain support for the changes. In addition, the IT department will need to develop a communication process that is directed at the end users to update them on the changes, the impact and benefit of the changes, and provide a method for obtaining user feedback and involvement.

User feedback should be obtained throughout the project, including validation of the business requirements and training on and testing of the new or changed functionality.

2.9.4 FINANCIAL MANAGEMENT PRACTICES

Financial management is a critical element of all business functions. In a cost-intensive computer environment, it is imperative that sound financial management practices are in place.

The user-pays scheme, a form of chargeback, can improve application and monitoring of IS expenses and available resources. In this scheme the costs of IS services—including staff time, computer time and other relevant costs—are charged back to the end users based on a standard (uniform) formula or calculation.

Chargeback provides all involved parties with a "marketplace" measure of the effectiveness and efficiency of the service provided by the information processing facility. Where implemented, the chargeback policy shall be set forth by the board and jointly implemented by the CFO, user management and IS management.

IS Budgets

IS management, like all other departments, must develop a budget.

A budget allows for forecasting, monitoring and analyzing financial information. The budget allows for an adequate allocation of funds, especially in an IS environment where expenses can be cost-intensive. The IS budget should be linked to short- and long-range IT plans.

Software Development

In the United States and in countries using International Accounting Standards Board (IASB) guidance, accounting standards require that companies have a detailed understanding of their development efforts, including time spent on specific projects and activities. An IS auditor should understand these requirements and the practices used by companies to track software development costs.

In the United States, the AICPA details these requirements in their Accounting Statement of Position (SOP) 98-1, Accounting for the Costs of Computer Software Developed or Obtained for Internal Use. This SOP explains that companies should capitalize certain internal-use software costs. Internal-use software is software that an entity has no substantive plans to market externally.

International Accounting Standard 38 (IAS 38) outlines six criteria that must be met if development costs are to be capitalized. Of these, an organization should demonstrate, according to IAS 38.57.d, "how the intangible asset will generate probable future economic benefits." Intangible assets include web sites, software, etc., if they satisfy these criteria. Interpretations of what "demonstrating the usefulness of the intangible asset" means vary. Therefore, the IS auditor working with organizations following International Financial Reporting Standards (IFRS) will need to obtain the guidance from the chartered accountants responsible for financial reporting.

2.9.5 QUALITY MANAGEMENT

Quality management is one of the means by which IT department-based processes are controlled, measured and improved. Processes in this context are defined as a set of tasks that, when properly performed, produce the desired results. Areas of control for quality management may include the following:

- Software development, maintenance and implementation
- Acquisition of hardware and software
- Day-to-day operations
- Service management
- Security
- HR management
- General administration

The development and maintenance of defined and documented processes by the IT department is evidence of effective governance of information resources. Insistence on the observance of processes and related process management techniques is key to the effectiveness and efficiency of the IT organization. Various standards have emerged to assist IT organizations in achieving these results. Quality standards are increasingly being used to assist IT organizations in achieving an operational environment that is predictable, measurable, repeatable and certified for their IT resources.

Note: The IS auditor should be aware of quality management. However, the CISA exam does not test specifics on any ISO standards.

A prominent standard receiving wide recognition and acceptance is ISO 9001:2008, which replaces earlier ISO standards governing the management of quality. Other standards, such as the ISO/IEC 27000 series, set the foundation to create quality information security programs. The standards explain the overall Plan-Do-Check-Act (PDCA) approach and provide detailed guidance for its implementation.

The introductory standard ISO/IEC 27000 defines the scope and vocabulary used throughout the information security management system (ISMS) standard and provides a directory of the publications that comprise the standard. ISO/IEC 27001 is the formal set of specifications against which organizations may seek independent certification of their ISMS. ISO/IEC 27002 contains a structured set of suggested controls that may be used by organizations as appropriate to address information security risk. Additional ISO/IEC 27000 series publications offer guidance for managing information security in specific industries and situations. The ISO/IEC 27000 series evolved from

ISO/IEC 17799, which was based on the 1995 United Kingdom BSI standard BS1799 for information security management good practices.

2.9.6 INFORMATION SECURITY MANAGEMENT

Information security management provides the lead role to ensure that the organization's information and the information processing resources under its control are properly protected. This would include leading and facilitating the implementation of an organizationwide information security program that includes the development of a BIA, a BCP and aDRP related to IT department functions in support of the organization's critical business processes. A major component in establishing such programs is the application of risk management principles to assess the risk to IT assets, mitigate the risk to an appropriate level as determined by management and monitor the remaining residual risk.

See chapter 5 Protection of Information Assets, for more details on information security management.

2.9.7 PERFORMANCE OPTIMIZATION

Performance is not how well a system works; performance is the service perceived by users and stakeholders. Performance optimization is the process of both improving perceived service performance along with improving information system productivity to the highest level possible without unnecessary, additional investment in the IT infrastructure.

Within the foundation of effective performance management approaches, measures are not just used for assigning accountabilities or to comply with reporting requirements. Measures are used to create and facilitate action to improve performance and, therefore, GEIT.

Effective performance measurement depends on two key aspects being addressed:
- The clear definition of performance goals
- The establishment of effective metrics to monitor achievement of goals

A performance measurement process is also required to help ensure that performance is monitored consistently and reliably. Effective governance significantly enables overall performance optimization and is achieved when:
- Goals are set from the top down and aligned with high-level, approved business goals.
- Metrics are established from the bottom up and aligned in a way that enables the achievement of goals at all levels to be monitored by each layer of management.

Critical Success Factors

Two critical governance success factors (enabling overall performance optimization) are:
- The approval of goals by stakeholders
- The acceptance of accountability for achievement of goals by management

IT is a complex and technical topic; therefore, it is important to achieve transparency by expressing goals, metrics and performance reports in language meaningful to the stakeholders so that appropriate actions can be taken.

Methodologies and Tools

A variety of improvement and optimization methodologies are available that complement simple, internally developed approaches. These include:
- Continuous improvement methodologies, such as the PDCA cycle
- Comprehensive best practices, such as ITIL
- Frameworks, such as COBIT

PDCA is an iterative four-step management method used in business for the control and continuous improvement of processes and products. The steps in each successive PDCA cycle are:
- **Plan**—Establish the objectives and processes necessary to deliver results in accordance with the expected output (the target or goals). By establishing output expectations, the completeness and accuracy of the specification is also a part of the targeted improvement. When possible, start on a small scale to test possible effects.
- **Do**—Implement the plan, execute the process and make the product. Collect data for charting and analysis in the following Check and Act steps.
- **Check**—Study the actual results (measured and collected in the Do step) and compare against the expected results (targets or goals from the Plan step) to ascertain any differences. Look for deviation in implementation from the plan, and also look for the appropriateness/completeness of the plan to enable the execution (i.e., the Do step). Charting data can make it much easier to see trends over several PDCA cycles and to convert the collected data into information. Information is needed for the next step, Act.
- **Act**—Request corrective actions on significant differences between actual and planned results. Analyze the differences to determine their root causes. Determine where to apply changes that will include improvement of the process or product. When a pass through these four steps does not result in the need to improve, the scope to which PDCA is applied may be refined to plan and improve with more detail in the next iteration of the cycle, or attention needs to be placed in a different stage of the process.

Using the PDCA following agile development allows for reassessment of the direction of the project at points throughout the development life cycle. This is done through "sprints" or "iterations," which require working groups to produce a functional product. This focus on abbreviated work cycles has led to the description of agile methodology as "iterative" and "incremental." As compared to a single opportunity to achieve each aspect of a project as in the waterfall method, agile development allows for each aspect to be continually revisited.

The *COBIT 5 for Assurance* guide explains how assurance professionals can provide independent assurance to boards of directors regarding IT performance.

Tools and Techniques

Tools and techniques that facilitate measurements, good communication and organizational change include:
- Six Sigma
- IT BSC
- KPIs
- Benchmarking
- Business process reengineering (BPR)
- Root cause analysis
- Life cycle cost-benefit analysis

Six Sigma and Lean Six Sigma are proven quantitative process analysis and improvement approach that easily translates to IT processes. Six Sigma's objective is the implementation of a measurement-oriented strategy focused on process improvement and defect reduction. A Six Sigma defect is defined as anything outside customer specifications.

Lean Six Sigma examines the measurement-oriented strategy focused on process improvement and defect reduction and the efficiency of these processes. Both Six Sigma and Lean Six Sigma use statistical data drive processes in defining process from data source, input, processes, output, and products and services provided by the process under review.

The **IT BSC** is a process management evaluation technique that can be applied to the GEIT process in assessing IT functions and processes. See section 2.3.3 The IT Balanced Scorecard for more information.

A **KPI** is a measure that determines how well the process is performing in enabling the goal to be reached. It is a lead indicator of whether a goal will likely be reached and a good indicator of capabilities, practices and skills. For example, a service delivered by IT is a goal for IT, but a performance indicator and a capability for the business. This is why performance indicators are sometimes referred to as performance drivers, particularly in BSCs.

As controls are selected for implementation, criteria should also be established to determine the operational level and effectiveness of the controls. These criteria will often be based on KPIs that indicate whether a control is functioning correctly. For example, a KPI for the implementation process measures the relative success of the changeover compared to desired performance objectives. Success of a changeover is often measured as a percentage of errors, number of trouble reports, duration of system outage or degree of customer satisfaction. The use of the KPI indicates to management whether the change control process was managed correctly, with sufficient levels of quality and testing.

Benchmarking is a systematic approach to comparing enterprise performance against peers and competitors in an effort to learn the best ways of conducting business. Examples include benchmarking of quality, logistic efficiency and various other metrics.

BPR is the thorough analysis and significant redesign of business processes and management systems to establish a better performing structure, more responsive to the customer base and market conditions, while yielding material cost savings. For more information on BPR, see section 3.12.1 Business Process Reengineering and Process Change Projects.

IT performance measurement and reporting may be a statutory or contractual requirement. Appropriate performance measurement practices for the enterprise include outcome measures for business value, competitive advantage and defined performance metrics that show how well IT performs. Incentives, such as rewards, compensation and recognition should be linked to performance measures. It is also important to share results and progress with employees, customers and stakeholders.

Root cause analysis is the process of diagnosis to establish the origins of events (root causes). Once identified, the root causes can then be used to develop needed controls to accurately address these root causes that lead to system failures and deficiencies. Furthermore, root cause analysis also enables an organization to learning from consequences, typically from errors and problems, in the effort to not repeat undesired actions or results.

Life cycle cost-benefit analysis is the assessment of the following element to determine strategic direction for IT enterprise systems and overall IT portfolio management. Key terms for this process include the following:
- Life cycle (LC): A series of stages that characterize the course of existence of an organizational investment (e.g., product, project, program)
- Life cycle cost (LCC): The estimated costs of maintenance/updates, failure, and maintaining interoperability with mainstream and emerging technologies
- Benefit analysis (BA): The user costs (or benefits) and business operational costs (or benefits) derived from the information system(s)

2.10 IT ORGANIZATIONAL STRUCTURE AND RESPONSIBILITIES

An IT department can be structured in different ways. One such format is shown in **figure 2.12**. The organizational chart depicted includes functions related to security, applications development and maintenance, technical support for network and systems administration, and operations. The organizational structure shows the IT department typically headed by an IT manager/director or, in large organizations, by a CIO.

> **Note:** The CISA exam does not test specific job responsibilities because they may vary within organizations. However, universally known responsibilities such as business owners, information security functions and executive management might be tested, especially when access controls and data ownership are tested. A CISA should be familiar with segregation of duties (SoD).

2.10.1 IT ROLES AND RESPONSIBILITIES

An organizational chart is an important item for all employees to know, because it provides a clear definition of the department's hierarchy and authorities. Additionally, job descriptions, RACI charts and swimlane workflow diagrams provide IT department employees a more complete and clear direction regarding their

Figure 2.12—IT Department Organization

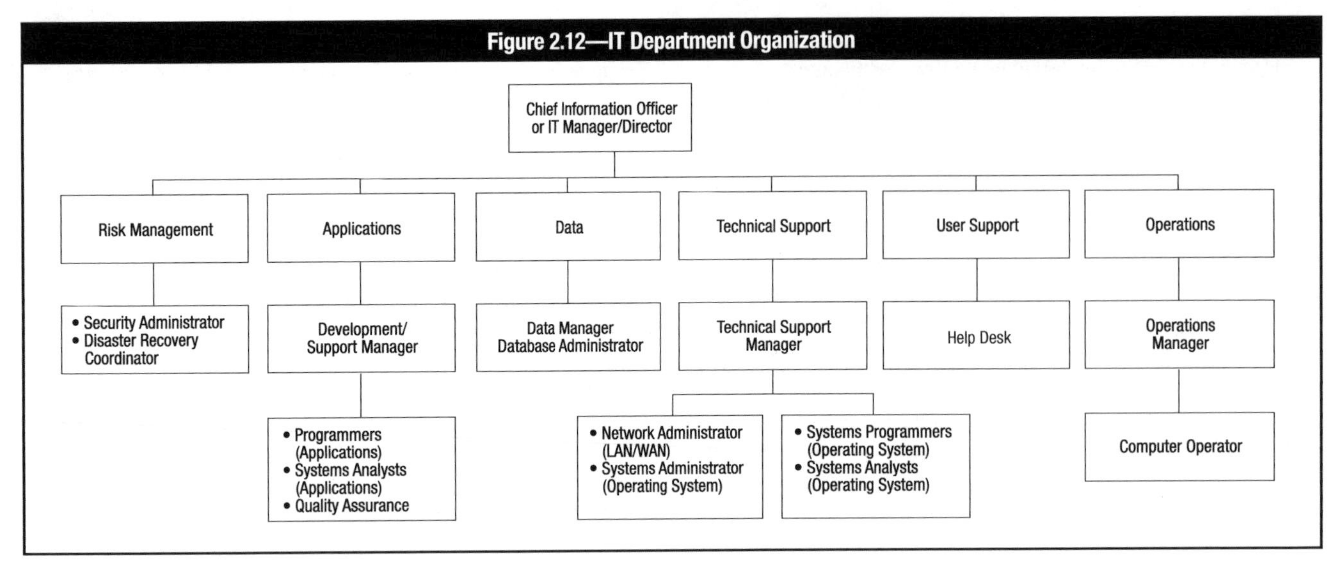

(and others') roles and responsibilities. The IS auditor should spend time in an auditee's area to observe and determine whether the formal job description and structures coincide with real ones and are adequate. Generally, the following IT functions should be reviewed:

- **Systems development manager**—Systems development managers are responsible for programmers and analysts who implement new systems and maintain existing systems
- **Project management**—Project managers are responsible for planning and executing IS projects and may report to a project management office or to the development organization. Project management staff utilize budgets assigned to them for the delivery of IS initiatives and report on project progress to the IT steering committee. Project managers play a central role in executing the vision of the IT strategy and IT steering committee by planning, coordinating and delivering IT projects to the enterprise.
- **Help desk (service desk)**—More and more companies find it important to have a help desk function for their IT departments. A help desk is a unit within an organization that responds to technical questions and problems faced by users. Most software companies have help desks. Questions and answers can be delivered by telephone, fax, email or instant messaging. Help desk personnel may use third-party help desk software that enables them to quickly find answers to common questions. A procedure to record the problems reported, solved and escalated should be in place for analysis of the problems/questions. It helps in monitoring the user groups and improving the software/information processing facility (IPF) services.
  - Help desk/support administration includes the following activities:
    - Acquiring hardware/software (HW/SW) on behalf of end users
    - Assisting end users with HW/SW difficulties
    - Training end users to use HW/SW and databases
    - Answering end-user queries
    - Monitoring technical developments and informing end users of pertinent developments
    - Determining the source of problems with production systems and initiating corrective actions
    - Informing end users of problems with HW/SW or databases that could affect their control of the installation of HW/SW upgrades
    - Initiating changes to improve efficiency
- **End user**—End users are responsible for operations related to business application services; used to distinguish the person for whom the product (generally application level) was designed from the person who programs, services or installs applications. It is worth noting that there is a small distinction between the terms "end user" and "user." End user is slightly more specific and refers to someone who will access a business application, as stated above. The term user is broader and could refer to administrative accounts and accounts to access platforms.
- **End-user support manager**—The end-user support manager acts as a liaison between the IT department and the end users.
- **Data management**—Data management personnel are responsible for the data architecture in larger IT environments and tasked with managing data as a corporate asset.
- **Quality assurance (QA) manager**—The QA manager is responsible for negotiating and facilitating quality activities in all areas of information technology.
- **Information security management**—This is a function that generally needs to be separate from the IT department and headed by a CISO. The CISO may report to the CIO or have a dotted-line (indirect reporting) relationship to the CIO. Even when the security officer reports to the CIO there is a possibility of conflict because the goals of the CIO are to efficiently provide continuous IT services whereas the CISO may be less interested in cost reduction if this impacts the quality of protection.

Vendor and Outsourcer Management

With the increase in outsourcing, including the use of multiple vendors, dedicated staff may be required to manage the vendors and outsourcers including performing the following functions:

- Act as the prime contact for the vendor and outsourcer within the IT function.
- Provide direction to the outsourcer on issues and escalate internally within the organization and IT function.
- Monitor and report on the service levels to management.
- Review changes to the contract due to new requirements and obtain IT approvals.

Infrastructure Operations and Maintenance

An **operations manager** is responsible for computer operations personnel, including all the staff required to run the data center efficiently and effectively (e.g., computer operators, librarians, schedulers and data control personnel). The data center includes the servers and mainframe, peripherals such as high-speed printers, networking equipment, magnetic media and storage area networks. It constitutes a major asset investment and impacts the organization's ability to function effectively.

The **control group** is responsible for the collection, conversion and control of input, and the balancing and distribution of output to the user community. The supervisor of the control group usually reports to the IPF operations manager. The input/output control group should be in a separate area where only authorized personnel are permitted since they handle sensitive data. For more information, see section 3.13.1 Input/Origination Controls.

Media Management

Media management is required to record, issue, receive and safeguard all program and data files that are maintained on removable media. Depending on the size of the organization, this function may be assigned to a full-time individual or a member of operations who also performs other duties.

This is a crucial function. Therefore, many organizations provide additional support for this function through the use of software that assists in maintaining inventory and movement of media. The use of this software also helps to maintain version control and configuration management of the programs.

Data Entry

Data entry is critical to the information processing activity and includes batch entry or online entry.

In most organizations personnel in user departments do their own data entry online. In many online environments, data are captured from the original source (e.g., electronic data interchange [EDI] input documents, data captured from bar codes for time management, departmental store inventory). The user department and the system application must have controls in place to ensure that data are validated, accurate, complete and authorized.

Supervisory Control and Data Acquisition

With the advancement of technology and need to acquire data at its origination, automated systems for data acquisition are being deployed by organizations. These systems include barcode readers, or systems that are referred to as Supervisory Control and Data Acquisition (SCADA). The term SCADA usually refers to centralized systems that monitor and control entire sites, or complexes of systems spread out over large areas (on the scale of kilometers or miles). These systems are typical of industrial plants, steel mills, power plants, electrical facilities and similar. Most site control is performed automatically by remote terminal units (RTUs) or by programmable logic controllers (PLCs). Host control functions are usually restricted to basic site overriding or supervisory level intervention. An example of automated systems for data acquisition are those used on oil rigs to measure and control the extraction of oil and to control the temperature and flow of water.

Data acquisition begins at the RTU or PLC level and includes meter readings and equipment status reports that are communicated to SCADA as required. Data are then compiled and formatted in such a way that a control room operator using human machine interfacing (HMI) networks can make supervisory decisions to adjust or override normal RTU (PLC) controls. Data may also be fed to a history log, often built on a commodity database management system, to allow trending and other analytical auditing.

SCADA applications traditionally used dedicated communication lines. Currently there is a significant migration to the Internet. This trend has obvious advantages, among them easier integration in the company business applications. However, a disadvantage is that many such companies are nation-critical infrastructures and become easy prey to cyberattacks (see section 3.7.16 Industrial Control Systems for greater detail).

Systems Administration

The **systems administrator** is responsible for maintaining major multiuser computer systems, including local area networks (LANs), wireless local area networks (WLANs), wide area networks (WANs), virtual machine/server/network environments, personal area networks (PANs), storage area networks (SANs), intranets and extranets, and mid-range and mainframe systems. Typical duties include:

- Adding and configuring new workstations and peripherals
- Setting up user accounts
- Installing systemwide software
- Performing procedures to prevent/detect/correct the spread of viruses
- Allocating mass storage space

Small organizations may have one systems administrator, whereas larger enterprises may have a team of systems administrators. Some mainframe-centric organizations may refer to a systems administrator as a systems programmer.

Security Administration

Security administration begins with management's commitment. Management must understand and evaluate security risk and develop and enforce a written policy that clearly states the standards and procedures to be followed. The duties of the **security administrator** should be defined in the policy. To provide adequate SoD, this individual should be a full-time employee who may report directly to the infrastructure director. However, in a small organization, it may not be practical to hire a full-time individual for this position. The individual performing the function should ensure that the various users are complying with the corporate security policy and controls are adequate to prevent unauthorized access to the company assets (including data, programs and equipment). The security administrator's functions usually include:

- Maintaining access rules to data and other IT resources
- Maintaining security and confidentiality over the issuance and maintenance of authorized user IDs and passwords
- Monitoring security violations and taking corrective action to ensure that adequate security is provided
- Periodically reviewing and evaluating the security policy and suggesting necessary changes to management

- Preparing and monitoring the security awareness program for all employees
- Testing the security architecture to evaluate the security strengths and detect possible threats
- Working with compliance, risk management and audit functions to ensure that security is appropriately designed and updated based on audit feedback or testing

Quality Assurance

The terms "quality assurance" and "quality control" are often used interchangeably to refer to ways of ensuring the quality of a service or product. The terms, however, do have different meanings.

Quality assurance personnel usually perform two distinct tasks:
- **Quality assurance (QA)**—A planned and systematic pattern of all actions necessary to provide adequate confidence that an item or product conforms to established technical requirements. QA helps the IT department to ensure that personnel are following prescribed quality processes. For example, QA will set up procedures (e.g., ISO 9001-compliant) to facilitate widespread use of quality management/assurance practices.
- **Quality control (QC)**—The observation techniques and activities used to fulfill requirements for quality. QC is responsible for conducting tests or reviews to verify and ensure that software is free from defects and meets user expectations. This could be done at various stages of the development of an application system, but it must be done before the programs are moved into production. For example, QC will help to ensure that programs and documentation adhere to the standards and naming conventions.

The QA function within an organization is in charge of developing, promulgating and maintaining standards for the IT function. They also provide training in QA standards and procedures. The QC group assists by periodically checking the accuracy and authenticity of the input, processing and output of various applications.

To enable the QA function to play an effective role, the QA group should be independent within the organization. In some organizations this function may be a part of the larger control entity. In smaller organizations it may not be possible to have a separate QA function, in which case individuals may possess more than one role. However, under no circumstances should an individual review of his/her own work. Additionally, the review should not be performed by an individual whose role would create an SoD conflict (e.g., a database administrator performing quality review of application system changes that would impact the database).

Database Administration

The **database administrator (DBA)**, as custodian of an organization's data, defines and maintains the data structures in the corporate database system. The DBA must understand the organization, and user data and data relationship (structure) requirements. This position is responsible for the security of the shared data stored on database systems. The DBA is responsible for the actual design, definition and proper maintenance of the corporate databases. The DBA usually reports directly to the director of the IPF. The DBA's role includes:
- Specifying the physical (computer-oriented) data definition
- Changing the physical data definition to improve performance
- Selecting and implementing database optimization tools
- Testing and evaluating programmer and optimization tools
- Answering programmer queries and educating programmers in the database structures
- Implementing database definition controls, access controls, update controls and concurrency controls
- Monitoring database usage, collecting performance statistics and tuning the database
- Defining and initiating backup and recovery procedures

The DBA has the tools to establish controls over the database and the ability to override these controls. The DBA also has the capability of gaining access to all data, including production data. It is usually not practical to prohibit or completely prevent access to production data by the DBA. Therefore, the IT department must exercise close control over database administration through:
- Segregation of duties
- Management approval of DBA activities
- Supervisor review of access logs and activities
- Detective controls over the use of database tools

Systems Analyst

Systems analysts are specialists who design systems based on the needs of the user and are usually involved during the initial phase of the system development life cycle (SDLC). These individuals interpret the needs of the user and develop requirements and functional specifications as well as high-level design documents. These documents enable programmers to create a specific application.

Security Architect

Security architects evaluate security technologies; design security aspects of the network topology, access control, identity management and other security systems; and establish security policies and security requirements. One may argue that systems analysts perform the same role; however, the set of skills required are quite different. The deliverables (e.g., program specifications versus policies, requirements, architecture diagrams) are different as well. Security architects should also work with compliance, risk management and audit functions to incorporate their requirements and recommendations for security into the security policies and architecture.

System Security Engineer

The **system security engineer**, as defined under *ISO/IEC 21827:2008: Information technology—Security techniques—Systems Security Engineering—Capability Maturity Model*, provides technical information system security engineering support to the organization that encompasses:
- Project life cycles, including development, operation, maintenance and decommissioning activities
- Entire organizations, including management, organizational and engineering activities
- Concurrent interactions with other disciplines, such as system software and hardware, human factors, test engineering, system management, operation and maintenance
- Interactions with other organizations, including acquisition, system management, certification, accreditation and evaluation

Applications Development and Maintenance

Applications staff is responsible for developing and maintaining applications. Development can include developing new code or changing the existing setup or configuration of the application. Staff develop the programs or change the application setup that will ultimately run in a production environment. Therefore, management must ensure that staff cannot modify production programs or application data. Staff should work in a test-only environment and turn their work to another group to move programs and application changes into the production environment.

Infrastructure Development and Maintenance

Infrastructure staff is responsible for maintaining the systems software, including the OS. This function may require staff to have broad access to the entire system. IT management must closely monitor activities by requiring that electronic logs capture this activity and are not susceptible to alteration. Infrastructure staff should only have access to the system libraries of the specific software that they maintain. Usage of domain administration and superuser accounts should be tightly controlled and monitored.

Network Management

Today many organizations have widely dispersed IPFs. They may have a central IPF, but they also make extensive use of:
- **LANs** at branches and remote locations
- **WANs**, where LANs may be interconnected for ease of access by authorized personnel from other locations
- **Wireless networks** established through mobile devices

Network administrators are responsible for key components of this infrastructure (routers, switches, firewalls, network segmentation, performance management, remote access, etc.). Because of geographical dispersion, each LAN may need an administrator. Depending on the policy of the company, these administrators can report to the director of the IPF or, in a decentralized operation, may report to the end-user manager, although at least a dotted line to the director of the IPF is advisable. This position is responsible for technical and administrative control over the LAN. This includes ensuring that transmission links are functioning correctly, backups of the system are occurring, and software/hardware purchases are authorized and installed properly. In smaller installations this person may be responsible for security administration over the LAN. The LAN administrator should have no application programming responsibilities but may have systems programming and end-user responsibilities.

2.10.2 SEGREGATION OF DUTIES WITHIN IT

Actual job titles and organizational structures vary greatly from one organization to another depending on the size and nature of the business. However, an IS auditor shall obtain enough information to understand and document the relationships among the various job functions, responsibilities and authorities, and assess the adequacy of the SoD. SoD avoids the possibility that a single person could be responsible for diverse and critical functions in such a way that errors or misappropriations could occur and not be detected in a timely manner and in the normal course of business processes. SoD is an important means by which fraudulent and/or malicious acts can be discouraged and prevented.

Duties that should be segregated include:
- Custody of the assets
- Authorization
- Recording transactions

If adequate SoD does not exist, the following could occur:
- Misappropriation of assets
- Misstated financial statements
- Inaccurate financial documentation (i.e., errors or irregularities)
- Improper use of funds or modification of data could go undetected
- Unauthorized or erroneous changes or modification of data and programs may not be detected

When duties are segregated, access to the computer, production data library, production programs, programming documentation, and OS and associated utilities can be limited, and potential damage from the actions of any one person is, therefore, reduced. The IS and end-user departments should be organized to achieve adequate SoD. See **figure 2.13** for a guideline of the job responsibilities that should not be combined.

Note: The SoD control matrix (**figure 2.13**) is not an industry standard but a guideline indicating which positions should be separated and which require compensating controls when combined. The matrix illustrates potential SoD issues and should not be viewed or used as an absolute; rather, it should be used to help identify potential conflicts so that proper questions may be asked to identify compensating controls.

In actual practice, functions and designations may vary in different enterprises. Further, depending on the nature of the business processes and technology deployed, risk may vary. However, it is important for an IS auditor to understand the functions of each of the designations specified in the manual. IS auditors need to understand the risk of combining functions as indicated in the SoD control matrix. In addition, depending on the complexity of the applications and systems deployed, an automated tool may be required to evaluate the actual access a user has against the SoD matrix. Most tools come with a predefined SoD matrix that must be tailored to an organization's IT and business processes, including any additional functions or risk that are not included in the delivered SoD matrix.

Regarding privileged users of the system, remote logging (sending system logs to separate log server) should be enabled, so that the privileged users do not have access to their own logs. For example, the activities of the DBA may be remotely logged to another server where an official in the IT department can review/audit the DBA's actions. The activities of system administrators may be similarly monitored via separation of log review duties on an independent log server.

Compensating controls are internal controls that are intended to reduce the risk of an existing or potential control weakness when duties cannot be appropriately segregated. The organization structure and roles should be taken into account when determining the appropriate controls for the relevant environment. For example, an organization may not have all the positions described in the matrix or one person may be responsible for many of the roles

Figure 2.13—Segregation of Duties Control Matrix

	Control Group	Systems Analyst	Application Programmer	Help Desk and Support Manager	End User	Data Entry	Computer Operator	Database	Network	Systems	Security Administrator	Systems Programmer	Quality Assurance
Control Group		X	X	X		X	X	X	X	X		X	
Systems Analyst	X			X	X		X				X		X
Application Programmer	X			X	X	X	X	X	X	X	X	X	X
Help Desk and Support Manager	X	X	X		X	X		X	X	X		X	
End User		X	X	X			X	X	X			X	X
Data Entry	X		X	X			X	X	X	X	X	X	
Computer Operator	X	X	X		X	X		X	X	X	X	X	
Database Administrator	X		X	X	X	X	X		X	X		X	
Network Administrator	X		X	X	X	X	X	X					
System Administrator	X		X	X		X	X	X				X	
Security Administrator		X	X			X	X					X	
Systems Programmer	X		X	X	X	X	X	X		X	X		X
Quality Assurance		X	X		X							X	

X—Combination of these functions may create a potential control weakness.

described. The size of the IT department may also be an important factor that should be considered (i.e., certain combinations of roles in an IT department of a certain size should never be used). However, if for some reason combined roles are required, then compensating controls should be developed and put in place.

2.10.3 SEGREGATION OF DUTIES CONTROLS

Several control mechanisms can be used to strengthen SoD. The controls are described in the following sections.

Transaction Authorization

Transaction authorization is the responsibility of the user department. Authorization is delegated to the degree that it relates to the particular level of responsibility of the authorized individual in the department. Periodic checks must be performed by management and audit to detect the unauthorized entry of transactions.

Custody of Assets

Custody of corporate assets must be determined and assigned appropriately. The data owner usually is assigned to a particular user department, and his/her duties should be specific and in writing. The owner of the data has responsibility for determining authorization levels required to provide adequate security, while the administration group is often responsible for implementing and enforcing the security system.

Access to Data

Controls over access to data are provided by a combination of physical, system and application security in the user area and the IPF. The physical environment must be secured to prevent unauthorized personnel from accessing the various tangible devices connected to the central processing unit, thereby permitting access to data. System and application security are additional layers that may prevent unauthorized individuals from gaining access to corporate data. Access to data from external connections is a growing concern since the advent of the Internet. Therefore, IT management has added responsibilities to protect information assets from unauthorized access.

Access control decisions are based on organizational policy and two generally accepted standards of practice—SoD and least privilege. Controls for effective use must not disrupt the usual work flow more than necessary or place too much burden on administrators, auditors or authorized users. Further access must be conditional and access controls must adequately protect all of the organization's resources.

Policies establish levels of sensitivity—such as top secret, secret, confidential and unclassified—for data and other resources. These levels should be used for guidance on the proper procedures for handling information resources. The levels may be also used as a basis for access control decisions. Individuals are granted access to only those resources at or below a specific level of sensitivity. Labels are used to indicate the sensitivity level of electronically stored documents. Policy-based controls may be characterized as either mandatory or discretionary.

AUTHORIZATION FORMS

System owners must provide IT with formal authorization forms (either hard copy or electronic) that define the access rights of each individual. In other words, managers must define who should have access to what. Authorization forms must be evidenced properly with management-level approval. Generally, all users should be authorized with specific system access via formal request of management. In large companies or in those with remote sites, signature authorization logs should be maintained and formal requests should be compared to the signature log. Access privileges should be reviewed periodically to ensure that they are current and appropriate to the user's job functions.

USER AUTHORIZATION TABLES

The IT department should use the data from the authorization forms to build and maintain user authorization tables. These will define who is authorized to update, modify, delete and/or view data. These privileges are provided at the system, transaction or field level. In effect, these are user access control lists. These authorization tables must be secured against unauthorized access by additional password protection or data encryption. A control log should record all user activity and appropriate management should review this log. All exception items should be investigated.

Compensating Controls for Lack of Segregation of Duties

In a small business where the IT department may only consist of four to five people, compensating control measures must exist to mitigate the risk resulting from a lack of SoD. Before relying on system generated reports or functions as compensating controls, the IS auditor should carefully evaluate the reports, applications and related processes for appropriate controls, including testing and access controls to make changes to the reports or functions. Compensating controls include the following:

- **Audit trails** are an essential component of all well-designed systems. Audit trails help the IT and user departments as well as the IS auditor by providing a map to retrace the flow of a transaction. Audit trails enable the user and IS auditor to recreate the actual transaction flow from the point of origination to its existence on an updated file. In the absence of adequate SoD, good audit trails may be an acceptable compensating control. The IS auditor should be able to determine who initiated the transaction, time of day and date of entry, type of entry, what fields of information it contained, and what files it updated.
- **Reconciliation** is ultimately the responsibility of the user department. In some organizations limited reconciliation of applications may be performed by the data control group with the use of control totals and balance sheets. This type of independent verification increases the level of confidence that the application processed successfully and the data are in proper balance.
- **Exception reporting** should be handled at the supervisory level and should require evidence, such as initials on a report, noting that the exception has been handled properly. Management should also ensure that exceptions are resolved in a timely manner.
- **Transaction logs** may be manual or automated. An example of a manual log is a record of transactions (grouped or batched) before they are submitted for processing. An automated transaction log provides a record of all transactions processed and is maintained by the computer system.
- **Supervisory reviews** may be performed through observation and inquiry or remotely.
- **Independent reviews** are carried out to compensate for mistakes or intentional failures in following prescribed procedures. These reviews are particularly important when duties in a small organization cannot be appropriately segregated. Such reviews will help detect errors or irregularities.

2.11 AUDITING IT GOVERNANCE STRUCTURE AND IMPLEMENTATION

While many conditions concern the IS auditor when auditing the IT function, some of the more significant indicators of potential problems include:

- Unfavorable end-user attitudes
- Excessive costs
- Budget overruns
- Late projects
- High staff turnover
- Inexperienced staff
- Frequent HW/SW errors
- An excessive backlog of user requests
- Slow computer response time
- Numerous aborted or suspended development projects
- Unsupported or unauthorized HW/SW purchases
- Frequent HW/SW upgrades
- Extensive exception reports
- Exception reports that were not followed up
- Poor motivation
- Lack of succession plans
- A reliance on one or two key personnel
- Lack of adequate training

2.11.1 REVIEWING DOCUMENTATION

The following documents should be reviewed:

- **IT strategies, plans and budgets**—They provide evidence of planning and management's control of the IT environment and alignment with the business strategy.
- **Security policy documentation**—This documentation provides the standard for compliance. The documentation should state the position of the organization with regard to any and all security risk. The documentation should identify who is responsible for the safeguarding of company assets, including programs and data, and it should state the preventive measures to be taken to provide adequate protection and actions to be taken against violators. For this reason, this part of the policy document should be treated as confidential.

- **Organization/functional charts**—These charts provide the IS auditor with an understanding of the reporting line within a particular department or organization. The charts illustrate a division of responsibility and give an indication of the degree of SoD within the organization.
- **Job descriptions**—These descriptions define the functions and responsibilities of positions throughout the organization. Job descriptions provide an organization with the ability to group similar jobs in different grade levels to ensure fair compensation for the workforce. Furthermore, job descriptions give an indication of the degree of SoD within the organization and may help identify possible conflicting duties. Job descriptions should identify the position to which these personnel report. The IS auditor should then verify that the levels of reporting relationships are based on sound business concepts and do not compromise the SoD.
- **IT steering committee reports**—These reports provide documented information regarding new system projects. The reports are reviewed by upper management and disseminated among the various business units.
- **System development and program change procedures**—These procedures provide a framework within which to undertake system development or program change.
- **Operations procedures**—These procedures describe the responsibilities of the operations staff. Performance measurement procedures are generally embedded in operational procedures and are periodically reported to senior management/steering committees. IS auditors should ensure that these procedures are embedded in operational procedures.
- **HR manuals**—These manuals provide the rules and regulations (determined by an organization) that specify how employees should conduct themselves. HR manuals will also contain the rules relating to the taking of annual leave, which helps to protect the organization against the risk of fraudulent or inappropriate activity and over dependence on key staff.
- **QA procedures**—These procedures provide framework and standards that can be followed by the IT department.

The various documents reviewed should be further assessed to determine whether:
- They were created as management authorized and intended
- They are current and up to date

2.11.2 REVIEWING CONTRACTUAL COMMITMENTS

There are various phases to computer hardware, software and IT service contracts, including:
- Development of contract requirements and service levels
- Contract bidding process
- Contract selection process
- Contract acceptance
- Contract maintenance
- Contract compliance

Each of these phases should be supported by legal documents, subject to the authorization of management. The IS auditor should verify management participation in the contracting process and ensure a proper level of timely contract compliance review. The IS auditor may wish to perform a separate compliance review on a sample of such contracts.

In reviewing a sample of contracts, the IS auditor should evaluate the adequacy of the following terms and conditions:
- Service levels
- Right to audit or third-party audit reporting
- Software escrow
- Penalties for noncompliance
- Adherence to security policies and procedures
- Protection of customer information
- Ownership of intellectual property (IP)
- Contract change process
- Contract termination and any associated penalties

> **Note:** An IS auditor should be familiar with the request for proposal (RFP) process and know what needs to be reviewed in an RFP. It is also important to note that a CISA should know, from a governance perspective, the evaluation criteria and methodology of an RFP, and the requirements to meet organizational standards.

2.12 BUSINESS CONTINUITY PLANNING

The purpose of business continuity/disaster recovery is to enable a business to continue offering critical services in the event of a disruption and to survive a disastrous interruption to activities. Rigorous planning and commitment of resources is necessary to adequately plan for such an event.

The first step in preparing a new BCP, or in updating an existing one, is to identify the business processes of strategic importance—those key processes that are responsible for both the permanent growth of the business and for the fulfillment of the business goals. Ideally, the BCP/DRP should be supported by a formal executive policy that states the organization's overall target for recovery and empowers those people involved in developing, testing and maintaining the plans.

Based on the key processes, the risk management process should begin with a risk assessment. The risk is directly proportional to the impact on the organization and the probability of occurrence of the perceived threat. Thus, the result of the risk assessment should be the identification of the following:
- The human resources, data, infrastructure elements and other resources (including those provided by third parties) that support the key processes
- A list of potential vulnerabilities—the dangers or threats to the organization
- The estimated probability of the occurrence of these threats
- The efficiency and effectiveness of existing risk mitigation controls (risk countermeasures)

BCP is primarily the responsibility of senior management, as they are entrusted with safeguarding the assets and the viability of the organization, as defined in the BCP/DRP policy. The BCP is generally followed by the business and supporting units, to provide a reduced but sufficient level of functionality in the business operations immediately after encountering an interruption, while recovery is taking place. The plan should address all functions and assets required to continue as a viable organization. This includes continuity procedures determined necessary to survive and minimize the consequences of business interruption.

BCP takes into consideration:
- Those critical operations that are necessary to the survival of the organization
- The human/material resources supporting them

Besides the plan for the continuity of operations, the BCP includes:
- The DRP that is used to recover a facility rendered inoperable, including relocating operations into a new location
- The restoration plan that is used to return operations to normality whether in a restored or new facility

Depending on the complexity of the organization, there could be one or more plans to address the various aspects of business continuity and disaster recovery. These plans do not necessarily have to be integrated into one single plan. However, each has to be consistent with other plans to have a viable BCP strategy.

It is highly desirable to have a single integrated plan to ensure that:
- There is proper coordination among various plan components.
- Resources committed are used in the most effective way, and there is reasonable confidence that, through its application, the organization will survive a disruption.

Even if similar processes of the same organization are handled at a different geographic location, the BCP and DRP solutions may be different for different scenarios. Solutions may be different due to contractual requirements (e.g., the same organization is processing an online transaction for one client and the back office is processing for another client. A BCP solution for the online service will be significantly different than one for the back office processing.)

2.12.1 IT BUSINESS CONTINUITY PLANNING

In the case of IT business continuity planning, the approach is the same as in BCP with the exception being that the continuity of IT processing is threatened. IT processing is of strategic importance—it is a critical component because most key business processes depend on the availability of key systems infrastructure components and data.

The IT business continuity plan should be aligned with the strategy of the organization. The criticality of the various application systems deployed in the organization depends on the nature of the business as well as the value of each application to the business.

The value of each application to the business is directly proportional to the role of the information system in supporting the strategy of the organization. The components of the information system (including the technology infrastructure components) are then matched to the applications (e.g., the value of a computer or a network is determined by the importance of the application system that uses it).

Therefore, the information system BCP/DRP is a major component of an organization's overall business continuity and disaster recovery strategy. If the IT plan is a separate plan, it must be consistent with and support the corporate BCP.

Throughout the IT business continuity (sometimes referred to as IT service continuity) planning process, the overall BCP of the organization should be taken into consideration; again, this should be supported by the executive policy. All IT plans must be consistent with and support the corporate BCP. This means that alternate processing facilities that support key operations must be ready, be compatible with the original processing facility and have up-to-date plans regarding their use.

Again, all possible steps must be taken to reduce or remove the likelihood of a disruption using the method described in other sections of this manual. Examples include:
- Minimizing threats to the data center by considering location:
  - Not on a flood plain
  - Not on or near an earthquake fault line
  - Not close to an area where explosive devices or toxic materials are regularly used
- Making use of resilient network topographies such as Loop or Mesh with alternative processing facilities already built into the network infrastructure

Developing and testing an information system BCP/DRP is a major component of an organization's overall business continuity and disaster recovery strategy. The plan is based on the coordinated use of whatever risk countermeasures are available for the organization (i.e., duplicate processing facility, redundant data networks, resilient hardware, backup and recovery systems, data replication, etc.). If the IT plan is a separate plan (or multiple separate plans), it must be consistent with and support the corporate BCP.

Establishing dependencies among critical business processes, applications, the information system and IT infrastructure components is a subject of risk assessment. The resulting dependencies map with threats to and vulnerabilities of the components/dependencies (along with the key applications grouped by their criticality) are the outcomes of the risk assessment.

After the risk assessment identifies the importance of the IS components to the organization, and the threats to and vulnerabilities of those components, a remedial action plan can be developed for establishing the most appropriate methods to protect the components. There is always a choice of risk mitigation measures (risk countermeasures)—either to remove the threat and/or fix the vulnerability.

The risk can be either estimated in a qualitative way (assigning qualitative values to the impact of the threat and its probability) or calculated in a quantitative way (assigning a monetary value to the impact [i.e., loss] and assigning a probability).

> **Note:** The CISA candidate will not be tested on the actual calculation of risk analysis; however, the IS auditor should be familiar with risk analysis calculation.

If the organization is willing to investigate the extent of the losses that the business will suffer from the disruption, the organization may conduct a business impact analysis (BIA), which is discussed in a separate section of this manual. The BIA allows

the organization to determine the maximum downtime possible for a particular application and how much data could be lost. The BIA also allows the organization to quantify the losses as they grow after the disruption, thus allowing the organization to make a decision on the technology (and facilities) used for protection and recovery of its key information assets (information system, IT components, data, etc.).

The results of risk assessment and BIA are fed into the IS business continuity strategy, which outlines the main technology and principles behind IT protection and recovery as well as the road map to implement the technology and principles.

As the IT business continuity strategy and its overarching IT strategy are executed, the IT infrastructure of the organization changes. New risk countermeasures are introduced and old ones become obsolete. The information system BCP must be changed accordingly and retested periodically to ensure that these changes are satisfactory.

Similar to any BCP, an information system BCP is much more than just a plan for information systems. A BCP identifies what the business will do in the event of a disaster. For example, where will employees report to work, how will orders be taken while the computer system is being restored, which vendors should be called to provide needed supplies? A subcomponent of the BCP is the IT disaster recovery plan (DRP). This typically details the process IT personnel will use to restore the computer systems, communications, applications and their data. DRPs may be included in the BCP or as a separate document altogether, depending on the needs of the business.

Not all systems will require a recovery strategy. Based upon the results of the risk assessment and BIA, management may not see a tangible cost benefit for restoring certain applications in the event of a disaster. An overriding factor when determining recovery options is that the cost should never exceed the benefit (this usually becomes clear after completing a BIA). One of the important outcomes of BIA, apart from the RTO and recovery point objective (RPO), is a way to group information systems according to their recovery time. This usually guides the selection of the technological solutions (i.e., controls) supporting business continuity and IT disaster recovery.

The IT disaster recovery usually happens in unusual, stressful circumstances (e.g., fire, flood, hurricane devastation). Often, the security controls (both physical and IS) may not be functioning. It is, therefore, recommended that the organization implement an ISMS to maintain the integrity, confidentiality and availability of IS, and not only under normal conditions.

2.12.2 DISASTERS AND OTHER DISRUPTIVE EVENTS

Disasters are disruptions that cause critical information resources to be inoperative for a period of time, adversely impacting organizational operations. The disruption could be a few minutes to several months, depending on the extent of damage to the information resource. Most important, disasters require recovery efforts to restore operational status.

A disaster may be caused by natural calamities—such as earthquakes, floods, tornados, severe thunderstorms and fire—which cause extensive damage to the processing facility and the locality in general. Other disastrous events causing disruptions may occur when expected services, such as electrical power, telecommunications, natural gas supply or other delivery services are no longer supplied to the company due to a natural disaster or other cause.

Not all critical disruptions in service or disasters are due to natural causes. A disaster could also be caused by events precipitated by human beings, such as terrorist attacks, hacker attacks, viruses or human error. Disruption in service is sometimes caused by system malfunctions, accidental file deletions, untested application releases, loss of backup, network denial of service (DoS) attacks, intrusions and viruses. These events may require action to recover operational status in order to resume service. Such actions may necessitate restoration of hardware, software or data files.

Many disruptions start as mere incidents. Normally, if the organization has a help desk, it would act as the early warning system to recognize the first signs of an upcoming disruption. Often, such disruptions (e.g., gradually deteriorating database performance) go undetected. Until these "creeping disasters" strike (the database halts), they cause only infrequent user complaints.

Based on risk assessment, worst-case scenarios and short- and long-term fallback strategies are formulated in the IS business continuity strategy for later incorporation into the BCP (or other plan). In the short term, an alternate processing facility may be needed to satisfy immediate operational needs (as in the case of a major natural disaster). In the long term, a new permanent facility must be identified for disaster recovery and equipped to provide for continuation of IS processing services on a regular basis.

Pandemic Planning

Pandemics can be defined as epidemics or outbreaks of infectious diseases in humans that have the ability to spread rapidly over large areas, possibly worldwide. Several pandemics have occurred throughout history, and recently pandemic threats such as the avian or swine flu outbreaks have further raised awareness regarding this issue. There are distinct differences between pandemic planning and traditional business continuity planning, and therefore, the IS auditor should evaluate an organization's preparedness for pandemic outbreaks. Pandemic planning presents unique challenges; unlike natural disasters, technical disasters, malicious acts or terrorist events, the impact of a pandemic is much more difficult to determine because of the anticipated difference in scale and duration.

Dealing With Damage to Image, Reputation or Brand

Damaging rumors may rise from many sources (even internal). They may or may not be associated with a serious incident or crisis. Whether they are "spontaneous" or a side effect of a business continuity or disaster recovery problem, their consequences may be devastating. One of the worst consequences of crises is the loss of trust.

Effective public relations (PR) activities in an organization may play an important role in helping to contain the damage to the image and ensure that the crisis is not made worse. Certain industries (e.g., banks, health care organizations, airlines, petroleum refineries, chemical, transportation, or nuclear power plants or other organizations with relevant social impact) should have elaborate protocols for dealing with accidents and catastrophes.

A few basic good practices should be considered and applied by an organization experiencing a major incident. Irrespective of the resultant objective consequences of an incident (delay or interruption in service, economic losses, etc.), a negative public opinion or negative rumors can be costly. Reacting appropriately in public (or to the media) during a crisis is not simple. A properly trained spokesperson should be appointed and prepared beforehand. Normally, senior legal counsel or a PR officer is the best choice. No one, irrespective of his/her rank in the organizational hierarchy, except for the spokesperson, should make any public statement.

As part of the preparation, the spokesperson should draft and keep on file a generic announcement with blanks to be filled in with the specific circumstances. This should not be deviated from because of improvisation or time pressure. The announcement should not state the causes of the incident but rather indicate that an investigation has been started and results will be reported. Liability should not be assumed. The system or the process should not be blamed.

Unanticipated/Unforeseeable Events

Management should consider the possible impacts of unforeseeable (black swan) events on the business of the organization. Black swan events are those events that are a surprise (to the observer), have a major effect and after the fact are often inappropriately rationalized with the benefit of hindsight.

One example of a black swan event is the Fukushima nuclear disaster in Japan in March 2012. An earthquake triggered a tsunami that disabled the back-up power for generators that were essential to pump in water for cooling of the nuclear reactors, which ultimately led to the nuclear disaster. Prior to this event, a contingency plan would not have considered or contemplated such a interlinkage of events by any stretch of imagination. While these events are few and far between, once they occur, they have such a crippling impact on the organization that, based on the criticality of the process or industry or activity, management should start thinking about contingency planning to meet such events. Senior executives who have shared responsibilities being forbidden from traveling together is another example where management is proactive, ensuring that, should a common disaster occur, the organization would not be left headless.

2.12.3 BUSINESS CONTINUITY PLANNING PROCESS

The BCP process can be divided into the life cycle phases depicted in **figure 2.14**.

2.12.4 BUSINESS CONTINUITY POLICY

A business continuity policy is a document approved by top management that defines the extent and scope of the business continuity effort (a project or an ongoing program) within the organization. The business continuity policy can be broken into two parts: public and internal. The business continuity policy serves several other purposes:

- Its internal portion is a message to internal stakeholders (i.e., employees, management, board of directors) that the company is undertaking the effort, committing its resources and expecting the rest of the organization to do the same.
- Its public portion is a message to external stakeholders (shareholders, regulators, authorities, etc.) that the organization is treating its obligations (e.g., service delivery, compliance) seriously.

Figure 2.14—Business Continuity Planning Life Cycle

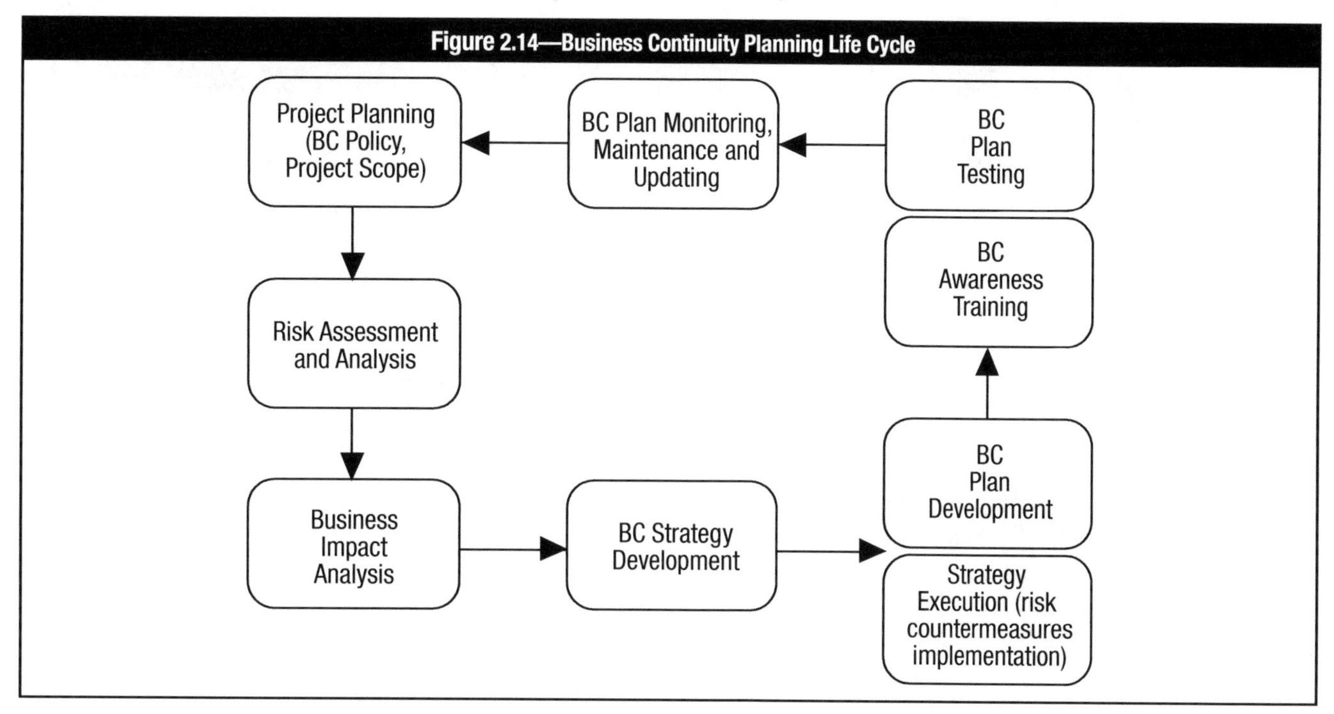

- It is a statement to the organization, empowering those who are responsible for business continuity.
- It may broadly state the general principles on which business continuity will be based.

A business continuity policy should be proactive. The message delivered to the organization must be that all possible controls to detect and prevent disruptions should be used and, if disruption still occurs, to have the controls necessary to mitigate the consequences. This is later reflected in the IT business continuity strategy and its execution. There are preventive and detective controls to reduce the likelihood of a disruption and corrective controls to mitigate the consequences.

The BCP (or IT DRP) is the most critical corrective control. It depends on other controls being effective; in particular, it depends upon incident management and backup and recovery solutions.

Incidents and their impacts can, to some extent, be mitigated through preventive controls. These relationships are depicted in **figure 2.15**.

This requires that the incident management group (help desk) be adequately staffed, supported and trained in crisis management, and that the BCP be well designed, documented, drill tested, funded and audited.

2.12.5 BUSINESS CONTINUITY PLANNING INCIDENT MANAGEMENT

Incidents and crises are dynamic by nature. They evolve, change with time and circumstances, and are often rapid and unforeseeable. Because of this, their management must be dynamic, proactive and well documented. An incident is any unexpected event, even if it causes no significant damage. See section 5.2.13 Security Incident Handling and Response for more information.

Depending on an estimation of the level of damage to the organization, all types of incidents should be categorized. A classification system could include the following categories: negligible, minor, major and crisis. Classification can dynamically change while the incident is resolved. These levels can be broadly described as follows:

- **Negligible** incidents are those causing no perceptible or significant damage, such as very brief OS crashes with full information recovery or momentary power outages with uninterruptible power supply (UPS) backup.
- **Minor** events are those that, while not negligible, produce no negative material (of relative importance) or financial impact.
- **Major** incidents cause a negative material impact on business processes and may affect other systems, departments or even outside clients.
- **Crisis** is a major incident that can have serious material (of relative importance) impact on the continued functioning of the business and may also adversely impact other systems or third parties. The severity of the impact depends on the industry and circumstances but is generally directly proportional to the time elapsed from the inception of the incident to incident resolution.

Minor, major and crisis incidents should be documented, classified and revisited until corrected or resolved. This is a dynamic process because a major incident may decrease in extent momentarily and later expand to a crisis incident.

Negligible incidents can be analyzed statistically to identify any systemic or avoidable causes.

Figure 2.16 provides an example of an incident classification system and reaction protocol.

The security officer (SO) or other designated individual should be notified of all relevant incidents as soon as any triggering event occurs. This person should then follow a pre-established escalation protocol (e.g., calling in a spokesperson, alerting top

Figure 2.15—Incident and Impact Relationship Diagram

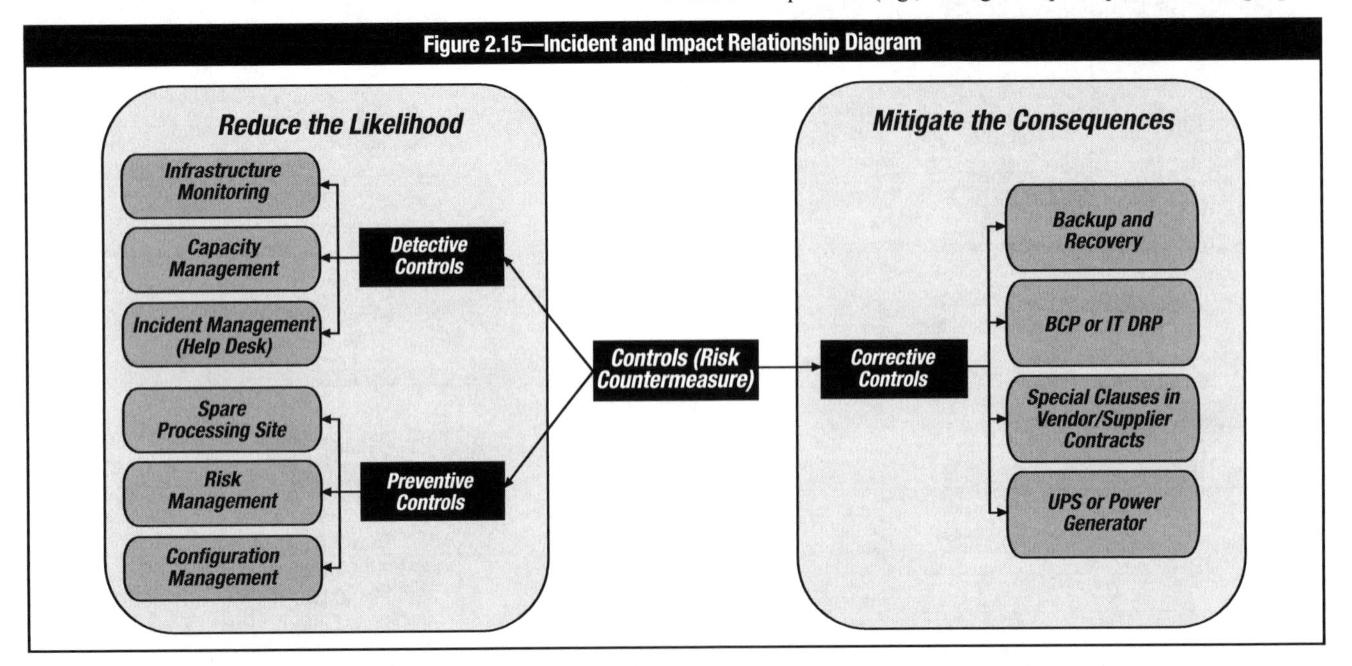

management and involving regulatory agencies) that may be followed by invoking a recovery plan such as the IT DRP.

Service can be defined as including commitments with clients that can be either external customers or internal departments. Often, the service delivery is regulated by SLAs which may state the maximum downtime and recovery estimates. Although not always true, severity is usually driven to a large extent by the estimated downtime. Other criteria may include the impact on data or platforms and the degree to which the functioning of the organization is adversely impacted. A conservative fail-safe approach would be to assign any nonnegligible incident a starting, provisional severity level 3 (shown in **figure 2.16**). As the incident evolves, this level should be reevaluated regularly by the person or team in charge, often referred to as an incident response or firecall team.

2.12.6 BUSINESS IMPACT ANALYSIS

BIA is a critical step in developing the business continuity strategy and the subsequent implementation of the risk countermeasures and BCP in particular.

BIA is used to evaluate the critical processes (and IT components supporting them) and to determine time frames, priorities, resources and interdependencies. Even if an extensive risk assessment was done prior to BIA, and the criticality and risk are input into BIA, the rule of thumb is to double-check. Often, the BIA uncovers some less visible, but nonetheless vital, component that supports the critical business process. Where IT activities have been outsourced to third-party service providers, the contractual commitments (in a BCP context) should also be considered.

To perform this phase successfully, one should obtain an understanding of the organization, key business processes and IT resources used by the organization to support the key business processes. Often, this may be obtained from the risk assessment results. BIA requires a high level of senior management support/sponsorship and extensive involvement of IT and end-user personnel. The criticality of the information resources (e.g., applications, data, networks, system software, facilities) that support an organization's business processes must be approved by senior management.

For the BIA, it is important to include all types of information resources and to look beyond traditional information resources (i.e., database servers).

Information systems consist of multiple components. Some of the components (e.g., database servers or storage arrays) are quite visible. Other components (e.g., gateways, transport servers, network devices) may fall out of scope and remain "invisible." For instance, a banking application may not perform its services if the payment gateways are down. Often, the vital parts of the application or the critical data may reside on the user workstations. Ideally, upon completion of the BIA, these "hidden" components must be uncovered and included in the business continuity program (project) scope for further inclusion in the BCP.

Note: The IS auditor should be able to evaluate the BIA. Task statement T2.10 in the CISA job practice states "Evaluate the organization's business continuity plan (BCP), including the alignment of the IT disaster recovery plan (DRP) with the BCP, to determine the organization's ability to continue essential business operations during the period of an IT disruption." The auditor needs to know what is involved in developing a BIA so that he/she can properly evaluate it. However, a CISA candidate will not be tested on how a BIA is performed or what method is used to perform a BIA.

Figure 2.16—Incident/Crisis Levels

	1 LEVEL	MAIN CRITERION (hours) SERVICE DOWNTIME FORECAST >=	ACTUAL >=	COMPLEMENTARY CRITERIA DATA	PLATFORMS
	7		24		
CRISIS	6	24	12		
	5	12	6	Database loss of integrity	Hacked or Denial of Service Attack
	4	6	4		Viruses, worms. Hardware failure.
MAJOR INC'T	3	4	2		
	2	2	1	Lost transactions	
MINOR INC'T	1	1	0.5		
NEGLIGIBLE	0				

	LEVEL	2 ACTIONS	
	7	Follow Business Continuity Plan	Alert SM and eventually Reg. Agencies
CRISIS	6	Follow Business Continuity Plan	Alert SM and eventually Reg. Agencies
	5	Prepare for Business Continuity Plan	Alert SM
	4	Correct/Clean/Restore/Replace	Alert SM
MAJOR	3	Correct	If confirmed, alert SO
	2	Correct	
MINOR	1	Correct	
NEGLIGIBLE	0	Log	(Analyze logs regularly)

SM = Senior Management
SO = Security Officer

Source: Personas & Técnicas Multimedia SL © 2007. All rights reserved. Used by permission.

Information is collected for the BIA from different parts of the organization that own critical processes/applications. To evaluate the impact of downtime for a particular process/application, the impact bands are developed (i.e., high, medium, low) and, for each process, the impact is estimated in time (hours, days, weeks). The same approach is used when estimating the impact of data loss. If necessary, the financial impact may be estimated using the same techniques, assigning the financial value to the particular impact band.

In addition, data for the BIA may be collected on the time frames needed to supply vital resources—how long the organization may run if a supply is broken or when the replacement has arrived. For example, how long will the bank run without plastic cards with chips to be personalized into credit cards or when will IT need to have the desktop workstations shipped in after a disaster?

There are different approaches for performing a BIA. One of the popular approaches is the questionnaire approach. This approach involves developing a detailed questionnaire and circulating it to key users in IT and end-user areas. The information gathered is tabulated and analyzed. If additional information is required, the BIA team would contact the relevant users for additional information. Another popular approach is to interview groups of key users. The information gathered during these interview sessions is tabulated and analyzed for developing a detailed BIA plan and strategy. A third approach is to bring relevant IT personnel and end users (i.e., those owning the critical processes) together in a room to come to a conclusion regarding the potential business impact of various levels of disruptions. The latter method may be used after all the data are collected. Such a mixed group will quickly decide on the acceptable downtime and vital resources.

Wherever possible, the BCP team should analyze past transaction volume in determining the impact to the business if the system were to be unavailable for an extended period of time. This would substantiate the interview process that the BCP team conducts for performing a BIA.

The three main questions that should be considered during the BIA phase are depicted in **figure 2.17.**

To make decisions, there are two independent cost factors to consider as shown in **figure 2.18**. One is the downtime cost of the disaster. This component, in the short run (e.g., hours, days, weeks), grows quickly with time, where the impact of a disruption increases the longer it lasts. At a certain moment, it stops growing, reflecting the moment or point when the business can no longer function. The cost of downtime (increasing with time) has many components (depending on the industry and the specific company and circumstances), among them: cost of idle resources (e.g., in production), drop in sales (e.g., orders), financial costs (e.g., not invoicing nor collecting), delays (e.g., procurement) and indirect costs (e.g., loss of market share, image and goodwill).

Figure 2.17—BIA Considerations

1. What are the different business processes? Each process needs to be assessed to determine its relative importance. Indications of criticality include, for example:
   - The process supporting health and safety, such as hospital patient records and air traffic control systems
   - Disruption of the process causing a loss of income to the organization or exceptional unacceptable costs
   - The process meeting legal or statutory requirements
   - The number of business segments or number of users that are affected

   A process can be critical or noncritical depending on factors such as time of operation and mode of operation (e.g., business hours or ATM operations).

2. What are the critical information resources related to an organization's critical business processes? This is the first consideration because disruption to an information resource is not a disaster in itself, unless it is related to a critical business process (e.g., an organization losing its revenue-generating business processes due to an IS failure).

   Other examples of potential critical business processes may include:
   - Receiving payments
   - Production
   - Paying employees
   - Advertising
   - Dispatching of finished goods
   - Legal and regulatory compliance

3. What is the critical recovery time period for information resources in which business processing must be resumed before significant or unacceptable losses are suffered? In large part, the length of the time period for recovery depends on the nature of the business or service being disrupted. For instance, financial institutions, such as banks and brokerage firms, usually will have a much shorter critical recovery time period than manufacturing firms. Also, the time of year or day of week may affect the window of time for recovery. For example, a bank experiencing a major outage on Saturday at midnight has a longer time in which to recover than on Monday at midnight, assuming that the bank is not processing on Sunday.

Figure 2.18—Disruption Costs vs. Recovery Costs

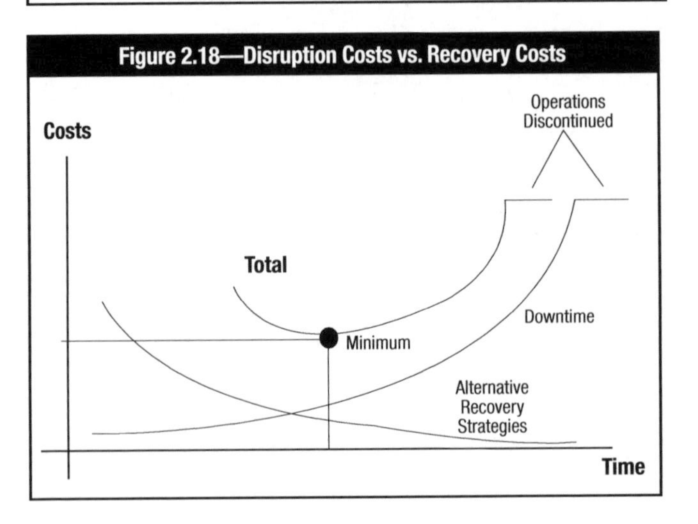

The other factor is the cost of the alternative corrective measures (i.e., the implementation, maintenance and activation of the BCP). This cost decreases with the target chosen for recovery time. The recovery cost also has many components (most of them rigid-inelastic). This includes the costs of preparing and periodically testing the BCP, offsite backup premises, insurance coverage, alternative site arrangements, etc. The cost of alternative recovery strategies may be plotted as discrete points on the time and cost coordinates and a curve drawn joining the points (**figure 2.18**). The curve as a whole is representative of all possible strategies. Each possible strategy has a fixed-base cost (i.e., does not change with time until an eventual disaster happens). Note that the fixed-base cost of each possible strategy will normally differ. If the business continuity strategy aims at a longer recovery time, it will be less expensive than a more stringent requirement but may be more susceptible to downtime costs spiraling out of control. Normally, the shorter the target recovery time, the higher the fixed cost. The organization pays for the cost of planning and implementation even if no disaster takes place.

If there is a disaster, variable costs will significantly increase (e.g., a warm site contract may consist of a flat annual fee plus a daily fee for actual occupation; extra staff, overtime, transportation and other logistics (e.g.. staff *per diem*, new communication lines, etc.) need to be considered. Variable costs will depend on the strategy implemented.

Having plotted the two curves—downtime costs and costs of alternative recovery strategies—**figure 2.18** shows the curve of total cost (the sum of the other two cost curves). An organization would choose the point at which those total costs are minimal.

In summary, the sum of all costs—downtime and recovery—should be minimized. The first group (downtime costs) increases with time, and the second (recovery costs) decreases with time; the sum usually is a U curve. At the bottom of the U curve, the lowest cost can be found.

Note: The CISA candidate will not be tested on calculations of costs.

Classification of Operations and Criticality Analysis

What is the system's risk ranking? It involves a determination of risk based upon the impact derived from the critical recovery time period, as well as the likelihood that an adverse disruption will occur. Many organizations will use a risk of occurrence to determine a reasonable cost of being prepared. For example, they may determine that there is a 0.1 percent risk (or 1 in 1,000) that over the next five years the organization will suffer a serious disruption. If the assessed impact of a disruption is US $10 million, then the maximum reasonable cost of being prepared might be US $10 million × 0.1 percent = US $10,000 over five years. Such a method is called the annual loss expectancy (ALE). From this risk-based analysis process, prioritizing critical systems can take place in developing recovery strategies. The risk ranking procedure should be performed in coordination with IS processing and end-user personnel.

A typical risk ranking system may contain the classifications as found in **figure 2.19**.

Figure 2.19—Classification of Systems

Classification	Description
Critical	These functions cannot be performed unless they are replaced by identical capabilities. Critical applications cannot be replaced by manual methods. Tolerance to interruption is very low; therefore, cost of interruption is very high.
Vital	These functions can be performed manually, but only for a brief period of time. There is a higher tolerance to interruption than with critical systems and, therefore, somewhat lower costs of interruption, provided that functions are restored within a certain time frame (usually five days or less).
Sensitive	These functions can be performed manually, at a tolerable cost and for an extended period of time. While they can be performed manually, it usually is a difficult process and requires additional staff to perform.
Nonsensitive	These functions may be interrupted for an extended period of time, at little or no cost to the company, and require little or no catching up when restored.

The next phase in continuity management is to identify the various recovery strategies and available alternatives for recovering from an interruption and/or disaster. The selection of an appropriate strategy based on the BIA and criticality analysis is the next step for developing BCPs and DRPs. The two metrics that help in determining the recovery strategies are the RPO and RTO.

Recovery strategies are described in greater detail in chapter 4 Information Systems Operations, Maintenance and Service Management.

2.12.7 DEVELOPMENT OF BUSINESS CONTINUITY PLANS

Based on the inputs received from the BIA, criticality analysis and recovery strategy selected by management, a detailed BCP and DRP should be developed or reviewed. They should address all the issues included in the business continuity scope that are involved in interruption to business processes, including recovering from a disaster. The various factors that should be considered while developing/reviewing the plan are:

- Predisaster readiness covering incident response management to address all relevant incidents affecting business processes
- Evacuation procedures
- Procedures for declaring a disaster (rating and escalation procedures)
- Circumstances under which a disaster should be declared. Not all interruptions are disasters, but a small incident if not addressed in a timely or proper manner may lead to a disaster. For example, a virus attack not recognized and contained in time may bring down the entire IT facility.
- The clear identification of the responsibilities in the plan
- The clear identification of the persons responsible for each function in the plan
- The clear identification of contract information
- The step-by-step explanation of the recovery process
- The clear identification of the various resources required for recovery and continued operation of the organization

The plan should be documented and written in simple language, understandable to all.

It is common to identify teams of personnel who are made responsible for specific tasks in case of disasters. Some important teams should be formed, and their responsibilities are explained in the next section. Copies of the plan should be maintained offsite. The plan must be structured so that its parts can easily be handled by different teams.

2.12.8 OTHER ISSUES IN PLAN DEVELOPMENT

The personnel who must react to the interruption/disaster are those responsible for the most critical resources. Therefore, management and user involvement is vital to the success of the execution of the BCP. User management involvement is essential to the identification of critical systems, their associated critical recovery times and the specification of needed resources. The three major divisions that require involvement in the formulation of the BCP are support services (who detect the first signs of incident/disaster), business operations (who may suffer from the incident) and information processing support (who are going to run the recovery).

Because the underlying purpose of BCP is the recovery and resumption of business operations, it is essential to consider the entire organization, not just IS processing services, when developing the plan. Where a uniform BCP does not exist for the entire organization, the plan for IS processing should be extended to include planning for all divisions and units that depend on IS processing functions.

When formulating the plan, the following items should also be included:
- A list of the staff, with redundant contact information (backups for each contact), required to maintain critical business functions in the short, medium and long term
- The configuration of building facilities, desks, chairs, telephones, etc., required to maintain critical business functions in the short, medium and long term
- The resources required to resume/continue operations (not necessarily IT or even technology resources), such as company letterhead stationery)

2.12.9 COMPONENTS OF A BUSINESS CONTINUITY PLAN

Depending on the size and/or requirements of an organization, a BCP may consist of more than one plan document.

This should include:
- Continuity of operations plan
- DRP
- Business resumption plan

It may also include:
- Continuity of support plan/IT contingency plan
- Crisis communications plan
- Incident response plan
- Transportation plan
- Occupant emergency plan
- Evacuation and emergency relocation plan

One example of the components of a BCP as suggested by *NIST Special Publication 800-34 Revision 1: Contingency Planning Guide for Federal Information Systems*, is shown in **figure 2.20**.

For the planning, implementation and evaluation phase of the BCP, the following should be agreed on:
- The policies that will govern all of the continuity and recovery efforts
- The goals/requirements/products for each phase
- Alternate facilities to perform tasks and operations
- Critical information resources to deploy (e.g., data and systems)

Figure 2.20—Components of a Business Continuity Plan

Plan	Purpose	Scope	Plan Relationship
Business continuity plan (BCP)	Provides procedures for sustaining mission/business operations while recovering from a significant disruption.	Address mission/business processes at a lower or expanded level from COOP MEFs.	Mission/business process focused plan that may be activated in coordination with a COOP plan to sustain non-MEFs.
Continuity of operations (COOP) plan	Provides procedures and guidance to sustain an organization's MEFs at an alternate site for up to 30 days; mandated by federal directives.	Addresses MEFs at a facility; information systems are addressed based only on their support of the mission essential functions.	MEF focused plan that may also activate several business unit-level BCPs, ISCPs, or DRPs, as appropriate.
Crisis communications Plan	Provides procedures for disseminating internal and external communications; means to provide critical status information and control rumors.	Addresses communications with personnel and the public; not information system-focused.	Incident-based plan often activated with a COOP or BCP, but may be used alone during a public exposure event.
Critical Infrastructure Protection (CIP) Plan	Provides policies and procedures for protection of national critical infrastructure components as defined in the National Infrastructure Protection Plan.	Addresses critical infrastructure components that are supported or operated by an agency or organization.	Risk management plan that supports COOP plans for organizations with critical infrastructure and key resource assets.
Cyber incident response plan	Provides procedures for mitigating and correcting a cyber attack, such as a virus, worm, or Trojan horse.	Address mitigation and isolation of affected systems, cleanup, and minimizing loss of information.	Information system-focused plan that may activate an ISCP or DRP depending on the extent of the attack.

Figure 2.20—Components of a Business Continuity Plan (cont.)

Plan	Purpose	Scope	Plan Relationship
Disaster recovery plan (DRP)	Provides procedures for relocating information systems operations to an alternate location.	Activated after major system disruptions with long-term effects.	Information system-focused plan that activates one or more ISCPs for recovery of individual systems.
Information System Contingency Plan (ISCP)	Provides procedures and capabilities for recovering an information system.	Addresses single information system recovery at the current or, if appropriate alternate location.	Information system-focused plan that may be activated independent from other plans or as part of a larger recovery effort coordinated with a DRP, COOP, and/or BCP.
Occupant emergency plan (OEP)	Provides coordinated procedures for minimizing loss of life or injury and protecting property damage in response to a physical threat.	Focuses on personnel and property particular to the specific facility; not mission/business process or information system-based.	Incident-based plan that is initiated immediately after an event, preceding a COOP or DRP activation.

Source: National Institute of Standards and Technology, *NIST Special Publication 800-34 Revision 1: Contingency Planning Guide for Federal Information Systems*, USA, 2010. Reprinted courtesy of the National Institute of Standards and Technology, U.S. Department of Commerce. Not copyrightable in the United States.

- Persons responsible for completion
- Available resources to aid in deployment (including human)
- The scheduling of activities with priorities established

Most BCPs are created as procedures that accommodate recovery of information systems (i.e., data storage, servers, etc.), user workstations, other selected equipment (card readers, barcode scanners, printers, etc.) and the network (channels, equipment). Copies of the plan should be kept offsite—at the recovery facility, at the media storage facility and possibly at the homes of key decision-making personnel. More and more frequently, an organization places the electronic version of the plan on a mirrored web site.

Key Decision-making Personnel

The plan should contain a telephone list or "call tree" (i.e., a notification directory, of key decision-making IT and end-user personnel required to initiate and carry out recovery efforts). This is usually a telephone directory of people who should be notified in the event of an incident/disaster or catastrophe. Points to remember when preparing the list are:

- In the event of a widespread disaster or a fire/explosion during normal business hours that heavily damages the organization's offices, many team leaders may not be available
- The telephone list or "call tree" should be highly redundant and updated on a regular basis.

This directory should contain the following information:

- A prioritized list of contacts (i.e., who gets called first?)
- Primary and emergency telephone numbers and addresses for each critical contact person. These usually will be key team leaders responsible for contacting the members of their team.
- Phone numbers and addresses for representatives of equipment and software vendors
- Phone numbers of contacts within companies that have been designated to provide supplies and equipment or services
- Phone numbers of contact persons at recovery facilities, including hot-site representatives and predefined network communications rerouting services
- Phone numbers of contact persons at offsite media storage facilities and the contact persons within the company who are authorized to retrieve media from the offsite facility
- Phone numbers of insurance company agents
- Phone numbers of contacts at contract personnel services
- Phone numbers and contacts of legal/regulatory/governmental agencies, if required
- A procedure to ascertain how many people were reached while using the call tree

Backup of Required Supplies

The plan should have provisions for all supplies necessary for the continuation of normal business activities in the recovery effort. This includes detailed, up-to-date hard copy procedures that can be followed easily by staff and contract personnel who are unfamiliar with the standard and recovery operations. Also, a supply of special forms, such as check stock, invoice forms and order forms, should be secured at an offsite location.

If the data entry function depends on certain hardware devices and/or software programs, these programs and equipment should be provided at the hot site. The same would apply to cryptographic equipment, including electronic keys (e.g., RSA tokens, universal serial bus [USB] keys, etc.).

Insurance

The plan should contain key information about the organization's insurance. The IT processing insurance policy is usually a multiperil policy designed to provide various types of IT coverage. It should be constructed in modules so it can be adapted to the insured's particular IT environment.

> **Note:** Specifics on insurance policies are not tested on the CISA exam because they differ from country to country. The test covers what should be included in policies and third-party agreements but would not test the specific types of coverage.

Specific types of coverage available are:

- **IT equipment and facilities**—Provides coverage for physical damage to the IPF and owned equipment. (Insurance of leased equipment should be obtained when the lessee is responsible for hazard coverage.) The IS auditor is cautioned to review these policies because many policies obligate insurance vendors to replace nonrestorable equipment only with "like kind and

quality," not necessarily with new equipment by the same vendor as the damaged equipment.

- **Media (software) reconstruction**—Covers damage to IT media that is the property of the insured and for which the insured may be liable. Insurance is available for on-premises, off-premises or in-transit situations and covers the actual reproduction cost of the property. Considerations in determining the amount of coverage needed are programming costs to reproduce the media damaged; backup expenses; and physical replacement of media devices such as tapes, cartridges and disks.
- **Extra expense**—Designed to cover the extra costs of continuing operations following damage or destruction at the IPF. The amount of extra-expense insurance needed is based on the availability and cost of backup facilities and operations. Extra expense can also cover the loss of net profits caused by computer media damage. This provides reimbursement for monetary losses resulting from suspension of operations due to the physical loss of equipment or media. An example of a situation requiring this type of coverage is if the information processing facilities were on the sixth floor and the first five floors were burned out. In this case, operations would be interrupted even though the IPF remained unaffected.
- **Business interruption**—Covers the loss of profit due to the disruption of the activity of the company caused by any malfunction of the IT organization
- **Valuable papers and records**—Covers the actual cash value of papers and records (not defined as media) on the insured's premises against direct physical loss or damage
- **Errors and omissions**—Provides legal liability protection in the event that the professional practitioner commits an act, error or omission that results in financial loss to a client. This insurance was originally designed for service bureaus but it is now available from several insurance companies for protecting systems analysts, software designers, programmers, consultants and other IS personnel.
- **Fidelity coverage**—Usually takes the form of bankers blanket bonds, excess fidelity insurance and commercial blanket bonds and covers loss from dishonest or fraudulent acts by employees. This type of coverage is prevalent in financial institutions operating their own IPF.
- **Media transportation**—Provides coverage for potential loss or damage to media in transit to off-premises IPFs. Transit coverage wording in the policy usually specifies that all documents must be filmed or otherwise copied. When the policy does not state specifically that data be filmed prior to being transported and the work is not filmed, management should obtain from the insurance carrier a letter that specifically describes the carrier's position and coverage in the event data are destroyed.

Several key points are important to remember about insurance. Most insurance covers only financial losses based on the historical level of performance and not the existing level of performance. The IS auditor will also be concerned with ensuring that the valuation of insured items, such as technical equipment and infrastructure and data, is appropriate and up to date. Also, insurance does not compensate for loss of image/goodwill.

2.12.10 PLAN TESTING

Most business continuity tests fall short of a full-scale test of all operational portions of the organization, if they are in fact tested at all. This should not preclude performing full or partial testing because one of the purposes of the business continuity test is to determine how well the plan works or which portions of the plan need improvement.

The test should be scheduled during a time that will minimize disruptions to normal operations. Weekends are generally a good time to conduct tests. It is important that the key recovery team members be involved in the test process and allotted the necessary time to put their full effort into it. The test should address all critical components and simulate actual primetime processing conditions, even if the test is conducted in off hours.

Specifications

The test should strive to accomplish the following tasks:

- Verify the completeness and precision of the BCP.
- Evaluate the performance of the personnel involved in the exercise.
- Appraise the training and awareness of employees who are not members of a business continuity team.
- Evaluate the coordination among the business continuity team and external vendors and suppliers.
- Measure the ability and capacity of the backup site to perform prescribed processing.
- Assess the vital records retrieval capability.
- Evaluate the state and quantity of equipment and supplies that have been relocated to the recovery site.
- Measure the overall performance of operational and IT processing activities related to maintaining the business entity.

> **Note:** Assessing the results and the value of the BCP and the DRP tests is an important part of the IS auditor's responsibility.

Test Execution

To perform testing, each of the following test phases should be completed:

- **Pretest**—The set of actions necessary to set the stage for the actual test. This ranges from placing tables in the proper operations recovery area to transporting and installing backup telephone equipment. These activities are outside the realm of those that would take place in the case of a real emergency, in which there is no forewarning of the event and, therefore, no time to take preparatory actions.
- **Test**—This is the real action of the business continuity test. Actual operational activities are executed to test the specific objectives of the BCP. Data entry, telephone calls, information systems processing, handling orders, and movement of personnel, equipment and suppliers should take place. Evaluators review staff members as they perform the designated tasks. This is the actual test of preparedness to respond to an emergency.
- **Posttest**—The cleanup of group activities. This phase comprises such assignments as returning all resources to their proper place, disconnecting equipment, returning personnel, and deleting all company data from third-party systems. The post-test cleanup also includes formally evaluating the plan and implementing indicated improvements.

In addition, the following types of tests may be performed:
- **Desk-based evaluation/paper test**—A paper walk-through of the plan, involving major players in the plan's execution who reason out what might happen in a particular type of service disruption. They may walk through the entire plan or just a portion. The paper test usually precedes the preparedness test.
- **Preparedness test**—Usually a localized version of a full test, wherein actual resources are expended in the simulation of a system crash. This test is performed regularly on different aspects of the plan and can be a cost-effective way to gradually obtain evidence about how good the plan is. It also provides a means to improve the plan in increments.
- **Full operational test**—This is one step away from an actual service disruption. The organization should have tested the plan well on paper and locally before endeavoring to completely shut down operations. For purposes of the BCP testing, this is the disaster.

Documentation of Results

During every phase of the test, detailed documentation of observations, problems and resolutions should be maintained. Each team should have a diary form, with specific steps and information to be recorded, which can be used as documentation. This documentation serves as important historical information that can facilitate actual recovery during a real disaster. Additionally, the insurance company or the local authorities may ask for it. The documentation also aids in performing detailed analysis of both the strengths and weaknesses of the plan.

Results Analysis

It is important to have ways to measure the success of the plan and test against the stated objectives. Therefore, results must be quantitatively gauged as opposed to an evaluation based only on observation.

Specific measurements vary depending on the test and the organization; however, these general measurements usually apply:
- **Time**—Elapsed time for completion of prescribed tasks, delivery of equipment, assembly of personnel and arrival at a predetermined site
- **Amount**—Amount of work performed at the backup site by clerical personnel and information systems processing operations
- **Count**—The number of vital records successfully carried to the backup site versus the required number and the number of supplies and equipment requested versus actually received. Also, the number of critical systems successfully recovered can be measured with the number of transactions processed.
- **Accuracy**—Accuracy of the data entry at the recovery site versus normal accuracy (as a percentage). Also, the accuracy of actual processing cycles can be determined by comparing output results with those for the same period processed under normal conditions.

Plan Maintenance

Plans and strategies for business continuity should be reviewed and updated on a scheduled basis to reflect continuing recognition of changing requirements or extraordinarily (unscheduled revisions) when there is an important change affecting the plans and strategies. The following factors, and others, may impact business continuity requirements and the need for the plan to be updated:
- A strategy that is appropriate at one point in time may not be adequate as the needs of the organization change (business processes, new departments, changes in key personnel)
- New resources/applications may be developed or acquired.
- Changes in business strategy may alter the significance of critical applications or deem additional applications as critical.
- Changes in the software or hardware environment may make current provisions obsolete or inappropriate.
- New events or a change in the likelihood of events may cause disruption.
- Changes are made to key personnel or their contact details.

An important step in maintaining a BCP is to update and test it whenever relevant changes take place within the organization. It is also desirable to include BCP as part of the SDLC process.

The responsibility for maintaining the BCP often falls on the BCP coordinator. Specific plan maintenance responsibilities include:
- Developing a schedule for periodic review and maintenance of the plan advising all personnel of their roles and the deadline for receiving revisions and comments
- Calling for unscheduled revisions when significant changes have occurred
- Reviewing revisions and comments and updating the plan within a certain number days (e.g., 30 days, 2 weeks) of the review date
- Arranging and coordinating scheduled and unscheduled tests of the BCP to evaluate its adequacy
- Participating in the scheduled plan tests, which should be performed at least once per year on specific dates. For scheduled and unscheduled tests, the coordinator will write evaluations and integrate changes to resolve unsuccessful test results into the BCP within a certain number of days (e.g., 30 days, 2 weeks)
- Developing a schedule for training recovery personnel in emergency and recovery procedures as set forth in the BCP. Training dates should be scheduled within 30 days of each plan revision and scheduled plan test.
- Maintaining records of BCP maintenance activities—testing, training and reviews
- Periodically updating, at least quarterly (shorter periods are recommended), the notification directory of all personnel changes including phone numbers, responsibilities or status within the company

A software tool for administering continuity and recovery plans may be useful to track and follow-up on maintenance tasks.

Business Continuity Management Good Practices

The need to continually and periodically revisit and improve on the business continuity program is critical to the development of successful and robust recovery strategy for an organization, irrespective of whether the organization is at the initial stage of developing a BCP. In an effort to enhance business continuity management capabilities (and to comply with regulatory guidelines), some organizations have started adopting good practices from industry-independent and industry-specific entities and regulatory agencies.

Some of these entities or practices/regulations/standards are:
- Business Continuity Institute (BCI)—Provides good practices for business continuity management
- Disaster Recovery Institute International (DRII)—Provides professional practices for business continuity professionals
- US Federal Emergency Management Association (FEMA)—Provides business and industry guidance for emergency management
- ISACA—The COBIT standard provides guidance on IT controls that are relevant to the business.
- US National Institute of Standards and Technology (NIST)
- US Federal Financial Institutions Examination Council (FFIEC)
- US Health and Human Services (HHS)—The Health Insurance Portability and Accountability Act (HIPAA) describes the requirements for managing health information.
- *ISO 22301:2012: Societal security—Business continuity management systems—Requirements*

Note: The CISA candidate will not be tested on specific practices/regulations/standards.

2.12.11 SUMMARY OF BUSINESS CONTINUITY

To ensure continuous service, a BCP should be written to minimize the impact of disruptions. This plan should be based on the long-range IT plan and should support and be aligned with the overall business continuity strategy. Therefore, the process of developing and maintaining an appropriate DRP/BCP would be to:
- Conduct a risk assessment.
  - Identify and prioritize the systems and other resources required to support critical business processes in the event of a disruption.
  - Identify and prioritize threats and vulnerabilities.
- Prepare BIA of the effect of the loss of critical business processes and their supporting components.
- Choose appropriate controls and measures for recovering IT components to support the critical business processes.
- Develop the detailed plan for recovering IS facilities (DRP).
- Develop a detailed plan for the critical business functions to continue to operate at an acceptable level (BCP).
- Test the plans.
- Maintain the plans as the business changes and systems develop.

2.13 AUDITING BUSINESS CONTINUITY

The IS auditor's tasks include:
- Understanding and evaluating business continuity strategy and its connection to business objectives
- Reviewing the BIA findings to ensure that they reflect current business priorities and current controls
- Evaluating the BCPs to determine their adequacy and currency, by reviewing the plans and comparing them to appropriate standards and/or government regulations including the RTO, RPO, etc., defined by the BIA
- Verifying that the BCPs are effective, by reviewing the results from previous tests performed by IT and end-user personnel
- Evaluating cloud-based mechanisms
- Evaluating offsite storage to ensure its adequacy, by inspecting the facility and reviewing its contents and security and environmental controls
- Verifying the arrangements for transporting backup media to ensure that they meet the appropriate security requirements
- Evaluating the ability of personnel to respond effectively in emergency situations, by reviewing emergency procedures, employee training and results of their tests and drills
- Ensuring that the process of maintaining plans is in place and effective and covers both periodic and unscheduled revisions
- Evaluating whether the business continuity manuals and procedures are written in a simple and easy to understand manner. This can be achieved through interviews and determining whether all the stakeholders understand their roles and responsibilities with respect to business continuity strategies.

2.13.1 REVIEWING THE BUSINESS CONTINUITY PLAN

When reviewing the developed plan, IS auditors should verify that basic elements of a well-developed plan are evident. Audit procedures to address basic elements are discussed in the following sections.

Review the Document
- Obtain a copy of the current business continuity policy and strategy.
- Obtain a current copy of the BCP or manual.
- Obtain a copy of the most recent BIA findings and identify the RTO, RPO and other key strategic directives.
- Sample the distributed copies of the manual and verify that they are current.
- Verify whether the BCP supports the overall business continuity strategy.
- Evaluate the effectiveness of the documented procedures for the invocation of the BCP execution.
- Evaluate the procedure for updating the manual. Are updates applied and distributed in a timely manner? Are specific responsibilities documented for maintenance of the manual?

Review the Applications Covered by the Plan
- Review the identification, priorities and planned support of critical applications, both server-based and workstation-based applications.
- Determine whether all applications have been reviewed for their level of tolerance in the event of a disaster.
- Determine whether all critical applications (including PC applications) have been identified.
- Determine whether the secondary site has the correct versions of all system software. Verify that all of the software is compatible; otherwise, the system will not be able to process production data during recovery.

Review the Business Continuity Teams
- Obtain a member list for each recovery/continuity/response team.
- Obtain a copy of agreements relating to use of backup facilities
- Review the list of business continuity personnel, emergency hot-site contacts, emergency vendor contacts, etc., for appropriateness and completeness.
- Call a sample of the people indicated and verify that their phone numbers and addresses are correct, as indicated, and that they possess a current copy of the business continuity manual.
- Interview them for an understanding of their assigned responsibilities in case of interruption/disaster situation.

Plan Testing
- Evaluate the procedures for documenting the tests.
- Review the backup procedures followed for each area covered by the DRP.
- Determine whether the backup and recovery procedures are being followed.

In addition to the above steps:
- Evaluate whether all written emergency procedures are complete, appropriate, accurate, current and easy to understand.
- Identify whether the transactions reentered in the system through recovery process need to be separately identified from the normal transactions.
- Determine whether all recovery/continuity/response teams have written procedures to follow in the event of a disaster.
- Determine whether a suitable procedure exists for updating the written emergency procedures.
- Determine whether user recovery procedures are documented.
- Determine whether the plan adequately addresses movement to the recovery site.
- Determine whether the plan adequately addresses recovering from the recovery site.
- Determine whether items necessary for the reconstruction of the information processing facility are stored offsite, such as blueprints, hardware inventory and wiring diagrams.

Questions to consider include:
- Who is responsible for administration or coordination of the plan?
- Is the plan administrator/coordinator responsible for keeping the plan up to date?
- Where is the DRP stored?
- What critical systems are covered by the plan?
- What systems are not covered by the plan? Why not?
- What equipment is not covered by the plan? Why not?
- Does the plan operate under any assumptions? What are they?
- Does the plan identify rendezvous points for the disaster management committee or emergency management team to meet and decide if business continuity should be initiated?
- Are the documented procedures adequate for successful recovery?
- Does the plan address disasters of varying degrees?
- Are telecommunication's backups (both data and voice line backups) addressed in the plan?
- Where is the backup facility site?
- Does the plan address relocation to a new information processing facility in the event that the original center cannot be restored?
- Does the plan include procedures for merging master file data, automated tape management system data, etc., into predisaster files?
- Does the plan address loading data processed manually into an automated system?
- Are there formal procedures that specify backup procedures and responsibilities?
- What training has been given to personnel in using backup equipment and established procedures?
- Are the restoration procedures documented?
- Are regular and systematic backups of required sensitive and/or crucial applications and data files, being taken?
- Who determines the methods and frequency of data backup for critical information stored?
- What type of media is being used for backups?
- Is offsite storage used to maintain backups of critical information required for processing either onsite or offsite operations?
- Is there adequate documentation to perform a recovery in case of disaster or loss of data?
- Is there a schedule for testing and training on the plan?

2.13.2 EVALUATION OF PRIOR TEST RESULTS

The BCP coordinator should maintain historical documentation of the results of prior business continuity tests. The IS auditor should review these results and determine whether actions requiring correction have been incorporated into the plan. Also, the IS auditor should evaluate BCP/DRP prior tests for thoroughness and accuracy in accomplishing their objectives. Test results should be reviewed to determine whether the appropriate results were achieved and to determine problem trends and appropriate resolutions of problems.

2.13.3 EVALUATION OF OFFSITE STORAGE

The offsite storage facility should be evaluated to ensure the presence, synchronization and currency of critical media and documentation. This includes data files, applications software, applications documentation, systems software, systems documentation, operations documentation, necessary supplies, special forms and a copy of the BCP. To verify the conditions mentioned above, the IS auditor should perform a detailed inventory review. This inventory includes testing for correct dataset names, volume serial numbers, accounting periods and bin locations of media. The IS auditor should also review the documentation, compare it for currency with production documentation, evaluate the availability of the facility and ensure it conforms with management's requirements.

The IS auditor should also review the method of transporting backup data to and from the offsite storage facility to ensure it does not represent a weakness in the information security management system.

2.13.4 INTERVIEWING KEY PERSONNEL

The IS auditor should interview key personnel required for the successful recovery of business operations. All key personnel should have an understanding of their assigned responsibilities as well as up-to-date detailed documentation describing their tasks.

2.13.5 EVALUATION OF SECURITY AT OFFSITE FACILITY

The security of the offsite facility should be evaluated to ensure that it has the proper physical and environmental access controls. These controls include the ability to limit access to only authorized users of the facility, raised flooring, humidity controls, temperature controls, specialized circuitry, uninterruptible power supply, water detection devices, smoke detectors and an appropriate fire extinguishing system. The IS auditor should examine the equipment for current inspection and calibration tags. This review should also consider the security requirements of media transportation.

2.13.6 REVIEWING ALTERNATIVE PROCESSING CONTRACT

The IS auditor should obtain a copy of the contract with the vendor of the alternative processing facility. The vendor's references should be checked to ensure reliability, and all vendor promises should be verified in writing. The contract should be reviewed against the following guidelines:

- Ensure that the contract is written clearly and is understandable
- Legal review for required terms and condition to meet all applicable laws and regulations
- Reexamine and confirm the organization's agreement with the rules that apply to sites shared with other subscribers.
- Ensure that insurance coverage ties in with and covers all (or most) expenses of the disaster.
- Ensure that tests can be performed at the hot site at regular intervals.
- Review and evaluate communications requirements for the backup site.
- Ensure that enforceable source code escrow is reviewed by a lawyer specializing in such contracts.
- Determine the limitation recourse tolerance in the event of a breached agreement.

2.13.7 REVIEWING INSURANCE COVERAGE

It is essential that insurance coverage reflect the actual cost of recovery. Taking into consideration the insurance premium (cost), the coverage for media damage, business interruption, equipment replacement and business continuity processing should be reviewed for adequacy. The specific areas of risk should be found within the BIA, customer contracts and SLAs along with regulatory impacts due to a break in business operations.

Note: The CISA candidate should know what critical provisions need to be included within insurance policies to safeguard the organization.

2.14 CASE STUDIES

The following case studies are included as a learning tool to reinforce the concepts introduced in this chapter.

2.14.1 CASE STUDY A

An IS auditor has been asked to review the draft of an outsourcing contract and SLA and recommend any changes or point out any concerns prior to these documents being submitted to senior management for final approval. The agreement includes outsourcing support of Windows and UNIX server administration, and network management to a third party. Servers will be relocated to the outsourcer's facility that is located in another country, and connectivity will be established using the Internet. OS software will be upgraded on a semiannual basis, but it will not be escrowed. All requests for addition or deletion of user accounts will be processed within three business days. Intrusion detection software will be continuously monitored by the outsourcer and the customer notified by email if any anomalies are detected. Employees hired within the last three years were subject to background checks. Prior to that time there was no policy in place. A right to audit clause is in place but 24-hour notice is required prior to an onsite visit. If the outsourcer is found to be in violation of any of the terms or conditions of the contract, the outsourcer will have 10 business days to correct the deficiency. The outsourcer does not have an IS auditor but is audited by a regional public accounting firm.

CASE STUDY A QUESTIONS	
A1.	Which of the following should be of **MOST** concern to the IS auditor? A. User account changes are processed within three business days. B. Twenty-four hour notice is required prior to an onsite visit. C. The outsourcer does not have an IS audit function. D. Software escrow is not included in the contract.
A2.	Which of the following would be the **MOST** significant issue to address if the servers contain personally identifiable customer information that is regularly accessed and updated by end users? A. The country in which the outsourcer is based prohibits the use of strong encryption for transmitted data. B. The outsourcer limits its liability if it took reasonable steps to protect the customer data. C. The outsourcer did not perform background checks for employees hired over three years ago. D. System software is only upgraded once every six months.

See answers and explanations to the case study questions at the end of the chapter (page 134).

2.14.2 CASE STUDY B

An organization has implemented an integrated application for supporting business processes. It has also entered into an agreement with a vendor for application maintenance and providing support to the users and system administrators. This support will be provided by a remote vendor support center using a privileged user ID with OS level super user authority having read and write access to all files. The vendor will use this special user ID to log on to the system for troubleshooting and implementing application updates (patches). Due to the volume of transactions, activity logs are only maintained for 90 days.

CASE STUDY B QUESTIONS	
B1.	Which of the following is a **MAJOR** concern for the IS auditor? A. User activity logs are only maintained for 90 days. B. The special user ID will access the system remotely. C. The special user ID can alter activity log files. D. The vendor will be testing and implementing patches on servers.
B2.	Which of the following actions would be **MOST** effective in reducing the risk that the privileged user account may be misused? A. The special user ID should be disabled except when maintenance is required. B. All usage of the special user account should be logged. C. The agreement should be modified so that all support is performed onsite. D. All patches should be tested and approved prior to implementation.

See answers and explanations to the case study questions at the end of the chapter (page 134).

2.14.3 CASE STUDY C

An IS auditor was asked to review alignment between IT and business goals for a small financial institution. The IS auditor requested various information including business goals and objectives and IT goals and objectives. The IS auditor found that business goals and objectives were limited to a short bulleted list, while IT goals and objectives were limited to slides used in meetings with the CIO (the CIO reports to the CFO). It was also found in the documentation provided that over the past two years, the risk management committee (composed of senior management) only met on three occasions, and no minutes of what was discussed were kept for these meetings. When the IT budget for the upcoming year was compared to the strategic plans for IT, it was noted that several of the initiatives mentioned in the plans for the upcoming year were not included in the budget for that year.

CASE STUDY C QUESTIONS	
C1.	Which of the following should be of **GREATEST** concern to the IS auditor? A. Strategy documents are informal and incomplete. B. The risk management committee seldom meets and does not keep minutes. C. Budgets do not appear adequate to support future IT investments. D. The CIO reports to the CFO.
C2.	Which of the following would be the **MOST** significant issue to address? A. The prevailing culture within IT. B. The lack of information technology policies and procedures. C. The risk management practices as compared to peer organizations. D. The reporting structure for IT.
See answers and explanations to the case study questions at the end of the chapter (page 135).	

2.14.4 CASE STUDY D

An IS auditor is auditing the IT governance practices for an organization. During the course of the work, it is noted that the organization does not have a full time CIO. The organization chart of the entity provides for an IS manager reporting to the CFO, who in turn reports to the board of directors. The board plays a major role in monitoring IT initiatives in the entity and the CFO communicates on a frequent basis the progress of IT initiatives. From reviewing the SoD matrix, it is apparent that application programmers are only required to obtain approval from the DBA to directly access production data. It is also noted that the application programmers have to provide the developed program code to the program librarian, who then migrates it to production. IS audits are carried out by the internal audit department, which reports to the CFO at the end of every month, as part of business performance review process; the financial results of the entity are reviewed in detail and signed off by the business managers for correctness of data contained therein.

CASE STUDY D QUESTIONS	
D1.	Given the circumstances described, what would be of **GREATEST** concern from an IT governance perspective? A. The organization does not have a full-time CIO. B. The organization does not have an IT steering committee. C. The board of the organization plays a major role in monitoring IT initiatives. D. The information systems manager reports to the CFO.
D2.	Given the case, what would be of **GREATEST** concern from a segregation of duties perspective? A. Application programmers are required to obtain approval only from the DBA for direct write access to data. B. Application programmers are required to turn over the developed program code to the program librarian for migration to production. C. The internal audit department reports to the CFO. D. Business performance reviews are required to be signed off only by the business managers.
D3.	Which of the following would **BEST** address data integrity from a mitigating control standpoint? A. Application programmers are required to obtain approval from DBA for direct access to data. B. Application programmers are required to hand over the developed program codes to the program librarian for transfer to production. C. The internal audit department reports to the CFO. D. Business performance results are required to be reviewed and signed off by the business managers.
See answers and explanations to the case study questions at the end of the chapter (page 135).	

2.14.5 CASE STUDY E

An organization is developing revised BCPs and DRPs for its headquarters facility and network of 16 branch offices. The current plans have not been updated in more than eight years, during which time the organization has grown by over 300 percent. At the headquarters facility, there are approximately 750 employees. These individuals connect over a LAN to an array of more than 60 application, database and file print servers located in the corporate data center and over a frame relay network to the branch offices. Traveling users access corporate systems remotely by connecting over the Internet using virtual private networking. Users at both headquarters and the branch offices access the Internet through a firewall and proxy server located in the data center. Critical applications have a RTO of between three and five days. Branch offices are located between 30 and 50 miles from one another, with none closer to the headquarters' facility than 25 miles. Each branch office has between 20 and 35 employees plus a mail server and a file/print server. Backup media for the data center are stored at a third-party facility 35 miles away. Backups for servers located at the branch offices are stored at nearby branch offices using reciprocal agreements between offices. Current contracts with a third-party hot site provider include 25 servers, work area space equipped with desktop computers to accommodate 100 individuals, and a separate agreement to ship up to two servers and 10 desktop computers to any branch office declaring an emergency. The contract term is for three years, with equipment upgrades occurring at renewal time. The hot site provider has multiple facilities throughout the country in case the primary facility is in use by another customer or rendered unavailable by the disaster. Senior management desires that any enhancements be as cost effective as possible.

CASE STUDY E QUESTIONS	
E1.	On the basis of the above information, which of the following should the IS auditor recommend concerning the hot site? A. Desktops at the hot site should be increased to 750. B. An additional 35 servers should be added to the hot site contract. C. All backup media should be stored at the hot site to shorten the RTO. D. Desktop and server equipment requirements should be reviewed quarterly.
E2.	On the basis of the above information, which of the following should the IS auditor recommend concerning branch office recovery? A. Add each of the branches to the existing hot site contract. B. Ensure branches have sufficient capacity to back each other up. C. Relocate all branch mail and file/print servers to the data center. D. Add additional capacity to the hot site contract equal to the largest branch.
See answers and explanations to the case study questions at the end of the chapter (page 135).	

2.15 ANSWERS TO CASE STUDY QUESTIONS

ANSWERS TO CASE STUDY A QUESTIONS

A1. **A** Three business days to remove the account of a terminated employee would create an unacceptable risk to the organization. In the intervening time significant damage could be done. In contrast, some degree of advance notice prior to an onsite visit is generally accepted within the industry. Also, not every outsourcer will have its own internal audit function or IS auditor. Software escrow is primarily of importance when dealing with custom application software where there is a need to store a copy of the source code with a third party. OS software for generally available commercial OSs would not require software escrow.

A2. **A** Because connectivity to the servers is over the Internet, the prohibition against strong encryption will place any transmitted data at risk. The limitation of liability is a standard industry practice. Although the failure to perform background checks for employees hired more than three years ago is of importance, it is not as significant an issue. Upgrading system software once every six months does not present any significant exposure.

ANSWERS TO CASE STUDY B QUESTIONS

B1. **C** Because the super user ID has read and write access to all files, there is no way to ensure that the activity logs are not modified to hide unauthorized activity by the vendor. Remote access is not a major concern as long as the connection is made over an encrypted line, and testing and implementing patches on servers is part of vendor-provided support. Although 90-day retention of logs may not be sufficient in some business situations, it is not as major a concern as is the fact that the vendor has the ability to alter the activity logs.

B2. **A** The **MOST** effective and practical control in this situation is to lock the special user account when it is not needed. The account should be opened only when vendor needs access for support and closed immediately after use. All activities should be logged and reviewed for appropriateness. The other choices are not as effective or practical in reducing the risk.

ANSWERS TO CASE STUDY C QUESTIONS

C1. **B** The fact that the risk management committee seldom meets and when it does meet, no minutes are taken, is the greatest concern. Because senior management is not meeting regularly to discuss key risk issues, and minutes are not captured which would provide for follow up, analysis and commitment, this indicates a serious lack of governance. The other options are not as serious in their potential impact on the organization.

C2. **B** The absence of policies and procedures makes it difficult if not impossible to implement effective IT governance. Other issues are secondary by comparison.

ANSWERS TO CASE STUDY D QUESTIONS

D1. **D** The information systems manager should ideally report to the board of directors or the CEO to provide a sufficient degree of independence. The reporting structure that requires the information systems manager to report to the CFO is not a desirable situation and could lead to the compromise of certain controls.

D2. **A** The application programmers should obtain approval from the business owners before accessing data. DBAs are only custodians of the data and should only provide access that is authorized by the data owner.

D3. **D** Sign-off on data contained in the financial results by the business managers at the end of the month would detect any significant discrepancies that would result from tampering of data through inappropriate direct access of data without the approval or knowledge of the business managers.

ANSWERS TO CASE STUDY E QUESTIONS

E1. **D** As equipment needs in a rapidly growing business are subject to frequent change, quarterly reviews are necessary to ensure that the recovery capability keeps pace with the organization. Because not all employee job functions are critical during a disaster, it is not necessary to contact the same number of desktops at a recovery facility as the number of employees. Similarly, not every server is critical to the continued operation of the business. In both cases, only a subset will be required. Because there is no assurance that the hot site will not already be occupied, it would not be advisable to store backup media at the facility. These facilities are generally not designed to provide extensive media storage, and frequent testing by other customers could compromise the security of the media.

E2. **B** The most cost-effective solution is to recommend that branches have sufficient capacity to accommodate critical personnel from another branch. Because critical job functions would represent only perhaps 20 percent of the staff from the affected branch, accommodations for only four to seven critical staff members would be needed. Adding each of the branches to the hot site contract would be far more expensive, while adding capacity to the hot site contract would not provide coverage as hot site contracts base their pricing on each location covered. Finally, relocating branch servers to the data center could result in performance issues, and would not address the question of where to locate displaced employees.

Page intentionally left blank

Chapter 3:

Information Systems Acquisition, Development and Implementation

Section One: Overview

Section Two: Content

Section One: Overview

DEFINITION

This chapter on information systems acquisition, development and implementation provides an overview of key processes and methodologies used by organizations when creating and changing application systems and infrastructure components.

OBJECTIVES

The objective of this domain is to ensure that the CISA candidate understands and can provide assurance that the practices for the acquisition, development, testing and implementation of information systems meet the organization's strategies and objectives.

This domain represents 18 percent of the CISA examination (approximately 27 questions).

TASK AND KNOWLEDGE STATEMENTS

TASKS

There are seven tasks within the domain covering information systems acquisition, development and implementation:

- T3.1 Evaluate the business case for the proposed investments in information systems acquisition, development, maintenance and subsequent retirement to determine whether it meets business objectives.
- T3.2 Evaluate IT supplier selection and contract management processes to ensure that the organization's service levels and requisite controls are met.
- T3.3 Evaluate the project management framework and controls to determine whether business requirements are achieved in a cost-effective manner while managing risks to the organization.
- T3.4 Conduct reviews to determine whether a project is progressing in accordance with project plans, is adequately supported by documentation and has timely and accurate status reporting.
- T3.5 Evaluate controls for information systems during the requirements, acquisition, development and testing phases for compliance with the organization's policies, standards, procedures and applicable external requirements.
- T3.6 Evaluate the readiness of information systems for implementation and migration into production to determine whether project deliverables, controls and organization's requirements are met.
- T3.7 Conduct post-implementation reviews of systems to determine whether project deliverables, controls and organization's requirements are met.

KNOWLEDGE STATEMENTS

The CISA candidate must have a good understanding of each of the topics or areas delineated by the knowledge statements. These statements are the basis for the examination.

There are 14 knowledge statements within the domain covering information systems acquisition, development and implementation:

- K3.1 Knowledge of benefits realization practices, (e.g., feasibility studies, business cases, total cost of ownership [TCO], return on investment [ROI])
- K3.2 Knowledge of IT acquisition and vendor management practices (e.g., evaluation and selection process, contract management, vendor risk and relationship management, escrow, software licensing) including third-party outsourcing relationships, IT suppliers and service providers.
- K3.3 Knowledge of project governance mechanisms (e.g., steering committee, project oversight board, project management office)
- K3.4 Knowledge of project management control frameworks, practices and tools
- K3.5 Knowledge of risk management practices applied to projects
- K3.6 Knowledge of requirements analysis and management practices (e.g., requirements verification, traceability, gap analysis, vulnerability management, security requirements)
- K3.7 Knowledge of enterprise architecture related to data, applications, and technology (e.g., web-based applications, web services, n-tier applications, cloud services, virtualization)
- K3.8 Knowledge of system development methodologies and tools including their strengths and weaknesses (e.g., agile development practices, prototyping, rapid application development [RAD], object-oriented design techniques, secure coding practices, system version control)
- K3.9 Knowledge of control objectives and techniques that ensure the completeness, accuracy, validity and authorization of transactions and data
- K3.10 Knowledge of testing methodologies and practices related to the information system development life cycle (SDLC)
- K3.11 Knowledge of configuration and release management relating to the development of information systems
- K3.12 Knowledge of system migration and infrastructure deployment practices and data conversion tools, techniques and procedures
- K3.13 Knowledge of project success criteria and project risk
- K3.14 Knowledge of post-implementation review objectives and practices (e.g., project closure, control implementation, benefits realization, performance measurement)

Relationship of Task to Knowledge Statements

The task statements are what the CISA candidate is expected to know how to perform. The knowledge statements delineate each of the areas in which the CISA candidate must have a good understanding in order to perform the tasks. The task and knowledge statements are mapped in **figure 3.1** insofar as it is possible to do so. Note that although there is often overlap, each task statement will generally map to several knowledge statements.

Note: Hardware and software are integrated components of the technical infrastructure that make it possible for the business applications to operate. Considering them as separate audit items is an intrinsic scope limitation that requires a cautious definition of the audit objectives because these must generally be framed around their efficiency and effectiveness in supporting the business. The need to separately evaluate a component, or a class of components, must be determined as a part of the overall system assessment. A specific audit would be necessary, for example, if a preceding higher-level review concluded that a component, or a class of components, weakens the overall system performance. Furthermore, IS auditors must be aware of other synergies and restrictions that affect a specific component's functionality in the overall system because these have a direct impact on the significance of the audit results.

Figure 3.1—Task and Knowledge Statements Mapping

Task Statement	Knowledge Statements
T3.1 Evaluate the business case for the proposed investments in information systems acquisition, development, maintenance and subsequent retirement to determine whether it meets business objectives.	K3.1 Knowledge of benefits realization practices, (e.g., feasibility studies, business cases, total cost of ownership [TCO], return on investment [ROI]) K3.3 Knowledge of project governance mechanisms (e.g., steering committee, project oversight board, project management office) K3.5 Knowledge of risk management practices applied to projects K3.7 Knowledge of enterprise architecture related to data, applications, and technology (e.g., web-based applications, web services, n-tier applications, cloud services, virtualization) K3.13 Knowledge of project success criteria and project risk
T3.2 Evaluate IT supplier selection and contract management processes to ensure that the organization's service levels and requisite controls are met.	K3.2 Knowledge of IT acquisition and vendor management practices (e.g., evaluation and selection process, contract management, vendor risk and relationship management, escrow, software licensing) including third-party outsourcing relationships, IT suppliers and service providers.
T3.3 Evaluate the project management framework and controls to determine whether business requirements are achieved in a cost-effective manner while managing risks to the organization.	K3.1 Knowledge of benefits realization practices, (e.g., feasibility studies, business cases, total cost of ownership [TCO], return on investment [ROI]) K3.3 Knowledge of project governance mechanisms (e.g., steering committee, project oversight board, project management office) K3.4 Knowledge of project management control frameworks, practices and tools K3.5 Knowledge of risk management practices applied to projects K3.6 Knowledge of requirements analysis and management practices (e.g., requirements verification, traceability, gap analysis, vulnerability management, security requirements) K3.8 Knowledge of system development methodologies and tools including their strengths and weaknesses (e.g., agile development practices, prototyping, rapid application development [RAD], object-oriented design techniques, secure coding practices, system version control) K3.13 Knowledge of project success criteria and project risk
T3.4 Conduct reviews to determine whether a project is progressing in accordance with project plans is adequately supported by documentation and has timely and accurate status reporting.	K3.1 Knowledge of benefits realization practices, (e.g., feasibility studies, business cases, total cost of ownership [TCO], return on investment [ROI]) K3.3 Knowledge of project governance mechanisms (e.g., steering committee, project oversight board, project management office) K3.4 Knowledge of project management control frameworks, practices and tools K3.5 Knowledge of risk management practices applied to projects K3.8 Knowledge of system development methodologies and tools including their strengths and weaknesses (e.g., agile development practices, prototyping, rapid application development [RAD], object-oriented design techniques, secure coding practices, system version control) K3.13 Knowledge of project success criteria and project risk

Figure 3.1—Task and Knowledge Statements Mapping *(cont.)*

Task Statement	Knowledge Statements
T3.5 Evaluate controls for information systems during the requirements, acquisition, development and testing phases for compliance with the organization's policies, standards, procedures and applicable external requirements.	K3.2 Knowledge of IT acquisition and vendor management practices (e.g., evaluation and selection process, contract management, vendor risk and relationship management, escrow, software licensing) including third-party outsourcing relationships, IT suppliers and service providers. K3.4 Knowledge of project management control frameworks, practices and tools K3.5 Knowledge of risk management practices applied to projects K3.6 Knowledge of requirements analysis and management practices (e.g., requirements verification, traceability, gap analysis, vulnerability management, security requirements) K3.7 Knowledge of enterprise architecture related to data, applications, and technology (e.g., web-based applications, web services, n-tier applications, cloud services, virtualization) K3.8 Knowledge of system development methodologies and tools including their strengths and weaknesses (e.g., agile development practices, prototyping, rapid application development [RAD], object-oriented design techniques, secure coding practices, system version control) K3.9 Knowledge of control objectives and techniques that ensure the completeness, accuracy, validity, and authorization of transactions and data K3.10 Knowledge of testing methodologies and practices related to the information system development life cycle (SDLC) K3.11 Knowledge of configuration and release management relating to the development of information systems K3.12 Knowledge of system migration and infrastructure deployment practices and data conversion tools, techniques and procedures
T3.6 Evaluate the readiness of information systems for implementation and migration into production to determine whether project deliverables, controls and organization's requirements are met.	K3.5 Knowledge of risk management practices applied to projects K3.6 Knowledge of requirements analysis and management practices (e.g., requirements verification, traceability, gap analysis, vulnerability management, security requirements) K3.8 Knowledge of system development methodologies and tools including their strengths and weaknesses (e.g., agile development practices, prototyping, rapid application development [RAD], object-oriented design techniques, secure coding practices, system version control) K3.9 Knowledge of control objectives and techniques that ensure the completeness, accuracy, validity, and authorization of transactions and data K3.10 Knowledge of testing methodologies and practices related to the information system development life cycle (SDLC) K3.11 Knowledge of configuration and release management relating to the development of information systems K3.12 Knowledge of system migration and infrastructure deployment practices and data conversion tools, techniques, and procedures K3.13 Knowledge of project success criteria and project risk
T3.7 Conduct post-implementation reviews of systems to determine whether project deliverables, controls and organization's requirements are met.	K3.1 Knowledge of benefits realization practices, (e.g., feasibility studies, business cases, total cost of ownership [TCO], return on investment [ROI]) K3.4 Knowledge of project management control frameworks, practices and tools K3.9 Knowledge of control objectives and techniques that ensure the completeness, accuracy, validity and authorization of transactions and data K3.10 Knowledge of testing methodologies and practices related to the information system development life cycle (SDLC) K3.13 Knowledge of project success criteria and project risk K3.14 Knowledge of post-implementation review objectives and practices (e.g., project closure, control implementation, benefits realization, performance measurement)

Knowledge Statement Reference Guide

Each knowledge statement is explained in terms of underlying concepts and relevance of the knowledge statement to the IS auditor. It is essential that the exam candidate understand the concepts. The knowledge statements are what the IS auditor must know in order to accomplish the tasks. Consequently, only the knowledge statements are detailed in this section.

The sections identified in K3.1 through K3.14 are described in greater detail in section two of this chapter.

K3.1 Knowledge of benefits realization practices (e.g., feasibility studies, business cases, total cost of ownership [TCO], return on investment [ROI])

<table>
<tr><th>Explanation</th><th>Key Concepts</th><th>Reference in Manual</th></tr>
<tr><td rowspan="2">The premise of benefits realization is driven from a well-founded concern at board and senior management levels that the large expenditures on IT-related initiatives are not realizing the business benefits they promise. The objective of benefits realization is to ensure that IT and the business fulfill their value management responsibilities, particularly that:
• IT-enabled business investments achieve the promised benefits and deliver measurable business value.
• Required capabilities (solutions and services) are delivered:
– On time, both with respect to schedule and time-sensitive market, industry and regulatory requirements
– Within budget
• IT services and other IT assets continue to contribute to business value.

Benefits realization of both systems and software projects are a compromise of major factors such as cost, quality, development/delivery time, reliability and dependability. Prior to initiating these types of projects, senior management performs a comprehensive study and evaluates which factors are “qualifying” or “winning.” These results are then compared to strengths, weaknesses and competencies of services available to complete and maintain systems or software to be delivered by the project. Organizations should employ structured project management principles to support changes to their information systems environment. The IS auditor should understand how the business defines business cases, processes used during feasibility studies and resultant determinations with regard to ROI) for development-related projects. If companies fail to consistently meet their ROI objectives, this may suggest weakness in their system development life cycle (SDLC) and related project management practices.</td><td>Understanding the business case development approach for program management and SDLC processes</td><td>3.2 Benefits Realization
3.2.2 Business Case Development and Approval
3.2.3 Benefits Realization Techniques
3.5.2 Description of Traditional SDLC Phases
3.15.2 Feasibility Study</td></tr>
<tr><td>Understanding benefits realization techniques</td><td>3.2.3 Benefits Realization Techniques</td></tr>
</table>

K3.2 Knowledge of IT acquisition and vendor management practices (e.g., evaluation and selection process, contract management, vendor risk and relationship management, escrow, software licensing) including third-party outsourcing relationships, IT suppliers and service providers

Explanation	Key Concepts	Reference in Manual
Based on the system or software requirements defined and feasibility studies, the project management team must evaluate if the organization will use internal development resources, a combination of internal resources supplemented by vendor-provided capabilities and products or completely outsource the project system/software development. Outsourced or vended capabilities, services and products must be evaluated for the risk these outside entities bring to the organization and the impact on benefits-realization goals.	Understanding system and software project acquisition evaluation and selection process, contract management, vendor risk and relationship management, escrow, software licensing	3.4.2 Project Planning 3.9 Infrastructure Development/ Acquisition Practices 3.9.4 Hardware Acquisition 3.9.5 System Software Acquisition
The IS auditor must understand the variety of vendor provided services (commercial off-the-shelf hardware/software products, outsourced services to include cloud offerings, managed services, etc.) and the functional requirements these services are addressing. Furthermore, the IS auditor needs to understand the vendors' service level agreements that are in place to address system/software operational and technical support requirements. Additional considerations also include suppliers' financial viability, licensing scalability and provisions for software escrow. Use of vendors can speed a project and potentially reduce total costs. However, use of vendors adds risk, especially if the vendor is a single- or sole-source provider. Proper vendor management can reduce or prevent problems caused when a vendor is chosen that is unable to achieve the required solution or time frame. Proper vendor management can also ensure that contracts address business needs and do not expose the business to unnecessary risk. Contract management processes may include many activities (a complete contract life cycle) and involve various organizational units (user department, IT development, finance, legal). Standards, frameworks and certifications play an important role because they are cost-effective ways of assessing and transmitting trust and they save effort, time and paperwork. The overall process may include specifications development, contract solicitation, vendor evaluation and selection, contract drafting, negotiation and execution, change orders, continuous vendor rating and monitoring, and contract close out. Although they are not legal or "contract auditors," IS auditors must understand the importance of requirements specification that forms the request for proposal (RFP). The IS auditor must understand the need for required security and other controls to be specified, the essential elements of vendor selection to ensure that a reliable and professional vendor is chosen and the essential contents of the contract—most notably, the need, as appropriate, for an escrow agreement to be in place. The right to audit must also be addressed in the contract.	Understanding the key information that should be included in a vendor contract	3.9.4 Hardware Acquisition 3.9.5 System Software Acquisition

K3.3 Knowledge of project governance mechanisms (e.g., steering committee, project oversight board, project management office)

Explanation	Key Concepts	Reference in Manual
All projects require governance structures, policies and procedures, and specific control mechanisms assure strategic and tactical alignment with the organization's respective goals and objectives and management of associated risk. Without proper governance oversight, all aspects of a project may be endangered. The IS auditor needs to understand program management governance concepts and how to evaluate the program office and/or project steering committee integration within the organization. With this understanding, the IS auditor can perform more proactive evaluations of top level system/software development activities within the organization at both an enterprise and operational level that can have dramatic impacts, both positive and negative, on the organization.	Understanding project portfolio management and project oversight structure and processes	3.2.1 Portfolio/Program Management 3.2.2 Business Case Development and Approval 3.4.4 Project Controlling 3.5.4 Risk Associated With Software Development

K3.4 Knowledge of project management control frameworks, practices and tools

Explanation	Key Concepts	Reference in Manual
Effective and efficient project management requires that the project manager's skill set be commensurate with the project at hand. Without the application of project management practices, tools and control frameworks, it is virtually impossible to manage all the relevant parameters of a large project. Projects need to be managed based on hard factors such as deliverables, quality, costs and deadlines; on soft factors such as team dynamics, conflict resolution, leadership issues, cultural differences and communication; and, finally, on environmental factors such as the political and power issues in the sponsoring organization, managing the expectations of stakeholders, and the larger ethical and social issues that may surround a project. While the CISA exam will not test knowledge of any particular methodology, the IS auditor must understand the need for an established development management framework within the organization, the constituent elements of a standard methodology, and the contents and deliverables of each phase in order to ascertain the degree of necessary audit involvement.	Understanding project management practices, tools and control frameworks	3.3 Project Management Structure 3.3.1 General Aspects 3.3.3 Project Organizational Forms 3.3.5 Project Objectives 3.3.6 Roles and Responsibilities of Groups and Individuals 3.4 Project Management Practices 3.4.1 Initiation of a Project 3.4.2 Project Planning 3.4.3 Project Execution 3.4.4 Project Controlling 3.4.5 Closing a Project
	Understanding the system development life cycle (SDLC) process as it relates to project management	3.5.2 Description of Traditional SDLC Phases 3.8 Development Methods 3.8.1 Use of Structured Analysis, Design and Development Techniques 3.9 Infrastructure Development/ Acquisition Practices 3.12 Process Improvement Practices 3.12.1 Business Process Reengineering and Process Change Projects 3.12.2 ISO/IEC 25010:2011 3.12.3 Capability Maturity Model Integration 3.12.4 ISO/IEC 330xx Series

K3.5 Knowledge of risk management practices applied to projects

Explanation	Key Concepts	Reference in Manual
Risk is defined as a possible negative event or condition that would disrupt relevant aspects of the project. There are two main categories of project risk: the category that impacts the business benefits (and, therefore, endangers the reasons for the project's very existence) and the category that impacts the project itself. The project *sponsor* is responsible for mitigating the first category of risk and the project *manager* is responsible for mitigating the second category. The IS auditor needs to understand how risk management processes are integrated throughout program management processes and system and software development activities.	Understanding of the key concepts and processes used to mitigate risk with regards to program and project management	3.4.4 Project Controlling 3.7 Business Application Systems 3.15.1 Project Management

K3.6 Knowledge of requirements analysis and management practices (e.g., requirements verification, traceability, gap analysis, vulnerability management, security requirements)

Explanation	Key Concepts	Reference in Manual
Requirements capture is one of the most critical and difficult activities of the development life cycle. It is critical because it is an initial phase and the basis for every other activity. It is difficult because it involves subjective decisions and extensive interaction between units, functions and people (mainly between users and developers) with very different functions, interests and skills. Certain techniques and tools (interviews, protocol analysis, etc.) facilitate the requirements capture. Requirements should be prudent, feasible, cost-effective and aligned with an enterprise's strategy, plans and policies. Requirements should be duly documented to facilitate the understanding of the developers and formally approved and frozen (baselined) to prevent scope creep. It is incumbent that the IS auditor understand the life cycle of program, project and unique system and software development requirements. This understanding covers the processes of discovery of key functional, operational and security requirements through the maintenance, verification, traceability, gap analysis and vulnerability management controls implementation over requirements baseline.	Understanding the requirements analysis during SDLC phases	3.5 Business Application Development 3.5.1 Traditional SDLC Approach 3.5.2 Description of Traditional SDLC Phases 3.5.3 Integrated Resource Management Systems
	Understanding the use of the V-model to reduce the risk of not meeting requirements	3.5 Business Application Development
	Understanding the sponsor's role in developing success criteria	3.3.6 Roles and Responsibilities of Groups and Individuals
	Understanding the reasons to use an object breakdown structure	3.3.5 Project Objectives

K3.7 Knowledge of enterprise architecture related to data, applications and technology (e.g., web-based applications, web services, n-tier applications, cloud services, virtualization)

Explanation	Key Concepts	Reference in Manual
An enterprise architecture is a systematic description of an organization's structure, including business processes, information systems, human resources and organizational units. There are various enterprise architecture models; knowledge of any specific model is not essential for the IS auditor and will not be tested on the CISA exam. Enterprise architectures are supported or served by IT architectures (e.g., n-tier, client-server, web-based and distributed components). The IS auditor must understand the role of these components and how control objectives are met across all components to determine whether risk is sufficiently mitigated by these controls.	Understanding the components of an enterprise architecture	3.6 Virtualization and Cloud Computing Environments 3.7 Business Application Systems 3.7.1 E-commerce 3.7.16 Industrial Control Systems 3.9 Infrastructure Development/ Acquisition Practices

K3.8 Knowledge of system development methodologies and tools including their strengths and weaknesses (e.g., agile development practices, prototyping, rapid application development [RAD], object-oriented design techniques, secure coding practices, system version control)

Explanation	Key Concepts	Reference in Manual
In the face of increasing system complexity and the need to implement new systems more quickly to achieve benefits before the business changes, system and software development practitioners have adopted new ways of organizing system and software projects that vary, or in some cases radically depart from, the traditional waterfall model. In addition, there has been continued evolution in the thinking about how best to analyze, design and construct software systems and in the information technologies available to perform these activities. The IS auditor's understanding of these differing methodologies enables them to better evaluate the existence and effectiveness of critical system development controls and to look at the selection of the methodologies and whether the process will properly address the project requirements (time, budget, resources). The CISA exam will not test the detailed knowledge of any specific methodology or project framework nor of any particular tool, but an understanding of the relative merits of and risk associated with recognized, standard tools is expected.	Understanding the different types of system/software development methodologies and tools	3.8 Development Methods
	How adopting these system/software development methodologies and tools impacts programs and respective projects	

K3.9 Knowledge of control objectives and techniques that ensure the completeness, accuracy, validity and authorization of transactions and data

Explanation	Key Concepts	Reference in Manual
Poor controls over data input, processing, storage or output increase the risk of loss to an enterprise. The IS auditor must be aware of the need for controls to ensure the authorization, accuracy and completeness of data input to, processing by and output from computer applications. The IS auditor also must know what types of control techniques are available at each level and how each may be evidenced in the form of reports, logs and audit trails.	Understanding authorization and editing, processing and output controls	3.7 Business Application Systems 3.13.1 Input/Origination Controls 3.13.2 Processing Procedures and Controls 3.13.3 Output Controls

K3.10 Knowledge of testing methodologies and practices related to the information system development life cycle (SDLC)

Explanation	Key Concepts	Reference in Manual
Integral to the system and software development project success is the proper selection of testing methodologies, development of testing plans fully traceable to requirements and acquisition of essential resources to successfully complete testing. Once completed, testing provides confidence to stakeholders that a system or system component operates as intended and delivers the benefits realization as required at the start of project. The IS auditor should also be aware of other forms of testing, such as unit, integration, sociability and regression testing. Furthermore, the IS auditor needs to understand the role quality assurance (QA) monitoring and evaluation play within quality of an enterprise's internal processes (e.g., project management, software development process or IT service) and the quality of the final products produced by these processes (e.g., the system implemented or software developed).	Understanding types of testing in the system development process	3.14.6 Test Application Systems
	Understanding the goals of a quality assurance program within the context of the SDLC	3.3.6 Roles and Responsibilities of Groups and Individuals 3.5.2 Description of Traditional SDLC Phases
	Understanding quality assurance testing	3.14.4 Data Integrity Testing

K3.11 Knowledge of configuration and release management relating to the development of information systems

Explanation	Key Concepts	Reference in Manual
Effective and efficient development and maintenance of complicated IT systems requires that rigorous configuration, change and release management processes be implemented and adhered to within the organization. These processes provide systematic, consistent and unambiguous control on attributes of IT components comprising the system (hardware, software, firmware, and network connectivity including physical connecting media wire, fiber, radio frequency [RF]). Knowledge of the current configuration status of computing environments is critical to system reliability, availability and security along with achieving timely maintenance of these systems. Changes to IT systems must be carefully assessed, planned, tested, approved, documented and communicated to minimize any undesirable consequences to the business processes. The IS auditor should be aware of the tools available for managing configuration, change and release management and of the controls in place to ensure segregation of duties (SoD) between development staff and the production environment.	Understanding the configuration, change and release management processes and ways to examine the accomplishment of these processes	3.10 Information Systems Maintenance Practices 3.10.1 Change Management Process Overview 3.10.2 Configuration Management

K3.12 Knowledge of system migration and infrastructure deployment practices and data conversion tools, techniques and procedures

Explanation	Key Concepts	Reference in Manual
New software applications tend to be more comprehensive and integrated than older applications. Furthermore, organizations rely increasingly on data warehouses, models and simulation for decision making; thus, importing data from old (and legacy) systems into the new application is crucial. Data format, coding, structure and integrity are to be preserved or properly translated. A migration scenario must be set up, and a rollback plan needs to be in place. There are many direct (old to new application) and indirect (using interim repositories) strategies and tools. Data conversion is a one-time task in many development projects. The importance of correct results is critical, and success depends on the use of good practices by the development team as the programmed input checks under development will not be available for the conversion. Source data must be correctly characterized, and the destination database must accommodate all existing data values. Resulting data should be carefully tested. Steps for the conversion that are developed in the test environment must be recorded so that they can be repeated on the production system. The IS auditor should ensure that any tools and techniques selected for the process are adequate and appropriate, that data conversion achieves the necessary objectives without data loss or corruption, and that any loss of data is both minimal and formally accepted by user management.	Understanding the criteria for successful data conversion during system migration	3.5.2 Description of Traditional SDLC Phases
	Understanding computer-aided software engineering and how it can be leveraged for data conversion	3.11.2 Computer-aided Software Engineering
	Understanding process improvement practices	3.12 Process Improvement Practices 3.12.1 Business Process Reengineering and Process Change Projects 3.12.2 ISO/IEC 25010:2011 3.12.3 Capability Maturity Model Integration 3.12.4 ISO/IEC 330xx Series

K3.13 Knowledge of project success criteria and project risk

Explanation	Key Concepts	Reference in Manual
From the start, program and project managers must identify and document the criteria that clearly indicate successfully meeting stakeholder requirements ranging from form, fit, function, schedule, budget and interoperability. As part of the program and project planning and overall management activities, risk to meeting the identified success criteria are also identified, analyzed and evaluated to enable managers to take preventive and corrective actions as needed. The IS auditor needs to understand the specific success criteria for the system/software programs and respective projects and how management is evaluating progress toward meeting these criteria along with addressing deficiencies that put the program and respective projects at risk.	Understanding methods to develop, monitor, measure and attain critical program and project success criteria	3.2.3 Benefits Realization Techniques 3.4 Project Management Practices 3.4.2 Project Planning 3.4.5 Closing a Project 3.15.8 Postimplementation Review

K3.14 Knowledge of post-implementation review objectives and practices (e.g., project closure, control implementation, benefits realization, performance measurement)

Explanation	Key Concepts	Reference in Manual
Projects should be formally closed to provide accurate information on project results, improve future projects and allow an orderly release of project resources. The closure process should determine whether project objectives were met or excused and should identify lessons learned to avoid mistakes and encourage repetition of good practices. In contrast to project closure, a postimplementation review typically is carried out in several weeks or months after project completion, when the major benefits and shortcomings of the solution implemented will be realized. The review is part of a benefits realization process and includes an estimate of the project's overall success and impact on the business.	Understanding the methods and benefits of postimplementation reviews following the completion of a project	3.2.3 Benefits Realization Techniques 3.4.5 Closing a Project 3.5.2 Description of Traditional SDLC Phases 3.15.8 Postimplementation Review

SUGGESTED RESOURCES FOR FURTHER STUDY

Bonham, Stephen S.; *IT Project Portfolio Management*, Artech House Inc., USA, 2005

Maizlish, Bryan; Robert Handler; *IT Portfolio Management Step-by-Step: Unlocking the Business Value of Technology*, John Wiley & Sons, USA, 2005

Natan, Ron Ben; *Implementing Database Security and Auditing*, Elsevier Digital Press, USA, 2005

Project Management Institute (PMI), *www.pmi.org*

Project Management Institute (PMI), *A Guide to the Project Management Body of Knowledge (PMBOK® Guide), 5th Edition*, USA, 2013

Wysocki, Robert K.; *Effective Project Management: Traditional, Agile, Extreme, 6th Edition*, Wiley Publishing Inc., USA, 2011

*Note: **Publications in bold are stocked in the ISACA Bookstore.***

SELF-ASSESSMENT QUESTIONS

CISA self-assessment questions support the content in this manual and provide an understanding of the type and structure of questions that have typically appeared on the exam. Questions are written in a multiple-choice format and designed for one best answer. Each question has a stem (question) and four options (answer choices). The stem may be written in the form of a question or an incomplete statement. In some instances, a scenario or a description problem may also be included. These questions normally include a description of a situation and require the candidate to answer two or more questions based on the information provided. Many times question will require the candidate to choose the **MOST** likely or **BEST** answer among the options provided.

In each case, the candidate must read the question carefully, eliminate known incorrect answers and then make the best choice possible. Knowing the format in which questions are asked, and how to study and gain knowledge of what will be tested, will help the candidate correctly answer the questions.

3-1 To assist in testing a core banking system being acquired, an organization has provided the vendor with sensitive data from its existing production system. An IS auditor's **PRIMARY** concern is that the data should be:

A. sanitized.
B. complete.
C. representative.
D. current.

3-2 Which of the following is the **PRIMARY** purpose for conducting parallel testing?

A. To determine whether the system is cost-effective
B. To enable comprehensive unit and system testing
C. To highlight errors in the program interfaces with files
D. To ensure the new system meets user requirements

3-3 When conducting a review of business process reengineering (BPR), an IS auditor found that a key preventive control had been removed. In this case the IS auditor should:

A. inform management of the finding and determine whether management is willing to accept the potential material risk of not having that preventive control.
B. determine if a detective control has replaced the preventive control during the process and, if it has not, report the removal of the preventive control.
C. recommend that this and all control procedures that existed before the process was reengineered be included in the new process.
D. develop a continuous audit approach to monitor the effects of the removal of the preventive control.

3-4 Which of the following data validation edits is effective in detecting transposition and transcription errors?

A. Range check
B. Check digit
C. Validity check
D. Duplicate check

3-5 Which of the following weaknesses would be considered the **MOST** serious in enterprise resource planning (ERP) software used by a bank?

A. Access controls have not been reviewed.
B. Limited documentation is available.
C. Two-year-old backup tapes have not been replaced.
D. Database backups are performed once a day.

3-6 When auditing the requirements phase of a software acquisition, the IS auditor should:

A. assess the feasibility of the project timetable.
B. assess the vendor's proposed quality processes.
C. ensure that the best software package is acquired.
D. review the completeness of the specifications.

3-7 An organization decides to purchase a software package instead of developing it. In such a case, the design and development phases of a traditional system development life cycle (SDLC) would be replaced with:

A. selection and configuration phases.
B. feasibility and requirements phases.
C. implementation and testing phases.
D. nothing; replacement is not required.

3-8 User specifications for a project using the traditional system development life cycle (SDLC) methodology have not been met. An IS auditor looking for a cause should look in which of the following areas?

A. Quality assurance
B. Requirements
C. Development
D. User training

3-9 When introducing thin client architecture, which of the following types of risk regarding servers is significantly increased?

A. Integrity
B. Concurrency
C. Confidentiality
D. Availability

3-10 Which of the following procedures should be implemented to help ensure the completeness of inbound transactions via electronic data interchange (EDI)?

A. Segment counts built into the transaction set trailer
B. A log of the number of messages received, periodically verified with the transaction originator
C. An electronic audit trail for accountability and tracking
D. Matching acknowledgment transactions received to the log of EDI messages sent

ANSWERS TO SELF-ASSESSMENT QUESTIONS

3-1
A. Test data should be sanitized to prevent sensitive data from leaking to unauthorized persons.
B. Although it is important that the data set be complete, the primary concern is that test data should be sanitized to prevent sensitive data from leaking to unauthorized persons.
C. Although it is important to encompass a representation of the transactional data, the primary concern is that test data should be sanitized to prevent sensitive data from leaking to unauthorized persons.
D. Although it is important that the data set represent current data being processed, the primary concern is that test data should be sanitized to prevent sensitive data from leaking to unauthorized persons.

3-2
A. Parallel testing may show that the old system is more cost-effective than the new system, but this is not the primary reason.
B. Unit and system testing are completed before parallel testing.
C. Program interfaces with files are tested for errors during system testing.
D. The purpose of parallel testing is to ensure that the implementation of a new system will meet user requirements.

3-3
A. Management should be informed immediately to determine whether they are willing to accept the potential material risk of not having that preventive control in place.
B. The existence of a detective control instead of a preventive control usually increases the risk that a material problem may occur.
C. Often during business process reengineering (BPR), many nonvalue-added controls will be eliminated. This is good, unless they increase the business and financial risk.
D. The IS auditor may wish to monitor or recommend that management monitor the new process, but this should be done only after management has been informed and accepts the risk of not having the preventive control in place.

3-4
A. A range check is checking data that matches a predetermined range of values.
B. A check digit is a numeric value that is calculated mathematically and is appended to data to ensure that the original data have not been altered (e.g., an incorrect, but valid, value substituted for the original). This control is effective in detecting transposition and transcription errors.
C. Avalidity check is programmed checking of the data validity in accordance with predetermined criteria.
D. In a duplicate check, new or fresh transactions are matched to those previously entered to ensure that they are not already in the system.

3-5
A. A lack of review of access controls in a financial organization could have serious consequences.
B. A lack of documentation may not be as serious as not having reviewed access controls.
C. It may not even be possible to retrieve data from two-year-old backup tapes.
D. It may be acceptable to the business to perform database backups once a day, depending on the volume of transactions.

3-6
A. A project timetable normally would not be found in a requirements document.
B. Assessing the vendor's quality processes would come after the requirements have been completed.
C. The decision to purchase a package from a vendor would come after the requirements have been completed.
D. The purpose of the requirements phase is to specify the functionality of the proposed system; therefore, the IS auditor would concentrate on the completeness of the specifications.

3-7
A. With purchase packages becoming more common, the design and development phases of the traditional life cycle have become replaceable with selection and configuration phases. A request for proposal (RFP) from the supplier of packaged systems is called for and evaluated against predefined criteria for selection, before a decision is made to purchase the software. Thereafter, it is configured to meet the organization's requirement.
B. The other phases of the system development life cycle (SDLC) such as feasibility study, requirements definition, implementation and postimplementation remain unaltered.
C. The other phases of the SDLC such as feasibility study, requirements definition, implementation and postimplementation remain unaltered.
D. In this scenario, the design and development phases of the traditional life cycle have become replaceable with selection and configuration phases.

3-8

A. Quality assurance has its focus on formal aspects of software development such as adhering to coding standards or a specific development methodology.

B. To fail at user specifications implies that requirements engineering has been done to describe the users' demands. Otherwise, there would not be a baseline of specifications to check against.

C. Obviously, project management has failed to either set up or verify controls that provide for software or software modules under development that adhere to those specifications made by the users. Functionality issues in terms of business usability are out of scope and are, in this case, part of the development phase.

D. Depending on the chosen approach (traditional waterfall, rapid application development, etc.), these discrepancies might show up during user training or acceptance testing—whatever occurs first.

3-9

A. Because the other elements do not need to change, the integrity risk is not increased.

B. Because the other elements do not need to change, the concurrency risk is not increased.

C. Because the other elements do not need to change, the confidentiality risk is not increased.

D. The main change when using thin client architecture is making the servers critical to the operation; therefore, the probability that one of them fails is increased, and as a result, the availability risk is increased.

3-10

A. Control totals built into the trailer record of each segment is the only option that will ensure all individual transactions sent are received completely.

B. A log of the number of messages received provides supporting evidence, but their findings are either incomplete or not timely.

C. An electronic audit trail provides supporting evidence, but their findings are either incomplete or not timely.

D. Matching acknowledgment transactions received to the log of electronic data interchange (EDI) messages sent provides supporting evidence, but their findings are either incomplete or not timely.

Section Two: Content

3.1 QUICK REFERENCE

Quick Reference Review

Chapter 3 addresses the need for systems and infrastructure life cycle management. For an IS auditor to provide assurance that an enterprise's objectives are being met by the management practices of its systems and infrastructure, it is important to understand how an organization evaluates, develops, implements, maintains and disposes of its IT systems and related components. CISA candidates should have a sound understanding of the following items, not only within the context of the present chapter, but also to correctly address questions in related subject areas. It is important to keep in mind that it is not enough to know these concepts from a definitional perspective. The CISA candidate must also be able to identify which elements may represent the greatest risk and which controls are most effective at mitigating this risk. Examples of key topics in this chapter include the following:

- The management practice of benefits realization is the process by which an organization evaluates technology solutions to business problems. A feasibility study scopes the problem, outlines possible solutions and makes a recommendation. A business case is normally derived from the feasibility study and contains information for the decision-making process on whether it should be undertaken; if so, it becomes the basis for a project.
- A project portfolio is defined as all of the projects (related or unrelated) being carried out in an organization at a given point in time. A program can be seen as a group of projects that are linked together through common objectives, strategies, budgets and schedules. Portfolios, programs and projects are controlled by a project management office, which governs the processes of project management but are not typically involved in the management of their content.
- A project may be initiated from within any part of the organization, including the IT department. A project is time bound, with specific start and end dates, a specific objective and a set of deliverables. Throughout the project life cycle, the business case should be reviewed at predefined decision points. The project management processes of project initiation, planning, execution, controlling and project closing are management practices that outline how an organization works to implement projects.
- Using object and work breakdown structures, the project activities are identified and sized using a variety of techniques, such as estimating lines of code (LOC) to function and feature point analysis. The project manager then determines the optimal schedule by using the critical path method. The schedule can be depicted in several forms, such as the Gantt chart or Program Evaluation Review Technique (PERT) diagrams.
- A project must take into account the duration, costs and deliverables (quality) of the project. These elements are managed throughout the controlling and execution phases of the project. At project closing, a postproject review is conducted to assess lessons learned of the project management process.
- The management practices of specific IT processes are used for managing and controlling IT resources and activities. These IT processes form a system development life cycle (SDLC). A project may follow various forms of an SDLC, such as the classical waterfall method, the iterative method, the agile and verification and validation development models or other alternative methods of development, testing and implementation.
- The major phases of an SDLC include the release planning phase, definition phase, development phase, validation phase and deployment phase throughout which feasibility study, requirements definition, design, development, testing, training, implementation, certification and postimplementation review will be accomplished. The phases for each project depend on whether a solution is developed or acquired.

Quick Reference Review *(cont.)*

- Development processes may differ based on the use of system development tools and productivity aids including code generators, computer-aided software engineering (CASE) applications and collaborative code sharing and messaging services along with evolving fourth-generation languages and related utilities.
- It is essential to identify risk to the project. Business risk includes the new system not meeting the users' business requirements, whereas project risk relates to activities, such as the project exceeding the limits of the budget or time frame commonly caused by requirements and related scope creep.
- As part of its IT processes, an organization uses management practices to identify and analyze the needs for IT infrastructure. The physical architecture analysis, the definition of a new architecture and the necessary road map to move from one architecture to another is a critical task for an IT department.
- As an organization's systems and infrastructure change, management practices are used to manage that change (change management, release management and configuration management). Change management processes include the governance on how a change is submitted, assigned, tested, documented and approved.
- The IS auditor should understand the systems development, acquisition and maintenance methodology in use and to identify potential vulnerabilities and points requiring control. It is the IS auditor's role to advise the project team and senior management of the deficiencies and best practices within each of these processes.
- Other key impacts and influencers on an organization's management practices for its systems and architecture include e-commerce, business intelligence and cloud computing driven by market, industry and regulatory factors.
- In IT practices, input control procedures are evaluated to determine that they permit only valid, authorized and unique transactions. In an integrated application environment, controls are embedded and designed into the application that supports the processes. Business process control assurance involves evaluating controls at the process and activity level. These controls may be a combination of programmed and manual controls.

3.2 BENEFITS REALIZATION

The objective of benefits realization is to ensure that IT and the business fulfill their value management responsibilities, particularly that:

- IT-enabled business investments achieve the promised benefits and deliver measurable business value.
- Required capabilities (solutions and services) are delivered:
  - On time, both with respect to schedule and time-sensitive market, industry and regulatory requirements
  - Within budget
- IT services and other IT assets continue to contribute to business value.

The premise of benefits realization is that there is strong concern at board and senior management levels that the high expenditures on IT-related initiatives are not realizing the business benefits they promise. Studies and surveys also indicate high levels of loss from ill-planned and ill-executed initiatives. The focus on value has become more prevalent while the capability maturity of value management practices in most enterprises has remained low.

Benefits realization of projects is a compromise among major factors such as cost, quality, development/delivery time, reliability and dependability. Strategy-makers perform a

comprehensive study and evaluate which factors are "qualifying" or "winning" and then compare those factors with strengths, weaknesses and competencies of services available to complete and maintain systems. Most large organizations employ structured project management principles to support changes to their information systems environment. As a starting point, the IS auditor should understand how the business defines value or a return on investment (ROI) for development-related projects. If companies fail to consistently meet their ROI objectives, this may suggest weakness in their system development life cycle (SDLC) and related project management practices.

Example 1

Note: This is not an exhaustive example, but an illustrative example to explain the concept. Other factors may be applicable to each situation.

An entity is planning to invest in an application that would enable customers to manager their orders online. The following would be relevant for the ROI calculation:

A. Costs
1. Cost of developing the application
2. Cost of controls to ensure integrity of data at rest and in process, while ensuring nonrepudiation.

B. Benefits
1. Increase in operating profits attributable to expected spike in business driven by customer satisfaction (percent of revenue)
2. Reduction in operating costs (in terms of dedicated personnel who previously interacted with customers and executed changes)

ROI may be measured as value of benefit/costs, which then can be compared with the entity's cost of funds, to make a go/no-go decision. This ROI framework can then be used as a benchmark to evaluate the progress of the project and identify causes, if the actual ROI is not comparing with the planned ROI.

3.2.1 PORTFOLIO/PROGRAM MANAGEMENT

A program can be seen as a group of projects and time-bound tasks that are closely linked together through common objectives, a common budget, and intertwined schedules and strategies. Like projects, programs have a limited time frame (i.e., a defined start and end date) and organizational boundaries. A differentiator is that programs are more complex, usually have a longer duration, a higher budget and higher risk associated with them, and are of higher strategic importance.

A typical IS-related program may be the implementation of a large-scale enterprise resource planning (ERP) system that includes projects that address technology infrastructure, operations, organizational realignment, business process reengineering (BPR) and optimization, training, and development. Mergers and acquisitions (M&As) may serve as an example of a non–IS-related program that impacts both the gaining and/or divesting organizations' IS architectures and system, organizational structure, and business processes.

The objective of program management is the successful execution of programs including, but not limited to, management of:
- Program scope, program financials (costs, resources, cash flow, etc.), program schedules, and program objectives and deliverables
- Program context and environment
- Program communication and culture
- Program organization

To make autonomous projects possible while making use of synergies between related projects in the program, a specific program organization is required. Typical program roles are the program owner, the program manager and the program team. The program owner role is distinct from the project owner role. Typical communication structures in a program are program owner's meetings and program team meetings.

Methodology and processes used in program management are very similar to those in project management and run in parallel to each other. However, they must not be combined and have to be handled and carried out separately.

To formally start a program, some form of written assignment from the program sponsor (owner) to the program manager and the program team is required. Because programs most often emerge from projects, such an assignment is of paramount importance to set the program context and boundaries as well as formal management authority.

A project portfolio is defined as all the projects being carried out in an organization at a given point in time (snapshot). In contrast to program management in which all relevant projects are closely coupled, this is not a requirement in a project portfolio. Projects of a program belong to the company's project portfolio as do projects that are not associated with a program.

To manage portfolios, programs and projects, an organization requires specific and well-designed structures such as expert pools, a project management office and project portfolio groups. Specific integrative tools such as project management guidelines, standard project plans and project management marketing instruments are employed.

The project management office, as an owner of the project management and program management process, must be a permanent structure and adequately staffed to provide professional support in these areas to maintain current and develop new procedures and standards.

The objective of the project management office is to improve project and program management quality and secure project success, but it can only focus on activities and tasks and not on project or program content. Therefore, an IS auditor has to differentiate between auditing project content and/or procedural aspects of a program or project.

The objectives of project portfolio management are:
- Optimization of the results of the project portfolio (not of the individual projects)
- Prioritizing and scheduling projects

- Resource coordination (internal and external)
- Knowledge transfer throughout the projects

A project portfolio database is mandatory for project portfolio management. It must include project data such as owner, schedules, objectives, project type, status, cost, etc. Project portfolio management requires specific project portfolio reports. Typical project portfolio reports are a project portfolio bar chart, a profit versus risk matrix, a project portfolio progress graph, etc.

Example 2

An entity is carrying out the following activities:

A. An entity is migrating from legacy applications to an ERP system and has the strategic goal of delivering cutting-edge computers and maintaining high cash flow to continue to fund research and development.
   1. The entity is using its internal pool of application developers to code its strategic business process of the manufacture of newly designed computers to deliver finished goods and sales order to cash receipts considering the sensitivity of its business model.
   2. The entity is using vendors to code the non-strategic business processes of procure to pay and financial accounting.

B. The entity is also pursuing a program for outsourcing transactional processes to a third-party service provider for online sales and a payment portal.

In this context, activities A.1, A.2 and B, individually, are projects. Activities A.1 and A.2 represent a single program, as they are part of the single larger activity of migrating from legacy applications to ERP, and activity B is part of another larger program to outsource non-core manufacturing processes. Activities A.1, A.2 and B (assuming these are the only activities underway in the entity) represent the portfolio for the entity.

3.2.2 BUSINESS CASE DEVELOPMENT AND APPROVAL

An important consideration in any IT project, whether it is the development of a new system or an investment in new infrastructure, is the business case. It has been increasingly recognized that the achievement of business benefits should drive projects.

A business case provides the information required for an organization to decide whether a project should proceed. Depending on the organization and often on the size of the investment, the development of a business case is either the first step in a project or a precursor to the commencement of a project.

The initial business case would normally derive from a feasibility study undertaken as part of project initiation/planning. This is an early study of a problem to assess if a solution is practical, meets requirements within established budgets and schedule requirements. The feasibility study will normally include the following six elements:

1. The **project scope** defines the business problem and/or opportunity to be addressed. It should be clear, concise and to the point.
2. The **current analysis** defines and establishes an understanding of a system, a software product, etc. Based on this analysis, it may be determined that the current system or software product is working correctly, some minor modifications are needed, or a complete upgrade or replacement is required. At this point in the process, the strengths and weaknesses of the current system or software product are identified.
3. **Requirements** are defined based upon stakeholder needs and constraints. Defining requirements for software differs from defining requirements for systems. The following are examples of needs and constraints used to define requirements:
   - Business, contractual and regulatory processes
   - End-user functional needs
   - Technical and physical attributes defining operational and engineering parameters
4. The **approach** is the recommended system and/or software solution to satisfy the requirements. This step clearly identifies the alternatives that were considered and the rationale as to why the preferred solution was selected. This is the process wherein the use of existing structures and commercial alternatives are considered (e.g., "build versus buy" decisions).
5. **Evaluation** is based upon the previously completed elements within the feasibility study. The final report addresses the cost-effectiveness of the approach selected. Elements of the final report include:
   - The estimated total cost of the project if the preferred solution is selected along with the alternates to provide a cost comparison, including:
     - Estimate of employee hours required to complete
     - Material and facility costs
     - Vendors and third-party contractors costs
     - Project schedule start and end dates
     - A cost and evaluation summary encompassing cost-benefit analysis, ROI, etc.
6. A formal **review** of the feasibility study report is conducted with all stakeholders. This review will both validate the completeness and accuracy of the feasibility study and render a decision to either approve or reject the project or ask for corrections before making a final decision. If the feasibility study is approved, all key stakeholders sign the document. Rationale for rejection of the feasibility study should be explained and attached to the document as part of a lessons learned list for use in future project studies.

Part of the work in developing options is to calculate and outline a business case for each solution in order to allow a comparison of costs and business benefits.

The business case should be of sufficient detail to describe the justification for setting up and continuing a project. The business case should provide the reasons for the project and answer the question, "Why should this project be undertaken?"

The business case should also be a key element of the decision process throughout the life cycle of any project. If at any stage the business case is thought to be no longer be valid, through increased costs or reduction in the anticipated benefits, the project sponsor or IT steering committee should consider whether the project should proceed. In a well-planned project, there will be decision points, sometimes called "stage gates" or "kill points," at which the business case is formally reviewed to ensure

that it is still valid. If the business case changes during the course of an IT project, the project should be reapproved through the departmental planning and approval process.

3.2.3 BENEFITS REALIZATION TECHNIQUES

COBIT 5 provides the industry accepted framework under which IT governance goals and objectives are derived from stakeholder drivers with the intent of enterprise IT generating business value from IT-enabled investments (i.e., achieve strategic goals and realize business benefits through effective and innovative use of IT). As indicated in **figure 3.2**, principle 1, Meeting Stakeholder Needs, stresses the need for enterprise IT to create value for stakeholders by maintaining a balance between the realization of benefits and the optimization of risk and use of resources. COBIT 5 addresses the required processes and other enablers to support business value creation through the use of IT. Because every enterprise has different objectives, an enterprise can customize COBIT 5 to suit its own context through the goals cascade—translating high-level enterprise goals into manageable, specific, IT-related goals and mapping these to specific processes and practices.

Figure 3.2—COBIT 5 Principles

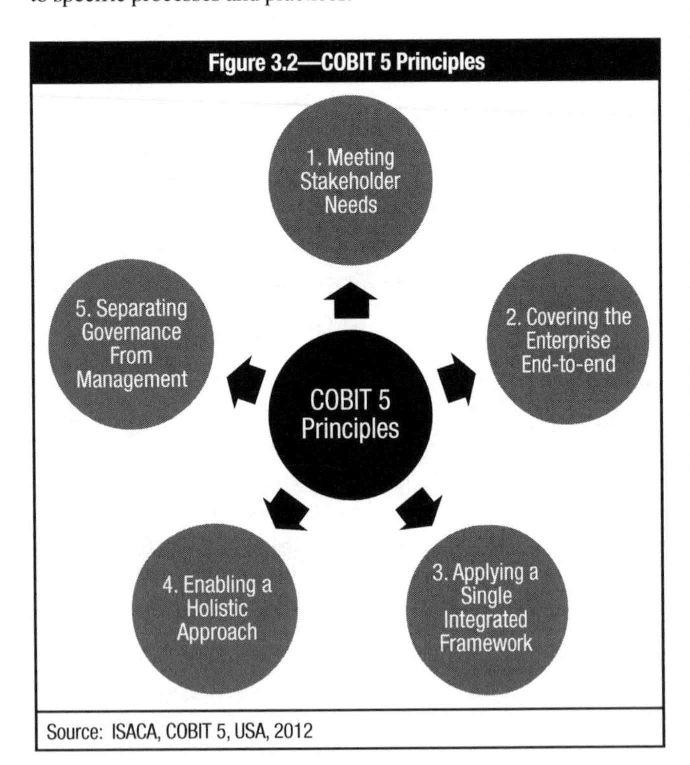

Source: ISACA, COBIT 5, USA, 2012

The majority of business benefits are obtained through changes enabled by technology. With the evolution of the application of IT from straight automation of work, through information management to business transformation, the realization of benefits has become a more complex task. Benefits do not just happen when the new technology is delivered; they occur throughout the business cycle.

A planned approach to benefits realization is required, looking beyond project cycles to longer term cycles that consider the total benefits and total costs throughout the life of the new system. Benefits rarely come about exactly as envisioned in plans.

Organizations have to keep checking and adjusting strategies. This concept is called benefits management or benefits realization, and key elements are as follows:
- Describing benefits management or benefits realization
- Assigning a measure and target
- Establishing a tracking/measuring regimen
- Documenting the assumption
- Establishing key responsibilities for realization
- Validating the benefits predicted in the business
- Planning the benefit that is to be realized

Generally, benefits realization at the project level encompass four phases, as indicated in the **figure 3.3**.

Figure 3.3—Benefits Realization Phases—Project Level

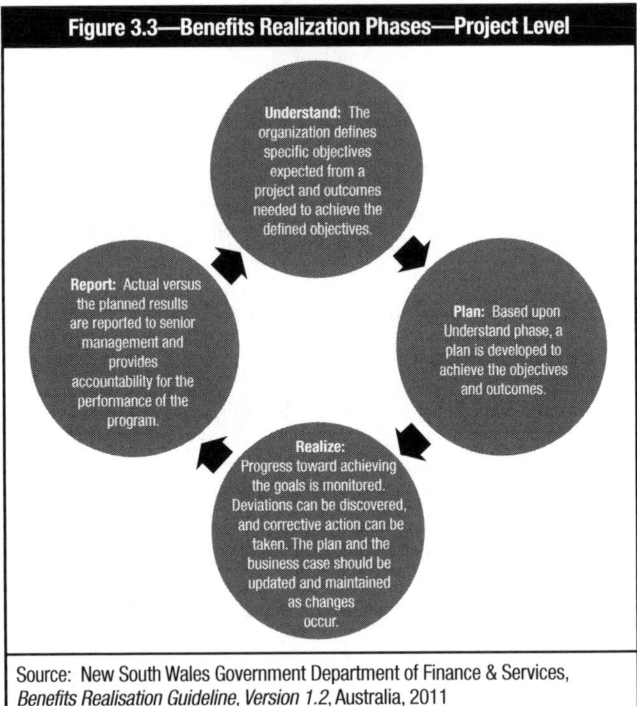

Understand: The organization defines specific objectives expected from a project and outcomes needed to achieve the defined objectives.
Plan: Based upon Understand phase, a plan is developed to achieve the objectives and outcomes.
Realize: Progress toward achieving the goals is monitored. Deviations can be discovered, and corrective action can be taken. The plan and the business case should be updated and maintained as changes occur.
Report: Actual versus the planned results are reported to senior management and provides accountability for the performance of the program.

Source: New South Wales Government Department of Finance & Services, *Benefits Realisation Guideline, Version 1.2,* Australia, 2011

Benefits realization is a continuous process that must be managed just like any business process. Assessment of the benefits realization processes and the business case should be a key element of benefits realization processes. Enterprisewide benefits realization studies should be aggregated and lessons learned should fine-tune the enterprise benefit realization process.

Benefits realization often includes a postimplementation review six to eighteen months after the implementation of systems. Time must be allowed for initial technical problems to be resolved and for the project benefits to accrue as users become familiar with the new processes and procedures.

Benefits realization must be part of governance and management of projects. There must be business sponsorship of projects. Project governance structures should involve the project and the functional line organization, all governance decisions about the project should be driven through the business case, and there must be periodic review of benefits.

3.3 PROJECT MANAGEMENT STRUCTURE

Today, many approaches to project management exist. Some are focused on software development, others have a more general approach; some concentrate on a holistic and systemic view, others provide a very detailed workflow including templates for document creation. Some of the most prominent *de facto* standards and organizations are the Project Management Body of Knowledge (PMBOK®) (i.e., IEEE standard 1490 from the Project Management Institute [PMI], Projects in a Controlled Environment [PRINCE2™] from the Office of Government Commerce (OGC) in the UK, and the International Project Management Association [IPMA]). Because there are significant differences in scope, content and wording in each of these standards, an IS auditor has to become familiar with the standard in use prior to involvement in specific projects.

Although each project management approach has its own pros and cons, several elements are common across all project management methodologies.

Project management structures are dependent on the size of the organization and complexity of business/operations. Accordingly, some roles and responsibilities may be grouped or replaced. Under such circumstances, the role of the IS auditor is to ensure that rules of system development, as they relate to segregation of duties (SoD) and responsibilities, are not compromised.

3.3.1 GENERAL ASPECTS

IS projects may be initiated from within any part of the organization, including the IT department.

A project is always a time-bound effort. A project can be complex and involve an element of risk. A project has specific objectives, deliverables and start and end dates. Most IS projects are divisible into explicit phases (e.g., SDLC). A project can be perceived as a group of complex tasks as well as a social system and/or a temporary organization (in contrast to the relatively stable structures of the permanent organization).

Project management is a business process in a project-oriented organization. The project management process begins with the project charter and ends with the completion of the project.

The complexity of project management requires a careful and explicit design of the project management process, which is normally done for a business process but sometimes neglected for the project management process. Thus, all design issues applicable for business process engineering should also be applied for the project management process.

3.3.2 PROJECT CONTEXT AND ENVIRONMENT

A project context can be divided into a time and a social context. In the analysis of the content's dimension of the context, the following must be taken into account:
- Importance of the project in the organization
- Connection between the organization's strategy and the project
- Relationship between the project and other projects within the same program as well as other projects under different program(s)
- Connection between the project and the underlying business case

Because there are normally several projects running at the same time, the relationships between those projects have to be investigated to identify common objectives for the business organization, identify and manage risk, and identify resource connections. A common approach to address these issues is to establish a project portfolio management and/or a program management structure.

A project, by definition, has a specific start date and end date. The project's time context, however, should consider the phase before project startup and the phase after project closing. During the preproject phase, key activities and objectives that must be considered in project planning are identified. A thorough transfer of information about the negotiations and decisions as well as necessary knowledge from the preproject phase is critical. The expectations in regard to the postproject phase also influence the tasks to be fulfilled and the strategy for designing the project environment relationships.

Because a project represents a social system, it is also necessary to consider its relationships to its own social environments. The design of the project environment relationships is a project management activity. The objective is to determine all relevant environments for the project, which will have a significant influence on overall project planning and project success.

3.3.3 PROJECT ORGANIZATIONAL FORMS

Three major forms of organizational alignment for project management within the business organization can be observed:
1. Influence project organization
2. Pure project organization
3. Matrix project organization

In influence project organization, the project manager has only a staff function without formal management authority. The project manager is only allowed to advise peers and team members as to which activities should be completed.

In a pure project organization, the project manager has formal authority over those taking part in the project. Often, this is bolstered by providing a special working area for the project team that is separated from their normal office space.

In a matrix project organization, management authority is shared between the project manager and the department heads.

For an IS auditor, it is important to understand these organizational forms and their implications on controls in project management activities.

Requests for major projects should be submitted to and prioritized by the IT steering committee. A project manager should be identified and appointed by the IT steering committee. The project manager, who need not be an IT staff member, should be given complete operational control over the project and be

allocated the appropriate resources, including IS professionals and other staff from user departments, for the successful completion of the project. IS auditors may be included in the project team as control experts. They may also provide an independent, objective review to ensure that the level of involvement (commitment) of the responsible parties is appropriate. In such cases, IS auditors are not performing an audit, but are participating on the project in an advisory role; depending on the level of their involvement, they may become ineligible to perform audits of the application when it becomes operational. An example of a project's organizational structure is shown in **figure 3.4**.

3.3.4 PROJECT COMMUNICATION AND CULTURE

Depending on the size and complexity of the project and the affected parties, communication when initiating the project management process may be achieved by:

- One-on-one meetings
- Kick-off meetings
- Project start workshops
- A combination of the three

One-on-one meetings and a project start workshop help to facilitate two-way communication between the project team members and the project manager. A kick-off meeting may be used by the project manager to inform the team of what has to be done for the project. Communications involving significant project events should be documented as part of the project artifacts (i.e., project charter meeting, kick-off meeting, gate reviews, stakeholder meetings, etc.).

A preferred method to ensure that communication is open and clear among the project team is to use a project start workshop to obtain cooperation from all team members and buy-in from stakeholders. This helps develop a common overview of the project and communicates the project culture early in the project.

As an independent social system, each project has its own culture that defines its norms and rules of engagement. The project culture cannot be described but manifests itself in the applied project management techniques, including project planning, forms of communication, etc. The dynamics of the project culture provide for a common understanding of what is seen as desirable and serves as an orientation for the team.

A project culture is comprised of shared norms, beliefs, values and assumptions of the project team. A key success factor for establishing the correct project culture is defining and adapting the unique characteristics of a project, which enables the correct culture to flourish that is a match to complexity of that project.

Methods for developing a project culture include the establishment of a project mission statement, project name and logo, project office or meeting place, project intranet, project team meeting rules and communication protocol, and project-specific social events.

3.3.5 PROJECT OBJECTIVES

Project objectives are the specific action statements that support the road map to obtain established project goals. All project goals will have one or more objectives identified as the actions needed to reach that goal. The project objective is the means to meet the project goal. When a project objective is defined, it always starts with an action verb. This syntax ensures that the objective is measurable and can be used to verify that the project outcome is directly traceable to the action of the objective.

A project needs clearly defined results that are specific, measurable, attainable, realistic and timely (SMART). A comprehensive project view ensures the consideration and consolidation of all closely coupled objectives. These objectives are broken down into main objectives, additional objectives and nonobjectives.

- **Main objectives** are the primary reason for the project and will always be directly coupled with business success (see section 3.5 Business Application Development).
- **Additional objectives** are objectives that are not directly related to the main results of the project but may contribute to project success (e.g., business unit reorganization in a software development project).

Figure 3.4—Project Management, Sample Organization Chart

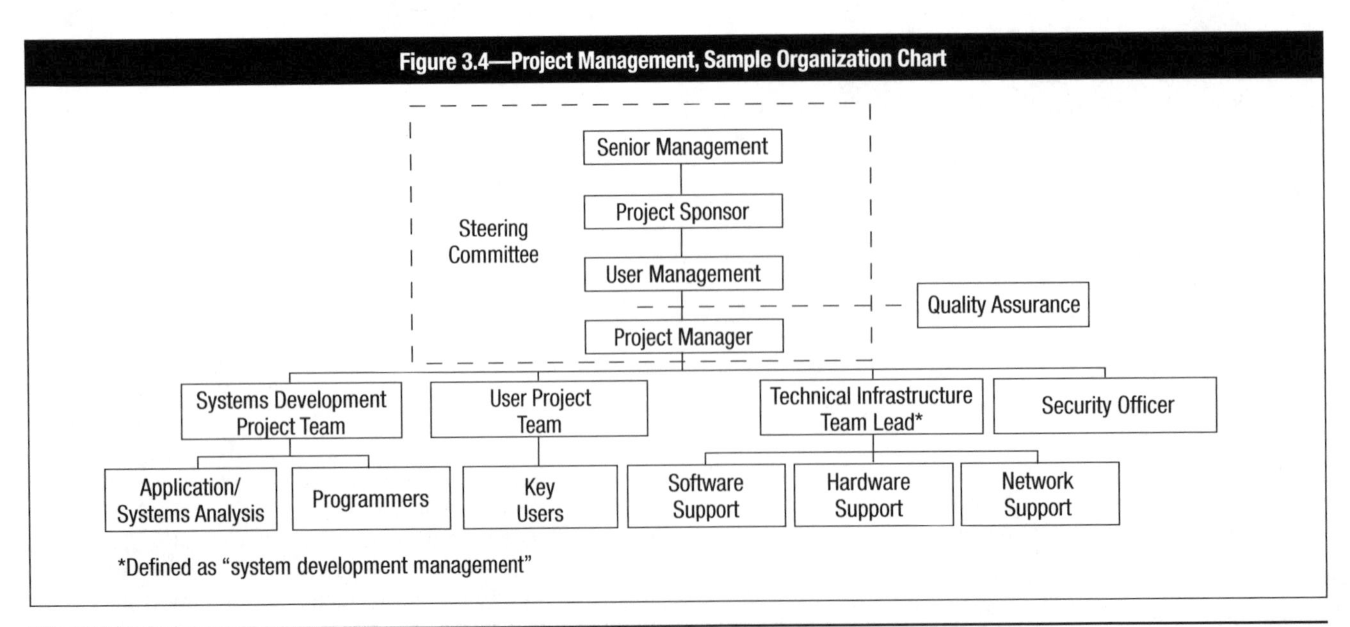

- **Nonobjectives** are the results that are not to be expected on completion of the project. Defining the nonobjectives brings clarity to scope, and project boundaries become clearer. Nonobjectives shape the confines of the project deliverables and help all parties to gain a clear understanding of what results are within expectations and clearly not within scope of the project to avoid any ambiguities.

A commonly accepted approach to define project objectives is to start off with an object breakdown structure (OBS). It represents the individual components of the solution and their relationships to each other in a hierarchical manner, either graphically or in a table. An OBS can help, especially when dealing with nontangible project results such as organizational development, to ensure that a material deliverable is not overlooked.

After the OBS has been compiled or a solution is defined, a work breakdown structure (WBS) is designed to structure all the tasks that are necessary to build up the elements of the OBS during the project. The WBS represents the project in terms of manageable and controllable units of work, serves as a central communications tool in the project, and forms the baseline for cost and resource planning.

In contrast to the OBS, the WBS does not include basic elements of the solution to build but shows individual work packages (WPs) instead. The structuring of the WBS is process-oriented and in phases. The level of detail of the WBS serves as the basis for the negotiations of detailed objectives between the project sponsor, project manager and project team members.

Figure 3.5 shows an example of this process.

Detailed specifications regarding the WBS can be found in WPs. Each WP must have a distinct owner and a list of main objectives, and may have a list of additional objectives and nonobjectives. The WP specifications should include dependencies on other WPs and a definition of how to evaluate performance and goal achievement. An example of a WBS is shown in **figure 3.6.**

Figure 3.5—Defining Project Objectives

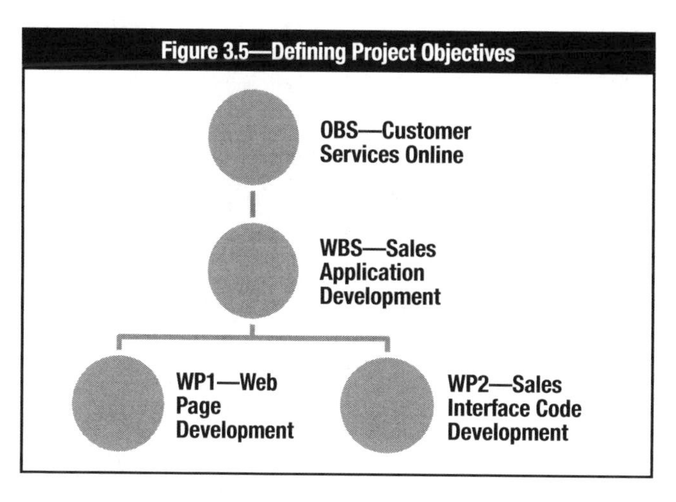

Key things to remember with WBS and respective WPs include:
- The top WBS level represents the final deliverable or project.
- Sub-deliverables contain work packages that are assigned to an organization's department or unit.
- All elements of the WBS do not need to be defined to the same level.
- WPs define the work, duration, and costs for the tasks required to produce the sub-deliverable.
- WPs should not exceed duration of 10 days.
- WPs need to be independent of each other in the WBS.
- WPs are unique; do not duplicate a WP across the WBS.

To support communications, task lists are often used. A task list is a list of actions to be carried out in relation to work packages and includes assigned responsibilities and deadlines. The task list aids the individual project team members in operational planning and in making agreements.

These task lists are most typically compiled into a project schedule at the planning phase of a project and are used in the controlling phase of the project to monitor and track the progress and completion of the WPs. Project schedules are living documents and should indicate the tasks for a WP, the start and finish dates, percentage completed, task dependencies and

Figure 3.6—Sample Work Breakdown Structure

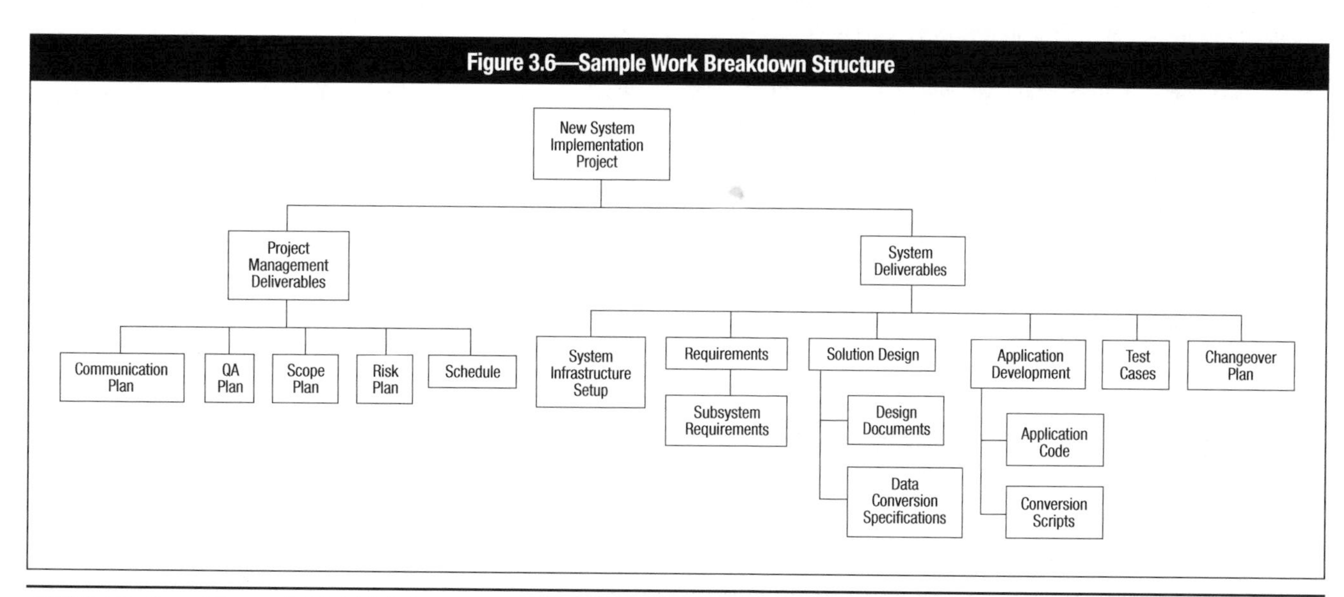

resource names of individuals planned to work on those tasks. A project schedule will also indicate the stage boundaries explained in section 3.2.2 Business Case Development and Approval.

3.3.6 ROLES AND RESPONSIBILITIES OF GROUPS AND INDIVIDUALS

To achieve a successful completion and implementation of any new system, it is advisable that the audit function has an active part, where appropriate, in the life cycle development of the business application. This will facilitate efforts to ensure that proper controls are designed and implemented in the new system (e.g., continuous concurrent controls for paperless e-commerce systems).

Additionally, there are other key stakeholders who should be involved in the system's design, development and implementation. The various roles and responsibilities of groups/individuals that may be involved in the development process are summarized as follows:

- **Senior management**—Demonstrates commitment to the project and approves the necessary resources to complete the project. This commitment from senior management helps ensure involvement by those needed to complete the project.
- **User management**—Assumes ownership of the project and resulting system, allocates qualified representatives to the team, and actively participates in business process redesign, system requirements definition, test case development, acceptance testing and user training. User management should review and approve system deliverables as they are defined and implemented. User management is concerned particularly with the following questions:
  - Are the required functions available in the software?
  - How reliable is the software?
  - How efficient is the software?
  - Is the software easy to use?
  - How easy is it to transfer or adapt old data from preexisting software to this environment?
  - How easy is it to transfer the software to another environment?
  - Is it possible to add new functions?
  - Does it meet regulatory requirements?
- **Project steering committee**—Provides overall direction and ensures appropriate representation of the major stakeholders in the project's outcome. The project steering committee is ultimately responsible for all deliverables, project costs and schedules. This committee should be comprised of a senior representative from each business area that will be significantly impacted by the proposed new system or system modification. Each member must have the authority to make decisions related to system designs that will affect their respective departments. Generally, a project sponsor who would assume the overall ownership and accountability of the project will chair the steering committee. The project manager should also be a member of this committee. The project steering committee performs the following functions:
  - Reviews project progress regularly (for example, semimonthly or monthly) and holds emergency meetings when required.
  - Serves as coordinator and advisor. Members of the committee should be available to answer questions and make user-related decisions about system and program design.
  - Takes corrective action. The committee should evaluate progress and take action or make recommendations regarding personnel changes on the project team, managing budgets or schedules, changes in project objectives, and the need for redesign. The committee should be available to address risk and issues that are escalated and cannot be resolved at the project level. The project manager should have the ability to escalate such matters and rely on the project steering committee to resolve risk and project issues for the benefit of the organization. In some cases, the committee may recommend that the project be halted or discontinued.
- **Project sponsor**—Provides funding for the project and works closely with the project manager to define the critical success factors (CSFs) and metrics for measuring the success of the project. It is crucial that success is translated to measurable and quantifiable terms. Data and application ownership are assigned to a project sponsor. A project sponsor is typically the senior manager in charge of the primary business unit that the application will support.
- **Systems development management**—Provides technical support for hardware and software environments by developing, installing and operating the requested system. This area also provides assurance that the system is compatible with the organization's computing environment and strategic IT direction, and assumes operating support and maintenance activities after installation.
- **Project manager**—Provides day-to-day management and leadership of the project, ensures that project activities remain in line with the overall direction, ensures appropriate representation of the affected departments, ensures that the project adheres to local standards, ensures that deliverables meet the quality expectations of key stakeholders, resolves interdepartmental conflicts, and monitors and controls costs and the project timetable. The project manager may also facilitate the definition of the scope of the project, manage the budget of the project, and control the activities via a project schedule. This person can be an end user, a member of the systems development team or a professional project manager. Where projects are staffed by personnel dedicated to the project, the project manager will have a line responsibility for such personnel.
- **Systems development project team**—Completes assigned tasks, communicates effectively with users by actively involving them in the development process, works according to local standards and advises the project manager of necessary project plan deviations.
- **User project team**—Completes assigned tasks, communicates effectively with the systems developers by actively involving themselves in the development process as subject matter experts (SMEs), works according to local standards and advises the project manager of expected and actual project plan deviations.
- **Security officer**—Ensures that system controls and supporting processes provide an effective level of protection, based on the data classification set in accordance with corporate security policies and procedures; consults throughout the life cycle on appropriate security measures that should be incorporated into the system; reviews security test plans and reports prior to implementation; evaluates security-related documents developed in reporting the system's security effectiveness for accreditation; and periodically monitors the security system's effectiveness during its operational life.
- **Information system security engineer**—Applies scientific and engineering principles to identify security vulnerabilities and minimize or contain risk associated with these

vulnerabilities (International Organization for Standardization [ISO]/International Electrotechnical Commission [IEC] 21827:2008 specifies the Systems Security Engineering–Capability Maturity Model® [SSE-CMM®]). Key to meeting this role is defining the needs, requirements, architectures and designs to construct network, platform and application constructs according to the principles of both defense-in-breadth and security-in-depth.
- **Quality assurance (QA)**—Reviews results and deliverables within each phase and at the end of each phase and confirms compliance with requirements. The objective of this group is to ensure the quality of the project by measuring the adherence of the project staff to the organization's SDLC, advise on deviations, and propose recommendations for process improvements or greater control points when deviations occur. The points where reviews occur depend on the SDLC methodology used, the structure and magnitude of the system and the impact of potential deviations. Additionally, focus may include a review of appropriate, process-based activities related to either project management or the use of specific software engineering processes within a particular life cycle phase. Such a focus is crucial to completing a project on schedule and within budget and in achieving a given software process maturity level (see section 3.12.4 ISO/IEC 330xx Series). Specific objectives of the QA function include:
  - Ensuring the active and coordinated participation by all relevant parties in the revision, evaluation, dissemination, and application of standards, management guidelines and procedures
  - Ensuring compliance with the agreed-on systems development methodology
  - Reviewing and evaluating large system projects at significant development milestones and making appropriate recommendations for improvement
  - Establishing, enhancing and maintaining a stable, controlled environment for the implementation of changes within the production software environment
  - Defining, establishing and maintaining a standard, consistent and well-defined testing methodology for computer systems
  - Reporting to management on systems that are not performing as defined or designed

It is essential for the IS auditor to understand the systems development, acquisition and maintenance methodology in use and to identify potential vulnerabilities and points requiring control. If controls are lacking (either as a result of the organizational structure or of the software methods used) or the process is disorderly, it is the IS auditor's role to advise the project team and senior management of the deficiencies. It may also be necessary to advise those engaged in development and acquisition activities of appropriate controls or processes to implement and follow.

> **Note:** The CISA candidate should be familiar with general roles and responsibilities of groups or individuals involved in the systems development process.

3.4 PROJECT MANAGEMENT PRACTICES

Project management is the application of knowledge, skills, tools and techniques to a broad range of activities to achieve a stated objective such as meeting the defined user requirements, budget and deadlines for an IS project. Project management knowledge and practices are best described in terms of their component processes of initiating, planning, executing, controlling and closing a project. Overall characteristics of successful project planning are that it is a risk-based management process and iterative in nature. Project management techniques also provide systematic quantitative and qualitative approaches to software size estimating, scheduling, allocating resources and measuring productivity.

There are numerous project management techniques and tools available to assist the project manager in controlling the time and resources utilized in the development of a system. The techniques and tools may vary from a simple manual effort to a more elaborate computerized process. The size and complexity of the project may require different approaches. These tools typically provide assistance in areas described in the following sections.

There are various elements of a project that should always be taken into account. Their relationship is shown in **figure 3.7**.

Project management should pay attention to three key intertwining elements: deliverables, duration and budget (**figure 3.7A**). Their relationship is very complex but is shown in an oversimplified and schematized manner in the figure. Project duration and budget must be commensurate with the nature and characteristics of the deliverables. In general, there will be a positive correlation between highly demanding deliverables, a long duration and a high budget.

Budget is deduced (**figure 3.7B**) from the resources required to carry out the project by multiplying fees or costs by the amount of each resource. Resources required by the project are estimated at the beginning of the project using techniques of software/project size estimation.

Size estimation yields a "total resources" calculation. Project management decides on the resources allocated at any particular moment in time. In general, it is convenient to assign an (almost) fixed amount of resources, thus probably minimizing costs (direct and administration). **Figure 3.7C** is simplified in its assumption that there is a fixed amount of resources during the entire project. The curve shows resources assigned (R) × duration (D) = total resources (TR, a constant quantity); which is the classic "man × month" dilemma curve. Any point along the curve meets the condition R × D = TR. If we choose any point O on the curve, the area of the rectangle will be TR, proportional to the budget. If few resources are used, the project will take a long time (a point close to L_R); if many resources are used, the project will take a shorter time (a point close to L_D). L_R and L_D are two practical limits: a duration that is too long may not seem possible; use of too many (human) resources at once would be unmanageable.

Figure 3.7—Relationships Between Project Management Elements

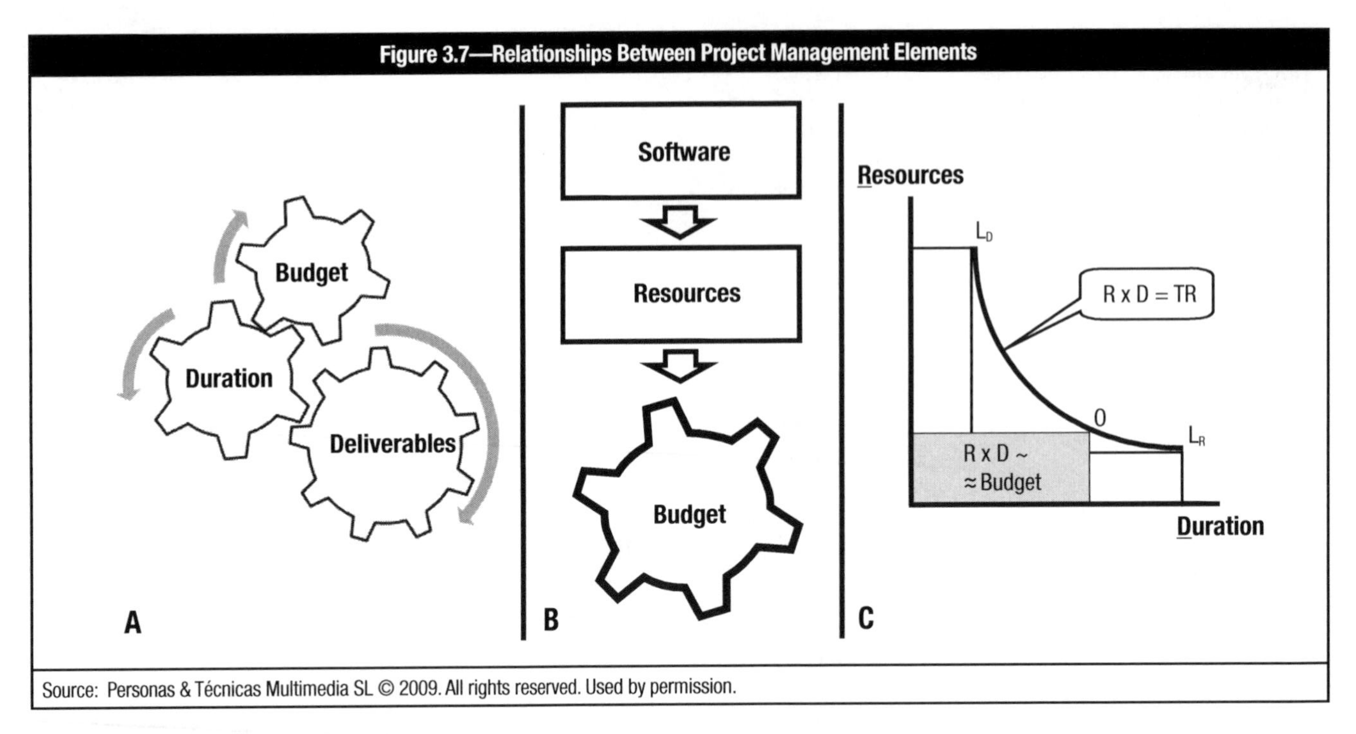

Source: Personas & Técnicas Multimedia SL © 2009. All rights reserved. Used by permission.

3.4.1 INITIATION OF A PROJECT

A project will be initiated by a project manager or sponsor gathering the information required to gain approval for the project to be created. This will often be compiled into terms of reference or a project charter that states the objective of the project, the stakeholders in the system to be produced, and the project manager and sponsor. Approval of a project initiation document (PID) or a project request document (PRD) is the authorization for a project to begin.

3.4.2 PROJECT PLANNING

System and software development/acquisition or maintenance projects have to be planned and controlled.

The project manager needs to determine:

- The various tasks that need to be performed to produce the expected business application system
- The sequence or the order in which these tasks need to be performed
- The duration or the time window for each task
- The priority of each task
- The IT and non-IT supporting resources, which are available and required to perform these tasks
- Budget or costing for each of these tasks
- Source and means of funding for labor, services, materials, and plant and equipment resources involved in the project

Realistically, on other than small projects, detailed task-level planning can only occur toward the end of a phase or iteration for the upcoming phase or iteration, or through to the next decision gate. Until work products and decisions are produced during the current phase or iteration, insufficient information is available to understand in detail what will occur in the next phase/iteration.

There are some techniques that are useful in creating a project plan and monitoring its progress throughout the execution of the project to provide continual support for making it successful. Some of these techniques are addressed in the following sections.

To measure the development effort, metrics are required. The first step is to identify resources (e.g., people with requisite skills, development tools, facilities) for system and software development. This will help in estimating and budgeting system and software development resources. Several different sizing and measurement techniques are discussed below.

System Development Project Cost Estimation

Normally much larger in scope and size, the system development project focuses on a more complete and integrated solution (hardware, software, facilities, services, etc.) Therefore, these types of projects require much greater planning with regard to estimating and budgeting.

Four commonly used methodologies to estimate the cost of a system development project are:

- **Analogous estimating**—By using estimates from prior projects, the project manager can develop the estimated cost for a new project. This is the quickest estimation technique.
- **Parametric estimating**—The project manager looks at the same past data that were used in analogous estimating and leverages statistical data (estimated employee hours, materials costs, technology, etc.) to develop the estimate. This approach is more accurate than analogous estimation.
- **Bottom-up estimating**—In this method, the cost of each activity in the project is estimated to the greatest detail (i.e., starting at the bottom), and then all the costs are added to arrive at the cost estimate of the entire project. While the most accurate estimate, this is the most time-consuming approach.

- **Actual costs**—Like analogous estimation, this approach takes an extrapolation from the actual costs that were incurred on the same system during past projects.

Software Size Estimation

Software size estimation relates to methods of determining the relative physical size of the application software to be developed. Estimates can be used to guide to the allocation of resources and to judge the time and cost required for its development, and to compare the total effort required by the resources. See **figure 3.7**.

Traditionally, software sizing has been performed using single-point estimations (based on a single parameter) such as source lines of code (SLOC). For complex systems, single-point estimation techniques have not worked because they do not support more than one parameter in different types of programs, which in turn affects the cost, schedule and quality metrics. To overcome this limitation, multiple-point estimations have been designed.

Current technologies now take the form of more abstract representations such as diagrams, objects, spreadsheet cells, database queries and graphical user interface (GUI) widgets. These technologies are more closely related to "functionality" deliverables rather than "work" or lines that need to be created.

Function Point Analysis

The function point analysis (FPA) technique has evolved over the years to become a multiple-point technique widely used for estimating complexity in developing large business applications.

The results of FPA are a measure of the size of an information system based on the number and complexity of the inputs, outputs, files, interfaces and queries with which a user sees and interacts. This is an indirect measure of software size and the process by which it is developed versus direct size-oriented measures such as SLOC counts.

Function points (FPs) are computed by first completing a table (**figure 3.8**) to determine whether a particular entry is simple, average or complex. Five FP count values are defined, including the number of user inputs, user outputs, user inquiries, files and external interfaces.

Upon completion of the table entries, the count total in deriving the function point is computed through an algorithm that takes into account complexity adjustment values (i.e., rating factors) based on responses to questions related to issues such as reliability, criticality, complexity, reusability, changeability and portability. Function points derived from this equation are then used in a manner analogous to SLOC counts as a measure for cost, schedule, productivity and quality metrics (e.g., productivity = FP/person-month, quality = defects/FP, and cost = $/FP).

FPA is an indirect measurement of the software size. It is based on the number and complexity of inputs, outputs, files, interfaces and queries.

> **Note:** The CISA candidate should be familiar with the use of FPA; however, the exam does not test the specifics on how to perform calculations.

FPA Feature Points

In most standard applications, lists of functions are identified and the corresponding effort is estimated. In web-enabled applications, the development effort depends on the number of screens (forms), number of images, type of images (static or animated), features to be enabled, interfaces and cross-referencing that is required. Thus, from the point of view of web applications, the effort would include all that is mentioned under function point estimation, plus the features that need to be enabled for different types of user groups. The measurement would involve identification or listing of features, access rules, links, storage, etc.

FPA behaves reasonably well in estimating business applications but not as well for other types of software (such as OS, process control, communications and engineering). Other estimation methods are more appropriate for such software and include the constructive cost model (COCOMO) and FPA Feature Points of De Marco and Watson-Felix.

Cost Budgets

A system development project should be analyzed with a view toward estimating the amount of effort that will be required to carry out each task. The estimates for each task should contain some or all of the following elements:
- Personnel hours by type (e.g., system analyst, programmer, clerical)
- Machine hours (predominantly computer time as well as duplication facilities, office equipment and communication equipment)

Figure 3.8—Computing Function Point Metrics

Measurement Parameter	Count	Weighting Factor			
		Simple	Average	Complex	Results
Number of user inputs		× 3	4	6	=________
Number of user outputs		× 4	5	7	=________
Number of user inquiries		× 3	4	6	=________
Number of files		× 7	10	15	=________
Number of external interfaces		× 5	7	10	=________
Count total:					

Note: Organizations that use FP methods develop criteria for determining whether a particular entry is simple, average or complex.

- Other external costs such as third-party software, licensing of tools for the project, consultant or contractor fees, training costs, certification costs (if required), and occupation costs (if extra space is required for the project)

Having established a best estimate of expected work efforts by task (i.e., actual hours, minimum/maximum) for personnel, costs budgeting now becomes a two-step process to:

1. Obtain a phase-by-phase estimate of human and machine effort by summing the expected effort for the tasks within each phase
2. Multiply the effort expressed in hours by the appropriate hourly rate to obtain a phase-by-phase estimate of systems development expenditure

Other costs may require tenders or quotes.

Software Cost Estimation

Cost estimation is a consequence of software size estimation. This is a necessary step in properly scoping a project. Alternatively, there are automated techniques for cost estimation of projects at each phase of system development. To use these products, a system is usually divided into main components, and a set of cost drivers is established. Components include:

- Source code language
- Execution time constraints
- Main storage constraints
- Data storage constraints
- Computer access
- The target machine used for development
- The security environment
- Staff experience

After all the drivers are defined, the program will develop cost estimates of the system and total project.

Scheduling and Establishing the Time Frame

While budgeting involves totaling the human and machine effort involved in each task, scheduling involves establishing the sequential relationship among tasks. This is achieved by arranging tasks according to:

- Earliest start date by considering the logical sequential relationship among tasks and attempting to perform tasks in parallel, wherever possible
- Latest expected finish date by considering the estimate of hours per the budget and the expected availability of personnel or other resources, and allowing for known, elapsed-time considerations (e.g., holidays, recruitment time, full-time/part-time employees)

The schedule can be graphically represented using various techniques such as Gantt charts, the Critical Path Method (CPM) or Program Evaluation Review Technique (PERT) diagrams. During the project execution, the budget and schedule should be revisited to verify compliance and identify variances at key points and milestones. Any variances to the budget and schedule should be analyzed to determine the cause and corrective action to take in minimizing or eliminating the total project variance. Variances and the variance analysis should be reported to management on a timely basis.

Critical Path Methodology

All project schedules have a critical path. Because the activities of a project are ordered and independent, a project can be represented as a network where activities are shown as branches connected at nodes immediately preceding and immediately following activities.

A path through the network is any set of successive activities that go from the beginning to the end of the project. Associated with each activity in the network is a single number that best estimates the amount of time that the activity will consume. There are different ways to estimate the activity duration.

Regardless of how the activity duration is estimated, the critical path is the sequence of activities whose sum of activity time is longer than that for any other path through the network. All project schedules have (at least) one critical path, usually only one in nonmanipulated project schedules. Critical paths are important because, if everything goes according to schedule, their duration gives the shortest possible completion time for the overall project. Activities that are not in the critical path have time slack. This is the difference between the latest possible completion time of each activity that will not delay the completion of the overall project and the earliest possible completion time based on all predecessor activities. Activities on a critical path have zero slack time, and conversely, activities with zero slack time are on a critical path.

The critical path(s) and the slack times for a project are computed by simply working forward through the network (forward pass), computing the earliest possible completion time for each activity, until the earliest possible completion time for the total project is found. Then by working backward through the network, the latest completion time for each activity is found, the slack time computed and the critical path identified. This procedure is computer-supported for easy calculation and what-if scenarios.

Within limits, activities can be "crashed" (reduced in time by payment of a premium for early completion) or relaxed (with associated cost reductions). In this way, total duration and budget can be managed (within limits).

Most CPM packages facilitate the analysis of resource utilization per time unit (e.g., day, week, etc.) and resource leveling, which is a way to level off resource peaks and valleys. Resource peaks and valleys are expensive due to management, hiring, firing, and/or overtime and idle resource costs. A constant, base resource utilization is preferable.

There are few, if any, scientific (algorithmic) resource-leveling methods available, but there is a battery (which CPM packages offer) of efficient heuristic methods that yield satisfactory results.

Gantt Charts

Gantt charts (**figure 3.9**) can be constructed to aid in scheduling the activities (tasks) needed to complete a project. The charts show when an activity should begin and when it should end along a timeline. The charts also show which activities can be in progress concurrently and which activities must be completed sequentially. Gantt charts can also reflect the resources assigned to each task and by what percent allocation. The charts aid in

Figure 3.9—Sample Gantt Chart

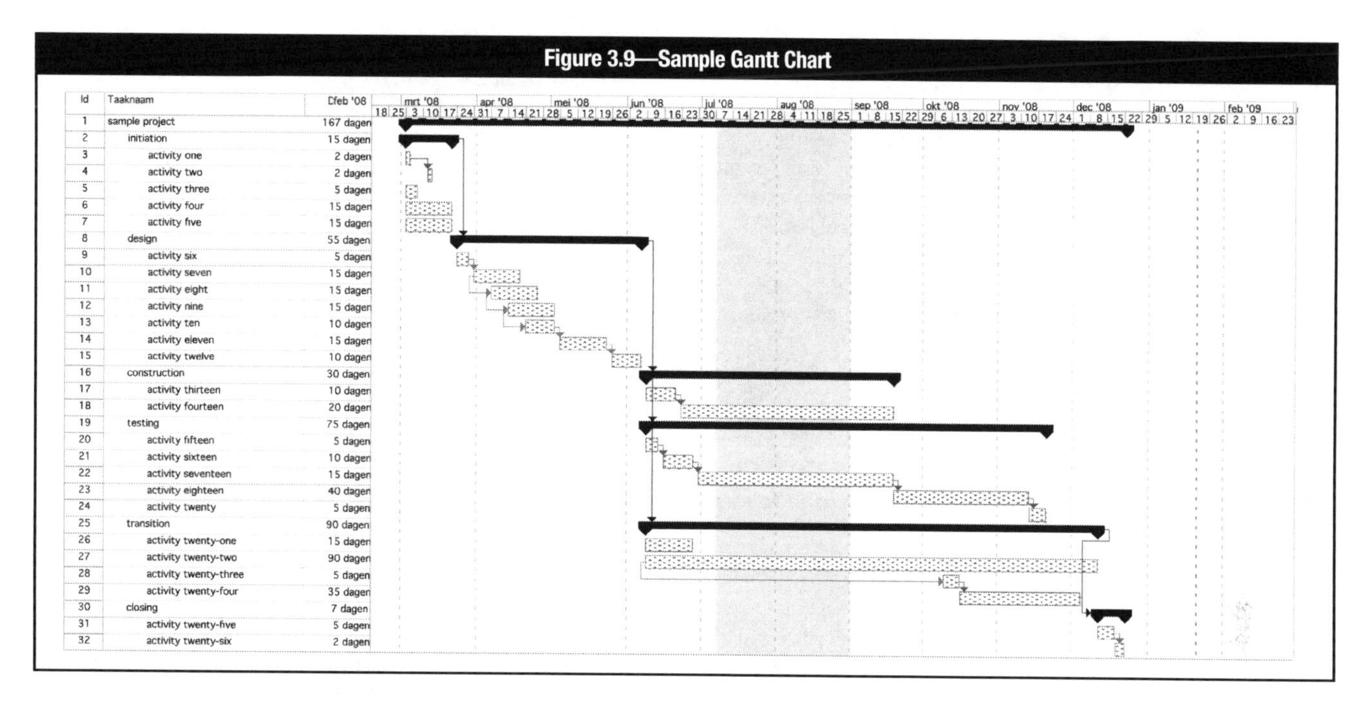

identifying activities that have been completed early or late by comparison to a baseline. Progress of the entire project can be read from the Gantt chart to determine whether the project is behind, ahead or on schedule compared to baseline project plan. Gantt charts can also be used to track the achievement of milestones or significant accomplishments for the project such as the end of a project phase or completion of a key deliverable.

Program Evaluation Review Technique

PERT is a CPM-type technique which uses three different estimates of each activity duration in lieu of using a single number for each activity duration (as used by CPM). The three estimates are then reduced (applying a mathematical formula) to a single number and then the classic CPM algorithm is applied. PERT is often used in system development projects with uncertainty about the duration (e.g., pharmaceutical research or complex software development). A diagram illustrating use of the PERT network management technique is shown in **figure 3.10**, where events are points in time or milestones for starting and completing activities (arrows). To determine a task's completion, three estimates are shown for completing each activity. The first is the most optimistic time (if everything went well) and the third is the pessimistic or worst-case scenario. The second is the most likely scenario. This estimate is based on experience attained from projects similar in size and scope. To calculate the PERT time estimate for each given activity, the following calculation is applied:

$$[\text{Optimistic} + \text{Pessimistic} + 4(\text{most likely})]/6$$

Using PERT, a critical path is also derived. The critical path is the longest path through the network (only one critical path in a network). The critical path is the route along which the project is shortened (accelerated) or lengthened (delayed). In **figure 3.10**, the critical path is A, C, E, F, H and I.

The advantage of PERT over CPM is that the formula above is based on the reasonable assumption that the three time estimates follow a Beta statistical distribution and, accordingly, probabilities (with associated confidence levels) can be associated with the total project duration.

Figure 3.10—PERT Network-based Chart

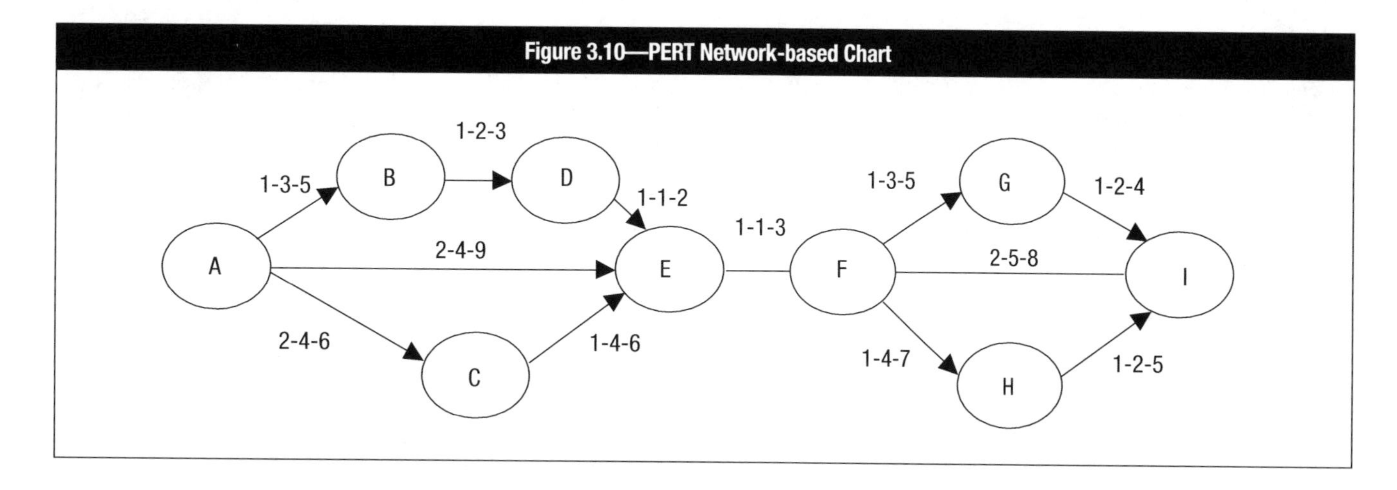

When designing a PERT network for system development projects, the first step is to identify all the activities and related events/milestones of the project and their relative sequence. For example, an event or result may be the completion of the operational feasibility study or the point at which the user accepts the detailed design. The analyst must be careful not to overlook any activity. Additionally, some activities such as analysis and design must be preceded by others before program coding can begin. The list of activities determines the detail of the PERT network. The analyst may prepare many diagrams that provide increasingly more detailed time estimates.

Timebox Management

Timebox management is a project management technique for defining and deploying software deliverables within a relatively short and fixed period of time, and with predetermined specific resources. There is a need to balance software quality and meet the delivery requirements within the timebox or time window. The project manager has some degree of flexibility and uses discretion in scoping requirements. Timebox management can be used to accomplish prototyping or rapid application development-type approaches in which key features are to be delivered in a short time frame. Key features include interfaces for future integrations. The major advantage of this approach is that it prevents project cost overruns and delays from scheduled delivery. The project does not necessarily eliminate the need for a quality process. The design and development phase is shortened due to the use of newer developmental tools and techniques. Preparation of test cases and testing requirements are easily written down as a result of end-user participation. System test and user acceptance testing are normally performed together.

3.4.3 PROJECT EXECUTION

Once planning efforts have been completed, the program manager, in coordination with the program office, starts the actual execution of the planned tasks as described in the OBS, WBS and WPs (the plans, processes and procedures). The program and project management team initiates monitoring of internal team production and quality metrics and monitors these metrics from contractors and vendors. A key success factor is the project's oversight of the integrated team in the IT system requirements, architecture, design, development, testing, implementation and transitioning to production operations.

3.4.4 PROJECT CONTROLLING

The controlling activities of a project include management of scope, resource usage and risk. It is important that new requirements for the project are documented and, if approved, allocated appropriate resources. Control of change during a project ensures that projects are completed within stakeholder requirements of time, use of funds and quality objectives. Stakeholder satisfaction should be addressed with effective and accurate requirements capture, proper documentation, baselining and skilled steering committee activity.

Management of Scope Changes

Managing the scope of projects requires careful documentation in the form of a work breakdown structure. This documentation forms part of the project plan or the project baseline. For complex deliverables, it is best to document the WBS in a component management database (CMDB). Changes to the scope almost invariably lead to changes in required activities and impact deadlines and budget. Therefore, it is necessary to have a change management process. This process starts with a formal change request that contains a clear description of the requested change and the reasons for the change. Change requests must be submitted to the project manager. Obviously, only stakeholders are allowed to submit change requests. Copies of all change requests should be archived in the project file. The project manager judges the impact of each change request on project activities (scope), schedule and budget. The change advisory board then evaluates the change request (on behalf of the sponsor) and decides whether to recommend the change. If the change is accepted, the project manager is instructed to update the project plan to reflect the requested change. The updated project plan must be formally confirmed by the project sponsor—accepting or rejecting the recommendation of the change advisory board.

Management of Resource Usage

Resource usage is the process by which the project budget is being spent. To determine whether actual spending is in line with planned spending, resource usage must be measured and reported. It is not sufficient to only monitor actual spending. Every budget and project plan presupposes a certain "productivity" of resources (e.g., if a task is planned to take 24 man-hours, then it is implicitly supposed that the resource being deployed is capable of finishing that task in 24 man-hours and, at the same time, delivering results at a satisfactory quality level). Whether this is actually happening can be checked with a technique called earned value analysis (EVA).

EVA consists of comparing the following metrics at regular intervals during the project: budget to date, actual spending to date, estimate to complete and estimate at completion. If a single-task project is planned to take three working days, with eight hours spent each day, the resource will have spent eight hours after the first day. This is according to budget, but the following question must be answered: Is this project on track? The answer cannot be known unless the estimate to complete is known. If productivity of the resource is as it is supposed to be, the resource would need another 16 hours to complete the task. Rather than assuming this to be the case, the worker must be asked how much more time is needed to complete the task. The answer might well be that another 22 hours are needed to complete the task. The estimate at completion is then 30 hours for the project. This results in an overrun of 25 percent. On the first day of this small project, the resource spent is according to budget, but the "earned value" of the eight hours spent is only two hours (at planned productivity). Obviously, reporting via time sheet and management are at the basis of all of these processes.

Management of Risk

Risk is defined as a possible negative event or condition that would disrupt relevant aspects of the project. There are two main categories of project risk: the category that impacts the business benefits (and, therefore, endangers the reasons for the project's very existence) and the category that impacts the project itself. The project *sponsor* is responsible for mitigating the first category of risk and the project *manager* is responsible for mitigating the second category.

The risk management process consists of five steps that are repeatedly executed during a project. Phase-end milestones are a good anchor point in time at which to review and update the initial risk assessments and the related mitigations. The risk management process steps are:
- **Identify risk**—Perform a brainstorming session with the team and create an inventory of possible risk.
- **Assess and evaluate risk**—Quantify the likelihood (expressed as a percentage) and the impact of the risk (expressed as an amount of money). The "insurance policy" that needs to be in the project budget is calculated as the likelihood multiplied by the impact.
- **Manage risk**—Create a risk management plan, describing the strategy adopted and measures to deal with the risk. Generally, the more important the risk, the more budget should be made available for countermeasures. Countermeasures include prevention, detection and damage control/reconstruction activities. Any risk can be mitigated, avoided, transferred or accepted depending on its severity, likelihood and cost of countermeasures and the enterprise's policy.
- **Monitor risk**—Discover risk that materializes, and act accordingly.
- **Evaluate the risk management process**—Review and evaluate the effectiveness and costs of the risk management process.

3.4.5 CLOSING A PROJECT

A project should have a finite life so, at some point, it is closed and the new or modified system is handed over to the users and/or system support staff. At this point, any outstanding issues will need to be assigned. The project sponsor should be satisfied that the system produced is acceptable and ready for delivery. Key areas to consider include:
- When will the project manager issue the final project closure notification?
- Who will the final project notification come from?
- How will the project manager assist the project team transition to new projects or release them to their regular assigned duties?
- What will the project manager do for actions, risk and issues that remain open? Who will pick up these actions and how will these be funded?

Custody of contracts may need to be assigned, and documentation archived or passed on to those who will need it. It is also common at this stage to survey the project team, development team, users and other stakeholders to identify any lessons learned that can be applied to future projects including content-related criteria such as performance fulfillment, fulfillment of additional objectives, adherence to the schedule and costs, and process-related criteria such as quality of the project teamwork and relationships to relevant environments.

Review may be done in a formal process such as a postproject review in which lessons learned and an assessment of project management processes used are documented to allow reference, in the future, by other project managers or users working on projects of similar size and scope.

A postimplementation review, in contrast, is typically completed after the project has been in use (or in "production") for some time—long enough to realize its business benefits and costs, and measure the project's overall success and impact on the business units. Metrics used to quantify the value of the project include: total cost of ownership (TCO) and ROI.

Note: Project management practice descriptions and related concepts and theories behind best practices have been brought together in "body of knowledge" (BoK) reference libraries. Certification schemes have subsequently been based upon such BoKs. There are three relevant professional organizations that have such certification schemes: PMI, IPMA and PRINCE2 Foundation.

3.5 BUSINESS APPLICATION DEVELOPMENT

Companies often commit significant IT resources (e.g., people, applications, facilities and technology) to develop, acquire, integrate and maintain application systems that are critical to the effective functioning of key business processes. These systems, in turn, often control critical information assets and should be considered assets that need to be effectively managed and controlled. IT processes for managing and controlling these IT resources and other such activities are part of a life cycle process with defined phases applicable to business application development, deployment, maintenance and retirement. In this process, each step or phase in the life cycle is an incremental step that lays the foundation for the next phase, for effective management control in building and operating business application systems.

The implementation process for business applications follows the project planning and management processes as outlined previously. Normally, the business application development project begins when an individual application feasibility study is initiated as a result of one or more of the following situations:
- A new opportunity that relates to a new or existing business process
- A problem that relates to an existing business process
- A new opportunity that will enable the organization to take advantage of technology
- A problem with the current technology
- Alignment of business applications with business partners/ industry standard systems and respective interfaces

All of these situations are tightly coupled with key business drivers. Key business drivers, in this context, can be defined as the attributes of a business function that drive the behavior and implementation of that business function to achieve the strategic business goals of the company.

Thus, all critical business objectives (as a breakdown of the corporate strategy) have to be translated into key business drivers for all parties involved in business operations during an SDLC project. Objectives should follow the SMART quality guidelines mentioned in section 3.3.5 Project Objectives, so that general requirements will be expressed in scorecard form, which allows objective evidence to be collected in order to measure the business value of an application and to prioritize requirements.

Benefits of using this approach are that all affected parties will have a common and clear understanding of the objectives and how they contribute to business support. Additionally, conflicting key business drivers (e.g., costversus functionality) and mutually dependent key business drivers can be detected and resolved in early stages of an SDLC project.

Business application projects should be initiated using well-defined procedures or activities as part of a defined process to communicate business needs to management. These procedures often require detailed documentation identifying the need or problem, specifying the desired solution and relating the potential benefits to the organization. All internal and external factors affected by the problem and their effect on the corporation should be identified.

A risk in any software development project is that the final outcome may not meet all requirements. Problems due to translation errors arise when initially defining the requirements for interim products. The waterfall model and variants of the model normally involve a life cycle verification approach that ensures that potential mistakes are corrected early and not solely during final acceptance testing. The verification and validation model, sometimes called the V-model, also emphasizes the relationship between development phases and testing levels (**figure 3.11**). The most granular testing—the unit test—occurs immediately after programs have been written. Following this model, testing occurs to validate the detailed design. System testing relates to the architectural specification of the system while final user-acceptance testing references the requirements.

From an IS auditor's perspective, the V-model's defined life cycle phases and specific points for review and evaluation provides the following advantages:
- The IS auditor's influence is significantly increased when there are formal procedures and guidelines identifying each phase in the business application life cycle and the extent of auditor involvement.
- The IS auditor can review all relevant areas and phases of the systems development project and report independently to management on the adherence to planned objectives and company procedures.
- The IS auditor can identify selected parts of the system and become involved in the technical aspects on the basis of his/her skills and abilities.
- The IS auditor can provide an evaluation of the methods and techniques applied through the development phases of the business application life cycle.

Any business application system developed will fall under one of two major categories:
- **Organizationcentric** (management information system [MIS], ERP, customer relationship management [CRM], supply chain management [SCM], etc.)—The objective of organizationcentric applications is to collect, collate, store, archive and share information with business users and various applicable support functions on a need-to-know basis. Thus, sales data are made available to accounts, administration, governmental levy payment departments, etc. Regulatory levy fulfillment (i.e., tax compliance) is also addressed by the presence of organizationcentric applications. Organizationcentric application projects usually use the SDLC or other more detailed software engineering approaches for development.
- **End-user-centric computing**—The objective of an end-user-centric application is to provide different views of data for their performance optimization. This objective includes DSS, geographic information systems (GIS), techniques, etc. Most of these applications are developed using alternative development approaches.

3.5.1 TRADITIONAL SDLC APPROACH

Over the years, business application development has occurred largely through the use of the traditional SDLC phases shown in **figure 3.12**. Also referred to as the waterfall technique, this life cycle approach is the oldest and most widely used for developing business applications. The approach is based on a systematic, sequential approach to software development (largely of business applications) that begins with a feasibility study and progresses through requirements definition, design, development, implementation and postimplementation. The series of steps or phases have defined goals and activities to perform with assigned responsibilities, expected outcomes and target completion dates.

Figure 3.11—Verification and Validation

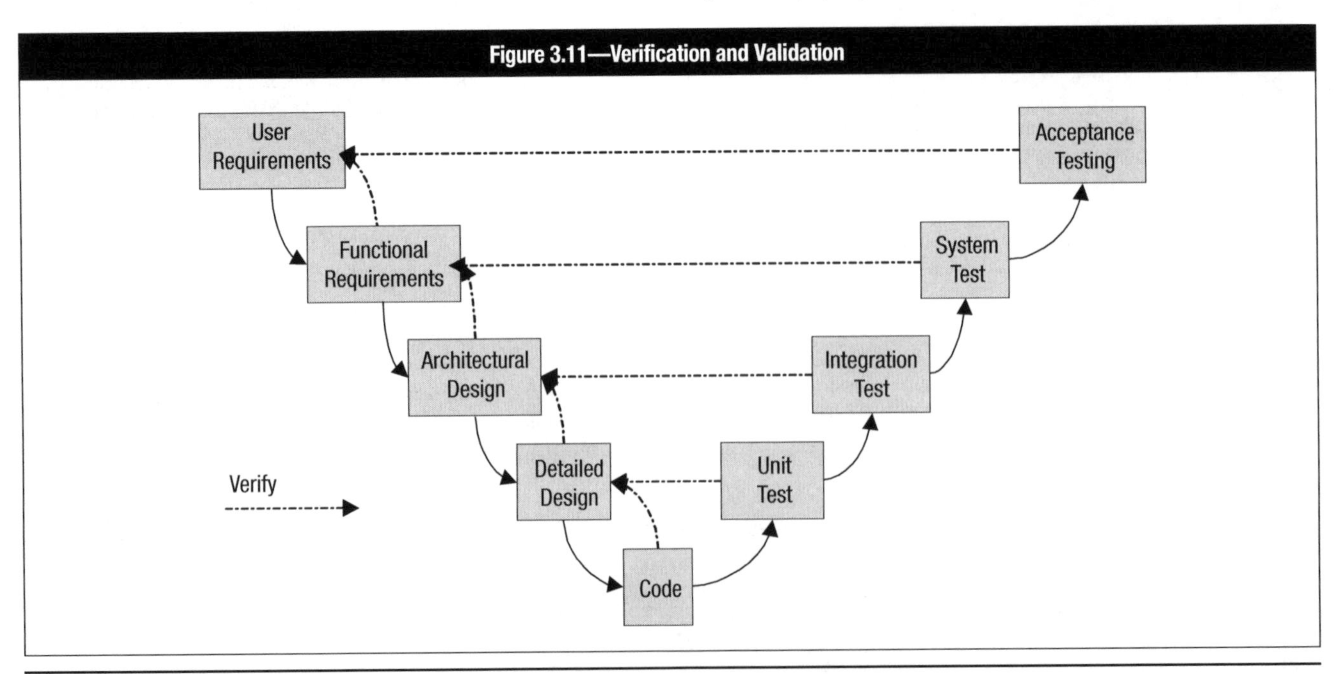

Figure 3.12—Traditional System Development Life Cycle Approach	
SDLC Phase	**General Description**
Phase 1—Feasibility Study	Determine the strategic benefits of implementing the system either in productivity gains or in future cost avoidance, identify and quantify the cost savings of a new system, and estimate a payback schedule for costs incurred in implementing the system. Further, intangible factors such as readiness of the business users and maturity of the business processes will also be considered and assessed. This business case provides the justification for proceeding to the next phase.
Phase 2—Requirements Definition	Define the problem or need that requires resolution and define the functional and quality requirements of the solution system. This can be either a customized approach or vendor-supplied software package, which would entail following a defined and documented acquisition process. In either case, the user needs to be actively involved.
Phase 3A—Software Selection and Acquisition (*purchased systems*)	Based on requirements defined, prepare a request for proposal outlining the entity requirements to invite bids from prospective suppliers, in respect of those systems that are intended to be procured from vendors or solution providers.
Phase 3B—Design (*in-house development*)	Based on the requirements defined, establish a baseline of system and subsystem specifications that describe the parts of the system, how they interface, and how the system will be implemented using the chosen hardware, software and network facilities. Generally, the design also includes program and database specifications and will address any security considerations. Additionally, a formal change control process is established to prevent uncontrolled entry of new requirements into the development process.
Phase 4A—Configuration (*purchased systems*)	Configure the system, if it is a packaged system, to tailor it to the organization's requirements. This is best done through the configuration of system control parameters, rather than changing program code. Modern software packages are extremely flexible, making it possible for one package to suit many organizations simply by switching functionality on or off and setting parameters in tables. There may be a need to build interface programs that will connect the acquired system with existing programs and databases.
Phase 4B—Development (*in-house development*)	Use the design specifications to begin programming and formalizing supporting operational processes of the system. Various levels of testing also occur in this phase to verify and validate what has been developed. This generally includes all unit and system testing and several iterations of user acceptance testing.
Phase 5—Final Testing and Implementation	Establish the actual operation of the new information system, with the final iteration of user acceptance testing and user sign-off conducted in this phase. The system also may go through a certification and accreditation process to assess the effectiveness of the business application in mitigating risk to an appropriate level and providing management accountability over the effectiveness of the system in meeting its intended objectives and in establishing an appropriate level of internal control.
Phase 6—Postimplementation	Following the successful implementation of a new or extensively modified system, implement a formal process that assesses the adequacy of the system and projected cost-benefit or ROI measurements *vis-à-vis* the feasibility stage findings and deviations. In so doing, IS project and end-user management can provide lessons learned and/or plans for addressing system deficiencies as well as recommendations for future projects regarding system development and project management processes followed.

This approach works best when a project's requirements are likely to be stable and well defined. It facilitates the determination of a system architecture relatively early in the development effort. Another type of software development approach is the iterative approach where business requirements are developed and tested in iterations until the entire application is designed, built and tested. The traditional approach is useful in web applications in which prototypes of screens are necessary to aid in the completion of requirements and design.

As purchased packages have become more common, the design and development phases of the traditional life cycle are being replaced with the selection and configuration phases.

Note: The CISA candidate should be familiar with the SDLC phases and should be aware of what the IS auditor should look for when reviewing the feasibility study.

Some examples of measurements of critical success factors (CSFs) for an SDLC project could include productivity, quality, economic value and customer service, as shown in **figure 3.13**.

Figure 3.13—Measurements of Critical Success Factors	
Productivity	Dollars spent per user Number of transactions per month Number of transactions per user
Quality	Number of discrepancies Number of disputes Number of occurrences of fraud/misuse detection
Economic value	Total processing time reduction Monetary value of administration costs
Customer service	Turnaround time for customer question handling Frequency of useful communication to users

The primary advantage of the traditional approach is that it provides a template into which methods for the requirements (i.e., definition, design, programming, etc.) can be placed. However, some of the problems encountered with this approach include:
- Unanticipated events. Because real projects rarely follow the sequential flow prescribed, iteration always occurs and creates problems in implementing the approach.
- The difficulty of obtaining an explicit set of requirements from the customer/user as the approach requires
- Managing requirements and convincing the user about the undue or unwarranted requirements in the system functionality, which may lead to conflict in the project
- The necessity of customer/user patience, which is required because under this approach a working version of the system's programs will not be available until late in the project's life cycle
- A changing business environment that alters or changes the customer/user requirements before they are delivered

Moreover, the actual phases for each project may vary depending on whether a developed or acquired solution is chosen. For example, system maintenance efforts may not require the same level of detail or number of phases as new applications. The phases and deliverables should be decided during the early planning stages of the project.

3.5.2 DESCRIPTION OF TRADITIONAL SDLC PHASES

A traditional SDLC approach is made up of a number of distinct phases (six are shown in **figure 3.12**), each with a defined set of activities and outcomes. (There are other interpretations that use a slightly different number of phases with different names; for example, a seventh phase—maintenance—is often added.)

The following section describes in detail each phase's purpose and relationship to prior phases, the general activities performed and expected outcomes.

Phase 1—Feasibility Study

After the initial approval has been given to move forward with a project, an analysis begins to clearly define the need and to identify alternatives for addressing the need. This analysis is known as the feasibility study.

A feasibility study is concerned with analyzing the benefits and solutions for the identified problem area. This study includes development of a business case, which states the strategic benefits of implementing the system either in productivity gains or in future cost avoidance; identifies and quantifies the cost savings of the new system; and estimates a payback schedule for the cost incurred in implementing the system or shows the projected ROI. Intangible benefits such as improved morale may also be identified; however, benefits should be quantified wherever possible.

Within the feasibility study, the following typically are addressed:
- Define a time frame for the implementation of the required solution.
- Determine an optimum alternative risk-based solution for meeting business needs and general information resource requirements (e.g., whether to develop or acquire a system). Information resources include people, applications, technology, facilities and data organized and managed through a given set of IT "natural" processes. Such processes can easily be mapped to SDLC and, with some more effort, to rapid application development (RAD).
- Determine whether an existing system can correct the situation with slight or no modification (e.g., workaround).
- Determine whether a vendor product offers a solution to the problem. Vendor products include services such as cloud infrastructure; platform and software as service; managed security services; and commercial off-the-shelf applications.
- Determine the approximate cost to develop the system to correct the situation.
- Determine whether the solution fits the business strategy.

Factors impacting whether to develop or acquire a system include:
- The date the system needs to be functional
- The cost to develop the system as opposed to buying it
- The resources, staff (availability and skill sets) and hardware required to develop the system or implement a vendor solution
- In a vendor system, the license characteristics (e.g., yearly renewal, perpetual) and maintenance costs
- The other systems needing to supply information to or use information from the vendor system that will need the ability to interface with the system
- Compatibility with strategic business plans
- Compatibility with risk appetite and regulatory compliance needs
- Compatibility with the organization's IT infrastructure
- Likely future requirements for changes to functionality offered by the system

The result of the completed feasibility study should be some type of a comparative report that shows the results of criteria analyzed (e.g., costs, benefits, risk, resources required and organizational impact) and recommends one of the alternatives/solutions and a course of action.

Closely related to a feasibility study is the development of an impact assessment. An impact assessment is a study of the potential future effects of a development project on current projects and resources. The resulting document should list the pros and cons of pursuing a specific course of action.

Phase 2—Requirements Definition

Requirements definition is concerned with identifying and specifying the business requirements of the system chosen for development during the feasibility study. Requirements include descriptions of what a system should do, how users will interact with a system, conditions under which the system will operate and the information criteria the system should meet. The COBIT framework defines information criteria that should be incorporated in system requirements to address issues associated with effectiveness, efficiency, confidentiality, integrity, availability, compliance and reliability. The requirements definition phase also deals with overarching issues that are sometimes called nonfunctional requirements (e.g., access control). There is heightened focus in today's marketplace about ensuring that security-related considerations are addressed early in the SDLC. Many IT security weaknesses can be corrected with a more critical focus on this topic within the context of the SDLC and in particular during the requirements definition.

To successfully complete the requirements definition phase, the project team should:

- Identify and consult stakeholders to determine their requirements.
- Analyze requirements to detect and correct conflicts (mainly, differences between requirements and expectations) and determine priorities.
- Identify system bounds and how the system should interact with its environment.
- Convert user requirements into system requirements (e.g., an interactive user interface prototype that demonstrates the screen look and feel).
- Record requirements in a structured format. Historically, requirements have been recorded in a written requirements specification, possibly supplemented by some schematic models. Commercial requirements management tools now are available that allow requirements and related information to be stored in a multiuser database.
- Verify that requirements are complete, consistent, unambiguous, verifiable, modifiable, testable and traceable. Because of the high cost of rectifying requirements' problems in downstream development phases, effective requirements reviews have a large payoff.
- Resolve conflicts between stakeholders.
- Resolve conflicts between the requirements set and the resources that are available.

The users in this process specify their information resource needs, nonautomated as well as automated, and how they wish to have them addressed by the system (e.g., access controls, regulatory requirements, management information needs and interface requirements).

The IS auditor should pay close attention to the degree the organization system security engineering team is involved in the development of security controls throughout the data life cycle within the business application. This means the controls are in place regarding applicable confidentiality, integrity and availability need of data from creation/receipt, processing, storage, transmission and ultimately destruction.

From this interactive process, a general preliminary design of the system may be developed and presented to user management for their review, modification, approval and endorsement. A project schedule is created for developing, testing and implementing the system. Also, commitments are obtained from the system's developers and affected user departments to contribute the necessary resources to complete the project. It is important to note that all concerned management and user groups must be actively involved in the requirements definition phase to prevent problems such as expending resources on a system that will not satisfy the business requirements. User involvement is necessary to obtain commitment and full benefit from the system. Without management sponsorship, clearly defined requirements and user involvement, the benefits may never be realized.

IS auditors are involved at this stage to determine whether adequate security requirements have been defined to address, at a minimum, the confidentiality, integrity and availability requirements of the system. This includes whether adequate audit trails are defined as part of the system because these affect the auditor's ability to identify issues for proper follow-up.

Phase 3A—Software Selection and Acquisition

At this point in the project, it may be appropriate to evaluate the risk and benefits of developing a new system versus acquiring from a vendor a suitable system that is complete, tested and proven. Consideration should be given to the ability of the organization to undertake the proposed development project, the costs and risk of doing so, and the benefits of having total ownership and control over the new system rather than becoming dependent on a vendor. Software acquisition is not a phase in the standard SDLC. However, if a decision was reached to acquire rather than develop software, software acquisition is the process that should occur after the requirements definition phase. The decision is generally based on various factors such as the cost differential between development and acquisition, availability of generic software, and the time gap between development and acquisition. Please note that if the result of the decision to develop/acquire is to buy a vendor-supplied software package, the user must be actively involved in the package evaluation and selection process.

The feasibility study should contain documentation that supports the decision to acquire the software. Depending on the software required there could be four cases:

1. Software is required for a generic business process for which vendors are available and software can be implemented without customization.
2. The vendor's software needs to be customized to suit business processes.
3. Software needs to be developed by the vendor.
4. Software is available as a service through the cloud, software as a service (SaaS). This is generally available for generic processes.

A project team with participation by technical support staff and key users should be created to write a request for proposal (RFP) or invitation to tender (ITT). An RFP needs to be prepared separately for each case referred to previously.

The invitation to respond to an RFP should be widely distributed to appropriate vendors and, if possible, posted via a public procurement medium (Internet or newspaper). This process allows the business to determine which of the responding vendors' products offers the best solution at the most cost-effective price.

The RFP should include the areas shown in **figure 3.14**.

Figure 3.14—RFP Contents

Item	Description
Product vs. system requirements	The chosen vendor's product should come as close as possible to meeting the defined requirements of the system. If no vendor's product meets all of the defined requirements, the project team, especially the users, will have to decide whether to accept the deficiencies. An alternative to living with a product's deficiencies is for the vendor or the purchaser to make customized changes to the product.
Product scalability and interoperability	The project management should not only look at vendor's product ability to meet the existing requirements for the project but also the ability of the product to grow and/or contract with the organization's business processes. Vendor products should be assessed as to the applications' ability to interconnect with other systems whose interconnections are currently out of the project's scope but may be needed in the future.
Customer references	Project management should check vendor-supplied references to validate the vendor's claims of product performance and completion of work by the vendor.
Vendor viability/financial stability	The vendor supplying or supporting the product should be reputable and able to provide evidence of financial stability. A vendor may not be able to prove financial stability; if the product is new, the vendor presents a substantially higher risk to the organization.
Availability of complete and reliable documentation	The vendor should be willing and able to provide a complete set of system documentation for review prior to acquisition. The level of detail and precision found in the documentation may be an indicator of the detail and precision utilized within the design and programming of the system itself.
Vendor support	The vendor should have available a complete line of support products for the software package. This may include a 24-hour, seven-day-a-week help line, onsite training during implementation, product upgrades, automatic new version notification and onsite maintenance, if requested.
Source code availability	The source code should be received either from the vendor initially or there should be provisions for acquiring the source code in the event that the vendor goes out of business. Usually, these clauses are part of a software escrow agreement in which a third party holds the software in escrow should such an event occur. The acquiring company should ensure that product updates and program fixes are included in the escrow agreement.
Number of years of experience in offering the product	More years indicate stability and familiarity with the business that the product supports.
A list of recent or planned enhancements to the product, with dates	A short list suggests the product is not being kept current.
Number of client sites using the product with a list of current users	A larger number suggests wide acceptance of the product in the marketplace.
Acceptance testing of the product	Such testing is crucial in determining whether the product really satisfies the system requirements. This is allowed before a purchasing commitment must be made.

It is worth noting that a distinction can sometimes be drawn between ITT and RFP with respect to their applicability. When the product and related services are known in advance, a user organization will prefer an ITT so they can obtain the best combination of price and services. This is more applicable where procurement of hardware, network, database, etc., is involved. When the requirement is more toward a solution and related support and maintenance, an organization generally prefers an RFP, where the capability, experience and approach can be measured against the requirement. This is more applicable in system integration projects such as ERP and SCM that involve delivery or escrowing of source code.

Often, prior to the development of an RFP, organizations will develop a request for information (RFI) to solicit software development vendors for advice in addressing problems with existing systems. Information obtained in this manner may be used to develop an RFP.

Note: The CISA candidate should be familiar with the RFP process.

The project team needs to carefully examine and compare the vendors' responses to the RFP. This comparison should be done using an objective method such as a scoring and ranking methodology. After the RFP responses have been examined, the project team may be able to identify a single vendor whose product satisfies most or all of the stated requirements in the RFP. Other times, the team may narrow the list to two or three acceptable candidates (i.e., short list of vendors). In evaluating the best-fit solution and vendor against the given set of business requirements and conditions, a suitable methodology of evaluation should be adopted. The methodology should ensure objective, equitable and fair comparison of the products/vendors (e.g., a gap analysis to find out the differences between requirements and software, the parameters required to modify, etc.).

It is important to keep in mind the minimum and recommended requirements to use the software, including:

- Required hardware such as memory, disk space and server or client characteristics

- Operating system (OS) versions and patch levels supported
- Additional tools such as import and export tools
- Databases supported

Often it is likely that more than one product/vendor fits the requirements with advantages and disadvantages with respect to each other. To resolve such a situation, agenda-based presentations should be requested from the short-listed vendors. The agenda-based presentations are scripted business scenarios that are designed to show how the vendor will perform certain critical business functions. Vendors are typically invited to demonstrate their product and follow the sample business scenarios given to them to prepare. It is highly recommended that adequate participation from various user groups is included when evaluating the product/vendor's fit and the system's ease of use. The project team thus has an opportunity to check the intangible issues such as the vendor's knowledge of the product and the vendor's ability to understand the business issue at hand. With each short-listed vendor demonstrating their product following a scripted document, this also enables the project team to evaluate and finalize the product/vendor with knowledge and objectivity built into the process. The finalist vendor candidate is then requested to organize site visits to confirm the findings from the agenda-based presentations and check the system in a live environment. Once the finalist is confirmed, a conference room pilot needs to be conducted. The conference room pilot will enable the project team to understand the system with a hands-on session and identify the areas that need certain customizations or workarounds.

Additionally, for the short list of vendors, it can be beneficial for the project team to talk to current users of each of the potential products. If it can be arranged and can be cost-justified, an onsite visit can be even more beneficial. Whenever possible, the companies chosen should be those that use the products in a manner similar to the way the company will use the products. The IS auditor should encourage the project team to contact current users. The information obtained from these discussions or visits validates statements made in the vendor's proposal and can determine which vendor is selected.

The discussions with the current users should concentrate on each vendor's:

- **Reliability**—Are the vendor's deliverables (enhancements or fixes) dependable?
- **Commitment to service**—Is the vendor responsive to problems with its product? Does the vendor deliver on time?
- **Commitment to providing training, technical support and documentation for its product**—What is the level of customer satisfaction?

Upon completing the activities cited, vendor presentations and final evaluations, the project team can make a product selection. The reasons for making a particular choice should be documented.

The last step in the acquisition process is to negotiate and sign a contract for the chosen product. Appropriate legal counsel should review the contract prior to its signing.

The contract should contain the following items:

- Specific description of deliverables and their costs
- Commitment dates for deliverables
- Commitments for delivery of documentation, fixes, upgrades, new release notifications and training
- Commitments for data migration
- Allowance for a software escrow agreement, if the deliverables do not include source code
- Description of the support to be provided during installation/customization
- Criteria for user acceptance
- Provision for a reasonable acceptance testing period, before the commitment to purchase is made
- Allowance for changes to be made by the purchasing company
- Maintenance agreement
- Allowance for copying software for use in business continuity efforts and for test purposes
- Payment schedule linked to actual delivery dates
- Confidentiality clauses
- Data protection clauses

Managing the contract should also involve a major level of effort to ensure that deployment efforts are controlled, measured and improved on, where appropriate. This may include regular status reporting requirements. Additionally, the milestones and metrics to be reported against should be agreed on with the vendor.

IS auditors are involved in the software acquisition process to determine whether an adequate level of security controls has been considered prior to any agreement being reached. If security controls are not part of the software, it may become difficult to ensure data integrity for the information that will be processed through the system. Risk involved with the software package includes inadequate audit trails, password controls and overall security of the application. Because of the risk, the IS auditor should ensure that these controls are built into the software application.

Phase 3B—Design

Based on the general preliminary design and user requirements defined in the requirements definition phase, a detailed design should be developed. Generally, a programming and analyst team is assigned the tasks of defining the software architecture depicting a general blueprint of the system and then detailing or decomposing the system into its constituent parts such as modules and components. This approach is an enabler for effective allocation of resources to design and for defining how the system will satisfy all its information requirements. Depending on the complexity of the system, several iterations in defining system-level specifications may be needed to get down to the level of detail necessary to start development activities such as coding.

Key design phase activities include:

- Developing system flowcharts and entity relationship models to illustrate how information will flow through the system
- Determining the use of structured design techniques (which are processes to define applications through a series of data or process flow diagrams) that show various relationships from the top level down to the details
- Describing inputs and outputs such as screen designs and reports. If a prototyping tool is going to be used, they most often are used in the screen design and presentation process (via online programming facilities) as part of an integrated development environment.

- Determining processing steps and computation rules when addressing functional requirement needs
- Determining data file or database system file design
- Preparing program specifications for various types of requirements or information criteria defined
- Developing test plans for the various levels of testing:
  - Unit (program)
  - Subsystem (module)
  - Integration (system)
  - Interface with other systems
  - Loading and initializing files
  - Stress
  - Security
  - Backup and recovery
- Developing data conversion plans to convert data and manual procedures from the old system to the new system. Detailed conversion plans will alleviate implementation problems that arise due to incompatible data, insufficient resources or staff who are unfamiliar with the operations of the new system.

ENTITY RELATIONSHIP DIAGRAMS

An important tool in the creation of a general preliminary design is the use of entity relationship diagrams (ERDs). An ERD depicts a system's data and how these data interrelate. An ERD can be used as a requirements analysis tool to obtain an understanding of the data a system needs to capture and manage. In this case, the ERD represents a logical data model. An ERD can also be used later in the development cycle as a design tool that helps document the actual database schema that will be implemented. Used in this way, the ERD represents a physical data model.

As the name suggests, the essential components of an ERD are entities and relationships.

Entities are groupings of like data elements or instances that may represent actual physical objects or logical constructs. An entity is described by attributes, which are properties or characteristics common to all or some of the instances of the entity. Particular attributes, either singularly or in combination, form the keys of an entity. An entity's primary key uniquely identifies each instance of the entity. Entities are represented on ERDs as rectangular boxes with an identifying name.

Relationships depict how two entities are associated (and, in some cases, how instances of the same entity are associated). The classic way of depicting a relationship is a diamond with connecting lines to each related entity. The name in the diamond describes the nature of the relationship. The relationship may also specify the foreign key attributes that achieve the association among the entities. A foreign key is one or more attributes held in one entity that map to the primary key of a related entity.

SOFTWARE BASELINING

The software design phase represents the optimum point for software baselining to occur. The term software baseline means the cutoff point in the design and is also referred to as design freeze. User requirements are reviewed, item by item, and considered in terms of time and cost. The changes are undertaken after taking into account various types of risk, and change does not occur without undergoing formal strict procedures for approval based on a cost-benefit impact analysis. Failure to adequately manage the requirements for a system through baselining can result in a number of types of risk. Foremost among these types of risk is scope creep—the process through which requirements change during development. Empirical studies have shown that a typical project can experience at least a 25 percent change in requirements throughout development, resulting in an increase in the effort and costs required for development.

Software baselining also relates to the point when formal establishment of the software configuration management process occurs. At this point, software work products are established as configuration baselines with version numbers. This would include, for example, functional requirements, specifications and test plans. All of these work products are configuration items and are identified and brought under formal change management control. This process will be used throughout the application system's life cycle, where SDLC procedures for analysis, design, development, testing and deployment are enforced on new requirements or changes to existing requirements.

USER INVOLVEMENT IN THE DESIGN

After business processes have been documented and it is understood how those processes might be executed in the new system, involvement of users during the design phase is limited. Given the technical discussion that usually occurs during a design review, end-user participation in the review of detailed design work products is normally not appropriate. However, developers should be able to explain how the software architecture will satisfy system requirements and outline the rationale for key design decisions. Choices of particular hardware and software configurations may have cost implications of which stakeholders need to be aware and control implications that are of interest to the IS auditor.

END OF DESIGN PHASE

After the detailed design has been completed, including user approvals and software baselining, the design is distributed to the system developers for coding.

IS AUDITOR INVOLVEMENT

The IS auditor involvement is primarily focused on whether an adequate system of controls is incorporated into system specifications and test plans, and whether continuous online auditing functions are built into the system (particularly for e-commerce applications and other types of paperless environments). Additionally, the IS auditor is interested in evaluating the effectiveness of the design process itself (such as in the use of structured design techniques, prototyping and test plans, and software baselining) to establish a formal software change process that effectively freezes the inclusion of any changes to system requirements without a formal review and approval process.

The key documents coming out of this phase include system, subsystem, program and database specifications, test plans, and a defined and documented formal software change control process.

Phase 4A—Configuration

System configuration, as it relates to the SDLC, consists of defining, tracking and controlling changes in a purchased system to meet the needs of the business. For ERP systems, the task often involves the modification of configuration tables as well as some development, primarily to ensure that the ERP system is integrated into the existing IT architecture. System configuration is supported by the change management policies and processes, which define:

- Roles and responsibilities
- Classification and prioritization of all changes based on business risk
- Assessment of impact
- Authorization and approval of all changes by the business process owners and IT
- Tracking and status of changes
- Impact on data integrity (e.g., all changes to data files being made under system and application control rather than by direct user intervention)

Phase 4B—Development

The development phase uses the detailed design developed in Phase 3B—Design to begin coding, moving the system one step closer to a final software product. Responsibilities in this phase rest primarily with programmers and systems analysts who are building the system. Key activities performed in a test/development environment include:

- Coding and developing program and system-level documents
- Debugging and testing the programs developed
- Developing programs to convert data from the old system for use on the new system
- Creating user procedures to handle transition to the new system
- Training selected users on the new system because their participation will be needed
- Ensuring modifications are documented and applied accurately and completely to vendor-acquired software to ensure that future updated versions of the vendor's code can be applied

PROGRAMMING METHODS AND TECHNIQUES

To enhance the quality of programming activities and future maintenance capabilities, program coding standards should be applied. Program coding standards are essential to writing, reading and understanding code, simply and clearly, without having to refer back to design specifications. Elements of program coding standards include methods and techniques for internal (source code level) documentation, methods for data declaration, and an approach to statement construction and techniques for input/output (I/O). The programming standards applied are an essential control because they serve as a method of communicating among members of the program team, and between the team and users during system development. Program coding standards minimize system development setbacks resulting from personnel turnover, provide the material needed to use the system effectively, and are required for efficient program maintenance and modifications.

Additionally, traditional structured programming techniques should be applied in developing quality and easily maintained software products. They are a natural progression from the top-down structuring design techniques previously described. Like the design specifications, structured application programs are easier to develop, understand and maintain because they are divided into subsystems, components, modules, programs, subroutines and units. Generally, the greater extent to which each software item described performs a single, dedicated function (cohesion) and retains independence from other comparable items (coupling), the easier it is to maintain and enhance a system because it is easier to determine where and how to apply a change, and reduce the chances of unintended consequences.

ONLINE PROGRAMMING FACILITIES (INTEGRATED DEVELOPMENT ENVIRONMENT)

To facilitate effective use of structured programming methods and techniques, an online programming facility should be available as part of an integrated development environment (IDE). This allows programmers to code and compile programs interactively with a remote computer or server from a terminal or a client's PC workstation. Through this facility, programmers can enter, modify and delete programming codes as well as compile, store and list programs (source and object) on the development computer. The online facilities can also be used by non-IS staff to update and retrieve data directly from computer files.

Online programming facilities are used on PC workstations. The program library is on a server, such as a mainframe library management system, but the modification/development and testing are performed on the workstation. This approach can lower the development costs, maintain rapid response time and expand the programming resources and aids available (e.g., editing tools, programming languages, debugging aids). From the perspective of control, this approach introduces the potential weaknesses of:

- The proliferation of multiple versions of programs
- Reduced program and processing integrity through the increased potential for unauthorized access and updating
- The possibility that valid changes could be overwritten by other changes

In general, an online programming facility allows faster program development and helps to enforce the use of standards and structured programming techniques. Online systems improve the programmer's problem-solving abilities, but online systems create vulnerabilities resulting from unauthorized access. Access control software should be used to help reduce the risk.

PROGRAMMING LANGUAGES

Application programs must first be coded in statements, instructions or a programming language that is easy for a programmer to write and that can be read by the computer. These statements (source code) will then be translated by the language translator/compiler into a binary machine code or machine language (object code) that the computer can execute.

Programming languages commonly used for developing application programs are:

- High-level, general-purpose programming languages such as COBOL and the C programming language
- Object-oriented languages for business purposes such as C++, Eiffel and Java

- IDEs such as Visual Studio or JBuilder, which provide coding templates automatically
- Hypertext Markup Language (HTML)
- Scripting languages such as shell, Perl, Tcl, Python, JavaScript and VBScript. In web development, scripting languages are used commonly to write common gateway interface (CGI) scripts that are used to extend the functionality of web server application software (e.g., to interface with search engines, create dynamic web pages and respond to user input).
- Low-level assembler languages designed for a specific processor type that are usually used for embedded applications (e.g., slot machines, vending machines, aerospace devices)
- Fourth-generation, high-level programming languages (4GLs), which consist of a database management system (DBMS), embedded database manager, and a nonprocedural report and screen generation facility. 4GLs provide fast iteration through successive designs. Examples of 4GLs include FOCUS, Natural and dBase.
- Decision support or expert systems languages (EXPRESS, Lisp and Prolog)

PROGRAM DEBUGGING

Many programming bugs are detected during the system development process, after a programmer runs a program in the test environment. The purpose of debugging programs during system development is to ensure that all program abends (unplanned ending of a program due to programming errors) and program coding flaws are detected and corrected before the final program goes into production. A debugging tool is a program that will assist a programmer in debugging, fixing or fine-tuning the program under development. Compilers have some potential to provide feedback to a programmer, but they are not considered debugging tools. These tools fall into three main categories:

- **Logic path monitors**—Report on the sequence of events performed by the program, thus providing the programmer with clues on logic errors
- **Memory dumps**—Provide a picture of the internal memory's content at one point in time. This is often produced at the point where the program fails or is aborted, providing the programmer with clues on inconsistencies in data or parameter values. A variant, called a trace, will do the same at different stages in the program execution to show changes in machine-level structures such as counters and registers.
- **Output analyzers**—Help check results of program execution for accuracy. This is achieved by comparing expected results with the actual results.

TESTING

Testing is an essential part of the development process that verifies and validates that a program, subsystem or application performs the functions for which it has been designed. Testing also determines whether the units being tested operate without any malfunction or adverse effect on other components of the system.

The variety of development methodologies and organizational requirements provide for a large range of testing schemes or levels. Each set of tests is performed with a different set of data and under the responsibility of different people or functions. The IS auditor can play a preventive or detective role in the testing process.

ELEMENTS OF A SOFTWARE TESTING PROCESS

To guide the testing process and help ensure that all facets of the system function as expected, basic elements for application software testing activities have been defined and are discussed below.

Developed early in the life cycle and refined until the actual **testing phase**, test plans identify the specific portions of the system to be tested. Test plans may include a categorization of types of deficiencies that can be found during the test. Categories of such deficiencies may be system defects, incomplete requirements, designs, specifications, or errors in the test case itself. Test plans also specify severity levels of problems found as well as guidelines on identifying the business priority. The tester determines the severity of the problem found during testing. Based on the severity level, the problem may be fixed prior to implementation or may be noted for correction following implementation. Often, cosmetic problems with the interface are classified as a lower severity and may not be fixed if time constraints become an issue for the project manager. This would be an example of a low severity, low priority defect. However, a defect in a report output such as spelling, look and feel, or specific content may also be a low-severity defect for the tester, but a high-priority defect for the business and thus would warrant resolution within the time constraints of the project. The project sponsor, end-user management and the project manager decide early in the test phase on the severity definitions.

Test plans also identify test approaches, such as the two reciprocal approaches, to software testing:

- **Bottom up**—Begin testing of atomic units, such as programs or modules, and work upward until a complete system testing has taken place. The advantages are:
  - No need for stubs or drivers
  - Can be started before all programs are complete
  - Errors in critical modules are found early
- **Top down**—Follow the opposite path, either in depth-first or breadth-first search order. The advantages are:
  - Tests of major functions and processing are conducted early
  - Interface errors can be detected sooner
  - Confidence is raised in the system because programmers and users actually see a working system

Generally, most application testing of large systems follows a bottom-up testing approach that involves ascending levels of integration and testing (e.g., unit or program, subsystem/integration, system, etc.):

- **Conduct and report test results**—Describe resources implied in testing, including personnel involved and information resources/facilities used during the test as well as actual versus expected test results. Results reported, along with the test plan, should be retained as part of the system's permanent documentation.
- **Address outstanding issues**—Identify errors and irregularities from the actual tests conducted. When such problems occur, the specific tests in question have to be redesigned in the test plan until acceptable conditions occur when the tests are redone.

TESTING CLASSIFICATIONS

The following tests relate, to varying degrees, to the above approaches that can be performed based on the size and complexity of the modified system:

- **Unit testing**—The testing of an individual program or module. Unit testing uses a set of test cases that focus on the control structure of the procedural design. These tests ensure that the internal operation of the program performs according to specification.
- **Interface or integration testing**—A hardware or software test that evaluates the connection of two or more components that pass information from one area to another. The objective is to take unit-tested modules and build an integrated structure dictated by design. The term integration testing is also used to refer to tests that verify and validate the functioning of the application under test with other systems, where a set of data is transferred from one system to another.
- **System testing**—A series of tests designed to ensure that modified programs, objects, database schema, etc., which collectively constitute a new or modified system, function properly. These test procedures are often performed in a nonproduction test/development environment by software developers designated as a test team. The following specific analyses may be carried out during system testing:
  - Recovery testing—Checking the system's ability to recover after a software or hardware failure
  - Security testing—Making sure the modified/new system includes provisions for appropriate access controls and does not introduce any security holes that might compromise other systems
  - Load testing—Testing an application with large quantities of data to evaluate its performance during peak hours
  - Volume testing—Studying the impact on the application by testing with an incremental volume of records to determine the maximum volume of records (data) that the application can process
  - Stress testing—Studying the impact on the application by testing with an incremental number of concurrent users/services on the application to determine the maximum number of concurrent users/services the application can process
  - Performance testing—Comparing the system's performance to other equivalent systems using well-defined benchmarks
- **Final acceptance testing**—After the system staff is satisfied with their system tests, the new or modified system is ready for the acceptance testing, which occurs during the implementation phase. During this testing phase, the defined methods of testing to apply should be incorporated into the organization's QA methodology. QA activities should proactively encourage that adequate levels of testing be performed on all software development projects. Final acceptance testing has two major parts: quality assurance testing (QAT) focusing on technical aspects of the application, and user acceptance testing (UAT) focusing on functional aspect of the application. QAT and UAT have different objectives and, therefore, should not be combined.

QAT focuses on the documented specifications and the technology employed. It verifies that the application works as documented by testing the logical design and the technology itself. It also ensures that the application meets the documented technical specifications and deliverables. QAT is performed primarily by the IT department. The participation of the end user is minimal and on request. QAT does not focus on functionality testing.

> **Note:** The CISA candidate should be familiar with the need for coding standards and with details on QA activities, software QA plan and the application QA function.

UAT supports the process of ensuring that the system is production-ready and satisfies all documented requirements. The methods include:

- Definition of test strategies and procedures
- Design of test cases and scenarios
- Execution of the tests
- Utilization of the results to verify system readiness

Acceptance criteria are defined criteria that a deliverable must meet to satisfy the predefined needs of the user. A UAT plan must be documented for the final test of the completed system. The tests are written from a user's perspective and should test the system in a manner as close to production as possible. For example, tests may be based around typical predefined, business process scenarios. If new business processes have been developed to accommodate the new or modified system, they should also be tested at this point. A key aspect of testing should also include testers seeking to verify that supporting processes integrate into the application in an acceptable fashion. Successful completion would generally enable a project team to hand over a complete integrated package of application and supporting procedures.

Ideally, UAT should be performed in a secure testing or staging environment. A secure testing environment where both source and executable code are protected helps to ensure that unauthorized or last-minute changes are not made to the system without going through the standard system maintenance process. The nature and extent of the tests will be dependent on the magnitude and complexity of the system change.

Even though packaged systems are tested by the vendor prior to distribution, these systems and any subsequent changes should be tested thoroughly by the end user and the system maintenance staff. These supplemental tests will help ensure that programs function as designed by the vendor and the changes do not interact adversely with existing systems.

In the case of acquired software, after attending to the changes during testing by the vendor, the accepted version should be controlled and used for implementation. In the absence of controls, the risk of introducing malicious patches/Trojan horse programs is very high.

Some organizations rely on integrated test facilities (ITFs). Test data usually are processed in production-like systems. This confirms the behavior of the new application or modules in real-life conditions. These conditions include peak volume and other resource-related constraints. In this environment, IS will perform their tests with a set of fictitious data whereas client representatives use extracts of production data to cover the most possible scenarios as well as some fictitious data for scenarios that would not be tested by the production data. In some organizations in which a subset of production data is used in a test environment, such production data may be altered to scramble the

data so that the confidential nature of data is obscured from the tester. This is often the case when the acceptance testing is done by team members who, under usual circumstances, would not have access to such production data.

On completion of acceptance testing, the final step is usually a certification and accreditation process. Certification/accreditation should only be performed after the system is implemented and in operation for some time to produce the evidence needed for certification/accreditation processes. This process includes evaluating program documentation and testing effectiveness. The process will result in a final decision for deploying the business application system. For information security issues, the evaluation process includes reviewing security plans, the risk assessments performed and test plans, and the evaluation process results in an assessment of the effectiveness of the security controls and processes to be deployed. Generally involving security staff and the business owner of the application, this process provides some degree of accountability to the business owner regarding the state of the system that he/she will accept for deployment.

Note: The CISA candidate should be familiar with details on independent certification, evaluating confidentiality for security certification and benchmarking.

When the tests are completed, the IS auditor should issue an opinion to management as to whether the system meets the business requirements, has implemented appropriate controls, and is ready to be migrated to production. This report should specify the deficiencies in the system that need to be corrected and should identify and explain the risk that the organization is taking by implementing the new system.

OTHER TYPES OF TESTING

Other types of testing include:

- **Alpha and beta testing**—An alpha version is an early version of the application system (or software product) submitted to internal users for testing. The alpha version may not contain all of the features that are planned for the final version. Typically, software goes through two stages of testing before it is considered finished. The first stage, called alpha testing, is often performed only by users within the organization developing the software (i.e., systems testing). The second stage, called beta testing, a form of user acceptance testing, generally involves a limited number of external users. Beta testing is the last stage of testing, and normally involves real-world exposure, sending the beta version of the product to independent beta test sites or offering it free to interested users.
- **Pilot testing**—A preliminary test that focuses on specific and predetermined aspects of a system. It is not meant to replace other testing methods, but rather to provide a limited evaluation of the system. Proofs of concept are early pilot tests—usually over interim platforms and with only basic functionalities.
- **White box testing**—Assesses the effectiveness of software program logic. Specifically, test data are used in determining procedural accuracy or conditions of a program's specific logic paths (i.e., applicable to unit and integration testing). However, testing all possible logic paths in large information systems is not feasible and would be cost-prohibitive and, therefore, is used on a select basis only.
- **Black box testing**—An integrity-based form of testing associated with testing components of an information system's "functional" operating effectiveness without regard to any specific internal program structure. Applicable to integration (interface) and user acceptance testing processes.
- **Function/validation testing**—It is similar to system testing but is often used to test the functionality of the system against the detailed requirements to ensure that the software that has been built is traceable to customer requirements (i.e., Are we building the right product?).
- **Regression testing**—The process of rerunning a portion of a test scenario or test plan to ensure that changes or corrections have not introduced new errors. The data used in regression testing should be the same as the data used in the original.
- **Parallel testing**—This is the process of feeding test data into two systems—the modified system and an alternative system (possibly the original system)—and comparing the results.
- **Sociability testing**—The purpose of these tests is to confirm that the new or modified system can operate in its target environment without adversely impacting existing systems. This should cover not only the platform that will perform primary application processing and interfaces with other systems but, in a client server or web development, changes to the desktop environment. Multiple applications may run on the user's desktop, potentially simultaneously, so it is important to test the impact of installing new dynamic link libraries (DLLs), making OS registry or configuration file modifications, and possibly extra memory utilization.

AUTOMATED APPLICATION TESTING

Automated testing techniques are used in this process. For example, test data generators can be used to systematically generate random data that can be used to test programs. The generators work by using the field characteristics, layout and values of the data. In addition to test data generators, there are interactive debugging aids and code logic analyzers available to assist in the testing activities.

Phase 5—Final Testing and Implementation

During the implementation phase, the actual operation of the new information system is established and tested. Final UAT is conducted in this environment. The system may also go through a certification and accreditation process to assess the effectiveness of the business application at mitigating risk to an appropriate level, and provide management accountability over the effectiveness of the system in meeting its intended objectives and establishing an appropriate level of internal control.

After a successful full-system testing, the system is ready to migrate to the production environment. The programs have been tested and refined; program procedures and production schedules are in place; all necessary data have been successfully converted and loaded into the new system; and the users have developed procedures and been fully trained in the use of the new system. A date (or dates) for system migration is determined and production turnover takes place. In the case of large organizations and complex systems, this may involve a project in itself and require a phased approach.

Planning for the implementation should commence well in advance of the actual implementation date, and a formal implementation plan should be constructed in the design phase and revised accordingly as development progresses. Each step in setting up the production environment should be stipulated, including who will be responsible, how the step will be verified and the backout procedure if problems are experienced. If the new system will interface with other systems or is distributed across multiple platforms, some final commissioning tests of the production environment may be desirable to prove end-to-end connectivity. If such tests are run, care will be needed to ensure test transactions do not remain in production databases or files.

In the case of acquired software, the implementation project should be coordinated by user management with the help of IS management, if required. The total process should not be delegated to the vendor—to avoid possible unauthorized changes or introduction of malicious code by the vendor's employees/representatives.

After operations are established, the next step is to perform site acceptance testing, which is a full-system test conducted on the actual operations environment. UAT supports the process of ensuring that the system is production-ready and satisfies all documented requirements.

IMPLEMENTATION PLANNING

Once developed and ready for operation, the new system delivered by the project will need an efficient support structure. It is not enough to set up roles for a support structure and naming people to fulfill these roles. Support personnel will need to acquire new skills. Workload has to be distributed, in order for the right people to support the right issues, thus new processes have to be developed while respecting the specificities of IT department requirements. Additionally, an infrastructure dedicated for support staff has to be made available.

For these and other reasons, setting up a support structure normally is a project in itself and requires planning, a methodology and good practices adaptation from past experiences.

The objective of such a project is to develop and establish the to-be support structure for the new technical infrastructure.

The main goals are to:
- Provide appropriate support structures for first-, second- and third-line support teams
- Provide a single point of contact (SPOC)
- Provide roles and skills definitions with applicable training plans

Often the project sponsor's company operates and supports a legacy solution and will implement a new system environment based on new system architecture. The existing support procedures and the organizational units will have to maintain the future system to provide the appropriate level of support for the new platform as well as for the old one.

One of the major challenges, therefore, is to manage the phases from build to integrate, to migrate, and for the phasing-out of the existing system and the phasing-in of the new one.

The migration cannot be accomplished via a single event. Instead, a step-by-step transition of the affected services must take place. Further, the implemented processes for a legacy environment might be different from what may be implemented with the new platform and any changes must be communicated to users and system support staff.

To achieve significant success in updating staff on changes to the business process and introducing new software, it is necessary to address some important questions such as:
- How can the existing support staff be involved in the setup of the new project without neglecting the currently running system?
- What is the gap of knowledge/skills that must be addressed in the training plan?
- How large is the difference from the current legacy environment operation to the operation of the new platform?

Generally, a transition project should conform to the following guidelines:
- There should be a smooth transition from the existing platform to the new platform, without any negative effect on users of the system.
- There should be maximum employment of the existing support staff to operate the new system environment and keep the effort of new hires at a minimum level.

Step 1—Develop To-be Support Structures

Step 1 includes the development of:
- **Gap analysis:** One possible approach to determine the gap—the differences between the current support organization and the future one—is to schedule workshops with the appropriate staff members who are dealing with system operational tasks or support processes at present. This should also include representatives of the current help desk unit.
  - The gap analysis should be based on the results of those workshops and on the data that can be gathered from self-assessment where representatives of all present support units will take part. The topics of the gap analysis should include processes, skills, tools, headcount, services of the help desk and interfaces to other organizational units.
- **Role definitions:** The definitions for the required roles must be provided in detail.

Step 2—Establish Support Functions

Step 2 includes the development of service level agreements (SLAs) and implementation and training plans.

Service Level Agreement

The SLA should at least consider the following key attributes:
- Operating time
- Support time
- Mean time between failures (MTBF)
- Mean time to repair (MTTR)
- Technical support response time

All attributes of the SLA must be measurable with a reasonable effort.

Implementation Plan/Knowledge Transfer Plan

In accordance with good practices, the transfer should follow the shadowing and relay-baton method. Shadowing gives staff the opportunity to become accustomed to the system by observation. The relay-baton approach is the best suitable concept to transfer knowledge and also to transfer responsibility in a transparent way. The metaphor of the relay-baton expresses exactly what must be achieved—that is, knowledge is transferred in small portions.

Training Plans

After the roles and responsibilities are defined, they will be documented in the form of a chart to allow for a clear and easy-to-read overview.

The training plans for the staff should show all of the required training in terms of:

- Content
- Scheduling information
- Duration
- Delivery mechanism (classroom and/or web-based)
- Train-the-trainer concept

The plan should consider the role definitions and skill profiles for the new to-be structure, and the results of the gap analysis. The plan takes into account that the staff who need to be trained must still run the current system, so a detailed coordination with the daily business tasks is maintained.

The following list gives an example of work tasks defined to fulfill the overall project goal:

- Collate existing support structure documentation.
- Review the existing IT organization model.
- Define the new support organization structure.
- Define the new support processes.
- Map the new process to the organization model.
- Execute the new organization model.
- Establish support functions.
- Develop communications material for support staff.
- Conduct briefing and training sessions.
- Review mobilization progress.
- Transfer to new organization structure.
- Review of items above.

END-USER TRAINING

The goal of a training plan is to ensure that the end user can become self-sufficient in the operation of the system.

One of the most important keys in end-user training is to ensure that training is considered and a training project plan is created early in the development process. A strategy can be developed that would take into consideration the timing, extent and delivery mechanisms. The training should be piloted using a cross-section of users to determine how best to customize the training to the different user groups. Following the pilot, the training approach can be adjusted as necessary, based on the feedback received from the pilot group. Separate classes should be developed for individuals who will assist in the training process. These train-the-trainer classes also provide useful feedback for improving the content of the training program. The timing of the delivery of training is very important. If training is delivered too early, users will forget much of the training by the time the system actually goes into production. If training is delivered too late, there will not be enough time to obtain feedback from the pilot group and implement the necessary changes into the main training program.

Training classes should be customized to address skill level and needs of users based on their role within the organization.

To develop the training strategy, the organization must name a training administrator. The training administrator will identify users who need to be trained with respect to their specific job functions. Consideration should be given to the following format and delivery mechanisms:

- Case studies
- Role-based training
- Lecture and breakout sessions
- Modules at different experience levels
- Practical sessions on how to use the system
- Remedial computer training (if needed)
- Online sessions on the web or on a CD-ROM

It is important to have a library of cases or tests, including user errors and the system response to those errors.

The training administrator needs to record student information in a database or spreadsheet, including student feedback for improving the training course.

DATA MIGRATION

A data conversion (also known as data porting) is required if the source and target systems utilize different field formats or sizes, file/database structures, or coding schemes. For example, a number may be stored as text, floating point or as binary-coded-decimal. Another example is the colors red, green and blue being represented using the codes "R," "G" and "B" on the existing system and "RD," "GN" and "BL" in the new system. Note that this conversion would also address a change in the field size. Conversions are often necessary when the source and target systems are on different hardware and/or OS platforms, and where different file or database structures (e.g., relational database, flat files, VSAM, etc.) are used. Another example of different coding schemes requiring conversion would be the use of different character representation schemes on the two systems (e.g., American standard code for information interchange [ASCII] vs. extended binary-coded decimal interchange code [EBCDIC]).

The objective of data conversion is to convert existing data into the new required format, coding and structure while preserving the meaning and integrity of the data. The data conversion process must provide some means, such as audit trails and logs, which allow for the verification of the accuracy and completeness of the converted data. This verification of accuracy and completeness may be performed through a combination of manual processes, system utilities, vendor tools and one-time-use special applications.

A large-scale data conversion can potentially become a project within a project as considerable analysis, design and planning will be required. Among the steps necessary for a successful data conversion are:

- Determining what data should be converted using programs and what, if any, should be converted manually
- Performing any necessary data cleansing ahead of conversion
- Identifying the methods to be used to verify the conversion, such as automated file comparisons, comparing record counts and control totals, accounting balances, and individual data items on a sample basis
- Establishing the parameters for a successful conversion. For example, is 100 percent consistency between the old and new systems necessary, or will some differences within defined ranges be acceptable?
- Scheduling the sequence of conversion tasks
- Designing audit trail reports to document the conversion, including data mappings and transformations
- Designing exception reports that will record any items that cannot be converted automatically
- Establishing responsibility for verifying and signing off on individual conversion steps and accepting the overall conversion (typically, the responsibility of the system owner)
- Developing and testing conversion programs, including functionality and performance
- Performing one or more conversion dress rehearsals to familiarize persons with the sequence of events and their roles, and test the conversion process end-to-end with real data
- Outsourcing the conversion process should be controlled with a proper agreement covering nondisclosure, data privacy, data destruction and other warranties
- Running the actual conversion with all necessary personnel onsite or able to be contacted

A successful data migration delivers the new system on time, on budget and with the required quality. The data migration project should be carefully planned and utilize appropriate methodologies and tools to minimize the risk of:
- Disruption of routine operations
- Violation of the security and confidentiality of data
- Conflicts and contention between legacy and migrated operations
- Data inconsistencies and loss of data integrity during the migration process

The data model and the new application model should be stored in an enterprise repository. Using a repository allows a simulation of the migration scenario and traceability during the project. An enterprise repository enables an overview of the reengineering and data migration process (e.g., which modules and entities are in which stage such as in service or already migrated). These models will be modified in the course of the processes described in the following sections.

Refining the Migration Scenario

In order to determine the scope of the implementation project, module analysis should be undertaken to identify the affected functional modules and data entities. The plan of the implementation project should be refined based on this information and an analysis of business requirements.

The next step is to develop a migration screenplay. This is a detailed listing of tasks for the production deployment of the new system. Within this plan decision points are defined to make "go" or "no-go" decisions. The following processes require decision points:

- **Support migration process**—A support process to administer the enterprise repository must be implemented. Because this repository should be used on completion of the project to manage the software components of the new architecture, this process should be capable of supporting future development processes. The enterprise repository administration and report generation supports the migration by supporting the reverse-engineering of changes in the legacy architecture and facilitating the creation of impact analysis reports.
- **Migration infrastructure**—The project develops specifications for the infrastructure of the migration project. This approach ensures consistency and increases confidence in the functionality of the fallback scenario. The migration project team completes a high-level analysis of the legacy and new data models to establish links between them that will be refined later. The migration infrastructure is the basis for specifying the following components:
  - Data redirector (temporary adapters)—Good practices suggest the staged deployment of applications to minimize the end-user impact of their implementation and limit the risk by having a fallback scenario with minimum impact. For this reason, an infrastructure component is needed to handle distributed data on different platforms within distributed applications. The design of a data redirector on the new architecture corresponds to service-oriented architectures and should cover features such as access to the not-yet-migrated legacy data during run time (e.g., EBCDIC–ASCII conversion), data consistency due to the usage of standards such as X/Open XA interface, and a homogeneous new architecture.
  - Data conversion components—The need to create an enterprise data model to eliminate data redundancies and inconsistencies often is identified. For this reason, infrastructure components to transform the legacy data model to the new data model must be provided. These components can be described as follows:
    - Unload components to copy the data (either "as is" or suitably modified to align with data format of target system) in legacy database that have been identified for migration
    - Transfer components to execute the data transfer from the legacy system to the new system
    - Load components to execute the load of the data into the new database

Software packages that support data migration, such as ERP and document management software, should be acquired as soon as the software evaluation is done. The data conversion plan should be based on the available databases and migration tools provided by the selected vendor(s).

Fallback (Rollback) Scenario

Not all new system deployments go as planned. To mitigate the risk of downtime for mission-critical systems, good practices dictate that the tools and applications required to reverse the migration are available prior to attempting the production cutover. Some or all of these tools and applications may need to be developed as part of the project.

Components have to be delivered that can back out all changes and restore data to the original applications in the case of nonfunctioning new applications. Two types of components should be considered as part of a fallback contingency plan:

1. The first consists of: (i) unload components to execute the unloading of the data from the new data structures, (ii) transfer components for the data conversion, and (iii) load components to execute the loading of the data into the legacy data structures.
2. The second consists of: (i) a log component to log the data modifications within the new data model during runtime within the service layer, (ii) transfer components for the data conversion, and (iii) load components to execute the load of the data into the legacy data structures.

The decision on which method to use for data conversion has to be made as part of the implementation project and should be based on the following criteria:

- Transaction volume
- Change degree of the data model

In summary, data migration projects begin with planning and preparation. The first step is to understand the new system's data structure. This is done by reading the software user guides, analyzing the entity relationship diagrams, understanding the relationships between data elements and reviewing definitions of key terms (such as entity and record) in the new system.

The next step in data conversion is to review the decisions on how business processes should be conducted in the new system. Changes are identified and the output of this exercise is a table of new data terminology against current definitions of data elements. In this step, the project team identifies how current data are defined in the new system. Following this step, a data cleanup is completed to eliminate inconsistencies in the current database, if possible, and duplications of data sets are discovered and resolved. The rules of conversion are defined and documented with the objective of ensuring the business processes executed in the new system yield results that maintain data integrity and relationships.

Data conversion rules are programmed by the software development team. Data conversion scripts are created to convert the data from the old database to the new database. The data conversion scripts are tested on a discrete selection of data that is carefully selected to include all cases. This is referred to as program or unit testing. Following the sign-off of data conversion scripts by programmers, the scripts are run on a test copy of the production database. The values of data are tested by executing tests including business process tests. Users and developers complete cycles of testing until conversion scripts are fine-tuned. After testing has been completed, the next step is to promote the converted database to production. Data conversion can also be done manually if the conversion of data requires interpretation or if the conditions or data conversion rules are too complex for programming.

The key points to be taken into consideration in a data conversion project are to ensure:

- Completeness of data conversion (i.e., the total number of records from the source database is transferred to the new database [assuming the number of fields is the same])
- Integrity of data (i.e., the data are not altered manually, mechanically or electronically by person, program, substitution or overwriting in the new system): Integrity problems also include errors due to transposition, transcription errors, and problems transferring particular records, fields, files and libraries.
- Storage and security of data under conversion (i.e., data are backed up before conversion for future reference or any emergency that might arise out of data conversion program management): Unauthorized copy or too many copies can lead to misuse, abuse or theft of data from the system.
- Consistency of data (i.e., the field/record called for from the new application should be consistent with that of the original application): This should enable consistency in repeatability of the testing exercise.
- Continuity (i.e., the new application should be able to continue with newer records as addition [append] and help in ensuring seamless business continuity)
- The last copy of the data before conversion from the old platform and the first copy of the data after conversion to the new platform should be maintained separately in the archive for any future reference

CHANGEOVER (GO-LIVE OR CUTOVER) TECHNIQUES

Changeover refers to an approach to shift users from using the application from the existing (old) system to the replacing (new) system. This is appropriate only after testing the new system with respect to its program and relevant data. This is sometimes called the go-live technique because it enables the start of the new system. This approach is also called the cutover technique because it helps in cutting out from the older system and moving over to the newer system.

This technique can be achieved in three different ways. See **figures 3.15, 3.16** and **3.17**.

Figure 3.15—Parallel Changeover

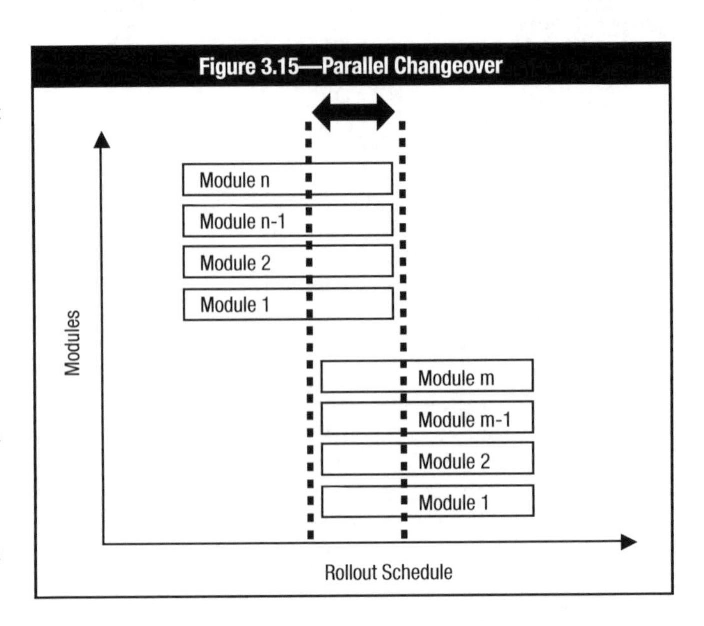

Figure 3.16—Phased Changeover

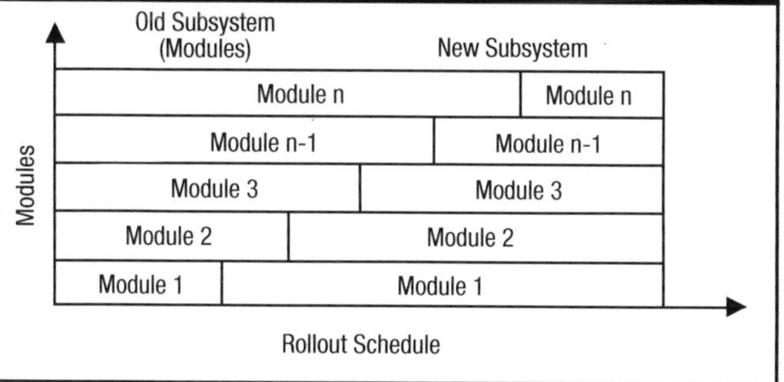

Figure 3.17—Abrupt Changeover

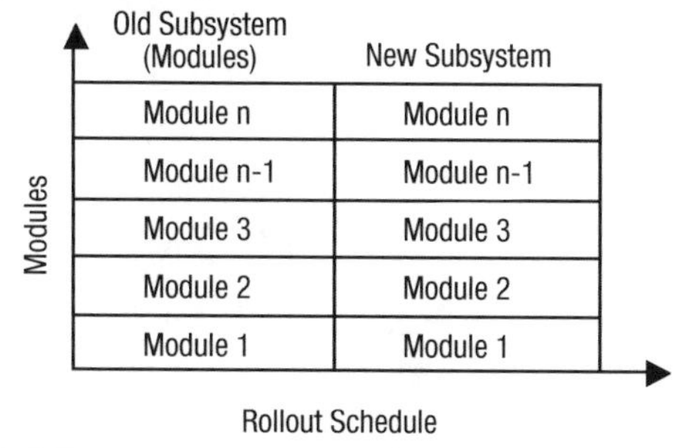

Parallel Changeover

This technique includes running the old system, then running both the old and new systems in parallel, and finally, fully changing over to the new system after gaining confidence in the working of the new system. With this approach, the users will have to use both systems during the period of overlap. This will minimize the risk of using the newer system and, at the same time, help in identifying problems, issues or any concerns that the user comes across in the newer system in the beginning. After a period of overlap, the user gains confidence and assurance in relying on the newer system. At this point, the use of the older system is discontinued and the new system becomes totally operational. Note in **figure 3.15** that the number (m, n, respectively) of modules in the new and old systems may be different.

Phased Changeover

In this approach the older system is broken into deliverable modules. Initially, the first module of the older system is phased out using the first module of the newer system. Then, the second module of the older system is phased out, using the second module of the newer system and so forth until reaching the last module. Thus, the changeover from the older system to the newer system takes place in a preplanned, phased manner.

Some of the risk areas that may exist in the phased changeover include:

- Resource challenges (both on the IT side—to be able to maintain two unique environments such as hardware, OSs, databases and code; and on the operations side—to be able to maintain user guides, procedures and policies, definitions of system terms, etc.)
- Extension of the project life cycle to cover two systems
- Change management for requirements and customizations to maintain ongoing support of the older system

Abrupt Changeover

In this approach the newer system is changed over from the older system on a cutoff date and time, and the older system is discontinued once changeover to the new system takes place.

Changeover to the newer system involves four major steps or activities:

1. Conversion of files and programs; test running on test bed
2. Installation of new hardware, OS, application system and the migrated data
3. Training employees or users in groups
4. Scheduling operations and test running for go-live or changeover

Some of the risk areas related to changeover include:

- Asset safeguarding
- Data integrity
- System effectiveness
- System efficiency
- Change management challenges (depending on the configuration items considered)
- Duplicate or missing records (duplicate or erroneous records may exist if data cleansing is not done correctly)

CERTIFICATION/ACCREDITATION

Certification is a process by which an assessor organization performs a comprehensive assessment against a standard of management and operational and technical controls in an information system. The assessor examines the level of compliance in meeting certain requirements such as standards, policies, processes, procedures, work instructions and guidelines—requirements made in support of accreditation. The goal is to determine the extent to which controls are implemented correctly, operating as intended and producing the desired outcome with respect to meeting the system's security requirements. The results of a certification are used to reassess the risk and update the system security plan, thus providing the factual basis for an authorizing official to render an accreditation decision.

Accreditation is the official management decision (given by a senior official) to authorize operation of an information system and to explicitly accept the risk to the organization's operations, assets or individuals based on the implementation of an agreed-upon set of requirements and security controls. Security accreditation provides a form of quality control and challenges managers and technical staff at all levels to implement the most effective security controls possible in an information system, given mission requirements, and technical, operational and cost/schedule constraints.

By accrediting an information system, a senior official accepts responsibility for the security of the system and is fully accountable for any adverse impact to the organization if a breach of security occurs. Thus, responsibility and accountability are core principles that characterize accreditation.

> **Note:** The CISA candidate should be familiar with the auditor's role in the certification process.

Phase 6—Postimplementation Review

Following the successful implementation of a new or extensively modified system, it is beneficial to verify the system has been properly designed and developed and that proper controls have been built into the system. A postimplementation review should meet the following objectives:

- Assess the adequacy of the system.
  - Does the system meet user requirements and business objectives?
  - Have access controls been adequately defined and implemented?
- Evaluate the projected cost benefits or ROI measurements.
- Develop recommendations that address the system's inadequacies and deficiencies.
- Develop a plan for implementing the recommendations.
- Assess the development project process.
  - Were the chosen methodologies, standards and techniques followed?
  - Were appropriate project management techniques used?

It is important to note that, for a postimplementation review to be effective, the information to be reviewed should be identified during the project feasibility and design phase, and collected during each stage of the project. For instance, the project manager might establish certain checkpoints to measure effectiveness of software development processes and accuracy of software estimates during the project execution. Business measurements should also be established up front and collected before the project begins and after the project is implemented (for examples of critical success factor measurements, see **figure 3.13**).

It is also important to allow a sufficient number of business cycles to be executed in the new system to realize the new system's actual return on investment.

A postproject review should be performed jointly by the project development team and appropriate end users. Typically, the focus of this type of internal review is to assess and critique the project process, whereas a postimplementation review has the objective of assessing and measuring the value the project has on the business (benefits realization).

Alternatively, an independent group not associated with the project implementation (internal or external audit) can perform a postimplementation review. The IS auditors performing this review should be independent of the system development process. Therefore, IS auditors involved in consulting with the project team on the development of the system should not perform this review. Unlike internal project team reviews, postimplementation reviews performed by IS auditors have a tendency to concentrate on the control aspects of the system development and implementation processes.

It is important that all audit involvement in the development project be thoroughly documented in the audit work papers to support the IS auditor's findings and recommendations. This audit report and documentation should be reused during maintenance and changes to validate, verify and test the impact of any changes made to the system. The system should periodically undergo a review to ensure the system is continuing to meet business objectives in a cost-effective manner and that control integrity still exists.

> **Note:** The CISA candidate should be familiar with issues related to dual control as they apply to authorization within the postimplementation review and with those related to reviewing results of live processing.

3.5.3 INTEGRATED RESOURCE MANAGEMENT SYSTEMS

A growing number of organizations—public and private—are shifting from separate groups of interrelated applications to a fully integrated corporate solution. Such solutions are often marketed as ERP solutions. Many vendors, mainly from Europe and the USA, have been focusing on this market and offer packages with commercial names such as SAP®, Oracle® Financials or SSG (Baan).

An integrated solutions implementation is a very large software acquisition project. The acquisition and implementation of an ERP system impacts the way the corporation does business, its entire control environment, technological direction and internal resources. Generally, a corporation that adopts an integrated solution is required to convert management philosophies, policies and practices to those of the integrated software solution providers, notwithstanding the numerous customization options. In this respect, such a solution will either impair or enhance IT's ability to support the organization's mission and goals. When considering a change of this magnitude, it is imperative that a thorough impact and risk assessment be conducted.

When implementing an ERP solution or any off-the-shelf software, the business unit has the option of implementing and configuring the new system in the simplest configuration possible: as-is, out-of-the-box and not developing any additional functionality or customization to bridge the gaps in the corporation's specific business processes. The business opts to change business processes to suit the industry standard as dictated by the software solution. While this decision results in less software design, development and testing work than does customization, it does require greater change in the business units to work differently. Due to the large costs in software development, maintenance and continuous upgrading and patching, customization is not usually recommended by software vendors.

Because of the magnitude of the risk involved, it is imperative that senior management assess and approve all plans and changes in the system's architecture, technological direction, migration strategies and IS budgets.

3.5.4 RISK ASSOCIATED WITH SOFTWARE DEVELOPMENT

There are many potential types of risk that can occur when designing and developing software systems.

One type of risk is business risk (or benefit risk), relating to the likelihood that the new system may not meet the users' business needs, requirements and expectations. For example, the business requirements that were to be addressed by the new system are still unfulfilled, and the process has been a waste of resources. In such a case, even if the system is implemented, it will most likely be underutilized and not maintained, making it obsolete in a short period of time.

Another risk is project risk (or delivery risk), where the project activities to design and develop the system exceed the limits of the financial resources set aside for the project and, as a result, it may be completed late, if ever. There are many potential types of risk that can occur when designing and developing software systems. Software project risk exists at multiple levels:
- Within the project (e.g., risk associated with not identifying the right requirements to deal with the business problem or opportunity that the system is meant to address and not managing the project to deliver within time and cost constraints)
- With suppliers (e.g., risk associated with a failure to clearly communicate requirements and expectations, resulting in suppliers delivering late, at over expected cost and/or with deficient quality)
- Within the organization (e.g., risk associated with stakeholders not providing needed inputs or committing resources to the project, and changing organizational priorities and politics
- With the external environment (e.g., risk associated with impacts on the projects caused by the actions and changing preferences of customers, competitors, government/regulators and economic conditions)
- With the technology chosen (e.g., sudden displacement of technology chosen by a one more cost efficient; insufficient compatibility in the marketplace, resulting in barriers to potential clients' use of the new system)

The foremost cause of these problems is a lack of discipline in managing the software development process or the use of a methodology inappropriate to the system being developed. In such instances, organizations are not providing the infrastructure and support necessary to help projects avoid these problems. In such cases, successful projects, if occurring, are not repeatable, and SDLC activities are not defined and followed adequately (i.e., insufficient maturity). However, with effective management, SDLC management activities can be controlled, measured and improved.

The IS auditor should be aware that merely following an SDLC management approach does not ensure the successful completion of a development project. The IS auditor should also review the management discipline over a project related to the following:
- The project meets cooperative goals and objectives
- Project planning is performed, including effective estimates of resources, budget and time
- Scope creep is controlled and there is a software baseline to prevent requirements from being added into the software design or having an uncontrolled development process
- Management is tracking software design and development activities
- Senior management support is provided to the software project's design and development efforts
- Periodic review and risk analysis is performed in each project phase

(See section 3.12 Process Improvement Practices for more information.)

3.6 VIRTUALIZATION AND CLOUD COMPUTING ENVIRONMENTS

The need for greater scalability and agility in the market place and efforts to lower operating costs have driven the growth in both system virtualization and its use within cloud service providers (CSPs). To develop effective audit programs, the IS auditor must obtain a clear understanding of both virtualization and CSP architectures supporting the organization business applications and processes.

3.6.1 VIRTUALIZATION

As the sophistication and complexity of business processes grow, so does the supporting network and computing infrastructure. The concept of virtualization is generally believed to have its origins in the late 1960s and early 1970s when the mainframe-enabled robust time-sharing option (TSO) environments with several users accessing the OS concurrently, without impacting others, are also accessing the same OS—resulting in more effective use of central processing unit (CPU) processing, memory and storage space. However, the high cost of these resources at that time did not enable the ability to deploy these computing capabilities on the scale as seen in today's virtualization model. The same rationale for the TSO model is being applied today via virtualization because the majority of current server technologies are underutilized and capable of supporting much more than one or two server functions.

Data centers and many other organizations use virtualization techniques to create an abstraction of the physical hardware and make large pools of logical resources consisting of CPUs, memory, disks, file storage, applications and networking. This approach enables greater availability of these resources to the user base. The main focus of virtualization is to enable a single physical computing environment to run multiple logical, yet independent, systems at the same time.

The most common use for full virtualization is operational efficiency, which uses existing hardware more efficiently by placing greater loads on each computer. Second, using full virtualization of desktops enables end users to have one computer hosting multiple OSs if needed to support various OS-dependent applications. Furthermore, the IT organization can better control deployed OSs to ensure that they meet the organization's security requirements, security threat and respective control requirements are dynamic, and the virtual desktop images can be changed to respond to new threats.

Elements of the virtualized computing environment are normally comprised of the following:
- Server or other hardware product
- Virtualization hypervisor: A piece of computer software, firmware or hardware that creates and runs virtual machine environment—normally called the "host."
- Guest machine: Virtual environment elements (e.g., OS, switches, routers, firewalls, etc.) residing on the computer on which a hypervisor host machine has been installed

There are two methods of deploying a fully virtualized environment. **Figure 3.18** compares these architectures.

- **Bare metal/native virtualization** occurs when the hypervisor runs directly on the underlying hardware, without a host OS.
- **Hosted virtualization** occurs when the hypervisor runs on top of the host OS (Windows, Linux or MacOS). The hosted virtualization architectures usually have an additional layer of software (the virtualization application) running in the guest OS that provides utilities to control the virtualization while in the guest OS, such as the ability to share files with the host OS.

Key Risk Areas

Overall, migrating computing resources to a virtualized environment does not change the threat plane for most of the systems' vulnerabilities and threats. If a service has inherent vulnerabilities on a physical server or network product and it is migrated to a virtualized server, the service remains vulnerable to exploitation. However, the use of virtualization may also provide additional virtual environment attack vectors (e.g., hypervisor misconfiguration or security flaws, memory leakage, etc.), thus increasing the likelihood of successful attacks. The following types of high-level risk are representative of the majority of virtualized systems in use:

- Rootkits on the host installing themselves as a hypervisor below the OS, enabling the interception of any operations of the guest OS (i.e., logging password entry, etc.): Antivirus software may not detect this, because the malware runs below the entire OS.
- Default and/or improper configuration of the hypervisor partitioning resources (CPU, memory, disk space and storage): This can lead to unauthorized access to resources, one guest OS injecting malware into another or placing malware code into another guest OS's memory.
- On hosted virtualization, mechanisms called guest tools enable a guest OS to access files, directories, the copy/paste buffer, and other resources on the host OS or another guest OS: This functionality can inadvertently provide an attack vector for malware or allow an attacker to gain access to particular resources.
- Snapshots/images of guests' environments contain sensitive data (such as passwords, personal data, etc.) like a physical hard drive: These snapshots pose a greater risk than images because snapshots contain the contents of random access memory (RAM) at the time that the snapshot was taken, and this might include sensitive information that was not stored on the drive itself.
- In contrast to bare metal installations, hosted virtualization products rarely have hypervisor access controls: Therefore, anyone who can launch an application on the host OS can run the hypervisor. The only access control is whether someone can log into the host OS.

Typical Controls

The IS auditor will need to understand the following concepts:

- Hypervisors and guest images (OS and networks) are securely configured according to industry standards. Apply hardening to these virtual components as closely as one would to a physical server, switch, router, firewall or other computing device.
- Hypervisor management communications should be protected on a dedicated management network. Management communications carried on untrusted networks should be encrypted, and encryption should encapsulate the management traffic.
- The hypervisor should be patched as the vendor releases the fixes.
- The virtualized infrastructure should be synchronized to a trusted authoritative timeserver.
- Unused physical hardware should be disconnected from the host system.
- All hypervisor services, such as clipboard- or file-sharing between the guest OS and the host OS, should be disabled unless they are needed.
- Host inspection capabilities should be enabled to monitor the security of each guest OS. Hypervisor security services can allow security monitoring even when the guest OS is compromised.
- Host inspection capabilities should be enabled to monitor the security of activity occurring between guest OSs. Of special focus is communications in a non-virtualized environment carried and monitored over networks by network security

Figure 3.18—Full Virtualization Architectures

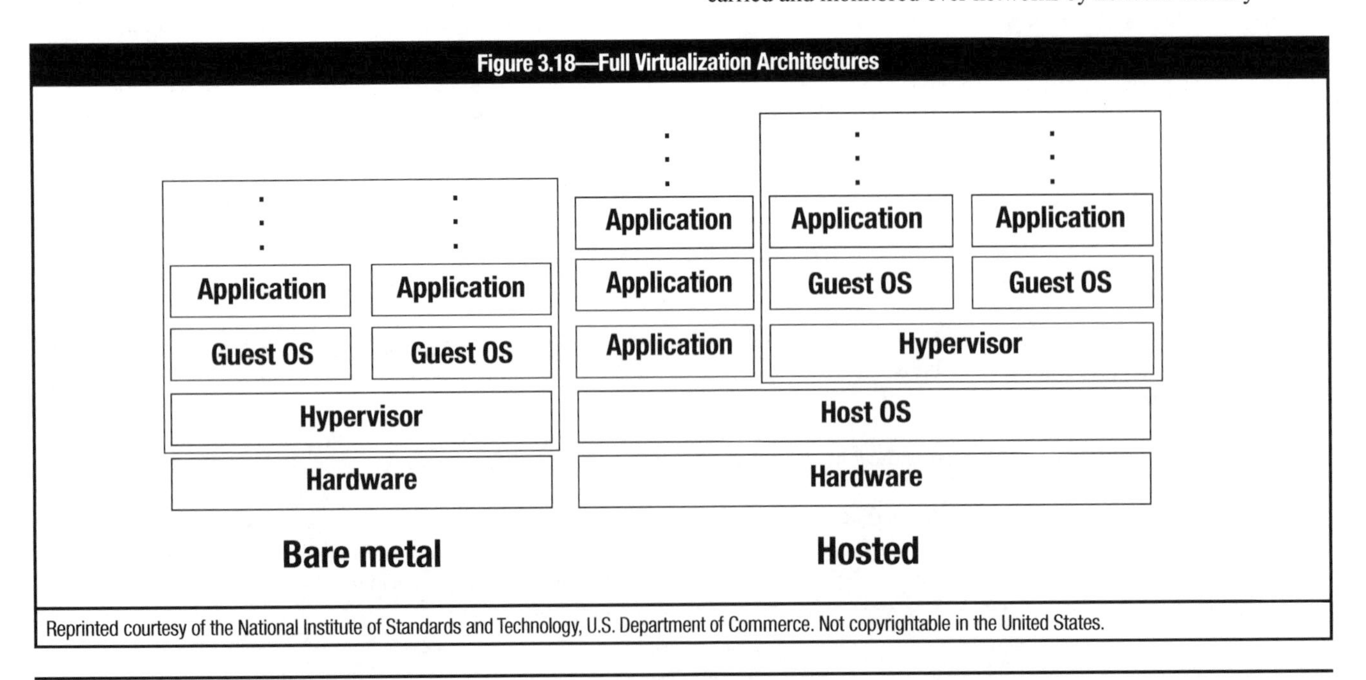

Reprinted courtesy of the National Institute of Standards and Technology, U.S. Department of Commerce. Not copyrightable in the United States.

controls (such as network firewalls, security appliances, and network IDPS sensors).
- File integrity monitoring of the hypervisor should be used to monitor for signs of compromise.

See sections 5.4.1 LAN Security and 5.13 Cloud Computing for details regarding the risk and controls over virtual computing environments.

3.7 BUSINESS APPLICATION SYSTEMS

To develop effective audit programs, the IS auditor must obtain a clear understanding of the application system under review. Some types of application systems and the related processes are described in the following sections.

Numerous financial and operational functions are computerized for the purpose of improving efficiency and increasing the reliability of information. These applications range from traditional (including general ledger, accounts payable and payroll) to industry-specific (such as bank loans, trade clearing and material requirements planning). Given their unique characteristics, computerized application systems add complexity to audit efforts. These characteristics may include limited audit trails, instantaneous updating and information overload. Application systems may reside in the various environments that follow.

3.7.1 E-COMMERCE

E-commerce is the buying and selling of goods online, usually via the Internet. Typically, a web site will advertise goods and services, and the buyer will fill in a form on the web site to select the items to be purchased and provide delivery and payment details or banking services such as transfers and payment orders. The web site may gather details about customers and offer other items that may be of interest. The cost of a brick-and-mortar store is avoided, and the savings are often a benefit to the customers, sometimes leading to spectacular growth. The term e-business includes buying and selling online as well as other aspects of online business such as customer support or relationships between businesses.

E-commerce, as a general model, uses technology to enhance the processes of commercial transactions among a company, its customers and business partners. The used technology can include the Internet, multimedia, web browsers, proprietary networks, automatic teller machines (ATMs) and home banking, and the traditional approach to electronic data interchange (EDI). However, the primary area of growth in e-commerce is through the use of the Internet as an enabling technology.

E-commerce Models

E-commerce models include the following:
- **Business-to-consumer (B-to-C) relationships**—The greatest potential power of e-commerce comes from its ability to redefine the relationship with customers in creating a new convenient, low-cost channel to transact business. Companies can tailor their marketing strategies to an individual customer's needs and wants. As more of its business shifts online, a company will have an enhanced ability to track how its customers interact with it.
- **Business-to-business (B-to-B) relationships**—The relationship among the selling services of two or more businesses opens up the possibility of reengineering business processes across the boundaries that have traditionally separated external entities from each other. Because of the ease of access and the ubiquity of the Internet; for example, companies can build business processes (order processing, payments and after-sale service) that combine previously separated activities. The result is a faster, higher-quality and lower-cost set of transactions. The market has even created a subdivision of B-to-B called business-to-small business (B-to-SB) relationships.
- **Business-to-employee (B-to-E) relationships**—Web technologies also assist in the dissemination of information to and among an organization's employees.
- **Business-to-government (B-to-G) relationships**—Covers all transactions between companies and government organizations. Currently this category is expanding quite rapidly as governments use their own operations to promote awareness and growth of e-commerce. In addition to public procurement, administrations may also offer the option of electronic interchange for such transactions as VAT returns and the payment of corporate taxes.

E-COMMERCE ARCHITECTURES

There are a large number of choices to be made in determining an appropriate e-commerce architecture. Initially, e-commerce architectures were either two-tiered (i.e., client browser and web server) or three-tiered (i.e., client browser, web server and database server). With increasing emphasis on integrating the web channel with a business' internal legacy systems and the systems of its business partners, company systems now typically will run on different platforms, running different software and with different databases. It is also the case that, in addition to supporting browser connections, companies eventually may move in the direction of supporting connections from active content clients, mobile phones or other wireless devices, and host-to-host connections. Systems that rely on multiple computer platforms are described as "*n*-tiered" also referred to as multitiered architectures. The *n*-tiered architecture separates presentation, application processing and data management functions, thereby allowing modification to be made to any of these functions instead of reworking all as one application. Types of architectures include:
- Single-tier architecture is a client-based application running on a single computer.
- Two-tier architecture is composed of the client and server.
- Three-tier architecture is comprised of the following:
  - The presentation tier displays information which users can access directly such as a web page, or an OS's GUI.
  - The application tier (business logic/applications) controls an application's functionality by performing detailed processing.
  - The data tier is usually comprised of the database servers, file shares, etc.) and the data access layer that encapsulates the persistence mechanisms and exposes the data.

The challenge of integrating diverse technologies within and beyond the business has increasingly led companies to move to component-based systems that utilize a middleware infrastructure based around an application server. This supports current trends in the evolution of software development: build systems from

proven quality and cataloged components, just as hardware is built. While this is yet to be fully realized, component models—notably Microsoft Component Object Model (COM) and Oracle Enterprise JavaBeans (EJB)—are widely used and fall under the grouping of "mobile code." Mobile code is software transferred between systems (i.e., transferred across a network) and executed on a local system using cross-platform code without explicit installation by the recipient computer (e.g., Adobe® Flash®, Shockwave®, Java applets, VBScripts, ActiveX, etc.). The continued adoption of mobile code brings another vector for the spread of malware via ever-evolving delivery vectors ranging from email, malicious web sites and mobile device applications.

E-components often seen in a B-to-C system include marketing, sales and customer service components (e.g., personalization, membership, product catalog, customer ordering, invoicing, shipping, inventory replacement, online training and problem notification).

Application servers support a particular component model and provide services (such as data management, security and transaction management) either directly or through connection to another service or middleware product such as MQSeries. A number of major software vendors offer application servers.

Application servers in conjunction with other middleware products provide for multitiered systems (i.e., a business transaction can span multiple platforms and software layers). For example, a system's presentation layer typically will consist of a browser or other client application. A web server will be used to manage web content and connections; business logic and other services will be provided by the application server; and one or more database(s) will be used for data storage.

Databases play a key role in most e-commerce systems, maintaining data for web site pages, accumulating customer information and storing click-stream data for analyzing web site usage. To provide full functionality and achieve back-end efficiencies, an e-commerce system may involve connections to in-house legacy systems—accounting, inventory management or an ERP system—or business partner systems. Thus, further business logic and data persistence tiers are added.

For security reasons, persistent customer data should not be stored on web servers that are exposed directly to the Internet. Extensible Markup Language (XML) is also likely to form an important part of an organization's overall e-commerce architecture. While originally conceived as a technique to facilitate electronic publishing, XML was quickly seized on as a medium that could store and enclose any kind of structured information so it could be passed between different computing systems. XML has emerged as a key means of exchanging a wide variety of data on the web and elsewhere. In addition to basic XML, a variety of associated standards has been and is continuing to be developed. Some of these include:

- **Extensible Stylesheet Language (XSL)**—Defines how an XML document is to be presented (e.g., on a web page)
- **XML query (XQuery)**—Deals with querying XML format data
- **XML encryption**—Deals with encrypting, decrypting and digitally signing XML documents

A particularly important offshoot of XML is web services. Web services represent a way of using XML format information to remotely invoke processing. Because a web services message can contain both an XML document and a corresponding schema defining the document, in theory, it is self-describing and assists in achieving the goal of "loose coupling." If the format of a web services message changes, the receiving web services will still work, provided the accompanying schema is updated. This advantage, combined with the support for web services that has emerged from major software industry players such as IBM and Microsoft, means web services is now the key middleware to connect distributed web systems.

Cooperation between the different software vendors means that web services can interoperate, irrespective of hardware, OS and programming language.

It is necessary to reach some agreement on metadata definitions for web services to serve as a means of enabling cooperative processing across organizational boundaries. (Metadata are data about data, and the term is referred to in web services' standards as ontology). Web services may be successfully called and the resulting XML data may be successfully parsed by the calling program, but to use these data effectively it is necessary to understand the business meaning of the data. This is similar to previous attempts at interorganizational computing (such as EDI), where it was necessary to agree in advance on electronic document formats and meanings. The Object Management Group (OMG) and *ISO/IEC 19510:2013: Information technology—Object Management Group Business Process Model and Notation* have established a recognized industry-standard XML documents and standard business process definitions that can be represented in XML.

E-COMMERCE RISK

E-commerce, as any other form of commerce, depends on the existence of a level of trust between two parties. For example, the Internet presents a challenge between the buyer and seller, similar to those a catalog or direct-mail retailer faces. The challenges are proving to the buyer that the seller is who they say they are, proving to the buyer that their personal information, such as credit card numbers (and other personally identifiable information), remains confidential and that the seller cannot later refute the occurrence of a valid transaction. Some of the most important elements at risk are:

- **Confidentiality**—Potential consumers are concerned about providing unknown vendors with personal (sometimes sensitive) information for a number of reasons including the possible theft of credit card information from the vendor following a purchase. Connecting to the Internet via a browser requires running software on the computer that has been developed by someone unknown to the organization. Moreover, the medium of the Internet is a broadcast network, which means that whatever is placed on it is routed over wide-ranging and essentially uncontrolled paths. The current trend of outsourcing and hosting services on the cloud expands the risk perimeter beyond the boundaries of the transacting entity.
- **Integrity**—Data, both in transit and in storage, could be susceptible to unauthorized alteration or deletion (i.e., hacking or the e-business system itself could have design or configuration problems).

- **Availability**—The Internet holds out the promise of doing business on a 24-hour, seven-day-a-week basis. Hence, high availability is important, with any system's failure becoming immediately apparent to customers or business partners.
- **Authentication and nonrepudiation**—The parties to an electronic transaction should be in a known and trusted business relationship, which requires that they prove their respective identities before executing the transaction in preventing man-in-the-middle attacks (i.e., preventing the seller from being an impostor). Then, after the fact, there must be some manner of ensuring that the transacting parties cannot deny that the transaction was entered into and the terms on which it was completed.
- **Power shift to customers**—The Internet gives consumers unparalleled access to market information and generally makes it easier to shift between suppliers. Firms participating in e-business need to make their offerings attractive and seamless in terms of service delivery. This will involve not only system design, but also reengineering of business processes. Back-end support processes need to be as efficient as possible because, in many cases, doing business over the Internet forces down prices (e.g., online share brokering). To avoid losing their competitive advantage of doing business online, firms need to enhance their services, differentiate from the competition and build additional value. Hence, the drive to personalize web sites by targeting content based on analyzed customer behavior and allowing direct contact with staff through instant messaging technology and other means.

It is important to take into consideration the importance of security issues that extend beyond confidentiality objectives.

E-COMMERCE REQUIREMENTS

Some e-commerce requirements include:
- Build a business case (IT as an enabler).
- Develop a clear business purpose.
- Use technology to first improve costs.
- Build business case around the four C's: customers, costs, competitors and capabilities.

Other requirements for e-commerce include the following:
- **Top-level commitment**—Because of the breadth of changes required (i.e., business processes, company culture, technology and customer boundaries), e-commerce cannot succeed without a clear vision and strong commitment from the top of the organization.
- **Business process reconfiguration**—Technology is not the key innovation needed to make e-commerce work, but it is the ingenuity needed to envision how that technology can enable the company to fundamentally reconfigure some of its basic business processes. This requires thinking that is outside-the-box and outside-the-walls (i.e., looking outside of the organization and understanding what customers are doing and how changes in the overall process can create new value for them).
- **Links to legacy systems**—Organizations must take seriously the requirement to accelerate response times, provide real interaction to customers and customize responses to individual customers. Specifically, in applying enterprise application integration (EAI), organizations must create online interfaces and make sure those interfaces communicate with existing databases and systems for customer service and order processing. A term often referred to in establishing this communication is middleware, which is defined as independent software and services that distributed business applications use to share computing resources across heterogeneous technologies. A range of middleware technologies—message brokers, gateways, process managers, data transformation software and file transfer—are likely to be deployed to create an integration infrastructure. Increasingly, integration will be viewed not as a responsibility of an individual application development team, but as something to be managed across the organization using a standard approach and technologies.

E-COMMERCE AUDIT AND CONTROL ISSUES (GOOD PRACTICES)

When reviewing the adequacy of contracts in e-commerce applications, audit and control professionals should assess applicable use of the following items:
- For B-to-C, B-to-B, B-to-E and B-to-G, interconnection agreements need to be established and documented prior to engaging in the e-commerce agreement. This can be as simple as accepting terms of use for B-to-C and B-to-E systems to detailed terms and conditions to be in place before the e-commerce interconnections can be established.
- Security mechanisms and procedures that, taken together, constitute a security architecture for e-commerce (e.g., Internet firewalls, public key infrastructure [PKI], encryption, certificates, PCI DSS compliance and password management)
- Firewall mechanisms that are in place to mediate between the public network (the Internet) and an organization's private network
- A process whereby participants in an e-commerce transaction can be identified uniquely and positively (e.g., process of using some combination of public and private key encryption and certifying key pairs)
- Digital signatures so the initiator of an e-commerce transaction can be uniquely associated with it. Attributes of digital signatures include:
  - The digital signature is unique to the person using it.
  - The signature can be verified.
  - The mechanism for generating and affixing the signature is under the sole control of the person using it.
  - The signature is linked to data in such a manner that if the data are changed, the digital signature is invalidated.
- Infrastructure to manage and control public key pairs and their corresponding certificates which include:
  - Certificate authority (CA)—Attests, as trusted provider of the public/private key pairs, to the authenticity of the owner (entity or individual) to whom a public/private key pair has been given. The process involves a CA who makes a decision to issue a certificate based on evidence or knowledge obtained in verifying the identity of the recipient. On verifying the identity of the recipient, the CA signs the certificate with its private key for distribution to the user. On receipt, the user will decrypt the certificate with the CA's public key (e.g., commercial CAs such as Verisign® provide public keys on web browsers). The ideal CA is authoritative (someone that the user trusts) for the name or key space it represents.
  - Registration authority (RA)—An optional entity separate from a CA that would be used by a CA with a very large customer

base. CAs use RAs to delegate some of the administrative functions associated with recording or verifying some or all of the information needed by a CA to issue certificates or certification revocation lists (CRLs) and to perform other certificate management functions. However, with this arrangement, the CA retains sole responsibility for signing either digital certificates or CRLs. If an RA is not present in the established PKI structure, the CA is assumed to have the same set of capabilities as defined for an RA.
  - Certification revocation list (CRL)—Instrument for checking the continued validity of the certificates. If a certificate is compromised, if the holder is no longer authorized to use the certificate or if there is a fault in binding the certificate to the holder, the certificate must be revoked and taken out of circulation as rapidly as possible and all parties in the trust relationship must be informed. The CRL is usually a highly controlled online database through which subscribers and administrators may determine the status of a target partner's certificate.
  - Certification practice statement (CPS)—A detailed set of rules governing the certificate authority's operations. The CPS provides an understanding of the value and trustworthiness of certificates issued by a given CA, the terms of the controls that an organization observes, the method used to validate the authenticity of certificate applicants and the CA's expectations of how its certificates may be used.
- Procedures in place to control changes to an e-commerce presence
- E-commerce application logs, which are monitored by responsible personnel. This includes OS logs and console messages, network management messages, firewall logs and alerts, router management messages, intrusion detection alarms, application and server statistics, and system integrity checks.
- Methods and procedures to recognize security breaches when they occur (network and host-based intrusion detection systems [IDSs])
- Features in e-commerce applications to reconstruct the activity performed by the application
- Protection in place to ensure that data collected about individuals are not disclosed without the individuals' consent nor used for purposes other than that for which they are collected
- Means to ensure confidentiality of data communicated between customers and vendors (safeguarding resources such as through encrypted Secure Sockets Layer [SSL])
- Mechanisms to protect the presence of e-commerce and supporting private networks from computer viruses and to prevent them from propagating viruses to customers and vendors
- Features within the e-commerce architecture to keep all components from failing and allow them to repair themselves, if they should fail
- Plan and procedure to continue e-commerce activities in the event of an extended outage of required resources for normal processing
- Commonly understood set of practices and procedures to define management's intentions for the security of e-commerce
- Shared responsibility within an organization for e-commerce security
- Communications from vendors to customers about the level of security in an e-commerce architecture
- Regular program of audit and assessment of the security of e-commerce environments and applications to provide assurance that controls are present and effective

3.7.2 ELECTRONIC DATA INTERCHANGE

EDI replaces the traditional paper document exchange, such as medical claims and records, purchase orders, invoices, or material release schedules. Therefore, the proper controls and edits need to be built within each company's application system to allow this communication to take place.

General Requirements

An EDI system requires communications software, translation software and access to standards. Communications software moves data from one point to another, flags the start and end of an EDI transmission and determines how acknowledgments are transmitted and reconciled. Translation software helps build a map and shows how the data fields from the application correspond to elements of an EDI standard. Later, it uses this map to convert data back and forth between the application and EDI formats.

To build a map, an EDI standard appropriate for the kind of EDI data to be transmitted is selected. For example, there are specific standards for medical claims, patient records, invoices, purchase orders, advance shipping notices, etc.

The final step is to write a partner profile that tells the system where to send each transaction and how to handle errors and exceptions.

In summary, components of an EDI process include system software and application systems. EDI system software includes transmission, translation and storage of transactions initiated by or destined for application processing. EDI is also an application system in that the functions it performs are based on business needs and activities. The applications, transactions and trading partners supported will change over time, and the intermixing of transactions, purchase orders, shipping notices, invoices and payments in the EDI process makes it necessary to include application processing procedures and controls in the EDI process.

In reviewing EDI, IS auditors need to be aware of the two approaches related to EDI: the traditional proprietary version of EDI used by large companies and government parties, and the development of EDI through the publicly available commercial infrastructure offered through the Internet. The difference between the approaches relates to cost, where use of a public commercial infrastructure such as the Internet provides significantly reduced costs versus development of a customized proprietary approach. From a security standpoint, risk associated with not having a completely trustworthy relationship arise in addressing Internet security and risk.

Traditional EDI

Moving data in a batch transmission process through the traditional EDI process generally involves three functions within each trading partner's computer system:

1. **Communications handler**—Process for transmitting and receiving electronic documents between trading partners via dial-up lines, public-switched network, multiple dedicated

lines or a value-added network (VAN). VANs use computerized message switching and storage capabilities to provide electronic mailbox services similar to a post office. The VAN receives all the outbound transactions from an organization, sorts them by destination and passes them to recipients when they log on to check their mailbox and receive transmissions. VANs may also perform translation and verification services. VANs specializing in EDI applications also provide technical support, help desk and troubleshooting assistance for EDI and telecommunications problems. VANs help in configuration of software, offer upgrades to telecommunications connectivity, provide data and computer security, audit and trace transactions, recover lost data, and confirm service reliability and availability.

2. **EDI interface**—Interface function that manipulates and routes data between the application system and the communications handler. The interface consists of two components:
   - EDI translator—This device translates the data between the standard format (ANSI X12) and a trading partner's proprietary format.
   - Application interface—This interface moves electronic transactions to or from the application systems and performs data mapping. Data mapping is the process by which data are extracted from the EDI translation process and integrated with the data or processes of the receiving company. The EDI interface may generate and send functional acknowledgments, verify the identity of partners and check the validity of transactions by checking transmission information against a trading partner master file. Functional acknowledgments are standard EDI transactions that tell the trading partners that their electronic documents were received. Different types of functional acknowledgments provide various levels of detail and can, therefore, act as an audit trail for EDI transactions.
3. **Application system**—The programs that process the data sent to, or received from, the trading partner. Although new controls should be developed for the EDI interface, the controls for existing applications, if left unchanged, are usually unaffected.

Application-initiated transactions (such as purchase orders from the purchasing system) are passed to a common application interface for storage and interpretation. All outbound transactions are formatted according to an externally defined standard and batched by destination and transaction type by the translator. The batches of transactions, like functional groups, are routed to the communications processor for transmission. This entire process is reversed for inbound transactions, including invoices destined for the purchasing and accounts payable systems. Controls need to recognize and deal with error conditions and provide feedback on the process for the EDI system to be considered well managed.

Web-based EDI

Web-based EDI has come into prominence because of the following:

- Internet-through-Internet service providers offer a generic network access (i.e., not specific to EDI) for all computers connected to the Internet, whereas VAN services have typically used a proprietary network or a network gateway linked with a specific set of proprietary networks. The result is a substantially reduced cost to EDI applications.
- Its ability to attract new partners via web-based sites to exchange information, take orders and link the web site to back-end order processing and financial systems via EDI
- New security products available to address issues of confidentiality, authentication, data integrity and nonrepudiation of origin and return
- Improvements in the X12 EDI formatting standard

Web-based EDI trading techniques aim to improve the interchange of information between trading partners, suppliers and customers by bringing down the boundaries that restrict how they interact and do business with each other. For example, the use of Internet service provider (ISP)-related services can provide functions similar to those of the more traditional VANs, but with a much broader array of available services (i.e., processing transactions of all types through the Internet). This is beneficial particularly for smaller organizations wanting to enter the e-commerce EDI market because ISPs have a ready network infrastructure of servers offering email, web services and the network of routers, and modems attached to a permanent, high-speed Internet "backbone" connection.

3.7.3 EDI RISK AND CONTROLS

The hybrid nature of EDI adds a new dimension to the design and auditing of the EDI process. The traditional procedures for managed and controlled implementation of system software—such as requirements definition, version and release identification, testing and limited implementation with a fallback strategy—apply to software used for EDI. In addition, there are issues and risk unique to EDI.

Transaction authorization is the biggest EDI risk. Because the interaction between parties is electronic, there is no inherent authentication occurring. Computerized data can look the same no matter what the source and do not include any distinguishing human element or signature.

Where responsibilities of trading partners are not clearly defined by a trading partner agreement, there could be uncertainty related to specific, legal liability. Therefore, it is important that, to protect both parties, any agreement is codified legally in what is known as a trading partner agreement. Another risk is the loss of business continuity. Corruption of EDI applications, whether done innocently or deliberately, could affect every EDI transaction undertaken by a company. This would have a negative impact on both customer and vendor relations. In an extreme situation, it could ultimately affect the ability of a company to stay in business.

Additional security types of risk include:

- Unauthorized access to electronic transactions
- Deletion or manipulation of transactions prior to or after establishment of application controls
- Loss or duplication of EDI transmissions
- Loss of confidentiality and improper distribution of EDI transactions while in the possession of third parties

3.7.4 CONTROLS IN THE EDI ENVIRONMENT

Security risk can be addressed by enforcing general controls and establishing an added layer of application control procedures over the EDI process that can take over where traditional application controls leave off. These controls need to secure the current EDI activity as well as historical activities that may be called on to substantiate business transactions should a dispute arise.

To protect EDI transmissions, the EDI process should include the following electronic measures:

- Standards should be set to indicate that the message format and content are valid to avoid transmission errors.
- Controls should be in place to ensure that standard transmissions are properly converted for the application software by the translation application.
- The receiving organization must have controls in place to test the reasonableness of messages received. This should be based on a trading partner's transaction history or documentation received that substantiates special situations.
- Controls should be established to guard against manipulation of data in active transactions, files and archives. Attempts to change records should be recorded by the system for management review and attention.
- Procedures should be established to determine messages are only from authorized parties and transmissions are properly authorized.
- Direct or dedicated transmission channels among the parties should exist to reduce the risk of tapping into the transmission lines.
- Data should be encrypted using algorithms agreed on by the parties involved.
- Electronic signatures should be in the transmissions to identify the source and destination.
- Message authentication codes should exist to ensure that what is sent is received.

The EDI process needs the ability to detect and deal with transactions that do not conform to the standard format or are from/to unauthorized parties. Options for handling detected errors include requesting retransmissions or manually changing the data.

The critical nature of many EDI transactions, such as orders and payments, requires that there be positive assurances that the transmissions were complete. The transactions need to be successfully passed from the originating computer application to the destination organization. Methods for providing these assurances include internal batch total checking, run-to-run and transmission record count balancing, and use of special acknowledgment transactions for functional acknowledgments.

Organizations desiring to exchange transactions using EDI are establishing a new business relationship. This business relationship needs to be defined so both parties can conduct business in a consistent and trusting manner. This relationship usually is defined in a legal document called a trading partner agreement. The document should define the transactions to be used, responsibilities of both parties in handling and processing the transactions, as well as the written business terms and conditions associated with the transactions.

The evolving nature of EDI means that the transaction standards are also evolving, particularly with use of the Internet as an available, low-cost primary business transaction delivery system. Specifically, not all trading partners desire or need to use the current standard. As a result, the EDI process needs to adapt to changes in standards and be able to support multiple versions of the standard as ISPs offer more EDI-related services (vs. VANs).

Other issues relate to many organizations with a large, installed base of applications that need to be retrofit to accommodate EDI. In addition, not all transactions will be processed through EDI. Some transactions will continue to be processed in the traditional way. The application processing control procedures must be modified to include the EDI transaction processing and the dual sources/destinations.

Receipt of Inbound Transactions

Controls should ensure that all inbound EDI transactions are accurately and completely received (communication phase), translated (translation phase) and passed to an application (application interface phase) as well as processed only once.

The control considerations for receipt of inbound transactions are as follows:

- Use appropriate encryption techniques when using public Internet infrastructures for communication in assuring confidentiality, authenticity and integrity of transactions.
- Perform edit checks to identify erroneous, unusual or invalid transactions prior to updating an application.
- Perform additional computerized checking to assess transaction reasonableness, validity, etc. (Consider expert system front ends for complex comparisons.)
- Log each inbound transaction on receipt.
- Use control totals on receipt of transactions to verify the number and value of transactions to be passed to each application; reconcile totals between applications and with trading partners.
- Segment count totals built into the transaction set trailer by the sender.
- Control techniques in the processing of individual transactions such as check digits on control fields, loop or repeat counts.
- Ensure the exchange of control totals of transactions sent and received between trading partners at predefined intervals.
- Maintain a record of the number of messages received/sent and validate these with the trading partners periodically.
- Arrange for security over temporary files and data transfer to ensure that inbound transactions are not altered or erased between time of transaction receipt and application updates.

Outbound Transactions

Controls should ensure that only properly authorized outbound transactions are processed. This includes the objectives that outbound EDI messages are initiated on authorization, that they contain only preapproved transaction types and that they are sent only to valid trading partners.

The control considerations for outbound transactions are as follows:

- Control the set up and change of trading partner details.
- Compare transactions with trading partner transaction profiles.
- Match the trading partner number to the trading master file (prior to transmission).
- Limit the authority of users within the organization to initiate specific EDI transactions.
- Segregate initiation and transmission responsibilities for high-risk transactions.
- Document management sign-off on programmed procedures and subsequent changes.
- Log all payment transactions to a separate file, which is reviewed for authorization before transmission.

- Segregate duties within the transaction cycle, particularly where transactions are automatically generated by the system.
- Segregate access to different authorization processes in a transaction cycle.
- Report large (value) or unusual transactions for review prior to or after transmission.
- Log outbound transactions in a secure temporary file until authorized and due for transmission.
- Require paperless authorization that would establish special access to authorization fields (probably two levels, requiring the intervention of different users) within the computer system.

Auditing EDI

The IS auditor must evaluate EDI to ensure that all inbound EDI transactions are received and translated accurately, passed to an application, and processed only once.

To accomplish this goal, IS auditors must review the following:
- Internet encryption processes put in place to ensure authenticity, integrity, confidentiality and nonrepudiation of transactions
- Edit checks to identify erroneous, unusual or invalid transactions prior to updating the application
- Additional computerized checking to assess transaction reasonableness and validity
- Each inbound transaction to ensure that it is logged on receipt
- The use of control totals on receipt of transactions to verify the number and value of transactions to be passed to each application and reconcile totals between applications and with trading partners
- Segment count totals built into transaction set trailers by the sender
- Transaction set count totals built into the functional group headers by the sender
- Batch control totals built into the functional group headers by the sender
- The validity of the sender against trading partner details by:
  - Using control fields within an EDI message at either the transaction, function, group or interchange level (often within the EDI header, trailer or control record)
  - Using VAN sequential control numbers or reports (if applicable)
  - Sending an acknowledgment transaction to inform the sender of message receipt. The sender should then match this against a file/log of EDI messages sent.

EDI audits also involve:
- **Audit monitors**—Devices can be installed at EDI workstations to capture transactions as they are received. Such transactions can be stored in a protected file for use by the auditor. Consideration should be given to storage requirements for voluminous amounts of data.
- **Expert systems**—Within the context of utilizing the computer system for internal control checks, consideration should be given to having audit monitors evaluate the transactions received. Based upon judgmental rules, the system can determine the audit significance of such transactions and provide a report for the auditor's use.

As use of EDI becomes more widespread, additional methods for auditing transactions will be developed. It is important to stay current with these developments.

3.7.5 EMAIL

Email may be the most heavily used feature of the Internet or LANs in an organization. At the most basic level, the email process can be divided into two principal components:
- **Mail servers**—Hosts that deliver, forward and store mail
- **Clients**—Interface with users and allow users to read, compose, send and store email messages

Email messages are sent in the same way as most Internet data. When a user sends an email message, it is first broken up by the Transmission Control Protocol (TCP) into Internet Protocol (IP) packets. Those packets are then sent to an internal router (a router that is inside the user's network) that examines the address. Based on the address, the router decides whether the mail is to be delivered to someone on the same network or to someone outside of the network. If the mail goes to someone on the same network, the mail is delivered to them. If the mail is addressed to someone outside the network, it may pass through a firewall, which is a security mechanism (hardware/software) that shields the network from the broader Internet so intruders cannot break into the network. The firewall keeps track of messages and data going into and out of the network—to and from the Internet. The firewall also can prevent certain packets from getting through it.

Once out on the Internet, the message is sent to an Internet router. The router examines the address, determines where the message should be sent and sends the message on its way. A gateway at the receiving network receives the email message. This gateway uses TCP to reconstruct the IP packets into a full message. The gateway then translates the message into the protocol the target network uses and sends it on its way. The message may be required to also pass through a firewall on the receiving network. The receiving network examines the email address and sends the message to a specific mailbox.

A user can also attach binary files such as pictures, videos, sounds and executable files to the email message. To do this, the user must encode the file in a way that will allow it to be sent across the network. The receiver will have to decode the file once it is received. There are a variety of different encoding schemes that can be used. Some email software packages automatically do the encoding for the user and the decoding on the receiving end.

When a user sends email to someone on the Internet or within a closed network, that message often has to travel through a series of networks before it reaches the recipient. These networks might use different email formats. Gateways perform the job of translating email formats from one network to another so the messages can make their way through all the networks. An email message is made up of binary data, usually in the ASCII text format. ASCII is a standard that allows any computer, regardless of its OS or hardware, to read the text. ASCII code describes the characters that users see on their computer screens.

There are several email protocols in use. The following protocols are the primary protocols the IS auditor will need to be familiar with during the review of email services:
- Outgoing email
  - Simple Mail Transport Protocol (SMTP): can only be used to send emails, not to receive them

- Incoming email
  - Post Office Protocol (POP)
  - Internet Message Access Protocol (IMAP)
  - Hypertext Transfer Protocol (HTTP)—also called "web-based email"
  - Messaging Application Programming Interface (MAPI)—used with Outlook in conjunction with a Microsoft Exchange Server mail server; very close to IMAP but has extended features to interact with other applications

More organizations are moving their email systems to the cloud. This essentially outsources many of the maintenance and security management issues associated with maintaining email servers and shifts expenditures from capital investments to operational expenditures. It also provides additional scalability and availability, which would be more difficult to achieve in smaller IT operations. However, the IS auditor must be mindful of the regulatory requirements of his or her organization. For example, for government agencies, if a foreign national uses a certain type of email that contains restricted data, this may be in violation of a law or regulation.

Security Issues of Email

According to J. Klensin and the Network Working Group, when reading the engineering standards established for email, these documents clearly indicate the following:

- SMTP mail is inherently insecure; a low level of complexity and skill is needed to perform a man-in-the-middle attack between receiving and relaying SMTP servers to then spoof legitimate email traffic.
- Real mail security lies only in end-to-end methods involving the message bodies, such as those that use digital signatures or integrity checks provided, at the transport level.

Some other security issues involved in emails are as follows:

- Phishing and spear phishing are electronic social engineering attacks that have become exceedingly sophisticated and can only be addressed through security awareness training.
- Flaws in the configuration of the mail server application may be used as the means of compromising the underlying server and the attached network.
- Denial-of-service (DoS) attacks may be directed to the mail server, denying or hindering valid users from using the mail server.
- Sensitive information transmitted unencrypted between mail server and email client may be intercepted.
- Information within the email may be altered at some point between the sender and recipient.
- Viruses and other types of malicious code may be distributed throughout an organization via email.
- Users may send inappropriate, proprietary or other sensitive information via email leading to a legal exposure.

Standards for Email Security

To improve email security, organizations should:

- Address the security aspects of the deployment of a mail server through maintenance and administration standards
- Ensure that the mail server application is deployed, configured and managed to meet the security policy and guidelines instituted by management
- Consider the implementation of encryption technologies to protect user authentication and mail data

As authorizers in many organizations use email to communicate approvals for business transactions (e.g., payroll run, journal entry posting, payment authorization), it is important to adopt a calibrated approach for ensuring the authenticity of emails based on an informed risk assessment. For example, organizations may require emails communicating authorizations to be digitally signed by the sender before they can be acted upon. In email security, a digital signature authenticates a transmission from a user in an untrusted network environment. A digital signature is a sequence of bits appended to a digital document. Like a handwritten signature, its authenticity can be verified. Unlike a handwritten signature, it is unique to the document being signed. Digital signatures are another application of public key cryptography. Digital signatures are a good method of securing email transmissions because:

- The signature cannot be forged.
- The signature is authentic and encrypted.
- The signature cannot be reused (a signature on one document cannot be transferred to another document).
- The signed document cannot be altered; any alteration to the document (whether or not it has been encrypted) renders the signature invalid.

There are two different types of encryption techniques used to ensure security. Messages can be secured with a single, bidirectional (encrypt/decrypt), secret, symmetric key system using Advanced Encryption Standard (AES) or with an asymmetric key system using pairs of unidirectional (only encrypt or decrypt), complementary keys, such as the public key management system RSA, a public key cryptosystem, was developed by R. Rivest, A. Shamir and L. Adleman and is often used (see section 5.4.5 Encryption).

If the email transmission is secured with use of a symmetric key on the receiver's end, the user needs to know the single secret key to decrypt the message. If the transmission is secured with an asymmetric key system using a public key, the user at the receiving end needs to use the private key to decrypt the message, as well as a digital signature verification program to verify the signature. Digital signatures are based on a procedure called message digesting, which computes a short, fixed-length number called a digest for any message of any length. Several different messages may have the same digest, but it is extremely difficult to produce any of them from the digest. A message digest is a cryptographically strong, one-way hash function of the message. It is similar to a checksum in that it compactly represents the message and is used to detect changes in the message. The message digest authenticates the user's message in such a way that if it were altered, the message would be considered corrupted.

Organizations should employ their network infrastructure to protect their mail server(s) through appropriate use of firewalls, routers and IDSs. (See chapter 5 Protection of Information Assets, for more information.)

3.7.6 POINT-OF-SALE SYSTEMS

Point-of-sale (POS) systems enable the capture of data at the time and place that sales transactions occur. The most common payment instruments to operate with POS are credit and debit cards, which are associated with bank accounts. POS terminals may have attached peripheral equipment—such as optical scanners to read bar codes and magnetic card readers for credit or debit cards or electronic readers for smart cards—to improve the efficiency and accuracy of the transaction recording process.

POS systems may be online to a central computer owned by a financial institution or a third-party administrator or may use local processors/microcomputers owned by a business to hold the transactions for a specified period, after which they are sent to the main computer for batch processing.

It is most important for the IS auditor to determine whether any cardholder data is stored on the local POS system such as primary account numbers (PANs), personal identification numbers (PINs), etc. Any such information, if stored on the POS system, should be encrypted using strong encryption methods. Certain data can never be stored on these devices such as card verification value (CVV) numbers.

Further information on standards for POS systems can be obtained at *www.pcisecuritystandards.org*, which addresses information on Payment Card Industry Data Security Standards (PCI DSS), and at *www.emvco.com*, which addresses standards for smart cards using embedded microprocessor chips.

3.7.7 ELECTRONIC BANKING

Banking organizations have been remotely delivering electronic services to consumers and businesses for years. Electronic funds transfer (EFT) (including small payments and corporate cash management systems), publicly accessible automated machines for currency withdrawal and retail account management are global fixtures.

Continuing technological innovation and competition among existing banking organizations and new market entrants has allowed for a much wider array of electronic banking products and services for retail and wholesale banking customers. However, the increased worldwide acceptance of the Internet as a delivery channel for banking products and services provides new business opportunities as well as new risk.

Major risk associated with banking activities includes strategic, reputational, operational (including security—sometimes called transactional—and legal risk), credit, price, foreign exchange, interest rate and liquidity. Electronic banking (e-banking) activities do not raise risk that was not already identified in traditional banking, but e-banking increases and modifies some of types of traditional risk. The core business and the IT environment are tightly coupled, thereby influencing the overall risk profile of e-banking.

In particular, from the perspective of the IS auditor, the main issues are strategic, operational and reputational risk because these are directly related to threats to reliable data flow and operational risk and are certainly heightened by the rapid introduction and underlying technological complexity of electronic banking.

Risk Management Challenges in E-banking

E-banking presents a number of risk management challenges:

- The speed of change relating to technological and service innovation in e-banking is unprecedented. Currently, banks are experiencing competitive pressure to roll out new business applications in very compressed time frames. This competition intensifies the management challenge to ensure that adequate strategic assessment, risk analysis and security reviews are conducted prior to implementing new e-banking applications.
- Transactional e-banking web sites and associated retail and wholesale business applications are typically integrated as much as possible with legacy computer systems to allow more straight-through processing of electronic transactions. Such straight-through automated processing reduces opportunities for human error and fraud inherent in manual processes, but it also increases dependence on sound system design and architecture as well as system interoperability and operational scalability.
- E-banking increases banks' dependence on information technology, thereby increasing the technical complexity of many operational and security issues and furthering a trend toward more partnerships, alliances and outsourcing arrangements with third parties such as ISPs, telecommunication companies and other technology firms.
- The Internet is ubiquitous and global by nature. It is an open network accessible from anywhere in the world by unknown parties. Messages are routed through unknown locations and via fast-evolving wireless devices. Therefore, the Internet significantly magnifies the importance of security controls, customer authentication techniques, data protection, audit trail procedures and customer privacy standards.

Risk Management Controls for E-banking

Effective risk management controls for electronic banking include the following 15 controls divided among three categories:

- **Board and management oversight:**
  1. Effective management oversight of e-banking activities
  2. Establishment of a comprehensive security control process
  3. Comprehensive due diligence and management oversight process for outsourcing relationships and other third-party dependencies
- **Security controls:**
  4. Authentication of e-banking customers
  5. Nonrepudiation and accountability for e-banking transactions
  6. Appropriate measures to ensure SoD
  7. Proper authorization controls within e-banking systems, databases and applications
  8. Data integrity of e-banking transactions, records and information
  9. Establishment of clear audit trails for e-banking transactions
  10. Confidentiality of key bank information
- **Legal and reputational risk management:**
  11. Appropriate disclosures for e-banking services
  12. Privacy of customer information
  13. Capacity, business continuity and contingency planning to ensure availability of e-banking systems and services
  14. Incident response planning
  15. Compliance to banking sector directives (e.g., Basel Accords)

3.7.8 ELECTRONIC FINANCE

Electronic finance (e-finance) is an integral element of the financial services industry and enables providers to emerge within and across countries, including online banks, brokerages and companies that allow consumers to compare financial services such as mortgage loans and insurance policies. Nonfinancial entities are also entering the market, including telecommunication and utility companies that offer payment and other services.

Advantages of this approach to consumers are:
- Lower costs
- Increased breadth and quality
- Widening access to financial services
- Asynchrony (time-decoupled)
- Atopy (location-decoupled)

By using credit scoring and other data mining techniques, providers can create and tailor products over the Internet without much human input and at a very low cost. Providers can better stratify their customer base through analysis of Internet-collected data and allow consumers to build preference profiles online. This not only permits personalization of information and services, it also allows more personalized pricing of financial services and more effective identification of credit risk. At the same time, the Internet allows new financial service providers to compete more effectively for customers because it does not distinguish between traditional brick-and-mortar providers of financial services and those without physical presence. All these forces are delivering large benefits to consumers at the retail and commercial levels. These mechanisms should be used within privacy law statements (regarding confidentiality and authorization) to gather diverse user information and set up profiles.

3.7.9 PAYMENT SYSTEMS

There are two types of parties involved in all payment systems—the issuers and the users. An issuer is an entity that operates the payment service. An issuer holds the items that the payments represent (e.g., cash held in regular bank accounts). The users of the payment service perform two main functions—making payments and receiving payments—and, therefore, can be described as a payer or a payee, respectively.

Electronic Money Model

The objective of electronic money systems is to emulate physical cash. An issuer attempts to do this by creating digital certificates, which are then purchased (or withdrawn) by the users who redeem (deposit) them with the issuer at a later date. In the interim, certificates can be transferred among users to trade for goods or services.

For the certificates to take on some of the attributes of physical cash, certain techniques are used so that when a certificate is deposited, the issuer cannot determine the original withdrawer of the certificate. This provides the electronic certificates with unconditional untraceability.

Electronic money systems can be hard to implement in practice due to the overheads of solving the "double spending" problem (i.e., preventing a user from depositing the same money twice).

Some advantages of electronic money systems are:
- The payer does not need to be online (generally) at the time of the purchase (because electronic money can be stored on the payer's computer).
- The payer can have unconditional untraceability (albeit at the expense of lost interest on deposits).

Electronic Checks Model

Electronic check systems model real-world checks quite well and are thus relatively simple to understand and implement. A user writes an electronic check, which is a digitally signed instruction to pay. This is transferred (in the course of making a purchase) to another user, who then deposits the electronic check with the issuer. The issuer will verify the payer's signature on the payment and transfer the funds from the payer's account to the payee's account.

Some advantages of electronic check systems are:
- Easy to understand and implement
- The availability of electronic receipts, allowing users to resolve disputes without involving the issuer
- No need for payer to be online to create a payment

These systems are usually fully traceable, which is an advantage for certain law enforcement, tax collection and marketing purposes, but a disadvantage for those concerned about privacy.

Electronic Transfer Model

Electronic transfer systems are the simplest of the three payment models. The payer simply creates a payment transfer instruction, signs it digitally and sends it to the issuer. The issuer then verifies the signature on the request and performs the transfer. This type of system requires the payer to be online, but not the payee.

Some advantages of electronic transfer systems are:
- Easy to understand and implement
- The payee does not need to be online—a considerable advantage in some circumstances (e.g., paying employee wages)

3.7.10 INTEGRATED MANUFACTURING SYSTEMS

Integrated manufacturing systems (IMS) have a long history and, accordingly, there has been quite a diverse group of models and approaches.

Some of the integrated manufacturing systems include bill of materials (BOM), BOM processing (BOMP), manufacturing resources planning (MRP), computer-assisted design (CAD), computer-integrated (or computer-intensive) manufacturing (CIM) and manufacturing accounting and production (MAP).

Original IMSs were based on BOM and BOMP and usually supported by a hierarchical DBMS.

Evolution toward further integration with other business functions (e.g., recording of raw materials, work-in-process and finished goods transactions, inventory adjustments, purchases, supplier management, sales, accounts payable, accounts receivable, goods received, inspection, invoices, cost accounting, maintenance, etc.) led to MRP (initially standing for material requirements processing, now for manufacturing resources planning), which is a family of widely used standards and standard-based packages.

MRP is a typical module of most ERP packages such as SAP or Oracle Financials, and is usually integrated in modern customer relationship management (CRM) and supply chain management (SCM) systems.

CAD, computer-assisted engineering (CAE) and CAM—the latter including computerized numeric control (CNC)—have led to CIM. CIM is frequently used to run huge lights-out plants, with a significant portion of consumer goods being manufactured in these environments.

The importance for the IS auditor lies in the high number of systems and applications using these technologies. The larger the scale of integration, the more auditor attention is required.

Highly integrated CIM projects require the same attention from the auditor as the ERPs previously mentioned in this chapter. They are major undertakings that should be based on comprehensive feasibility studies and subject to top management approval and close supervision.

Continuity planning is also a primary area that should be reviewed by the IS auditor.

3.7.11 ELECTRONIC FUNDS TRANSFER

The method of payment plays a significant role in the relationship between seller and buyer. The underlying goal of the automated environment is to wring out costs inherent in the business processes. EFT is the exchange of money via telecommunications without currency actually changing hands. In other words, EFT is the electronic transfer of funds between a buyer, seller and his/her respective financial institution. EFT refers to any financial transaction that transfers a sum of money from one account to another electronically. EFT allows parties to move money from one account to another account, replacing traditional check writing and cash collection procedures. EFT services have been available for two decades. With the increased interest in Internet business, more and more consumers and businesses have begun to utilize EFT services. In the settlement between parties, EFT transactions usually function via an internal bank transfer from one party's account to another or via a clearinghouse network. Usually, transactions originate from a computer at one institution (location) and are transmitted to a computer at another institution (location) with the monetary amount recorded in the respective organization's accounts. Because of the potential high volume of money being exchanged, these systems may be in an extremely high-risk category. Therefore, access security and authorization of processing are important controls. Regarding EFT transactions, central bank requirements should be reviewed for application in these processes.

Controls in an EFT Environment

Because of the potential high volume of money being exchanged, these systems may be in an extremely high-risk category and security in an EFT environment becomes extremely critical.

Security includes the methods used by the customer to gain access to the system, the communications network, and the host or application processing site. Individual consumer access to the EFT system may be controlled by a plastic card and a PIN or by other means that bypass the need for a card. The IS auditor should review the physical security of unissued plastic cards, the procedures used to generate PINs, the procedures used to issue cards and PINs, and the conditions under which the consumer uses the access devices.

Security in an EFT environment ensures that:
- All the equipment and communication linkages are tested to effectively and reliably transmit and receive data
- Each party uses security procedures that are reasonably sufficient for affecting the authorized transmission of data and for protecting business records and data from improper access
- There are guidelines set for the receipt of data and to ensure that the receipt date and time for data transmitted are the date and time the data have been received
- On receipt of data, the receiving party will immediately transmit an acknowledgment or notification to communicate to the sender that a successful transmission occurred
- Data encryption standards are set
- Standards for unintelligible transmissions are set
- Regulatory requirements for enforceability of electronic data transmitted and received are explicitly stated

The IS auditor should ensure that reasonable authentication methods are required for access to EFT systems. The communications network should be designed to provide maximum security. Data encryption is recommended for all transactions; however, the IS auditor should determine any conditions under which the PIN might be accessible in a clear mode.

An EFT switch involved in the network is also an audit concern. An EFT switch is the facility that provides the communication linkage for all equipment in the network. The IS auditor should review the contract with the switch and the third-party audit of the switch operations. If a third-party audit has not been performed, the auditor should consider visiting the switch location.

At the application processing level, the IS auditor should review the interface between the EFT system and the application systems that process the accounts from which funds are transferred. Availability of funds or adequacy of credit limits should be verified before funds are transferred. Unfortunately, this is not always the case. Because of the penalties for failure to make a timely transfer, the IS auditor should review backup arrangements or other methods used to ensure continuity of operations. Because EFT reduces the flow of paper and consequently reduces normal audit trails, the IS auditor should determine that alternative audit trails are available.

3.7.12 AUTOMATED TELLER MACHINE

An ATM is a specialized form of the POS terminal that is designed for the unattended use by a customer of a financial institution. These machines customarily allow a range of banking and debit operations—especially financial deposits and cash withdrawals. ATMs are usually located in uncontrolled areas to facilitate easy access to customers after hours. This facility can be within a bank, across local banks and in banks outside a region. They are becoming known as retail EFT networks, transferring information and money over communication lines. Therefore the

system must provide high levels of logical and physical security for both the customer and the machinery. The ATM architecture has a physical network layer, a switch and a communication layer connecting the various ATM POS terminals.

Recommended internal control guidelines for ATMs, apart from what has been provided for any EFT, include the following:
- Written policies and procedures covering personnel, security controls, operations, disaster recovery credit and check authorization, floor limits, override, settlement, and balancing
- Reconciliation of all general ledger accounts related to retail EFTs and review of exception items and suspense accounts
- Procedures for PIN issuance and protection during storage
- Procedures for the security of PINs during delivery and the restriction of access to a customer's account after a small number of unsuccessful attempts
- Systems should be designed, tested and controlled to preclude retrieval of stored PINs in any nonencrypted form. Application programs and other software containing formulas, algorithms and data used to calculate PINs must be subject to the highest level of access for security purposes.
- Controls over plastic card procurement should be adequate with a written agreement between the card manufacturer and the bank that details control procedures and methods of resolution to be followed if problems occur.
- Controls and audit trails of the transactions that have been made in the ATM. This should include internal registration in the ATM, either in internal paper or digital media, depending on regulation or laws in each country and on the hosts that are involved in the transaction.

Audit of ATMs

To perform an audit of ATMs, the IS auditor should:
- Review physical security to prevent introduction of malware (such as Tyupkin)
- Review measures to establish proper customer identification and maintenance of their confidentiality
- Review file maintenance and retention system to trace transactions
- Review exception reports to provide an audit trail
- Review daily reconciliation of ATM transactions including:
  - Review SoD in the opening of ATM and recount of deposit
  - Review the procedures made for the retained cards
- Review encryption key change management procedures
  - Physical security measures to ensure security of the ATM and the money contained in the ATM
  - Review of ATM card slot, key pad and enclosure to prevent skimming of card data and capture of PIN during entry

3.7.13 INTERACTIVE VOICE RESPONSE

In telephony, interactive voice response (IVR), is a phone technology that allows a computer to detect voice and touch tones using a normal phone call. The caller uses the telephone keypad to select from preset menu choices provided by the IVR. The IVR system then responds with pre-recorded or dynamically generated audio to further direct callers or route the caller to a customer service representative. IVR systems can be used to control almost any function where the interface can be broken down into a series of simple menu choices. IVR systems generally scale well to handle large call volumes. Controls over such systems must be in place to prevent unauthorized individuals from entering system—level commands that may permit them to change or rerecord menu options.

3.7.14 PURCHASE ACCOUNTING SYSTEM

Financial transactions frequently go through more than one system when processed. In a department store, a sale is first processed in the sales accounting system, then processed by the accounts receivable system (if the purchase was by credit card) and, for either cash or credit sales, through the inventory system (when they are linked). That same sale might trigger the purchase accounting system to replace depleted inventory. Eventually the transactions become part of the general ledger system because all transactions are recorded somewhere in that system. For the integration of systems to be effective, processing of transactions must be complete, accurate and timely. If it is not, a ripple effect impairs the integrity of the data.

Purchase accounting systems process the data for purchases and payments. Because purchases automatically lead to payments, if purchases are properly contracted, partial control over payments exists. Additional controls over payments are still needed to ensure that each payment was made for goods and services received, that the same purchases were not paid for twice and that they were, indeed, paid. Most purchase accounting systems perform three basic accounting functions:
1. **Accounts payable processing**—Recording transactions in the accounts payable records
2. **Goods received processing**—Recording details of goods received but not yet invoiced
3. **Order processing**—Recording goods ordered but not yet received

The computer may be involved in each of these activities, and the extent to which they are computerized determines the complexity of the purchase accounting system.

3.7.15 IMAGE PROCESSING

Some of the many algorithms used in image processing include convolution (on which many others are based), fast Fourier transform (FFT), discrete cosine transform (DCT), thinning (or skeletonization), edge detection and contrast enhancement. These are usually implemented in software but may also use special-purpose hardware for speed.

An imaging system stores, retrieves and processes graphic data, such as pictures, charts and graphs, instead of or in addition to text data. The storage capacities must be enormous and most image systems include optical disk storage. In addition to optical disks, the systems include high-speed scanning, high-resolution displays, rapid and powerful compression, communications functions, and laser printing. The systems include techniques that can identify levels of shades and colors that cannot be differentiated by the human eye. These systems are expensive and companies do not invest in them lightly.

Most businesses that perform image processing obtain benefits from using the imaging system. Examples of potential benefits are:
- Item processing (e.g., signature storage and retrieval)
- Immediate retrieval via a secure optical storage medium
- Increased productivity
- Improved control over paper files
- Reduced deterioration due to handling
- Enhanced disaster recovery procedures

Imaging systems are the fastest growth area of the micrographics industry and are an outgrowth of microfilm and microfiche, which have, in the past, been heavily used in paper-intensive fields such as insurance and banking. Not surprisingly, these same fields were the first to incorporate imaging systems into their standard operations. The replacement of paper documents with electronic images can have a significant impact on the way that an institution does business. Many of the traditional audit and security controls for paper-based systems may be reduced or absent in electronic documentation workflow. New controls must be developed and designed into the automated process to ensure that information image files cannot be altered, erased or lost.

Risk areas that management should address when installing imaging systems and that IS auditors should be aware of when reviewing an institution's controls over imaging systems include:
- **Planning**—The lack of careful planning in selecting and converting paper systems to document imaging systems can result in excessive installation costs, the destruction of original documents and the failure to achieve expected benefits. Critical issues include converting existing paper storage files and integration of the imaging system into the organization workflow and electronic media storage to meet audit and document retention legal requirements.
- **Audit**—Imaging systems may change or eliminate the traditional controls as well as the checks and balances inherent in paper-based systems. Audit procedures may have to be redesigned and new controls designed into the automated process.
- **Redesign of workflow**—Institutions generally redesign or reengineer workflow processes to benefit from imaging technology.
- **Scanning devices**—Scanning devices are the entry point for image documents and a significant risk area in imaging systems. Scanning operations can disrupt workflow if the scanning equipment is not adequate to handle the volume of documents or the equipment breaks down. The absence of controls over the scanning process can result in poor quality images, improper indexing, and incomplete or forged documents being entered into the system. Factors that should be considered in an imaging system are quality control over the scanning and indexing process, the scanning rate of the equipment, the storage of images, equipment backup, and the experience level of personnel scanning the document. Procedures should be in place to ensure that original documents are not destroyed before determining that a good image has been captured.
- **Software security**—Security controls over image system documents are critical to protect institutions and customer information from unauthorized access and modifications. The integrity and reliability of the imaging system database is related directly to the quality of controls over access to the system.
- **Training**—Inadequate training of personnel scanning the documents can result in poor-quality document images and indexes, and the early destruction of original documents. The installation and use of imaging systems can be a major change for department personnel. They must be trained adequately to ensure quality control over the scanning and storage of imaging documents as well as the use of the system to maximize the benefits of converting to imaging systems.

3.7.16 INDUSTRIAL CONTROL SYSTEMS

Industrial control system (ICS) is a general term that encompasses several types of control systems, including supervisory control and data acquisition (SCADA) systems, distributed control systems (DCS), and other control system configurations such as programmable logic controllers (PLC), often found in the industrial sectors and critical infrastructures.

Figure 3.19 provides a high-level overview of typical ICS process flows.

Figure 3.19—ICS Operation

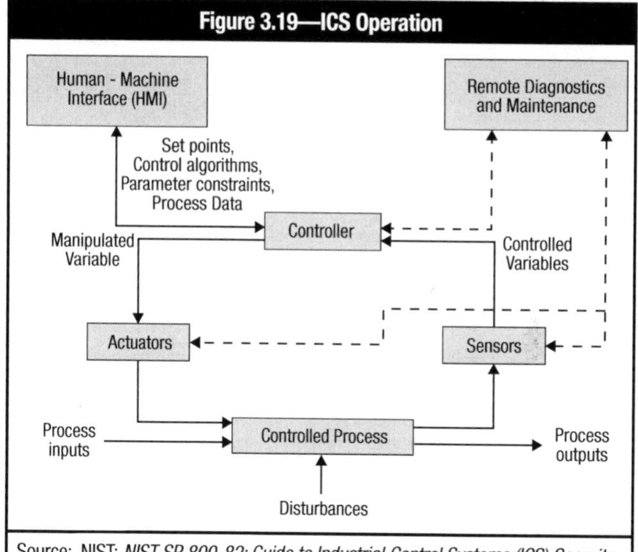

Source: NIST; *NIST SP 800-82: Guide to Industrial Control Systems (ICS) Security*, USA, 2011

Risk Factors

Based on the criticality that ICSs have on manufacturing, chemical processes and, more important, critical infrastructure (energy generation, transmission and control, water treatment, etc.), there are key risk factors that the IS auditor must consider:
- Blocked or delayed flow of information through ICS networks, which could disrupt ICS operation
- Unauthorized changes to instructions, commands or alarm thresholds, which could damage, disable or shut down equipment, create environmental impacts and/or endanger human life
- Inaccurate information sent to system operators, either to disguise unauthorized changes or to cause the operators to initiate inappropriate actions, which could have various negative effects
- ICS software or configuration settings modified or ICS software infected with malware, which could have various negative effects
- Interference with the operation of safety systems, which could endanger human life

Typical Controls

To address risk, preventive, detective and corrective controls need to be considered within the administrative, operational and technical implementation of ICS:

- Restricting logical access to the ICS network and network activity. This includes using a demilitarized zone (DMZ) network architecture with firewalls to prevent network traffic from passing directly between the corporate and ICS networks and having separate authentication mechanisms and credentials for users of the corporate and ICS networks. The ICS should also use a network topology that has multiple layers, with the most critical communications occurring in the most secure and reliable layer.
- Restricting physical access to the ICS network and devices. Unauthorized physical access to components could cause serious disruption of the ICS's functionality. A combination of physical access controls should be used, such as locks, card readers and/or guards.
- Protecting individual ICS components from exploitation. This includes deploying security patches in as expeditious a manner as possible, after testing them under field conditions; disabling all unused ports and services; restricting ICS user privileges to only those that are required for each person's role; tracking and monitoring audit trails; and using security controls such as antivirus software and file integrity checking software, where technically feasible, to prevent, deter, detect and mitigate malware.
- Maintaining functionality during adverse conditions. This involves designing the ICS so that each critical component has a redundant counterpart. Additionally, if a component fails, it should fail in a manner that does not generate unnecessary traffic on the ICS or other networks, or does not cause another problem elsewhere, such as a cascading event.
- Restoring system after an incident. Incidents are inevitable and an incident response plan is essential. A major characteristic of a good security program is how quickly a system can be recovered after an incident has occurred.

3.7.17 ARTIFICIAL INTELLIGENCE AND EXPERT SYSTEMS

Artificial intelligence (AI) is the study and application of the principles by which:

- Knowledge is acquired and used.
- Goals are generated and achieved.
- Information is communicated.
- Collaboration is achieved.
- Concepts are formed.
- Languages are developed.

AI fields include, among others:

- Expert systems
- Natural and artificial (such as programming) languages
- Neural networks
- Intelligent text management
- Theorem proving
- Abstract reasoning
- Pattern recognition
- Voice recognition
- Problem solving
- Machine translation of foreign languages

Two main programming languages that have been developed for AI are Lisp and Prolog.

Expert systems are an area of AI and perform a specific function or are prevalent in certain industries. An expert system allows the user to specify certain basic assumptions or formulas and then uses these assumptions or formulas to analyze arbitrary events. Based on the information used as input to the system, a conclusion is produced.

The use of expert systems has many potential benefits within an organization including:

- Capturing the knowledge and experience of individuals
- Sharing knowledge and experience
- Enhancing personnel productivity and performance
- Automating highly (statistically) repetitive tasks (help desk, score credits, etc.)
- Operating in environments where a human expert is not available (e.g., medical assistance on board of a ship, satellites, etc.)

Expert systems are comprised of the primary components shown in **figure 3.20**, called shells, when they are not populated with particular data, and the shells are designed to host new expert systems.

Figure 3.20—Artificial Intelligence Expert Systems

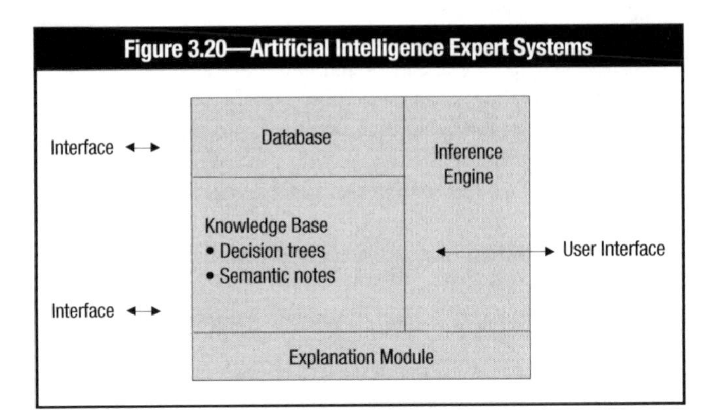

Key to the system is the knowledge base (KB), which contains specific information or fact patterns associated with particular subject matter and the rules for interpreting these facts. The KB interfaces with a database in obtaining data to analyze a particular problem in deriving an expert conclusion. The information in the KB can be expressed in several ways:

- **Decision trees**—Using questionnaires to lead the user through a series of choices, until a conclusion is reached. Flexibility is compromised because the user must answer the questions in an exact sequence.
- **Rules**—Expressing declarative knowledge through the use of if-then relationships. For example, if a patient's body temperature is over 39°C (102.2°F) and his/her pulse is under 60, then the patient might be suffering from a certain disease.
- **Semantic nets**—Consist of a graph in which the nodes represent physical or conceptual objects and the arcs describe the relationship between the nodes. Semantic nets resemble a data flow diagram and make use of an inheritance mechanism to prevent duplication of data.

Additionally, the inference engine shown is a program that uses the KB and determines the most appropriate outcome based on the information supplied by the user. In addition, an expert system includes the following components:
- **Knowledge interface**—Allows the expert to enter knowledge into the system without the traditional mediation of a software engineer
- **Data interface**—Enables the expert system to collect data from nonhuman sources, such as measurement instruments in a power plant

An explanation module that is user-oriented in addressing the problem is analyzed, and the expert conclusion reached is also provided.

Expert systems are gaining acceptance and popularity as audit tools. Specifically related to IS auditing, expert systems have been developed to facilitate the IS auditor in auditing such areas as OSs, online software environments, access control products and microcomputer environments. These tools can take the form of a series of well-designed questionnaires or actual software that integrates and reports on system parameters and data sets. Other accounting- and auditing-related applications for expert systems include audit planning, internal control analysis, account attribute analysis, quality review, accounting decisions, tax planning and user training.

Consistent with standard systems development methodologies, stringent change control procedures should be followed because the basic assumptions and formulas may need to be changed as more expertise is gained. As with other systems, access should be on a need-to-know basis.

The IS auditor should be knowledgeable about the various AI and expert system applications used within the organization.

The IS auditor needs to be concerned with the controls relevant to these systems when used as an integral part of an organization's business process or mission critical functions, and the level of experience or intelligence used as a basis for developing the software. This is critical because errors produced by AI systems may have a more severe impact than those produced by traditional systems. This is true especially of intelligent systems that facilitate health care professionals in the diagnosis and treatment of injuries and illnesses. Error loops/routines should be designed into these systems.

Specifically, the IS auditor should:
- Understand the purpose and functionality of the system.
- Assess the system's significance to the organization and related businesses processes as well as the associated potential risk.
- Review the adherence of the system to corporate policies and procedures.
- Review the decision logic built into the system to ensure that the expert knowledge or intelligence in the system is sound and accurate. The IS auditor should ensure that the proper level of expertise was used in developing the basic assumptions and formulas.
- Review procedures for updating information in the KB.
- Review security access over the system, specifically the KB.
- Review procedures to ensure that qualified resources are available for maintenance and upgrading.

3.7.18 BUSINESS INTELLIGENCE

Business intelligence (BI) is a broad field of IT that encompasses the collection and analysis of information to assist decision making and assess organizational performance.

Investments in BI technology can be applied to enhance understanding of a wide range of business questions. Some typical areas in which BI is applied for measurement and analysis purposes include:
- Process cost, efficiency and quality
- Customer satisfaction with product and service offerings
- Customer profitability, including determination of which attributes are useful predictors of customer profitability
- Staff and business unit achievement of key performance indicators
- Risk management (e.g., by identifying unusual transaction patterns and accumulation of incident and loss statistics)

The interest in BI as a distinct field of IT activity is being spurred by a number of factors:
- **The increasing size and complexity of modern organizations**—The result is that even fundamental business questions cannot be properly answered without establishing serious BI capability.
- **Pursuit of competitive advantage**—Most organizations have, for many years, automated their basic, high-volume activities. Significant organizationwide IT investment such as ERP systems is now common place. Many companies have or are now investing in Internet technology as a means of distributing product/service and supply chain integration. However, utilization of IT to maintain and extend a firm's knowledge capital represents a new opportunity to use technology to gain an advantage over competitors.
- **Legal requirements**—Legislation such as the US Sarbanes-Oxley Act and the US Patriot Act exist to enforce the need for companies to have an understanding of the "whole of business." Financial institutions must now be able to report on all accounts/instruments that their customers have and all transactions against those accounts/instruments, including any suspicious transaction patterns.

To deliver effective BI, organizations need to design and implement (progressively, in most cases) a data architecture. A complete data architecture consists of two components:
- The enterprise data flow architecture (EDFA)
- A logical data architecture

An example of optimized enterprise data flow architecture is depicted in **figure 3.21**. Explanations of the various layers/components of this data flow architecture follow:
- **Presentation/desktop access layer**—This is where end users directly deal with information. This layer includes familiar desktop tools such as spreadsheets, direct querying tools, reporting and analysis suites offered by vendors such as Cognos and Business Objects, and purpose-built applications such as balanced scorecards (BSCs) and digital dashboards. Power users will have the ability to build their own queries and reports

Figure 3.21—Data Flow Architecture Sample

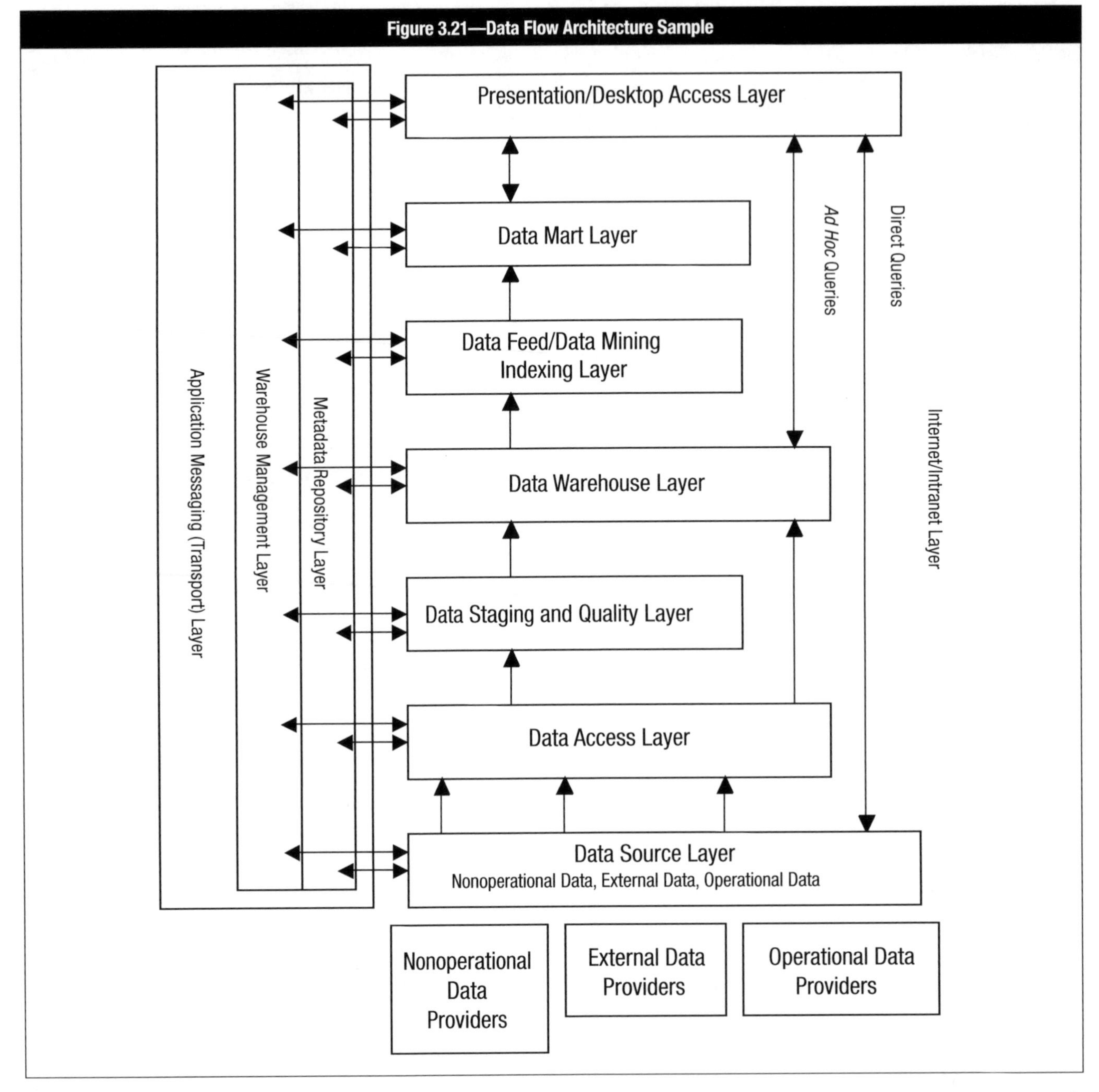

while other users will interact with the data in predefined ways. Increasingly, users are being provided with more than static reporting capabilities by enhancing reports with parameterization and drill-down capabilities and presenting data in visual formats as a supplement or replacement of textual/tabular data presentation.

- **Data source layer**—Enterprise information derives from a number of sources:
  - Operational data—Data captured and maintained by an organization's existing systems and usually held in system-specific databases or possibly flat files
  - External data—Data provided to an organization by external sources. This could include data such as customer demographics and market share information.
  - Nonoperational data—Information needed by end users that is not currently maintained in a computer-accessible format
- **Core data warehouse**—A core data warehouse (DW) is where all (or at least the majority of) the data of interest to an organization is captured and organized to assist reporting and analysis. DWs are normally instituted as large relational databases. While there is not unanimous agreement, many pundits suggest the warehouse should hold fully normalized data to give it the flexibility to deal with complex and changing business structures. A properly constituted DW should support three basic forms of inquiry:
  - Drilling up and drilling down—Using dimensions of interest to the business, it should be possible to aggregate data (e.g., sum store sales to get region sales and ultimately national sales) as well as drill down (e.g., break store sales down to counter sales). Attributes available at the more granular levels of the warehouse can also be used to refine the analysis (e.g., analyze sales by product).

  - Drill across—Use common attributes to access a cross-section of information in the warehouse such as sum sales across all product lines by customer and groups of customers according to length of association with the company (and/or other attribute of interest)
  - Historical analysis—The warehouse should support this by holding historical, time-variant data. An example of historical analysis would be to report monthly store sales and then repeat the analysis using only customers who were preexisting at the start of the year in order to separate the effect of new customers from the ability to generate repeat business with existing customers.
- **Data mart layer**—Data marts represent subsets of information from the core DW selected and organized to meet the needs of a particular business unit or business line. Data marts may be relational databases or some form of online analytical processing (OLAP) data structure (also known as a data cube). OLAP technologies and some variants (e.g., relational OLAP [ROLAP]), allow users to "slice and dice" data presented in terms of standardized measures (i.e., numerical facts) and dimensions (i.e., business hierarchies). Data marts have a simplified structure compared to the normalized DW.
- **Data staging and quality layer**—This layer is responsible for data copying, transformation into DW format and quality control. It is particularly important that only reliable data get loaded to the core DW. This layer needs to be able to deal with problems periodically thrown up by operational systems such as changes to account number formats and reuse of old account and customer numbers (when the DW still holds information on the original entity).
- **Data access layer**—This layer operates to connect the data storage and quality layer with data stores in the data source layer and, in the process, avoiding the need to know exactly how these data stores are organized. Technology now permits structured query language (SQL) access to data even if it is not stored in a relational database.
- **Data preparation layer**—This layer is concerned with the assembly and preparation of data for loading into data marts. The usual practice is to precalculate the values that are loaded into OLAP data repositories to increase access speed. Specialist data mining also normally requires preparation of data. Data mining is concerned with exploring large volumes of data to determine patterns and trends of information. Data mining often identifies patterns that are counterintuitive due to the number and complexity of data relationships. Data quality needs to be very high to not corrupt the results.
- **Metadata repository layer**—Metadata are data about data. The information held in the metadata layer needs to extend beyond data structure names and formats to provide detail on business purpose and context. The metadata layer should be comprehensive in scope, covering data as they flow between the various layers, including documenting transformation and validation rules. Ideally, information in the metadata layer can be directly sourced by software operating in the other layers, as required.
- **Warehouse management layer**—The function of this layer is the scheduling of the tasks necessary to build and maintain the DW and populate data marts. This layer is also involved in the administration of security.
- **Application messaging layer**—This layer is concerned with transporting information between the various layers. In addition to business data, this layer encompasses generation, storage and targeted communication of control messages.
- **Internet/intranet layer**—This layer is concerned with basic data communication. Included here are browser-based user interfaces and TCP/IP networking.

The construction of the logical data architecture for an enterprise is a major undertaking that would normally be undertaken in stages. One reason for separating logical data model determination by business domain is that different parts of large business organizations will often deal with different transaction sets, customers and products.

Ultimately, the data architecture needs to be structured to accommodate the needs of the organization in the most efficient manner. Factors to consider include the types of transactions in which the organization engages, the entities that participate in or form part of these transactions (e.g., customers, products, staff and communication channels), and the dimensions (hierarchies) that are important to the business (e.g., product and organization hierarchies).

With modern DWs, storage capacity is not really an issue. Therefore, the goal should be to obtain the most granular or atomic data possible. The lowest level data are most likely to have attributes that can be used for analysis purposes that would be lost if summarized data are loaded.

Various analysis models used by data architects/analysts follow:
- **Context diagrams**—Outline the major processes of an organization and the external parties with which the business interacts.
- **Activity or swim-lane diagrams**—Deconstruct business processes.
- **Entity relationship diagrams**—Depict data entities and how they relate. These data analysis methods obviously play an important part in developing an enterprise data model. However, it is also crucial that knowledgeable business operatives are involved in the process. This way proper understanding can be obtained of the business purpose and context of the data. This also mitigates the risk of the replication of suboptimal data configurations from existing systems and databases into the DW.

Business Intelligence Governance

To maximize the value an organization obtains from its BI initiatives, an effective BI governance process needs to be in place.

An important part of the governance process involves determining which BI initiatives to fund, what priority to assign to initiatives and how to measure their ROI. This is particularly important because the investment needed to build BI infrastructure, such as a DW, is considerable. Additionally, the scope and complexity of an organizationwide DW means that, realistically, it must be built in stages.

A recommended practice in the area of BI funding governance is to establish a business/IT advisory team that allows different functional perspectives to be represented, recommends investment priorities and establishes cross-organizational benefit measures. Final funding decisions should rest with a technology steering committee that comprises senior management.

A further important part of overall BI governance is data governance. Aspects to be considered here include establishing standard definitions for data, business rules and metrics, identifying approved data sources, and establishing standards for data reconciliation and balancing.

3.7.19 DECISION SUPPORT SYSTEM

A decision support system (DSS) is an interactive system that provides the user with easy access to decision models and data from a wide range of sources in order to support semistructured decision-making tasks typically for business purposes. It is an informational application that is designed to assist an organization in making decisions through data provided by business intelligence tools (in contrast to an operational application that collects the data in the course of normal business operations). Typical information that a decision support application might gather and present would be:

- Comparative sales figures between one week and the next
- Projected revenue figures based on new product sales assumptions
- The consequences of different decision alternatives given past experience in the described context

A DSS may present information graphically and may include an expert system or AI. Further, it may be aimed at business executives or some other group of knowledge workers.

Characteristics of a DSS are:

- Aims at solving less-structured, underspecified problems that senior managers face
- Combines the use of models or analytic techniques with traditional data access and retrieval functions
- Emphasizes flexibility and adaptability to accommodate changes in the environment and the decision-making approach of the users

Efficiency vs. Effectiveness

A principle of DSS design is to concentrate less on efficiency (i.e., performing tasks quickly and reducing costs) and more on effectiveness (i.e., performing the right task). Therefore DSSs are often developed using 4GL tools that are less efficient but allow for flexible and easily modified systems.

Decision Focus

A DSS is often developed with a specific decision or well-defined class of decisions to solve; therefore, some commercial software packages that claim to be DSS are nothing more than a DSS generator (tools with which to construct a DSS).

DSS Frameworks

Frameworks are generalizations about a field that help put many specific cases and ideas into perspective. The G. Gorry-M.S. Morton framework is the most complete knowledge- and system-control-related IS model, and it is based on problem classification into structured and unstructured types as well as the time horizon of the decisions. This framework characterizes DSS activities along two dimensions:

1. The degree of structure in the decision process being supported
2. The management level at which decision making takes place

This framework also portrays all IS efforts as addressing distinct types of problems, depending on the above two factors.

The management-level dimension is broken into three parts:

1. Operational control
2. Management control
3. Strategic planning

The decision-structure dimension is also broken into three parts:

1. Structured
2. Semistructured
3. Unstructured

The degree to which a problem or decision is structured corresponds roughly to the extent to which it can be automated or programmed.

Another DSS framework is the Sprague-Carlson framework that is initiated with an effort to create family trees—which is a generalization of the structure of a DSS. This framework suggests that every DSS has data, a model and a dialog generator subsystem. This framework emphasizes the importance of data management in DSS work. This framework also stresses the importance of interactive user interfaces in a DSS. The generation and management of these interfaces require appropriate software and hardware—the dialog management system. In general, a system must offer more than one interface and might need to provide a tailored interface for each user.

Design and Development

Prototyping is the most popular approach to DSS design and development. Prototyping usually bypasses the usual requirement definition. System requirements evolve through the user's learning process. The benefits of prototyping include the following:

- Learning is explicitly incorporated into the design process because of the iterative nature of the system design.
- Feedback from design iterations is rapid to maintain an effective learning process for the user.
- The user's expertise in the problem area helps the user suggest system improvements.
- The initial prototype must be inexpensive to create.

Implementation and Use

It is difficult to implement a DSS because of its discretionary nature. Using a DSS to solve a problem represents a change in behavior on the part of the user. Implementing a DSS is an exercise in changing an organization's behavior. The main challenge is to get the users to accept the use of software. The following are the steps involved in changing behavior:

- **Unfreezing**—This step alters the forces acting on individuals such that the individuals are distracted sufficiently to change. Unfreezing is accomplished either through increasing the pressure for change or by reducing some of the threats of or resistance to change.
- **Moving**—This step presents a direction of change and the actual process of learning new attitudes.
- **Refreezing**—This step integrates the changed attitudes into the individual's personality.

Risk Factors

Developers should be prepared for eight implementation risk factors:
1. Nonexistent or unwilling users
2. Multiple users or implementers
3. Disappearing users, implementers or maintainers
4. Inability to specify purpose or usage patterns in advance
5. Inability to predict and cushion impact on all parties
6. Lack or loss of support
7. Lack of experience with similar systems
8. Technical problems and cost-effectiveness issues

Implementation Strategies

To plan for risk and prevent it from occurring:
- Divide the project into manageable pieces.
- Keep the solution simple.
- Develop a satisfactory support base.
- Meet user needs and institutionalize the system.

Assessment and Evaluation

The true test of a DSS lies in whether it improves a manager's decision making, which is something not easily measured. A DSS also rarely results in cost displacements such as a reduction in staff or other expenses. In addition, because a DSS is evolutionary in nature, it lacks neatly defined completion dates.

Using an incremental approach to DSS development reduces the need for evaluation. By developing one step at a time and achieving tangible results at the end of each step, the user does not need to make extensive commitments of time and money at the beginning of the development process.

The DSS designer and user should use broad evaluation criteria. These criteria should include:
- Traditional cost-benefit analysis
- Procedural changes, more alternatives examined and less time consumed in making the decision
- Evidence of improvement in decision making
- Changes in the decision process

Some common trends in DSS usage include:
- The need for more accurate information by managers is an important motivator for DSS development.
- Few DSS evaluations use the traditional cost-benefit analysis.
- End users are usually the motivators in developing a DSS.
- Development staff members are drawn largely from functional area staff or the planning department, not from the IT department.
- Users perceive flexibility as the most important factor influencing the system's success.
- Few DSS projects are being developed today for third-generationsoftware facilities. User-oriented, 4GLs and planning languages predominate.
- Planning, evaluation and training for DSS projects traditionally have been performed quite poorly.

DSS Common Characteristics

Some of the common characteristics of DSSs include:
- Oriented toward decision making
- Usually based on 4GL
- Surfable
- Linkable
- Drill-down
- Semaphores (signals to automatically alert when a decision needs to be made)
- Time series analysis
- What if (refers to scenario modeling; i.e., determining what is the end result of a changing variable or variables)
- Sensitivity analysis
- Goal-seeking
- Excellent graphic presentations
- Dynamic graphic, data editing
- Simulation

DSS Trends

DSS trends include:
- Gradual improvement and sharpening of skills in the development and implementation of a traditional DSS
- Advances in database and graphics capabilities for microcomputers
- Exploratory work in such fields as expert systems, a DSS to support group decision making and visual interactive modeling

3.7.20 CUSTOMER RELATIONSHIP MANAGEMENT

The customer-driven business trend is to be focused on the wants and needs of the customers. With the customer's expectations constantly increasing, these objectives are becoming more difficult to achieve. This emphasizes the importance of focusing on information relating to transaction data, preferences, purchase patterns, status, contact history, demographic information and service trends of customers rather than on products.

All these factors lead CRM, which is an optimum combination of strategy, tactics, processes, skill sets and technology. CRM has become a strategic success factor for all types of business, and its proficiency has a significant impact on profitability.

The customer expectations are increasing tremendously and that, in turn, raises the expectation of service levels. Therefore, the customer-centered applications focus on CRM processes emphasizing the customer, rather than marketing, sales or any other function. The new business model will have an integration of telephony, web and database technologies, and interenterprise integration capabilities. Also, this model spreads to the other business partners who can share information, communicate and collaborate with the organization with the seamless integration of web-enabled applications and without changing their local network and other configurations.

It is possible to distinguish between operational and analytical CRM. Operational CRM is concerned with maximizing the utility of the customer's service experience while also capturing useful data about the customer interaction. Analytical CRM seeks to analyze information captured by the organization about its customers and their interactions with the organization into information that allows greater value to be obtained from the customer base. Among uses of analytical CRM are increasing customer product holdings or "share of customer wallet," moving customers into higher margin products, moving customers to lower-cost service channels, increasing marketing success rates, and making pricing decisions.

3.7.21 SUPPLY CHAIN MANAGEMENT

Supply chain management (SCM) is linking the business processes between the related entities such as the buyer and the seller. The link is provided to all the connected areas such as managing logistics and the exchange of information, services and goods among supplier, consumer, warehouse, wholesale/retail distributors and the manufacturer of goods.

SCM has become a focal point and is seen as a new area in strategic management because of the shift in the business scenario at the advent of global competition, proliferation of the Internet, the instantaneous transmission of information and web presence in all spheres of business activities. SCM is all about managing the flow of goods, services and information among suppliers, manufacturers, wholesalers, distributors, stores, consumers and end users.

SCM shifts the focus; all the entities in the supply chain can work collaboratively and in a real-time mode, thus reducing, to a great extent, the required available inventory. The just-in-time (JIT) concept becomes possible, and the cycle time is reduced with an objective toward reducing the unneeded inventory. Seasonal (e.g., both availability and demand) and regional (e.g., preferences in size, shape, quantity, etc.) factors are addressed.

Stock levels of nonmoving items are significantly reduced, and there is an automated flow of supply and demand. Also, the intrinsic costs and errors associated with the manual means such as fax, input of data, delay and inaccurate orders, can be avoided.

3.8 DEVELOPMENT METHODS

In the face of increasing system complexity and the need to implement new systems more quickly to achieve benefits before the business changes, software development practitioners have adopted new ways of organizing software projects that vary, or in some cases radically depart from, the traditional waterfall model previously described. In addition, there has been continued evolution in the thinking about how best to analyze, design and construct software systems and in the information technologies available to perform these activities.

This section describes different techniques of understanding, designing and constructing a software system. The choice of a particular method will be driven by considerations such as organizational policy, developer knowledge and preference, and the technology being used.

Please note that the selection of one of the methods described in this section is generally independent of the selection of a project organization model. An object-oriented approach to design and coding could be utilized on a project organized into distinct phases as in the waterfall model of software development just as it could be with an agile project where each short iteration delivers working software.

3.8.1 USE OF STRUCTURED ANALYSIS, DESIGN AND DEVELOPMENT TECHNIQUES

The use of structured analysis, design and development techniques is closely associated with the traditional, classic SDLC approach to software development. These techniques provide a framework for representing the data and process components of an application using various graphic notations at different levels of abstraction, until the abstraction level that enables programmers to code the system is reached. Early on, for example, the following activities occur in defining the requirements for a new system:

- Develop system context diagrams (e.g., high-level business process flow schema).
- Perform hierarchical data flow/control flow decomposition.
- Develop control transformations.
- Develop minispecifications.
- Develop data dictionaries.
- Define all external events—inputs from external environment.
- Define single transformation data flow diagrams (DFDs) from each external event.

The next level of design provides greater detail for building the system, including developing system flowcharts, inputs/outputs, processing steps and computations, and program and data file or database specifications. It should be noted that representation of functions is developed in a modularized top-down fashion. This enables programmers to systematically develop and test modules in a linear fashion.

Auditors should be particularly concerned with whether the processes under a structured approach are well defined, documented and followed when using the traditional SDLC approach to business application development.

3.8.2 AGILE DEVELOPMENT

The term "agile development" refers to a family of similar development processes that espouse a nontraditional way of developing complex systems. One of the first agile processes, Scrum (a rugby analogy), emerged in the early 1990s.

Scrum aims to move planning and directing tasks from the project manager to the team, leaving the project manager to work on removing the obstacles to the team, achieving their objectives. Scrum is a project management approach that fits well with other agile techniques. Other agile processes have since emerged such as Extreme Programming (XP), Crystal, Adaptive Software Development, Feature Driven Development and Dynamic Systems Development Method. These processes are termed "agile" because they are designed to flexibly handle changes to the system being developed or the project that is performing the development.

Agile development processes have a number of common characteristics:

- The use of small, time-boxed subprojects or iterations, as shown in **figure 3.22**. In this instance, each iteration forms the basis for planning the next iteration.
- Replanning the project at the end of each iteration (referred to as a "sprint" in Scrum), including reprioritizing requirements, identifying any new requirements and determining within which release delivered functionality should be implemented
- Relatively greater reliance, compared to traditional methods, on tacit knowledge—the knowledge in people's heads—as opposed to external knowledge that is captured in project documentation

- A heavy influence on mechanisms to effectively disseminate tacit knowledge and promote teamwork. Therefore, teams are kept small in size, comprise both business and technical representatives, and are located physically together. Team meetings to verbally discuss progress and issues occur daily, but with strict time limits.
- At least some of the agile methods stipulate pair-wise programming (two persons code the same part of the system) as a means of sharing knowledge and as a quality check.
- A change in the role of the project manager, from one primarily concerned with planning the project, allocating tasks and monitoring progress to that of a facilitator and advocate. Responsibility for planning and control is delegated to the team members.

Agile development does not ignore the concerns of traditional software development, but approaches them from a different perspective:
- Agile development only plans for the next iteration of development in detail, rather than planning subsequent development phases far out in time.
- Agile development's adaptive approach to requirements does not emphasize managing a requirements baseline.
- Agile development's focus is to quickly prove an architecture by building actual functionality versus formally defining, early on, software and data architecture in increasingly more detailed models and descriptions.
- Agile development assumes limits to defect testing but attempts to validate functions through a frequent-build test cycle and correct problems in the next subproject before too much time and cost are incurred.
- Agile development does not emphasize defined and repeatable processes but instead performs and adapts its development based on frequent inspections.

3.8.3 PROTOTYPING-EVOLUTIONARY DEVELOPMENT

Prototyping, also known as heuristic or evolutionary development, is the process of creating a system through controlled trial and error procedures to reduce the level of risk in developing the system. That is, it enables the developer and customer to understand and react to risk at each evolutionary level (using prototyping as a risk reduction mechanism). It combines the best features of classic SDLC by maintaining the systematic stepwise approach and incorporates it into an iterative framework that more realistically reflects the real world.

The initial emphasis during the development of the prototype is usually placed on the reports and screens, which are the system aspects most used by the end users. This allows the end user to see a working model of the proposed system within a short time. There are two basic methods or approaches to prototyping:
1. Build the model to create the design (i.e., the mechanism for defining requirements). Then, based on that model, develop the system design with all the performance, quality and maintenance features needed.
2. Gradually build the actual system that will operate in production using a 4GL that has been determined to be appropriate for the system being built.

The problem with the first approach is that there can be considerable pressure to implement an early prototype. Often, users observing a working model cannot understand why the early prototype must be refined further. The fact that the prototype needs to be expanded to handle transaction volumes, client-server network connectivity, and backup and recovery procedures, and provide for security, auditability and control is not often understood.

The second approach typically works with small applications using 4GL tools. However, for larger efforts, it is necessary to

Figure 3.22—Data Flow Architecture Sample

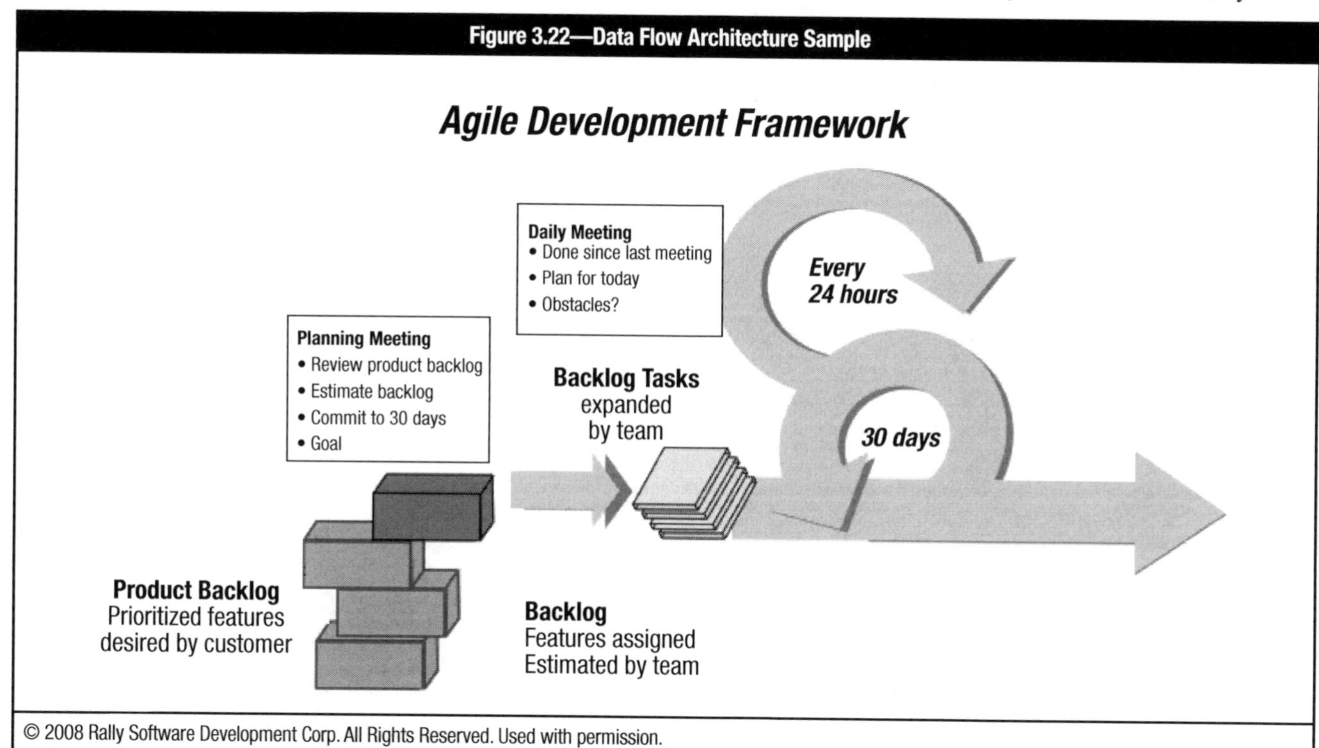

develop a design strategy for the system even if a 4GL is used. The use of 4GL techniques alone will cause the same difficulties (e.g., poor quality, poor maintainability and low user acceptance) encountered when developing business applications using conventional approaches.

Another overall disadvantage of prototyping is that it often leads to functions or extras being added to the system that are not included in the initial requirements document. All major enhancements beyond the initial requirements document should be reviewed to ensure that they meet the strategic needs of the organization and are cost-effective. Otherwise, the final system may be functionally rich but inefficient.

A potential risk with prototyped systems is that the finished system will have poor controls. By focusing mainly on what the user wants and what the user sees, system developers may miss some of the controls that come out of the traditional system development approach such as backup recovery, security and audit trails.

Change control often becomes much more complicated with prototyped systems. Changes in designs and requirements happen so quickly that they are seldom documented or approved, and the system can escalate to a point of not being maintainable.

Although the IS auditor should be aware of the risk associated with prototyping, the IS auditor should also be aware that this method of system development can provide the organization with significant time and cost savings.

3.8.4 RAPID APPLICATION DEVELOPMENT

RAD is a methodology that enables organizations to develop strategically important systems quickly while reducing development costs and maintaining quality. This is achieved by using a series of proven application development techniques within a well-defined methodology. These techniques include the use of:
- Small, well-trained development teams
- Evolutionary prototypes
- Integrated power tools that support modeling, prototyping and component reusability
- A central repository
- Interactive requirements and design workshops
- Rigid limits on development time frames

RAD supports the analysis, design, development and implementation of individual application systems. However, RAD does not support the planning or analysis required to define the information needs of the enterprise as a whole or of a major business area of the enterprise. RAD provides a means for developing systems faster while reducing cost and increasing quality. This is done by automating large portions of the SDLC, imposing rigid limits on development time frames and reusing existing components. The RAD methodology has four major stages:
1. The **concept definition stage** defines the business functions and data subject areas that the system will support and determines the system scope.
2. The **functional design stage** uses workshops to model the system's data and processes and build a working prototype of critical system components.
3. The **development stage** completes the construction of the physical database and application system, builds the conversion system and develops user aids and deployment work plans.
4. The **deployment stage** includes final-user testing and training, data conversion and the implementation of the application system.

RAD uses prototyping as its core development tool no matter which underlying technology is used. In contrast, object-oriented software development (OOSD) and data-oriented system development (DOSD) use continuously developing models but have a focus on content solution space (e.g., how to best address the problem to make the code reusable and maintainable) and can be applied using a traditional waterfall approach. It should also be noted that business process reengineering (BPR) attempts to convert an existing business process rather than make dynamic changes.

3.8.5 OBJECT-ORIENTED SYSTEM DEVELOPMENT

OOSD is the process of solution specification and modeling where data and procedures can be grouped into an entity known as an object. An object's data are referred to as its attributes and its functionality is referred to as its methods. This contrasts with the traditional (structured SDLC) approach which considered data separately from the procedures that act on them (e.g., program and database specifications). Proponents of OOSD claim the combination of data and functionality is aligned with how humans conceptualize everyday objects.

OOSD is a programming technique, not a software development methodology. One can do OOSD while following any of the widely diverse set of software methodologies: waterfall, iterative, software engineering, agile and even pure hacking (and prototyping). A particular programming language, or use of a particular programming technique, does not imply or require use of a particular software development methodology.

Objects usually are created from a general template called a class. The template contains the characteristics of the class without containing the specific data that need to be inserted into the template to form the object. Classes are the basis for most design work in objects. Classes are either superclasses (i.e., root or parent classes) with a set of basic attributes or methods or subclasses which inherit the characteristics of the parent class and may add (or remove) functionality as required. In addition to inheritance, classes may interact through sharing data, referred to as aggregate or component grouping, or sharing objects. Aggregate classes interact through messages, which are requests for services from one class (called a client) to another class (called a server). The ability of two or more objects to interpret a message differently at execution, depending on the superclass of the calling object, is termed polymorphism.

The first objected-oriented language, Simula67, was released in 1967. Smalltalk emerged during the 1970s as the first commercial object-oriented language. Then followed a series of languages that were either object-oriented from inception (e.g., Eiffel) or had been modified to include object-oriented capabilities (e.g., C++, Object Pascal, and Ada95). The emergence of Java during the late 1990s provided a significant boost to the acceptance of object technology.

To realize the full benefits of using object-oriented programming, it is necessary to employ object-oriented analysis and design approaches. Dealing with objects should permit analysts, developers and programmers to consider larger logical chunks of a system and clarify the programming process.

The major advantages of OOSD are as follows:
- The ability to manage an unrestricted variety of data types
- Provision of a means to model complex relationships
- The capacity to meet the demands of a changing environment

A significant development in OOSD has been the decision by some of the major players in object-oriented development to join forces and merge their individual approaches into a unified approach using the Unified Modeling Language (UML). UML is a general-purpose notational language for specifying and visualizing complex software for large object-oriented projects. This signals a maturation of the object-oriented development approach. While object-orientation is not yet pervasive, it can accurately be said to have entered the computing mainstream. Applications that use object-oriented technology are:
- Web applications
- E-business applications
- Computer-aided software engineering (CASE) for software development
- Office automation for email and work orders
- Artificial intelligence
- CAM for production and process control

3.8.6 COMPONENT-BASED DEVELOPMENT

Component-based development can be regarded as an outgrowth of object-oriented development. Component-based development means assembling applications from cooperating packages of executable software that make their services available through defined interfaces (i.e., enabling pieces of programs called objects, to communicate with one another regardless of which programming language they were written in or what OS they are running). The basic types of components are:
- **In-process client components**—These components must run from within a container of some kind such as a web browser; they cannot run on their own.
- **Stand-alone client components**—Applications that expose services to other software can be used as components. Well-known examples are Microsoft's Excel and Word.
- **Stand-alone server components**—Processes running on servers that provide services in standardized ways can be components. These are initiated by remote procedure calls or some other kind of network call. Technologies supporting this include Microsoft's Distributed Component Object Model (DCOM), Object Management Group's Common Object Request Broker Architecture (CORBA) and Sun's Java through Remote Method Invocation (RMI) are sometimes referred to as distributed object technologies
- **In-process server components**—These components run on servers within containers. Examples include Microsoft's Transaction Server (MTS) and Oracle's JavaBeans.

> **Note:** The CISA candidate will not be tested on vendor-specific products or services in the CISA exam.

A number of different component models have emerged. Microsoft has its Component Object Model (COM). MTS when combined with COM allows developers to create components that can be distributed in the Windows environment. COM is the basis for ActiveX technologies, with ActiveX Controls being among the most widely used components. Alternative component models include the CORBA Component Model and JavaBeans.

Please note that COM/DCOM, CORBA (itself a standard, not a specific product) and RMI are sometimes referred to as distributed object technologies. As the name suggests, they allow objects on distributed platforms to interact. These technologies, among others, are also termed middleware. Middleware is a broad term, but a basic definition is software that provides run-time services whereby programs/objects/components can interact with one another.

Tool developers are supporting one or another of these standards with powerful, visual tools now available for designing and testing component-based applications. Industry "heavyweights" such as Microsoft and IBM are supporting component-based development. Additionally, a growing number of commercially available application servers now support MTS or EJB. There is a growing market for third-party components. A primary benefit of component-based development is the ability to buy proven, tested software from commercial developers. The range of components available has increased. The first components were simple in concept (e.g., buttons and list boxes). Components now provide much more diverse functionality. Databases are now available on the web to search for commercial components.

Components play a significant role in web-based applications. Applets are required to extend static HTML, ActiveX controls or Java. Both technologies are compatible with component development. Component-based development:
- **Reduces development time**—If an application system can be assembled from prewritten components and only code for unique parts of the system needs to be developed, then this should prove faster than writing the entire system from scratch.
- **Improves quality**—Using prewritten components means a significant percentage of the system code has been tested already.
- **Allows developers to focus more strongly on business functionality**—An outcome of component-based development and its enabling technologies is to further increase abstraction already achieved with high-level languages, databases and user interfaces. Developers are shielded from low-level programming details.
- **Promotes modularity**—By encouraging or forcing impassable interfaces between discrete units of functionality, it encourages modularity.
- **Simplifies reuse**—It avoids the need to be conversant with procedural or class libraries, allowing cross-language combination and allowing reusable code to be distributed in an executable format (i.e., no source is required). (To date, large-scale reuse of business logic has not occurred.)
- **Reduces development cost**—Less effort needs to be expended on design and build. Instead, the cost of software components can be spread across multiple users.
- **Supports multiple development environments**—Components written in one language can interact with components written in other languages or running on other machines.

- **Allows a satisfactory compromise between build and buy options**—Instead of buying a complete solution, which perhaps does not entirely fit requirements, it could be possible to purchase only needed components and incorporate these into a customized system.

To realize these advantages, attention to software integration should be provided early and continuously during the development process. No matter how efficient component-based development is, if system requirements are poorly defined or the system fails to adequately address business needs, the project will not be successful.

3.8.7 WEB-BASED APPLICATION DEVELOPMENT

Web-based application development is an important emerging software development approach designed to achieve easier and more effective integration of code modules within and between enterprises. Historically, software written in one language on a particular platform has used a dedicated application programming interface (API). The use of specialized APIs has caused difficulties in integrating software modules across platforms. Technologies such as CORBA and COM that use remote procedure calls (RPCs) have been developed to allow real-time integration of code across platforms. However, using these RPC approaches for different APIs still remains complex. Web-based application development and associated XML technologies are designed to further facilitate and standardize code module and program integration.

The other problem that web-based application development seeks to address is to avoid the need to perform redundant computing tasks with the inherent need for redundant code. One obvious example of this is a change of address notification from a customer. Instead of having to update details separately in multiple databases (e.g., contact management, accounts receivable and credit control), it is preferable that a common update process updates the multiple places required. Web services are intended to make this relatively easy to achieve.

Web application development is different than traditional third- or fourth-generation program developments in many ways—from the languages and programming techniques used, to the methodologies (or lack thereof) used to control the development work, to the way the users test and approve the development work. The risk of application development remains the same. For example, buffer overflows had been a risk since computer programming was invented (for example, truncation issues with first generation computer programs), but they are widely known when they could be exploited by almost anyone, almost anywhere in the world, courtesy of the Internet.

As with traditional program development, a risk-based approach should be taken in the assessment of web application vulnerabilities: identify the business goals and supporting IT goals related to the development, then identify what can go wrong. One's previous experience can be used to identify risk related to inadequate specifications, poor coding techniques, inadequate documentation, inadequate QC and QA (including testing inadequacies), lack of proper change control and controls over promotion into production, and so on, and put these in the context of the web application languages, development processes and deliverables (perhaps with the support of best practice material/literature on web applications development). The focus should be on application development risk, the associated business risk and technical vulnerabilities, and how these could materialize and be controlled/addressed. Some controls will look the same for all application development activity, but many will need to reflect the way the development activity is taking place in the area under review.

With web-based application development, an XML language known as Simple Object Access Protocol (SOAP) is used to define APIs. SOAP will work with any OS and programming language that understands XML. SOAP is simpler than using the more complex RPC-based approach, with the advantage that modules are coupled loosely so that a change to one component does not normally require changes to other components.

The second key component of web development is the Web Services Description Language (WSDL), which is also based on XML. WSDL is used to identify the SOAP specification that is to be used for the code module API and the formats of the SOAP messages used for input and output to the code module. The WSDL is also used to identify the particular web service accessible via a corporate intranet or across the Internet by being published to a relevant intranet or Internet web server.

The final component of web services is another XML-based language—Universal Description, Discovery and Integration (UDDI). UDDI is used to make an entry in a UDDI directory, which acts as an electronic directory accessible via a corporate intranet or across the Internet, and allows interested parties to learn of the existence of available web services.

Standards for SOAP, WSDL and UDDI have been accepted by the World Wide Web Consortium. A number of current software products and development environments, including Microsoft's .Net family of products, support web services. However, some important standards, such as those addressing security and transaction management, are yet to be defined. Other issues, such as charging for use of commercially developed web services, also need to be addressed.

3.8.8 SOFTWARE REENGINEERING

Reengineering is a process of updating an existing system by extracting and reusing design and program components. This process is used to support major changes in the way an organization operates. A number of tools are now available to support this process. Typical methodologies used in software reengineering generally fall into the following categories:

- Business process reengineering (BPR) is the thorough analysis and significant redesign of business processes and management systems to establish a better performing structure, more responsive to the customer base and market conditions, while yielding material cost savings.
- The service-oriented software reengineering methodology is based upon the service-oriented computer architecture, and the reengineering processes apply many concepts of RAD development (see section 3.7.4 Controls in the EDI Environment) leveraging RACI (responsible, accountable, consulted and informed) charts and UML modeling.

3.8.9 REVERSE ENGINEERING

Reverse engineering is the process of studying and analyzing an application, a software application or a product to see how it functions and to use that information to develop a similar system. This process can be carried out in several ways:

- Decompiling object or executable code into source code and using it to analyze the program
- Black box testing the application to be reverse-engineered to unveil its functionality

The major advantages of reverse engineering are:

- Faster development and reduced SDLC duration
- The possibility of introducing improvements by overcoming the reverse-engineered application drawbacks

The IS auditor should be aware of the following risk:

- Software license agreements often contain clauses prohibiting the licensee from reverse engineering the software so that any trade secrets or programming techniques are not compromised.
- Decompilers are relatively new tools with functions that depend on specific computers, OSs and programming languages. Any change in one of these components may require developing or purchasing a new decompiler.

3.9 INFRASTRUCTURE DEVELOPMENT/ ACQUISITION PRACTICES

The physical architecture analysis, the definition of a new one and the necessary road map to move from one to the other is a critical task for an IT department. Its impact is not only economic but also technological because it decides many other choices downstream, such as operational procedures, training needs, installation issues and TCO.

Conflicting requirements such as evolving toward a services-based architecture, legacy hardware considerations, secure data access independent of data location, zero data loss, 24/7 availability and many others ensure that no single platform satisfies all these requirements equally. Thus, physical architecture analysis cannot be based solely on price or isolated features. A formal, reasoned choice must be made.

In section 3.9.1 Project Phases of Physical Architecture Analysis, steps are listed to arrive at the choice of a good physical architecture and the way to define a possible road map for supporting the migration of the technical architecture to a new one to reach the following goals:

- To successfully analyze the existing architecture (including data flow analysis defining all data being received, processed, stored and transmitted)
- To design a new architecture that takes into account the existing architecture and a company's particular constraints/requirements, such as:
  - Scalability and Interoperability to handle all data currently and having the potential of being received, processed, stored and transmitted
  - Reduced costs
  - Increased functionality
  - Minimum impact on daily work
  - Security and confidentiality issues
  - Progressive migration to the new architecture
- To write the functional requirements of this new architecture
- To develop a proof of concept (POC) based on these functional requirements:
  - To characterize price, functionality and performance
  - To identify additional requirements that will be used later

The resulting requirements will be documented in specifications and drawings describing the reference infrastructure that will be used by all projects downstream. With these requirements in hand, these projects can begin to start implementation.

The requirements are validated using a proof of concept. The proof of concept is a test-bed implementation of the physical architecture. It saves money because any problems are detected and corrected early, when they are cheaper to correct, and it gives confidence to the teams that the requirements correctly instruct potential vendors on the requirements they have to meet.

The main objective of section 3.9.2 Planning Implementation of Infrastructure, is to plan the physical implementation of the required technical infrastructure to set up the future environment (normally production, test and development environment). This task will cover the procurement activities such as contracting partners, setting up the SLAs, and developing installation plans and installation test plans. A well-designed selection process must be ensured, taking analytical results and intuition into account and guaranteeing alignment and commitment for implementation success. Due to the possible heterogeneous nature of the infrastructure found, it is necessary to develop a clear implementation plan (including deliverables, delivery times, test plans, etc.). It is also necessary to plan the coexistence of the old and new system, to avert possible mistakes during the installation and go-live phase (**figure 3.23**).

Thus, information and communication technologies (ICT) departments often face these requirements. The suggested solution must:

- Ensure alignment of the ICT with corporate standards
- Provide appropriate levels of security
- Integrate with current IT systems
- Consider IT industry trends
- Provide future operational flexibility to support business processes
- Allow for projected growth in infrastructure without major upgrades
- Include technical architecture considerations for information security, secure storage, etc.
- Ensure cost-effective, day-to-day operational support
- Foster the usage of standardized hardware and software
- Maximize ROI, cost transparency and operational efficiency

3.9.1 PROJECT PHASES OF PHYSICAL ARCHITECTURE ANALYSIS

Figure 3.24 shows the project phases to physical architecture analysis and, in the background, the time at which the vendor selection process may take place.

Figure 3.23—Risks of Current System

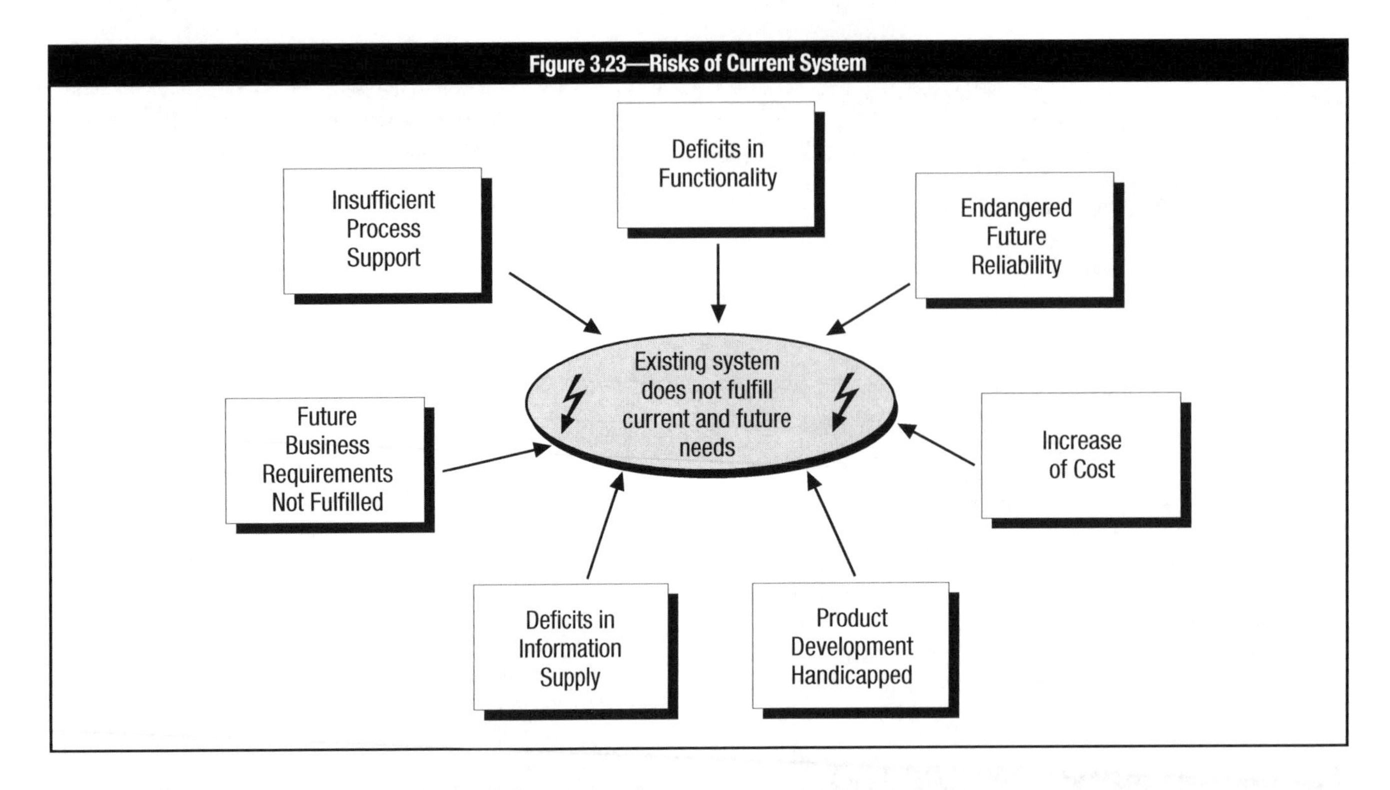

Figure 3.24—Project Phases of Physical Architecture Analysis

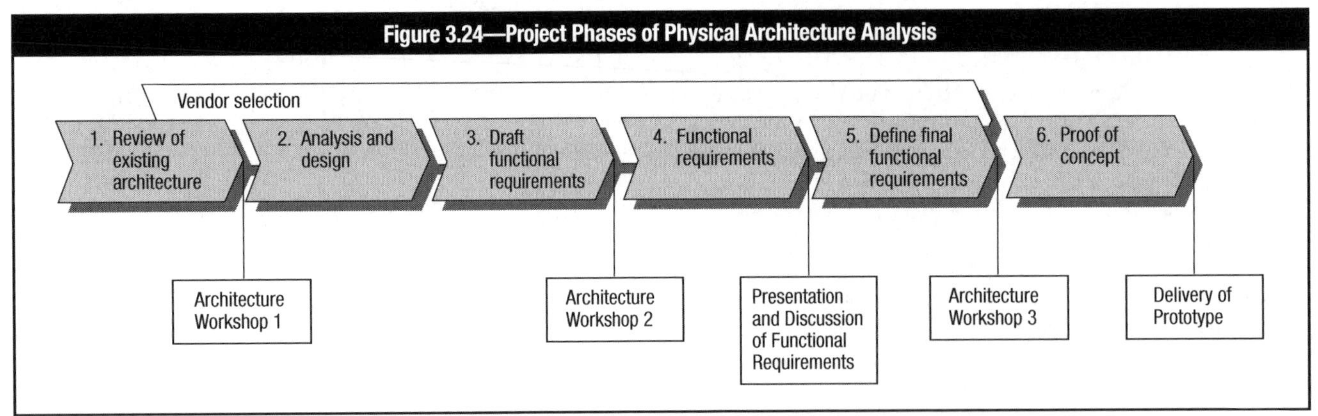

Review of Existing Architecture

To start the process, the latest documents about the existing architecture must be reviewed. Participants of the first workshop will be specialists of the ICT department in all areas directly impacted by physical architecture. Examples are server, storage, security and overall IT infrastructure.

Special care must be taken in characterizing all the operational constraints that impact physical architecture such as:

- Ground issues
- Size limits
- Weight limits
- Current power supply
- Environmental operating limitations (temperature and humidity minimum and maximum)
- Physical security issues

The output of the first workshop is a list of components of the current infrastructure and constraints defining the target physical architecture.

Analysis and Design

After reviewing the existing architecture, the analysis and design of the actual physical architecture has to be undertaken, adhering to good practices and meeting business requirements.

Draft Functional Requirements

With the first physical architecture design in hand, the first (draft) of functional requirements is composed. This material is the input for the next step and the vendor selection process.

Vendor and Product Selection

While the draft functional requirements are written, the vendor selection process proceeds in parallel. This process is described in detail later in this chapter.

Writing Functional Requirements

After finishing the draft functional requirements and feeding the second part of this project, the functional requirements document is written, which will be introduced at the second architecture workshop with staff from all affected parties. The results will be discussed and a list of the requirements that need to be refined or added will be composed.

This is the last checkpoint before the sizing and the POC starts, although the planning of the POC starts after the second workshop. With the finished functional requirements, the proof of concept phase begins.

Proof of Concept

Establishing a POC is highly recommended to prove that the selected hardware and software are able to meet all expectations, including security requirements. The deliverable of the POC should be a running prototype, including the associated document and test protocols describing the tests and their results.

To start, the POC should be based on the results of the procurement phase described below in this section. For this purpose, a representative subset of the target hardware is used. The software to run the POC can be either test versions or software already supplied by the vendor; therefore, additional costs are expected to be minimal.

To keep costs low, most elements of the framework are implemented in a simplified form. They will be extended to their final form in later phases.

The prototype should demonstrate the following features:
- The basic setup of the core security infrastructure
- Correct functionality of auditing components
- Basic but functional implementation of security measures as defined
- Secured transactions
- Characterization in terms of installation constraints and limits (server size, server current consumption, server weight, server room physical security)
- Performance
- Resiliency to include basic fail-over to a trusted operational state
- Funding and costing model

Related implementation projects that prepare the ground for deployment should also be part of the POC because they will be used in the same way as they are used in the production physical architecture. At the end of this phase, a last workshop is held where the production sizing and layout is adapted to include POC conclusions.

Additional considerations may apply if the entity goes in for an outsourcing/offshoring model for deployment and operation of applications. Also, the platform for operation of the IT environment (i.e., owned, cloud-based, virtualization) can give rise to additional considerations. For example, if the entity operates in a highly regulated industry or an industry that demands high levels of availability, adequate redundancy and safeguards for ensuring data privacy and confidentiality may have to be factored in while testing the POC.

3.9.2 PLANNING IMPLEMENTATION OF INFRASTRUCTURE

To ensure the quality of the results, it is necessary to use a phased approach to fit the entire puzzle together. It is also fundamental to set up the communication processes to other projects like those described earlier. Through these different phases the components are fit together, and a clear understanding of the available and contactable vendors is established by using the selection process during the procurement phase and beyond. Furthermore, it is necessary to select the scope of key business and technical requirements to prepare the next steps, which include the development of the delivery, installation and test plans. Moreover, to ensure a future proven solution, it is crucial to choose the right partners with the right skills.

As shown in **figure 3.25**, the requirements analysis is not part of this process but constantly feeds results into the process. If a Gantt chart is produced with these phases, most likely some phases overlap; therefore, the different phases must be considered an iterative process.

During the four different phases, it is necessary to fit all the components together to prepare for projects downstream (e.g., data migration).

Procurement Phase

During the procurement phase, the communication processes is established with the analysis project to get an overview of the chosen solution and determine the quantity structure of the deliverables. The requirements statements are also produced.

Additionally, the procurement process begins the service-level management process. During these activities, the preferred partners are invited to the negotiations process and the deliverables, contracts and SLAs are signed (**figure 3.26**).

Delivery Time

During the delivery time phase, the delivery plan is developed (**figure 3.27**). This phase overlaps in some parts with the procurement phase.

The delivery plan should include topics such as priorities, goals and nongoals, key facts, principles, communication strategies, key indicators, progress on key tasks, and responsibilities.

Installation Plan

During the installation planning phase, the installation plan is developed in cooperation with all affected parties (**figure 3.28**).

An additional step is to review the plan with the involved parties and, of course, with those responsible for the integration projects. This is an iterative process.

Figure 3.25—Project Phases of Planning the Implementation of Infrastructure

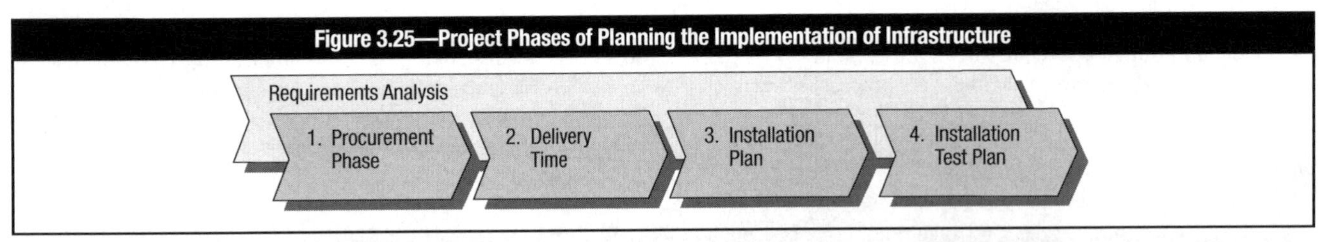

Figure 3.26—Procurement Phases

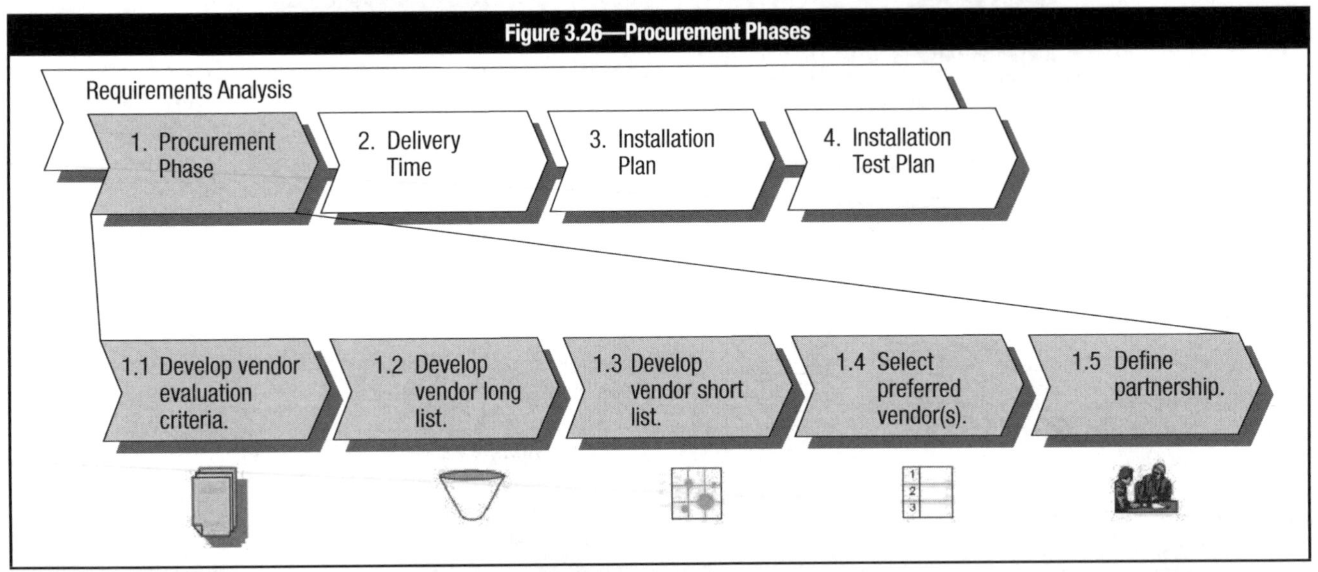

Figure 3.27—Delivery Time

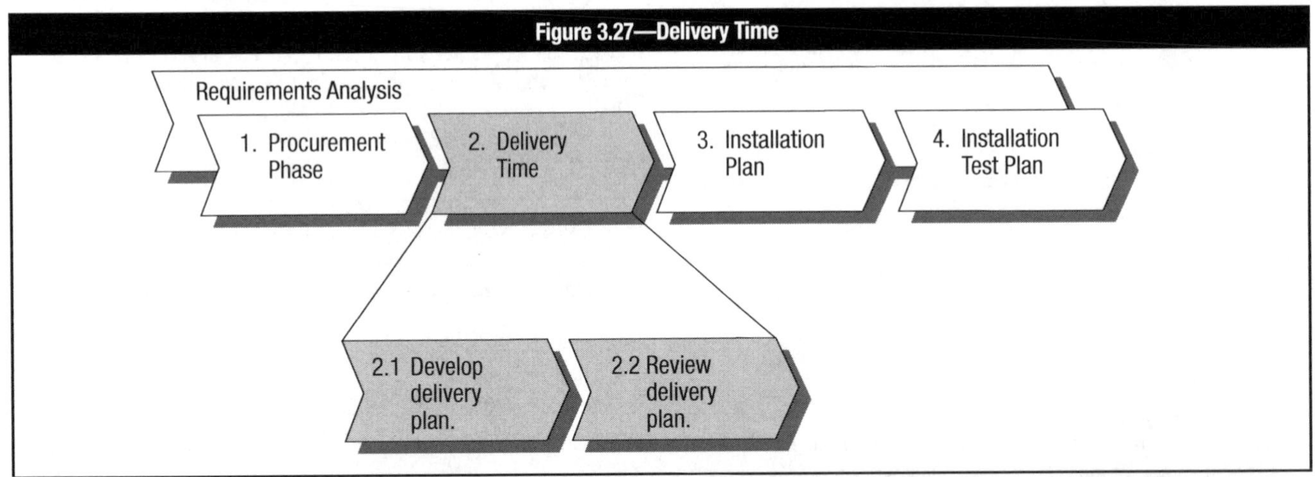

Figure 3.28—Installation Plan

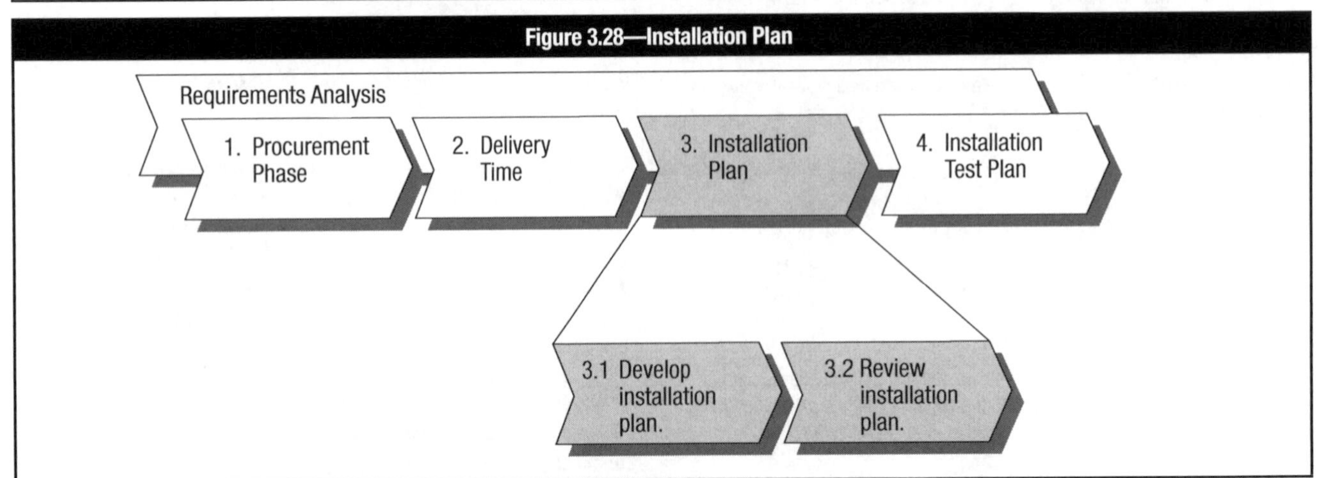

Installation Test Plan

Based on the known dependencies of the installation plan, the test plan is developed (**figure 3.29**).

The test plan includes test cases, basic requirements' specifications, definition of the processes and, as far as possible, measurement information for the applications and the infrastructure.

3.9.3 CRITICAL SUCCESS FACTORS

Critical success factors of planning the implementation include:
- To avoid delays, the appropriate skilled staff must attend workshops and participate for the entire project duration.
- The documentation needed for carrying out the work needs to be ready at project initiation.
- Decision makers must be involved at all steps to ensure all necessary decisions can be made quickly.
- Part one of the project (Analysis of Physical Architecture) must be completed, and the needed infrastructure decisions must be made.

3.9.4 HARDWARE ACQUISITION

Selection of a computer hardware and software environment frequently requires the preparation of specifications for distribution to hardware/software (HW/SW) vendors and criteria for evaluating vendor proposals. The specifications are sometimes presented to vendors in the form of an ITT, also known as a RFP.

The specifications must define, as completely as possible, the usage, tasks and requirements for the equipment needed and must include a description of the environment where that equipment will be used.

When acquiring a system, the specifications should include the following:
- Organizational descriptions indicating whether the computer facilities are centralized or decentralized, distributed, outsourced, manned or lights-out
- HW/SW evaluation assurance levels (EALs) for security robustness, based on *ISO/IEC 15408:2009: Information technology—Security techniques—Evaluation criteria for IT security*; developing HW/SW requirements using common criteria–evaluated IT products provides a level of confidence (based in independent laboratory testing) that the security functionality of these IT products meet stated security specifications.
- Information processing requirements such as:
  - Major existing application systems and future application systems
  - Workload and performance requirements
  - Processing approaches (e.g., online/batch, client-server, real-time databases, continuous operation).
- Hardware requirements such as:
  - CPU speed
  - Disk space requirements
  - Memory requirements
  - Number of CPUs required
  - Peripheral devices (e.g., sequential devices such as tape drives; direct access devices such as magnetic disk drives, printers, compact disc drives, digital video disc drives, universal serial bus [USB] peripherals and secure digital multimedia cards [SD/MMC]) required or to be excluded (usually for security reasons)
  - Data preparation/input devices that accept and convert data for machine processing
  - Direct entry devices (e.g., terminals, point-of-sale terminals or automated teller machines)
  - Networking capability (e.g., Ethernet connections, modems and integrated services digital network [ISDN] connections)
  - Number of terminals or nodes the system needs to support
- System software applications such as:
  - OS software (current version and any required upgrades)
  - Utilities
  - Compilers
  - Program library software
  - Database management software and programs
  - Communications software
  - Access control software
  - Job scheduling software
- Support requirements such as:
  - System maintenance (for preventive, detective [fault reporting] or corrective purposes)
  - Training (user and technical staff)
  - Backups (daily and disaster backups)
- Adaptability requirements such as:
  - Hardware and software upgrade capabilities
  - Compatibility with existing hardware and software platforms
  - Changeover to other equipment capabilities

Figure 3.29—Installation Test Plan

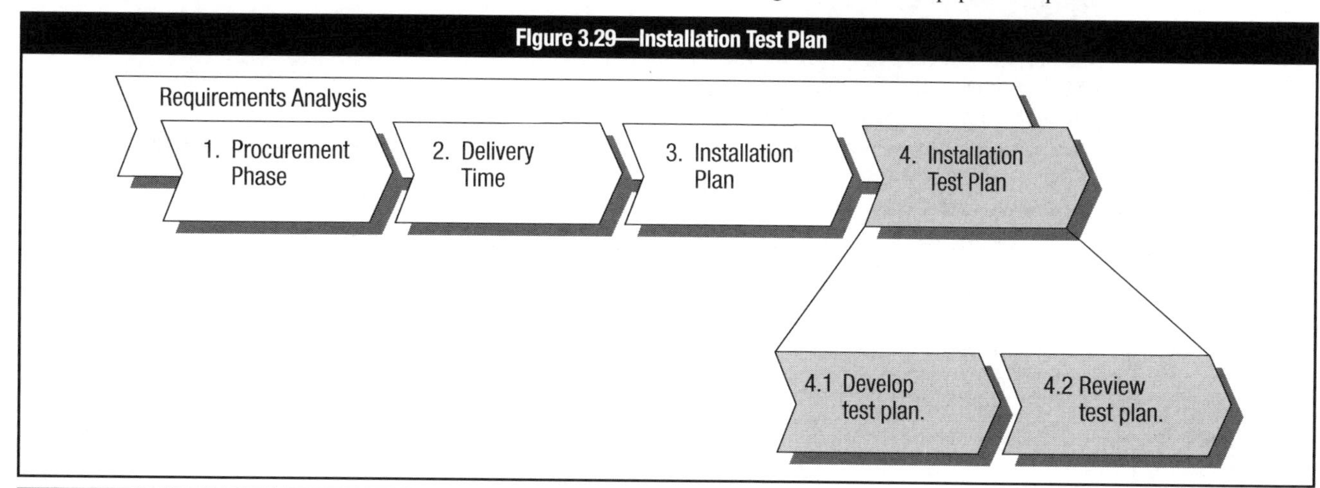

- Constraints such as:
  - Staffing levels
  - Existing hardware capacity
  - Delivery dates
- Conversion requirements such as:
  - Test time for the hardware and software
  - System conversion facilities
  - Cost/pricing schedule

Acquisition Steps

When purchasing (acquiring) hardware and software from a vendor, consideration should be given to the following:

- Testimonials or visits with other users
- Provisions for competitive bidding
- Analysis of bids against requirements
- Comparison of bids against each other using predefined evaluation criteria
- Analysis of the vendor's financial condition
- Analysis of the vendor's capability to provide maintenance and support (including training)
- Review of delivery schedules against requirements
- Pedigree of the hardware to verify it is not sourced from "grey market" supply sources (through distribution sources that are legal but are unofficial, unauthorized or unintended by the original manufacturer) that can increase the risk of malware and other unknown operability of the product
- Analysis of hardware and software upgrade capability
- Analysis of security and control facilities
- Evaluation of performance against requirements
- Review and negotiation of price
- Review of contract terms (including warranties, penalties and right to audit clauses)
- Preparation of a formal written report summarizing the analysis for each of the alternatives and justifying the selection based on benefits and cost

The criteria and data used for evaluating vendor proposals should be properly planned and documented. The following are some of the criteria that should be considered in the evaluation process:

- **Turnaround time**—The time that the help desk or vendor takes to fix a problem from the moment it is logged in
- **Response time**—The time a system takes to respond to a specific query by the user
- **System reaction time**—The time taken for logging into a system or getting connected to a network
- **Throughput**—The quantity of useful work made by the system per unit of time. Throughput can be measured in instructions per second or some other unit of performance. When referring to a data transfer operation, throughput measures the useful data transfer rate and is expressed in kilobits per second (Kbps), megabits per second (Mbps), and gigabits per second (Gbps).
- **Workload**—The capacity to handle the required volume of work or the volume of work that the vendor's system can handle in a given time frame
- **Compatibility**—The capability of an existing application to run successfully on the newer system supplied by the vendor
- **Capacity**—The capability of the newer system to handle a number of simultaneous requests from the network for the application and the volume of data that it can handle from each of the users
- **Utilization**—The system availability time versus the system downtime

When performing an audit of this area, the IS auditor should:

- Determine if the acquisition process began with a business need and whether the hardware requirements for this need were considered in the specifications.
- Determine if several vendors were considered and whether the comparison between them was done according to the aforementioned criteria.

3.9.5 SYSTEM SOFTWARE ACQUISITION

Every time a technological development has allowed for increased computing speeds or new capabilities, these have been absorbed immediately by the demands placed on computing resources by more ambitious applications. Consequently, improvements have led to decentralized, interconnected open systems through functions bundled in OS software to meet these needs. For example, network management and connectivity are features now found in most OSs.

It is IS management's responsibility to be aware of HW/SW capabilities because they may improve business processes and provide expanded application services to businesses and customers in a more effective way. Short- and long-term plans should document IS management's plan for migrating to newer, more efficient and more effective OSs and related systems software.

When selecting new system software, a number of business and technical issues must be considered including:

- Business, functional and technical needs and specifications
- Cost and benefit(s)
- Obsolescence
- Compatibility with existing systems
- Security
- Demands on existing staff
- Training and hiring requirements
- Future growth needs
- Impact on system and network performance
- Open source code versus proprietary code

3.9.6 SYSTEM SOFTWARE IMPLEMENTATION

System software implementation involves identifying features, configuration options and controls for standard configurations to apply across the organization. Additionally, implementation involves testing the software in a nonproduction environment and obtaining some form of certification and accreditation to place the approved OS software into production.

3.10 INFORMATION SYSTEMS MAINTENANCE PRACTICES

System maintenance practices refer primarily to the process of managing change to application systems while maintaining the integrity of both the production hardware (e.g., network and related server management products) and application source and executable code.

After a system is moved into production, it seldom remains static. Change is expected in all systems regardless of whether they are vendor-supplied or internally developed. Reasons for change in normal operations include internal IT/business changes, new external regulations, changes in classification related to either sensitivity or criticality, audits, and adverse incidents such as intrusions and viruses.

To control the ongoing maintenance of the system, a standard process for performing and recording changes is necessary. This process, which will be an integral part of the organization's overall SDLC process, should include steps to ensure that the system changes are appropriate to the needs of the organization, appropriately authorized, documented, thoroughly tested and approved by management. The process typically is established in the design phase of the application when application system requirements are baselined.

3.10.1 CHANGE MANAGEMENT PROCESS OVERVIEW

The change management process begins with authorizing changes to occur. For this purpose, a methodology should exist for prioritizing and approving system change requests. Change requests are initiated from end users as well as operational staff and system development/maintenance staff. In any case, authorization needs to be obtained from appropriate levels of end-user and systems management (e.g., a change control group, configuration control boards). For acquired systems, a vendor may distribute periodic updates, patches or new release levels of the software. User and systems management should review such changes. Determination should be made as to whether the changes are appropriate for the organization or will negatively affect the existing system.

Users should convey system change requests to the system management using some type of formal correspondence such as a standard change request form, memo or email message. At a minimum, the user request should include the requestor's name, date of request, date the change is needed, priority of the request, a thorough description of the change request, a description of any anticipated effects on other systems or programs, and fallback procedures in case the changes cause the failure of the system. The user could also provide a reason for the change, a cost justification analysis and the expected benefits of the change. In addition, the request should provide evidence that it has been reviewed and authorized by user management. A signature on the request form or memo typically provides this evidence.

Change requests should be in a format that ensures all changes are considered for action and allows the system management staff to easily track the status of the request. This is usually done by assigning a unique control number to each request and entering the change request information into a computerized system. This can also be performed manually. With detailed information regarding each request, management can identify those requests that have been completed and are still in progress or have not been addressed yet. Management can also use this information to help ensure that user requests are addressed in a timely manner. See **figure 3.30**.

All requests for changes and related information should have a security impact analysis performed to validate that change's impact on the system's overall security posture. The system maintenance staff as part of the system's permanent documentation should maintain the completed change documentation and supporting information.

Maintenance records of all program changes should exist either manually or automatically. Several library management software products provide this type of audit trail. The maintenance information usually consists of the programmer ID, time and date of change, project or request number associated with the change, and before and after images of the lines of code that were changed.

This process becomes even more important when the programmer who creates the program is also the operator. In this case, it is assumed that the IT department is either small or there are few applications being processed. Special change management procedures must be closely followed because SoD cannot be established in this environment and compensating controls are required. It requires user management to pay more attention to changes and upgrades made by the programmer, and proper authorization must be given to the programmer before putting any change into production. In lieu of the manual process of management approving changes before the programmer can submit them into production, management could have automated change control software installed to prevent unauthorized program changes. By doing this, the programmer is no longer responsible for migrating changes into production. The change control software becomes the operator that migrates programmer changes into production based on approval by management.

Programmers should not have write, modify or delete access to production data. Depending on the type of information in production, programmers may not even have read-only access (or access to customer credit card numbers, US Social Security numbers/national ID numbers or other sensitive information that may require added security).

Deploying Changes

After the end user is satisfied with the system test results and the adequacy of the system documentation, approval should be obtained from user management. Users should convey system change requests to the system management using some type of formal correspondence such as a standard change request form, memo or email message.

Documentation

To ensure the effective utilization and future maintenance of a system, it is important that all relevant system documentation be updated. Due to tight time constraints and limited resources, thorough updates to documentation are often neglected. Documentation requiring revision may consist of program and/or system flowcharts, program narratives, data dictionaries, entity relationship models, data flow diagrams (DFDs), operator run books and end-user procedural manuals. Keeping the internal coherence of all these items is a challenge; software configuration management packages can be a valuable tool.

Procedures should be in place to ensure that documentation stored offsite for disaster recovery purposes is also updated. This documentation is often overlooked and may be out of date.

Figure 3.30—Sample Change Request Form

Request for Change (RFC) Document

1. Contents

This document shall ensure that, at a minimum, every major change will be applied to the customer's mission or business critical systems in a controlled environment. It shall support all affected parties in gaining a more reliable and resilient infrastructure. Thus the usage of this document is mandatory for all involved personnel and—during normal operations—has to be distributed in a timely manner to provide the setup of a Forward Schedule of Change. In the rare condition of emergency changes it must be belatedly completed for documentation purpose. **In either case there must always be a formal approval for the RFC covered in this document.**

2. Usage Guidance

The following figure shall illustrate the usage guideline for this document.

The area shaded in red symbolizes the mandatory usage of the formal RFC procedure depicted in this document.

If the proposed change has an impact either on critical/vital systems (criticality is high) or on a massive amount of systems (wide penetration) then it is—by definition—a major change and the RFC procedure has to be executed.

In any other case (area left blank on the figure) it is up to the requestor to choose a formal approach or to bypass the RFC procedure.

To follow best practices a list shall be compiled which itemizes all systems classified as either critical or vital so the requestor will know what to head for (criticality). Additionally a percentage rate of affected systems shall be published as a threshold so that it is clearly defined when to execute the RFC procedure (penetration).

3. General RFC data

RFC title:			ID:	
Standard/emergency change:	☐/☐			
Issued by:		Intended recipient:		
Issued on:		Scheduled for:		
Checked by:		Checked on:		Status:
Postimplementation review done by:		Postimplementation review done on:		Status:

Has to be filled out by the requestor
Has to be filled out by the approver

4. Scope of Change

Abstract
Detailed description (please be as specific as you can)
Expected benefit

Figure 3.30—Sample Change Request Form (cont.)

Implementation checklist/release management		
Task	Description	Responsible Party

Affected configuration items (CIs) and impact on CIs

	Yes/No	Description
Performance degradation	☐ ☐	
Redundancy loss	☐ ☐	
Service disruption	☐ ☐	
Reboot	☐ ☐	
Downtime	☐ ☐	
Fallback/backout implemented	☐ ☐	

Associated costs (hardware, software, manpower, etc.)

5. **Approval/Rejection**

Comments

6. **Postimplementation Review**

Summary

	Yes/No	Description
Does the implementation meet your expectations?	☐ ☐	
Are there any deviations from the outlined procedure above?	☐ ☐	
Is the documentation/configuration management database up to date?	☐ ☐	
Are the stakeholders informed of the change?	☐ ☐	

Testing Changed Programs

Changed programs should be tested and also eventually certified with the same discipline as newly developed systems to ensure that the changes perform the intended functions. In addition, if the risk analysis determines it is necessary, additional testing would be required to ensure:
- Existing functionality is not damaged by the change
- System performance is not degraded because of the change
- No security exposures have been created because of the change

Auditing Program Changes

In evaluating whether procedures for program changes are adequate, the IS auditor should ensure that controls are in place to protect production application programs from unauthorized changes. The control objectives are as follows:
- Access to program libraries should be restricted.
- Supervisory reviews should be conducted.
- Change requests should be approved and documented.
- Potential impact of changes should be assessed.
- The change request should be documented on a standard form, paying particular attention to the following:
  - The change specifications should be adequately described, a cost analysis developed and a target date established.
  - The change form should be signed by the user to designate approval.
  - The change form should be reviewed and approved by programming management.
  - The work should be assigned to an analyst, programmer and programming group leader for supervision.
- A sample of program changes made during the audit period should be selected and traced to the maintenance form to determine whether the changes are authorized, check that the form has appropriate approvals, and compare the date on the form with the date of production update for agreement.
- If an independent group updates the program changes in production, the IS auditor should determine before the update whether procedures exist to ensure possession of the change request form. (This is accomplished by watching the groups perform their jobs.)

Emergency Changes

There may be times when emergency changes are required to resolve system problems and enable critical "production job" processing to continue. Procedures should primarily exist in the application's operations manual to ensure emergency fixes can be performed without compromising the integrity of the system. This typically involves the use of special logon IDs (i.e., emergency IDs) that grant a programmer/analyst temporary access to the production environment during these emergency situations. The use of emergency IDs should be logged and carefully monitored because their use grants someone powerful privileges. Emergency fixes should be completed using after-the-fact, follow-up procedures that ensure that all normal change management controls are retroactively applied. Changes completed in this fashion are held in a special emergency library from where they should be moved through the change management process into normal production libraries in an expeditious manner. IS auditors need to pay particular attention that emergency changes are handled in an appropriate and transparent manner.

Deploying Changes Back Into Production

Once user management has approved the change, the modified programs can be moved into the production environment. A group that is independent of computer programming should perform the migration of programs from test to production. Groups such as computer operations, QA or a designated change control group should perform this function.

To ensure that only authorized individuals have the ability to migrate programs into production, proper access restrictions must be in place. Such restrictions can be implemented through the use of OS security or an external security package.

Distributed systems, such as point-of-sale systems, offer an additional challenge in ensuring that changed programs are rolled out to all nodes. The rollout may be performed over a significant period of time to enable:
- Controls to be exercised over conversion of data
- Training of staff who will be using the changed software
- Support to be provided to users of the changed system
- Reduction of the risk associated with changing all nodes at the same time, and having to back them all out if something goes wrong

In view of the time it may take to roll out changes to the whole distributed system, controls must ensure that all nodes are eventually updated. If changes are made on a regular basis, it may be necessary to check regularly that all nodes are running the same versions of all software.

Change Exposures (Unauthorized Changes)

An unauthorized change to application system programs can occur for several reasons:
- The programmer has access to production libraries containing programs and data including object code.
- The user responsible for the application was not aware of the change (no user signed the maintenance change request approving the start of the work).
- A change request form and procedures were not formally established.
- The appropriate management official did not sign the change form approving the start of the work.
- The user did not sign the change form signifying acceptance before the change was updated into production.
- The changed source code was not properly reviewed by the appropriate programming personnel.
- The appropriate management official did not sign the change form approving the program for update to production.
- The programmer put in extra code for personal benefit (i.e., committed fraud).
- Changes received from the acquired software vendor were not tested or the vendor was allowed to load the changes directly into production/site. This happens in cases of distributed processing sites, such as POS, banking applications, ATM networks, etc.

3.10.2 CONFIGURATION MANAGEMENT

Because of the difficulties associated with exercising control over both system and programming maintenance activities, more and more organizations implement configuration management systems. In fact in many cases, regulatory requirements mandate these levels of control to provide a high degree of reliability and repeatability in all associated system processes (government systems, critical infrastructure, ICS, etc.). In a configuration management system, maintenance requests must be formally documented and approved by a change control group (e.g., configuration control boards). In addition, careful control is exercised over each stage of the maintenance process via checkpoints, reviews and sign-off procedures. From an audit perspective, effective use of this software provides important evidence of management's commitment to careful control over the maintenance process.

Configuration management involves procedures throughout the system hardware and software life cycle (from requirements analysis to maintenance) to identify, define and baseline software items in the system and thus provide a basis for problem management, change management and release management.

The process of checking out also prevents or manages simultaneous code edits, with hardware, network and system architects reviewing and approving the changes or updates to both the hardware asset and inventory tracking systems.

Checking in is the process of moving an item to the controlled environment. When a change is required (and supported by a change control form), the configuration manager will check out the item. Once the change is made, it can be checked using a different version number. The process of checking out also prevents or manages simultaneous code edits. With hardware, network and system architects review and approve the changes or updates to both the hardware asset and the inventory tracking systems.

For configuration management to work, management must support the concept of configuration management. The configuration management process is implemented by developing and following a configuration management plan and operating procedures. This plan should not be limited to just the software developed but should also include all system documentation, test plans and procedures. As part of the software configuration management task, the maintainer performs the following task steps:

1. Develop the configuration management plan.
2. Baseline the hardware, code, network physical and logical connections, ports protocols and services and associated engineering, operational, and administrative documents.
3. Analyze and report on the results of configuration control.
4. Develop the reports that provide configuration status information.
5. Develop release procedures.
6. Perform configuration control activities such as identification and recording of the request.
7. Update the configuration status accounting database.

In many cases, commercial software products will be used to automate the manual processes. Such tools should allow control to be maintained for applications software from the outset of system analysis and design to running live.

Configuration management tools will support change management and release management by providing automated support for the following:

1. Identification of items affected by a proposed change to assist with impact assessment (functional, operational and security)
2. Recording configuration items affected by authorized changes
3. Implementation of changes in accordance with authorization records
4. Registering of configuration item changes when authorized changes and releases are implemented
5. Recording of baselines that are related to releases (with known consequences) to which an organization would revert if an implemented change fails
6. Preparing a release to avoid human errors and resource costs

A new version of the system (or builds) should only be built from the baselined items. The baseline becomes the trusted recovery source for these systems and applications.

3.11 SYSTEM DEVELOPMENT TOOLS AND PRODUCTIVITY AIDS

System development tools and productivity aids include code generators, CASE applications and 4GL. These tools and aids are addressed in the following sections.

3.11.1 CODE GENERATORS

Code generators are tools, often incorporated with CASE products, that generate program code based on parameters defined by a systems analyst or on data/entity flow diagrams developed by the design module of a CASE product. These products allow most developers to implement software programs with efficiency. The IS auditor should be aware of source code generated by such tools.

3.11.2 COMPUTER-AIDED SOFTWARE ENGINEERING

Application development efforts require collecting, organizing and presenting a substantial amount of data at the application, systems and program levels. A substantial amount of the application development effort involves translating this information into program logic and code for subsequent testing, modification and implementation. This often is a time consuming process but it is necessary to develop, use and maintain computer applications.

CASE is the use of automated tools to aid in the software development process. Their use may include the application of software tools for software requirements capture and analysis, software design, code production, testing, document generation and other software development activities.

CASE products are generally divided into three categories:

1. **Upper CASE**—Products used to describe and document business and application requirements. This information includes data object definitions and relationships, and process definitions and relationships.
2. **Middle CASE**—Products used for developing the detailed designs. These include screen and report layouts, editing criteria, data object organization and process flows. When elements or relationships change in the design, it is necessary to make only minor alterations to the automated design and all other relationships are automatically updated.

3. **Lower CASE**—Products involved with the generation of program code and database definitions. These products use detailed design information, programming rules and database syntax rules to generate program logic, data file formats or entire applications.

Some CASE products embrace two of these categories or all three of them.

CASE tools are available for mainframe, minicomputer and microcomputer environments. These tools can provide higher quality systems more quickly. CASE products enforce a uniform approach to system development, facilitate storage and retrieval of documents, and reduce the manual effort in developing and presenting system design information. This power of automation changes the nature of the development process by eliminating or combining some steps and altering the means of verifying specifications and applications.

The IS auditor needs to recognize the changes in the development process brought on by CASE. Some CASE systems allow a project team to produce a complete system from the DFDs and data elements without any traditional source code. In these situations, the DFDs and data elements become the source code.

The IS auditor should gain assurances that approvals are obtained for the appropriate specifications, users continue to be involved in the development process, and investments in CASE tools yield benefits in quality and speed. Other key issues the IS auditor needs to consider with CASE include the following:
- CASE tools help in the application design process but do not ensure that the design, programs and system are correct or that they fully meet the needs of the organization.
- CASE tools should complement and fit into the application development methodology, but there needs to be a project methodology in place for CASE to be effective. The methodology should be understood and used effectively by the organization's software developers.
- The integrity of data moved between CASE products or between manual and CASE processes needs to be monitored and controlled.
- Changes to the application should be reflected in stored CASE product data.
- Just like a traditional application, application controls need to be designed.
- The CASE repository (the database that stores and organizes the documentation, models and other outputs from the different phases) needs to be secured on a need-to-know basis. Strict version control should be maintained on this database.

The IS auditor may also become a user of CASE tools as several features facilitate the audit process. DFDs, which may be the product of upper and middle CASE tools, may be used as an alternative to other flowcharting techniques. IS auditors whose IS departments are moving into CASE are using CASE-generated documentation as part of the audit. Some are even experimenting with the use of CASE tools to create audit documentation. In addition, CASE tools can be used to develop interrogation software and embedded audit modules (EAMs). Repository reports should be used to gain an understanding of the system and to review controls over the development process.

3.11.3 FOURTH-GENERATION LANGUAGES

While a standard definition of a 4GL does not exist, the common characteristics of 4GLs are the following:
- **Nonprocedural language**—Most 4GLs do not obey the procedural paradigm of continuous statement execution and subroutine call and control structures. Instead, they are event-driven and make extensive use of object-oriented programming concepts such as objects, properties and methods.
  - For example, a COBOL programmer who wants to produce a report sorted in a given sequence must first open and read the data file, then sort the file and finally produce the report. A typical 4GL treats the report as an object with properties, such as input file name and sort order, and methods such as sort file and print report.
  - Care should be taken when using 4GLs. Unlike traditional languages, the 4GLs can lack the lower level detail commands necessary to perform certain types of data intensive or online operations. These operations are usually required when developing major applications. For this reason, the use of 4GLs as development languages should be weighed carefully against traditional languages already discussed.
- **Environmental independence (portability)**—Many 4GLs are portable across computer architectures, OSs and telecommunications monitors. Some 4GLs have been implemented on mainframe processors and microcomputers.
- **Software facilities**—These facilities include the ability to design or paint retrieval screen formats, develop computer-aided training routines or help screens, and produce graphical outputs.
- **Programmer workbench concepts**—The programmer has access through the terminal to easy filing facilities, temporary storage, text editing and OS commands. This type of a workbench approach is closely associated with the CASE application development approach. It is often referred to as an integrated development environment (IDE).
- **Simple language subsets**—4GLs generally have simple language subsets that can be used by less-skilled users in an information center.

4GLs are often classified in the following ways:
- **Query and report generators**—These specialized languages can extract and produce reports (audit software). Recently, more powerful languages have been produced that can access database records, produce complex online outputs and be developed in an almost natural language.
- **Embedded database 4GLs**—These depend on self-contained database management systems. This characteristic often makes them more user-friendly but also may lead to applications that are not integrated well with other production applications. Examples include FOCUS, RAMIS II and NOMAD 2.
- **Relational database 4GLs**—These high-level language products are usually an optional feature on a vendor's DBMS product line. These allow the applications developer to make better use of the DBMS product, but they often are not end-user-oriented. Examples include SQL+, MANTIS and NATURAL.

- **Application generators**—These development tools generate lower-level programming languages (3GLs) such as COBOL and C. The application can be further tailored and customized. Data processing development personnel, not end users, use application generators.

3.12 PROCESS IMPROVEMENT PRACTICES

Business processes require improvements, which are accomplished with practices and techniques addressed in the following sections.

3.12.1 BUSINESS PROCESS REENGINEERING AND PROCESS CHANGE PROJECTS

The generic model shown in **figure 3.31** is a very basic description of a process (some form of information enters the process, is processed, and the outcome is measured against the goal or objective of the process). The level of detail needed (e.g., breakdown of subprocesses to activities) depends highly on the complexity of the process, the knowledge of the affected staff and the company's requirements regarding audit functionality (performance and compliance) of the process and shall fit into an existing quality management system (e.g., ISO 9001:2000).

Any output produced by a process must be bound to a business objective and adhere to defined corporate standards. Monitoring of effectiveness (goal achievement), efficiency (minimum effort) and compliance must be done on a regular basis and shall be included in management reports for review under the plan-do-check-act (PDCA) cycle.

BPR is the process of responding to competitive and economic pressures, and customer demands to survive in the current business environment. This is usually done by automating system processes so that there are fewer manual interventions and manual controls. BPR achieved with the help of implementing an ERP system is often referred to as package-enabled reengineering (PER). Advantages of BPR are usually experienced where the reengineering process appropriately suits the business needs. BPR has increased in popularity as a method for achieving the goal of cost savings through streamlining operations.

The steps in a successful BPR are:
- Define the areas to be reviewed.
- Develop a project plan.
- Gain an understanding of the process under review.
- Redesign and streamline the process.
- Implement and monitor the new process.
- Establish a continuous improvement process.

As a reengineering process takes hold, new results begin to emerge:
- New business priorities based on value and customer requirements
- A concentration on process as a means of improving product, service and profitability
- New approaches to organizing and motivating people inside and outside the enterprise
- New approaches to the use of technologies in developing, producing and delivering goods and services
- New approaches to the use of information as well as powerful and more accessible information technologies
- Refined roles for suppliers including outsourcing, joint development, quick response, JIT inventory and support
- Redefined roles for clients and customers, providing them with more direct and active participation in the enterprise's business process

A successful BPR/process change project requires the project team to perform the following for the existing processes:
- Process decomposition to the lowest level required for effectively assessing a business process (typically referred to as an elementary process), which is a unit of work performed with a definitive input and output

Figure 3.31—Generic Process Model

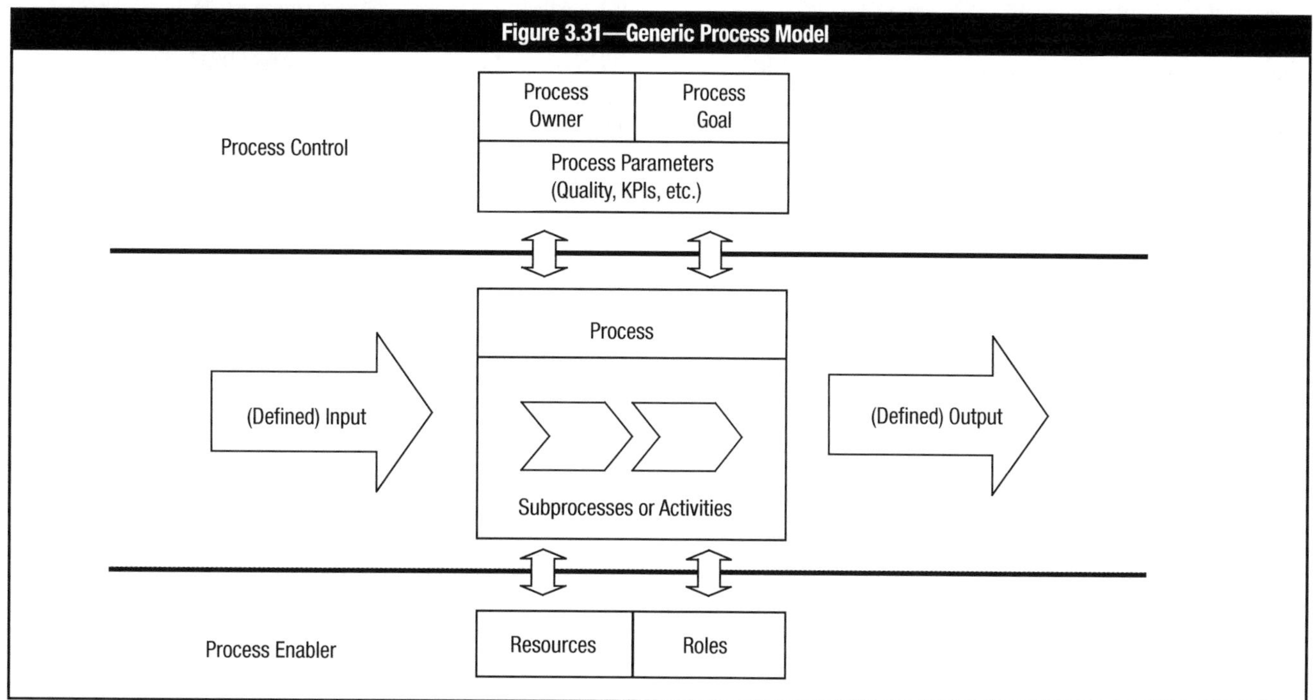

- Identification of customers, process-based managers or process owners responsible for processes from beginning to end
- Documentation of the elementary process-related profile information including:
  - Duration
  - Trigger (which triggers the process to act)
  - Frequency
  - Effort
  - Responsibility (process owner)
  - Input and output
  - External interfaces
  - System interaction
  - Risk and control information
  - Performance measurement information
  - Identified problematic areas and their root causes

The existing baseline processes must be documented—preferably in the form of flowcharts and related profile documents—so the baseline processes can be compared to the processes after reengineering.

The newly designed business processes inevitably involve changes in the way(s) of doing business and could impact the finances, philosophy and personnel of the organization, its business partners and customers.

Throughout the change process, the BPR team must be sensitive to organization culture, structure, direction and the components of change. Management must also be able to predict and/or anticipate issues and problems, and offer appropriate resolutions that will accelerate the change process.

BPR teams can be used to facilitate and assist the staff in transitioning into the reengineered business processes. BPR professionals are valuable in monitoring progress toward the achievement of the strategic plan of the organization.

A major concern in BPR is that key controls may be reengineered out of a business process. The IS auditor's task is to identify the existing key controls and evaluate the impact of removing these controls. If the controls are key preventive controls, the IS auditor must ensure that management is aware of the removal of the control and management is willing to accept the potential material risk of not having that preventive control.

BPR Methods and Techniques

Applying BPR methods and techniques to a process creates an immediate environment for change and provides consistency of results.

BENCHMARKING PROCESS

Benchmarking is about improving business processes. It is defined as a continuous, systematic process for evaluating the products, services or work processes of organizations recognized as a world-class "reference" in a globalized world. Reference products services or processes are systematically analyzed for one or more of the following purposes:

- Comparing and ranking
- Strategic planning, SWOT (strengths, weaknesses, opportunities and threats) analysis
- Investment decisions, company takeovers, mergers
- Product or process design or redesign/reengineering
- BPR

The steps listed below are followed generally by the benchmarking team in a benchmarking exercise:

1. **Plan**—In the planning stage, critical processes are identified for the benchmarking exercise. The benchmarking team should identify the critical processes and understand how they are measured, the kinds of data that are needed and how the data need to be collected.
2. **Research**—The team should collect baseline data about the processes of its own organization before collecting these data about other organizations. The next step is to identify the reference products or companies through sources such as business newspapers and magazines, quality award winners, trade journals, consultants, etc. Depending on the team's own preferences and resources, and on the marketplace, several scenarios may result:
   - Benchmarks that satisfy the organization's interest already exist at no charge from professional associations, journals or analysis firms.
   - The organization may join or promote a survey launched by a single or multi-industry specialized web portal (e.g., a bookmark portal).
   - The organization may conduct or subcontract business intelligence.
   - The organization may enter into an agreement with one or more "benchmark partners" who agree to share information.
   - Depending on the aims of the exercise and the resulting scenario above, the next steps will be skipped or adapted.
3. **Observe**—The next step is to collect data and visit the benchmarking partner. There should be an agreement with the partner organization, a data collection plan and a method to facilitate proper observation.
4. **Analyze**—This step involves summarizing and interpreting the data collected, and analyzing the gaps between an organization's process and its partner's process. Converting key findings into new operational goals will be the goal of this stage.
5. **Adopt**—Adopting the results of benchmarking can be the most difficult step. In this step, the team needs to translate the findings into a few core principles and work down from principles to strategies to action plans.
6. **Improve**—Continuous improvement is the key focus in a benchmarking exercise. Benchmarking links each process in an organization with an improvement strategy and organizational goals.

BPR Audit and Evaluation

When reviewing an organization's business process change (reengineering) efforts, IS auditors must determine whether:

- The organization's change efforts are consistent with the overall culture and strategic plan of the organization.
- The reengineering team is making an effort to minimize any negative impact the change might have on the organization's staff.
- The BPR team has documented lessons to be learned after the completion of the BPR/process change project.

The IS auditor would also provide a statement of assurance or conclusion with respect to the objectives of the audit.

3.12.2 ISO/IEC 25010:2011

ISO/IEC 25010:2011 is an international standard to assess the quality of software products. It provides the definition of the characteristics and associated quality evaluation process to be used when specifying the requirements for, and evaluating the quality of, software products throughout their life cycle. Attributes evaluated include:

- **Functionality**—The set of attributes that bears on the existence of a set of functions and their specified properties. The functions are those that satisfy stated or implied needs.
- **Reliability**—The set of attributes that bears on the capability of software to maintain its level of performance under stated conditions for a stated period of time.
- **Usability**—The set of attributes that bears on the effort needed for use and on the individual assessment of such use by a stated or implied set of users.
- **Efficiency**—The set of attributes that bears on the relationship between the level of performance of the software and the amount of resources used under stated conditions.
- **Maintainability**—The set of attributes that bears on the effort needed to make specified modifications.
- **Portability**—The set of attributes that bears on the ability of software to be transferred from one environment to another.

3.12.3 CAPABILITY MATURITY MODEL INTEGRATION

Following the release and successful adoption of capability maturity model (CMM) for Software, other models were developed for disciplines such as systems engineering, integrated product development, etc. The Capability Maturity Model Integration (CMMI) was conceived as a means of combining the various models into a set of integrated models. CMMI also describes five levels of maturity, although the descriptions of what constitutes each level differ from those used in the original CMM. CMMI is considered less directly aligned with the traditional waterfall approach toward development and better aligned with contemporary software development practices including:

- Iterative development
- Early definition of architecture
- Model-based design notation
- Component-based development
- Demonstration-based assessment of intermediate development products
- Use of scalable, configurable processes

Maturity models, such as CMMI, are useful to evaluate management of a computer center, the development function management process, and implement and measure the IT change management process.

See **figure 3.32** for characteristics of the maturity levels.

3.12.4 ISO/IEC 330XX SERIES

ISO/IEC 3300xx is a series of standards that provide guidance on process assessment. It supersedes and extends parts of the ISO/IEC 15504 series of standards. ISO/IEC 33002:2015 lists requirements for effective process assessment. ISO/IEC 33004:2015 provides requirements for process reference, assessment and maturity models. ISO/IEC 33020:2015 describes a process measurement framework for the assessment of process capability. Other ISO/IEC 330xx documents further outline the relationship of this series to the ISO/IEC 15504 series and cover other topics such as process improvement.

Figure 3.32—Characteristics of the Maturity Levels (CMMI)

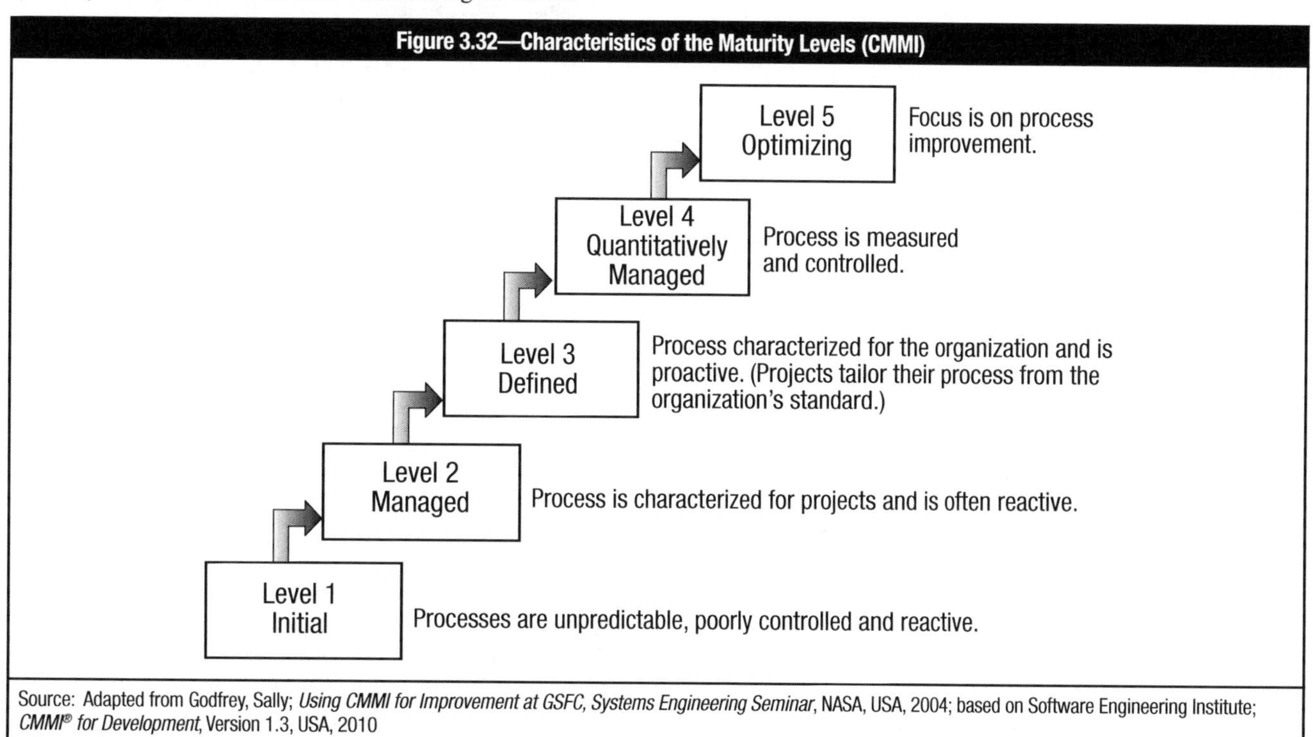
Level 5
Optimizing
Focus is on process improvement.
Level 4
Quantitatively Managed
Process is measured and controlled.
Level 3
Defined
Process characterized for the organization and is proactive. (Projects tailor their process from the organization's standard.)
Level 2
Managed
Process is characterized for projects and is often reactive.
Level 1
Initial
Processes are unpredictable, poorly controlled and reactive.

Source: Adapted from Godfrey, Sally; *Using CMMI for Improvement at GSFC, Systems Engineering Seminar*, NASA, USA, 2004; based on Software Engineering Institute; *CMMI® for Development*, Version 1.3, USA, 2010

3.13 APPLICATION CONTROLS

Application controls are controls over input, processing and output functions. They include methods for ensuring that:
- Only complete, accurate and valid data are entered and updated in a computer system
- Processing accomplishes the correct task
- Processing results meet expectations
- Data are maintained

Application controls may consist of edit tests, totals, reconciliations and identification and reporting of incorrect, missing or exception data. Automated controls should be coupled with manual procedures to ensure proper investigation of exceptions.

These controls help ensure data accuracy, completeness, validity, verifiability and consistency, thus achieving data integrity and data reliability. Implementation of these controls helps ensure system integrity, that applicable system functions operate as intended, and that information contained by the system is relevant, reliable, secure and available when needed.

The IS auditor's tasks include the following:
- Identifying the significant application components and the flow of transactions through the system and gaining a detailed understanding of the application by reviewing the available documentation and interviewing appropriate personnel
- Identifying the application control strengths, and evaluating the impact of the control weaknesses
- Developing a testing strategy
- Testing the controls to ensure their functionality and effectiveness by applying appropriate audit procedures
- Evaluating the control environment by analyzing the test results and other audit evidence to determine that control objectives were achieved
- Considering the operational aspects of the application to ensure its efficiency and effectiveness by comparing the system with efficient system design standards, analyzing procedures used and comparing them to management's objectives for the system

3.13.1 INPUT/ORIGINATION CONTROLS

Input control procedures must ensure that every transaction to be processed is entered, processed and recorded accurately and completely. These controls should ensure that only valid and authorized information is input and that these transactions are only processed once. In an integrated systems environment, output generated by one system is the input for another system.

Therefore, the system receiving the output of another system as input/origination must in turn apply edit checks, validations and access controls to those data.

Input Authorization

Input authorization verifies that all transactions have been authorized and approved by management.

Authorization of input helps ensure that only authorized data are entered for processing by applications. Authorization can be performed online at the time when the data are entered into the system. A computer-generated report listing the items requiring manual authorization may also be generated. It is important that controls exist throughout processing to ensure that the authorized data remain unchanged. This can be accomplished through various accuracy and completeness checks incorporated into an application's design.

Types of authorization include:
- **Signatures on batch forms or source documents**—Provide evidence of proper authorization.
- **Online access controls**—Ensure that only authorized individuals may access data or perform sensitive functions.
- **Unique passwords**—Necessary to ensure that access authorization cannot be compromised through use of another individual's authorized data access. Individual passwords also provide accountability for data changes. (See chapter 5 Protection of Information Assets, for more information.)
- **Terminal or client workstation identification**—Used to limit input to specific terminals or workstations as well as to individuals. Terminals or client workstations in a network can be configured with a unique form of identification such as serial number or computer name that is authenticated by the system.
- **Source documents**—The forms used to record data. A source document may be a piece of paper, a turnaround document or an image displayed for online data input. A well-designed source document achieves several purposes. It increases the speed and accuracy with which data can be recorded, controls work flow, facilitates preparation of the data in machine-readable form for pattern recognition devices, increases the speed and accuracy with which data can be read, and facilitates subsequent reference checking.

Ideally, source documents should be preprinted or electronic forms to provide consistency, accuracy and legibility. Source documents should include standard headings, titles, notes and instructions. Source document layouts should:
- Emphasize ease of use and readability
- Group similar fields together to facilitate input
- Provide predetermined input codes to reduce errors
- Contain appropriate cross-reference numbers or a comparable identifier to facilitate research and tracing
- Use boxes to prevent field size errors
- Include an appropriate area for management to document authorization

All source documents should be appropriately controlled. Procedures should be established to ensure that all source documents have been input and taken into account. Prenumbering source documents facilitates this control.

Batch Controls and Balancing

Batch controls group input transactions to provide control totals. The batch control can be based on total monetary amount, total items, total documents or hash totals.

Batch header forms are a data preparation control. All input forms should be clearly identified with the application name and transaction codes. Where possible, preprinted and prenumbered forms, with transaction identification codes and other constant data items, are recommended. This would help ensure that all pertinent data have been recorded on the input forms and can reduce data recording/entry errors.

Types of batch controls include:
- **Total monetary amount**—Verification that the total monetary value of items processed equals the total monetary value of the batch documents. For example, the total monetary value of the sales invoices in the batch agrees with the total monetary value of the sales invoices processed. This provides assurance on the completeness and accuracy of the sales value processed for the batch.
- **Total items**—Verification that the total number of items included on each document in the batch agrees with the total number of items processed. For example, the total number of units ordered in the batch of invoices agrees with the total number of units processed. This provides assurance on the completeness and accuracy of the units ordered in the batch processed.
- **Total documents**—Verification that the total number of documents in the batch equals the total number of documents processed. For example, the total number of invoices in a batch agrees with the total number of invoices processed. This provides assurance on the completeness of the number of invoices processed.
- **Hash totals**—Verification that total in a batch agrees with the total calculated by the system. Hash total is the total of non-value numeric fields in the batch (like total of dates or customer number fields, which by themselves, do not have informative value). This provides assurance on the completeness and accuracy of data entered for the numeric fields in the batch.

Batch balancing can be performed through manual or automated reconciliation. Batch totaling must be combined with adequate follow-up procedures. Adequate controls should exist to ensure that each transaction creates an input document, all documents are included in a batch, all batches are submitted for processing, all batches are accepted by the computer, batch reconciliation is performed, procedures for the investigation and timely correction of differences are followed, and controls exist over the resubmission of rejected items.

Types of batch balancing include:
- **Batch registers**—These registers enable recording of batch totals and subsequent comparison with system reported totals.
- **Control accounts**—Control account use is performed through an initial edit file to determine batch totals. The data are then processed to the master file, and a reconciliation is performed between the totals processed during the initial edit file and the master file.
- **Computer agreement**—Computer agreement with batch totals is performed through the input of batch header details that record the batch totals; the system compares these to calculated totals, either accepting or rejecting the batch.

Error Reporting and Handling

Input processing requires that controls be identified to verify that only correct data are accepted into the system and input errors are recognized and corrected.

Data conversion error corrections are needed during the data conversion process. Errors can occur due to duplication of transactions and inaccurate data entry. These errors can, in turn, impact the completeness and accuracy of the data. Corrections to data should be processed through normal data conversion processes and should be verified, authorized and reentered into the system as a part of the normal processing.

Input error handling can be processed by:
- **Rejecting only transactions with errors**—Only transactions containing errors would be rejected; the rest of the batch would be processed.
- **Rejecting the whole batch of transactions**—Any batch containing errors would be rejected for correction prior to processing.
- **Holding the batch in suspense**—Any batches containing errors would not be rejected; however, the batch would be held in suspense, pending correction.
- **Accepting the batch and flagging error transactions**—Any batch containing errors would be processed; however, those transactions containing errors would be flagged for identification, enabling subsequent error correction.

Input control techniques include:
- **Transaction log**—Contains a detailed list of all updates. The log can be either manually maintained or provided through automatic computer logging. A transaction log can be reconciled to the number of source documents received to verify that all transactions have been input.
- **Reconciliation of data**—Controls whether all data received are properly recorded and processed
- **Documentation**—Written evidence of user, data entry and data control procedures
- **Error correction procedures**—These include:
  - Logging of errors
  - Timely corrections
  - Upstream resubmission
  - Approval of corrections
  - Suspense file
  - Error file
  - Validity of corrections
- **Anticipation**—The user or control group anticipates the receipt of data
- **Transmittal log**—Documents transmission or receipt of data
- **Cancellation of source documents**—Procedures to cancel source documents such as by punching with holes or marking them to avoid duplicate entry

3.13.2 PROCESSING PROCEDURES AND CONTROLS

Processing procedures and controls are meant to ensure the reliability of application program processing. IS auditors need to understand the procedures and controls that can be exercised over processing to evaluate what exposures are covered by these controls and what exposures remain.

Data Validation and Editing Procedures

Procedures should be established to ensure that input data are validated and edited as close to the time and point of origination as possible. Preprogrammed input formats ensure that data are input to the correct field in the correct format. If input procedures allow supervisor overrides of data validation and editing, automatic logging should occur. A manager who did not initiate the override should review this log.

Data validation is meant to identify data errors, incomplete or missing data and inconsistencies among related data items. Front-end data editing and validation can be performed if intelligent terminals are used.

Edit controls are preventive controls that are used in a program before data are processed. If not in place or not working effectively, the preventive controls are not effective. This may cause processing of inaccurate data. **Figure 3.33** describes various types of data validation edits.

Processing Controls

Processing controls are meant to ensure the completeness and accuracy of accumulated data. They would ensure that data in a file/database remain complete and accurate until changed as a result of authorized processing or modification routines. The following are processing control techniques that can be used to address the issues of completeness and accuracy of accumulated data:

- **Manual recalculations**—A sample of transactions may be recalculated manually to ensure that processing is accomplishing the anticipated task.
- **Editing**—An edit check is a program instruction or subroutine that tests the accuracy, completeness and validity of data. It may be used to control input or later processing of data.
- **Run-to-run totals**—Run-to-run totals provide the ability to verify data values through the stages of application processing. Run-to-run total verification ensures that data read into the computer were accepted and then applied to the updating process.
- **Programmed controls**—Software can be used to detect and initiate corrective action for errors in data and processing. For example, if the incorrect file or file version is provided for processing, the application program could display messages instructing that the proper file and version be used.
- **Reasonableness verification of calculated amounts**—Application programs can verify the reasonableness of calculated amounts. The reasonableness can be tested to ensure appropriateness to predetermined criteria. Any transaction that is determined to be unreasonable may be rejected pending further review.
- **Limit checks on amounts**—An edit check can provide assurance, through the use of predetermined limits, that amounts have been keyed or calculated correctly. Any transaction exceeding the limit may be rejected for further investigation.

Figure 3.33—Data Validation Edits and Controls

Edits	Description
Sequence check	The control number follows sequentially and any sequence or duplicated control numbers are rejected or noted on an exception report for follow-up purposes. For example, invoices are numbered sequentially. The day's invoices begin with 12001 and end with 15045. If any invoice larger than 15045 is encountered during processing, that invoice would be rejected as an invalid invoice number.
Limit check	Data should not exceed a predetermined amount. For example, payroll checks should not exceed US $4,000. If a check exceeds US $4,000, the data would be rejected for further verification/authorization.
Range check	Data should be within a predetermined range of values. For example, product type codes range from 100 to 250. Any code outside this range should be rejected as an invalid product type.
Validity check	Programmed checking of the data validity in accordance with predetermined criteria. For example, a payroll record contains a field for marital status and the acceptable status codes are M or S. If any other code is entered, the record should be rejected.
Reasonableness check	Input data are matched to predetermined reasonable limits or occurrence rates. For example, a widget manufacturer usually receives orders for no more than 20 widgets. If an order for more than 20 widgets is received, the computer program should be designed to print the record with a warning indicating that the order appears unreasonable.
Table lookups	Input data comply with predetermined criteria maintained in a computerized table of possible values. For example, the input clerk enters a city code of 1 to 10. This number corresponds with a computerized table that matches the code to a city name.
Existence check	Data are entered correctly and agree with valid predetermined criteria. For example, a valid transaction code must be entered in the transaction code field.
Key verification	The keying process is repeated by a separate individual using a machine that compares the original keystrokes to the repeated keyed input. For example, the worker number is keyed twice and compared to verify the keying process.
Check digit	A numeric value that has been calculated mathematically is added to data to ensure that the original data have not been altered or an incorrect, but valid, value substituted. This control is effective in detecting transposition and transcription errors. For example, a check digit is added to an account number so it can be checked for accuracy when it is used.
Completeness check	A field should always contain data rather than zeros or blanks. A check of each byte of that field should be performed to determine that some form of data, not blanks or zeros, is present. For example, a worker number on a new employee record is left blank. This is identified as a key field and the record would be rejected, with a request that the field be completed before the record is accepted for processing.
Duplicate check	New transactions are matched to those previously input to ensure that they have not already been entered. For example, a vendor invoice number agrees with previously recorded invoices to ensure that the current order is not a duplicate and, therefore, the vendor will not be paid twice.
Logical relationship check	If a particular condition is true, then one or more additional conditions or data input relationships may be required to be true and consider the input valid. For example, the hire date of an employee may be required to be more than 16 years past his/her date of birth.

- **Reconciliation of file totals**—Reconciliation of file totals should be performed on a routine basis. Reconciliations may be performed through the use of a manually maintained account, a file control record or an independent control file.
- **Exception reports**—An exception report is generated by a program that identifies transactions or data that appear to be incorrect. These items may be outside a predetermined range or may not conform to specified criteria.

Data File Control Procedures

File controls should ensure that only authorized processing occurs to stored data. Types of controls over data files are shown in **figure 3.34**.

Contents of data files, or indeed database tables, generally fall into four categories:

- **System control parameters**—The entries in these files change the workings of the system and may alter controls exercised by the system; for example, the tolerance allowed before an exceptional transaction is reported or blocked. Any change to these files should be controlled in a similar way to program changes.
- **Standing data**—These "master files" include data, such as supplier/customer names and addresses, that do not frequently change and are referred to during processing. These data should be authorized before entry or maintenance. Input controls may include a report of changed data that is checked and approved. Audit trails may log all changes.

Figure 3.34—Data File Controls

Method	Description
Before and after image reporting	Computer data in a file prior to and after a transaction is processed can be recorded and reported. The before and after images make it possible to trace the impact transactions have on computer records.
Maintenance error reporting and handling	Control procedures should be in place to ensure that all error reports are properly reconciled and corrections are submitted on a timely basis. To ensure SoD, error corrections should be reviewed properly and authorized by personnel who did not initiate the transaction.
Source documentation retention	Source documentation should be retained for an adequate time period to enable retrieval, reconstruction or verification of data. Policies regarding the retention of source documentation should be enforced. Originating departments should maintain copies of source documentation and ensure that only authorized personnel have access. When appropriate, source documentation should be destroyed in a secure, controlled environment.
Internal and external labeling	Internal and external labeling of removable storage media is imperative to ensure that the proper data are loaded for processing. External labels provide the basic level of assurance that the correct data medium is loaded for processing. Internal labels, including file header records, provide assurance that the proper data files are used and allow for automated checking.
Version usage	It is critical that the proper version of a file be used as well as the correct file, for processing to be correct. For example, transactions should be applied to the most current database, while restart procedures should use earlier versions.
Data file security	Data file security controls prevent unauthorized access by unauthorized users that may have access to the application to alter data files. These controls do not provide assurances relating to the validity of data but ensure that unauthorized users who may have access to the application cannot alter stored data improperly.
One-for-one checking	Individual documents agree with a detailed listing of documents processed by the computer. It is necessary to ensure that all documents have been received for processing.
Prerecorded input	Certain information fields are preprinted on blank input forms to reduce initial input errors.
Transaction logs	All transaction input activity is recorded by the computer. A detailed listing, including date of input, time of input, user ID and terminal location, can then be generated to provide an audit trail. It also permits operations personnel to determine which transactions have been posted. This will help to decrease the research time needed to investigate exceptions and decrease recovery time if a system failure occurs.
File updating and maintenance authorization	Proper authorization for file updating and maintenance is necessary to ensure that stored data are safeguarded adequately, correct and up to date. Application programs may contain access restrictions in addition to the overall system access restrictions. The additional security may provide levels of authorization as well as an audit trail of file maintenance.
Parity checking	Data transfers in a computer system are expected to be made in a relatively error-free environment. However, when programs or vital data are transmitted, additional controls are needed. Transmission errors are controlled primarily by error-detecting or correcting codes. The former is used more often because error-correcting codes are costly to implement and are unable to correct all errors. Generally, error detection methods such as a check bit and redundant transmission are adequate. Redundancy checking is a common error-detection routine. A transmitted block of data containing one or more records or messages is checked for the number of characters or patterns of bits contained in it. If the numbers or patterns do not conform to predetermined parameters, the receiving device ignores the transmitted data and instructs the user to retransmit. Check bits are often added to the transmitted data by the telecommunications control unit and may be applied either horizontally or vertically. These checks are similar to the parity checks normally applied to data characters within on-premises equipment. A parity check on a single character generally is referred to as a vertical or column check, and a parity check on all the equivalent bits is known as a horizontal, longitudinal or row check. Use of both checks greatly improves the possibilities of detecting a transmission error, which may be missed when either of those checks is used alone.

- **Master data/balance data**—Running balances and totals that are updated by transactions should not be capable of adjustment except under strict approval and review controls. Audit trails are important here since there may be financial reporting implications for the change.
- **Transaction files**—These are controlled using validation checks, control totals, exception reports, etc.

It should be noted that the controls built into the application represent the management design of controls on how a business process (procurement or sales or payroll) should be run. While the applications contain the rules for the business, the data that are the outcome of the processing are stored in the database. An entity may have the best controls built into the application, but if management personnel directly update data in the database, then the benefit of the best controls in the application will be overridden.

However, in real-world production operations, in some situations, entities may have to carry out direct updates to database. For example, if due to a systems outage, the transactions could not be processed in real time for a few days, it is not practical to insist that once the system availability is restored, the backlog should be entered through the application (front end) before the transactions of the subsequent days is entered or processed. In such cases, management may take a decision to catch-up on the backlog by directly updating the transactions in the database (back end). Therefore, the IS auditor should ensure that there are controls in place to ensure that such direct back-end data fixes are supported by authorization of the business for completeness and accuracy and are processed subject to computer operations controls. The important point to remember is that the quality of application controls is only as good as the quality of controls around direct back-end data fixes, in any entity.

3.13.3 OUTPUT CONTROLS

Output controls are meant to provide assurance that the data delivered to users will be presented, formatted and delivered in a consistent and secure manner.

Output controls include:

- **Logging and storage of negotiable, sensitive and critical forms in a secure place**—Negotiable, sensitive or critical forms should be properly logged and secured to provide adequate safeguards against theft, damage or disclosure. The form log should be routinely reconciled to have inventory on hand, and any discrepancies should be properly researched.
- **Computer generation of negotiable instruments, forms and signatures**—The computer generation of negotiable instruments, forms and signatures should be properly controlled. A detailed listing of generated forms should be compared to the physical forms received. One should properly account for all exceptions, rejections and mutilations.
- **Report accuracy, completeness and timeliness**—Often reports are generated using third-party data analysis and reporting applications (ESSbase, etc.). Even with the most reliable and accurate data sources, improperly configured, constructed and prepared reports are still a significant risk. Report design and generation specifications, templates and creation/change request processes are critical system output controls.
- **Reports generated from the system**—These represent the data that management relies upon for business decisions and review of business results. Therefore, ensuring the integrity of data in reports is key for the reliability of information in information systems. The IS auditor should validate that the reports are accurate and correct representation of the source data

Example 3

In a trial balance report (where the debit side and credit side is expected to total, if the application had processed transactions accurately), the programmer was found to have coded the report in such a manner that the report simply reproduced the credit total under the debit column to force the tie out of the debit and credit totals.

The IS auditor needs to apply an assessment approach in validating reports depending on the situation (more evaluation when the entity has undergone a system change or evaluating customized reports as against standard reports of a widely used application).

- **Report distribution**—Output reports should be distributed according to authorized distribution parameters, which may be automated or manual. Operations personnel should verify that output reports are complete and delivered according to schedule. All reports should be logged prior to distribution. In most environments, processing output is spooled to a buffer or print spool on completion of job processing, where it waits for an available printer. Controls over access to the print spools are important to prevent reports from being deleted accidentally from print spools or directed to a different printer. In addition, changes to the output print priority can delay printing of critical jobs. Access to distributed reports can compromise confidentiality. Therefore, physical distribution of reports should be controlled adequately. Reports containing sensitive data should be printed under secure, controlled conditions. Secure output drop-off points should be established. Output disposal should also be secured adequately to ensure that no unauthorized access can occur. Reports that are distributed electronically through the computer system also need to be considered. Logical access to these reports should also be controlled carefully and subject to authorization. When distributed manually, assurance should be provided that sensitive reports are properly distributed. Such assurance should include the recipient signing a log as evidence of receipt of output (i.e., manual nonrepudiation).
- **Balancing and reconciling**—Data processing application program output should be balanced routinely to the control totals. Audit trails should be provided to facilitate the tracking of transaction processing and the reconciliation of data.
- **Output error handling**—Procedures for reporting and controlling errors contained in the application program output should be established. The error report should be timely and delivered to the originating department for review and error correction.
- **Output report retention**—A record retention schedule should be adhered to firmly. Any governing legal regulations should be included in the retention policy.
- **Verification of receipt of reports**—To provide assurance that sensitive reports are properly distributed, the recipient should sign a log as evidence of receipt of output.

The IS auditor should be aware of existing concerns regarding record-retention policies for the organization and address legal requirements. Output can be restricted to particular IT resources or devices (e.g., a particular printer)

3.13.4 BUSINESS PROCESS CONTROL ASSURANCE

In an integrated application environment, controls are embedded and designed into the application that supports the processes. Business process control assurance involves evaluating controls at the process and activity level. These controls may be a combination of management, programmed and manual controls. In addition to evaluating general controls that affect the processes, business process owner-specific controls—such as establishing proper security and SoD, periodic review and approval of access, and application controls within the business process—are evaluated.

Specific matters to consider in the business process control assurance are:
- Process and data flow mapping
- Process controls
- Assessing business risks within the process
- Benchmarking with best practices
- Roles and responsibilities
- Activities and tasks
- Data restrictions

3.14 AUDITING APPLICATION CONTROLS

The IS auditor's tasks include the following:
- Identifying the significant application components and the flow of information through the system, and gaining a detailed understanding of the application by reviewing the available documentation and interviewing appropriate personnel. Developing a data flow diagram can help visualize the flow of information.
- Understanding and evaluation of interfaces, including APIs, in case the application connects with other applications
- Identifying the application control strengths and evaluating the impact of the control weaknesses to develop a testing strategy by analyzing the accumulated information
- Reviewing application system documentation to provide an understanding of the functionality of the application. In many cases—mainly in large systems or packaged software—it is not feasible to review the whole application documentation. Thus, a selective review should be performed. If an application is vendor supplied, technical and user manuals should be reviewed. Any changes to applications should be documented properly.

The following documentation should be reviewed to gain an understanding of an application's development:
- **System development methodology documents**—These documents include cost-benefit analysis and user requirements.
- **Functional design specifications**—This document provides a detailed explanation of the application. An understanding of key control points should be noted during review of the design specifications.
- **Program changes**—Documentation of any program change should be available for review. Any change should provide evidence of authorization and should be cross-referenced to source code.
- **User manuals**—A review of the user manuals provides the foundation for understanding how the user is utilizing the application. Often control weaknesses can be noted from the review of this document.
- **Technical reference documentation**—This documentation includes any vendor-supplied technical manuals for purchased applications in addition to any in-house documentation. Access rules and logic usually are included in these documents.

3.14.1 FLOW OF TRANSACTIONS THROUGH THE SYSTEM

A transaction flowchart provides information regarding key processing controls. Points where transactions are entered, processed and posted should be reviewed for control weaknesses.

3.14.2 RISK ASSESSMENT MODEL TO ANALYZE APPLICATION CONTROLS

Risk assessment, as discussed in chapter 1, provides information relating to the inherent risk of an application.

A risk assessment model can be based on many factors, which may include a combination of the following:
- The quality of internal controls
- Economic conditions (impacts on organizational resource availability, capacity and capability)
- Recent accounting system changes
- Time elapsed since last audit
- Complexity of operations
- Changes in operations/environment
- Recent changes in key positions
- Time in existence
- Competitive environment
- Assets at risk
- Prior audit results
- Staff turnover
- Transaction volume
- Regulatory agency impact
- Monetary volume
- Sensitivity of transactions
- Impact of application failure

3.14.3 OBSERVING AND TESTING USER PERFORMING PROCEDURES

Some of the user procedures that should be observed and tested include:
- **Segregation of duties**—SoD ensures that no individual has the capability of performing more than one of the following processes: origination, authorization, verification or distribution. Observation and review of job descriptions and review of authorization levels and procedures may provide information regarding the existence and enforcement of SoD.
- **Authorization of input**—Evidence of input authorization can be achieved via written authorization on input documents or with the use of unique passwords. One may test this by looking through a sampling of input documents for proper authorization or reviewing computer-access rules. Supervisor overrides of data validation and editing should be reviewed to ensure that automatic logging occurs. This override activity report should

be tested for evidence of managerial review. Excessive overrides may indicate the need for modification of validation and editing routines to improve efficiency.
- **Balancing**—This is performed to verify that run-to-run control totals and other application totals are reconciled on a timely basis. This may be tested by independent balancing or reviewing past reconciliations.
- **Error control and correction**—This is completed in the form of reports that provide evidence of appropriate review, research, timely correction and resubmission. Input errors and rejections should be reviewed prior to resubmission. Managerial review and authorization of corrections should be evidenced. Testing of this effort can be achieved by retabulating or reviewing past error corrections.
- **Distribution of reports**—Critical output reports should be produced and maintained in a secure area and distributed in an authorized manner. The distribution process can be tested by observation and review of distribution output logs. Access to online output reports should be restricted. Online access may be tested through a review of the access rules or by monitoring user output.
- **Review and testing of access authorizations and capabilities**—Access control tables provide information regarding access levels by individuals. Access should be based on job descriptions and should provide for SoD. Testing can be performed through the review of access rules to ensure that access has been granted as management intended.
- **Activity reports**—These provide details, by user, of activity volume and hours. Activity reports should be reviewed to ensure that activity occurs only during authorized hours of operation.
- **Violation reports**—These indicate any unsuccessful and unauthorized access attempts. Violation reports should indicate the terminal location, date and time of attempted access. These reports should evidence managerial review. Repeated unauthorized access violations may indicate attempts to circumvent access controls. Testing may include review of follow-up activities.

3.14.4 DATA INTEGRITY TESTING

Data integrity testing is a set of substantive tests that examines accuracy, completeness, consistency and authorization of data presently held in a system. It employs testing similar to that used for input control. Data integrity tests will indicate failures in input or processing controls. Controls for ensuring the integrity of accumulated data in a file can be exercised by regularly checking data in the file. When this checking is done against authorized source documentation, it is common to check only a portion of the file at a time. Because the whole file is regularly checked in cycles, the control technique is often referred to as cyclical checking.

Two common types of data integrity tests are relational and referential integrity tests:
- **Relational integrity tests**—Performed at the data element and record-based levels. Relational integrity is enforced through data validation routines built into the application or by defining the input condition constraints and data characteristics at the table definition in the database stage. Sometimes it is a combination of both.
- **Referential integrity tests**—Define existence relationships between entities in different tables of a database that needs to be maintained by the DBMS. It is required for maintaining interrelation integrity in the relational data model. Whenever two or more relations are related through referential constraints (primary and foreign key), it is necessary that references be kept consistent in the event of insertions, deletions and updates to these relations. Database software generally provides various built-in automated procedures for checking and ensuring referential integrity. Referential integrity checks involve ensuring that all references to a primary key from another table (i.e., a foreign key) actually exist in their original table. In nonpointer databases (e.g., relational), referential integrity checks involve making sure that all foreign keys exist in their original table.

3.14.5 DATA INTEGRITY IN ONLINE TRANSACTION PROCESSING SYSTEMS

In multiuser transaction systems, it is necessary to manage parallel user access to stored data typically controlled by a DBMS, and deliver fault tolerance. Of particular importance are four online data integrity requirements known collectively as the ACID principle:
- **Atomicity**—From a user perspective, a transaction is either completed in its entirety (i.e., all relevant database tables are updated) or not at all. If an error or interruption occurs, all changes made up to that point are backed out.
- **Consistency**—All integrity conditions in the database are maintained with each transaction, taking the database from one consistent state into another consistent state.
- **Isolation**—Each transaction is isolated from other transactions, and hence, each transaction only accesses data that are part of a consistent database state.
- **Durability**—If a transaction has been reported back to a user as complete, the resulting changes to the database survive subsequent hardware or software failures.

This type of testing is vital in today's vast array of online Internet-accessible, multiuser DBMSs.

3.14.6 TEST APPLICATION SYSTEMS

Testing the effectiveness of application controls involves analyzing computer application programs, testing computer application program controls, or selecting and monitoring data process transactions. Testing controls by applying appropriate audit procedures is important to ensure their functionality and effectiveness. Methods and techniques for each category are described in **figure 3.35**.

To facilitate the evaluation of application system tests, an IS auditor may also want to use generalized audit software (GAS), also known as computer-assisted audit tools (CAATs). This is particularly useful when specific application control weaknesses are discovered that affect, for example, updates to master file records and certain error conditions on specific transaction records. Additionally, GAS can be used to perform certain application control tests, such as parallel simulation, in comparing expected outcomes to live data.

Figure 3.35—Testing Application Systems

Analyzing Computer Application Programs			
Technique	**Description**	**Advantages**	**Disadvantages**
Snapshot	• Records flow of designated transactions through logic paths within programs	• Verifies program logic	• Requires extensive knowledge of the IS environment
Mapping	• Identifies specific program logic that has not been tested and analyzes programs during execution to indicate whether program statements have been executed	• Increases efficiency by identifying unused code • Identifies potential exposures	• Cost of software
Tracing and tagging	• Tracing shows the trail of instructions executed during an application. • Tagging involves placing an indicator on selected transactions at input and using tracing to track them.	• Provides an exact picture of sequence of events, and is effective with live and simulated transactions	• Requires extensive amounts of computer time, an intimate knowledge of the application program and additional programming to execute trace routines
Test data/deck	• Simulates transactions through real programs	• May use actual master files or dummies • Source code review unnecessary • Can be used on a surprise basis • Provides objective review and verification of program controls and edits • Initial use can be limited to specific program functions minimizing scope and complexity • Requires minimal knowledge of the IS environment	• Difficult to ensure that the proper program is checked • Risk of not including all transaction scenarios • Requires good knowledge of application systems • Does not test master file and master file records
Base-case system evaluation	• Uses test data sets developed as part of a comprehensive testing of programs • Verifies correct system operations before acceptance, as well as periodic revalidation	• Comprehensive testing verification and compliance testing	• Extensive effort to maintain data sets • Close cooperation is required among all parties
Parallel operation	• Processes actual production data through existing and newly developed programs at the same time and compares results and is used to verify changed production prior to replacing existing procedures	• Verifies new system before discontinuing the old one	• Added processing costs
Integrated testing facility	• Creates a fictitious file in the database with test transactions processed simultaneously with live data	• Periodic testing does not require separate test process.	• Need for careful planning • Need to isolate test data from production data
Parallel simulation	• Processes production data using computer programs that simulate application program logic	• Eliminates need to prepare test data	• Programs must be developed
Transaction selection programs	• Use audit software to screen and select transactions input to the regular production cycle	• Independent of production system • Controlled by the auditor • Requires no modification to production systems	• Cost of development and maintenance

Figure 3.35—Testing Application Systems *(cont.)*			
Analyzing Computer Application Programs			
Technique	**Description**	**Advantages**	**Disadvantages**
Embedded audit data collection	• Software embedded in host computer applications screens. It selects input transactions and generates transactions during production. Usually, it is developed as part of system development. Types include: – Systems control audit review file (SCARF): Auditor determines reasonableness of tests incorporated into normal processing. It provides information for further review. – Sample audit review file (SARF): Randomly selects transactions to provide representative file for analysis	• Provides sampling and productions statistics	• High cost of development and maintenance • Auditor independence issues
Extended records	• Gathers all data that have been affected by a particular program	• Records are put into one convenient file.	• Adds to data storage costs and overhead, and to system development costs

3.14.7 CONTINUOUS ONLINE AUDITING

Continuous online auditing is becoming increasingly important in the e-business world because it provides a method for the IS auditor to collect evidence on system reliability while normal processing takes place. The approach allows IS auditors to monitor the operation of such a system on a continuous basis and gather selective audit evidence through the computer. If the selective information collected by the computer technique is not deemed serious or material enough to warrant immediate action, the information is stored in separate audit files for verification by the IS auditor at a later time. The continuous audit approach cuts down on needless paperwork and leads to the conduct of an essentially paperless audit. In such a setting, an IS auditor can report directly through the microcomputer on significant errors or other irregularities that may require immediate management action. This approach reduces audit cost and time.

Continuous audit techniques are important IS audit tools, particularly when they are used in time-sharing environments that process a large number of transactions but leave a scarce paper trail. By permitting IS auditors to evaluate operating controls on a continuous basis without disrupting the organization's usual operations, continuous audit techniques improve the security of a system. When a system is misused by someone withdrawing money from an inoperative account, a continuous audit technique will report this withdrawal in a timely fashion to the IS auditor. Thus the time lag between the misuse of the system and the detection of that misuse is reduced. The realization that failures, improper manipulation and lack of controls will be detected on a timely basis by the use of continuous audit procedures gives IS auditors and management greater confidence in a system's reliability.

Continuous audit very often relies on calls to GAS/CAAT services.

3.14.8 ONLINE AUDITING TECHNIQUES

There are five types of automated evaluation techniques applicable to continuous online auditing:

1. **Systems Control Audit Review File and Embedded Audit Modules (SCARF/EAM)**—The use of this technique involves embedding specially written audit software in the organization's host application system so the application systems are monitored on a selective basis.
2. **Snapshots**—This technique involves taking what might be termed pictures of the processing path that a transaction follows, from the input to the output stage. With the use of this technique, transactions are tagged by applying identifiers to input data and recording selected information about what occurs for the auditor's subsequent review.
3. **Audit hooks**—This technique involves embedding hooks in application systems to function as red flags and to induce IS security and auditors to act before an error or irregularity gets out of hand.
4. **Integrated test facility (ITF)**—In this technique, dummy entities are set up and included in an auditee's production files. The IS auditor can make the system either process live transactions or test transactions during regular processing runs and have these transactions update the records of the dummy entity. The operator enters the test transactions simultaneously with the live transactions that are entered for processing. The auditor then compares the output with the data that have been independently calculated to verify the correctness of the computer-processed data.
5. **Continuous and intermittent simulation (CIS)**—During a process run of a transaction, the computer system simulates the instruction execution of the application. As each transaction is entered, the simulator decides whether the transaction meets certain predetermined criteria and, if so, audits the transaction. If not, the simulator waits until it encounters the next transaction that meets the criteria.

In **figure 3.36**, the relative advantages and disadvantages of the various concurrent audit tools are presented.

The use of each of the continuous audit techniques has advantages and disadvantages. Their selection and implementation depends, to a large extent, on the complexity of an organization's computer systems and applications, and the IS auditor's ability to understand and evaluate the system with and without the use of continuous audit techniques. In addition, IS auditors must recognize that continuous audit techniques are not a cure for all control problems and that the use of these techniques provides only limited assurance that the information processing systems examined are operating as they were intended to function.

3.15 AUDITING SYSTEMS DEVELOPMENT, ACQUISITION AND MAINTENANCE

The IS auditor's tasks in system development, acquisition and maintenance may take place once the project is finished or during the project itself. Most tasks in the following list cover both scenarios and the IS auditor is expected to determine which task applies. Tasks generally include the following:

- Meet with key systems development and user project team members to determine the main components, objectives and user requirements of the system to identify the areas that require controls.
- Discuss the selection of appropriate controls with systems development and user project team members to determine and rank the major risks to and exposures of the system.
- Discuss references to authoritative sources with systems development and user project team members to identify controls to mitigate the risks to and exposures of the system.
- Evaluate available controls and participate in discussions with systems development and user project team members to advise the project team regarding the design of the system and implementation of controls.
- Periodically meet with systems development and user project team members, and review the documentation and deliverables to monitor the systems development process to ensure that controls are implemented, user and business requirements are met, and the systems development/acquisition methodology is being followed. Also review and evaluate the application system audit trails to ensure that documented controls are in place to address all security, edit and processing controls. Audit trails are tracking mechanisms that can help IS auditors ensure program change accountability. Tracking information in a change management system includes:
  - History of all work order activity (date of work order, programmer assigned, changes made and date closed)
  - History of logons and logoffs by programmers
  - History of program deletions
  - Adequacy of SoD and quality assurance activities
- Identify and test existing controls to determine the adequacy of production library security to ensure the integrity of the production resources.
- Participate in postimplementation reviews.
- Review and analyze test plans to determine if defined system requirements are being verified.
- Analyze test results and other audit evidence to evaluate the system maintenance process to determine whether control objectives were achieved.
- Review appropriate documentation, discuss with key personnel and use observation to evaluate system maintenance standards and procedures to ensure their adequacy.
- Discuss and examine supporting records to test system maintenance procedures to ensure that they are being applied as described in the standards.

3.15.1 PROJECT MANAGEMENT

Throughout the project management process the IS auditor should analyze the associated risk and exposures inherent in each phase of the SDLC and ensure that the appropriate control mechanisms are in place to minimize risk in a cost-effective manner. Caution should be exercised to avoid recommending controls that cost more to administer than the associated risk they are designed to minimize.

When reviewing the SDLC process, the IS auditor should obtain documentation from the various phases and attend project team meetings, offering advice to the project team throughout the system development process. The IS auditor should also assess the project team's ability to produce key deliverables by the promised dates.

Typically, the IS auditor should review the adequacy of the following project management activities:

- Levels of oversight by project committee/board
- Risk management methods within the project
- Issue management
- Cost management
- Processes for planning and dependency management
- Reporting processes to senior management
- Change control processes
- Stakeholder management involvement
- Sign-off process—At a minimum, signed approvals from systems development and user management responsible for the cost of the project and/or use of the system

Additionally, adequate and complete documentation of all phases of the SDLC process should be evident. Typical types of documentation include, but should not be limited to, the following:

- Objectives defining what is to be accomplished during that phase
- Key deliverables by phases with project personnel assigned direct responsibilities for these deliverables

Figure 3.36—Concurrent Audit Tools—Advantages and Disadvantages

	SCARF/EAM	Snapshots	Audit Hooks	ITF	CIS
Complexity	Very high	Medium	Low	High	Medium
Useful when:	Regular processing cannot be interrupted.	An audit trail is required.	Only select transactions or processes need to be examined.	It is not beneficial to use test data.	Transactions meeting certain criteria need to be examined.

- A project schedule with highlighted dates for the completion of key deliverables
- An economic forecast for that phase, defining resources and the cost of the resources required to complete the phase

3.15.2 FEASIBILITY STUDY

The IS auditor should perform the following functions:

- Review the documentation produced in this phase for reasonableness.
- Determine whether all cost justifications/benefits are verifiable and showing the anticipated costs and benefits to be realized.
- Identify and determine the criticality of the need.
- Determine if a solution can be achieved with systems already in place. If not, review the evaluation of alternative solutions for reasonableness.
- Determine the reasonableness of the chosen solution.

3.15.3 REQUIREMENTS DEFINITION

The IS auditor should perform the following functions:

- Obtain the detailed requirements definition document and verify its accuracy through interviews with the relevant user departments.
- Identify the key team members on the project team and verify that all affected user groups have/had appropriate representation.
- Verify that project initiation and cost have received proper management approval.
- Review the conceptual design specifications (e.g., transforms, data descriptions) to ensure that they address the needs of the user.
- Review the conceptual design to ensure that control specifications have been defined.
- Determine whether a reasonable number of vendors received a proposal covering the project scope and user requirements.
- Review the UAT specification.
- Determine whether the application is a candidate for the use of an embedded audit routine. If so, request that the routine be incorporated in the conceptual design of the system.

3.15.4 SOFTWARE ACQUISITION PROCESS

The IS auditor should perform the following functions:

- Analyze the documentation from the feasibility study to determine whether the decision to acquire a solution was appropriate (including consideration of common criteria evaluations).
- Review the RFP to ensure that it covers the items listed in this section.
- Determine whether the selected vendor is supported by RFP documentation.
- Attend agenda-based presentations and conference room pilots to ensure that the system matches the vendor's response to the RFP.
- Review the vendor contract prior to its signing to ensure that it includes the items listed.
- Ensure the contract is reviewed by legal counsel before it is signed.

3.15.5 DETAILED DESIGN AND DEVELOPMENT

The IS auditor should perform the following functions:

- Review the system flowcharts for adherence to the general design. Verify that appropriate approvals were obtained for any changes and all changes were discussed and approved by appropriate user management.
- Review the input, processing and output controls designed into the system for appropriateness.
- Interview the key users of the system to determine their understanding of how the system will operate and assess their level of input into the design of screen formats and output reports.
- Assess the adequacy of audit trails to provide traceability and accountability of system transactions.
- Verify the integrity of key calculations and processes.
- Verify that the system can identify and process erroneous data correctly.
- Review the quality assurance results of the programs developed during this phase.
- Verify that all recommended corrections to programming errors were made and the recommended audit trails or EAMs were coded into the appropriate programs.

3.15.6 TESTING

Testing is crucial in determining the user requirements have been validated, the system is performing as anticipated and internal controls work as intended. Therefore, it is essential that the IS auditor be involved in reviewing this phase and perform the following:

- Review the test plan for completeness; indicate evidence of user participation, such as user development of test scenarios and/or user sign-off of results; and consider rerunning critical tests.
- Reconcile control totals and converted data.
- Review error reports for their precision in recognizing erroneous data and resolution of errors.
- Verify cyclical processing for correctness (month-end, year-end processing, etc.).
- Verify accuracy of critical reports and output used by management and other stakeholders.
- Interview end users of the system for their understanding of new methods, procedures and operating instructions.
- Review system and end-user documentation to determine its completeness and verify its accuracy during the test phase.
- Review parallel testing results for accuracy.
- Verify that system security is functioning as designed by developing and executing access tests.
- Review unit and system test plans to determine whether tests for internal controls are planned and performed.
- Review the user acceptance testing and ensure that the accepted software has been delivered to the implementation team. The vendor should not be able to replace this version.
- Review procedures used for recording and following through on error reports.

3.15.7 IMPLEMENTATION PHASE

This phase is initiated only after a successful testing phase. The system should be installed according to the organization's change control procedures. The IS auditor should verify that appropriate sign-offs have been obtained prior to implementation and perform the following:

- Review the programmed procedures used for scheduling and running the system along with system parameters used in executing the production schedule.
- Review all system documentation to ensure its completeness and that all recent updates from the testing phase have been incorporated.
- Verify all data conversion to ensure that they are correct and complete before implementing the system in production.

3.15.8 POSTIMPLEMENTATION REVIEW

After the new system has stabilized in the production environment, a postimplementation review should be performed. Prior to this review, it is important that sufficient time be allowed for the system to stabilize in production. In this way, any significant problems will have had a chance to surface.

The IS auditor should perform the following functions:
- Determine if the system's objectives and requirements were achieved. During the postimplementation review, careful attention should be paid to the end users' utilization, trouble tickets, work orders and overall satisfaction with the system. This will indicate whether the system's objectives and requirements were achieved.
- Determine if the cost benefits identified in the feasibility study are being measured, analyzed and accurately reported to management.
- Review program change requests performed to assess the type of changes required of the system. The type of changes requested may indicate problems in the design, programming or interpretation of user requirements.
- Review controls built into the system to ensure that they are operating according to design. If an EAM was included in the system, use this module to test key operations.
- Review operators' error logs to determine if there are any resource or operating problems inherent within the system. The logs may indicate inappropriate planning or testing of the system prior to implementation.
- Review input and output control balances and reports to verify that the system is processing data accurately.

3.15.9 SYSTEM CHANGE PROCEDURES AND THE PROGRAM MIGRATION PROCESS

Following implementation and stabilization, a system enters into the ongoing development or maintenance stage. This phase continues until the system is retired. The phase involves those activities required to either correct errors in the system or enhance the capabilities of the system. In this regard, the IS auditor should consider the following:
- The existence and use of a methodology for authorizing, prioritizing and tracking system change requests from the user
- Whether emergency change procedures are addressed in the operations manuals
- Whether change control is a formal procedure for the user and the development groups
- Whether the change control log ensures all changes shown were resolved
- The user's satisfaction with the turnaround—timeliness and cost—of change requests
- The adequacy of the security access restrictions over production source and executable modules
- The adequacy of the organization's procedures for dealing with emergency program changes
- The adequacy of the security access restrictions over the use of the emergency logon IDs

For a selection of changes on the change control log:
- Determine whether changes to requirements resulted in appropriate change-development documents, such as program and operations documents.
- Determine whether changes were made as documented.
- Determine whether current documentation reflects the changed environment.
- Evaluate the adequacy of the procedures in place for testing system changes.
- Review evidence (test plans and test results) to ensure that procedures are carried out as prescribed by organizational standards.
- Review the procedures established for ensuring executable and source code integrity.
- Review production executable modules and verify there is one and only one corresponding version of the program source code.

Additionally, the IS auditor should review the overall change management process for possible improvements in acknowledgement, response time, response effectiveness and user satisfaction with the process.

3.16 CASE STUDIES

The following case studies are included as a learning tool to reinforce the concepts introduced in this chapter.

3.16.1 CASE STUDY A

A major retailer asked the IS auditor to review their readiness for complying with credit card company requirements for protecting cardholder information. The IS auditor subsequently learned the following information. The retailer uses wireless POS registers that connect to application servers located at each store. These registers use wired equivalent protection (WEP) encryption. The application server, usually located in the middle of the store's customer service area, forwards all sales data over a frame relay network to database servers located at the retailer's corporate headquarters, and using strong encryption over an Internet virtual private network (VPN) to the credit card processor for approval of the sale. Corporate databases are located on a protected screened subset of the corporate local area network. Additionally, weekly aggregate sales data by product line is copied from the corporate databases to magnetic media and mailed to a third party for analysis of buying patterns. It was noted that the retailer's database software has not been patched in over two years. This is because vendor support for the database package was dropped due to management's plans to eventually upgrade to a new ERP system.

CASE STUDY A QUESTIONS	
A1.	Which of the following would present the **MOST** significant risk to the retailer? A. Wireless POS registers use WEP encryption. B. Databases patches are severely out-of-date. C. Credit cardholder information is sent over the Internet. D. Aggregate sales data are mailed to a third party.
A2.	Based on the case study, which of the following controls would be the **MOST** important to implement? A. Store application servers should be located in a secure area. B. POS registers should use two-factor authentication. C. Wireless access points should use MAC address filtering. D. Aggregate sales data sent offsite should be encrypted.
See answers and explanations to the case study questions at the end of the chapter (page 237).	

3.16.2 CASE STUDY B

A large industrial concern has begun a complex IT project, with ERP, to replace the main component systems of its accounting and project control departments. Sizeable customizations were anticipated and are being carried out with a phased approach of partial deliverables. These deliverables are released to users for pilot usage on real data and actual projects. In the meanwhile, detailed design and programming of the next phase begins. After a period of initial adjustment, the pilot users start experiencing serious difficulties. In spite of positive test results, already stabilized functionalities began to have intermittent problems; transactions hang during execution, and more and more frequently, project data are corrupted in the database. Additional problems show up—errors already corrected started occurring again and functional modifications already tested tend to present other errors. The project, already late, is now in a critical situation. The IS auditor, after collecting the evidence, requests an immediate meeting with the head of the project steering committee to communicate findings and suggest actions capable to improve the situation.

CASE STUDY B QUESTIONS	
B1.	The IS auditor should indicate to the head of the project steering committee that: A. the observed project problems are a classic example of loss of control of project activities and loose discipline in following procedures and methodologies. A new project leader should be appointed. B. relays due to an underestimation of project efforts have led to failures in the versioning and modification control procedures. New programming and system resources must be added to solve the root problem. C. the problems are due to excessive system modifications after each delivery phase. The procedure for control of modifications must be tightened and made more selective. D. the nature of initial problems is such as to lead to doubts regarding the adequacy and reliability of the platform. An immediate technical review of the hardware and software platform (parameters, configuration) is necessary.
B2.	In order to contribute more directly to solve the situation, the IS auditor should: A. research the problems further to identify root causes and define appropriate countermeasures. B. review the validity of the functional project specifications as the basis for an improved software baselining definition. C. propose to be included in the project team as a consultant for the quality control of deliverables. D. contact the project leader and discuss the project plans and recommend redefining the delivery schedule using the PERT methodology.
See answers and explanations to the case study questions at the end of the chapter (page 237).	

3.17 ANSWERS TO CASE STUDY QUESTIONS

ANSWERS TO CASE STUDY A QUESTIONS

A1. **A** Use of WEP encryption would present the most significant risk because WEP uses a fixed secret key that is easy to break. Transmission of credit cardholder information by wireless registers would be susceptible to interception and would present a very serious risk. With regard to the unpatched database servers, since they are located on a screened subnet, this would mitigate the risk to the organization. Sending credit cardholder data over the Internet would be less of a risk because strong encryption is being utilized. Because the sales data being sent to the third party are aggregate data, no cardholder information should be included.

A2. **A** Locating application servers in an unsecured area creates a significant risk to the retailer because cardholder data may be retained on the server. Additionally, physical access to the server could potentially allow programs or devices to be installed that would capture sensitive cardholder information. Two-factor authentication for POS registers is not as critical. In the case of MAC address filtering, this can be spoofed. Because sales data are aggregate, no cardholder information should be at risk when the data are sent to a third party.

ANSWERS TO CASE STUDY B QUESTIONS

B1. **D** The IS auditor has only found symptoms but not the root causes of the severe problems, so attributing them to limited skills of the project leader is inappropriate. The sequence of negative events is to show that a fundamental problem is causing serious technical difficulties and, as a consequence of delays, negatively impacting the versioning and change control procedures. It is necessary, however, to ascertain immediately the nature of the suspect problems so if such a problem exists and is not found, it could severely affect the final result or kill the project. Also, the added technical programming resources are an inappropriate remedy because the development methodology is RAD, which is based on using a small group of skilled and experienced technical resources. The initial effects of the problem (intermittent block of transactions, corrupted data) lead to the suspicion that a problem might exist in the setup and configuration of the technical hardware/software environment or platform.

B2. **A** The only appropriate action is additional research, even if the apparently technical nature of the problem renders it unlikely that the auditor may find it alone. Functional project specifications should be executed by users and systems analysts, and not by the auditor. To propose to be project consultant for quality would not bring about an essential contribution since quality is a formal characteristic, whereas in the current case the problem is a substantial system instability. To contact the project leader and redesign the schedule of deliveries would not solve the problem. Furthermore, the definition of real causes may sensibly alter the project environmen

Page intentionally left blank

Chapter 4:

Information Systems Operations, Maintenance and Service Management

Section One: Overview

Section Two: Content

Section One: Overview

DEFINITION

Information systems operations, maintenance and service management are important to provide assurance to users as well as management that the expected level of service will be delivered. Service level expectations are derived from the organization's business objectives. IT service delivery includes IS operations, IT services and management of IS and the groups responsible for supporting them.

OBJECTIVES

The objective of this domain is to ensure that the CISA candidate understands and can provide assurance that the processes for information systems operations, maintenance and service management meet the organization's strategies and objectives.

This domain represents 20 percent of the CISA examination (approximately 30 questions).

TASK AND KNOWLEDGE STATEMENTS

TASKS

There are 10 tasks within the information systems operations, maintenance and service management domain:

- T4.1 Evaluate IT service management framework and practices (internal or third party) to determine whether the controls and service levels expected by the organization are being adhered to and whether strategic objectives are met.
- T4.2 Conduct periodic reviews of information systems to determine whether they continue to meet the organization's objectives within the enterprise architecture (EA).
- T4.3 Evaluate IT operations (e.g., job scheduling, configuration management, capacity and performance management) to determine whether they are controlled effectively and continue to support the organization's objectives.
- T4.4 Evaluate IT maintenance (patches, upgrades) to determine whether they are controlled effectively and continue to support the organization's objectives.
- T4.5 Evaluate database management practices to determine the integrity and optimization of databases.
- T4.6 Evaluate data quality and life cycle management to determine whether they continue to meet strategic objectives.
- T4.7 Evaluate problem and incident management practices to determine whether problems and incidents are prevented, detected, analyzed, reported and resolved in a timely manner to support organization's objectives.
- T4.8 Evaluate change and release management practices to determine whether changes made to systems and applications are adequately controlled and documented.
- T4.9 Evaluate end-user computing to determine whether the processes for end-user computing are effectively controlled and support the organization's objectives.
- T4.10 Evaluate IT continuity and resilience (backups/restores, disaster recovery plan [DRP]) to determine whether it is controlled effectively and continues to support the organization's objectives.

KNOWLEDGE STATEMENTS

The CISA candidate must have a good understanding of each of the topics or areas delineated by the knowledge statements. These statements are the basis for the examination.

There are 23 knowledge statements within the information systems operations, maintenance and service management domain:

- K4.1 Knowledge of service management frameworks
- K4.2 Knowledge of service management practices and service level management
- K4.3 Knowledge of techniques for monitoring third-party performance and compliance with service agreements and regulatory requirements
- K4.4 Knowledge of enterprise architecture (EA)
- K4.5 Knowledge of the functionality of fundamental technology (e.g., hardware and network components, system software, middleware, database management systems)
- K4.6 Knowledge of system resiliency tools and techniques (e.g., fault tolerant hardware, elimination of single point of failure, clustering)
- K4.7 Knowledge of IT asset management, software licensing, source code management and inventory practices
- K4.8 Knowledge of job scheduling practices, including exception handling
- K4.9 Knowledge of control techniques that ensure the integrity of system interfaces
- K4.10 Knowledge of capacity planning and related monitoring tools and techniques
- K4.11 Knowledge of systems performance monitoring processes, tools and techniques (e.g., network analyzers, system utilization reports, load balancing)
- K4.12 Knowledge of data backup, storage, maintenance and restoration practices
- K4.13 Knowledge of database management and optimization practices
- K4.14 Knowledge of data quality (completeness, accuracy, integrity) and life cycle management (aging, retention)
- K4.15 Knowledge of problem and incident management practices
- K4.16 Knowledge of change management, configuration management, release management and patch management practices
- K4.17 Knowledge of operational risks and controls related to end-user computing
- K4.18 Knowledge of regulatory, legal, contractual and insurance issues related to disaster recovery
- K4.19 Knowledge of business impact analysis (BIA) related to disaster recovery planning
- K4.20 Knowledge of the development and maintenance of disaster recovery plans (DRPs)
- K4.21 Knowledge of benefits and drawbacks of alternate processing sites (e.g., hot sites, warm sites, cold sites)
- K4.22 Knowledge of disaster recovery testing methods
- K4.23 Knowledge of processes used to invoke the disaster recovery plans (DRPs)

Relationship of Task to Knowledge Statements

The task statements are what the CISA candidate is expected to know how to do. The knowledge statements delineate each of the areas in which the CISA candidate must have a good understanding in order to perform the tasks. The task and knowledge statements are mapped in figure 4.1 insofar as it is possible to do so. Note that although there is often overlap, each task statement will generally map to several knowledge statements.

Figure 4.1—Task and Knowledge Statements Mapping

Task Statement	Knowledge Statements
T4.1 Evaluate IT service management framework and practices (internal or third party) to determine whether the controls and service levels expected by the organization are being adhered to and whether strategic objectives are met.	K4.1 Knowledge of service management frameworks K4.2 Knowledge of service management practices and service level management K4.3 Knowledge of techniques for monitoring third-party performance and compliance with service agreements and regulatory requirements K4.5 Knowledge of the functionality of fundamental technology (e.g., hardware and network components, system software, middleware, database management systems) K4.10 Knowledge of capacity planning and related monitoring tools and techniques K4.11 Knowledge of systems performance monitoring processes, tools and techniques (e.g., network analyzers, system utilization reports, load balancing) K4.14 Knowledge of data quality (completeness, accuracy, integrity) and life cycle management (aging, retention) K4.18 Knowledge of regulatory, legal, contractual and insurance issues related to disaster recovery
T4.2 Conduct periodic reviews of information systems to determine whether they continue to meet the organization's objectives within the enterprise architecture (EA).	K4.2 Knowledge of service management practices and service level management K4.3 Knowledge of techniques for monitoring third-party performance and compliance with service agreements and regulatory requirements K4.4 Knowledge of enterprise architecture (EA) K4.5 Knowledge of the functionality of fundamental technology (e.g., hardware and network components, system software, middleware, database management systems) K4.10 Knowledge of capacity planning and related monitoring tools and techniques K4.11 Knowledge of systems performance monitoring processes, tools and techniques (e.g., network analyzers, system utilization reports, load balancing) K4.14 Knowledge of data quality (completeness, accuracy, integrity) and life cycle management (aging, retention)
T4.3 Evaluate IT operations (e.g., job scheduling, configuration management, capacity and performance management) to determine whether they are controlled effectively and continue to support the organization's objectives.	K4.5 Knowledge of the functionality of fundamental technology (e.g., hardware and network components, system software, middleware, database management systems) K4.6 Knowledge of system resiliency tools and techniques (e.g., fault tolerant hardware, elimination of single point of failure, clustering) K4.7 Knowledge of IT asset management, software licensing, source code management and inventory practices K4.8 Knowledge of job scheduling practices, including exception handling K4.9 Knowledge of control techniques that ensure the integrity of system interfaces K4.10 Knowledge of capacity planning and related monitoring tools and techniques K4.11 Knowledge of systems performance monitoring processes, tools and techniques (e.g., network analyzers, system utilization reports, load balancing) K4.15 Knowledge of problem and incident management practices K4.16 Knowledge of change management, configuration management, release management and patch management practices
T4.4 Evaluate IT maintenance (patches, upgrades) to determine whether they are controlled effectively and continue to support the organization's objectives.	K4.5 Knowledge of the functionality of fundamental technology (e.g., hardware and network components, system software, middleware, database management systems) K4.7 Knowledge of IT asset management, software licensing, source code management and inventory practices K4.12 Knowledge of data backup, storage, maintenance and restoration practices K4.16 Knowledge of change management, configuration management, release management and patch management practices
T4.5 Evaluate database management practices to determine the integrity and optimization of databases.	K4.5 Knowledge of the functionality of fundamental technology (e.g., hardware and network components, system software, middleware, database management systems) K4.8 Knowledge of job scheduling practices, including exception handling K4.13 Knowledge of database management and optimization practices K4.16 Knowledge of change management, configuration management, release management and patch management practices

Figure 4.1—Task and Knowledge Statements Mapping *(cont.)*

Task Statement	Knowledge Statements
T4.6 Evaluate data quality and life cycle management to determine whether they continue to meet strategic objectives.	K4.5 Knowledge of the functionality of fundamental technology (e.g., hardware and network components, system software, middleware, database management systems) K4.7 Knowledge of IT asset management, software licensing, source code management and inventory practices K4.14 Knowledge of data quality (completeness, accuracy, integrity) and life cycle management (aging, retention) K4.17 Knowledge of operational risks and controls related to end-user computing
T4.7 Evaluate problem and incident management practices to determine whether problems and incidents are prevented, detected, analyzed, reported and resolved in a timely manner to support organization's objectives.	K4.5 Knowledge of the functionality of fundamental technology (e.g., hardware and network components, system software, middleware, database management systems) K4.8 Knowledge of job scheduling practices, including exception handling K4.9 Knowledge of control techniques that ensure the integrity of system interfaces K4.10 Knowledge of capacity planning and related monitoring tools and techniques K4.11 Knowledge of systems performance monitoring processes, tools, and techniques (e.g., network analyzers, system utilization reports, load balancing) K4.12 Knowledge of data backup, storage, maintenance and restoration practices K4.15 Knowledge of problem and incident management practices K4.16 Knowledge of change management, configuration management, release management and patch management practices K4.17 Knowledge of operational risks and controls related to end-user computing
T4.8 Evaluate change and release management practices to determine whether changes made to systems and applications are adequately controlled and documented.	K4.5 Knowledge of the functionality of fundamental technology (e.g., hardware and network components, system software, middleware, database management systems) K4.7 Knowledge of IT asset management, software licensing, source code management and inventory practices K4.9 Knowledge of control techniques that ensure the integrity of system interfaces K4.13 Knowledge of database management and optimization practices K4.14 Knowledge of data quality (completeness, accuracy, integrity) and life cycle management (aging, retention) K4.16 Knowledge of change management, configuration management, release management and patch management practices
T4.9 Evaluate end-user computing to determine whether the processes for end-user computing are effectively controlled and support the organization's objectives.	K4.4 Knowledge of enterprise architecture (EA) K4.9 Knowledge of control techniques that ensure the integrity of system interfaces K4.17 Knowledge of operational risks and controls related to end-user computing
T4.10 Evaluate IT continuity and resilience (backups/restores, disaster recovery plan [DRP]) to determine whether it is controlled effectively and continues to support the organization's objectives.	K4.4 Knowledge of enterprise architecture (EA) K4.5 Knowledge of the functionality of fundamental technology (e.g., hardware and network components, system software, middleware, database management systems) K4.6 Knowledge of system resiliency tools and techniques (e.g., fault tolerant hardware, elimination of single point of failure, clustering) K4.7 Knowledge of IT asset management, software licensing, source code management and inventory practices K4.8 Knowledge of job scheduling practices, including exception handling K4.12 Knowledge of data backup, storage, maintenance and restoration practices K4.15 Knowledge of problem and incident management practices K4.18 Knowledge of regulatory, legal, contractual and insurance issues related to disaster recovery K4.19 Knowledge of business impact analysis (BIA) related to disaster recovery planning K4.20 Knowledge of the development and maintenance of disaster recovery plans (DRPs) K4.21 Knowledge of benefits and drawbacks of alternate processing sites (e.g., hot sites, warm sites, cold sites) K4.22 Knowledge of disaster recovery testing methods K4.23 Knowledge of processes used to invoke the disaster recovery plans (DRPs)

Knowledge Statement Reference Guide

Each knowledge statement is explained in terms of underlying concepts and relevance of the knowledge statement to the IS auditor. It is essential that the exam candidate understand the concepts. The knowledge statements are what the IS auditor must know in order to accomplish the tasks. Consequently, only the knowledge statements are detailed in this section.

The sections identified in K4.1 through K4.23 are described in greater detail in section two of this chapter.

K4.1 Knowledge of service management frameworks

Explanation	Key Concepts	Reference in Manual
In order to provide the service that the organization needs to be effective, IT may leverage formal service management frameworks. The IS auditor should have awareness and knowledge of the major service management frameworks (e.g. IT Infrastructure Library, International Organization for Standardization [ISO] 20000), their contents and their objectives. The IS auditor should be able to determine whether the practices adopted meet the needs of the organization. The IS auditor should also be able to determine whether the service levels required by the organization have been implemented and are being met.	Understanding service management frameworks, their contents and their purpose	2.3.1 Good Practices for Governance of Enterprise IT 4.2.2 IT Service Management Frameworks
	Alignment of practices to the organization's requirements	

K4.2 Knowledge of service management practices and service level management

Explanation	Key Concepts	Reference in Manual
Service level management ensures that IT services meet the customer's expectations and that service level agreements (SLAs) are continuously maintained and improved as needed. SLAs are generally separate documents from the contracts with external vendors. Although generally associated with outsourced functions, the IS auditor should be aware that SLAs may also be created internally to assure key process owners of the level of service that the IT organization has agreed to provide. SLAs may include technical support elements such as expected response times; system availability (e.g., 08:00 to 18:00, Monday through Friday); help desk responses and escalation procedures, etc. Therefore, SLAs specify the underlying operational specifics for agreed-upon services which, if measured and managed, will deliver the commitments that meet the customer's expectations.	Understanding good practices for service level management	4.2.3 IT Service Management

K4.3 Knowledge of techniques for monitoring third-party performance and compliance with service agreements and regulatory requirements

Explanation	Key Concepts	Reference in Manual
With the increasing trend of outsourcing IT infrastructure to third-party service providers, it is essential to know the latest approaches in contracting strategies, processes and contract management practices. Outsourcing IT (and related solutions such as process management and infrastructure management) can help reduce costs and/or complement an enterprise's own expertise; however, outsourcing also may introduce additional risk. Thus, it is essential for the IS auditor to understand the latest approaches in contracting strategies, processes and contract management practices, such as which critical concepts must be included in an outsourcing contract and business case requirements.	Impact of sourcing practices on IT governance	2.9.2 Sourcing Practices
	Relationship between vendor management and IT governance of the outsourcing entity	2.10.1 IT Roles and Responsibilities
	Contractual terms and their impact on driving IT governance of the outsourcing entity	2.11.2 Reviewing Contractual Commitments 4.2.3 IT Service Management

K4.4 Knowledge of enterprise architecture (EA)

Explanation	Key Concepts	Reference in Manual
Enterprise architectures (EAs) are supported or served by IT architectures (e.g., n-tier, client-server, web-based and distributed components). The IS auditor must understand how EAs (e.g., Zachman, TOGAF) affect IT systems and how EAs may be leveraged when performing IT audits. The IS auditor should understand the current EA and identify potential assurance function coverage gaps.	Understanding the components, principles and concepts related to EA	2.3.5 Enterprise Architecture 4.7.1 Enterprise Architecture and Auditing
	Understand the objectives of EA	
	Relevance of different elements of EA and their impact on IT systems	

K4.5 Knowledge of the functionality of fundamental technology (e.g., hardware and network components, system software, middleware, database management systems)

Explanation	Key Concepts	Reference in Manual
The IS auditor must be familiar with the functionality of information system hardware and network components. This includes understanding the importance of the physical part of all IS/IT solutions that support the organizational objectives and goals. Although the CISA exam does not test technical knowledge of the working of individual components, an understanding of the risk associated with and possible control functions of each component is expected—for example, the risk that router access passwords may be shared but that, if properly programmed, passwords can make a major contribution to network resilience. The IS auditor should understand basic concepts related to system software. Application software resides within the environment controlled by the operating system, but other system software—such as utilities, security management, etc.—may have a material effect on the security, reliability, integrity and availability of both applications and data. System software issues are extremely important because all applications within the environment will be impacted and controls at the application level may be subject to circumvention at the system software level.	Understanding the key network and hardware components of a typical data center	4.4 Information Systems Hardware 4.6 IS Network Infrastructure 4.7.2 Hardware Reviews 4.7.5 Network Infrastructure and Implementation Reviews
	Understanding the key controls and risk involving system software and database management systems	3.9.5 System Software Acquisition 3.9.6 System Software Implementation 4.5.1 Operating Systems 4.5.2 Access Control Software 4.5.3 Data Communications Software 4.5.5 Database Management Systems 4.5.6 Utility Programs 4.7.3 Operating System Reviews 4.7.4 Database Reviews

K4.6 Knowledge of system resiliency tools and techniques (e.g., fault tolerant hardware, elimination of single point of failure, clustering)

Explanation	Key Concepts	Reference in Manual
System resiliency tools and techniques are important to ensure uninterrupted service. The IS auditor should be able to identify potential single points of failure within a process and understand related tools and techniques—such as high availability (HA), load balancing and clustering solutions—utilized to improve system resiliency.	Understanding good practices for ensuring system resiliency	4.8.3 Recovery Alternatives

K4.7 Knowledge of IT asset management, software licensing, source code management and inventory practices

Explanation	Key Concepts	Reference in Manual
The IS auditor should be familiar with asset management concepts including inventory practices and how these feed other processes such as software licensing. The IS auditor should be aware that IT asset management is key to information security. An asset cannot be protected if it is not identified. The IS auditor should be aware that the use of unlicensed software, also known as piracy, is regarded as unlawful throughout the world, although specific legislation may not be in force in every country. Software licensing should be subject to controls to ensure that the number of copies in circulation within an organization does not exceed the number purchased. The IS auditor should understand the different methods of software licensing (per seat, concurrent users, enterprise licenses, etc.) and the ways in which automated tools can be utilized to inventory the number of software products in use and to prevent and detect the use of unlicensed software. The IS auditor should be aware of the need to manage program source code. Source code may contain intellectual property and access should be restricted. Source code versioning should be controlled and always aligned with program objects. Source code management should be tightly aligned with change and release management.	Understanding the components, and purpose of IT asset management Understanding key controls for software licensing Understanding software version control systems	4.2.9 Quality Assurance 4.3 IT Asset Management 4.5.7 Software Licensing Issues 4.5.8 Source Code Management

K4.8 Knowledge of job scheduling practices, including exception handling

Explanation	Key Concepts	Reference in Manual
Operations management is critical in providing effective, efficient and appropriate technical solutions. The roles and responsibilities of operations management represent a high risk, not only to the day-to-day running of the IT organization, but to the protection of information assets both in the areas of restricting access to authorized people and the availability of IT. The IS auditor must understand operations management practices and controls to ensure the delivery of quality IT services to the business and to ensure the security of the information.	Understanding good practices for operations management	4.2 Information System Operations 4.7.5 Network Infrastructure and Implementation Reviews 4.7.6 IS Operations Reviews 4.7.7 Scheduling Reviews

K4.9 Knowledge of control techniques that ensure the integrity of system interfaces

Explanation	Key Concepts	Reference in Manual
System interfaces—including middleware, application program interfaces (APIs) and other similar software—present special risk because they may not be subject to the same security and control rigor that is found in large-scale application systems. The IS auditor needs to understand how these system interfaces are controlled and secured. Management should ensure that systems are properly tested and approved, modifications are adequately authorized and implemented, and appropriate version control procedures are followed.	Understanding the key controls and risk involving system interfaces	4.2.4 IS Operations 4.2.5 Incident and Problem Management 4.2.7 Change Management Process 4.2.8 Release Management 4.2.9 Quality Assurance 4.6.6 Application of the OSI Model in Network Architectures

K4.10 Knowledge of capacity planning and related monitoring tools and techniques

Explanation	Key Concepts	Reference in Manual
Capacity planning ensures that all the current and future capacity and performance aspects of business requirements are anticipated in advance, assessed and, as necessary, provided in a cost-effective manner. Capacity of information systems must be monitored on a continuous basis to meet business needs and should be planned using projections of future expected demands. Capacity includes the size and speed of the processor, internal system memory and storage and communications media. The IS auditor is expected to be aware of the concepts of capacity management and the essential information requirements of the task, such as technical performance reports and information on projected business needs. A detailed knowledge of the often complex mathematical models used by the process is not essential.	Capacity planning and monitoring	4.4.4 Capacity Management

K4.11 Knowledge of systems performance monitoring processes, tools and techniques (e.g., network analyzers, system utilization reports, load balancing)

Explanation	Key Concepts	Reference in Manual
IT performance monitoring of critical processes and assets should be conducted on a continuous basis to ensure reliable IT services that meet SLAs and achieve defined business objectives. Performance monitoring processes must be established with supporting tools and techniques, and although the CISA exam does not test knowledge of specific tools, the IS auditor should be aware of the importance of monitoring and of basic techniques that may be employed.	Understanding good practices for systems monitoring	4.2.5 Incident and Problem Management 4.4.3 Hardware Monitoring Procedures 4.4.4 Capacity Management 4.6.6 Application of the OSI Model in Network Architectures 4.8.3 Recovery Alternatives

K4.12 Knowledge of data backup, storage, maintenance and restoration practices

Explanation	Key Concepts	Reference in Manual
An IS auditor should understand the relationship between backup/recovery plans and business process requirements. It is essential that critical data be available in the event of data loss or contamination. Data must be backed up, available at a location that is not likely to be impacted by a disaster at the primary site and protected (i.e., physically secure and encrypted, if necessary). An organization should have documented policies, processes, procedures and standards that clearly explain data backup and recovery. The IS auditor is expected to understand that without backup, no disaster recovery plan (DRP) can work; that backup should be taken at appropriate intervals according to business need, as determined by the recovery point objective (RPO); and that backup must be securely transported for storage in an offsite location so backup will be available in the event of a seriously disruptive incident affecting its host site.	Understanding backup strategies including media rotation and proper storage, data protection, and relationship to recovery time objective (RTO)/RPO	2.13 Auditing Business Continuity 4.8.2 Recovery Strategies 4.8.6 Backup and Restoration

K4.13 Knowledge of database management and optimization practices

Explanation	Key Concepts	Reference in Manual
It is necessary for the IS auditor to understand the concepts of database design, database administration, relationships between database objects, potential problems in transaction processing and security issues associated with database management systems (DBMSs), especially when auditing such systems. The roles and responsibilities of key management, such as those of the database administrator (DBA), should be understood as should the control practices associated with those roles and responsibilities, and the technology managed by key personnel.	Understanding key control areas within database administration and security	4.5.5 Database Management System 4.7.4 Database Reviews

K4.14 Knowledge of data quality (completeness, accuracy, integrity) and life cycle management (aging, retention)

Explanation	Key Concepts	Reference in Manual
It is necessary for the IS auditor to understand the concepts of data quality and data life cycle management. The IS auditor should understand how these concepts are implemented in applications and database managements systems. The IS auditor should further be able to determine if the implementation meets the organizational objects.	Knowledge of data quality concepts Knowledge of data life cycle management	4.5.4 Data Management

K4.15 Knowledge of problem and incident management practices

Explanation	Key Concepts	Reference in Manual
An incident is any event that causes temporary disruption to the business. A problem may develop when such incidents are unresolved. The underlying cause of an incident may also be identified as a problem and addressed as such. All incidents or problems must be detected, reported, managed and resolved in a timely manner. A problem management tool should be used that can be checked by the IS auditor for evidence of satisfactory problem resolution and for the ability to identify trends in incidents and root cause analysis, which may point to an underlying problem.	Understanding good practices for incident and problem management	4.2.5 Incident and Problem Management 4.7.8 Problem Management Reporting Reviews

K4.16 Knowledge of change management, configuration management, release management and patch management practices

Explanation	Key Concepts	Reference in Manual
All changes to the production system or infrastructure should be approved according to an established change management process. Adequate segregation of duties (SoD) should be enforced—for example, ensure that the person making the change is not the same person approving the change. The IS auditor should also be aware of the need for established procedures to control changes made to systems in an emergency situation—such as when a programmer has been called in to address issues following a system stoppage. In such circumstances, it is often necessary for the programmer to have access to production systems, which then breaches the control of "division of duties." Logging of activity, together with management verification and post amendment approval, is an essential requirement.	Good practices for change management, release management and patch management	3.10.1 Change Management Process Overview 3.10.2 Configuration Management 4.2.7 Change Management Process 4.2.8 Release Management

K4.17 Knowledge of operational risks and controls related to end-user computing

Explanation	Key Concepts	Reference in Manual
It is necessary for the IS auditor to understand the risk associated with end-user computing (e.g., Microsoft® Excel, Access, etc.). This IS auditor should understand that these tools can be used to create key applications that are relied upon by the organization but not controlled by the IT department. This, in turn, means that they may not be backed up, liable to change management, etc.	Understanding the risk associated with end-user computing	4.5.9 End-user Computing

K4.18 Knowledge of regulatory, legal, contractual and insurance issues related to disaster recovery

Explanation	Key Concepts	Reference in Manual
An IS auditor should know how to analyze the degree to which the business continuity plan (BCP)/disaster recovery plan (DRP) is aligned with regulatory, legal, contractual and insurance requirements. Business continuity and disaster recovery strategies often depend, to varying degrees, on third-party service providers. Contractual terms determine the obligations of third-party vendors who are part of the DRP/BCP solution. BCP may also be mandatory depending on various regulatory or legal requirements. Additionally, insurance is an important component of the risk mitigation strategy, in terms of transfer of risk, and the IS auditor must be aware of the need to maintain an insurance valuation commensurate with the enterprise technology infrastructure.	Understanding BCP regulatory requirements, third-party contract provisions and insurance	2.12.1 IS Business Continuity Planning 4.8 Disaster Recovery Planning

K4.19 Knowledge of business impact analysis (BIA) related to disaster recovery planning

Explanation	Key Concepts	Reference in Manual
An IS auditor must be able to determine whether a business impact analysis (BIA) and business continuity plan (BCP) are suitably aligned. To be effective and efficient, BCP should be based on a well-documented BIA. A BIA drives the focus of the BCP efforts of an organization and helps in balancing costs to be incurred with the corresponding benefits to the organization. A good understanding of the BIA concept is essential for the IS auditor to audit the effectiveness and efficiency of a BCP.	Understanding the BIA as a key driver of the BCP/disaster recovery planning (DRP) process	2.12.6 Business Impact Analysis 4.8.7 Disaster Recovery Testing Methods

K4.20 Knowledge of the development and maintenance of disaster recovery plans (DRPs)

Explanation	Key Concepts	Reference in Manual
An IS auditor should be well-versed in the practices and techniques followed for development and maintenance of business continuity plans (BCPs)/DRPs, including the need to coordinate recovery plans across the organization. Plans should be tailored to fit the individual needs of organizations because differences in industry, size and scope of an organization, and even geographic location, can affect the contents of the plans. The size and nature of the selected recovery facility for technology will materially depend on the financial risk associated with disruption. In essence, the faster the required recovery, as determined by the recovery time objective (RTO), the greater the potential cost. Once established, recovery plans must be kept up to date with changes in the organization and with associated risk.	Understanding the life cycle of BCP/DRP development and maintenance	2.12.1 IS Business Continuity Planning 2.12.3 Business Continuity Planning Process 2.12.4 Business Continuity Policy 2.12.5 Business Continuity Planning Incident Management 2.12.7 Development of Business Continuity Plans 2.12.8 Other Issues in Plan Development 2.12.11 Summary of Business Continuity 4.8.2 Recovery Strategies 4.8.3 Recovery Alternatives 4.8.4 Development of Disaster Recovery Plans

K4.21 Knowledge of benefits and drawbacks of alternate processing sites (e.g., hot sites, warm sites, cold sites)

Explanation	Key Concepts	Reference in Manual
An IS auditor should be able to analyze whether an enterprise's selection of an alternate processing facility is appropriate, given the company's recovery requirements. The company should make provision for alternate processing facilities to sustain critical information systems in the event that the primary information systems become unavailable. The alternate processing site should meet the defined business requirements. An IS auditor should understand the differences among alternate processing site types and be able to evaluate whether the type selected is aligned with and adequate to meet the defined business requirements, as established in the business continuity plan (BCP)/disaster recovery plan (DRP). Solutions may involve, in descending order of recovery speed, a duplicate facility; a hot site; a warm site; a cold site; a contracted site, including the provision of a mobile facility; and reliance on vendors and a reciprocal agreement. The IS auditor should understand how contracts can be structured to accommodate the differing requirements of the business, depending on the intensity of the disaster.	Understanding alternate processing options, the advantages and disadvantages of each, and the methods used to monitor the contractual agreements with a third-party provider	4.8.2 Recovery Strategies 4.8.3 Recovery Alternatives

K4.22 Knowledge of disaster recovery testing methods

Explanation	Key Concepts	Reference in Manual
An IS auditor should know the testing approaches and methods for business continuity plan (BCP)/disaster recovery plan (DRP) to evaluate the effectiveness of the plans. To ensure that the BCP/DRP will work in the event of a disaster, it is important to periodically test the BCP/DRP, and ensure that the testing effort is efficient. The role of the IS auditor is to observe tests, ensure that all lessons learned are properly recorded and reflected in a revised plan, and review write-ups documenting previous tests. Key items to look for include the degree to which the test leverages resources or extensive preplanning meetings that would not be available during an actual disaster. The objective of a test should be to identify gaps that can be improved, rather than to have a flawless test. Another important aspect of DRP/BCP testing is to provide training for management and staff who may be involved in the recovery process.	Understanding the types of disaster recovery tests, factors to consider when choosing the appropriate test scope, methods for observing recovery tests and analyzing test results	2.13.4 Interviewing Key Personnel 4.8.7 Disaster Recovery Testing Methods

K4.23 Knowledge of processes used to invoke the disaster recovery plans (DRPs)

Explanation	Key Concepts	Reference in Manual
An IS auditor should understand the concepts behind the decision to declare a disaster and to invoke a BCP/DRP and should understand the impact of the decision on an organization, remembering that invocation of the BCP/DRP can, in itself, be a disruption. Key elements that the IS auditor should ensure are in place include clear instructions to individuals who have the authority to declare a disaster, identification of staff who will step into a decision-making role if the primary decision maker should be incapacitated or otherwise unavailable, and steps to ensure that the disaster declaration is properly communicated.	Understanding good practices for communicating the declaration of a disaster	2.12.5 Business Continuity Planning Incident Management 2.12.7 Development of Business Continuity Plans 2.13.6 Reviewing Alternative Processing Contract 4.8.5 Organization and Assignment of Responsibilities 4.8.8 Invoking Disaster Recovery Plans

SUGGESTED RESOURCES FOR FURTHER STUDY

Hiles, Andrew; *Business Continuity: Best Practices—World-class Business Continuity Management, 2nd Edition*, Rothstein Associates Inc., USA, 2003

Hobbs, Martyn; *IT Asset Management: A Pocket Survival Guide*, IT Governance Publishing, USA, 2011.

ISACA, COBIT 5, USA, 2012, *www.isaca.org/cobit*

ISACA, *COBIT 5: Enabling Information*, USA, 2013, *www.isaca.org/cobit*

ISACA, *COBIT 5: Enabling Processes*, USA, 2012, *www.isaca.org/cobit*

ISACA, *COBIT 5 for Assurance*, USA, 2013, *www.isaca.org/cobit*

International Organization for Standardization (ISO); *ISO/IEC 24762:2008: Information technology—Security techniques—Guidelines for information and communications technology disaster recovery services*, Switzerland, 2008

itSMF, the IT Service Management Forum; *Frameworks for IT Management*, Van Haren Publishing, Netherlands, 2006

Mullins, Craig S.; *Database Administration: The Complete Guide to DBA Practices and Procedures, 2nd Edition*, Addison-Wesley Professional, USA, 2012

National Institute of Standards and Technology (NIST), "Security Considerations for Voice Over IP Systems," USA, 2005, *http://csrc.nist.gov/publications/nistpubs/800-58/SP800-58-final.pdf*

Schneier, Bruce; *Secrets & Lies: Digital Security in a Networked World*, John Wiley & Sons, USA, 2004

Snedaker, Susan; *Business Continuity & Disaster Recovery for IT Professionals 2nd Edition*, Syngress Publishing Inc., USA, 2013

Wallace, Michael; Lawrence Webber; *The Disaster Recovery Handbook; A Step-by-Step Plan to Ensure Business Continuity and Protect Vital Operations, Facilities, and Assets, 2nd Edition*, AMACOM, USA, 2010

Wells, April; Charlyne Walker; Timothy Walker; David Abarca; *Disaster Recovery: Principles and Practices*, Pearson-Prentice Hall, USA, 2007

*Note: **Publications in bold are stocked in the ISACA Bookstore.***

SELF-ASSESSMENT QUESTIONS

CISA self-assessment questions support the content in this manual and provide an understanding of the type and structure of questions that have typically appeared on the exam. Questions are written in a multiple-choice format and designed for one best answer. Each question has a stem (question) and four options (answer choices). The stem may be written in the form of a question or an incomplete statement. In some instances, a scenario or a description problem may also be included. These questions normally include a description of a situation and require the candidate to answer two or more questions based on the information provided. Many times a question will require the candidate to choose the **MOST** likely or **BEST** answer among the options provided.

In each case, the candidate must read the question carefully, eliminate known incorrect answers and then make the best choice possible. Knowing the format in which questions are asked, and how to study and gain knowledge of what will be tested, will help the candidate correctly answer the questions.

4-1 Which one of the following provides the **BEST** method for determining the level of performance provided by similar information processing facility environments?

A. User satisfaction
B. Goal accomplishment
C. Benchmarking
D. Capacity and growth planning

4-2 For mission critical systems with a low tolerance to interruption and a high cost of recovery, the IS auditor would, in principle, recommend the use of which of the following recovery options?

A. Mobile site
B. Warm site
C. Cold site
D. Hot site

4-3 A university's IT department and financial services office (FSO) have an existing service level agreement (SLA) that requires availability during each month to exceed 98 percent. The FSO has analyzed availability and noted that it has exceeded 98 percent for each of the last 12 months, but has averaged only 93 percent during month-end closing. Which of the following options **BEST** reflects the course of action the FSO should take?

A. Renegotiate the agreement.
B. Inform IT that the agreement is not meeting the required availability standard.
C. Acquire additional computing resources.
D. Streamline the month-end closing process.

4-4 Which of the following is the **MOST** effective method for an IS auditor to use in testing the program change management process?

A. Trace from system-generated information to the change management documentation
B. Examine change management documentation for evidence of accuracy
C. Trace from the change management documentation to a system-generated audit trail
D. Examine change management documentation for evidence of completeness

4-5 The key objective of capacity planning procedures is to ensure that:

A. available resources are fully utilized.
B. new resources will be added for new applications in a timely manner.
C. available resources are used efficiently and effectively.
D. utilization of resources does not drop below 85 percent.

4-6 The **PRIMARY** benefit of database normalization is the:

A. minimization of redundancy of information in tables required to satisfy users' needs.
B. ability to satisfy more queries.
C. maximization of database integrity by providing information in more than one table.
D. minimization of response time through faster processing of information.

4-7 Which of the following would allow a company to extend its enterprise's intranet across the Internet to its business partners?

A. Virtual private network
B. Client-server
C. Dial-up access
D. Network service provider

4-8 The classification based on criticality of a software application as part of an IS business continuity plan is determined by the:

A. nature of the business and the value of the application to the business.
B. replacement cost of the application.
C. vendor support available for the application.
D. associated threats and vulnerabilities of the application.

4-9 When conducting an audit of client-server database security, the IS auditor should be **MOST** concerned about the availability of:

A. system utilities.
B. application program generators.
C. systems security documentation.
D. access to stored procedures.

4-10 When reviewing a network used for Internet communications, an IS auditor will **FIRST** examine the:

A. validity of password change occurrences.
B. architecture of the client-server application.
C. network architecture and design.
D. firewall protection and proxy servers.

4-11 An IS auditor should be involved in:

A. observing tests of the disaster recovery plan.
B. developing the disaster recovery plan.
C. maintaining the disaster recovery plan.
D. reviewing the disaster recovery requirements of supplier contracts.

4-12 The window of time for recovery of information processing capabilities is based on the:

A. criticality of the processes affected.
B. quality of the data to be processed.
C. nature of the disaster.
D. applications that are mainframe-based.

4-13 Data mirroring should be implemented as a recovery strategy when:

A. recovery point objective (RPO) is low.
B. RPO is high.
C. recovery time objective (RTO) is high.
D. disaster tolerance is high.

4-14 Which of the following components of a business continuity plan is **PRIMARILY** the responsibility of an organization's IS department?

A. Developing the business continuity plan
B. Selecting and approving the recovery strategies used in the business continuity plan
C. Declaring a disaster
D. Restoring the IT systems and data after a disaster

ANSWERS TO SELF-ASSESSMENT QUESTIONS

4-1 A. User satisfaction is the measure to ensure that an effective information processing operation meets user requirements.
B. Goal accomplishment evaluates effectiveness involved in comparing performance with predefined goals.
C. Benchmarking provides a means of determining the level of performance offered by similar information processing facility environments.
D. Capacity and growth planning are essential due to the importance of IT in organizations and the constant change in technology.

4-2 A. Mobile sites are specially designed trailers that can be quickly transported to a business location or to an alternate site to provide a ready-conditioned information processing facility (IPF).
B. Warm sites are partially configured, usually with network connections and selected peripheral equipment—such as disk drives, tape drives and controllers—but without the main computer.
C. Cold sites have only the basic environment to operate an IPF. Cold sites are ready to receive equipment, but do not offer any components at the site in advance of the need.
D. Hot sites are fully configured and ready to operate within several hours.

4-3 **A. The financial services office (FSO) agreed to an inadequate service level agreement (SLA). To meet business needs, the FSO should renegotiate as soon as possible.**
B. It is clear that IT is meeting the required availability standard.
C. Acquiring additional computing resources may be inefficient or cost prohibitive.
D. Streamlining month-end closing may not be possible and/or may not affect availability.

4-4 **A. When testing change management, the IS auditor should always start with system-generated information, containing the date and time a module was last updated, and trace from there to the documentation authorizing the change.**
B. Focusing exclusively on the accuracy of the documentation examined does not ensure that all changes were, in fact, documented.
C. To trace in the opposite direction would run the risk of not detecting undocumented changes.
D. Focusing exclusively on the completeness of the documentation examined does not ensure that all changes were, in fact, documented.

4-5 A. This does not mean that all resources must be fully utilized; full utilization (100 percent) is an indication that management should consider adding capacity.
B. New applications will not always require new resources since existing capacity may be sufficient to accommodate them.
C. Capacity management is the planning and monitoring of computer resources to ensure that available resources are used efficiently and effectively.
D. Utilization should routinely be between 85 and 95 percent, but occasional dips are also acceptable.

4-6 A. **The normalization means the elimination of redundant data. Therefore, the objective of normalization in relational databases is to minimize the quantum of information by eliminating redundant data in tables, quickly processing users' requests and maintaining data integrity.**

B. Maximizing the quantum of information is against the rules of normalization.

C. If particular information is provided in different tables, the objective of data integrity may be violated because one table may be updated and not others.

D. Normalization rules advocate storing data in only one table, therefore, minimizing the response time through faster processing of information.

4-7 A. **Virtual private network (VPN) technology allows external partners to securely participate in the extranet using public networks as a transport or shared private network. Because of low cost, using public networks (Internet) as a transport is the principal method. VPNs rely on tunneling/encapsulation techniques, which allow the Internet Protocol (IP) to carry a variety of different protocols (e.g., SNA, IPX, NETBEUI).**

B. Client-server does not address extending the network to business partners (i.e., client-servers refers to a group of computers within an organization connected by a communications network where the client is the requesting machine and the server is the supplying machine).

C. Although it may be technically possible for an enterprise to extend its intranet using dial-up access, it would not be practical or cost effective to do so.

D. A network service provider may provide services to a shared private network by providing Internet services, but it does not extend an organization's intranet.

4-8 **A. The criticality classification is determined by the role of the application system in supporting the strategy of the organization.**

B. The replacement cost of the application does not reflect the relative value of the application to the business.

C. Vendor support is not a relevant factor for determining the criticality classification.

D. The associated threats and vulnerabilities will get evaluated only if the application is critical to the business.

4-9 **A. System utilities may enable unauthorized changes to be made to data on the client-server database. In an audit of database security, the controls over such utilities would be the primary concern of the IS auditor.**

B. Application program generators are an intrinsic part of client-server technology, and the IS auditor would evaluate the controls over the generators access rights to the database rather than their availability.

C. Security documentation should be restricted to authorized security staff, but this is not a primary concern.

D. Access to stored procedures is not a primary concern.

4-10 A. Reviewing validity of password changes would be performed as part of substantive testing.

B. Understanding the network architecture and design is the starting point for identifying the various layers of information and the access architecture across the various layers such as client-server applications

C. The first step in auditing a network is to understand the network architecture and design. Understanding the network architecture and design provides an overall picture of the network and its connectivity.

D. Understanding the network architecture and design is the starting point for identifying the various layers of information and the access architecture across the various layers such as proxy servers and firewalls.

4-11 **A. The IS auditor should always be present when disaster recovery plans are tested to ensure that the tested recovery procedures meet the required targets for restoration, that recovery procedures are effective and efficient, and to report on the results, as appropriate.**

B. IS auditors may be involved in overseeing plan development, but they are unlikely to be involved in the actual development process.

C. Similarly, an audit of plan maintenance procedures may be conducted, but the IS auditor normally would not have any responsibility for the actual maintenance.

D. An IS auditor may be asked to comment upon various elements of a supplier contract, but, again, this is not always the case.

4-12 **A. The criticality of the processes affected by the disaster is the basis for defining the recovery window.**

B. The quality of the data to be processed is not the basis for determining the window of time.

C. The nature of the disaster is not the basis for determining the window of time.

D. Being a mainframe application does not, itself, provide a window-of-time basis.

4-13 **A. Recovery point objective (RPO) is the earliest point in time to which it is acceptable to recover the data. In other words, RPO indicates the "age" of the recovered data (i.e., how long ago the data were backed up). If RPO is very low, such as minutes, it means that the organization cannot afford to lose even a few minutes of data. In such cases, data mirroring (synchronous data replication) should be used as a recovery strategy.**

B. If RPO is high, such as hours, then other backup procedures—such as tape backup and recovery—could be used.

C. A high recovery time objective (RTO) will mean that the IT system may not be needed immediately after the disaster declaration/disruption (i.e., it can be recovered later).

D. RTO is the time from the disruption/declaration of disaster during which the business can tolerate nonavailability of IT facilities. If RTO is high, "slower" recovery strategies that bring up IT systems and facilities can be used.

4-14 A. Members of the organization's senior management are primarily responsible for overseeing the development of the business continuity plan for an organization and are accountable for the results.

B. Management is also accountable for selecting and approving the strategies used for disaster recovery.

C. IT may be involved in declaring a disaster, but is not primarily responsible.

D. The correct choice is restoring the IT systems and data after a disaster. The IT department of an organization is primarily responsible for restoring the IT systems and data after a disaster within the designated timeframes.

Section Two: Content

4.1 QUICK REFERENCE

Quick Reference Review

Chapter 4 addresses the need for IT service delivery and support. IT service management practices are important to provide assurance to users as well as to management that the expected level of service will be delivered. Service level expectations are derived from the organization's business objectives. IT service delivery includes IS operations, IT services and management of IS and the groups responsible for supporting them. IT services are built upon service management frameworks.

CISA candidates should have a sound understanding of the following items, not only within the context of the present chapter, but also to correctly address questions in related subject areas. It is important to keep in mind that it is not enough to know these concepts from a definitional perspective. The CISA candidate must also be able to identify which elements may represent the greatest risk and which controls are most effective at mitigating this risk. Examples of key topics in this chapter that CISA candidates should understand are:

- Service management frameworks and their purpose
- IS service delivery including service level, financial, capacity, service continuity, information security and availability management practices
- IT service delivery and support including the management of operations, architecture and software, network infrastructure and hardware
- The importance of service level agreements (SLAs) established for measuring performance
- Enterprise architecture and its relationship with auditing
- The process of incident handling as it relates to IT service management
  - Essential to this process is to prioritize items after determining their impact and urgency. Unresolved incidents should be escalated based on criteria set by management.
- IT asset management and the need to know what must be controlled. IT asset management is important for security, licensing etc.
- Wireless technologies, the methods for securing transmissions, and general issues and exposures related to wireless access
- Internet services such as uniform resource locator (URL), common gateway interface (CGI) scripts, cookies, applets and servlets
- The use of and risk associated with Telnet and File Transfer Protocol (FTP)
- Network administration and control, including the use of network performance metrics, management issues and tools
- Data quality and data life cycle management and how these concepts are implemented in applications and database management systems
- Client server technology within the context of thin clients, application servers, database servers, middleware and how the interaction of these elements may result in specific risk to the organization
  - The CISA candidate will be expected to exercise good judgment in determining which controls would be most effective in mitigating risk inherent in a client server environment.
- Steps to be conducted when performing reviews of operating systems (OSs), databases, network infrastructure and operations
  - This may include knowing which controls are most important or most effective in ensuring a controlled environment.

Quick Reference Review *(cont.)*

- Source code management including protection, versioning and alignment with change management
- The risk associated with end-user computing

This chapter also addresses the need for disaster recovery within an organization. Most organizations have some degree of disaster recovery plans (DRPs) in place for the recovery of IT infrastructure, critical systems and associated data. However, many organizations have not taken the next step and developed plans for how key business units will function during a period of IT disruption. CISA candidates should be aware of the components of disaster recovery and business continuity plans (see section 2.12 Business Continuity Planning), the importance of aligning one with the other, and aligning DRPs and business continuity plans (BCPs) with the organization's goals and risk tolerance. Also of importance are data backup, storage and retention, and restoration. Examples of key topics in this chapter that CISA candidates should understand include:

- DRPs: The recovery of IT must be aligned with BCPs, which address the recovery of key business processes and business units. Both must properly align with the goals and risk tolerance of the organization.
- Business impact analysis (BIA): For most organizations, it is not financially feasible to immediately recover all application systems and business processes. A BIA must be performed to understand the cost of interruption and identify which applications and processes are to be recovered first (those most critical to the continued functioning of the organization). The results of the BIA can then be used to decide which recovery strategies may be needed to achieve the agreed-upon recovery timetable.
- The difference between the recovery time objective (RTO) and the recovery point objective (RPO): The RTO is determined based on the acceptable downtime in case of a disruption of operations. It indicates the earliest point in time at which the business operations must resume their IT processing capacity after disaster. The RPO is determined based on the acceptable data loss in case of disruption of operations. It indicates the oldest age of the recovered data (i.e., to which point in time related to the disruption moment the recovered data must correspond).
- The differences between different recovery strategies—sites (hot, warm and cold), data storage (data replication and mirroring) applications (clustering), etc.—and which ones are appropriate given the needs of an organization: An organization needing the ability to recover rapidly would opt for a hot site, or in cases requiring very high availability/low RTO, a redundant site with redundant hardware, mirrored/replicated data and clustered applications.
- Familiarity with the different teams that are utilized in the recovery process and the components of a BCP
- Familiarity with the concepts of backup and recovery—tape backup, media rotation schemes and media expiration (grandfather-father-son)
- Familiarity with the concept of contract management of outsourced IT operations

4.2 INFORMATION SYSTEMS OPERATIONS

The information systems (IS) operations function is responsible for the ongoing support of an organization's computer and IS environment. This function plays a critical role in ensuring that computer operations processing requirements are met, end users are satisfied and information is processed securely. With the growth of cloud computing and the use of third parties, the IS operations function must also work closely with outside entities to meet the company's processing requirements. The IS auditor should understand the scope of the IS operations function when conducting an IS audit of this area. The organization of IS operations varies depending on the size of the computer environment and workload. **Figure 4.2** describes typical IS operation functional areas.

Figure 4.2—Typical IS Operations Functional Areas	
• Management of IS operations • Infrastructure support including computer operations • Technical support/help desk • Job scheduling • Quality assurance • IT asset management • Change control and release management • Configuration management	• Problem management procedures • Performance monitoring and management • Capacity monitoring and planning • Management of physical and environmental security • Information security management

4.2.1 MANAGEMENT OF IS OPERATIONS

The COBIT 5 framework makes a clear distinction between governance and management. These two disciplines encompass different types of activities, require different organizational structures and serve different purposes. COBIT 5's view on the key distinction between governance and management is as follows:

- **Governance**. Governance ensures that stakeholder needs, conditions and options are evaluated to determine balanced, agreed-on enterprise objectives to be achieved; setting direction through prioritization and decision making; and monitoring performance and compliance against agreed-on direction and objectives.
  - In most enterprises, overall governance is the responsibility of the board of directors under the leadership of the chairperson. Specific governance responsibilities may be delegated to special organizational structures at an appropriate level, particularly in larger, complex enterprises.
- **Management**. Management plans, builds, runs and monitors activities in alignment with the direction set by the governance body to achieve the enterprise objectives. In most enterprises, management is the responsibility of the executive management under the leadership of the chief executive officer (CEO).

IS management has the overall responsibility for all operations within the IT department. This area will involve allocation of resources (align, plan and organize [APO] domain in COBIT 5), adherence to standards and procedures (deliver, service and support [DSS] domain in COBIT 5) and monitoring of IS operation processes (monitor, evaluate and assess [MEA] domain in COBIT 5). Operations management functions include:

- Resource allocation—IS management is responsible for ensuring that the necessary resources are available to perform the planned activities within the IT function.
- Standards and procedures—IS management is responsible for establishing the necessary standards and procedures for all operations in accordance with the overall business strategies and policies.
- Process monitoring—IS management is responsible for monitoring and measuring the effectiveness and efficiency of IS operation processes, so that the processes will be improved over time.

Control Functions

Management control functions are listed in **figure 4.3**.

Figure 4.3—Management Control Functions	
IS Management	• Ensuring that adequate resources are allocated to support IS operations • Planning to ensure the most efficient and effective use of an operation's resources • Authorizing and monitoring IT resource usage based on corporate policy • Monitoring operations to ensure compliance with standards
IS Operations	• Ensuring that detailed schedules exist for each operating shift • Reviewing and authorizing changes to the operations schedules • Reviewing and authorizing changes to the network, system and applications • Ensuring that changes to hardware and software do not cause undue disruption to normal processing • Monitoring system performance and resource usage to optimize computer resource utilization • Monitoring service level agreements to ensure the delivery of quality IT services that meet business needs • Anticipating equipment replacement/capacity to maximize current job throughput and strategically plan future acquisitions • Maintaining job accounting reports and other audit records • Reviewing logs from all IT systems to detect critical system events and establish accountability of IS operations • Ensuring that all problems and incidents are handled in a timely manner • Ensuring that IS processing can recover in a timely manner from minor and major disruptions of operations
Information Security	• Ensuring the confidentiality, integrity and availability of the data • Monitoring the environment and security of the facility to maintain proper conditions for equipment performance • Ensuring that security vulnerabilities (internal and external) are identified and resolved in a timely manner • Ensuring that security patches are identified and installed in a timely manner • Detecting intrusion attempts • Resolving information security events, incidents and problems in a timely manner • Limiting logical and physical access to computer resources to those who require and are authorized to use it

4.2.2 IT SERVICE MANAGEMENT FRAMEWORKS

To manage IS operations, an organization may implement a service management framework. A framework is defined by Merriam-Webster as "a set of ideas or facts that provide support for something." IT service management (ITSM) is the implementation and management of IT services (people, process and information technology) to meet business needs. A service management framework is, therefore, a set of ideas or facts that provide support for the implementation of service management.

As noted in section 2.3.1 Good Practices for Governance of Enterprise IT, there are two main frameworks for ITSM: the IT Infrastructure Library (ITIL) and *ISO 20000-1:2011 Information technology – Service management – Part 1: Service management system requirements*.

ITIL is a reference body of knowledge for service delivery good practices. It is a comprehensive framework detailed over five volumes, which should be adopted for each business' needs. ITIL processes are interrelated, meaning one process may feed another. The main objective of ITIL is to improve service quality to the business. The five volumes of ITIL are Service Strategy (align organization strategy with IT strategy), Service Design (creates a service design from the strategy to meet stakeholders needs), Service Transition (creating IT services), Service Operations (maintaining IT services) and Continual Service Improvement (continually improve the quality of IT services).

Like most standards, ISO 20000 is primarily used as a demonstration of compliance to accepted good practice. In addition to the central elements of good ITSM practice, it also requires service providers to implement the plan-do-check-act (PDCA) methodology (Deming's quality circle) and apply it to their service management processes. This ensures continual service improvement by the service provider, so that the organization's processes develop, mature and adapt to their customers' requirements, errors and omissions are avoided, and those problems that have been dealt with do not recur. While the main objective of ISO 20000 is also to improve service quality, achievement of the standard certifies organizations as having passed auditable practices and processes in ITSM.

IT service delivery practices and processes have been well defined through ITIL and the ISO 20000 standard and continue to evolve. These practices and processes are applicable to IT service provider organizations, whether as an internal department or division or as an external service provider. The IS auditor should be aware of the how the content of the frameworks have been adopted and implemented in their organizations. The IS auditor should also ensure that the adopted practices meet the objectives of the organization.

4.2.3 IT SERVICE MANAGEMENT

Many organizations have leveraged ITIL and/or ISO 20000 to improve their ITSM.

The fundamental premise associated with ITSM is that IT can be managed through a series of discrete processes that provide "service" to the business. Although each process area may have separate and distinct characteristics, each process is also highly interdependent with other processes. The processes, once defined, can be better managed through service level agreements (SLAs) that serve to maintain and improve customer satisfaction (i.e., with the end business).

ITSM focuses on the business deliverables and covers infrastructure management of IT applications that support and deliver these IT services. This includes fine-tuning IT services to meet the changing demands of the enterprise as well as measuring and demonstrating improvements in the quality of IT services offered with a reduction in the cost of service in the long term. See **figure 4.4.**

Figure 4.4—IT Service Management (ITSM)

IT services support	• Help desk (service desk) • Incident management • Problem management • Configuration management • Change management • Release management	Although each management area is a separate process by itself, each process is highly interdependent with other processes.
IT service delivery	• Service-level management • IT financial management • Capacity management • IT service continuity management • Availability management	

IT services can be better managed with a SLA, and the services offered form a basis for such agreements. There is a possibility of a gap between customer expectations and the services offered, and this is narrowed by the SLA, which completely defines the nature, type, time and other relevant information for the services being offered. SLAs can also be supported by operational level agreements (OLAs), which are internal agreements covering the delivery of services that support the IT organization in its delivery of services.

For example, when a complaint is received, the help desk looks for an available solution from the Known Error Database (KEDB) after classifying and storing the complaint as an incident. Repeated incidents or major incidents may lead to problems that call for the problem management process. If changes are needed, the change management group of the process/program can provide a supporting role after consulting the configuration management group.

Any required change—whether it originated as a solution to a problem, an enhancement or for any other reason—goes through the change management process. The cost-benefit and feasibility studies are reviewed before the changes are accepted and approved. The risk of the changes should be studied, and a fallback plan should be developed. The change may be for one configuration item or for multiple items, and the change management process invokes the configuration management process.

For example, the software could comprise different systems, each containing different programs and each program having different modules. The configuration can be maintained at the system level, the program level or the module level. The organization may have a policy saying that any changes made at the system level will be released as a new version. It may also decide to release a new version, if it involves changes at the program level for yet another application.

The releases, whether major or minor, will have a unique identity. Sometimes, the minor or small fixes may trigger some other problem. Fully tested, major releases may not have such problems. Because of testing time, space and other constraints, it is also possible to have a partial release, which is known as a delta release. The delta release contains only those items that have undergone changes since the last release.

The releases are controlled, and in the event of any problems in the new release, one should be able to back out completely and restore the system to its previous state. Suitable contingency plans may also be developed, if it is not completely restorable. These plans are developed before the new release is implemented.

Service management metrics should be captured and appropriately analyzed so that this information can be used to enhance the quality of service.

Service Level

An SLA is an agreement between the IT organization and the customer. The SLA details the service(s) to be provided. The IT organization could be an internal IT department or an external IT service provider, and the customer is the business. The business may acquire IT services from an internal IT organization, such as email services, an intranet, an enterprise resource planning (ERP) system, etc. The business may acquire IT services from an external IT service provider, such as Internet connectivity, hosting of the public web site, etc.

The SLA describes the services in nontechnical terms, from the viewpoint of the customer. During the term of the agreement, it serves as the standard for measuring and adjusting the services.

Service-level management is the process of defining, agreeing upon, documenting and managing levels of service that are required and cost justified. Service-level management deals with more than the SLAs themselves; it includes the production and maintenance of the service catalog, service review meetings and service improvement plans (SIPs) for areas that are not achieving their SLAs.

The aim of service-level management is to maintain and improve customer satisfaction and to improve the service delivered to the customer. With clear definition of service level, the IT organization or service provider can design the service based on the service level, and the customer can monitor the performance of the IT services. If the services provided do not meet the SLA, the IT organization or service provider has to improve the services.

Characteristics of IT services are used to define the SLA. Characteristics that should be considered in the delivery of these services include accuracy, completeness, timeliness and security. Many tools are available to monitor the efficiency and effectiveness of services provided by IT personnel. These tools include:

- **Exception reports**—These automated reports identify all applications that did not successfully complete or otherwise malfunctioned. An excessive number of exceptions may indicate:
  - Poor understanding of business requirements
  - Poor application design, development or testing
  - Inadequate operation instructions
  - Inadequate operations support
  - Inadequate operator training or performance monitoring
  - Inadequate sequencing of tasks
  - Inadequate system configuration
  - Inadequate capacity management
- **System and application logs**—Logs generated from various systems and applications should be reviewed to identify all application problems. These logs would provide additional, useful information regarding activities performed on the computer because most abnormal system and application events will generate a record in the logs. Because of the size and complexity of the logs, it is difficult to manually review them. Programs have been developed which analyze the system log and report on specifically defined items. Using this software, the auditor can carry out tests to ensure that:
  - Only approved programs access sensitive data
  - Only authorized IT personnel access sensitive data
  - Software utilities that can alter data files and program libraries are used only for authorized purposes
  - Approved programs are run only when scheduled and, conversely, that unauthorized runs do not take place
  - The correct data file generation is accessed for production purposes
  - Data files are adequately protected
- **Operator problem reports**—These manual reports are used by operators to log computer operations problems and their resolutions. Operator responses should be reviewed by IS management to determine whether operator actions were appropriate or whether additional training should be provided to operators.
- **Operator work schedules**—These schedules are generally maintained manually by IS management to assist in human resource planning. By ensuring proper staffing of operation support personnel, IS management is assured that service requirements of end users will be met. This is especially important during critical or heavy computer usage periods. These schedules should be flexible enough to allow for proper cross-training and emergency staffing requirements.

Many IT departments define the level of service that they will guarantee to users of the IT services. This level of service is often documented in SLAs. It is particularly important to define service levels where there is a contractual relationship between the IT department and the end user or customer. SLAs are often tied to chargeback systems, in which a certain percentage of the cost is apportioned from the end-user department to the IT department. When functions of the IT department are performed by a third party, it is important to have an outsourcing SLA.

Service levels are often defined to include hardware and software performance targets (such as user response time and

hardware availability) but can also include a wide range of other performance measures. Such measures might include financial performance measures (such as year-to-year incremental cost reduction), human resources measures (such as resource planning, staff turnover, development or training) or risk management measures (compliance with control objectives). The IS auditor should be aware of the different types of measures available and should ensure that they are comprehensive and include risk, security and control measures as well as efficiency and effectiveness measures.

Monitoring of Service Levels

Defined service levels must be regularly monitored by an appropriate level of management to ensure that the objectives of IS operations are achieved. It is also important to review the impact on the customers and other stakeholders of the organization.

For example, a bank may be monitoring the performance and availability of its automated teller machines (ATMs). One of the metrics may be availability of ATM services at expected levels (99.9%); however, it may also be appropriate to monitor the impact on customer satisfaction due to nonavailability. Similar metrics may be defined for other services such as email, Internet, etc.

Monitoring of service levels is essential for outsourced services particularly if the third-party is involved in directly providing services to an organization's customers. Failure to achieve service levels will impact the organization more than the third party. For example, a fraud due to control weakness at a third party may result in reputation loss.

It is important to note that when service delivery is outsourced, only responsibility for serviced provision is outsourced—accountability is not and still rests with the organization. Where this is the case, the IS auditor should determine how management gains assurance that the controls at the third party are properly designed and operating effectively. Several techniques can be used by management, including questionnaires, onsite visits or an independent third-party assurance report such as a Statement on Standards for Attestation Engagements 16 (SSAE 16) (formerly SAS 70) Service Organization Control (SOC) 1 report or AT-101 (SOC 2 and SOC 3) report.

Service Levels and Enterprise Architecture

Defining and implementing an enterprise architecture (EA) helps an organization in aligning service delivery (see section 2.3.5 Enterprise Architecture). Organizations may use multiple service delivery channels such as mobile applications ("apps"), the Internet, service outlets, third-party service providers and automated kiosks. These channels use different technologies that are serviced by the same backend database.

When considering availability and recovery options, EA best helps in aligning operational requirements that can address the service delivery objectives. For example, an unacceptable recovery time may lead in choosing fault-tolerant, high-availability architecture for critical service delivery channels (see section 4.8.3 Recovery Alternatives).

4.2.4 IS OPERATIONS

IS operations are processes and activities that support and manage the entire IS infrastructure, systems, applications and data, focusing on day-to-day activities.

IS operations staff is responsible for the accurate and efficient operation of the network, systems and applications and for the delivery of high-quality IS services to business users and customers.

Tasks of the IS operations staff include:

- Executing and monitoring scheduled jobs
- Facilitating timely backup
- Monitoring unauthorized access and use of sensitive data
- Monitoring and reviewing the extent of adherence to IS operations procedures as established by IS and business management
- Participating in tests of disaster recovery plans (DRPs)
- Monitoring the performance, capacity, availability and failure of information resources
- Facilitating troubleshooting and incident handling

Procedures detailing instructions for operational tasks and procedures coupled with appropriate IS management oversight are necessary parts of the IS control environment.

This documentation should include:

- Operations procedures based on operating instructions and job flows for computer and peripheral equipment
- Procedures for monitoring systems and applications
- Procedures for detecting systems and applications errors and problems
- Procedures for handling IS problems and escalation of unresolved issues
- Procedures for backup and recovery

Job Scheduling

In complex IS environments, computer systems transfer hundreds and often thousands of data files daily. These files are often referred to as "batch jobs." A job schedule is typically created that lists the jobs that must be run and the order in which they are run, including any dependencies. Due to the inherent complexity of this process, automated job scheduling software provides control over the scheduling process. Job information is set up once, reducing the possibility of errors. Job dependencies can be defined and software can provide security over access to production data. In addition to the scheduling of batch jobs, job scheduling software can be used to schedule tape backups and other maintenance activities. Job scheduling is a major function within the IT department. The schedule includes the jobs that must be run, the sequence of job execution and the conditions that cause program execution. Low-priority jobs can also be scheduled, if time becomes available. Automated job scheduling software provides control over the scheduling process because job information is set up once—reducing the possibility of errors; job dependencies can be defined and software can provide security over access to production data.

High-priority jobs should be given optimal resource availability while maintenance functions such as backup and system reorganization should, if possible, be performed during nonpeak times. Schedules provide a means of keeping customer demand at a manageable level and permit unexpected or on-request jobs to be processed without unnecessary delay.

Job scheduling procedures are necessary to ensure that IS resources are utilized optimally based on processing requirements. Applications are increasingly required to be continually available; therefore, job scheduling (maintenance or long processing times) represents a greater challenge than before.

Job Scheduling Software

Job scheduling software is system software used by installations that process a large number of batch routines. The scheduling software sets up daily work schedules and automatically determines which jobs are to be submitted to the system for processing.

The advantages of using job scheduling software include:

- Job information is set up only once, reducing the probability of an error.
- Job dependencies are defined so that if a job fails, subsequent jobs relying on its output will not be processed.
- Records are maintained of all job successes and failures.
- Reliance on operators is reduced.

4.2.5 INCIDENT AND PROBLEM MANAGEMENT

Computer resources, like any other organizational asset, should be used in a manner that benefits the entire organization. This includes providing information to authorized personnel when and where it is needed, and at a cost that is identifiable and auditable. Computer resources include hardware, software, telecommunications, networks, applications and data.

Controls over these resources are sometimes referred to as general controls. Effective control over computer resources is critical because of the reliance on computer processing in managing the business.

Process of Incident Handling

Incident management is one of the critical processes in ITSM. IT needs to be attended to on a continuous basis to better serve the customer. Incident management focuses on providing increased continuity of service by reducing or removing the adverse effect of disturbances to IT services, and covers almost all nonstandard operations of IT services—thereby defining the scope to include virtually any nonstandard event. In addition to initiation, other steps in the incident life cycle include classification, assignment to specialists, resolution and closure.

It is essential for any incident handling process to prioritize items after determining the impact and urgency. For example, there could be a situation where a service request from the chief information officer (CIO) for a printer problem arrives at the same time as a request from the technology team to attend to a server crash. IS management should have parameters in place for assigning the priority of these incidents, considering both the urgency and impact.

Unresolved incidents are escalated based on the criteria set by IS management. Incident management is reactive and its objective is to respond to and resolve issues restoring normal service (as defined by the SLA) as quickly as possible. Formal SLAs are sometimes in place to define acceptable ranges for various incident management statistics.

Problem Management

Problem management aims to resolve issues through the investigation and in-depth analysis of a major incident or several incidents that are similar in nature in order to identify the root cause. Standard methodologies for root cause analysis include the development of fishbone/Ishikawa cause-and-effect diagrams, brainstorming and the use of the 5 Whys—an iterative question-asking technique used to explore the cause-and-effect relationships underlying a particular problem.

Once a problem is identified and analysis has identified a root cause, the condition becomes a "known error." A workaround can then be developed to address the error state and prevent future occurrences of the related incidents. This will then be added to the KEDB. The goal is to proactively prevent reoccurrence of the error elsewhere or, at a minimum, have a workaround that can be provided immediately should the incident reoccur.

Problem management and incident management are related but have different methods and objectives. Problem management's objective is to reduce the number and/or severity of incidents, while incident management's objective is to return the effected business process back to its "normal state" as quickly as possible, minimizing the impact on the business. Effective problem management can show a significant improvement in the quality of service of an IS organization.

Detection, Documentation, Control, Resolution and Reporting of Abnormal Conditions

Because of the highly complex nature of software, hardware and their interrelationships, a mechanism should exist to detect and document any abnormal conditions that could lead to the identification of an error. This documentation generally takes the form of an automated or manual log. See **figures 4.5** and **4.6**.

Figure 4.5—Typical Types of Errors That Are Logged	
• Application errors • System errors • Operator errors	• Network errors • Telecommunication errors • Hardware errors

Figure 4.6—Items to Appear in an Error Log Entry	
• Error date • Error resolution description • Error code • Error description • Source of error • Escalation date and time • Initials of the individual responsible for maintaining the log	• Initials of the individual responsible for closing the log entry • Department/center responsible for error resolution • Status code of problem resolution (i.e., problem open, problem closed pending some future specified date, or problem irresolvable in current environment) • Narrative of the error resolution status

For control purposes, the ability to add to the error log should not be restricted. The ability to update the error log, however, should be restricted to authorized individuals, and the updates should be traceable. Proper segregation of duties requires that the ability to close an error log entry be assigned to a different individual than the one responsible for maintaining or initiating the error log entry.

IS management should ensure that the incident and problem management mechanisms are properly maintained and monitored and that outstanding errors are being adequately addressed and resolved in a timely manner.

IS management should develop operations documentation to ensure that procedures exist for the escalation of unresolved problems to a higher level of IS management. While there are many reasons why a problem may remain outstanding for a long period of time, it should not be acceptable for a problem to remain unresolved indefinitely. The primary risk resulting from lack of attention to unresolved problems is the interruption of business operations. An unresolved hardware or software problem could potentially corrupt production data. Problem escalation procedures should be well documented. IS management should ensure that the problem escalation procedures are being adhered to properly. Problem escalation procedures generally include:

- Names/contact details of individuals who can deal with specific types of problems
- Types of problems that require urgent resolution
- Problems that can wait until normal working hours

Problem resolution should be communicated to appropriate systems, programming, operations and user personnel to ensure that problems are resolved in a timely manner. The IS auditor should examine problem reports and logs to ensure that they are resolved in a timely manner and are assigned to the individuals or groups most capable of resolving the problem.

The departments and positions responsible for problem resolution should be part of problem management documentation. This documentation must be maintained properly to be useful.

4.2.6 SUPPORT/HELP DESK

The responsibility of the technical support function is to provide specialist knowledge of production systems to identify and assist in system change/development and problem resolution. In addition, it is technical support's responsibility to apprise management of current technologies that may benefit overall operations.

Procedures covering the tasks to be performed by the technical support personnel must be established in accordance with an organization's overall strategies and policies. **Figure 4.7** illustrates common support functions.

Figure 4.7—Typical Support Functions

- Determining the source of computer incidents and taking appropriate corrective actions
- Initiating problem reports, as required, and ensuring that incidents are resolved in a timely manner
- Obtaining detailed knowledge of the network, system and applications
- Answering inquiries regarding specific systems
- Providing second- and third-tier support to business user and customer
- Providing technical support for computerized telecommunications processing
- Maintaining documentation of vendor software, including issuance of new releases and problem fixes, as well as documentation of utilities and systems developed in house
- Communicating with IS operations to signal abnormal patterns in calls or application behavior

Support is generally triaged when a help desk ticket/call is initiated and then escalated based on the complexity of the issue and the level of expertise required to resolve the problem.

The primary purpose of the help desk is to service the user. The help desk personnel must ensure that all hardware and software incidents that arise are fully documented and escalated based on the priorities established by management. In many organizations, the help desk function means different things. However, the basic function of the help desk is to be the first, single and central point of contact for users and to follow the incident management process.

4.2.7 CHANGE MANAGEMENT PROCESS

Change control procedures are a part of the more encompassing function referred to as change management and are established by IS management to control the movement of application changes (programs, jobs, configurations, parameters, etc.) from the test environment, where development and maintenance occurs, to the quality assurance (QA) environment, where thorough testing occurs, to the production environment. Typically, IS operations are responsible for ensuring the integrity of the production environment and often serve as the final approvers of any changes to production.

Change management is used when changing hardware, installing or upgrading to new releases of off-the-shelf applications, installing a software patch and configuring various network devices (firewalls, routers, switches).

The procedures associated with this process ensure that:

- All relevant personnel are informed of the change and when it is happening
- System, operations and program documentation are complete, up to date and in compliance with the established standards.
- Job preparation, scheduling and operating instructions have been established.
- System and program test results have been reviewed and approved by user and project management.
- Data file conversion, if necessary, has occurred accurately and completely as evidenced by review and approval by user management.

- System conversion has occurred accurately and completely as evidenced by review and approval by user management.
- All aspects of jobs turned over have been tested, reviewed and approved by control/operations personnel.
- Legal or compliance aspects have been considered.
- The risk of adversely affecting the business operation are reviewed and a rollback plan is developed to back out the changes, if necessary.

Apart from change control, standardized methods and procedures for change management are needed to ensure and maintain agreed-on levels in quality service. These methods are aimed at minimizing the adverse impact of any probable incidents triggered by change that may arise.

This is achieved by formalizing and documenting the process of change request, authorization, testing, implementation and communication to the users. Change requests are often categorized into emergency changes, major changes and minor changes, and may have different change management procedures in place for each type of change.

Patch Management

Patch management is an area of systems management that involves acquiring, testing and installing multiple patches (code changes) to an administered computer system in order to maintain up-to-date software and often to address security risk. Patch management tasks include the following:

- Maintaining current knowledge of available patches
- Deciding what patches are appropriate for particular systems
- Ensuring that patches are installed properly; testing systems after installation
- Documenting all associated procedures, such as specific configurations required

A number of products are available to automate patch management tasks. Patches can be ineffective and can cause more problems than they fix. Patch management experts suggest that system administrators take simple steps to avoid problems such as performing backups and testing patches on non-critical systems prior to installations. Patch management can be viewed as part of change management.

See chapter 3 Information Systems Acquisition, Development and Implementation, for details on program change controls.

4.2.8 RELEASE MANAGEMENT

Software release management is the process through which software is made available to users. The term "release" is used to describe a collection of authorized changes. The release will typically consist of a number of problem fixes and enhancements to the service.

A release consists of the new or changed software required.

Figure 4.8 presents some of the principal types of releases.

Figure 4.8—Types of Releases

Major releases	Normally contain a significant change or addition to new functionality. A major upgrade or release usually supersedes all preceding minor upgrades. Grouping together a number of changes facilitates more comprehensive testing and planned user training. Large organizations typically have a predefined timetable for implementing major releases throughout the year (e.g., quarterly). Smaller organizations may have only one release during the year or numerous releases if the organization is quickly growing.
Minor software releases	Upgrades, normally containing small enhancements and fixes. A minor upgrade or release usually supersedes all preceding emergency fixes. Minor releases are generally used to fix small reliability or functionality problems that cannot wait until the next major release. The entire release process should be followed for the preparation and implementation of minor releases, but it is likely to take less time because the development, testing and implementation activities do not require as much time as major releases do.
Emergency software releases	Normally containing the corrections to a small number of known problems. Emergency releases are fixes that require implementation as quickly as possible to prevent significant user downtime to business-critical functions. Depending upon the required urgency of the release, limited testing and release management activities are executed prior to implementation. Such changes should be avoided whenever possible because they increase the risk of errors being introduced.

Many new system implementations will involve phased delivery of functionality and thus require multiple releases. In addition, planned releases will offer an ongoing process for system enhancement.

The main roles and responsibilities in release management should be defined to ensure that everyone understands their role and level of authority and those of others involved in the process. The organization should decide the most appropriate approach, depending on the size and nature of the systems, the number and frequency of releases required, and any special needs of the users (for example, if a phased rollout is required over an extended period of time). All releases should have a unique identifier that can be used by configuration management.

Planning a release involves:

- Gaining consensus on the release's contents
- Agreeing to the release strategy (e.g., the phasing over time and by geographical location, business unit and customers)
- Producing a high-level release schedule
- Planning resource levels (including staff overtime)
- Agreeing on roles and responsibilities
- Producing back-out plans
- Developing a quality plan for the release
- Planning acceptance of support groups and the customer

While change management is the process whereby all changes go through a robust testing and approval process, release management is the process of actually putting the software changes into production.

4.2.9 QUALITY ASSURANCE

QA personnel verify that system changes are authorized, tested and implemented in a controlled manner prior to being introduced into the production environment according to a company's change and release management policies. With the assistance of source code management software (see section 4.5.8 Source Code Management), personnel also oversee the proper maintenance of program versions and source code to object integrity.

See chapter 3 Information Systems Acquisition, Development and Implementation, for more details on QA and on specific objectives of the QA function.

4.3 IT ASSET MANAGEMENT

An asset is something of either tangible or intangible value that is worth protecting and includes people, information, infrastructure, finances and reputation. However, you cannot effectively protect or manage an asset if you do not know that you have it. Likewise, it makes it more difficult to protect an asset if you do not know where it is or who is responsible for it.

According to COBIT 5 process BAI09 *Manage Assets*, assets should be managed as follows:

> *Manage IT assets through their life cycle to make sure that their use delivers value at optimal cost, they remain operational (fit for purpose), they are accounted for and physically protected, and those assets that are critical to support service capability are reliable and available. Manage software licenses to ensure that the optimal number are acquired, retained and deployed in relation to required business usage, and the software installed is in compliance with license agreements.*

The first step in IT asset management is the process of identifying and creating an inventory of IT assets. The inventory record of each information asset should include:

- Specific identification of the asset
- Relative value to the organization
- Loss implications and recovery priority
- Location
- Security/risk classification
- Asset group (where the asset forms part of a larger information system)
- Owner
- Designated custodian

Common methods to build the initial inventory include consulting the purchasing system, reviewing contracts and reviewing the software currently installed using tools, such as Microsoft® System Center Configuration Manager, Spiceworks and ManageEngine.

IT asset management is a fundamental prerequisite to developing a meaningful security strategy. Developing a list of assets is the first step in managing software licenses (see section 4.5.7 Software Licensing Issues) and in classifying and protecting information assets (see section 5.2.3 Classification of Information Assets).

IT asset management should be employed for both software and hardware assets. It is common to physically tag hardware assets.

4.4 INFORMATION SYSTEMS HARDWARE

This section provides an introduction to hardware platforms that make up the enterprise systems of today's organizations. The section describes the basic concepts of and history behind the different types of computers developed, and the advances in information technology that have occurred. Also discussed are the key audit considerations such as capacity management, system monitoring, maintenance of hardware and typical steps in the acquisition of new hardware.

> **Note:** Vendor-specific terminology is used within this manual for illustrative purposes only. Candidates will not be examined on the components of vendor-specific hardware offerings or on vendor-specific terminology unless this terminology has become generalized and is used globally.

4.4.1 COMPUTER HARDWARE COMPONENTS AND ARCHITECTURES

The hardware components of computer systems include differing interdependent components performing specific functions, which can be classified as either processing or input/output components.

Processing Components

The central component of a computer is the central processing unit (CPU). Computers may also:

- Have the CPU on a single chip (microprocessors)
- Have more than one CPU (multi-processor)
- Contain multiple CPUs on a single chip (multi-core processors)

The CPU consists of an arithmetic logic unit (ALU), a control unit and an internal memory. The control unit consists of electrical circuits that control/direct all operations in the computer system. The ALU performs mathematical and logical operations. The internal memory (i.e., CPU registers) is used for processing transactions.

Other key components of a computer include a motherboard, random access memory (RAM) and read-only memory (ROM). In order to operate, the computer requires permanent storage devices (hard disk drive or solid-state drive [SSD]) and a power supply unit. An SSD is nonvolatile storage device that stores persistent data on solid-state flash memory. SSDs have no moving components. This distinguishes them from hard disk drives, which contain spinning disks and movable read/write heads.

Input/Output Components

The input/output (I/O) components are used to pass instructions/information to the computer and to display or record the output generated by the computer. Some components, such as the keyboard and mouse, are input-only devices, while others, such as the touch screen, are both input and output devices. Printers are an example of an output-only device.

Types of Computers

Computers can be categorized following several criteria, mainly based on their processing power, size and architecture. These categories are illustrated in **figure 4.9**.

Figure 4.9—Common Types of Computers	
Supercomputers	Very large and expensive computers with the highest processing speed, designed to be used for specialized purposes or fields that require extensive processing power (e.g., complex mathematical or logical calculations). They are typically dedicated to a few specific specialized system or application programs.
Mainframes	Large, general-purpose computers that are made to share their processing power and facilities with thousands of internal or external users. Mainframes accomplish this by executing a large variety of tasks almost simultaneously. The range of capabilities of these computers is extensive. A mainframe computer often has its own proprietary OS that can support background (batch) and real-time (online) programs operating parallel applications. Mainframes have traditionally been the main data processing and data warehousing resource of large organizations and, as such, have long been protected by a number of the early security and control tools.
High-end and midrange servers	Multiprocessing systems capable of supporting thousands of simultaneous users. In size and power, they can be comparable to a mainframe. High-end/midrange servers have many of the control features of mainframes such as online memory and CPU management, physical and logical partitioning, etc. Their capabilities are also comparable to mainframes in terms of speed for processing data and execution of client programs, but they cost much less than mainframes. Their OSs and system software base components are often commercial products. The higher-end devices generally use UNIX and, in many cases, are used as database servers while smaller devices are more likely to utilize the Windows OS and be used as application servers and file/print servers.
Personal computers (PCs)	Small computer systems referred to as PCs or workstations that are designed for individual users, inexpensively priced and based on microprocessor technology. Their use includes office automation functions such as word processing, spreadsheets and email; small database management; interaction with web-based applications; and others such as personal graphics, voice, imaging, design, web access and entertainment. Although designed as single-user systems, these computers are commonly linked together to form a network.
Thin client computers	These are personal computers that are generally configured with minimal hardware features (e.g., diskless workstation) with the intent being that most processing occurs at the server level using software, such as Microsoft Terminal Services or Citrix Presentation Server, to access a suite of applications.
Laptop computers	Lightweight (under 10 pounds/5 kilograms) personal computers that are easily transportable and are powered by a normal AC connection or by a rechargeable battery pack. Similar to the desktop variety of personal computers in capability, they have similar CPUs, memory capacity and disk storage capacity, but the battery pack makes them less vulnerable to power failures. Being portable, these are vulnerable to theft. Devices may be stolen to obtain information contained therein and hijack connectivity, either within an internal local area network (LAN) or remotely.
Smartphones, tablets and other handheld devices	Handheld devices that enable their users to use a small computing device as a substitute for a laptop computer. Some of its uses include a scheduler, a telephone and address book, creating and tracking to-do lists, an expense manager, eReader, web browser, and an assortment of other functions. Such devices can also combine computing, telephone/fax and networking features together so they can be used anytime and anywhere. Handheld devices are also capable of interfacing with PCs to back up or transfer important information. Likewise, information from a PC can be downloaded to a handheld device.

Common Enterprise Back-end Devices

In a distributed environment, many different devices are used in delivering application services. The following are some of the most common devices encountered:

- **Print servers**—Businesses of all sizes require that printing capability be made available to users across multiple sites and domains. Generally, a network printer is configured based upon where the printer is physically located and who within the organization needs to use it. Print servers allow businesses to consolidate printing resources for cost savings.
- **File servers**—File servers provide for organizationwide access to files and programs. Document repositories can be centralized to a few locations within the organization and controlled with an access-control matrix. Group collaboration and document management are easier when a document repository is used, rather than dispersed storage across multiple workstations.
- **Application (program) servers**—Application servers typically host the software programs that provide application access to client computers, including the processing of the application business logic and communication with the application's database. Consolidation of applications and licenses in servers enables centralized management and a more secure environment.
- **Web servers**—Web servers provide information and services to external customers and internal employees through web pages. They are normally accessed by their universal resource locators (URLs).
- **Proxy servers**—Proxy servers provide an intermediate link between users and resources. As opposed to direct access, proxy servers will access services on a user's behalf. Depending on the services being proxied, a proxy server may render more secure and faster response than direct access.

- **Database servers**—Database servers store data and act as a repository. The servers concentrate on storing information rather than presenting it to be usable. Application servers and web servers use the data stored in database servers and process the data into usable information.
- **Appliances (specialized devices)**—Appliances provide a specific service and normally would not be capable of running other services. As a result, the devices are significantly smaller and faster, and very efficient. Capacity and performance demands require certain services to be run on appliances instead of generic servers. Examples of appliances are:
  - Firewalls—A firewall is a specific device that inspects all traffic going between segments and applies security policies to help ensure a secure network. An effective firewall implementation depends on the quality of the security policies written and their compliance with good practices.
  - Intrusion detection systems (IDSs)—An IDS listens to all incoming and outgoing traffic to deduce and warn of potentially malicious connections.
  - Intrusion prevention systems (IPSs)—An IPS actively attempts to prevent intrusion by monitoring traffic and identifying irregular usage patterns.
  - Switches—Switches are data link-level devices that can divide and interconnect network segments and help to reduce collision domains in Ethernet-based networks
  - Routers—Routers are devices used to link two or more physically separate network segments. The network segments linked by a router remain logically separate and can function as independent networks.
  - Virtual private networks (VPNs)—VPNs provide remote access to enterprise IT resources or can link two or more physically separate networks through a security tunnel. A secure sockets layer-virtual private network (SSL-VPN) provides clientless remote access only through an Internet browser.
  - Load balancers—A load balancer distributes traffic across several different devices to increase the performance and availability of IT services

Universal Serial Bus

The universal serial bus (USB) is a serial bus standard that interfaces devices with a host. USB was designed to allow connection of many peripherals to a single standardized interface socket and to improve the plug-and-play capabilities by allowing hot swapping, or allowing devices to be connected and disconnected without rebooting the computer or turning off the device. Other convenient features include providing power to low-consumption devices without the need for an external power supply and allowing many devices to be used without requiring installation of manufacturer-specific, individual device drivers.

USB ports overcome the limitations of the serial and parallel ports in terms of speed and the actual number of connections that can be made. USB 2.0 specifications support data transfer at up to 480 megabits per second (Mbps), while USB 3.0 can transfer data at up to ten times this speed.

USB ports can connect computer peripherals such as mice, keyboards, tablets, gamepads, joysticks, scanners, digital cameras, printers, personal media players, flash drives and external hard drives.

Most OSs recognize when a USB device is connected and load the necessary device drivers.

Memory Cards/Flash Drives

A memory card or flash drive is a solid-state electronic data storage device used with digital cameras, handheld and mobile computers, telephones, music players, video game consoles and other electronics. They offer high rerecordability, power-free storage, a small form factor, and rugged environmental specifications. Examples include a Memory Stick, CompactFlash, SD and flash drive.

RISK

Viruses and other malicious software—Users can bring infected documents from home to their place of employment or take home a business document to their infected PC, update the document and return the document to a corporate file server. USB drives present a vector for computer viruses that is very difficult to defend against.

Whenever files are transferred between two machines there is a risk that malware (viruses, spyware, keyloggers, etc.) will be transmitted, and USB drives are no exception. Some USB drives include a physical switch that can put the drive in read-only mode. When transferring files to an untrusted machine, a drive in read-only mode will prevent any data (including viruses) to be written to the device.

Data theft—Hackers, corporate spies and disgruntled employees steal data, and in many cases, these are crimes of opportunity. With a USB drive, any unattended and unlocked PC with a USB port provides an opportunity for criminal activity. Social engineering is a tool that can give a hacker physical access to a corporate PC in order to steal data or plant spyware.

Data and media loss—The portability of USB drives presents an increased risk for lost data and media. If an unencrypted USB device is lost, any individual who finds the device will be able to access the data on the drive.

Corruption of data—If the drive is improperly unplugged, then data loss can occur due to corruption. USB drives differ from other types of removable media, such as CD-ROM and DVD-ROM devices, because the computer is not automatically alerted when USB drives are removed. Users of USB drives must alert the computer when they intend to remove the device; otherwise, the computer will be unable to perform the necessary clean-up functions required to disconnect the device, especially if files from the device are currently open.

Loss of confidentiality—Because of its convenient small physical size and large logical size, a significant amount of data can be stored on a USB drive. Some stored information is confidential, and loss of data becomes a risk when the drive is lost, increasing the risk of the data falling into the hands of a competitor. Legal issues can also be associated with loss of confidentiality. For example, in the United States, lost or compromised patient data can indicate a breach of patient privacy, thus violating HIPAA.

SECURITY CONTROL

Encryption—An ideal encryption strategy allows data to be stored on the USB drive but renders the data useless without the required encryption key, such as a strong password or biometric data. Products are available to implement strong encryption and comply with the latest Federal Information Processing Standards (FIPS).

Encryption is a good method to protect information written to the device from loss or theft of the device. But unless the information is also encrypted on the network or local workstation hard drive, sensitive data still are exposed to theft.

Granular control—Products are available to provide centralized management of ports. Microsoft Active Directory (AD), within a group policy object, can be used to manage not only the USB and Firewire ports, but to also manage use of a CD-ROM drive. Because management is accomplished via AD, centralized management from the enterprise to the individual system is possible. As with all security issues a technological solution in isolation is insufficient. Strong policies, procedures, standards and guidelines must be put in place to ensure secure operation of memory card and USB drives. Further, an aggressive user awareness program is necessary to effect changes in employee behavior.

Security personnel education—Flash drives are so small and unobtrusive that they are easily concealed and removed from an enterprise. Physical security personnel should understand USB devices and the risk they present.

The "lock desktop" policy enforcement—In higher-risk environments, desktop computers should be configured to automatically lock after short intervals.

Antivirus policy—Antivirus software should be configured to scan all attached drives and removable media. Users should be trained to scan files before opening them.

Use of secure devices only—Enforce the use of encryption. Software is available to manage USBs, enforcing encryption or only accepting encrypted devices.

Inclusion of return information—In the event a USB drive is lost or misplaced, including a small, readable text file containing return information may help with device retrieval. It would be prudent to NOT include company details, but rather a phone number or post office box. It also would be prudent to include a legal disclaimer that clearly identifies the information on the drive as confidential and protected by law.

Radio Frequency Identification

Radio frequency identification (RFID) uses radio waves to identify tagged objects within a limited radius. A tag consists of a microchip and an antenna. The microchip stores information along with an ID to identify a product, while the antenna transmits the information an RFID reader.

The power needed to drive the tag can be derived in two modes. The first mode, used in passive tags, draws power from the incidental radiation arriving from the reader. The second and more expensive mode, used in active tags, derives its power from batteries and therefore is capable of utilizing higher frequencies and achieving longer communication distances. An active tag is reusable and can contain more data.

Tags can be used to identify an item based on either direct product identification or carrier identification. In the case of the latter, an article's ID is manually fed into the system (e.g., using a bar code) and is used along with strategically placed radio frequency readers to track and locate the item.

APPLICATIONS

Asset management—RFID-based asset management systems are used to manage inventory of any item that can be tagged. Asset management systems using RFID technology offer significant advantages over paper-based or bar-code systems, including the ability to read the identifiers of multiple items nearly simultaneously without optical line of sight or physical contact.

Tracking—RFID asset management systems are used to identify the location of an item or, more accurately, the location of the last reader that detected the presence of the tag associated with the item.

Authenticity verification—The tag provides evidence of the source of a tagged item. Authenticity verification often is incorporated into a tracking application.

Matching—Two tagged items are matched with each other and a signal (e.g., a light or tone) is triggered if one of the items is later matched with an incorrect tagged item.

Process control—This allows business processes to use information associated with a tag (or the item attached to the tag) and to take a customized action.

Access control—The system uses RFID to automatically check whether an individual is authorized to physically access a facility (e.g., a gated campus or a specific building) or logically access an information technology system.

Supply chain management (SCM)—SCM involves the monitoring and control of products from manufacture to distribution to retail sale. SCM typically bundles several application types, including asset management, tracking, process control and payment systems.

RISK

Business process risk—Direct attacks on RFID system components could undermine the business processes that the RFID system was designed to enable.

Business intelligence risk—An adversary or competitor could gain unauthorized access to RFID-generated information and use the information to harm the interests of the organization implementing the RFID system.

Privacy risk—Personal privacy rights or expectations may be compromised if an RFID system uses what is considered

personally identifiable information for a purpose other than originally intended or understood. The personal possession of functioning tags also is a privacy risk because possession could enable tracking of those tagged items.

Externality risk—RFID technology could represent a threat to non-RFID-networked or non-RFID-collocated systems, assets and people. An important characteristic of RFID that impacts the risk is that RF communication is invisible to operators and users.

SECURITY CONTROL

Management—A management control involves oversight of the security of the RFID system. For example, management staff of an organization may need to update existing policies to address RFID implementations, such as security controls needed for an RF subsystem.

Operational—An operational control involves the actions performed on a daily basis by the system's administrators and users. For example, RFID systems need operational controls that ensure the physical security of the systems and their correct use.

Technical—A technical control uses technology to monitor or restrict the actions that can be performed within the system. RFID systems need technical controls for several reasons such as protecting or encrypting data on tags, causing tags to self-destruct and protecting or encrypting wireless communications.

4.4.2 HARDWARE MAINTENANCE PROGRAM

To ensure proper operation, hardware must be routinely cleaned and serviced. Maintenance requirements vary based on complexity and performance workloads (e.g., processing requirements, terminals access and number of applications running). In any event, maintenance should be scheduled to closely coincide with vendor-provided specifications. Maintenance is also important for environmental hardware that controls temperature and humidity, fire protection and electrical power. The hardware maintenance program is designed to document the performance of this maintenance.

Information typically maintained by this program includes:
- Reputable service company information for each hardware resource requiring routine maintenance
- Maintenance schedule information
- Maintenance cost information
- Maintenance performance history information such as planned versus unplanned, executed and exceptional

IS management should monitor, identify and document any deviations from vendor maintenance specifications as well as provide supporting arguments for this deviation.

When performing an audit of this area, the IS auditor should:
- Ensure that a formal maintenance plan has been developed, approved by management and is being followed.
- Identify maintenance costs that exceed budget or are excessive. These overages may be an indication of a lack of adherence to maintenance procedures or of upcoming changes to hardware. Proper inquiry and follow-up procedures should be performed.

4.4.3 HARDWARE MONITORING PROCEDURES

The following are typical procedures and reports for monitoring the effective and efficient use of hardware:
- **Availability reports**—These reports indicate the time periods during which the computer is in operation and available for utilization by users or other processes. A key concern addressed by this report is excessive IS unavailability, referred to as downtime. This unavailability may indicate inadequate hardware facilities, excessive OS maintenance, the need for preventive maintenance, inadequate environmental facilities (e.g., power supply or air conditioning) or inadequate training for operators.
- **Hardware error reports**—These reports identify CPU, I/O, power and storage failures. These reports should be reviewed by IS operations management to ensure that equipment is functioning properly, to detect failures and to initiate corrective action. The IS auditor should be aware that a sure attribution of an error in hardware or software is not necessarily easy and immediate. Reports should be checked for intermittent or recurring problems, which might indicate difficulties in properly diagnosing the errors.
- **Asset management reports**—These reports provide an inventory of network-connected equipment such as PCs, servers, routers and other devices.
- **Utilization reports**—These automated reports document the use of the machine and peripherals. Software monitors are used to capture utilization measurements for processors, channels and secondary storage media (such as disk and tape drives). Depending on the OS, resource utilization for multiuser computing environments found in mainframe/large-scale computers should average in the 85 to 95 percent range, with allowances for utilization occasionally reaching 100 percent and falling below 70 percent. Trends from utilization reports can be used by IS management to predict whether more or fewer processing resources are required.

4.4.4 CAPACITY MANAGEMENT

Capacity management is the planning and monitoring of computing and network resources to ensure that the available resources are used efficiently and effectively. This requires that the expansion or reduction of resources takes place in parallel with the overall business growth or reduction. The capacity plan should be developed based on input from both user and IS management to ensure that business goals are achieved in the most efficient and effective way. This plan should be reviewed and updated at least annually.

Capacity planning should include projections substantiated by past experience, considering the growth of existing business as well as future expansions. The following information is key to the successful completion of this task:
- CPU utilization
- Computer storage utilization
- Telecommunications, local area network (LAN) and wide area network (WAN) bandwidth utilization
- I/O channel utilization
- Number of users
- New technologies
- New applications
- SLAs

The IS auditor must realize that the amount and distribution of these requirements has an intrinsic flexibility. Specialized resources of a given class may have an impact on the requirements for other classes. For example, the proper use of more "intelligent" terminals may consume less processor power and less communications bandwidth than other terminals. Consequently, the above information is strictly related to type and quality of used or planned system components.

An element in capacity management is deciding whether to host the organization's applications distributed across a number of small servers, consolidated onto a few large servers, in the cloud or combinations of the three. Consolidating applications on a few large servers (also known as application stacking) often allows the organization to make better overall use of the resources, but on the other hand, it increases the impact of a server outage and it affects more applications when the server has to be shut down for maintenance. Utilizing the cloud means that extra capacity may be purchased on demand but also brings the risk of relying on the supplier.

Larger organizations often have hundreds, if not thousands, of servers which are arrayed in groups referred to as server farms. Where virtual servers are utilized, these may be organized as private (also known as internal or corporate) clouds.

If an organization has put data storage hardware in place, the IS auditor should review the capacity management plans which involve both data storage utilization and storage area network (SAN) utilization.

Capacity management must also include network devices such as switches and routers which comprise physically and logically separated networks (virtual local area networks [VLANs]).

Capacity planning defines the business's requirements for IT capacity, in both business and technical terms, and presents the consequences of delivering the required volume of activity through the IT infrastructure and applications—at the right time and with optimal cost. Capacity management ensures that all current and future capacity and performance aspects of the business requirements are provided in a cost-effective manner.

Information system capacity is one of the key business requirements for IT systems. Business operations and processes can only be supported reliably when IT systems provide the required capacity. IT management should understand the capacity requirements prior to the design of their information systems, and verify the final design against the capacity requirements. IT management also must monitor capacity on an ongoing basis and provide additional capability as the business grows. For example, a file server may store all business files, but in two years, when the storage reaches the 80 percent threshold, an additional hard disk should be installed to keep up with the storage requirements.

IT capacity, as measured by CPU power and size of memory, hard disk or servers, is expensive. Organizations do not want to acquire more than what they need at the present time. Capacity planning is the process of ensuring that the resource provision can always meet business requirements. By continuously monitoring the threshold of the capacity utilization, additional capacity can be acquired and deployed before it no longer meets business requirements. With capacity management, expensive resources will only be provided when they are needed, thus resulting in a cost savings.

Capacity management monitors resource utilization and helps with resource planning. During procurement of the IT system, the capability management team will work with the architect to estimate resource requirements and to ensure that adequate, but not excessive, resources are provided to support the new solutions. The estimate is normally based on number of transactions, size of data being stored, transaction processing time and response time, etc. Estimates help determine capability requirements for the new solutions.

Capacity management aims to consistently provide the required IT resources—at the right time and cost and in alignment with current and future requirements of the business. Capacity management increases efficiency and cost savings by deferring the cost of new capacity to a later date and optimizing capacity to business needs. Capacity management reduces the risk of performance problems or failure by monitoring the resource utilization threshold and provision of new resources before a shortage occurs. Capacity management also provides accurate capacity forecasting through application sizing and modeling for new services.

Capacity planning and monitoring includes the elements listed in **figure 4.10**.

Figure 4.10—Capacity Planning and Monitoring Elements

Development	Develop a capacity plan that describes current and future requirements for capacity of IT resources.
Monitoring	Monitor IT components to ensure that agreed-upon service levels are achieved.
Analysis	Analyze data collected from monitoring activities to identify trends from which normal utilization and service level, or baseline, can be established.
Tuning	Optimize systems for actual or expected workload on the basis of analyzed and interpreted monitoring data.
Implementation	Introduce changes or new capacity to meet new capacity requirements.
Modeling	Model and forecast the behavior of IT resources to determine future capacity trends and requirements.
Application sizing	Take into consideration the predicted resources for new capacity. When designing the application, determine its size (no. of concurrent users that can be handled, no. of transactions, data storage requirements) and required server capability, memory size, processing power, etc.

4.5 IS ARCHITECTURE AND SOFTWARE

The architecture of most computers can be viewed as a number of layers of circuitry and logic, arranged in a hierarchical structure that interacts with the computer's OS. At the base of the hierarchy is the computer hardware, which includes some hard-coded

instructions (firmware). The next level up in the hierarchy comprises the nucleus functions. Functions of the nucleus relate to basic processes associated with the OS, which include:
- Interrupt handling
- Process creation/destruction
- Process state switching
- Dispatching
- Process synchronization
- Interprocess communication
- Support of I/O processes
- Support of the allocation and reallocation/release of memory

The nucleus is a highly privileged area where access by most users is restricted. Above the nucleus are various OS processes that support users. These processes, referred to as system software, are a collection of computer programs used in the design, processing and control of all computer applications used to operate and maintain the computer system. Comprised of system utilities and programs, the system software ensures the integrity of the system, controls the flow of programs and events in the computer, and manages the interfaces with the computer. Software developed for the computer must be compatible with its OS. Examples include:
- Access control software
- Data communications software
- Database management software
- Program library management systems
- Tape and disk management systems
- Network management software
- Job scheduling software
- Utility programs

Some or all of the above may be built into the OS.

4.5.1 OPERATING SYSTEMS

Before discussion of the various forms of system software, the most significant system software related to a computer—its OS—needs to be further addressed. The OS contains programs that interface between the user, processor and applications software. It is the control program that runs the computer and acts as a scheduler and traffic controller. It provides the primary means of managing the sharing and use of computer resources such as processors, real memory (e.g., RAM), auxiliary memory (e.g., disk storage) and I/O devices.

Most modern OSs have also expanded the basic OS functionalities to include capabilities for a more efficient operation of system and applications software. For example, all modern OSs possess a virtual storage memory capability which allows programs to use and reference a range of addresses greater than the real memory. This technique of mapping parts of a large slower memory to a faster and smaller working memory is used between various levels of "cached memory" within modern systems.

OSs vary in the resources managed, comprehensiveness of management and techniques used to manage resources. The type of computer, its intended use, and normal, expected attached devices and networks influence the OS requirements, characteristics and complexity. For example, a single-user microcomputer operating in stand-alone mode needs an OS capable of cataloging files and loading programs to be effective.

A mainframe computer handling large volumes of transactions for consolidation and distribution requires an OS capable of managing extensive resources and many concurrent operations, in terms of application input and output, with a very high degree of reliability. For example, the z/OS operating system from IBM has been engineered specifically to complement this environment.

A server with multiple users interacting with data and programs, from database servers and middleware connections to legacy mainframe applications, requires an OS that can accommodate multiprocessing, multitasking and multithreading. It must be able to share disk space (files) and CPU time among multiple users and system processes as well as manage connections to devices on the network. For example, the UNIX operating system is designed to specifically address this type of environment.

A microcomputer in a networked environment functioning as a server with specialized functions (applications, database management systems [DBMSs], directory/file storage, etc.) also has the ability to interact with data and programs of multiple users to provide services to client workstations throughout the network.

It is common for OSs to run on virtual servers. In a virtual environment, software is used to partition one physical server into multiple independent virtual servers. Each of these environments can then run its own (and if required different) OS. To the operator, the OS behaves as if it were running on a physical server.

Software Control Features or Parameters

Various OS software products provide parameters and options for the tailoring of the system and activation of features such as activity logging. Parameters are important in determining how a system runs because they allow a standard piece of software to be customized to diverse environments.

Software control parameters deal with:
- Data management
- Resource management
- Job management
- Priority setting

Parameter selections should be appropriate to the organization's workload and control environment structure. The most effective means of determining how controls are functioning within an OS is to review the software control features and/or parameters.

Improper implementation and/or monitoring of OSs can result in undetected errors and corruption of the data being processed as well as lead to unauthorized access and inaccurate logging of system usage.

Software Integrity Issues

OS integrity is a very important requirement and ability of the OS and involves utilizing specific hardware and software features to:
- Protect itself from deliberate and inadvertent modification

- Ensure that privileged programs cannot be interfered with by user programs
- Provide for effective process isolation to ensure that:
  - Multiple processes running concurrently will not interfere by accident or by design with each other and are protected from writing into each other's memory (e.g., changing instructions, sharing resources, etc.)
  - Enforcement of least privilege where processes have no more privilege than needed to perform functions and modules call on more privileged routines only if, and for as long as, needed.

To maintain system and data integrity, it is necessary to correctly and consistently define, enforce and monitor the operating environment and the granted permissions. IS management is responsible for the implementation of appropriate authorization techniques to prevent non-privileged users from gaining the ability to execute privileged instructions and thus take control of the entire machine.

For example, IBM mainframe z/OS systems are customized at system generation (SYSGEN) time. When these systems are started (initial program load), important options and parameters are read from information kept in a key system directory (referred to as the SYS1.PARMLIB partitioned data set). The directory specifies critical initialization parameters used to meet the data center's installation requirements (i.e., other system software activated for job scheduling, security, activity logging, etc.). These options, if uncontrolled, provide a nonprivileged user a way to gain access to the OS's supervisory state. The IS auditor should review system configuration directories/files in all OSs for control options used to protect the supervisory state.

Likewise, PC-based client-server Windows, UNIX and Linux OSs have special system configuration files and directories. The existence of program flaws or errors in configuring, controlling and updating the systems to the latest security patches makes them vulnerable to being compromised by perpetrators. Important Windows system options and parameters are set in special system configuration files, referred to as a registry. Therefore, the registry is an important aspect of IS auditing. Noting any changes that take place in the registry is crucial for maintaining the integrity, confidentiality and availability of the systems. In UNIX-based OSs, the same issues are present. Critical system configuration files and directories related to the nucleus (kernel) operations, system start-up, network file sharing and other remote services should be appropriately secured and checked for correctness.

Activity Logging and Reporting Options

Computer processing activity can be logged for analysis of system functions. The following are some of the areas that can be analyzed based on the activity log:

- Data file versions used for production processing
- Access to sensitive data
- Programs scheduled and run
- Utilities or service aids usage
- OS activities to ensure that the integrity of the OS has not been compromised due to improper changes to system parameters and libraries
- Databases to:
  - Evaluate the efficiency of the database structure
  - Assess database security
  - Validate the DBA's documentation
  - Determine whether the organization's standards have been followed
- Access control to:
  - Evaluate the access controls over critical data files/bases and programs
  - Evaluate security facilities that are active in communications systems, DBMSs and applications

Many intruders will attempt to alter logs to hide their activities. Secure logging is also needed to preserve evidence authenticity should the logs be required for legal/court use. It is, therefore, important that logs are protected against alteration. A common way to achieve this is to capture, centralize and analyze the logs on a secure server using security information and event management (SIEM) software.

4.5.2 ACCESS CONTROL SOFTWARE

Access control software is designed to prevent unauthorized access to data, unauthorized use of system functions and programs, and unauthorized updates/changes to data, and to detect or prevent unauthorized attempts to access computer resources. For more details on access control software, see chapter 5, Protection of Information Assets.

4.5.3 DATA COMMUNICATIONS SOFTWARE

Data communications software is used to transmit messages or data from one point to another either locally or remotely. For example, a database request from an end user is actually transmitted from that user's terminal to an online application, then to a DBMS in the form of messages handled by data communications software. Likewise, responses back to the user are handled in the same manner (i.e., from the DBMS to the online application and back to the user's terminal).

A typical simple data communications system has three components:

1. The transmitter (source)
2. The transmission path (channel or line)
3. The receiver

A one-way communication is said to exist when communication flows in one direction only. In a two-way communication, both ends may simultaneously operate as source and receiver, with data flowing over the same channel in both directions. The data communications system is concerned only with the correct transmission between two points. It does not operate on the content of the information.

A data communication system is divided into multiple functional layers. At each layer, software interfaces with hardware to provide a specific set of functions. All data communication systems have at least a physical layer and a data link layer. (See section 4.6.4 Network Standards and Protocols for a discussion regarding data communication layers.)

Communication-based applications operate in LAN and WAN environments to support:
- Electronic funds transfer (EFT) systems
- Database management systems
- Customer electronic services/electronic data interchange (EDI)
- Internet forums and email

The data communication system interfaces with the OS, application programs, database systems, telecommunication address method systems, network control system, job scheduling system and operator consoles.

4.5.4 DATA MANAGEMENT

The *Data Management Body of Knowledge* (DMBOK) defines data management as "the planning and execution of policies, practices, and projects that acquire, control, protect, deliver, and enhance the value of data and information assets."

Data management is a component of data architecture, which is a key part of enterprise architecture.

According to COBIT 5 (APO03.02 Define reference architecture), the reference architecture describes the current and target architectures for the business, information, **data**, application and technology domains. Further, one should "maintain an enterprise data dictionary that promotes a common understanding and a classification scheme that includes details about **data ownership**, definition of appropriate security levels, and **data retention** and **destruction requirements**."

Data Quality

Key to data management is data quality. There are three subdimensions of quality: intrinsic, contextual and security/accessibility. Each subdimension is divided further into several quality criteria, which are defined in **figure 4.11**.

Data Life Cycle

A life cycle describes a series of stages that characterize the course of existence of an organizational investment. Data life cycle management describes the stages that data go through in the course of existence in an organization. **Figure 4.12** shows how the COBIT 5 Information enabler distinguishes the life cycle phases:
- **Plan**—The phase in which the creation, acquisition and use of the information resource is prepared. Activities in this phase include understanding information use in the respective business processes, determining the value of the information asset and its associated classification, identifying objectives and planning the information architecture.
- **Design**—The phase in which more detailed work is done in specifying how the information will look and how systems processing the information will have to work. Activities in this phase may refer to the development of standards and definitions (e.g., data definitions, data collection, access, storage procedures and metadata characteristics).
- **Build/acquire**—The phase in which the information resource is acquired. Activities in this phase may refer to the creation of data records, the purchase of data and the loading of external files.
- **Use/operate**—This phase includes:
  - Store—The phase in which information is held electronically or in hard copy (or even just in human memory). Activities in this phase may refer to the storage of information in electronic form (e.g., electronic files, databases, data warehouses) or as hard copy (e.g., paper documents).
  - Share—The phase in which information is made available for use through a distribution method. Activities in this phase may refer to the processes involved in getting the information to places where it can be accessed and used (e.g., distributing documents by email). For electronically held information, this life cycle phase may largely overlap with the store phase (e.g., sharing information through database access, file/document servers).
  - Use—The phase in which information is used to accomplish (IT-related and thus enterprise) goals. Activities in this phase may refer to all kinds of information usage (e.g., managerial decision making, running automated processes), and also include activities such as information retrieval and converting information from one form to another. Information use as defined in the information model can be thought of as the purposes for which enterprise stakeholders need information when assuming their roles, fulfilling their activities and interacting with each other.
- **Monitor**—The phase in which it is ensured that the information resource continues to work properly (i.e., to be valuable). Activities in this phase may refer to keeping information up to date as well as other kinds of information management activities (e.g., enhancing, cleansing, merging, removing duplicate information data in data warehouses).
- **Dispose**—The phase in which the information resource is transferred or retained for a defined period, destroyed, or handled as part of an archive as needed. Activities in this phase may refer to information retention, archiving or destroying.

The IS auditor should ensure that the quality of the data allows the organization to meet its strategic objectives. Are the data being captured and processed to required standards? The IS auditor should also ensure that the configuration of the organization's applications and database management systems are in line with organizational objectives. For example, are data being archived, retained or destroyed in line with a data retention policy?

4.5.5 DATABASE MANAGEMENT SYSTEM

DBMS software aids in organizing, controlling and using the data needed by application programs. A DBMS provides the facility to create and maintain a well-organized database. Primary functions include reduced data redundancy, decreased access time and basic security over sensitive data.

DBMS data are organized in multilevel schemes, with basic data elements such as the fields (e.g., a Social Security number could be a field) at the lowest level. The levels above each field have differing properties depending on the architecture of the database.

The DBMS can include a data dictionary that identifies the fields, their characteristics and their use. Active data dictionaries require entries for all data elements and assist application processing of data elements such as providing validation characteristics or print formats. Passive dictionaries are only a repository of information that can be viewed or printed.

Figure 4.11—Information Goals/Quality Criteria

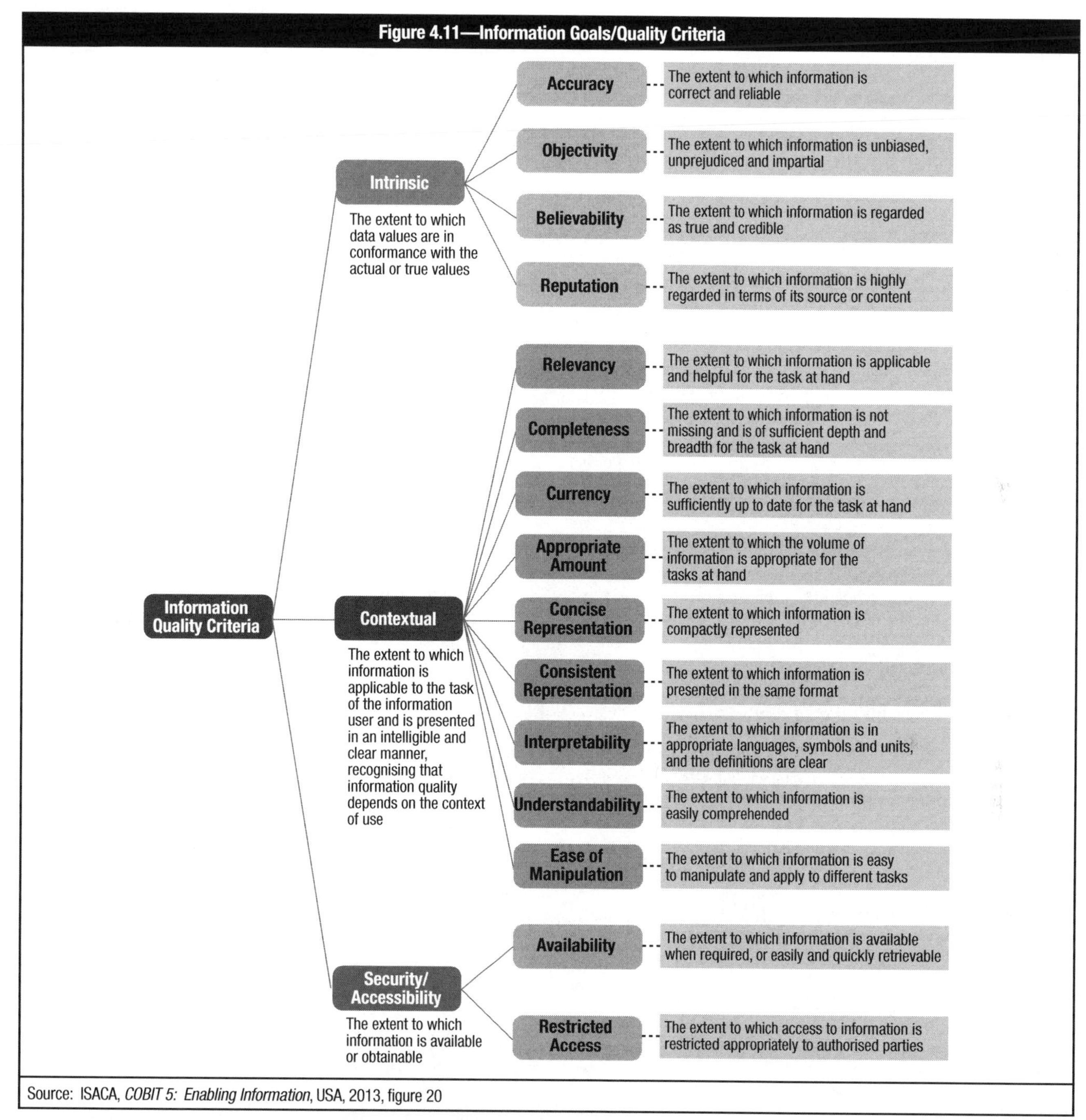
Information Quality Criteria
Intrinsic
The extent to which data values are in conformance with the actual or true values
Accuracy
The extent to which information is correct and reliable
Objectivity
The extent to which information is unbiased, unprejudiced and impartial
Believability
The extent to which information is regarded as true and credible
Reputation
The extent to which information is highly regarded in terms of its source or content
Contextual
The extent to which information is applicable to the task of the information user and is presented in an intelligible and clear manner, recognising that information quality depends on the context of use
Relevancy
The extent to which information is applicable and helpful for the task at hand
Completeness
The extent to which information is not missing and is of sufficient depth and breadth for the task at hand
Currency
The extent to which information is sufficiently up to date for the task at hand
Appropriate Amount
The extent to which the volume of information is appropriate for the tasks at hand
Concise Representation
The extent to which information is compactly represented
Consistent Representation
The extent to which information is presented in the same format
Interpretability
The extent to which information is in appropriate languages, symbols and units, and the definitions are clear
Understandability
The extent to which information is easily comprehended
Ease of Manipulation
The extent to which information is easy to manipulate and apply to different tasks
Security/ Accessibility
The extent to which information is available or obtainable
Availability
The extent to which information is available when required, or easily and quickly retrievable
Restricted Access
The extent to which access to information is restricted appropriately to authorised parties

Source: ISACA, *COBIT 5: Enabling Information*, USA, 2013, figure 20

Figure 4.12—Information Life Cycle

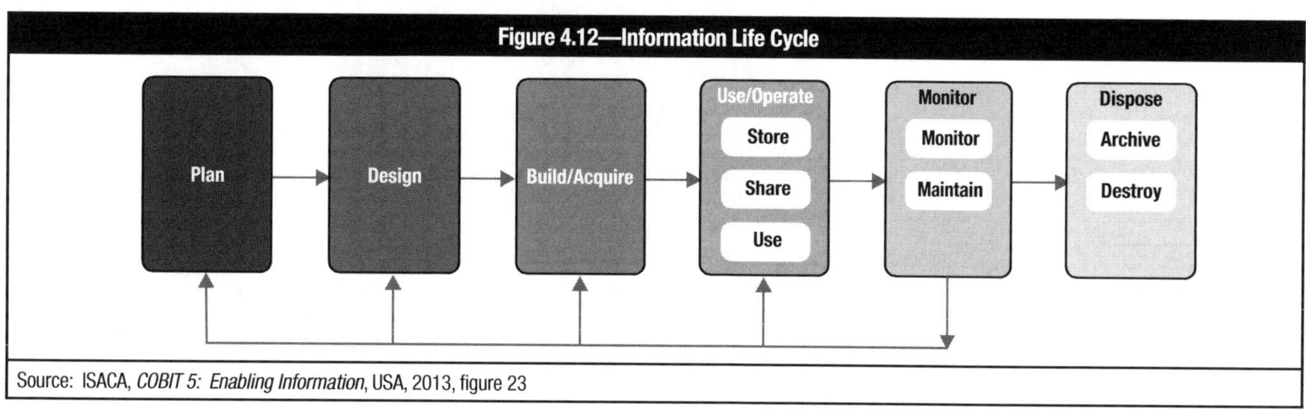
Plan
Design
Build/Acquire
Use/Operate
Store
Share
Use
Monitor
Monitor
Maintain
Dispose
Archive
Destroy

Source: ISACA, *COBIT 5: Enabling Information*, USA, 2013, figure 23

A DBMS can control user access at the following levels:
- User and the database
- Program and the database
- Transaction and the database
- Program and data field
- User and transaction
- User and data field

Some of the advantages of a DBMS include:
- Data independence for application systems
- Ease of support and flexibility in meeting changing data requirements
- Transaction processing efficiency
- Reduction of data redundancy
- Ability to maximize data consistency
- Ability to minimize maintenance cost through data sharing
- Opportunity to enforce data/programming standards
- Opportunity to enforce data security
- Availability of stored data integrity checks
- Facilitation of terminal users' *ad hoc* access to data, especially through designed query language/application generators

DBMS Architecture

Data elements required to define a database are called metadata. This includes data about data elements used to define logical and physical fields, files, data relationships, queries, etc. There are three types of metadata: conceptual schema, external schema and internal schema. If the schemas are not adjusted to smoothly work together, the DBMS may not be adequate to meet the users' needs.

Detailed DBMS Metadata Architecture

Within each level, there is a data definition language (DDL) component for creating the schema representation necessary for interpreting and responding to the user's request. At the external level, a DBMS will typically accommodate multiple DDLs for several application programming languages compatible with the DBMS. The conceptual level will provide appropriate mappings between the external and internal schemas. External schemas are location independent of the internal schema.

Data Dictionary/Directory System

A data dictionary/directory system (DD/DS) helps define and store source and object forms of all data definitions for external schemas, conceptual schemas, the internal schema and all associated mappings. The data dictionary contains an index and description of all of the items stored in the database. The directory describes the location of the data and the access method.

DD/DS provides the following functional capabilities:
- A data definition language processor, which allows the database administrator to create or modify a data definition for mappings between external and conceptual schemas
- Validation of the definition provided to ensure the integrity of the metadata
- Prevention of unauthorized access to, or manipulation of, the metadata
- Interrogation and reporting facilities that allow the DBA to make inquiries on the data definition

DD/DS can be used by several DBMSs; therefore, using one DD/DS could reduce the impact of changing from one DBMS to another DBMS. Some of the benefits of using DD/DS include:
- Enhancing documentation
- Providing common validation criteria
- Facilitating programming by reducing the needs for data definition
- Standardizing programming methods

Database Structure

There are three major types of database structure: hierarchical, network and relational. Most DBMSs have internal security features that interface with the OS access control mechanism/package. A combination of the DBMS security features and security package functions is often used to cover all required security functions. Types of DBMS structures are discussed below.

Hierarchical database model—In this model there is a hierarchy of parent and child data segments. To create links between them, this model uses parent-child relationships. These are 1:N (one-to-many) mappings between record types represented by logical trees, as shown in **figure 4.13**. A child segment is restricted to having only one parent segment, so data duplication is necessary to express relationships to multiple parents. Subordinate segments are retrieved through the parent segment. Reverse pointers are not allowed. When the data relationships are hierarchical, the database is easy to implement, modify and search. The registry in Microsoft Windows is an example of a hierarchical database. They are also used in geographic information systems.

Figure 4.13—Organization of a Hierarchical Database

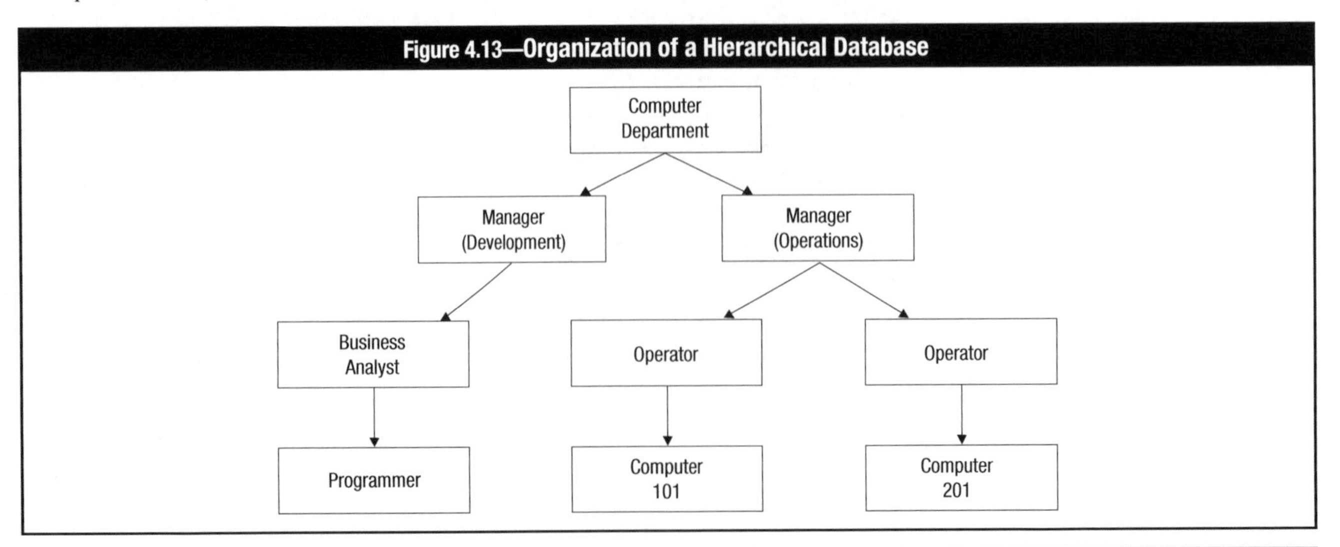

Network database model—In the network model, the basic data modeling construct is called a set. A set is formed by an owner record type, a member record type and a name. A member record type can have that role in more than one set, so a multiowner relationship is allowed. An owner record type can also be a member or owner in another set. Usually, a set defines a 1:N relationship, although one-to-one (1:1) is permitted. A disadvantage of the network model is that such structures can be extremely complex and difficult to comprehend, modify or reconstruct in case of failure. This model is rarely used in current environments. See **figure 4.14**. The hierarchical and network models do not support high-level queries. The user programs have to navigate the data structures.

Relational database model—An example of a relational database can be seen in **figure 4.15**. The relational model is based on the set theory and relational calculations. A relational database allows the definition of data structures, storage/retrieval operations and integrity constraints. In such a database, the data and relationships among these data are organized in tables. A table is a collection of rows, also known as tuples, and each tuple in a table contains the same columns. Columns, called domains or attributes, correspond to fields. Tuples are equal to records in a conventional file structure. Relational databases are used in most common enterprise resource planning (ERP) Systems. Common relational database management systems (RDBMS) include Oracle®, IBM® DB2® and Microsoft SQL Server.

Figure 4.14— Organization of a Network Database

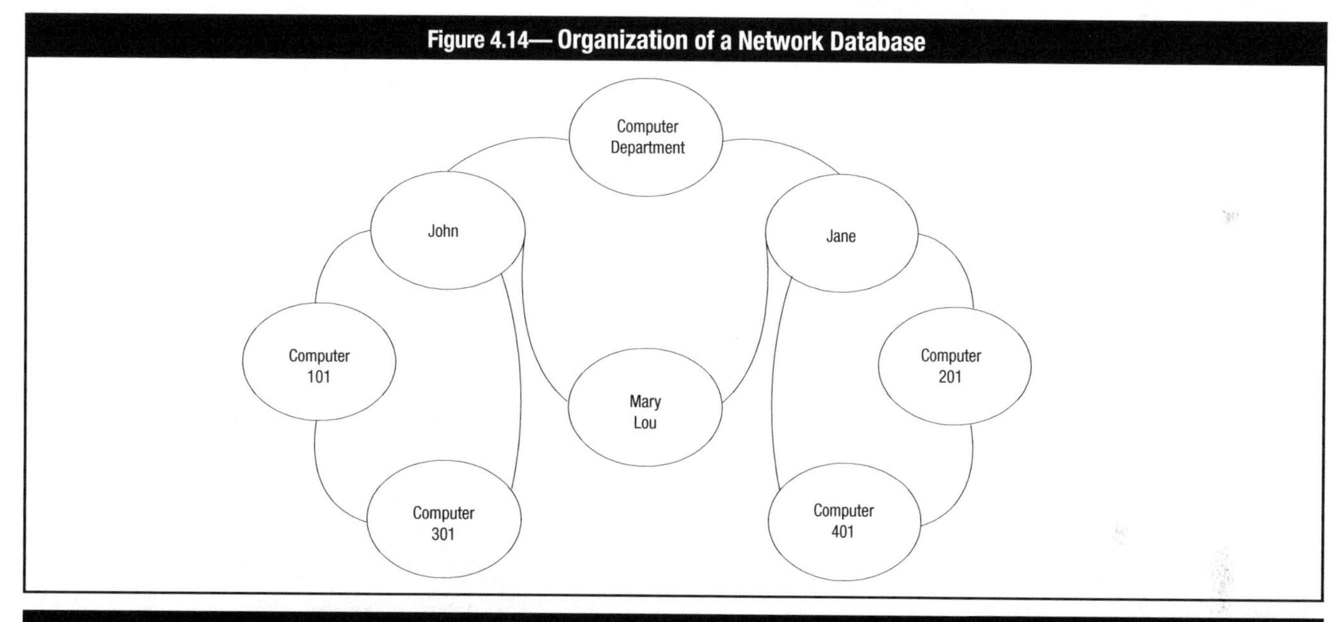

Figure 4.15—Organization of a Relational Database

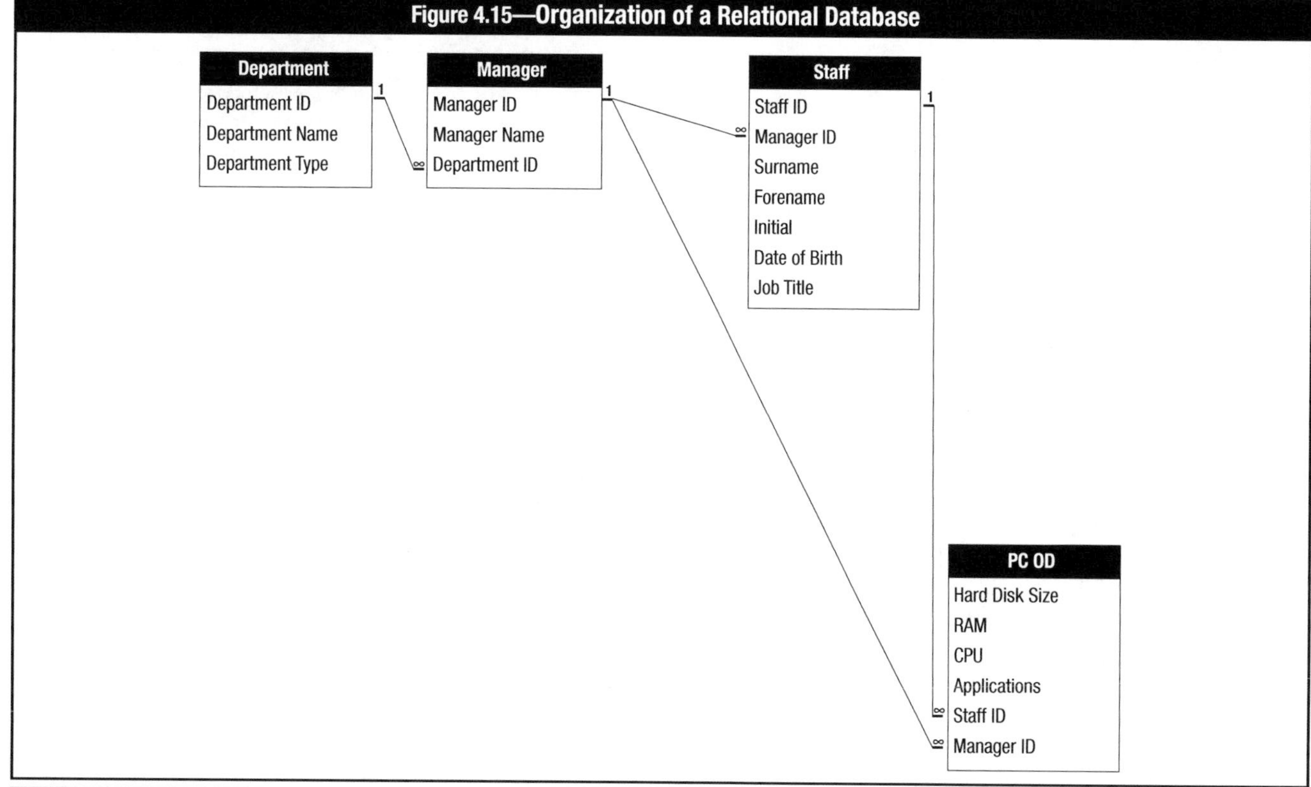

Relational tables have the following properties:
- Values are atomic.
- Each row is unique.
- Column values are of the same kind.
- The sequence of columns is insignificant.
- The sequence of rows is insignificant.
- Each column has a unique name.

Certain fields may be designated as keys, so searches for specific values of that field will be quicker because of the use of indexing. If fields in two different tables take their values from the same set, a join operation can be performed to select related records in the two tables by matching values in those fields. This can be extended to joining multiple tables on multiple fields. These relationships are only specified at retrieval time, so relational databases are dynamic. The relational model is independent from the physical implementation of the data structure, and has many advantages over the hierarchical and network database models. With relational databases, it is easier:
- For users to understand and implement a physical database system
- To convert from other database structures
- To implement projection and join operations (i.e., referencing groups of related data elements not stored together)
- To create new relations for applications
- To implement access control over sensitive data
- To modify the database

A key feature of relational databases is the use of "normalization" rules to minimize the amount of information needed in tables to satisfy the users' structured and unstructured queries to the database. Generally followed, normalization rules include:
- A given instance of a data object has one and only one value for each attribute.
- Attributes represent elementary data items; they should contain no internal structure.
- Each tuple (record) consists of a primary key that identifies some entity, together with a set of zero or more mutually independent attribute values that describes the entity in some way (fully dependent on primary key).
- Any foreign key should have a null value or should have an existing value linking to other tables; this is known as referential integrity.

Object-oriented Database Management Systems (OODBMS)—An example of an OODBMS can be seen in **figure 4.16**. In an OODBMS, information is stored as objects (as used in object-oriented programming) rather than data (as in rational databases). This means that all of the features related to object-oriented programming can be applied including encapsulation (i.e., the creation of data types or classes, including objects) and inheritance (i.e., classes inherit features from other classes). This results in objects that contain both executable code and data. The actual storage of the object in the database is achieved by assigning each object a unique identifier. These are loaded into virtual memory when referenced allowing them to be found quickly. OODBMS has found a niche in areas such as engineering, science and spatial databases. It is often used when the database is made up of graphics, diagrams or sound that cannot easily be defined or queried by relational databases.

Figure 4.16—Organization of an Object-oriented Database

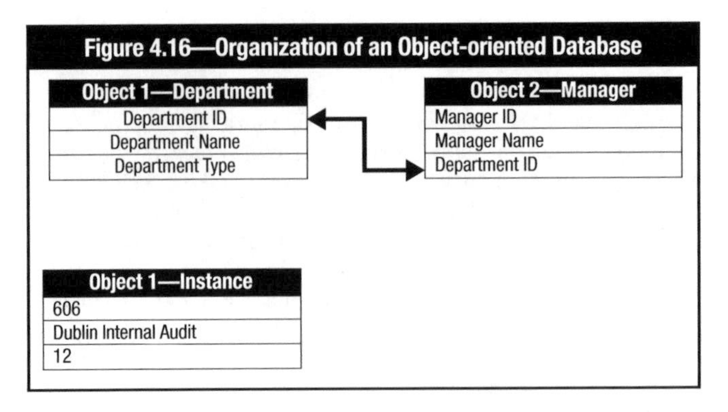

NoSQL—NoSQL databases were developed in response to a rise in the volume of data stored on the Internet commonly known as big data. Much of these data are unstructured being audio, video, tweets, logs, blogs, etc. These data cannot be broken out into components as required for a relational database; however, NoSQL databases may also support SQL, hence the term "Not only SQL." NoSQL databases may support object orientation (as per OODBMS) and other database technologies as seen in **figure 4.17**.

Figure 4.17—NoSQL Database Technologies

Data Model	Description
Key Value	All items in the database are stored as an attribute name (key) with its value.
Column-oriented	All of the values of a column are put together followed by all the values of the next column, then the values of the next column, etc.
Graph Database	Databases based on graph theory (mathematical models of the relationship between objects)
Document-oriented	Manages, stores and retrieves document-oriented information. This is achieved using storage methods such as XML and JSON.

The advantages of NoSQL databases include sharding—the ability to partition the database horizontally across database servers to spread the work load (important when dealing with big data)—and dynamic schemas—the schema does not have to be defined before you add data (as in relational databases). Common NoSQL databases include MongoDB and Cassandra.

Database Controls

It is critical that database integrity and availability be maintained. This is ensured through the following controls:
- Establish and enforce definition standards.
- Establish and implement data backup and recovery procedures to ensure database availability.
- Establish the necessary levels of access controls, including privileged access, for data items, tables and files to prevent inadvertent or unauthorized access.
- Establish controls to ensure that only authorized personnel can update the database.
- Establish controls to handle concurrent access problems such as multiple users desiring to update the same data elements at the same time (i.e., transaction commit, locking of records/files).
- Establish controls to ensure accuracy, completeness and consistency of data elements and relationships in the database. It

is important that these controls, if possible, be contained in the table/columns definitions. In this way, there is no possibility that these rules will be violated because of programming flaws or through the usage of utilities in manipulating data.
- Use database checkpoints at junctures in the job stream that minimize data loss and recovery efforts to restart processing after a system failure.
- Perform database reorganization to reduce unused disk space and verify defined data relationships.
- Follow database restructuring procedures when making logical, physical and procedural changes.
- Use database performance reporting tools to monitor and maintain database efficiency (e.g., available storage space, buffer size, CPU usage, disk storage configuration and deadlock conditions).
- Minimize the ability to use nonsystem tools or other utilities (i.e., those outside security control, to access the database).

4.5.6 UTILITY PROGRAMS

Utility programs are system software used to perform maintenance and routines that frequently are required during normal processing operations. Utility programs can be categorized by use, into five functional areas:

1. Understanding application systems (flowcharting software, transaction profile analyzer, executive path analyzer and data dictionary)
2. Assessing or testing data quality (data manipulation utilities, database dump utilities, data comparison utility and query facility)
3. Testing a program's ability to function correctly and maintain data integrity (test data generator, online debugging facility, output analyzer and network simulator)
4. Assisting in faster program development (visual display utility, library copy, text editor, online coding facility, report generators and code generators)
5. Improving operational efficiency (CPU and memory utilization monitors and communication line analyzers)

Smaller computer systems (i.e., PC and server OSs) are often equipped with specific utilities to:
- Operate verification, cleaning and defragmenting of hard disk and removable memory units
- Initialize removable data volumes and volumes of disk/removable memory
- Save/restore system images
- Reconstruct and restore (logically) cancelled files
- Test system units and peripherals

Many of these utility programs can perform outside the security system or can function without producing an audit trail of activity. As a result, access to and use of these sensitive and powerful utilities should be well controlled and restricted.

4.5.7 SOFTWARE LICENSING ISSUES

Software copyright laws must be followed to protect against the possibility of a company paying penalties over copyright infringements and the added reputational risk of being identified as a company that illegally uses software.

A software licensing agreement is a contract that establishes the terms and conditions under which a piece of software is being licensed (i.e., made legally available for use) from the software developer (owner) to the user. There are two different software licensing types: free (**figure 4.18**) and paid (**figure 4.19**).

Figure 4.18—Free Software Licensing Types

Type	Description
Open source	The software may be used, copied, studied, modified and redistributed as required. Open source is usually accompanied by the program source and a copy of the software license (for example, the GNU General Public License). A well-known example is Linux.
Freeware	The software is free, but the source code cannot be redistributed. A well-known example is Adobe Acrobat Reader®.
Shareware	The software may be free initially; however, this may only be on a trial basis or have limited functionality compared to the full, commercial version (may also be known as trial version, demo ware or an evaluation copy).

Figure 4.19—Paid Software Licensing Types

Type	Description
Per central processing unit (CPU)	Depends on the power of the server, specifically the number of the CPUs; could include the number of CPU cores
Per seat	Depends on the number of unique users of the system
Concurrent users	Depends on the total number of users using the software within a predefined period of time
Utilization	Depends on how busy the CPU is or the number of users that are active at any one time
Per workstation	Depends on the number of individual workstations (NOT users) that connect to the software
Enterprise	Usually allows unlimited use of the software throughout an organization without the need to apply any of the rules above, although there may be some restrictions

To detect software licensing violations, the IS auditor should:
- Review the listing of all standard, used and licensed application and system software.
- Obtain copies of all software contracts for these to determine the nature of the license agreements, be it an unlimited enterprise license, per-seat license or individual copies.
- Scan the entire network to produce a list of installed software.
- If required, review a list of server specifications including CPUs and cores.
- Compare the license agreements with the software that is actually installed noting any violations.

Options available to prevent software license violations include:
- A good software asset management process (see section 4.3 IT Asset Management)
- Centralizing control, distribution and installation of software (includes disabling the ability of users to install software, where possible)
- Requiring that all PCs be restricted workstations with disabled or locked down disk drives, USB ports, etc.
- Installing metering software on the LAN and requiring that all PCs access applications through the metered software
- Regularly scanning user networks endpoints to ensure that unauthorized copies of software have not been loaded (achieved by comparing actual software loaded to the list of software assets)
- Enforcing documented policies and procedures that require users to sign an agreement not to install software without management authorization and a software license agreement

Software licenses are primarily contractual compliance—that is, organizations agree to comply with the terms and conditions of the software publisher, with or without financial consideration. In certain circumstances, an IS auditor may need expert legal opinion to confirm compliance.

Note that some disaster recovery arrangements may require additional licenses and hosting of additional metering software. Refer to section 4.8 Disaster Recovery Planning for more information.

4.5.8 SOURCE CODE MANAGEMENT

Source code is the language in which a program is written. It is translated into object code by assemblers and compilers and tells the computer what to do. By its very nature, source code may contain intellectual property and should be protected, and access should be restricted.

Organizational access to source code may differ depending on the application and the nature of the agreement with the supplier. If no source code is supplied, it may be important to secure an escrow agreement. If the software is packaged, access to the source code may be granted under license to allow for customized modifications. If the software is bespoke or developed in house, the organization will have full access to the source code. In all instances source code is subject to the software development life cycle (see section 3.5.2 Description of Traditional SDLC Phases). Source code management is also tightly linked to change management, release management, quality assurance and information security management.

The actual source code should be managed using version control system (VCS), often called revision control software (RCS). These maintain a central repository, which allows programmers to check out a program source to make changes to it. Checking in the source creates a new revision of the program. A VCS provides the ability to synchronize source changes with changes from other developers, including conflict resolution when changes have been made to the same section of source. A VCS also allows for branching, a copy of the trunk (original main code) that exists independently to allow for customization for different customers, countries, locations etc.

An example of a popular VCS is Apache™ Subversion®. Git is a distributed version control system (DVCS). While Subversion manages a single centralized repository, a DVCS has multiple repositories. In a DVCS, the entire repository may be replicated locally with changes committed to the master repository when needed. This allows developers to work remotely, without a connection.

The advantages of VCSs include:
- Control of source code access
- Tracking of source code changes
- Allowing for concurrent development
- Allowing rollback to earlier versions
- Allowing for branching

The IS auditor should always be aware of the following:
- Who has access to source code
- Who can commit the code (push the code to production)
- Alignment of program source code to program objects
- Alignment with change and release management
- Backups of source code including those offsite and escrow agreements

4.5.9 END-USER COMPUTING

End-user computing (EUC) refers to the ability of end users to design and implement their own information system utilizing computer software products.

There are benefits to EUC as users can quickly build and deploy applications, taking the pressure off of the IT department. However, lack of IT department involvement also brings associated risk because the applications may not be subject to an independent review and, frequently, are not created in the context of a formal development methodology.

This can result in applications that:
- May contain errors and give incorrect results
- Are not subject to change management or release management, resulting in multiple, perhaps different, copies
- Are not secured
- Are not backed up

The IS auditor should ensure that policies for the use of EUC exist. An inventory (see section 4.3 IT Asset Management) of all such applications should exist with those deemed critical enough subject to the same controls as any other application.

4.6 IS NETWORK INFRASTRUCTURE

IS networks were developed from the need to share information resources residing on different computer devices, which enabled organizations to improve business processes and realize substantial productivity gains.

Generally, the telecommunication links or lines for networks are digital, although analog may still be used. They are classified according to the type of provider or the type of technology. Typically, they can be divided into dedicated circuit (also known as leased lines) and switched circuit.

A **dedicated circuit** is a symmetric telecommunications line connecting two locations. Each side of the line is permanently connected to the other. Dedicated circuits can be used for telephone, data or Internet services.

A **switched circuit** does not permanently connect two locations and can be set up on demand, based on the addressing method. There are two main types of switching mechanisms: circuit switching and packet switching.

The **circuit switching** mechanism is typically used over the telephone network (plain old telephone service [POTS], integrated services digital network [ISDN]). Switched circuits allow data connections that can be initiated when needed and terminated when communication is complete. This works much like a normal telephone line works for voice communication. ISDN is a good example of circuit switching. When a router has data for a remote site, the switched circuit is initiated with the circuit number of the remote network. In the case of ISDN circuits, the device places a call to the telephone number of the remote ISDN circuit. When the two networks are connected and authenticated, they can transfer data. When the data transmission is complete, the call can be terminated.

Packet switching is a technology in which users share common carrier resources. Because this allows the carrier to make more efficient use of its infrastructure, the cost to the customer is generally much lower than with leased lines. In a packet switching setup, networks have connections into the carrier's network, and many customers share the carrier's network. The carrier can then create virtual circuits between customers' sites by which packets of data are delivered from one to the other through the network. The section of the carrier's network that is shared is often referred to as a cloud. Some examples of packet-switching networks include asynchronous transfer mode (ATM), frame relay, Switched Multimegabit Data Services (SMDS) and X.25.

Methods for transmitting signals over analog telecommunication links or lines are either baseband or broadband, as described below:

- **Baseband**—The signals are directly injected on the communication link (no modulation or shift in the range of frequencies of the signal). Generally, only one communication channel is available at any a time (half-duplex), although full-duplex modems are now available.
- **Broadband network**—Different carrier frequencies defined within the available band, can carry analog signals, such as those generated by image processors or a data modem, as if they were placed on separate baseband channels. Interference is avoided by separating adjacent carrier frequencies with a gap that depends on the band requirements of the carried signals. The possibility of vectoring multiple independent channels on a single-carrier media enhances considerably the effectiveness of remote connections. The condition when simultaneous data or control transmission/reception takes place between two stations is called a full-duplex connection.

4.6.1 ENTERPRISE NETWORK ARCHITECTURES

Modern networks are part of a large, centrally managed, internetworked architecture solution of high-speed local- and wide-area computer networks serving organizations' client server–based environments. Such architectures include clustering common types of IT functions in network segments, each uniquely identifiable and specialized to a task. For example, network segments or blocks may include web-based front-end application servers (public or private), application and database servers, and mainframe servers using terminal emulation software to allow end users to access these back-end legacy-based systems. In turn, end users can be clustered together within their own network LANs, but with rapid access capabilities to incorporate information resources. Some organizations implement service-oriented architectures (SOA) in which web software components, using Simple Object Access Protocol (SOAP) and Extensible Markup Language (XML), interoperate in a loosely connected and distributed fashion across the network. Within this environment, information is highly accessible, available anytime and anywhere, and centrally managed for highly effective and efficient troubleshooting and performance management to achieve optimum use of network resources.

To understand the network architecture solutions offered from a business, performance and security design standpoint, an IS auditor must understand information technologies associated with the design and development of a telecommunications infrastructure (e.g., LAN and WAN specifications). Telecommunications is the electronic transmission of data, sound and images between connected end systems (two or more computers acting as sender and receiver). This process is enabled by a communications subsystem, such as a network interface card that interfaces each end user's computer to a common transmission medium, and network devices such as bridges, switches and routers, to connect computers residing on different networks.

4.6.2 TYPES OF NETWORKS

The types of networks common to all organizations are defined as follows:

- Personal area networks (PANs)—Generally, a PAN is a microcomputer network used for communications among computer devices (including telephones, tablets, printers, cameras, scanners, etc.) being used by an individual person. The extent of a PAN is typically within a range of 33 feet (about 10 meters). PANs can be used for communication among the personal devices themselves or to connect to a higher-level network and the Internet.
  - PANs may be wired with computer buses, such as USB, Firewire and other standards. If PANs are implemented without wires, they are called wireless PANs (WPANs), which can also be made possible with network technologies such as IrDA and Bluetooth.
  - A Bluetooth PAN is also called a piconet and is composed of up to eight active devices in a master-slave relationship. The first Bluetooth device in the piconet is the master, and all other devices are slaves that communicate with the master. A piconet typically has a range of 32.8 feet (10 meters), although ranges of up to 328 feet (100 meters) can be reached under ideal circumstances.

- **LANs**—LANs are computer networks that cover a limited area such as a home, office or campus. Characteristics of LANs are higher data transfer rates and smaller geographic range. Ethernet and Wi-Fi (WLANs) are the two most common technologies currently used.
- **SANs**—SANs are a variation of LANs and are dedicated to connecting storage devices to servers and other computing devices. SANs centralize the process for the storage and administration of data.
- **WANs**—WANs are computer networks that cover a broad area such as a city, region, nation or an international link. The Internet is the largest example of a WAN. WANs are used to connect LANs and other types of networks together so that users and computers in one location can communicate with users and computers in other locations. Many WANs are built for one particular organization and are private. Others, built by Internet service providers (ISPs), provide connections from an organization's LAN to the Internet. WANs may also be wireless (WWANs).
- **Metropolitan Area Networks (MANs)**—MANs are WANs that are limited to a city or region; usually, MANs are characterized by higher data transfer rates than WANs.

4.6.3 NETWORK SERVICES

Network services are functional features made possible by appropriate OS applications. They allow orderly utilization of the resources on the network. Instead of having a single OS that controls its own resources and shares them with the requesting programs, the network relies on standards and on a specific protocol or set of rules, enacted and operated through the basic system software of the various network devices that are capable of supporting the individual network services. Users and business applications can request network services through specific calls/interfaces. The following are network application services commonly used in organizations' networked environments:

- **Network file system**—Allows users to share files, printers and other resources in a network
- **Email services** —Provide the ability, via a terminal or PC connected to a communication network, to send an unstructured message to another individual or group of people
- **Print services**—Provide the ability, typically through a print server on a network, to manage and execute print request services from other devices on the network
- **Remote access services**—Provide remote access capabilities where a computing device appears, as if directly attached to the remote host
- **Directory services**—Store information about the various resources on a network and help network devices locate services, much like a conventional telephone directory
  - Directory services also help network administrators manage user access to network resources.
- **Network management**—Provides a set of functions to control and maintain the network
  - Network management provides detailed information about the status of all components in the network such as line status, active terminals, length of message queues, error rate on a line and traffic over a line.
  - It enables computers to share information and resources within a network and provides network reliability.
  - It provides the operator with an early warning signal of network problems before they affect network reliability, allowing the operator to take timely preventive or remedial actions.
- **Dynamic Host Configuration Protocol (DHCP)**—A protocol used by networked computers (clients) to obtain IP addresses and other parameters such as the default gateway, subnet mask and IP addresses of domain name systems (DNSs) from a DHCP server
  - The DHCP server ensures that all IP addresses are unique (e.g., no IP address is assigned to a second client while the first client's assignment is valid [its lease has not expired]). Thus, IP address pool management is performed by the server and not by a human network administrator.
- **DNS**—Translates the names of network nodes into network (IP) addresses

4.6.4 NETWORK STANDARDS AND PROTOCOLS

Network architecture standards facilitate the process of creating an integrated environment that applications can work within by providing a reference model that organizations can use for structuring intercomputer and network communication processes.

Besides the convenience of using compatible architectures, one major advantage of network standards is that they help organizations meet the challenge of designing and implementing an integrated, efficient, reliable, scalable and secure network of LANs and WANs with external connectivity (public Internet). This is a major challenge due to the requirements of the following:

- **Interoperability**—Occurs when connecting various systems to support communication among disparate technologies where different sites may use different types of media that may operate at differing speeds
- **Availability**—Means end users have continuous, reliable and secure service (24/7 access)
- **Flexibility**—Needed for network scalability to accommodate network expansion and requirements for new applications and services
- **Maintainability**—Means an organization provides centralized support and troubleshooting over heterogeneous, but highly integrated systems

To accomplish these tasks, organizations need to have the ability to define specifications for the types of networks to be established (e.g., LANs/WANs) when creating an integrated environment that their applications can work within. Organizations must also provide centralized support and troubleshooting over heterogeneous, but highly integrated systems.

4.6.5 OSI ARCHITECTURE

The purpose of network architecture standards is to facilitate this process by providing a reference model that organizations can use for building intercomputer and network communication processes, respectively.

The benchmark standard for this process, the Open Systems Interconnection (OSI) reference model, was developed by the ISO in 1984. The OSI is a proof-of-concept model composed of seven layers, each specifying particular specialized tasks or functions. Each layer is self-contained and relatively independent of the other layers in terms of its particular function. This enables solutions offered by one layer to be updated without adversely affecting the other layers.

Note: While it is beneficial for the IS auditor to know the OSI reference model, the CISA candidate will not be tested on the specifics of this standard in the exam.

The objective of the OSI reference model is to provide a protocol suite used to develop data-networking protocols and other standards to facilitate multivendor equipment interoperability. The OSI program was derived from a need for international networking standards and was designed to facilitate communication between hardware and software systems despite differences in underlying architectures.

It is important to note that in the OSI model each layer communicates not only with the layers above and below it in the local stack, but also with the same layer on the remote system. For example, the application layer on the local system appears to be communicating with the application layer on the remote system. All of the details of how the data are processed further down the stack are hidden from the application layer. This is true at every level of the model. Each layer appears to have a direct (virtual) connection to the same layer on the remote system.

The **application layer** provides a standard interface for applications that must communicate with devices on the network (e.g., print files on a network-connected printer, send an email or store data on a file server). Thus, the application layer provides an interface to the network. In addition, the application layer may communicate the computer's available resources to the rest of the network. The application layer should not be confused with application software. Application software uses the application layer interface to access network-connected resources.

The **presentation layer** transforms data to provide a standard interface for the application layer and provides common communication services such as encryption, text compression and reformatting (e.g., conversion of Extended Binary-coded for Decimal Interchange Code [EBCDIC] to ASCII code). The presentation layer converts the outgoing data into a format acceptable by the network standard and then passes the data to the session layer. Similarly, the presentation layer converts data received from the session layer into a format acceptable to the application layer.

The **session layer** controls the dialogs (sessions) between computers. It establishes, manages and terminates the connections between the local and remote application layers. All conversations, data exchanges and dialogs between the application layers are managed by the session layer.

The **transport layer** provides reliable and transparent transfer of data between end points, end-to-end error recovery and flow control. The transport layer ensures that all of the data sent to it by the session layer are successfully received by the remote system's transport layer. The transport layer is responsible for acknowledging every data packet received from the remote transport layer, ensuring that an acknowledgement is received from the remote transport layer for every packet sent. If an acknowledgement is not received for a packet, then that packet will be re-sent.

The **network layer** creates a virtual circuit between the transport layer on the local device and the transport layer on the remote device. This is the layer of the stack that understands IP addresses and is responsible for routing and forwarding. This layer prepares the packets for the data link layer.

The **data link layer** provides for the reliable transfer of data across a physical link. It receives packets of data from the network layer, encapsulates them into frames and sends them as a bit stream to the physical layer. These frames consist of the original data and control fields necessary to provide for synchronization, error detection and flow control. Error detection is accomplished through the use of a cyclic redundancy check (CRC) that is calculated for and then added to each frame of data. The receiving data link layer calculates the CRC value for the data portion of the received frame and discards the frame if the calculated and received values do not match. A CRC calculation will detect all single-bit and most multiple-bit errors.

A bit stream received from the physical layer is similarly converted to data packets and sent to the network layer. The data link layer logically connects to another device on the same network segment using a MAC address. Each device on the network has a unique MAC hardware address that is assigned to it at the time of manufacture. The MAC address can be overridden, but this practice is not recommended. The data link layer normally only listens to data intended for its MAC address. An important exception to this rule is that a network interface may be configured as a promiscuous interface, which will listen to all data that the physical layer sends it.

The **physical layer** provides the hardware that transmits and receives the bit stream as electrical, optical or radio signals over an appropriate medium or carrier. This layer defines the cables, connectors, cards and physical aspects of the hardware required to physically connect a device to the network. Error correction and detection is not usually implemented in the physical layer, with a few notable exceptions. Cell phones and digital microwave systems will typically implement some form of error correction code, not only detecting but actually correcting errors. The most sophisticated forms of these are used by the US National Aeronautics and Space Administration (NASA) program for communicating with their deep space probes.

ISO formulated the OSI model to establish standards for vendors developing protocols supporting open system architecture. The intent is to make different proprietary systems work seamlessly within the same network. The actual implementation of the functions defined in each layer is based on protocols developed for each layer. A protocol is an agreed-upon set of rules and procedures to follow when implementing the tasks associated with a given layer of the OSI model.

The intent of the OSI model is to provide a standard interface at each layer and to ensure that each layer does not have to be concerned with the details of how the other layers are implemented.

This approach supports system-to-system communication (peer-to-peer relationship) where each layer on the sender side provides information to its peer layer on the receiving side. The process also is characterized as a data traversal process with the following actions occurring:
- Data travels down through layers at the local end.
- Protocol-control information (headers/trailers) is used as an envelope at each layer to pick up control information.
- Data travels up through the layers at the receiving/destination end.
- Protocol-control information (headers/trailers) is removed as the information is passed up.

A traditional OSI model showing this process is depicted in **figure 4.20**.

4.6.6 APPLICATION OF THE OSI MODEL IN NETWORK ARCHITECTURES

The concepts of the OSI model are used in the design and development of organizations' network architectures. This includes LANs, WANs, MANs and use of the public Transmission Control Protocol/Internet Protocol (TCP/IP)-based global Internet. The following sections will provide a detailed technical discussion of each and will show how the OSI reference model applies to the various architectures. The discussion will focus on:
- Local area network (LAN)
- Wide area network (WAN)
- Wireless networks
- Public global Internet infrastructure
- Network administration and control
- Applications in a networked environment
- On-demand computing

Local Area Network

A LAN covers a small, local area—from a few devices in a single room to a network across a few buildings. The increase in reasonably priced bandwidth has reduced the design effort required to provide cost-effective LAN solutions for organizations of any size.

New LANs are almost always implemented using switched Ethernet (802.3). Twisted-pair cabling (100-Base-T or better and wireless LANs [WLANs]) connects floor switches to the workstations and printers in the immediate area. Floor switches can be connected to each other with 1000-Base-T or fiber-optic cabling. In larger organizations, the floor switches may be connected to larger, faster switches whose purpose is to properly route the switch-to-switch data.

As LANs get larger and traffic increases, the requirement to carefully plan the logical configuration of the network becomes more and more important. Network planners need to be highly skilled and very knowledgeable. Their tools include traffic monitors that allow them to monitor traffic volumes on critical links. Tracking traffic volumes, error rates and response times is every bit as important on larger LANs as it is on distributed servers and mainframes.

LAN DESIGN FUNDAMENTALS AND SPECIFICATIONS

To set up a LAN, an organization must assess cost, speed, flexibility and reliability. The issues include:
- Assessing media for physically transmitting data
- Assessing methods for the physical network medium
- Understanding from a performance and security standpoint how data will be transmitted across the network and how the actual LAN network is organized and structured in terms of optimizing the performance of the devices connected to it

Figure 4.20—Traditional OSI Model

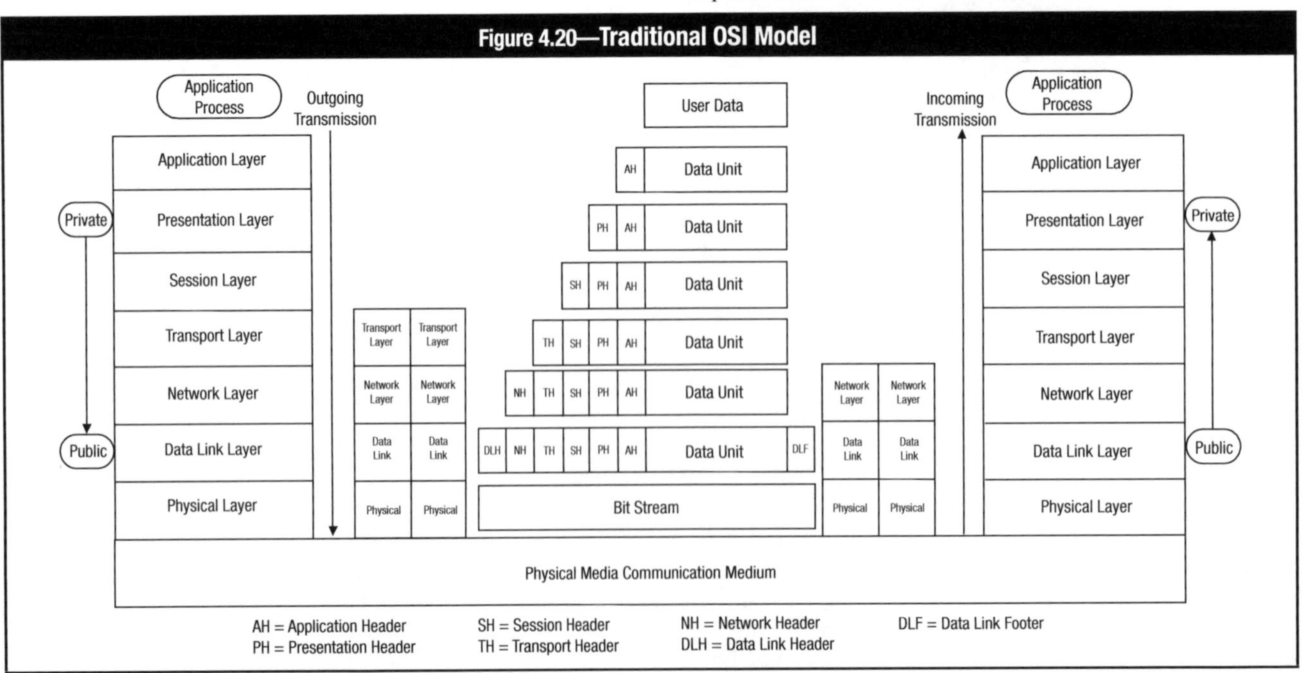

NETWORK PHYSICAL MEDIA SPECIFICATIONS

Physical media used to connect various types of computing devices together in a network include:
- Twisted pairs
- Fiber optics for high-capacity and specific architectures
- Infrared and radio (wireless)

Generally, twisted-pair cabling is still the most commonly used media for LANs; however, this is increasingly supplanted by wireless for LAN connectivity. The type and characteristics of physical media (e.g., speed, sensitivity to external disturbances, signal loss and propagation, security) not only affect the cost of implementation and support but also impact the capacity, flexibility and reliability of the network.

LANs can be implemented using various types of media including:
- **Copper (twisted-pair) circuits**—Two insulated wires are twisted around each other, with current flowing through them in opposite directions. This reduces the opportunity for cross talk between pairs in the same bundle and allows for lower sensitivity to electromagnetic disturbances (shielded twisted-pair circuits) within each individual pair. Twisted-pair circuits can also be used for some dedicated data networks. Today, the common standards for twisted-pair circuits are CAT5, CAT6 and CAT7. Organizations should buy certified cables from reputable suppliers and segment problem areas with switches. Additionally, assurance should be provided that maximum cabling lengths are not exceeded since this will produce intermittent failures. A disadvantage of unshielded twisted-pair cabling is that it is not immune to the effects of electromagnetic interference (EMI) and should be run in dedicated conduits, away from sources of potential interference such as fluorescent lights. Parallel runs of cable over long distances should also be avoided since the signals on one cable can interfere with signals on adjacent cables—an EMI condition known as cross talk.
- **Fiber-optic systems**—Glass fibers are used to carry binary signals as flashes of light. Fiber-optic systems have a low transmission loss as compared to twisted-pair circuits. Optical fibers do not radiate energy nor conduct electricity. In addition, they are not affected by EMI and present a significantly lower risk of security problems such as wiretaps. Optical fiber is a more fragile medium and is more attractive for applications where changes are infrequent. Optical fiber is smaller and lighter than metallic cables of the same capacity. Fiber is the preferred choice for high-volume, longer-distance runs. One example would be using fiber to connect floor switches to enterprise data switches. In addition, fiber-optic cable is often used to connect servers to SANs.
- **Radio systems (wireless)**—Data are communicated between devices using low-powered systems that broadcast (or radiate) and receive electromagnetic signals representing data.

LAN TOPOLOGIES AND PROTOCOLS

LAN topologies define how networks are organized from a physical standpoint, whereas protocols define how information transmitted over the network is interpreted by systems.

LAN physical topology was previously tied fairly tightly to the protocols that were used to transfer information across the wire. This is no longer true. For current technology, the physical topology is driven by ease of construction, reliability and practicality. Of the physical topologies that have been commonly used—bus, ring and star—only the star is used to any great extent in new construction. **Figure 4.21** illustrates commonly used physical topologies.

LAN MEDIA ACCESS TECHNOLOGIES

LAN media access technologies for accessing physical transmission media used are primarily either Ethernet or token passing. These technologies give devices shared access to the network, while also preventing a single device from monopolizing the network.

Ethernet has evolved from its original bus configuration, providing 10 Mbps speed with two coaxial cable versions (thin and thick), to star configurations initially using 10-Base-T (Ethernet using twisted-pair cabling) and now using today's more modern versions: Fast Ethernet (100 Mbps) and Gigabit Ethernet (1 Gbps).

Figure 4.21—Physical Topologies in Common Use

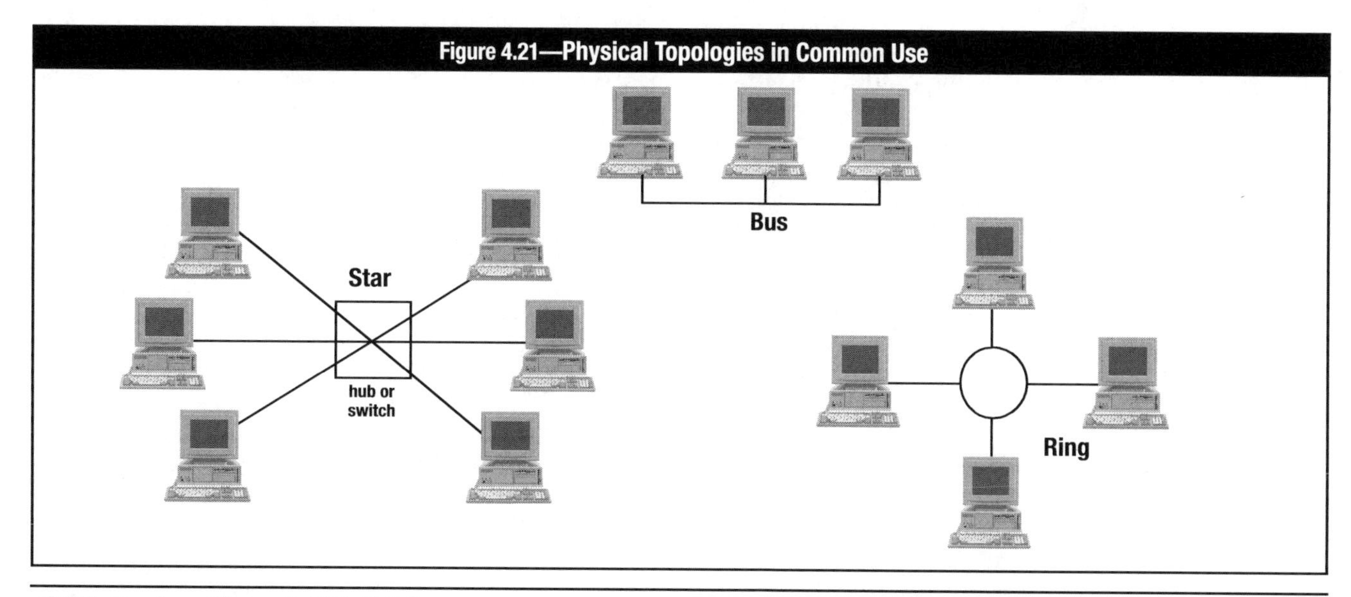

A critical aspect of any communication is determining the recipient of a message. At this level, considering Fast Ethernet and Gigabit Ethernet, a MAC address is used to specify the recipient. Every network interface that is manufactured has a unique MAC address, which is only used for the last hop of any communication (see TCP/IP and Its Relation to the OSI Reference Model to see how this fits in with real-world addresses, such as 192.168.4.5). Every network interface card (NIC) connected to the network listens to every conversation on the network. Normally, a NIC device driver (software) only collects the data with its address. A NIC that has been placed in promiscuous mode will read all data passing over the network (including user IDs and passwords).

The initial bus arrangement typically provides an effective throughput of 5 Mbps among all of the systems connected to a bus segment. Bus segments could be connected together with repeaters or bridges. Repeaters would regenerate signals—allowing a longer span for the network. Bridges would connect multiple buses together—blocking any traffic that could not be delivered on a given segment. Bridges also served another critical function—that of breaking the network into multiple collision domains.

Ethernet is a Carrier Sense Multiple Access/Collision Detection (CSMA/CD) protocol. This is analogous to a car attempting to turn into a street. The driver's view is limited to only the street in front of him/her. If nothing is visible, the driver attempts to turn on to the street. If the driver collides with another vehicle, he/she backs up and tries again later. It should be apparent that if the street is very busy, a lot of collisions would occur. Similarly, if all of the traffic is coming from one particular house, then many cars can be handled in an efficient manner. If there are cars coming from a lot of different houses, then the overall traffic volume that can be handled is much lower. This is the way in which Ethernet behaves in a bus arrangement.

The use of coaxial cable in this example is rather problematic. The cable itself is a single point of failure. Adding a new station would not solve the problem, and there exists a distinct lack of flexibility with such an implementation.

To alleviate this problem, a new physical implementation using a twisted-pair telephone cable was developed. This medium is much cheaper than coaxial cable and can be implemented using a star topology. The first implementation has all of the points of the star connected together using an unintelligent device called a hub—basically, a panel of connectors that allows all of the wires to be joined together. Circuitry within the hub electrically disconnects any branch that is not active. A problem on a single branch can still cause problems with the entire network, but the circuitry is simpler and a technician can easily isolate the problem at the hub. The traffic jam problem still exists, though.

Replacing hubs with switches was a significant advance in technology. A switch is an intelligent device that provides a private path for each pair of connections on the switch. If A is transferring data to B, it can do so without requiring C to transfer data to D. Further, transfers from A to D can be handled without fear of collision. This is analogous to a traffic light on a LAN. Collisions are then only an issue if more than one car is going to the same destination, and a traffic light can manage that problem.

While the traffic volume to or from any given device is still limited to the constraints set by the used technology (e.g., 10, 100, 1000 Mbps), this volume can be maintained between many pairs of devices. Additionally, the problem of collisions is eliminated; the switch ensures that they cannot happen. A packet may be delayed—while it waits for other traffic to clear the intersection—but it never encounters a delay caused by a collision or needs to be re-sent.

From a security perspective, switches provide another significant improvement. Each device on the network can only see traffic destined for its MAC address and cannot eavesdrop on network traffic intended for other destinations.

Today, switches are so inexpensive that there is little justification for continuing to use hubs. Switches that provide individual devices with 100 Mbps service and provide 1 Gbps connection to higher-level switches are in common use. Switches are increasingly providing additional functionality that can be used to implement corporate security policy.

Another media access technology used in LANs is the token ring medium access method which uses ring networks. Ring networks are usually implemented as a physical ring.

Devices using this method gain access to the network on the basis of a unique frame, called a token, that is passed around the network. The purpose of the token is to attach itself to a user or device when transmitting messages/data for its intended recipient. When unattached to a user or device, a free token's header, data field and trailer components are empty and are filled by devices needing to transmit. Token ring technologies have almost disappeared in today's networks.

LAN COMPONENTS

Components commonly associated with LANs are repeaters, hubs, bridges, switches and routers.

Repeaters are physical layer devices that extend the range of a network or connect two separate network segments together. Repeaters receive signals from one network segment and amplify (regenerate) the signal to compensate for signals (analog or digital) that are distorted due to a reduction of signal strength during transmission (i.e., attenuation).

Hubs are physical layer devices that serve as the center of a star-topology network or a network concentrator. Hubs can be active (if they repeat signals sent through them) or passive (if they merely split signals).

Bridges are data link layer devices that were developed to connect LANs or create two separate LAN or WAN network segments from a single segment to reduce collision domains. The two segments work as different LANs below the data link level of the OSI reference model, but from that level and above, they behave as a single logical network. Bridges act as store-and-forward devices in moving frames toward their destination. This is achieved by analyzing the MAC header of a data packet, which represents the hardware address of an NIC. Bridges can also filter frames based on Layer 2 information. For example, they can prevent frames sent from predefined MAC addresses from entering a particular network. Bridges are software-based, and they are less efficient than other similar hardware-based devices such as switches. Therefore, bridges are not major components in today's enterprise network designs.

Layer 2 switches are data link level devices that can divide and interconnect network segments and help to reduce collision domains in Ethernet-based networks. Furthermore, switches store and forward frames, filtering and forwarding packets among network segments, based on Layer 2 MAC source and destination addresses, as bridges and hubs do at the data link layer. Switches, however, provide more robust functionality than bridges through use of more sophisticated data link layer protocols that are implemented via specialized hardware called application-specific integrated circuits (ASICs). The benefits of this technology are performance efficiencies gained through reduced costs, low latency or idle time, and a greater number of ports on a switch with dedicated high-speed bandwidth capabilities (e.g., many ports on a switch are available with 10/100 Ethernet and/or Gigabit Ethernet speeds).

Switches are also applicable in WAN technology specifications.

Routers are similar to bridges and switches in that they link two or more physically separate network segments. The network segments linked by a router, however, remain logically separate and can function as independent networks. Routers operate at the OSI network layer by examining network addresses (i.e., routing information encoded in an IP packet). By examining the IP address, the router can make intelligent decisions to direct the packet to its destination. Routers differ from switches operating at the data link layer in that they use logically based network addresses, use different network addresses/segments off all ports, block broadcast information, block traffic to unknown addresses, and filter traffic based on network or host information.

Routers are often not as efficient as switches because they are generally software-based devices and they examine every packet coming through, which can create significant bottlenecks within a network. Therefore, careful consideration should be taken as to where routers are placed within a network. This should include leveraging switches in network design as well as applying load balancing principles with other routers for performance efficiency considerations.

Advances in switch technology have also provided switches with operating capabilities at Layer 3 and Layer 4 of the OSI reference model. A **Layer 3 switch** goes beyond the Layer 2–MAC addressing, acting at the network layer of the OSI model like a router. The Layer 3 switch looks at the incoming packet's networking protocol (e.g., IP). The switch compares the destination IP address to the list of addresses in its tables, to actively calculate the best way to send a packet to its destination. This creates a virtual circuit (i.e., the switch has the ability to segment the LAN within itself and will create a pathway between the receiving and the transmitting device to send the data). It then forwards the packet to the recipient's address. This provides the added benefit of reducing the size of network broadcast domains. A broadcast domain is the domain segment or segments where all connected devices may be simultaneously addressed by a message using a special common network address range, referred to as a broadcast address. This is needed for specific network management functions. As the broadcast domain grows larger, this may cause performance inefficiencies and major security concerns in terms of information leakage within a network (e.g., enumerating network domains, specific computers within a domain). Broadcast domains should be limited or aligned with business functional areas/workgroups within an organization to reduce the risk of information leakage to those without a need to know where systems can be targeted and their vulnerabilities exploited. The major difference between a router and a Layer 3 switch is that a router performs packet switching using a microprocessor, whereas a Layer 3 switch performs the switching using application ASIC hardware.

In creating separate broadcast domains, Layer 3 switches also enable the concept of establishing a virtual LAN (VLAN). A VLAN is a group of devices on one or more logically segmented LANs. A VLAN is set up by configuring ports on a switch, so devices attached to these ports may communicate as if they were attached to the same physical network segment, although the devices are located on different LAN segments. A VLAN is based on logical rather than physical connections and, thus, allow great flexibility. This flexibility enables administrators to restrict users' access of network resources to only those specified and segment network resources for optimal performance.

In **Layer 4 switching**, some application information is taken into account along with Layer 3 addresses. For IP, this information includes the port numbers from protocols such as User Datagram Protocol (UDP) and TCP. These devices, unlike Layer 3 switches, are more resource intensive since they have to store application-based protocol information. Only address information is stored at the Layer 2 and Layer 3 levels.

A Layer 4 (transport layer) switch allows for policy-based switching. With this functionality, Layer 4 switches can off-load a server by balancing traffic across a cluster of servers, based on individual session information and status.

Layer 4 through 7 switches are also known as content-switches, content services switches, web-switches or application-switches. They are typically used for load balancing among groups of servers. Load balancing can be based on Hypertext Transfer Protocol (HTTP), Secured Hypertext Transfer Protocol (HTTPS) and/or VPN, or for any application TCP/IP traffic using a specific port. Content switches can also be used to perform standard operations such as SSL encryption/decryption to reduce the load on the servers receiving the traffic, and to centralize the management of digital certificates.

Gateways are devices that are protocol converters. Typically, they connect and convert between LANs and the mainframe, or between LANs and the Internet, at the application layer of the OSI reference model. Depending on the type of gateway, the operation occurs at various OSI layers. The most common form of gateway is a systems network architecture (SNA) gateway, converting between a TCP/IP, NetBios or Inter-network Packet Exchange (IPX) session (terminal emulator) and the mainframe.

LAN TECHNOLOGY SELECTION CRITERIA

Some of the more relevant selection criteria are:

- What are the applications?
- What are the bandwidth needs?
- What is the area to be covered and what are the physical constraints?
- What is the budget?
- What are the remote management needs?
- What are the security needs?
- What network redundancy/resiliency is required?

Wide Area Network

A WAN is a data communications network that transmits information across geographically dispersed LANs such as among plant sites, cities and nations.

WAN characteristics include:

- They are applicable to the physical and data link layers of the OSI reference model.
- Data flow can be simplex (one-way flow), half duplex (one way at a time) or full duplex (both ways at one time without turnaround delay).
- Communication lines can be either switched or dedicated.

IMPLEMENTATION OF WANS

Fiber-optic cables are commonly used these days for most high-capacity network connections, both between buildings and between cities. Other systems that may be used include:

- **Microwave radio systems**—Microwave radio provides line-of-sight transmission of voice and data through the air. Historically, analog microwave circuits supplied the majority of long-haul low-speed data and voice transmission. This technology was used because it provided a lower-cost alternative to the low-capacity cable carrier systems of the time. Many, if not most, heavy route microwave systems have since been replaced by fiber-optic cable systems providing greatly increased capacity and greatly improved reliability at a cost per channel mile that is a tiny fraction of the cost for microwave circuits of similar capacity. All new microwave construction uses digital signals, providing greatly increased data rates and reduced error rates when compared with analog circuitry. Microwave radio circuits are still in common use on "light routes" where the economics do not favor installation of fiber. Most electrical utility companies will use microwave systems to connect their Supervisory Control and Data Acquisition (SCADA) systems together. Design of microwave circuits must take into account the physical topology of the area and the climate. Microwave antennae must be able to "see" each other. Climate conditions, such as rainfall, can adversely affect microwave links.
- **Satellite radio link systems**—These contain several receiver/amplifier/transmitter sections called transponders. Each transponder has a bandwidth of 36 megahertz (MHz), operates at a slightly different frequency, has individual transmitter sites and sends narrow beams of microwave signals to the satellite. Like microwaves, satellite signals can be affected by weather. Although satellite signals can carry large amounts of information at a time, the disadvantage is a bigger delay compared to all of the previous media, due to the "jump" from the earth to the satellite and back (estimated at about 300 milliseconds).

Figure 4.22 identifies the advantages and disadvantages of each physical layer medium available to networks. These physical specifications are applicable to WAN technologies.

WAN MESSAGE TRANSMISSION TECHNIQUES

WAN message transmission techniques include:

- **Message switching**—Sends a complete message to the concentration point for storage and routing to the destination point as soon as a communications path becomes available. Transmission cost is based on message length.
- **Packet switching**—A sophisticated means of maximizing transmission capacity of networks. This is accomplished by breaking a message into transmission units, called packets, and routing them individually through the network, depending on the availability of a channel for each packet. Passwords and all types of data can be included within the packet. The transmission cost is by packet and not by message, route or distance. Sophisticated error and flow control procedures are applied to each link by the network.
- **Circuit switching**—A physical communications channel is established between communicating equipment, through a circuit-switched network. This network can be, for instance, point-to-point (e.g., leased line) multipoint, a public-switched telephone network (PSTN) or an ISDN. The connection, once established, is used exclusively by the two subscribers for the duration of the call. The network does not provide any error or flow control on the transmitted data, so this task must be performed by the user.
- **Virtual circuits**—A logical circuit between two network devices that provides for reliable data communications. Two types are available—switched virtual circuits (SVCs) or permanent virtual circuits (PVCs). SVCs dynamically establish on-demand connectivity and PVCs establish an always-on connection.
- **WAN dial-up services**—Dial-up services using asynchronous and synchronous connectivity are widely available and well suited for organizations with a large number of mobile users. Their disadvantages are low bandwidth and limited performance.

WAN DEVICES

The following devices, typically operating at either the physical or data link layer of the OSI reference model, are specific to the WAN environment.

WAN switches are data link layer devices used for implementing various WAN technologies such as ATM, point-to-point frame relay and ISDN. These devices are typically associated with carrier networks providing dedicated WAN switching and router services to organizations via T-1/E-1 or T-3/E-3 connections.

Routers are devices that operate at the network layer of the OSI reference model and provide an interface between different network segments on an internal network or connects the internal network to an external network.

Modems (modulator/demodulator) are data communications equipment (DCE) devices that make it possible to use analog lines (generally, the public telephone network) as transmission media for digital networks. Modems convert computer digital

Figure 4.22—Transmission Media

Media	Use and Distance	Advantages	Disadvantages
Twisted Pair	• Used for short distances (< 200 feet [60.96 meters]) • Supports voice and data	• Cheap • Simple to install • Readily available • Simple to modify	• Easy to tap • Easy to splice • Cross talk • Interference • Noise
Coaxial cable	• Supports data and video	• Ease of installation • Straightforward • Readily available	• Thick • Expensive • Does not support many LANs • Distance sensitive • Difficult to modify
Fiber optics	• Used for long distances • Supports voice, data, image and video	• High bandwidth capabilities • Secure • Difficult to tap • No cross talk • Smaller and lighter than copper	• Expensive • Hard to splice • Difficult to modify
Radio systems	• Used for short distances	• Cheap	• Easy to tap • Interference • Noise
Microwave radio systems	• Line-of-sight carrier for voice and data signals	• Cheap • Simple to install • Available	• Easy to tap • Interference • Noise
Satellite radio link systems	• Uses transponders to send information	• High bandwidth and different frequencies	• Interference • Noise • Easy to tap

signals into analog data signals and analog data back to digital. When a link is established, modems operating at both ends of it automatically negotiate the fastest and safest standard that the line and the modems themselves can use, establishing speed, parity, cryptographic algorithm and compression.

For transmission purposes, modems disassemble bytes into a sequence of bits that are sent sequentially to the line. At the receiving end, these bits must be reassembled into bytes.

A main task of the modems at both ends is to maintain their synchronization so the receiving device knows when each byte starts and ends. Two methods can be used for this purpose:

- Synchronous transmission—Bits are transmitted without interruption at a constant speed. The sending modem uses a specific character when it starts transmitting a data block to "synchronize" the receiving device. This mode allows maximum efficiency, but only if blocks are not too short. Specific technical rules must be observed to maintain synchronization.
- Asynchronous transmission—The transmitting device marks the beginning and end of a byte by sending a "start" and a "stop" bit before and after each data byte. The efficiency of the line is lower, but the asynchronous standard is simpler and works well for character and block mode transmissions.

Communication links can be operated both ways. See **figure 4.23**.

Access servers provide centralized access control for managing remote access dial-up services.

Channel service unit/digital service unit (CSU/DSU) interfaces at the physical layer of the OSI reference model, data terminal equipment (DTE) to DCE, for switched carrier networks.

Multiplexors are physical layer devices used when a physical circuit has more bandwidth capacity than required by individual signals. The multiplexor can allocate portions of its total bandwidth and use each portion as a separate signal link. It can also link several low-speed lines to one high-speed line to enhance transmission capabilities.

Methods for multiplexing data include the following:

- **Time-division multiplexing (TDM)**—Information from each data channel is allocated bandwidth, based on preassigned time slots, regardless of whether there are data to transmit.
- **Asynchronous time division multiplexing (ATDM)**—Information from data channels is allocated bandwidth as needed via dynamically assigned time slots.
- **Frequency division multiplexing (FDM)**—Information from each data channel is allocated bandwidth, based on the signal frequency of the traffic.
- **Statistical multiplexing**—Bandwidth is allocated dynamically to any data channels that have the information to transmit.

WAN TECHNOLOGIES

Some common types of WAN technologies used to manage the communication links are described in the following sections.

Point-to-point Protocol
Point-to-point protocol (PPP) works in the data link layer. PPP provides a single, preestablished WAN communication path from the customer premises to a remote network, usually reached through a carrier network such as a telephone company. PPP is a widely available remote access solution that supports asynchronous and synchronous links, and operates over a wide range of media. Because PPP is more stable than the older Serial Line Internet Protocol (SLIP), PPP is the Internet standard for transmission of IP packets over serial lines. PPP makes use of two primary protocols for operation. The first, Link Control Protocol (LCP), is used when establishing, configuring and testing the data link connection. The second, Network Control Protocol (NCP), establishes and configures different network layer protocols (e.g., Internetwork packet exchange [IPX]). PPP features include address notification, authentication, support for multiple protocols and link monitoring.

X.25
As a packet-switched or virtual-circuit implementation, X.25 is a telecommunication standard (ITU-T) that defines how connections between data terminal equipment and data communications or circuit terminating equipment are maintained for remote terminal access and computer communications in public data networks (PDNs). Developed in 1976, X.25 operates at the lower three layers of the OSI reference model, but is no longer widely available today, primarily because it is resource-intensive in providing error control capabilities.

Frame Relay
As a packet-switched or virtual-circuit implementation, Frame Relay is a data link layer protocol for switch devices that uses a standard encapsulation technique to handle multiple virtual circuits between connected devices. The encapsulation method is high-level data link control (HDLC) for synchronous serial links using frame characters and checksums. Frame Relay is more efficient than X.25, the protocol for which it is generally considered a replacement. Contrary to X.25, Frame Relay relies more on upper layer protocols for significant error handling processes in data transmissions. Frame Relay is a low-cost, widely available LAN technology used in WAN point-to-point connections.

Integrated Services Digital Network
As a circuit-switched implementation, ISDN corresponds to integrated voice, data and video and is an architecture for worldwide telecommunications. This service integrates voice, data and video communication through digital switching and transmission over digital public carrier lines. The ISDN technologies now implemented are narrowband (basic-rate and primary-rate, not aggregated) ISDN; broadband ISDN has never been widely implemented. Separate channels are used for customer information (i.e., B, bearer channels—voice, data and video) and to send signals and control information (i.e., D, data channels). ISDN uses a packet-node layered protocol, based on the CCITT's X.25 standard. Unlike Frame Relay, it is moderately available to all.

Asynchronous Transfer Mode
As a packet-switched implementation operating at the data link layer, ATM is based on the use of a cell (a fixed-size data block) switching and multiplexing technology standard that combines the benefits of circuit switching (guaranteed capacity and constant transmission delay) with those of packet switching (flexibility and efficiency for intermittent traffic). Because ATM is asynchronous, time slots are available on demand with information identifying the source of the transmission contained in the header of each ATM cell. ATM is considered relatively expensive as a dedicated leased line option in comparison to other available WAN options.

Multiprotocol Label Switching
Multiprotocol label switching (MPLS) provides a mechanism for engineering network traffic patterns that is independent of routing tables. MPLS assigns short labels to network packets that describe how to forward them through the network. MPLS is independent of any routing protocol and can be used for unicast packets.

In traditional Level 3 forwarding, as a packet travels from one router to the next, an independent forwarding decision is made at each hop. The IP network layer header is analyzed, and the next hop is chosen based on this analysis and on the information in the routing table. In an MPLS environment, the analysis of the packet header is performed just once, when a packet enters the MPLS cloud. The packet is then assigned to a stream, which is identified by a label—a short (20-bit), fixed-length value at the front of the packet. Labels are used as lookup indexes into the label forwarding table. This table stores forwarding information for each label. Additional information, such as class-of-service (CoS) values, which can be used to prioritize packet forwarding, can be associated with a label.

Digital Subscriber Lines
Digital subscriber lines (DSL) is a network provider service using modem technology over existing twisted-pair telephone lines to transport high-bandwidth data such as multimedia and video. Characteristics of DSL include:
- Dedicated, point-to-point, public network access on the local loop. Local loops are generally the "last mile" between a network service provider's (NSP) central office and the customer site.
- Delivers high-bandwidth data rates to dispersed customers at low cost through the existing telecommunications infrastructure
- Always-on access, which eliminates call setup and makes it ideal for Internet/intranet and remote LAN access

DSL services vary in their speed and type of modulation:
- Asymmetric Digital Subscriber Line (ADSL)
- Symmetric Digital Subscriber Line (SDSL)
- High Bit-rate Digital Subscriber Line (HDSL)
- High Bit-rate Digital Subscriber Line version 2 (HDSL-2)
- Single-Pair High-speed Digital Subscriber Line (SHDSL)
- G.SHDSL (an international standard for symmetric DSL also known as G.001.2)
- Very High Speed Digital Subscriber Line(VDSL)

Virtual Private Networks

A VPN extends the corporate network securely via encrypted packets sent out via virtual connections over the public Internet to distant offices, home workers, salespeople and business partners. Rather than using expensive dedicated leased lines, VPNs take advantage of the public worldwide IP infrastructure, thereby enabling remote users to make a local call (versus dialing-in at long distance rates) or use an Internet cable modem or DSL connections for inexpensive public network connectivity.

VPNs are platform independent. Any computer system that is configured to run on an IP network can be connected through a VPN with no modifications, except for the installation of remote software.

There are three types of VPNs:
1. **Remote-access VPN**—Used to connect telecommuters and mobile users to the enterprise WAN in a secure manner; it lowers the barrier to telecommuting by ensuring that information is reasonably protected on the open Internet.
2. **Intranet VPN**—Used to connect branch offices within an enterprise WAN
3. **Extranet VPN**—Used to give business partners limited access to each other's corporate network; and example is an automotive manufacturer with its suppliers

The only difference between a traditional, intracompany VPN (intranet) and an intercompany VPN (extranet) is the way the VPN is managed. With an intranet VPN, all network and VPN resources are managed by a single organization. When an organization's VPN is used for an extranet, management control becomes weak. Therefore, it is recommended that in extranet VPN, each constituent company manage its own VPN and maintain control over it.

VPNs allow:
- Network managers to cost-efficiently increase the span of the corporate network
- Remote network users to securely and easily access their corporate enterprise
- Corporations to securely communicate with business partners
- Supply chain management to be efficient and effective
- Service providers to grow their businesses by providing substantial incremental bandwidth with value-added services

Determining which network resources should be linked via a VPN depends on the applications used on the various systems. Requirements often used to determine network connectivity include security policies, business models, intranet server access, application requirements, data sharing and application server access.

The process of encrypting packets, which makes VPN an effective protection scheme, uses the Internet Engineering Task Force's (IETF) IP Security (IPSec) standard. IPSec is implemented in two modes. The IPSec tunnel mode will encrypt the entire packet, including the header. The IPSec transport mode will encrypt only the data portion of the packet. A given VPN might use IPSec tunnel mode or might use IPSec transport mode with other encryption methods for the non-data parts of the packet.

Note: For the security implications of VPN and for information on IPSec encryption and VPN, see section 5.6.1 Auditing Remote Access.

Wireless Networks

Wireless technologies, in the simplest sense, enable one or more devices to communicate without physical connections (i.e., without requiring network or peripheral cabling). Wireless is a technology that enables organizations to adopt e-business solutions with tremendous growth potential. Wireless technologies use radio frequency transmissions/electromagnetic signals through free space as the means for transmitting data, whereas wired technologies use electrical signals through cables. Wireless technologies range from complex systems (such as wireless wide area networks [WWANs], wireless local area networks [WLANs] and cell phones) to simple devices (such as wireless headphones, microphones and other devices that do not process or store information). They also include Bluetooth devices with a miniradio frequency transceiver and infrared devices, such as remote controls, some cordless computer keyboards and mice, and wireless Hi-Fi stereo headsets, all of which require a direct line of sight between the transmitter and the receiver to close the link.

However, going wireless introduces new elements that must be addressed. For example, existing applications may need to be retrofit to make use of wireless interfaces. Also, decisions need to be made regarding general connectivity—to facilitate the development of completely wireless mobile applications or other applications that rely on synchronization of data transfer between mobile computing systems and corporate infrastructure. Other issues include narrow bandwidth, the lack of a mature standard, and unresolved security and privacy issues.

Wireless networks serve as the transport mechanism between devices, and among devices and the traditional wired networks. Wireless networks are many and diverse but are frequently categorized into four groups based on their coverage range:
- WANs
- LANs
- Wireless personal area networks (WPANs)
- Wireless *ad hoc* networks

WIRELESS WIDE AREA NETWORKS

Wireless wide area networking is the process of linking different networks over a large geographical area to allow wider IT resource sharing and connectivity. While computers are often connected to traditional WANs using cable networking solutions (such as telephone systems), wireless wide area networks are connected via radio, satellite and mobile phone technologies.

WWANs, using radio, satellite and mobile phone technologies, can complement and compete with more traditional systems of cable-based networking. These include wide coverage area technologies such as Long-term Evolution (LTE), Worldwide Interoperability for Microwave Access (WiMAX), Cellular Digital Packet Data (CDPD), global system for mobile communications (GSM) and Mobitex.

For some organizations, such as those in rural areas where laying cable is too expensive, wireless technology offers the only networking solution. For others, wireless wide area networking provides greater system flexibility, as well as the opportunity to control costs where the equipment is owned.

Implementing a WWAN requires careful attention to the planning and surveying of the network. The total cost of ownership involved in switching to this rapidly evolving system of networking should also be considered.

WIRELESS LOCAL AREA NETWORKS

WLANs allow greater flexibility and portability than traditional wired LANs. Unlike a traditional LAN, which requires a wire to connect a user's computer to the network, a WLAN connects computers, tablets, smartphones and other components to the network using an access point device. An access point, or wireless networking hub, communicates with devices equipped with wireless network adaptors within a specific range of the access point; it connects to a wired Ethernet LAN via an RJ-45 port. Access point devices typically have coverage areas of up to 300 feet (approximately 100 meters). This coverage area is called a cell or range. Users move freely within the cell with their laptop or other network devices. Access point cells can be linked together to allow users to roam within a building or between buildings. WLAN includes 802.11, HyperLAN, HomeRF and several others. WLANs are commonly referred to as Wi-Fi hotspots.

WLAN technologies conform to a variety of standards and offer varying levels of security features. The principal advantages of standards are to encourage mass production and to allow products from multiple vendors to interoperate. The most useful standard used currently is the IEEE 802.11 standard.

> **Note:** The CISA candidate will not be tested on these IEEE standards in the exam.

802.11 refers to a family of specifications for WLAN technology. 802.11 specifies an over-the-air interface between a wireless client and a base station or between two wireless clients.

WIRED EQUIVALENT PRIVACY AND WI-FI PROTECTED ACCESS (WPA/WPA2)

IEEE 802.11's Wired Equivalent Privacy (WEP) encryption uses symmetric, private keys, which means the end user's radio-based NIC and access point must have the same key. This leads to difficulties periodically involved with distributing new keys to each NIC. As a result, keys remain unchanged on networks for extended times. With static keys, several hacking tools easily break through the relatively weak WEP encryption mechanisms.

Because of the key reuse problem and other flaws, the current standardized version of WEP does not offer strong enough security for most corporate applications. Newer security protocols, such as 802.11i (WPA2) and Wi-Fi Protected Access (WPA), however, utilize public key cryptography techniques to provide effective authentication and encryption between users and access points.

WIRELESS PERSONAL AREA NETWORKS

WPANs are short-range wireless networks that connect wireless devices to one another. The most dominant form of WPAN technology is Bluetooth, which links wireless devices at very short distances. The oldest way to connect devices in a WPAN fashion is IR communications.

Bluetooth is an open source standard that borrows many features from existing wireless standards, such as IEEE 802.11, IrDA, Digital Enhanced Cordless Telecommunications (DECT), Motorola's Piano and TCP/IP, to connect portable devices without wires, via short-range radio frequencies (RF).

Bluetooth is a wireless protocol that connects devices within a range of up to 49 feet (15 meters) and has become a feature on some tablets, mobile phones, PC keyboards, mice, printers, etc. It is a system that changes frequencies from moment to moment using a technique called frequency-hopping. Bluetooth is used in computer systems, especially laptops, as a replacement for physical cables and for infrared connections, which are limited to line of sight. Bluetooth devices find one another when they are in range and automatically set up a background connection.

Bluetooth allows for high data speeds (between 1 Mbps and 2 Mbps), but is designed only for peer-to-peer data transfer. An alternative form of WPAN technology, called ZigBee, offers slower data speeds (250 Kbps) than Bluetooth, but is both cheaper than Bluetooth and requires far less energy to power.

AD HOC NETWORKS

Ad hoc networks are networks designed to dynamically connect remote devices such as mobile phones, laptops and tablets. These networks are termed *ad hoc* because of their shifting network topologies. Whereas WLANs or WPANs use a fixed network infrastructure, *ad hoc* networks maintain random network configurations, relying on a system of mobile routers connected by wireless links to enable devices to communicate. Bluetooth networks can behave as *ad hoc* networks, as mobile routers control the changing network topologies of these networks. The routers also control the flow of data between devices that are capable of supporting direct links to each other. As devices move about in an unpredictable fashion, these networks must be reconfigured to handle the dynamic topology. The routing protocol employed in Bluetooth allows the routers to establish and maintain these shifting networks.

The mobile router is commonly integrated in a handheld device. This mobile router, when configured, ensures that a remote, mobile device, such as a mobile phone, stays connected to the network. The router maintains the connection and controls the flow of communication.

INTERNET ACCESS ON MOBILE DEVICES

Smartphones and other mobile devices access the Internet by connecting to WLANs. These devices can also connect to the Internet over mobile networks.

Wireless Application Protocol (WAP) was the first protocol to enable this, connecting to the Internet over the second generation but first digital (2G) mobile network using Wireless Markup Language (WML). This has since been superseded by both third generation (3G) and fourth generation (4G) networks. 3G brought advances in Internet access times and download speeds. 4G is IP packet-switched network that adds Voice-over IP (VoIP) and mobile TV as well as further speed increases. These developments have also lead to changes in the way Internet content is accessed with applications (apps) being supported along with access through an Internet browser.

The following are general issues and exposures related to wireless and/or mobile access:

- **The interception of sensitive information**—Information is transmitted through the air, which increases the potential for unprotected information to be intercepted by unauthorized individuals.
- **The loss or theft of devices**—Devices tend to be relatively small, making them much easier to steal or lose.
- **The loss of data contained in the devices**—Theft or loss can result in the loss of data that has been stored on the device. This could be several gigabytes depending on the capacity of the device. If encryption is weak or not applied, a hacker may access the information because it may only be protected by a password or personal identification number (PIN).
- **The misuse of devices**—Devices can be used to gather information or intercept information that is being passed over wireless networks for financial or personal benefit.
- **Distractions caused by the devices**—The use of the devices may distract the user. If these devices are being used in situations where an individual's full attention is required (e.g., driving a car), they could result in an increase in the number of accidents.
- **Possible health effects of device usage**—The safety or health hazards have not, as yet, been identified. However, there are a number of concerns with respect to electromagnetic radiation, especially for those devices that must be held beside the head.
- **OS vulnerabilities**—The OS may contain vulnerabilities which allow access to the device. Vulnerabilities, for example, allow devices to be jail broken.
- **Applications**—Apps may contain vulnerabilities or malicious code which could allow access to data and the device itself. Jail broken devices may be more susceptible to this because the apps may not come from secure sources.
- **Wireless user authentication**—There is a need for stronger user authentication and authorization tools at the device level. The current technology is just emerging.
- **File security**—Wireless phones and tablets do not use the type of file access security that other computer platforms can provide.
- **WEP security encryption**—WEP security depends particularly on the length of the encryption key and on the usage of static WEP (many users on a WLAN share the same key) or dynamic WEP (per-user, per-session, dynamic WEP key tied to the network logon). The 64-bit encryption keys that are in use in the WEP standard encryption can be easily broken by the currently available computing power. Static WEP, used in many WLANs for flexibility purposes, is a serious security risk, because a static key can easily be lost or broken, and once this has occurred, all of the information is available for viewing and use. An attacker possessing the WEP key could also sniff packets being transmitted and decrypt them. WEP is rarely used today because it has been deprecated. WPA2 is the preferred solution for wireless networks.

Public "Global" Internet Infrastructure

The Internet is comprised of networks distributed over the entire world and interconnected via pathways that allow the exchange of information, data and files. Being connected to the Internet means to be logically part of it. By using these pathways, a connected computer can send or receive packets of data to/from any other Internet device.

Today, the Internet is a vast, global network of networks, ranging from university networks to corporate LANs to large online services. The Internet is not run or controlled by any single person, group or organization. The only thing that is centrally controlled is the availability and assignment of Internet addresses and the attached symbolic host names. Addresses and names are used for locating the source or destination networks.

Users can access the Internet through wireless enabled devices or an ISP. Routers, which connect networks, perform most of the work of directing traffic on the Internet. Networks are connected in different ways including telephone lines, ISDN telephone lines, leased lines, fiber-optic cables and satellite.

The networks in a particular geographic area are connected into a large regional network. Regional networks are connected to one another via high-speed backbones (connections that can send data at extremely high speeds). When data are sent from one regional network to another, they first travel to a network access point (NAP). NAPs then route the data to high-speed backbone network services (BNS). The data are then sent along the backbone to another regional network and then to a specific network and computer within that regional network.

TCP/IP AND ITS RELATION TO THE OSI REFERENCE MODEL

The protocol suite used as the *de facto* standard for the Internet is known as the TCP/IP. The TCP/IP suite includes both network-oriented protocols and application support protocols. **Figure 4.23** shows some of the standards associated with the TCP/IP suite and where these fit within the ISO model. It is interesting to note that the TCP/IP set of protocols was developed before the ISO/OSI framework; therefore, there is no direct match between the TCP/IP standards and the layers of the framework.

TCP/IP INTERNET WORLD WIDE WEB SERVICES

The most common way a user accesses a resource on the Internet is through the TCP/IP Internet World Wide Web (WWW) application service.

The **URL** identifies the address on the WWW where a specific resource is located. To access a web site, a user enters the site's location into their browser's URL space, or they click on the hypertext link that will send them to the location. The web browser looks up the IP address of the site, and sends a request for the URL via the HTTP. This protocol defines how the web browser and web server communicate with one another.

URLs contain several parts, as seen in **figure 4.24**.

Figure 4.23—OSI Association With the TCP/IP Suite

	OSI Model	TCP/IP Conceptual Layers	Protocol Data Unit (PDU)	TCP/IP Protocols	Equipment	Layer Functions	Layer Functions
7	Application	Application	Data	HTTP File Transport Protocol (FTP) Simple Mail Transport Protocol (SMTP) TFTP NFS Name Server Protocol (NSP) Simple Network Management Protocol (SNMP) Remote Terminal Control Protocol (Telnet) LPD X Windows DNS DHCP/BootP	Gateway	Provides user interface	File, print, message, database, and application services
6	Presentation					Presents data Handles processing such as encryption	Data encryption, compression and translation services
5	Session					Keeps separate the data of different applications	Dialog control
4	Transport	Transport	Segment	Transmission Control Protocol (TCP) User Datagram Protocol (UDP)	Layer 4 switch	Provide reliable or unreliable delivery	End-to-end connection
3	Network	Network interface	Packet	ICMP ARP RARP Internet Protocol (IP)	Route Layer 3 switch	Provides logical addressing which routers use for path determination	Routing
2	Data link	LAN or WAN interface	Frame	Ethernet Fast Ethernet FDDI Token Ring Point-to-point Protocol (PPP)	Layer 2 switch Bridge Wireless AP NIC	Combines packets into bytes and bytes into frames Provides access to media using MAC address Performs error detection, not error correction	Framing
1	Physical		Bits		Hub Repeater NIC	Moves bits between devices Specifies voltage, wire speed and pin-out of cables	Physical topology

Figure 4.24—Components of a URL

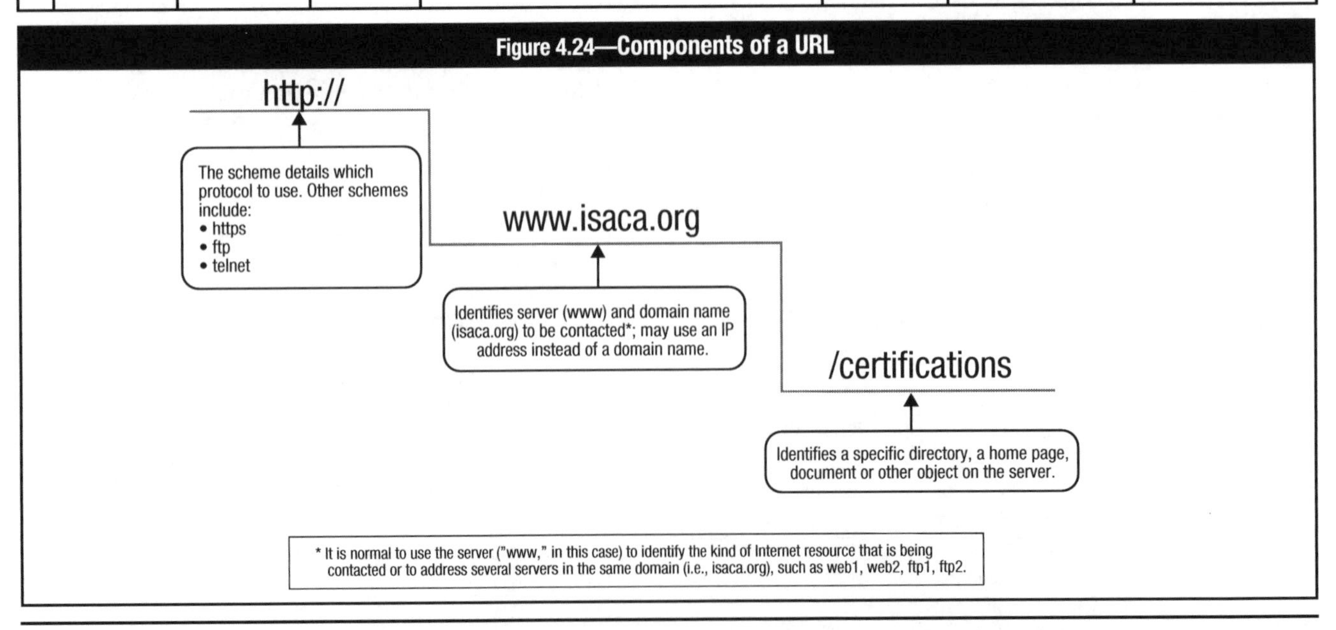

http://
The scheme details which protocol to use. Other schemes include:
• https
• ftp
• telnet
www.isaca.org
Identifies server (www) and domain name (isaca.org) to be contacted*; may use an IP address instead of a domain name.
/certifications
Identifies a specific directory, a home page, document or other object on the server.
* It is normal to use the server ("www," in this case) to identify the kind of Internet resource that is being contacted or to address several servers in the same domain (i.e., isaca.org), such as web1, web2, ftp1, ftp2.

A URL can also be used to access other TCP/IP Internet services:
- *ftp://isaca.org*
- *telnet://isaca.org*

The URL is the location of specific resources (e.g., pages, data) or services on the Internet. In the example, the resource is a web page called "certification" and is found on the web server of ISACA. This request is sent over the Internet and the routers transfer the request to the addressed web server, which activates the HTTP protocol and processes the request. When the server finds among its resources the requested home page, document or object, it sends the request back to the web browser. In the case of an HTML page, the information sent back contains data and formatting specifications. These are in the form of a program that is executed by the client web browser and produce the screen displayed for the user. After the page is sent by the server, the HTTP connection is closed and can be reopened. **Figure 4.25** displays the path.

Common gateway interface (CGI) scripts are an executable, machine-independent software program run on the server that can be called and executed by a web server. CGI scripts perform a specific set of tasks, such as processing input received from a client who typed information into a form on a web page. CGI scripts are coded in languages such as PERL or C. Note that CGI scripts need to be closely evaluated as they are run in the server; a bug in the scripts may allow a user to get unauthorized access to the server and, from there, eventually to the organization's network.

A **cookie** is a message stored by the web browser for the purpose of identifying users and possibly preparing customized web pages for them. Depending on the browser, the implementation may vary, but the process is as follows. When entering a web site that uses cookies for the first time, the user may be asked to go through a registration process such as filling out a form that provides information, including name and interests. The web server will send back a cookie with information (text message in HTTP header), which will be kept as a text message by the browser. Afterward, whenever the user's browser requests a page from that particular server, the cookie's message is sent back to the server so that the customized view, based on that user's particular interests and preferences, can be produced. Cookies are a very important functionality because the HTTP protocol does not natively support the concept of a session. Cookies allow the web server to discern whether a known or new user is connected and to keep track of information previously sent to that user. The browser's implementation of cookies has, however, brought several privacy and security concerns, allowing breaches of security and the theft of personal information (e.g., user passwords that validate the user's identity and enable restricted web services).

Applets are programs written in a portable, platform-independent computer language, such as Java, JavaScript or Visual Basic. Applets expose the user's machine to risk if the applets are not properly controlled by the browser. For example, the user's browser should be configured to not allow an applet to access a machine's information without prior authorization of the user.

Servlets are Java applets or small programs that run within a web server environment. A Java servlet is similar to a CGI program. Unlike a CGI program, once it is started, it stays in memory and can fulfill multiple requests, thereby saving server execution time and speeding up the services.

A **bookmark** is a marker or address that identifies a document or a specific place in a document.

GENERAL INTERNET TERMINOLOGY

The following terms are related to the use of the Internet:
- **Direct connection**—LANs or large computers, such as mainframes, that can be directly connected to the Internet. When a LAN is connected to the Internet, all the computers on the network can have full access to the Internet.

Figure 4.25—Components of a URL

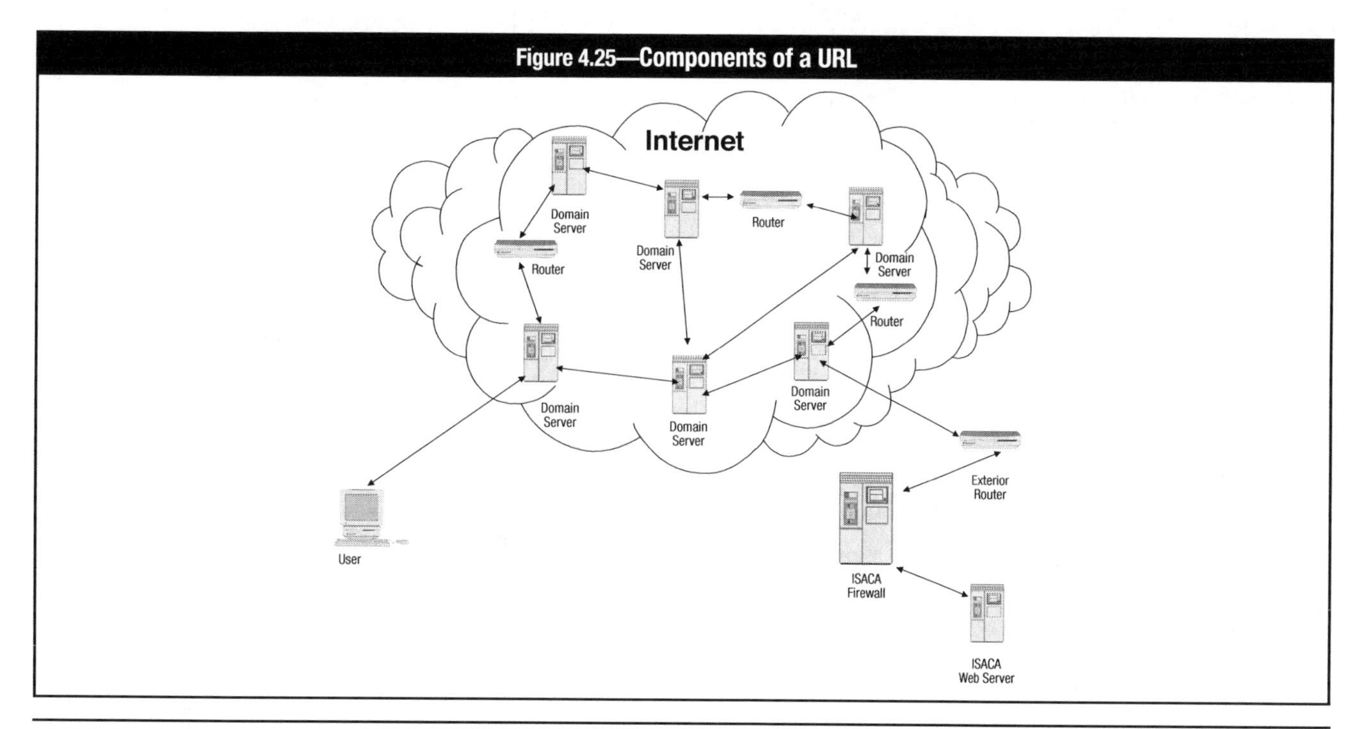

- **Domain name system (DNS)**—A hierarchical database that is distributed across the Internet that allows names to be resolved into IP addresses (and vice versa) to locate services such as web and email servers.
- **File Transfer Protocol (FTP)**—A protocol used to transfer files over a TCP/IP network (Internet, UNIX, etc.). These files can be of many types, including programs that the user can run on their computer—files with graphics, sounds and music, or text files that can be read. Most Internet files are downloaded using FTP. FTP can also be used to upload files from the computer to another computer on the Internet. To log onto an FTP site and download files, an account (or user name) and a password may need to be entered before the server or system allows the user to download or upload files. Some sites allow anyone to enter and download files. These sites are often referred to as anonymous FTP sites. As the definition suggests, anonymous FTP requires only a fictitious ID and password to transfer files. Anonymous FTP sites can be potentially dangerous if the network administrator setting up the site does not fully understand the risk associated with anonymous FTP. If file permissions have not been specified, the anonymous FTP user could also freely upload files to the server, introducing new files or changing existing files.
- **Internet link**—The connection between Internet users and the Internet service provider
- **Internet Service Provider (ISP)**—A third party that provides individuals and enterprises with access to the Internet and a variety of other Internet-related services.
- **Network Access Point (NAP)**—A traffic concentration spot, usually the point of convergence for Internet access by many Internet service providers
- **Online services**—All of the major online services allow users to tap the full power of the Internet. No special setup is required. When users dial into the online services, they are able to use the Internet resources, including browsing the World Wide Web.
- **Remote Terminal Control Protocol (Telnet)**—A standard terminal emulation protocol used for remote terminal connections, enabling users to log into remote systems and use resources as if they were local. An IS auditor should note that standard Telnet traffic is not encrypted by default, and consider this risk for any production Telnet use.
- **Secure Shell (SSH)**—Network protocol that uses cryptography to secure communication, remote command line login and remote command execution between two networked computers
- **Simple Mail Transport Protocol (SMTP)**— The standard email protocol on the Internet

TRANSBORDER DATA FLOW

Transborder data flow refers to data transmission between two countries. Information such as email, invoices, payment advice, etc., can be transmitted via sub-oceanic cables, telephone, television links and satellites. The selection of transmission alternatives should consider cost and possible transmission delays. The country of origin or the country of destination could have several laws applicable to transborder data flow that should be addressed. Legal compliance and protection, as well as data security and integrity, are a concern with transborder transmissions.

Privacy also is an issue because laws regarding protection and access to personal information may be different or conflicting between the source and destination countries.

Some countries also have laws concerning the encryption of data/information sent via transborder communications, thereby affecting the security and protection of data that may be exchanged between countries.

This is a particularly important issue in Internet communications, because the itinerary of the information is determined by the routers, is not fixed and, therefore, may cross a country border even while connecting two computers located in the same country.

Network Administration and Control

Network administration ensures that the network is functioning properly from a performance and security perspective. These duties include monitoring usage and throughput, load balancing, reacting to security violations and failure conditions, saving and restoring data, and making changes for scalability as the network usage grows. Therefore, an appropriate knowledge of network structure and topology, the protocols used, and the available administration tools is required.

The software used to monitor the network and enact changes should be accessible to the network administrator only. This software is the network OSs software associated with specific network devices, principally switches and routers.

The network OSs provide many functions aimed at shaping the network as a unified, controlled and uniform computing environment, including:
- Supporting local and remote terminal access to hosts and servers
- Supporting sharing of common network resources, such as file and print services
- Establishing links to hosts and servers

Network OSs have the following user-oriented features:
- Allow transparent access to the various resources of the network hosts.
- Check the user authorization to particular resources.
- Mediate and simplify the access to remote resources as easily as local resources.
- Establish uniform logon and logging procedures throughout the network.
- Make available up-to-the-minute online network documentation.
- Permit more reliable operation than possible on a single host or server, particularly when groups of equivalent hosts are used.

NETWORK PERFORMANCE METRICS

The major network performance metrics are latency and throughput. Network error counts and number of retransmissions are also measured to understand network performance.
- **Latency**—The delay that a message or packet will experience on its way from source to destination. Latency appears because the information needs to cross through different devices (switching and routing times) and, to a lesser extent, because

signals must travel some distance (propagation delay). When a network device is busy, the packets either must wait, be queued in a buffer or be dropped. A very easy way to measure latency in a TCP/IP network is to use the ping command.
- **Throughput**—The quantity of useful work made by the system per unit of time. In telecommunications, it is the number of bytes per second that are passing through a channel.

NETWORK MANAGEMENT ISSUES

It is much more common today to see WANs communicating with a mix of LAN and host systems network architecture (SNA) traffic, or pure LAN-oriented traffic. Almost all organizations are standardizing their telecommunications, infrastructure on TCP/IP and modern routers.

This trend to a different technical design approach is also made evident by the specific name (i.e., WAN) that designates telecommunication networks in a TCP/IP environment. A WAN needs to be monitored and managed similarly to a LAN. ISO, as part of its communications modeling effort (ISO/IEC 10040), has defined five basic tasks related to network management:
- **Fault management**—Detects the devices that present some kind of technical fault
- **Configuration management**—Allows users to know, define and change, remotely, the configuration of any device
- **Accounting resources**—Holds the records of the resource usage in the WAN (who uses what)
- **Performance management**—Monitors usage levels and sets alarms when a threshold has been surpassed
- **Security management**—Detects suspicious traffic or users, and generates alarms accordingly

NETWORK MANAGEMENT TOOLS

In an organization's modern inter-networking environment, all of the above tasks could be accomplished by a set of tools generically called network management tools.

Response time reports identify the time necessary for a command entered by a user at a terminal to be answered by the host system. Response time is important because end users experiencing slow response time will be reluctant to utilize IS resources to their fullest extent. These reports typically identify average, worst and best response times over a given time interval for individual telecommunication lines or systems. These reports should be reviewed by IS management and system support personnel to track potential problems. If response time is slow, all possible causes, such as I/O channel bottlenecks, bandwidth utilization and CPU capacity, should be investigated; various solutions should be analyzed; and an appropriate and cost-justified corrective action should be taken.

Downtime reports track the availability of telecommunication lines and circuits. Interruptions due to power/line failure, traffic overload, operator error or other anomalous conditions are identified in a downtime report. If downtime is excessive, IS management should consider the following remedies:
- Adding or replacing telecommunications lines
- Switching to a more dependable transmission link (such as dedicated lines versus shared lines)
- Installing backup power supplies
- Improving access controls
- Closely monitoring line utilization to better forecast user needs, both in the near and long term

Online monitors check data transmission accuracy and errors. Monitoring can be performed by echo checking (received data are bounced back to sender for verification) and status checking all transmissions, ensuring that messages are not lost or transmitted more than once.

Network monitors provide a real time display of network nodes and status.

Network (Protocol) analyzers are diagnostic tools attached to a network link that use network protocols' intelligence for monitoring the packets flowing along the link and produce network usage reports. Network analyzers are typically hardware-based and operate at the data link and/or network level. Output includes the following information:
- Protocol(s) in use
- The type of packets flowing along the monitored link
- Traffic volume analysis
- Hardware errors, noise and software problems
- Other performance statistics (e.g., percentage of used bandwidth)
- Problems and possible solutions

Simple Network Management Protocol (SNMP) is a TCP/IP-based protocol that monitors and controls different variables throughout the network, manages configurations, and collects statistics on performance and security. A master console polls all the network devices on a regular basis and displays the global status. SNMP software is capable of accepting, in real-time, specific operator requests. Based on the operator instructions, SNMP software sends specific commands to an SNMP-enabled device and retrieves the required information. To perform all of these tasks, each device (routers, switches, hubs, PCs, servers) needs to have a SNMP agent running. The actual SNMP communications occur between all the agents and the console.

Help desk reports are prepared by the help desk, which is staffed or supported by IT technicians trained to handle problems occurring during normal IS usage. If an end user encounters any problem, he/she can contact the help desk for assistance. Help desk facilities are critical to the telecommunication environment since they provide end users with an easy means of identifying and resolving problems quickly, before they have a major impact on IS performance and end-user resource utilization. Reports prepared by the help desk provide a history of the problems and their resolution.

Applications in a Networked Environment

There are different types of applications used in a networked architecture.

CLIENT-SERVER TECHNOLOGY

Client-server is a network architecture in which each computer or process on the network is either a server (a source of services and data) or a client (a user of these services and data that relies on servers to obtain them). In a client-server technology,

the available computing power can be distributed and shared among the client workstations. Use of client-server technology is one of the most popular trends in building applications aimed at networked environments. Often, in a client-server network environment, the server provides data distribution and security functions to other computers that are independently running various applications.

The client-server architecture has a number of advantages, such as distributing the work among servers and performing as much computational work as possible on the client workstation to save bandwidth and server computing power. Important tasks, such as manipulating and changing data, may be performed locally and without the need for controlling resources on the main processing unit. In this way, the applications may run more efficiently.

To achieve these advantages, client-server application systems are divided into separate pieces or tasks. The systems are split so that processing may take place on different machines (e.g., servers and clients). Each processing component is mutually dependent on the others. That tasks are performed on both client and server is the main difference between client-server processing and the traditional mainframe/distributed processing.

The typical client is a single PC or workstation. Presentation usually is provided by visually enhanced processing software, known as a graphical user interface (GUI). Clients may be thick or thin. A thin client (sometimes called a lean client) is a client computer or client software that depends primarily on the central server for processing activities and mainly focuses on conveying input and output between the user and the remote server. Many thin client devices run only web browsers or remote desktop software, meaning that all significant processing occurs on the server. In contrast, a thick or fat client does as much processing as possible and passes only data for communications and storage to the server.

The server is one or more multiuser computers. Server functions include any centrally supported role such as file sharing, printer sharing, database access and management, communication services, email services, and processing application logic. Multiple functions may be supported by a single server.

Client-server architecture can be two-tiered which is normally composed of:
- A thick client, focused on GUI tasks and running the application logic
- A group (one or more) of database servers

The main disadvantages of this model are the requirement to keep the programs on the clients synchronized (ensuring that they are running the same logic) and its scalability.

Client-server architecture is more normally based on (at least) three levels of computing tasks (i.e., three-tier architectures). A three-tier architecture is composed of:
- A thin client, focused on GUI tasks (most often but not always web browsers)
- A group (one or more) of application servers, focused on running the application logic
- A group (one or more) of database servers

This architecture does not have the limitations of two-tier applications and has other advantages such as:
- Thin clients, which are less complex and less costly to buy and maintain
- More scalability (up to several thousands of concurrent users) because the load is balanced among different servers. This, in turn, improves overall system performance and reliability since more of the processing load can be accommodated simultaneously.
- Can be implemented in applications for internal usage only or in e-business applications (in this case, there could be another tier represented by the web server)
- All of the program logic is separated from the rest of the code (via application servers)

Designs that contain more than two tiers are referred to as *multi*-tiered or *n*-tiered. *N*-tiered architecture applications are more complex to build and more difficult to maintain.

In an *n*-tired environment, each instance of the client software can send data requests to one or more connected servers. In turn, the servers can accept these requests, process them and return the requested information to the client. This concept can be applied to many different kinds of applications the architecture remaining fundamentally the same. The interaction between client and server is often described using sequence diagrams. Sequence diagrams are standardized in the Unified Modeling Language.

Note: Implicit in n-tiered architectures is the presence of middleware that supports not just the communications between clients and servers, but the more advanced features such as load balancing and fail over, dynamic location of components, and establishing synchronous connections or asynchronous queue-based messages.

MIDDLEWARE

Middleware is a client-server-specific term used to describe a unique class of software employed by client-server applications. Middleware serves as the glue between two otherwise distinct applications and provides services such as identification, authentication, authorization, directories and security. This software resides between an application and the network and manages the interaction between the GUI on the front end and data servers on the back end. Middleware facilitates the client-server connections over the network, and allows client applications to access and update remote databases and mainframe files.

Middleware is commonly used for:
- **Transaction processing (TP) monitors**—Programs that handle and monitor database transactions, and are used primarily for load balancing
- **Remote procedure calls (RPC)**—A protocol that enables a program on the client computer to execute another program on a remote computer (usually a server)
- **Object request broker (ORB) technology**—The use of shared, reusable business objects in a distributed computing environment
  - This provides the ability to support interoperability across languages and platforms, as well as enhance maintainability and adaptability of the system. Examples of such technologies are CORBA and Microsoft's COM/DCOM.

- **Messaging servers**—Programs which asynchronously prioritize, queue and/or process messages using a dedicated server

Risk and controls associated with middleware in a client-server environment are:
- **Risk**—System integrity may be adversely affected because of the very purpose of middleware, which is intended to support multiple operating environments interacting concurrently. Lack of proper software to control portability of data or programs across multiple platforms could result in a loss of data or program integrity.
- **Controls**—Management should implement compensating controls to ensure the integrity of the client-server networks. Management should ensure that systems are properly tested and approved, modifications are adequately authorized and implemented, and appropriate version control procedures are followed.

On-demand Computing

On-demand computing (ODC), also referred to as utility computing, is a computing model in which information system resources are allocated to users according to their current needs. The resources could be available within an organization or supplied by a third-party service provider. At any moment, a user (or organization) may need more bandwidth, CPU cycles, memory, application availability or other resource to a greater degree than another user. When that situation occurs, the resource can be made available to the user with the immediate need and taken away from the user with the lesser need.

A benefit of ODC is that an organization that is outsourcing its computing needs does not have to pay for excess computing capacity. A concern is the confidentiality of information maintained by the third-party provider.

4.7 AUDITING INFRASTRUCTURE AND OPERATIONS

The changing technological infrastructure and the manner in which to operate it have led to evolving ways to perform audits and specific reviews of hardware, OSs, databases, networks, IS operations and problem management reporting. The following sections enumerate important areas to be reviewed while performing an audit of these areas.

4.7.1 ENTERPRISE ARCHITECTURE AND AUDITING

Enterprise architecture (EA) involves documenting an organization's IT assets in a structured manner to facilitate understanding, management and planning for IT investments. An EA often involves both a current state and optimized future state representation (e.g., a road map).

EA for IT is a description of the fundamental underlying design of the IT components of the business, the relationships among them and the manner in which they support the enterprise's objectives.

When auditing infrastructure and operations, the IS auditor should follow the overall EA and use the EA as a main source of information. Further, the IS auditor should ensure that the systems are in line with the EA and meet the organization's objectives.

4.7.2 HARDWARE REVIEWS

When auditing infrastructure and operations, hardware reviews should include the areas shown in **figure 4.26**.

4.7.3 OPERATING SYSTEM REVIEWS

When auditing operating software development, acquisition or maintenance, the details shown in **figure 4.27** should be considered.

4.7.4 DATABASE REVIEWS

When auditing a database, an IS auditor should review the design, access, administration, interfaces, portability and database supported IS controls, as shown in **figure 4.28**.

4.7.5 NETWORK INFRASTRUCTURE AND IMPLEMENTATION REVIEWS

The IS auditor should review controls over network implementations to ensure that standards are in place for designing and selecting a network architecture, and for ensuring that the costs of procuring and operating the network do not exceed the benefits.

The unique nature of each network makes it difficult to define standard audit procedures. Modern networks are mixed with several kinds of devices and topologies (PANs, LANs, WANs, WPANs, WLANs, VLANs, etc.).

To effectively perform a review, the IS auditor should identify the following:
- Network topology and network design
- Significant network components (servers, routers, switches, hubs, modems, wireless devices, etc.)
- Interconnected boundary networks
- Network uses (including significant traffic types and main applications used over the network)
- Network gateway to the Internet
- Network administrator and operator
- Significant groups of network users
- Defined security standards or procedures

In addition, the IS auditor should gain an understanding of the following:
- Functions performed by the network administrators and operators
- The company division or department procedures and standards relating to network design, support, naming conventions and data security
- Network transmission media and techniques, including bridges, routers, gateways, switches and other relevant components

Understanding the above information should enable the IS auditor to make an assessment of the significant threats to the network, together with the potential impact and probability of occurrence of each threat. Having assessed the risk to the network, the IS auditor should evaluate the controls used to minimize the risk.

Figure 4.26—Hardware Reviews

Areas to Review	Questions to Consider
Hardware acquisition plan	• Is the plan aligned with business requirements? • Is the plan aligned with the enterprise architecture? • Is the plan compared regularly to business plans to ensure continued synchronization with business requirements? • Is the plan synchronized with IS plans? • Have criteria for the acquisition of hardware been developed? • Is the environment adequate to accommodate the currently installed hardware and new hardware to be added under the approved hardware acquisition plan? • Are the hardware and software specifications, installation requirements and the likely lead time associated with planned acquisitions adequately documented?
Acquisition of hardware	• Is the acquisition in line with the hardware acquisition plan? • Have the IS management staff issued written policy statements regarding the acquisition and use of hardware, and have these statements been communicated to the users? • Have procedures and forms been established to facilitate the acquisition approval process? • Are requests accompanied by a cost-benefit analysis? • Are purchases routed through the purchasing department to streamline the process, avoid duplications, ensure compliance with tendering requirements and legislation and to take advantage of quantity and quality benefits such as volume discounts?
IT asset management	• Has the hardware been tagged? • Has an owner been designated? • Where will the hardware be located? • Have we retained a copy of the contracts/SLAs?
Capacity management and monitoring	• Are criteria used in the hardware performance monitoring plan based on historical data and analysis obtained from the IS trouble logs, processing schedules, job accounting system reports, preventive maintenance schedules and reports? • Is continuous review performed of hardware and system software performance and capacity? • Is monitoring adequate for equipment that has been programmed to contact its manufacturer (without manual or human intervention) in the case of equipment failure?
Preventive maintenance schedule	• Is the prescribed maintenance frequency recommended by the respective hardware vendors being observed? • Is maintenance performed during off-peak workload periods? • Is preventive maintenance performed at times other than when the system is processing critical or sensitive applications?
Hardware availability and utilization reports	• Is scheduling adequate to meet workload schedules and user requirements? • Is scheduling sufficiently flexible to accommodate required hardware preventive maintenance? • Are IS resources readily available for critical application programs?
• Problem logs • Job accounting system reports	• Have IS management staff reviewed hardware malfunctions, reruns, abnormal system terminations and operator actions?

Figure 4.27—Operating Systems Reviews

Areas to Review	Questions to Consider
• System software selection procedures	• Do they align with the enterprise architecture? • Do they comply with short- and long-range IS plans? • Do they meet the IS requirements? • Are they properly aligned with the objectives of the business? • Do they include IS processing and control requirements? • Do they include an overview of the capabilities of the software and control options?
• Feasibility study • Selection process	• Are same selection criteria applied to all proposals? • Has the cost-benefit analysis of system software procedures addressed: – Direct financial costs associated with the product? – Cost of product maintenance? – Hardware requirements and capacity of the product? – Training and technical support requirements? – Impact of the product on processing reliability? – Impact on data security? – Financial stability of the vendor's operations?
• System software security	• Have procedures been established to restrict the ability to circumvent logical security access controls? • Have procedures been implemented to limit access to the system interrupt capability? • Have procedures been implemented to manage software patches and keep the system software up-to-date? • Are existing physical and logical security provisions adequate to restrict access to the master consoles? • Were vendor-supplied installation passwords for the system software changed at the time of installation?
• IT asset management	• Has an owner been designated? • Have we retained a copy of the contracts/SLAs? • What is the license agreement? Are we in compliance with it?
• System software implementation	• Are controls adequate in: – Change procedures? – Authorization procedures? – Access security features? – Documentation requirements? – Documentation of system testing? – Audit trails? – Access controls over the software in production?
• Authorization documentation	• Have additions, deletions or changes to access authorization been documented? • Does documentation exist of any attempted violations? If so, has there been follow-up?
• System documentation	• Are the following areas adequately documented: – Installation control statements? – Parameter tables? – Exit definitions? – Activity logs/reports?
• System software maintenance activities	• Is documentation available for changes made to the system software? • Are current versions of the software supported by the vendor? • Is there a defined patching process?
• System software change controls	• Is access to the libraries containing the system software limited to individual(s) needing to have such access? • Are changes to the software adequately documented and tested prior to implementation? • Is software authorized properly prior to moving from the test environment to the production environment?
• Controls over the installation of changed system software	• Have all appropriate levels of software been implemented? • Have predecessor updates taken place? • Are system software changes scheduled for times when the changes least impact IS processing? • Has a written plan been established for testing changes to system software? • Are test procedures adequate to provide reasonable assurance that changes applied to the system correct known problems and that they do not create new problems? • Are tests being completed as planned? • Have problems encountered during testing been resolved and were the changes retested? • Have fallback or restoration procedures been put in place in case of production failure?

Figure 4.28—Database Reviews

Areas to Review	Questions to Consider
• Logical schema	• Do all entities in the entity-relation diagram exist as tables or views? • Are all relations represented through foreign keys? • Are constraints specified clearly? • Are nulls for foreign keys allowed only when they are in accordance with the cardinality expressed in the entity-relation model?
• Physical schema	• Has allocation of initial and extension space (storage) for tables, logs, indexes and temporary areas been executed based on the requirements? • Are indexes by primary key or keys of frequent access present? • If the database is not normalized, is justification accepted?
• Access time reports	• Are indexes used to minimize access time? • Have indexes been constructed correctly? • If open searches not based on indexes are used, are they justified?
• Database security controls	• Are security levels for all users and their roles identified within the database and access rights for all users and/or groups of users justified? • Do referential integrity rules exist and are they followed? • How is a trigger created and when does it fire? • Is there a system for setting passwords? Does change of passwords exist and is it followed? • How many users have been given system administrator privileges? Do these users require the privilege to execute their job function? • Has an auditing utility been enabled? Are audit trails being monitored? • Can database resources be accessed without using DBMS commands and SQL statements? • Is system administrator authority granted to job scheduler? • Are actual passwords embedded into database utility jobs and scripts? • Has encryption been enabled where required? • Are copies of production data authorized? • Are copies of production data altered or masked to protected sensitive data?
• Interfaces with other programs/software	• Are integrity and confidentiality of data not affected by data import and export procedures? • Have mechanisms and procedures been put in place to ensure the adequate handling of consistency and integrity during concurrent accesses?
• Backup and disaster recovery procedures and controls	• Do backup and disaster recovery procedures exist to ensure the reliability and availability of the database? • Are there technical controls to ensure high availability and/or fast recovery of the database?
• Database-supported IS controls	• Is access to shared data appropriate? • Are adequate change procedures utilized to ensure the integrity of the database management software? • Is data redundancy minimized by the database management system? Where redundant data exist, is appropriate cross-referencing maintained within the system's data dictionary or other documentation? • Is the integrity of the database management system's data dictionary maintained?
• IT asset management	• Has an owner been designated? • Have we retained a copy of the contracts/SLAs? • What is the license agreement? Are we in compliance with it?

Physical controls should protect network components (hardware and software) and the access points by limiting access to those individuals authorized by management. Unlike most mainframes, the computers in a mixed network are usually decentralized. Company data stored on a file server are easier to damage or steal than those on a mainframe, and they should be physically protected. The IS auditor should review the areas as shown in **figure 4.29**.

Figure 4.29—Network Infrastructure and Implementation Reviews

Areas to Review	Questions to Consider
Physical controls	
• Network hardware devices • File server • Documentation	• Are network hardware devices located in a secure facility and restricted to the network administrator? • Is the housing of network file servers locked or otherwise secured to prevent removal of boards, chips or the computer itself? • Is the device tagged where appropriate?
• Key logs	• Are the keys to the network file server facilities controlled to prevent the risk of unauthorized access? • Are keys assigned only to the appropriate people (e.g., the network administrator and support staff)? • Select a sample of keys held by people without authorized access to the network file server facilities and wiring closet in order to determine that these keys do not permit access to these facilities.
• Network wiring closet and transmission wiring	• Is the wiring physically secured? • Is the wiring labeled where appropriate?
Environmental controls	
• Server facility	• Are temperature and humidity controls adequate? • Have static electricity guards been put in place? • Have electric surge protectors been put in place? • Has a fire suppression system been put in place and is it tested/inspected regularly? • Are fire extinguishers located nearby and inspected regularly? • Are the main network components equipped with an uninterruptible power supply (UPS) that will allow the network to operate in case of minor power fluctuations or to be brought down gracefully in case of a prolonged power outage? • Has electromagnetic insulation been put in place? • Is the network components power supply properly controlled to ensure that it remains within the manufacturer's specifications? • Are the backup media protected from environmental damage? • Is the server facility kept free of dust, smoke and other matter, particularly food?
Logical security control	
• Passwords	• Are users assigned unique passwords? • Are users required to change the passwords on a periodic basis? • Are passwords encrypted and not displayed on the computer screen when entered?
• Network user access	• Is network user access based on written authorization and given on a need-to-know/need-to-do basis and based on the individual's responsibilities? • Are network workstations automatically disabled after a short period of inactivity? • Is remote access to the system supervisor prohibited? • Are all logon attempts to the supervisor account captured in the computer system? • Are activities by supervisor or administrative accounts subject to independent review? • Is up-to-date information regarding all communication lines connected to the outside maintained by the network supervisor?
• Network access change requests	• Are network access change requests authorized by the appropriate manager? Are standard forms used? • Are requests for additions, changes and deletions of network logical access documented?
• Test plans	• Are appropriate implementation, conversion and acceptance test plans developed for the organization's distributed data processing network, hardware and communication links?

Figure 4.29—Network Infrastructure and Implementation Reviews *(cont.)*	
Areas to Review	**Questions to Consider**
Logical security control ***(cont.)***	
• Security reports	• Is only authorized access occurring? • Are security reports reviewed adequately and in a timely manner? • In the case of unauthorized users, are follow-up procedures adequate and timely?
• Security mechanisms	• Have all sensitive files/datasets in the network been identified and have the requirements for their security been determined? • Are all changes to the OS software used by the network and made by IS management (or at user sites) controlled? Can these changes be detected promptly by the network administrator or those responsible for the network? • Do individuals have access only to authorized applications, transaction processors and datasets? • Are system commands affecting more than one network site restricted to one terminal and to an authorized individual with an overall network control responsibility and security clearance? • Is encryption being used on the network to encode sensitive data? • Were procedures established to ensure effective controls over the hardware and software used by the departments served by the distributed processing network? • Are security policies and procedures appropriate to the environment: – Highly distributed?—Is security under the control of individual user management? – Distributed?—Is security under the direction of user management, but adheres to the guidelines established by IS management? – Mixed?—Is security under the direction of individual user management, but the overall responsibility remains with IS management? – Centralized?—Is security under the direction of IS management, with IS management staff maintaining a close relationship with user management? – Highly centralized?—Is security under the complete control of IS management?
• Network operation procedures	• Do procedures exist to ensure that data compatibility is applied properly to all the network's datasets and that the requirements for their security have been determined? • Have adequate restart and recovery mechanisms been installed at every user location served by the distributed processing network? • Has the IS distributed network been designed to ensure that failure of service at any one site will have a minimal effect on the continued service to other sites served by the network? • Are there provisions to ensure consistency with the laws and regulations governing transmission of data?
• Interview the person responsible for maintaining network security	• Is the person aware of the risk associated with physical and logical access that must be minimized? • Is the person aware of the need to actively monitor logons and to account for employee changes? • Is the person knowledgeable in how to maintain and monitor access?
• Interview users	• Are the users aware of management policies regarding network security and confidentiality?

4.7.6 IS OPERATIONS REVIEWS

Because processing environments vary among different installations, a tour of the information processing facility generally provides the IS auditor with a better understanding of operations tasks, procedures and control environment.

Audit procedures should include those shown in **figure 4.30**.

Figure 4.30—IS Operations Reviews	
Areas to Review	**Questions to Consider**
• Observation of IS personnel	• Have controls been put in place to ensure efficiency of operations and adherence to established standards and policies? • Is adequate supervision present? • Have controls been put in place regarding IS management review, data integrity and security?
• Operator access	• Is access to files and documentation libraries restricted to operators? • Are responsibilities for the operation of computer and related peripheral equipment limited? • Is access to correcting program and data problems restricted? • Should access to utilities that allow system fixes to software and/or data be restricted? • Is access to production source code and data libraries (including run procedures) limited?

Figure 4.30—IS Operations Reviews *(cont.)*	
Areas to Review	**Questions to Consider**
• Operator manuals	• Are instructions adequate to address: – The operation of the computer and its peripheral equipment? – Startup and shutdown procedures? – Actions to be taken in the event of machine/program failure? – Records to be retained? – Routine job duties and restricted activities?
• Access to the library	• Is the librarian prevented from accessing computer hardware? • Does the librarian have access only to the tape management system? • Is access to library facilities provided to authorized staff only? • Is removal of files restricted by production scheduling software? • Does the librarian handle the receipt and return of foreign media entering the library? • Are logs of the sign-in and sign-out of data files and media maintained?
• Contents and location of offline storage	• Are offline file storage media containing production system programs and data clearly marked with their contents? • Are offline library facilities located away from the computer room? • Are policies and procedures adequate for: – Administering the offline library? – Checking out/in media, including requirements for signature authorizations? – Identifying, labeling, delivering and retrieving offsite backup files? – Encryption of offsite backup files (especially if these physically move between locations)? – Inventorying the system for onsite and offsite media, including the specific storage locations of each tape? – Secure disposal/destruction of media, including requirements for signature authorizations?
• File handling procedures	• Have procedures been established to control the receipt and release of files and secondary storage media to/from other locations? • Are internal tape labels used to help ensure that the correct media are mounted for processing? • Are these procedures adequate and in accordance with management's intent and authorization? • Are these procedures being followed?
• Data entry	• Are input documents authorized and do the documents contain appropriate signatures? • Are batch totals reconciled? • Does segregation of duties exist between the person who keys the data and the person who reviews the keyed data for accuracy and errors? • Are control reports being produced? Are the reports accurate? Are the reports maintained and reviewed?
• Lights-out operations	• Remote access to the master console is often granted to standby operators for contingency purposes such as automated software failure. Is access to security sufficient to guard against unauthorized use? • Do contingency plans allow for the proper identification of a disaster in the unattended facility? • Are the automated operation software and manual contingency procedures documented and tested adequately at the recovery site? • Are proper program change controls and access controls present? • Are tests of the software performed on a periodic basis, especially after changes or updates are applied? • Do assurances exist that errors are not hidden by the software and that all errors result in operator notification?

4.7.7 SCHEDULING REVIEWS

Figure 4.31 describes an audit approach to be considered when reviewing workload job scheduling and personnel scheduling.

Figure 4.31—Scheduling Reviews	
Areas to Review	**Questions to Consider**
• Regularly scheduled applications • Input deadlines • Data preparation time • Estimated processing time • Output deadlines • Procedures for collecting, reporting and analyzing key performance indicators	• Are the items included in SLAs? • Are the items functioning according to the SLAs?

Figure 4.31—Scheduling Reviews *(cont.)*	
Areas to Review	**Questions to Consider**
• Job schedule	• Have critical applications been identified and the highest priority assigned to them? • Have processing priorities been established for other applications and are the assigned priorities justified? • Is scheduling of rush/rerun jobs consistent with their assigned priority? • Do scheduling procedures facilitate optimal use of computer resources while meeting service requirements? • Do operators record jobs that are to be processed and the required data files? • Do operators schedule jobs for processing on a predetermined basis and perform them using either automated scheduling software or a manual schedule?
• Daily job schedule	• Is the number of personnel assigned to each shift adequate to support the workload? • Does the daily job schedule serve as an audit trail? Does the schedule provide each shift of computer operators with the work to be carried out, the sequence in which programs are to be run and indication when lower-priority work can be performed? • At the end of a shift, does each operator pass to the work scheduler or the next shift of operators a statement of the work completed and the reasons any scheduled work was not finished?
• Console log	• Were jobs run and completed according to the schedule? • If not, are the reasons valid?
• Exception processing logs	• Do operators obtain written or electronic approval from owners when scheduling request-only jobs? • Do operators record all exception processing requests? • Do operators review the exception processing request log to determine the appropriateness of procedures performed?
• Reexecuted jobs	• Are all reexecution of jobs properly authorized and logged for IS management review? • Are procedures established for rerunning jobs to ensure that the correct input files are being used and subsequent jobs in the sequence also are rerun, if appropriate?
• Personnel	• Are personnel who are capable of assigning, changing job schedules or job priorities authorized to do so?

4.7.8 PROBLEM MANAGEMENT REPORTING REVIEWS

The audit approach shown in **figure 4.32** should be considered when reviewing problem management reporting.

Figure 4.32—Problem Management Reporting Reviews	
Areas to Review	**Questions to Consider**
• Interviews with IS operations personnel	• Have documented procedures been developed to guide IS operations personnel in logging, analyzing, resolving and escalating problems in a timely manner, in accordance with management's intent and authorization?
• Procedures used by the IT department • Operations documentation	• Are procedures for recording, evaluating, and resolving or escalating any operating or processing problems adequate? • Are procedures used by the IT department to collect statistics regarding online processing performance adequate and is the analysis accurate and complete? • Are all problems identified by IS operations being recorded for verification and resolution?
• Performance records • Outstanding error log entries • Help desk call logs	• Do problems exist during processing? • Are the reasons for delays in application program processing valid? • Are significant and recurring problems identified, and actions taken to prevent their recurrence? • Were processing problems resolved in a timely manner and was the resolution complete and reasonable? • Are there any reoccurring problems that are not being reported to IS management?

4.8 DISASTER RECOVERY PLANNING

Disaster recovery planning (DRP), in support of business operations/provisioning IT service, is an element of an internal control system established to manage availability and restore critical processes/IT services in the event of interruption. The purpose of this continuous planning process is to ensure that cost-effective controls to prevent possible IT disruptions and to recover the IT capacity of the organization in the event of a disruption are in place. The importance of the availability of individual applications/IT services depends on the importance of the business processes that they support. The importance and urgency of these business processes and corresponding IT services and applications can be defined through performing a business impact analysis (BIA) and assigning recovery point objectives (RPOs) and recovery time objectives (RTOs). The availability of business data and the ability to process and handle them are vital to the sustainable development and/or survival of any organization. Planning for disasters is, therefore, an important part of the risk management and business continuity planning (BCP) processes.

DRP is a continuous process. Once the criticality of business processes and supporting IT services, systems and data are defined, they are periodically reviewed and revisited. There are at least two important outcomes of DRP:

- Changes in IT infrastructure (servers, networks, data storage systems, etc.), changes in supporting processes (increasing the maturity), procedures and organizational structure (new headcount or new roles). These changes are combined into programs spanning three to five years, often called IT DR strategies.
- Disaster recovery plans developed as part of this process that direct the response to incidents ranging from simple emergencies to full-blown disasters. The plans range from departmental-level, simple procedures down to modular, multitiered plans that cover multiple locations and multiple lines of business.

The ultimate goal of the DRP process is to respond to incidents that may impact people and the ability of operations to deliver goods and services to the marketplace and to comply with regulatory requirements.

DRP may be subject to various compliance requirements depending upon geographic location, nature of business, and the legal and regulatory framework. Organizations engage third parties to perform the activities on their behalf, and these third parties are still subject to compliance. Most compliance requirements will focus on assuring continuity of service; however, human safety is the most essential aspect. For example, in case of fire, safe evacuation comes first; restoring service is a secondary activity.

This section focuses on the key activities that an organization must perform to proactively plan for, and manage, the consequences of a disaster.

4.8.1 RECOVERY POINT OBJECTIVE AND RECOVERY TIME OBJECTIVE

The RPO is determined based on the acceptable data loss in case of disruption of operations. It indicates the earliest point in time in which it is acceptable to recover the data. For example, if the process can afford to lose the data up to four hours before disaster, then the latest backup available should be up to four hours before disaster or interruption and the transactions that occurred during the RPO period and interruption need to be entered after recovery (known as catch-up data).

RPO effectively quantifies the permissible amount of data loss in case of interruption. It is almost impossible to recover the data completely. Even after entering incremental data, some data are still lost and are referred to as orphan data. The RPO directly affects the technology used to back up and recover data (see **figure 4.33**).

The RTO is determined based on the acceptable downtime in case of a disruption of operations. It indicates the earliest point in time at which the business operations (and supporting IT systems) must resume after disaster. **Figure 4.33** shows the relationship between the RTO and RPO and gives examples of technologies used to meet the RPOs and RTOs.

Both of these concepts are based on time parameters. The nearer the time requirements are to the center (0-1 hours), the higher the cost of the recovery strategies. If the RPO is in minutes (lowest possible acceptable data loss), then data mirroring or real-time replication should be implemented as the recovery strategy. If the RTO is in minutes (lowest acceptable time down), then a hot site, dedicated spare servers (and other equipment) and clustering must be used.

Disaster tolerance is the time gap within which the business can accept the unavailability of IT critical service; therefore, the lower the RTO, the lower the disaster tolerance.

RTO affects the technology used to make applications/IT systems available—what to use for recovery (i.e., warm site, hot site, clusters, etc.). RPO usually affects data protection solutions (backup and recovery, synchronous or asynchronous data replication).

> **Note:** The CISA candidate should be familiar with which recovery strategies would be best with different RTO and RPO parameters.

In addition to RTO and RPO, there are some additional parameters that are important in defining the recovery strategies. These include:

- **Interruption window**—The maximum period of time the organization can wait from the point of failure to the critical services/applications restoration. After this time, the progressive losses caused by the interruption are unaffordable.

Figure 4.33—Relationship Between RTO and RPO

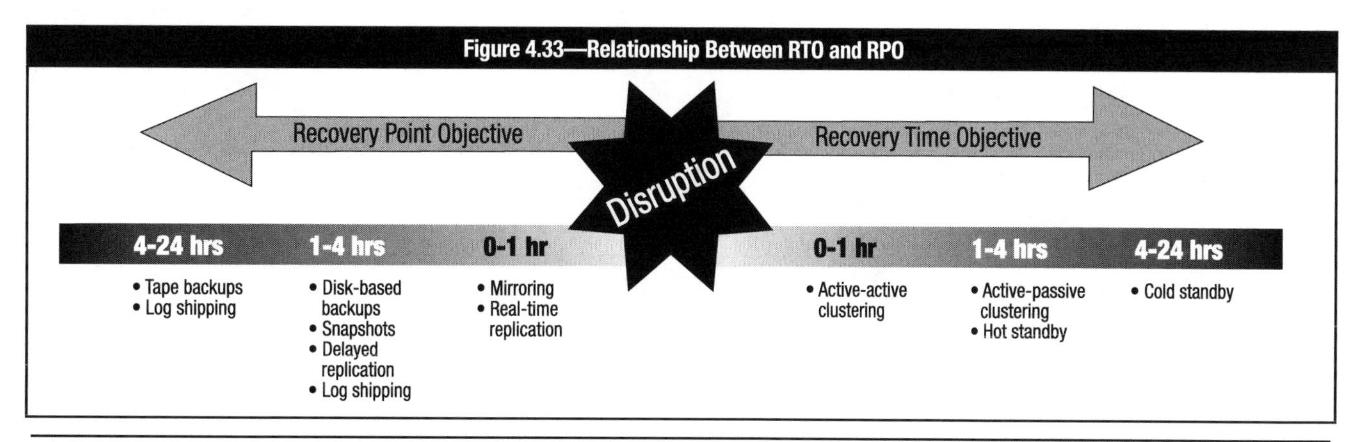

- **Service delivery objective (SDO)**—Level of services to be reached during the alternate process mode until the normal situation is restored. This is directly related to the business needs.
- **Maximum tolerable outages**—Maximum time the organization can support processing in alternate mode. After this point, different problems may arise, especially if the alternate SDO is lower than the usual SDO, and the information pending to be updated can become unmanageable.

4.8.2 RECOVERY STRATEGIES

A recovery strategy identifies the best way to recover a system (one or many) in case of interruption, including disaster, and provides guidance based on which detailed recovery procedures can be developed. Different strategies should be developed, and all alternatives should be presented to senior management. Senior management should select the most appropriate strategies from the alternatives provided and accept the inherent residual risk. The selected strategies should be used to further develop the detailed BCP.

The selection of a recovery strategy would depend on:

- The criticality of the business process and the applications supporting the processes
- Cost
- Time required to recover
- Security

There are various strategies for recovering critical information resources. The appropriate strategy is the one with a cost for an acceptable recovery time that is also reasonable compared to the impact and likelihood of occurrence as determined in the BIA. The cost of recovery is the cost of preparing for possible disruptions (e.g., the fixed costs of purchasing, maintaining and regularly testing redundant computers, and maintaining alternate network routing), as well as the variable costs of putting these into use in the event of a disruption. The latter costs can often be insured against, but the former generally cannot. However, the premiums for disaster insurance usually will be lower if there is a suitable plan.

Generally, each IT platform that runs an application supporting a critical business function will need a recovery strategy. There are many alternative strategies. The most appropriate alternative, in terms of cost to recover and impact cost, should be selected based on the relative risk level identified in the business impact analysis. Recovery strategies based on the risk level identified for recovery would include developing:

- Hot sites
- Warm sites
- Cold sites
- Duplicate information processing facilities
- Mobile sites
- Reciprocal arrangements with other organizations

> **Note:** The CISA candidate should know these recovery strategies and when to use them.

4.8.3 RECOVERY ALTERNATIVES

When the normal production facilities become unavailable, the business may utilize alternate facilities to sustain critical processing until the primary facilities can be restored. **Figure 4.34** lists the most common recovery alternatives.

Figure 4.34—Recovery Alternatives
Cold sites are facilities with the space and basic infrastructure adequate to support resumption of operations, but lacking any IT or communications equipment, programs, data or office support. A plan that specifies that a cold site will be utilized must also include provision to acquire and install the requisite hardware, software and office equipment to support the critical applications when the plan is activated. To use a sports analogy, a cold site is like having a substitute on the bench, ready to be called into the game.
Mobile sites are packaged, modular processing facilities mounted on transportable vehicles and kept ready to be delivered and set up at a location that may be specified upon activation. A plan to utilize mobile processing must specify the site locations that may be used. The plan must provide right-of-access to the selected site by the vendor and the company. The plan must also provide for any required ancillary infrastructure necessary to support the site such as access roads, water, waste disposal, power and communications.
Warm sites are complete infrastructures but are partially configured in terms of IT, usually with network connections and essential peripheral equipment such as disk drives, tape drives and controllers. The equipment may be less capable than the normal production equipment yet still be adequate to sustain critical applications on an interim basis. Typically, employees would be transferred to the warm site, and current versions of programs and data would need to be loaded before operations could resume at the warm site. Using the sports analogy, a warm site is a substitute warming up, getting ready to enter the game.
Hot sites are facilities with space and basic infrastructure and all of the IT and communications equipment required to support the critical applications, along with office furniture and equipment for use by the staff. Hot sites usually maintain installed versions of the programs required to support critical applications. Data may also be duplicated to the hot site in real or near real time. If this is not the case the most recent backup copies of data may need to be loaded before critical applications could be resumed. Although hot sites may have a small staff assigned, employees are usually transferred to the hot site from the primary site to support operations upon activation. Using the sports analogy, a hot site is a substitute on the sideline waiting to enter the game.
Mirrored sites are fully redundant sites with real-time data replication from the production site. They are fully equipped and staffed, and can assume critical processing with no interruption perceived by the users.
Reciprocal agreements are agreements between separate, but similar, companies to temporarily share their IT facilities in the event that one company loses processing capability. Reciprocal agreements are not considered a viable option due to the constraining burden of maintaining hardware and software compatibility between the companies, the complications of maintaining security and privacy compliance during shared operations, and the difficulty of enforcing the agreements should a disagreement arise at the time the plan is activated.
Reciprocal agreements with other organizations, although a less frequently used method, are agreements between two or more organizations with unique equipment or applications. Under the typical agreement, participants promise to provide assistance to each other when an emergency arises.

Alternatives which provide the fastest recovery time require the most dedicated resources on an ongoing basis, and thus incur the greatest ongoing cost to the company. By comparing the business costs associated with the interruption of critical processes (developed in the BIA) to the cost of the various alternative processing options, management will establish an optimal RTO and select an appropriate recovery alternative.

The alternate site should be selected with consideration that it will be located beyond the geographic area affected by any disruptive events considered in the plan. The impact and nature of the disruptive events should be considered in determining an adequate separation from the primary site rather than specifying a particular distance of separation.

Regardless of which type of alternative processing is utilized, the plan will need to include provision to establish network communication to the alternate site. The plan should provide for redundant solutions to ensure that communications can be established to the alternate site following interruption of normal processing by any anticipated cause.

The alternate processing facility can be provided by a third-party vendor or by the company using its own resources. When the facility is owned by the company, priority and conflicts can be prevented or quickly resolved by senior management. When the facility is provided by a third party, the company needs to have clearly stated contracts which ensure that the company will get access to the resources it needs without delay following a disaster. Consideration must be given to the likelihood that at the same time that the company needs to utilize the alternate processing facility, other companies in the area may also be trying to restore critical processing.

Contractual Provisions

Contractual provisions for the use of third-party sites should cover the following:

- **Configurations**—Are the hardware and software configurations for the facility adequate to meet company needs? Is there provision to update the configurations and conduct tests to ensure that the configurations remain adequate over time?
- **Disaster**—Is the definition of disaster broad enough to meet anticipated needs?
- **Access**—Is use of the facility exclusive or does the customer have to share the available space if multiple customers simultaneously declare a disaster? Does the company have guaranteed assurance that they will have adequate access to the site and the resources following a disaster? Does the agreement satisfactorily specify how access conflicts will be resolved?
- **Priority**—Does the agreement provide the company with satisfactory priority following a disaster? Does the agreement preclude the sharing of the needed resources with governmental entities that might preempt the company following a disaster?
- **Availability**—Will the facility be available to the company without delay when needed?
- **Speed of availability**—How soon after a disaster will facilities be available?
- **Subscribers per site**—Does the agreement limit the number of subscribers per site?
- **Subscribers per area**—Does the agreement limit the number of subscribers in a building or area?
- **Preference**—Who gets preference if there are common or regional disasters? Is there backup for the backup facilities? Is use of the facility exclusive or does the customer have to share the available space if multiple customers simultaneously declare a disaster? Does the vendor have more than one facility available for subscriber use?
- **Insurance**—Is there adequate insurance coverage for company employees at the backup site? Will existing insurance reimburse those fees?
- **Usage period**—How long is the facility available for use? Is this period adequate? What technical support will the site operator provide? Is this adequate?
- **Communications**—Are the communications adequate? Are the communication connections to the backup site sufficient to permit unlimited communication with the alternate site if needed?
- **Warranties**—What warranties will the vendor make regarding availability of the site and the adequacy of the facilities? Are there liability limitations (there usually are) and is the company willing to live with them?
- **Audit**—Is there a right-to-audit clause permitting an audit of the site to evaluate the logical, physical and environmental security?
- **Testing**—What testing rights are included in the contract? Check with the insurance company to determine any reduction of premiums that may be forthcoming due to the backup site availability.
- **Reliability**—Can the vendor attest to the reliability of the site(s) being offered? Ideally, the vendor should have an uninterruptible power supply (UPS), limited subscribers, sound technical management, and guarantees of computer hardware and software compatibility.
- **Security**—Can the site be adequately secured by the company to comply with the company's security policy?

Procuring Alternative Hardware

Companies planning to utilize a cold or warm site will need to include in their plan provision to acquire hardware and software to equip the sites upon activation. Companies can acquire and store the necessary equipment and software beforehand or can plan to acquire the hardware and software when it is needed. A key factor in the decision is whether standard systems are used that can be readily acquired when replacements are needed or are unique, specialized, outdated and therefore difficult to acquire on short notice. If companies depend on hardware that is not readily available to support critical business applications, plans must include provision to acquire the hardware in time to meet the RTO. This fact may dictate that the companies acquire the critical components beforehand and store them so they are available when required.

Additionally, part of the recovery of IT facilities will involve telecommunications, for which the strategies usually considered include:

- Network disaster prevention, which includes:
  - Alternative routing
  - Diverse routing
  - Long-haul network diversity
  - Protection of the local loop
  - Voice recovery
  - Availability of appropriate circuits and adequate bandwidth
- Server disaster recovery plans

Application Resiliency and Disaster Recovery Methods

Protecting an application against a disaster entails providing a way to restore it as quickly as possible. Clustering makes it possible to do so. A cluster is a type of software (agent) that is installed on every server (node) in which the application runs and includes management software that permits control of and tuning the cluster behavior. Clustering protects against single points of failure (a resource whose loss would result in the loss of service or production).

There are two major types of application clusters: active-passive and active-active. In active-passive clusters, the application runs on only one (active) node, while other (passive) nodes are used only if the application fails on the active node. In this case, cluster agents constantly watch the protected application and quickly restart it on one of the remaining nodes. This type of cluster does not require any special setup from the application side (i.e., the application does not need to be cluster-aware). Hence, it is one of the major ways to ensure application availability and DR. In active-active clusters, the application runs on every node of the cluster. With this setup, cluster agents coordinate the information processing between all of the nodes, providing load balancing and coordinating concurrent data access. When an application in such a cluster fails, users normally do not experience any downtime at all (possibly missing uncompleted transactions). Active-active clusters require that the application be built to utilize the cluster capabilities (for instance, if the transaction is not completed on the node that failed, some other remaining node will try to re-run the transaction). Such clusters are less common than active-passive and provide quick application recovery, load balancing and scalability. This type of cluster puts a greater demand on network latency. Very often, organizations use a combination of cluster setups; for instance, active-active for a particular processing site and active-passive between the sites. This combination protects applications against local software or hardware failure (active-active) and against site failure (active-passive). The clusters with a span of one city are called metro-clusters, while clusters spanning between cities, countries and continents are called geo-clusters. Although it is possible to develop cluster software in-house, generally, it is not economically viable, and there are a number of solutions available from major software vendors. Often, clustered applications require that the data are shared between all nodes of the cluster. Active-active clusters generally require that the same storage be available to all of the nodes; active-passive clusters are less demanding and require that the data are replicated from the active node to others.

Data Storage Resiliency and Disaster Recovery Methods

Redundant Array of Independent (or Inexpensive) Disks (RAID) is the most common, basic way to protect data against a single point of failure, in this instance, a disk failure. RAID provides performance improvements and fault-tolerant capabilities via hardware or software solutions, breaking up data and writing data to a series of multiple disks to simultaneously improve performance and/or save large files. These systems provide the potential for cost-effective mirroring offsite for data backup.

A variety of methods, categorized into 11 levels (the most popular being 0 [stripe], 1 [mirror], their combinations [0+1 or 1+0] and 5), is defined for combining several disk drives into what appears to the system as a single disk drive. RAID improves on the single-drive-only solution since it offers better performance and/or data redundancy.

Note: The CISA candidate will not be tested on the specifics of RAID levels.

Many vendors offer storage arrays—hardware that hides all the complexities of forming logical volumes from physical disks, thus completely removing the need for the low-level configuration. Typically, these storage arrays provide major RAID levels; however, that does not remove the need for responsible IT staff to understand the implications of the different RAID configurations.

To protect data against site failure and to ensure successful application recovery (with or without clusters), storage arrays provide data replication features, making sure that what data are saved to the disk on one site appear on the other site. Depending on the available network bandwidth and latency, this data replication may be synchronous (i.e., the local disk write is not confirmed until the data are written to the disk on the other site), asynchronous (data are replicated on a schedule basis) or adaptive (switching from one mode to another depending upon the network load).

The array-based (hardware) replication is absolutely transparent to the application (i.e., no special provisions are needed from the OS or the application side).

If there is no disk array, the data stored on local server volumes (RAID or not) can still be replicated to a remote site by using host-based data replication solutions. These act similarly to hardware-based solutions.

Telecommunication Networks Resiliency and Disaster Recovery Methods

The plan should contain the organization's telecommunication networks. Today, telecommunication networks are key to business processes in large and small organizations; therefore, the procedures to ensure continuous telecommunication capabilities should be given a high priority.

Telecommunication networks are susceptible to the same natural disasters as data centers but also are vulnerable to several disastrous events unique to telecommunications. These include central switching office disasters, cable cuts, communication software glitches and errors, security breaches connected to hacking (phone hackers are known as phreakers), and a host of other human mishaps. It is the responsibility of the organization and not the local exchange carriers to ensure constant communication capabilities. The local exchange carrier is not responsible for providing backup services, although many do back up main components within their systems. Therefore, the organization should make provisions for backing up its own telecommunication facilities.

To maintain critical business processes, the information processing facility's (IPF) BCP should provide for adequate telecommunications capabilities. Telecommunications capabilities to consider include telephone voice circuits, WANs (connections to distributed data centers), LANs (work group PC connections), and third-party EDI providers. The critical capacity requirements should be identified for the various thresholds of outage for each telecommunications capability such as two hours, eight hours or 24 hours. UPSs should be sufficient to provide backup to the telecommunication equipment as well as the computer equipment.

Methods for network protection are:
- **Redundancy**—This involves a variety of solutions, including:
  - Providing extra capacity with a plan to use the surplus capacity should the normal primary transmission capability not be available. In the case of a LAN, a second cable could be installed through an alternate route for use in the event the primary cable is damaged.
  - Providing multiple paths between routers
  - Dynamic routing protocols, such as Open Shortest Path First (OSPF) and Enhanced Interior Gateway Routing Protocol (EIGRP)
  - Providing for fail over devices to avoid single point of failures in routers, switches, firewalls, etc.
  - Saving configuration files for recovery in the event that network devices, such as those for routers and switches, fail. For example, organizations should utilize Trivial File Transport Protocol (TFTP) servers. Most network devices support TFTP for saving and retrieving configuration information.
- **Alternative routing**—The method of routing information via an alternate medium such as copper cable or fiber optics. This involves use of different networks, circuits or end points should the normal network be unavailable. Most local carriers are deploying counter-rotating, fiber-optic rings. These rings have fiber-optic cables that transmit information in two different directions and in separate cable sheaths for increased protection. Currently, these rings connect through one central switching office. However, future expansion of the rings may incorporate a second central office in the circuit. Some carriers are offering alternate routes to different points of presence or alternate central offices. Other examples include a dial-up circuit as an alternative to dedicated circuits; cellular phone and microwave communication as alternatives to land circuits; and couriers as an alternative to electronic transmissions.
- **Diverse routing**—The method of routing traffic through split cable facilities or duplicate cable facilities. This can be accomplished with different and/or duplicate cable sheaths. If different cable sheaths are used, the cable may be in the same conduit and, therefore, subject to the same interruptions as the cable it is backing up. The communication service subscriber can duplicate the facilities by having alternate routes, although the entrance to and from the customer premises may be in the same conduit. The subscriber can obtain diverse routing and alternate routing from the local carrier, including dual entrance facilities. However, acquiring this type of access is time-consuming and costly. Most carriers provide facilities for alternate and diverse routing, although the majority of services are transmitted over terrestrial media. These cable facilities are usually located in the ground or basement. Ground-based facilities are at great risk due to the aging infrastructures of cities. In addition, cable-based facilities usually share room with mechanical and electrical systems that can impose great risk due to human error and disastrous events.
- **Long-haul network diversity**—Many vendors of recovery facilities have provided diverse long-distance network availability, utilizing T1 circuits among the major long-distance carriers. This ensures long-distance access should any single carrier experience a network failure. Several of the major carriers now have installed automatic rerouting software and redundant lines that provide instantaneous recovery should a break in their lines occur. The IS auditor should verify that the recovery facility has these vital telecommunications capabilities.
- **Last-mile circuit protection**—Many recovery facilities provide a redundant combination of local carrier T1s or E1s, microwave, and/or coaxial cable access to the local communications loop. This enables the facility to have access during a local carrier communication disaster. Alternate local carrier routing also is utilized.
- **Voice recovery**—With many service, financial and retail industries dependent on voice communication, redundant cabling and VoIP are common approaches to deal with it.

4.8.4 DEVELOPMENT OF DISASTER RECOVERY PLANS

As part of a greater BCP process, IT DRP follows the same path. After conducting a BIA and risk assessment (or determining the risk and effectiveness of mitigation controls otherwise), the IT DR strategy is developed. Implementing this strategy means making changes to:
- IT systems
- Networks
- IT processing sites
- Organization structure (headcount, roles, positions)
- IT processes and procedures

An IT DRP is a well-structured collection of processes and procedures intended to make the disaster response and recovery effort swift, efficient and effective to achieve the synergy between recovery teams. The plan should be documented and written in simple language that is understandable to all.

IT DRP Contents

Typically the IT DRP contains:
- Procedures for declaring a disaster (escalation procedures)
- Criteria for plan activation (i.e., in which circumstances the disaster is declared, when the IT DRP is put to action, which scenarios are covered by the plan [loss of the IT system, loss of the processing site, loss of the office])
- Its linkage with the overarching plans (for instance, emergency response plan or crisis management plan or BCPs for different lines of business)
- The person (or people) responsible for each function in plan execution
- Recovery teams and their responsibilities
- Contact and notification lists (contact information for recovery teams, recovery managers, stakeholders, etc.)
- The step-by-step explanation of the whole recovery process (where and when the recovery should take place [the same site or backup site], what has to be recovered [IT systems, networks, etc.], the order of recovery)

- Recovery procedures (for each IT system or component). Note: the level of detail here greatly varies and depends on the practices used in the organization.
- Contacts for important vendors and suppliers
- The clear identification of the various resources required for recovery and continued operation of the organization

It is common to identify teams of personnel who are made responsible for specific tasks in case of disasters. Some important teams should be formed and their responsibilities are explained in section 4.8.5 Organization and Assignment of Responsibilities. Copies of the plan should be maintained offsite. The plan must be structured so that its parts can easily be handled by different teams.

IT DRP Scenarios

Although no two disasters are alike, the plan should outline which scenarios are covered, such as:

- Loss of network connectivity
- Loss of a key IT system
- Loss of the processing site (server room)
- Loss of critical data
- Loss of an office, etc.
- Loss of key service provider (e.g., cloud)

Normally, this section is quite short; however, it is important to remember that the best plan always accounts for the worst-case conditions (such as peak of sales, end of reporting period, etc.).

Recovery Procedures

Depending on the type of disaster, the sequence of the recovery effort may vary; however, the plan should contain a simple, high-level overview of the sequence for every major disaster scenario referring to the more detailed recovery procedures.

4.8.5 ORGANIZATION AND ASSIGNMENT OF RESPONSIBILITIES

The DRP should identify the teams with their assigned responsibilities in the event of an incident/disaster. IS and end-user personnel should be identified to go through the recovery procedures that have been developed for business/process recovery and key decision making. These individuals usually lead teams created in response to a critical function or task defined in the plan. Depending on the size of the business operation, these teams may be designated as single-person positions. The involvement of the following teams depends on the level of the disruption of service and the types of assets lost or damaged. It is a good idea to develop a matrix on the correlation between the teams needed to participate and the estimated recovery effort/level of disruption.

The recovery/continuity/response teams may include any of the following:

- **Incident response team**—This is a team that has been designated to receive the information about every incident that can be considered as a threat to assets/processes. This reporting can be useful for coordinating an incident in progress and or for postmortem analysis. The analysis of all incidents also provides input for updating the recovery plans.
- **Emergency action team**—They are first responders, designated fire wardens and bucket crews, whose function is to deal with fires or other emergency response scenarios. One of their primary functions is the orderly evacuation of personnel and the securing of human life.
- **Information security team**—The main mission of this team is to develop the needed steps to maintain a similar level of information and IT resource security as was in place in at the primary site before the contingency, and implement the needed security measures in the alternative procedures environment. Additionally, this team must continually monitor the security of system and communication links, resolve any security conflicts that impede the expeditious recovery of the system, and assure the proper installation and functioning of security software. The team is also responsible for the security of the organization's assets during the disorder following a disaster.
- **Damage assessment team**—This team assesses the extent of damage following the disaster. The team should be comprised of individuals who have the ability to assess damage and estimate the time required to recover operations at the affected site. This team should include staff skilled in the use of testing equipment, knowledgeable about systems and networks, and trained in applicable safety regulations and procedures. In addition, they have the responsibility to identify possible causes of the disaster and their impact on damage and predictable downtime.
- **Emergency management team**—This team is responsible for coordinating the activities of all other recovery/continuity/response teams and handling key decision making. They determine the activation of the BCP. Other functions entail arranging the finances of the recovery, handling legal matters evolving from the disaster, and handling public relations and media inquiries. This team functions as disaster overseers and is required to coordinate the following activities:
  - Retrieving critical and vital data from offsite storage
  - Installing and/or testing systems software and applications at the systems recovery site (hot site, cold site)
  - Identifying, purchasing, and installing hardware at the system recovery site
  - Operating from the system recovery site
  - Rerouting WAN communications traffic
  - Reestablishing the local area user/system network
  - Transporting users to the recovery facility
  - Restoring databases
  - Supplying necessary office goods (i.e., special forms, check stock, paper)
  - Arranging and paying for employee relocation expenses at the recovery facility
  - Coordinating systems use and employee work schedules
- **Offsite storage team**—This team is responsible for obtaining, packaging and shipping media and records to the recovery facilities, as well as establishing and overseeing an offsite storage schedule for information created during operations at the recovery site.
- **Software team**—This team is responsible for restoring system packs, loading and testing OSs software, and resolving system-level problems.
- **Applications team**—This team travels to the system recovery site and restores user packs and application programs on the backup system. As the recovery progresses, this team may have the responsibility of monitoring application performance and database integrity.

- **Emergency operations team**—This team consists of shift operators and shift supervisors who will reside at the systems recovery site and manage system operations during the entirety of the disaster and recovery projects. Another responsibility might be coordinating hardware installation, if a hot site or other equipment-ready facility has not been designated as the recovery center.
- **Network recovery team**—This team is responsible for rerouting wide-area voice and data communications traffic, reestablishing host network control and access at the system recovery site, providing ongoing support for data communications, and overseeing communications integrity.
- **Communications team**—This team travels to the recovery site where they work in conjunction with the remote network recovery team to establish a user/system network. This team also is responsible for soliciting and installing communications hardware at the recovery site and working with local exchange carriers and gateway vendors in the rerouting of local service and gateway access.
- **Transportation team**—This team serves as a facilities team to locate a recovery site, if one has not been predetermined, and is responsible for coordinating the transport of company employees to a distant recovery site. It also may assist in contacting employees to inform them of new work locations, and scheduling and arranging employee lodgings.
- **User hardware team**—This team locates and coordinates the delivery and installation of user terminals, printers, typewriters, photocopiers and other necessary equipment. This team also offers support to the communications team and to any hardware and facilities salvage efforts.
- **Data preparation and records team**—Working from terminals that connect to the user recovery site, the team updates the applications database. This team also oversees additional data-entry personnel and assists record salvage efforts in acquiring primary documents and other input information sources.
- **Administrative support team**—This team provides clerical support to the other teams and serves as a message center for the user recovery site. This team also may control accounting and payroll functions as well as ongoing facilities management.
- **Supplies team**—This team supports the efforts of the user hardware team by contacting vendors and coordinating logistics for an ongoing supply of necessary office and computer supplies.
- **Salvage team**—This team manages the relocation project. This team also makes a more detailed assessment of the damage to the facilities and equipment than was performed initially; provides the emergency management team with the information required to determine whether planning should be directed toward reconstruction or relocation; provides information necessary for filing insurance claims (insurance is the primary source of funding for the recovery efforts); and coordinates the efforts necessary for immediate records salvage, such as restoring paper documents and electronic media.
- **Relocation team**—This team coordinates the process of moving from the hot site to a new location or to the restored original location. This involves relocating the IS processing operations, communications traffic and user operations. This team also monitors the transition to normal service levels.
- **Coordination team**—This team is responsible for coordinating the recovery efforts across various offices located at different geographical locations. Where significant IT functions have been off-shored to distant geographical locations, this team acts as the focus for coordination between the organization and the third-party service providers.
- **Legal affairs team**—This team is responsible for handling the legal issues arising for various reasons due to any incident or unavailability of services (e.g., according to new laws enacted by many countries, the organization is responsible for securing its IT assets, and will be liable for damages to innocent parties in case of incidence).
- **Recovery test team**—This team is responsible for testing of various plans developed and analyzing the result.
- **Training team**—This team will provide training to the users for provisions of business continuity and disaster recovery procedures.

> **Note:** The IS auditor should have knowledge of these responsibilities; however, the CISA candidate will not be tested on these specific assignments as they vary from organization to organization.

4.8.6 BACKUP AND RESTORATION

To ensure that the critical activities of an organization (and supporting applications) are not interrupted in the event of a disaster, secondary storage media are used to store software application files and associated data for backup purposes. These secondary storage media are removable media (tape cartridges, CDs, DVDs) or mirrored disks (local or remote) or network storage. Typically, the removable media are recorded in one facility and stored in one or more remote physical facilities (referred to as offsite libraries). The number and locations of these remote storage facilities are based on availability of use and perceived business interruption risk. Maintaining the inventory (catalog) of the remote storage facility can be performed automatically (vaulting solutions) or manually. In the latter case, it is the offsite librarian's responsibility to maintain a continuous inventory of the contents of these libraries, to control access to library media and to rotate media between various libraries, as needed. As the amount of information increases, keeping manual inventories of tape backups (whether local or remote) becomes increasingly difficult and is gradually replaced by integrated backup and recovery solutions that handle the backup catalogs—remote and local.

Offsite Library Controls

When disaster strikes, the offsite storage library often becomes the only remaining copy of the organization's data. To ensure that these data are not lost, it is very important to implement strict controls over the data—both physical and logical. Unauthorized access, loss or tampering with this information (either onsite or while in transit) could impact the information system's ability to provide support for critical business processes, putting the very future of the organization at risk.

Controls over the offsite storage library include:
- Securing physical access to library contents, ensuring that only authorized personnel have access
- Encrypting backup media especially when it is in transit

- Ensuring that physical construction can withstand fire/heat/water
- Locating the library away from the data center, preferably in a facility that will not be subject to the same disaster event, to avoid the risk of a disaster affecting both facilities
- Ensuring that an inventory of all storage media and files stored in the library is maintained for the specified retention time
- Ensuring that a record of all storage media and files moved into and out of the library is maintained for the specified retention/expiration time
- Ensuring that a catalog of information regarding the versions and location of data files is maintained for the specified retention time and protecting this catalog against unauthorized disclosure

The retention time for the different records must be in accordance with the enterprise retention policy.

Security and Control of Offsite Facilities

The offsite IPF must be as secured and controlled as the originating site. This includes adequate physical access controls such as locked doors, no windows and active surveillance. The offsite facility should not be easily identified from the outside. This is to prevent intentional sabotage of the offsite facility should the destruction of the originating site be from a malicious attack. The offsite facility should not be subject to the same disaster event that affected the originating site.

The offsite facility should possess at least the same constant environmental monitoring and control as the originating site, or the ones that are dictated by business requirements. This includes monitoring the humidity, temperature and surrounding air to achieve the optimum conditions for storing optical and magnetic media, and, if applicable, servers, workstations, storage arrays and tape libraries. The proper environmental controls include a UPS, operating on a raised floor with proper smoke and water detectors installed, climate controls and monitoring for temperature and humidity, and a working/tested fire extinguishing system. Provisions for paper record storage should ensure that a fire hazard is not created. Additional controls should be implemented in case of specific legal, regulatory or business requirements.

Media and Documentation Backup

A crucial element of a DRP (on- or offsite) is the availability of adequate data. Duplication of important data and documentation, including offsite storage of such backup data and paper records, is a prerequisite for any type of recovery.

Where information is processed and stored in a confidential environment at the primary site and backup is to be stored in a similarly secure location, care should be exercised to ensure that the means of transporting data, whether in the form of physical backup media or via mirrored backups on the network, extend adequate protection to the information.

Types of Backup Devices and Media

The backup device and media must be chosen based on a variety of factors:

- **Standardization**—Very specific technologies require a lot of support for both the primary site and the offsite facility, increasing costs.
- **Capacity**—Backup media should have adequate capacity, in order to reduce the number of media necessary to implement a backup set.
- **Speed**—Processes to backup and restore should be completed in an acceptable time, to comply with business requirements.
- **Price**—Backup devices are only part of the costs; attention must be paid to media prices.

There are a lot of different devices and media types available. The technology chosen must be adequate to the business needs. **Figure 4.35** provides some examples.

Figure 4.35—Types of Media

Portability	Small Amounts, Few Changes	Large Amounts, Frequent Changes
Removable media	CDs, DVDs, removable hard drives or solid state drives	Tape-based backup systems (DDS, digital audio tape [DAT], DLT, AIT, LTO)
Nonremovable media		Disk-based backup (virtual tape libraries [VTLs]), disk snapshots, host-based or disk-array-based replication

Modern tape-based backup systems are libraries with up to hundreds of tape drives and up to several thousands of tape slots. These libraries may be equipped with robotic arms and barcode scanners. Barcode scanners are used to quickly determine the contents of backup tapes. Without a barcode scanner, the tape must be actually inserted into tape drive and its header must be read and compared to backup catalog to read the tape contents. Having a barcode scanner makes this process quicker—the backup catalog contains the tape numbers written on the barcode instead of reading it in the drive. The robotic arms make the process of scanning the barcode and transporting the tape to the tape drive significantly faster. Tape libraries are controlled by backup and recovery applications which are available from major software companies. These applications:

- Handle backup and recovery tasks according to backup and recovery policies
- Maintain backup catalog (local and remote)
- Control tape libraries

The most important feature of the tape drives is its data interface. Modern tape drives have fiber channel (FC) or serial attached SCSI (SAS) interfaces, conventional parallel SCSI is gradually coming out of use. Tape libraries are connected either to SAN (via FC) or attached to backup and recovery server through SAS or iSCSI connections. Typically, tape libraries have LAN interfaces for maintenance and diagnostics.

Disk-based backup systems exist in different types:

- **Virtual tape libraries (VTLs)**—These systems consist of disk storage (typically mid-range disk arrays) and software that control backup and recovery data sets. For an external user (backup and recovery software), VTLs behave like a conventional tape library; however, data are stored on a disk array. Often, for the disaster recovery purposes the contents of a VTL are replicated from primary site to a backup site using the hardware-based replication provided by a disk array.

- **Host-based replication**—This replication is executed at the host (server) level by a special software running on this server and on the target server. It can occur in real-time (synchronous mode, the data is not written to the primary site until the backup site sends the confirmation the replicated data has arrived and safely written to the disk) or with some delay (asynchronous mode, when data is transferred to the backup site with some delay). The software packages are available from major software vendors.
- **Disk-array-based replication**—The same as host-based replications, however the replication is performed at the disk array level, completely hidden from servers and applications. This feature is available from all major hardware vendors supplying mid-range and high-end disk arrays. The replication can be completed via SAN or LAN.
- **Snapshots**—This technology is very flexible, allowing making different types of momentary copies of volumes or file systems. Depending upon types of snapshots, either full copy is created each time or only the changed blocks of data or files are stored. This technology is especially efficient and effective while used in combination with backup and recovery software. For instance, a snapshot is taken and then mounted on a different server, full backup is performed, thus saving the production system from overhead load. Another example is replicating data to remote site, making snapshots on the remote site and using them for backup and recovery, thus utilizing the server equipment at the backup site.

In an environment where server virtualization is utilized, disk-based backup systems can provide an excellent disaster recovery solution because entire virtual servers may be replicated to the recovery site.

Copies of data taken for offsite backup must be given the same level of security as the original files. The offsite facility and transportation arrangements must, therefore, meet the security requirements for the most sensitive class of data on the backup media.

Periodic Backup Procedures

Both data and software files should be backed up on a periodic basis in accordance with the defined RPO. The time period in which to schedule the backup may differ per application program or software system. For instance, the locations (folders or volumes) where the application data are stored must be backed up regularly since the data are frequently changed by daily transactions. The locations where application configuration and software files (application or OS) are stored are updated less frequently—only when the configurations change or a patch is applied. Often, online/real-time systems that perform large-volume transaction processing require nightly or hourly backups or utilize data replication at a separate remote processing facility.

Scheduling the periodic backups can often be easily accomplished via an automated backup/media management system and automated job scheduling software. Using the integrated solution for backup/recovery procedures and media management will prevent erroneous or missed backup cycles due to operator error. Schedules describing backup of certain data are included in the backup procedures.

Modern backup and recovery solutions include special pieces of software called "agents" that are installed on the protected servers and workstations. These agents are collecting the data (data files, configuration files, software application files) and shipping it to the backup and recovery server(s) that convert data for subsequent storage on tape or disk. The same agents are used for data restoration.

Frequency of Rotation

Backup for data and software must allow for the continuing occurrence of change. A copy of the file or record, as of some point in time, is retained for backup purposes. All changes or transactions that occur during the interval between the copy and the current time also are retained.

Considerations for establishing file backup schedules include the following:

- The frequency of backup cycle and depth-of-retention generations must be determined for each application.
- The backup procedures must anticipate failure at any step of the processing cycle.
- For legacy systems, master files should be retained at appropriate intervals, such as at the end of an updating procedure, to provide synchronization between files and systems.
- Transaction files should be presented to coincide with master files so a prior generation of a master file can be brought completely up-to-date to recreate a current master file.
- DBMS require specialized backup, usually provided as an integral feature of the DBMS or the special part of the backup and recovery software (agent) designed especially for the particular make and version of the database.
- It may be necessary to secure the license to use certain vendor software at an alternate site; this should be arranged in advance of the need.
- Backup for custom-built software must include object-code and source-code libraries and provisions for maintaining program patches on a current basis at all backup locations.
- Backup hardware should be available at the offsite facility and should be compatible with backup media. Also, for long-term retention, it is necessary to have technical support and maintenance agreements to guarantee that the alternate backup hardware will work properly in case of restoration.

Likewise, any documentation required for the consistent and continual operation of the business should be preserved in an offsite backup facility. This includes source documents required for restoration of the production database. As with data files, the offsite copies should be kept up to date to ensure their usefulness. It is important to remember that adequate backup is a prerequisite to successful recovery.

Types of Media and Documentation Rotated

Without software, the computer hardware is of little value. Software, including OSs, programming languages, compilers, utilities and application programs, along with copies of paper documentation—such as operational guides, users manuals,

records, data files, databases, etc.—should be maintained and stored offsite in their current status. This information provides the raw materials and finished products for the IS processing cycle and should be stored offsite.

Figure 4.36 describes the documentation to be backed up and stored offsite.

Figure 4.36—Offsite Storage

Classification	Description
Operating procedures	Application run books, job stream control instructions, OS manuals and special procedures
System and program documentation	Flow charts, program source code listings, program logic descriptions, statements, error conditions and user manuals
Special procedures	Any procedures or instructions that are out of the ordinary such as exception processing, variations in processing and emergency processing
Input source documents, output documents	Duplicate copies, photocopies, microfiche, microfilm reports or summaries required for auditing, historical analysis, performance of vital work, satisfaction of legal requirements or expediting insurance claims
Business continuity plan	A copy of the correct plan for reference

Sensitive data that are stored offsite should be stored in a fire-resistant magnetic media container. When the data are shipped back to the recovery site, the data should be stored and sealed in the magnetic media container.

Every organization should have a written policy to govern what is stored and for how long. Backup schedules and rotation media to be used in an offsite location are important. This rotation of media can be performed via management software.

Backup Schemes

There are three main schemes for backup: full, incremental and differential. Each one has its advantages and disadvantages. Usually, the methods are combined, in order to complement each other.

FULL BACKUP

This type of backup scheme copies all files and folders to the backup media, creating one backup set (with one or more media, depending on media capacity). The main advantage is having a unique repository in case of restoration, but it requires more time and media capacity.

INCREMENTAL BACKUP

An incremental backup copies the files and folders that changed or are new since the last incremental or full backup. If you have a full backup on day 1, your incremental backup on day 2 will copy only the changes from day 1 to day 2. On day 3, it will copy only the changes from day 2 to day 3, and so on. Incremental backup is a faster method of backup and requires less media capacity, but it requires that all backup sets restore all changes since a full backup, and restoration will take more time.

Figure 4.37 provides an example of a full plus incremental backup scheme. On day 1 there was a full backup and all files were saved to backup media. On days 2 to 7, there were incremental backups. On day 2, file 1 changed. On day 3, file 2 changed. On day 4, file 3 changed. On day 5, file 4 changed. The X shows which files were backed up.

Figure 4.37—Full Plus Incremental Backup Scheme

	Day 1	Day 2	Day 3	Day 4	Day 5	Day 6	Day 7
File 1	X	X					
File 2	X		X				
File 3	X			X			
File 4	X				X		

DIFFERENTIAL BACKUP

A differential backup will copy all files and folders that have been added or changed since a full backup was performed. This type of backup is faster and requires less media capacity than a full backup and requires only the last full and differential backup sets to make a full restoration. It also requires less time to restore than incremental backups, but it is slower and requires more media capacity than incremental backups because data that are backed up are cumulative.

Figure 4.38 depicts an example of a full plus differential backup scheme. On day 1 there is a full backup. On days 2 to 7, there are differential backups. On day 2, file 1 changed. On day 3, file 2 changed. On day 4, file 3 changed. On day 5, file 4 changed. The X shows which files were backed up.

Note that, in differential backups, all files or folders that were changed since a full backup are repeatedly copied to the backup media.

Figure 4.38—Full Plus Differential Backup Scheme

	Day 1	Day 2	Day 3	Day 4	Day 5	Day 6	Day 7
File 1	X	X	X	X	X		
File 2	X		X	X	X		
File 3	X			X	X		
File 4	X				X		

Method of Rotation

Although there are various approaches for the rotation of media, one of the more accepted techniques is referred to as the Grandfather-Father-Son method. In this method, daily backups (son) are made over the course of a week. The final backup taken during the week becomes the backup for that week (father). The earlier daily backup media are then rotated for reuse as backup media for the second week. At the end of the month, the final weekly backup is retained as the backup for that month (grandfather). Earlier weekly backup media are then rotated for reuse in subsequent months. At the end of the year, the final monthly backup becomes the yearly backup. Normally, monthly and annual tapes/other media are retained and not subject to the rotation cycle. See **figures 4.39** and **4.40** for examples of typical rotation cycles.

Figure 4.39—Typical Rotation Cycle, Sample A

	Day 1	Day 2	Day 3	Day 4	Day 5	Day 6	Day 7
Week 1	Tape 1	Tape 2	Tape 3	Tape 4	Tape 5	Tape 6	Tape 7 (week tape)
Week 2	Tape 1	Tape 2	Tape 3	Tape 4	Tape 5	Tape 6	Tape 8 (week tape)
Week 3	Tape 1	Tape 2	Tape 3	Tape 4	Tape 5	Tape 6	Tape 9 (week tape)
Week 4	Tape 1	Tape 2	Tape 3	Tape 4	Tape 5	Tape 6	Tape 10 (week tape)
Week 5	Tape 1	Tape 2	Tape 3	Tape 4	Tape 5	Tape 6	Tape 7 (week tape)

Figure 4.40—Typical Rotation Cycle Sample B

	Mon	Tue	Wed	Thu	Fri
W1	S	S	S	S	F
W2	S	S	S	S	F
W3	S	S	S	S	F
W4	S	S	S	S	GF

S	Son
F	Father
GF	Grandfather

A key element to this approach is that backups rotated offsite should not be returned for reuse until their replacement has been sent offsite. As an example, the backup media for week 1 should not be returned from offsite storage until the month-end backup is safely stored offsite. Variations of this method can be used depending on whether quarterly backups are required and on the amount of redundancy an organization may wish to have.

Record Keeping for Offsite Storage

An inventory of contents at the offsite storage location should be maintained. This inventory should contain information such as:
- Data set name, volume serial number, date created, accounting period and offsite storage bin number for all backup media
- Document name, location, pertinent system and date of last update for all critical documentation

Automated media management systems usually have options that help in recording and maintaining this information—bar code stickers for magnetic tapes and robotic arms with bar code readers for tape libraries. If backup media are carried between facilities, then both receipt and shipment logs should be maintained to assist tracking in case of losses.

4.8.7 DISASTER RECOVERY TESTING METHODS

Based on the risk assessment and BIA, critical applications and infrastructure are identified for testing. These should be developed into a testing schedule.

Testing all aspects of the DRP is the most important factor in achieving success in an emergency situation. The main objective of testing is to ensure that executing the plans will result in the successful recovery of the infrastructure and critical business processes. Testing should focus on:
- Identifying gaps
- Verifying assumptions
- Testing time lines
- Effectiveness of strategies
- Performance of personnel
- Accuracy and currency of plan information

Testing promotes collaboration and coordination among teams and is a useful training tool. Many organizations require complete testing annually. In addition, testing should be considered on the completion or major revision of each draft plan or complementary plans and following changes in key personnel, technology or the business/regulatory environment.

Testing must be carefully planned and controlled to avoid placing the business at increased risk. To ensure that all plans are regularly tested, the IS auditor should be aware of the testing schedule and tests to be conducted for all critical functions.

All tests must be fully documented with pre-test, test and post-test reports. Test documentation should be reviewed by the IS auditor. Information security should also be validated during the test to ensure that it is not being compromised.

Recovery plans that have not been tested leave an organization with an unacceptable likelihood that plans will not work. As testing plans cost time and resources, an organization should carefully plan and develop test objectives to ensure that measurable benefits can be achieved. Once these objectives have been defined, an independent third party such as the IS auditor should be present to monitor and evaluate the test. A result of the evaluation step should be a list of recommendations to improve the plan.

In summary, testing should include:
- Developing test objectives
- Executing the test
- Evaluating the test
- Developing recommendations to improve the effectiveness of testing processes and recovery plans
- Implementing a follow-up process to ensure that the recommendations are implemented

It is extremely unlikely that no recommendations will result and that everything works as planned. If it does, it is likely that a more challenging test should have been planned.

Types of Tests

The types of disaster recovery tests include:
- **Checklist review**—This is a preliminary step to a real test. Recovery checklists are distributed to all members of a recovery team to review and ensure that the checklist is current.
- **Structured walk-through**—Team members physically implement the plans on paper and review each step to assess its effectiveness, identify enhancements, constraints and deficiencies.

- **Simulation test**—The recovery team role play a prepared disaster scenario without activating processing at the recovery site.
- **Parallel test**—The recovery site is brought to a state of operational readiness, but operations at the primary site continue normally.
- **Full interruption test**—Operations are shut down at the primary site and shifted to the recovery site in accordance with the recovery plan; this is the most rigorous form of testing but is expensive and potentially disruptive.

Testing should start simply and increase gradually, stretching the objectives and success criteria of previous tests so as to build confidence and minimize risk to the business. **Figure 4.41** shows how tests can become progressively more challenging.

Figure 4.41—Progression of Disaster Recovery Tests

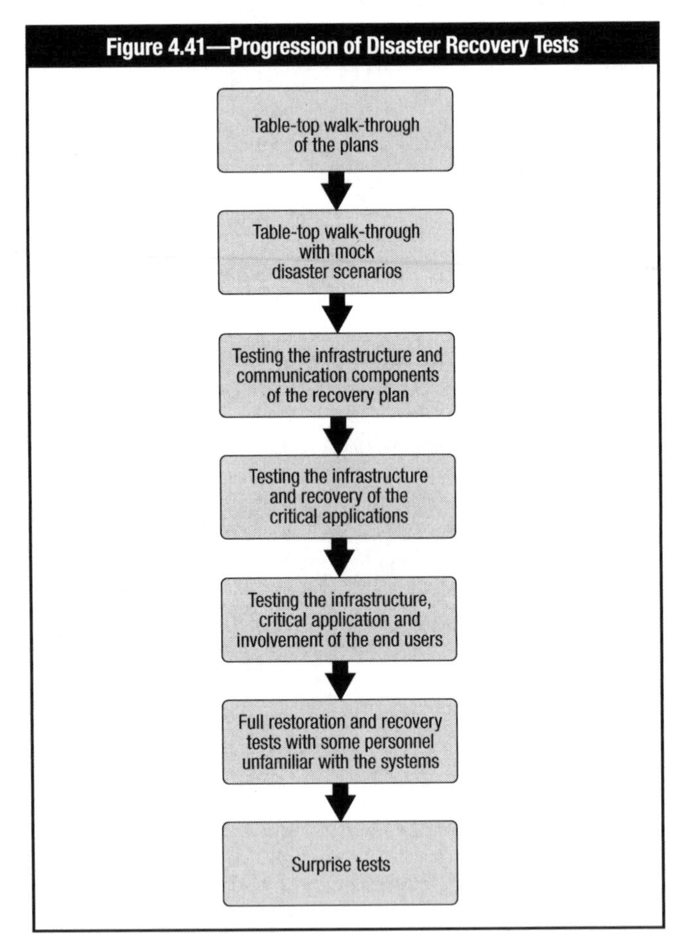
Table-top walk-through of the plans
Table-top walk-through with mock disaster scenarios
Testing the infrastructure and communication components of the recovery plan
Testing the infrastructure and recovery of the critical applications
Testing the infrastructure, critical application and involvement of the end users
Full restoration and recovery tests with some personnel unfamiliar with the systems
Surprise tests

Most recovery tests fall short of a full-scale test of all operational portions of the corporation. This should not preclude performing full or partial testing because one of the purposes of the disaster recovery test is to determine how well the plan works or which portions of the plan need improvement. Surprise tests are advantageous because they are similar to real-life incident response situations. However, they can be terribly disruptive to production and operations and can alienate individuals who are in some way disrupted by them.

The test should be scheduled during a time that will minimize disruptions to normal operations, such as long weekends. It is important that the key recovery team members are involved in the test process and are allotted the necessary time to devote their full effort. The test should address all critical components and simulate actual prime-time processing conditions, even if the test is conducted during off hours. Ideally, full-interruption tests should be performed annually after individual plans have been tested separately with satisfactory results.

Testing

The test should strive to accomplish the following tasks:

- Verify the completeness and precision of the response and recovery plan.
- Evaluate the performance of the personnel involved in the exercise.
- Appraise the demonstrated level of training and awareness of individuals who are not part of the recovery/response team.
- Evaluate the coordination among the team members and external vendors and suppliers.
- Measure the ability and capacity of the backup site to perform prescribed processing.
- Assess the vital records retrieval capability.
- Evaluate the state and quantity of equipment and supplies that have been relocated to the recovery site.
- Measure the overall performance of operational and information systems processing activities related to maintaining the business entity.

To perform testing, each of the following phases should be completed:

- **Pre-test**—The pre-test consists of the set of actions necessary to set the stage for the actual test, including transporting and installing required backup equipment, gaining access to the recovery site, accessing recovery documentation, etc.
- **Test**—The test is the real action of the disaster recovery test. Actual operational activities are executed to test the specific objectives of the plan. Applications are failed over; data entry and business processing should take place. Evaluators should review staff members as they perform the designated tasks. This is the actual test of preparedness to respond to an emergency.
- **Post-test**—The post-test is the cleanup of group activities. This phase comprises assignments such as restoring the applications back to the primary location and returning all resources to their proper place, disconnecting equipment, returning personnel to their normal locations, and deleting all company data from third-party systems. The post-test cleanup also includes formally evaluating the plan and implementing indicated improvements.

During every phase of the test, detailed documentation of observations, problems and resolutions should be maintained. Each team should have a diary with specific steps and information recorded. This documentation serves as important historical information that can facilitate actual recovery during a real disaster. The documentation also aids in performing detailed analysis of the strengths and weaknesses of the plan.

Test Results

Metrics should be developed and used in measuring the success of the plan and testing against the stated objectives. Results should be recorded and evaluated quantitatively, as opposed to an evaluation based only on verbal descriptions. The resulting metrics should be used not only to measure the effectiveness of the plan, but more importantly, to improve it. Although specific measurements vary depending on the test and the organization, the following types of metrics usually apply:

- **Time**—Elapsed time for completion of prescribed tasks. This is essential to refine the response time estimated for every task in the escalation process. Was the RTO met?
- **Data**—Were all data required data recovered? Was the RPO met? Was the recovery point aligned (where required) across all inter-connected applications?
- **Amount**—Amount of work performed at the backup site by clerical personnel and the amount of information systems processing operations. Does the recovery site allow the required throughput?
- **Percentage and/or number**—The number of critical systems successfully recovered can be measured with the number of transactions processed.
- **Accuracy**—Accuracy of the data entry at the recovery site versus normal accuracy (as a percentage). The accuracy of actual processing cycles can be determined by comparing output results with those for the same period processed under normal conditions.

4.8.8 INVOKING DISASTER RECOVERY PLANS

The BCP and DRP should be very closely aligned. As noted in section 2.12.5 Business Continuity Planning Incident Management, a designated individual should be notified of all relevant incidents as soon as any triggering event occurs. This person should then follow a pre-established escalation protocol (e.g., calling in a spokesperson, alerting top management and involving regulatory agencies), which may be followed by invoking a recovery plan such as the information technology DRP.

The required teams (see section 4.8.5 Organization and Assignment of Responsibilities) should be then be mobilized with the incident evaluated to confirm which of the tested scenarios it most closely resembles. Examples include:

- Loss of network connectivity
- Loss of a key IT system
- Loss of the processing site (server room)
- Loss of critical data
- Loss of an office, etc.
- Loss of key service provider (e.g., cloud)

Note that there may be more than one way to respond to a given incident. These should be evaluated with those most likely to deliver the required RPO and RTO selected. The documented recovery procedures should then be followed. It should be noted that recovery procedures may not include all required recovery steps as the testing may not have been comprehensive or the selected scenario an exact match. In such incidents the response teams may need to evaluate their options at each step. All decisions made should be documented and used to update the recovery procedures after normal service has been achieved.

4.9 CASE STUDIES

The following case studies are included as a learning tool to reinforce the concepts introduced in this chapter.

4.9.1 CASE STUDY A

The IS auditor has recently been asked to perform an external and internal network security assessment for an organization that processes health benefit claims. The organization has a complex network infrastructure with multiple local area and wireless networks and a Frame Relay network crosses international borders. Additionally, there is an Internet site that is accessed by doctors and hospitals. The Internet site has both open areas and sections containing medical claim information that requires an ID and password to access. An intranet site is also available that allows employees to check on the status of their personal medical claims and purchase prescription medications at a discount using a credit card. The frame relay network carries unencrypted, nonsensitive statistical data that are sent to regulatory agencies but do not include any customer identifiable information. The last review of network security was performed more than five years ago. At that time, numerous exposures were noted in the areas of firewall rule management and patch management for application servers. Internet applications were also found to be susceptible to SQL injection. It should be noted that wireless access as well as the intranet portal had not been installed at the time of the last review. Since the last review, a new firewall has been installed and patch management is now controlled by a centralized mechanism for pushing patches out to all servers. Internet applications have been upgraded to take advantage of newer technologies. Additionally, an intrusion detection system has been added, and reports produced by this system are monitored on a daily basis. Traffic over the network involves a mixture of protocols, as a number of legacy systems are still in use. All sensitive network traffic traversing the Internet is first encrypted prior to being sent. Traffic on the internal local area and wireless networks is encoded in hexadecimal so that no data appear in cleartext. A number of devices also utilize Bluetooth to transmit data between tablets and laptop computers.

CASE STUDY A QUESTIONS	
A1.	In performing an external network security assessment, which of the following should normally be performed **FIRST**? A. Exploitation B. Enumeration C. Reconnaissance D. Vulnerability scanning
A2.	Which of the following presents the **GREATEST** risk to the organization? A. Not all traffic traversing the Internet is encrypted. B. Traffic on internal networks is unencrypted. C. Cross-border data flow is unencrypted. D. Multiple protocols are being used.
See answers and explanations to the case study questions at the end of the chapter (page 316).	

4.9.2 CASE STUDY B

The IS auditor has been asked to represent the internal audit department on a task force to define the requirements for a new branch automation project for a community bank with 17 branches. This new system would handle deposit and loan information as well as other confidential customer information. The branches are all located within the same geographic area, so the director of branch operations has suggested the use of a microwave radio system to provide connectivity due to its low cost of operation and the fact that it is a private (and not a public) network. The director has also strongly suggested that it would be preferable to provide each branch with a direct coaxial connection to the Internet (using the local cable television provider) as a backup should the microwave system develop a fault. The direct Internet connection would also be connected to a wireless access point at each branch to provide free wireless access to customers. The director also asked that each branch be provided with mail and application servers that would be administered by the administrative manager of each branch. The IS auditor was informed by the IT manager for the bank that the cable service provider will encrypt all traffic sent over the direct coaxial connection to the Internet.

CASE STUDY B QUESTIONS	
B1.	In reviewing the information for this project, what would be the **MOST** important concern regarding the use of microwave radio systems based on the above scenario? A. Lack of encryption B. Lack of scalability C. Likelihood of a service outage D. Cost overruns in implementation
B2.	Which of the following would **BEST** reduce the likelihood of business systems being attacked through the wireless network? A. Scanning all connected devices for malware B. Placing the wireless network on a firewalled subnet C. Logging all access and issuing alerts for failed logon attempts D. Limiting access to regular business hours and standard protocols
See answers and explanations to the case study questions at the end of the chapter (page 316).	

4.10 ANSWERS TO CASE STUDY QUESTIONS

ANSWERS TO CASE STUDY A QUESTIONS

A1. **C** Information reconnaissance should be performed first to establish the "footprint" of the target organization (e.g., Internet-facing IP address ranges) and search for any "information leakage" that would inadvertently disclose technical details about the organization's network. Such leakage can occur as the result of Internet postings in which a network administrator asks a question regarding how to correct a network problem and identifies their organization, or when a job posting requests specific experience in a certain firewall or security package. Enumeration involves mapping the network services, protocols and devices and would normally occur after the initial reconnaissance. Vulnerability scanning and exploitation would occur in the later stages of the assessment.

A2. **B** The internal network is used to transmit sensitive information such as patient information and credit card numbers. Because the internal network also includes wireless, these factors create a major risk when such transmissions are not encrypted. With regard to the other choices, it is not necessary that all Internet traffic be encrypted. The fact that sensitive traffic traversing the Internet is encrypted should be sufficient. Because cross-border data flow does not include any sensitive information, this does not present a significant risk. The use of multiple protocols is typical and does not present a significant risk to the organization.

ANSWERS TO CASE STUDY B QUESTIONS

B1. **A** Lack of encryption is the most important concern since microwave radio systems are easy to tap. Lack of scalability and the likelihood of a service outage or cost overruns in implementation are important, but not as important as ensuring the confidentiality and integrity of customer data.

B2. **B** Isolating the wireless network by placing it on a firewalled subnet would best reduce the likelihood of attack. Scanning for malware would not detect the use of investigative tools designed to harvest passwords or reveal network vulnerabilities. Logging access and limiting access to normal business hours would not prevent a successful attack.

Chapter 5:

Protection of Information Assets

Section One: Overview

Section Two: Content

Section One: Overview

DEFINITION

This chapter addresses the key components that ensure confidentiality, integrity and availability (CIA) of information assets. The design, implementation and monitoring of logical and physical access controls are explained. Network infrastructure security, environmental controls, and processes and procedures used to classify, enter, store, retrieve, transport and dispose of confidential information assets are covered. The methods and procedures followed by organizations are described, focusing on the auditor's role in evaluating these procedures for suitability and effectiveness.

OBJECTIVES

The objective of this domain is to ensure that the CISA candidate understands and can provide assurance that the enterprise's security policies, standards, procedures and controls ensure the confidentiality, integrity and availability of information assets.

This area represents 25 percent of the CISA examination (approximately 38 questions).

TASK AND KNOWLEDGE STATEMENTS

TASKS

There are six tasks within the protection of information assets domain:

- T5.1 Evaluate the information security and privacy policies, standards and procedures for completeness, alignment with generally accepted practices and compliance with applicable external requirements.
- T5.2 Evaluate the design, implementation, maintenance, monitoring and reporting of physical and environmental controls to determine whether information assets are adequately safeguarded.
- T5.3 Evaluate the design, implementation, maintenance, monitoring and reporting of system and logical security controls to verify the confidentiality, integrity and availability of information.
- T5.4 Evaluate the design, implementation and monitoring of the data classification processes and procedures for alignment with the organization's policies, standards, procedures and applicable external requirements.
- T5.5 Evaluate the processes and procedures used to store, retrieve, transport and dispose of assets to determine whether information assets are adequately safeguarded.
- T5.6 Evaluate the information security program to determine its effectiveness and alignment with the organization's strategies and objectives.

KNOWLEDGE STATEMENTS

The CISA candidate must have a good understanding of each of the topics or areas delineated by the knowledge statements. These statements are the basis for the exam.

There are 26 knowledge statements within the protection of information assets domain:

- K5.1 Knowledge of generally accepted practices and applicable external requirements (e.g., laws, regulations) related to the protection of information assets
- K5.2 Knowledge of privacy principles
- K5.3 Knowledge of the techniques for the design, implementation, maintenance, monitoring and reporting of security controls
- K5.4 Knowledge of physical and environmental controls and supporting practices related to the protection of information assets
- K5.5 Knowledge of physical access controls for the identification, authentication and restriction of users to authorized facilities and hardware
- K5.6 Knowledge of logical access controls for the identification, authentication and restriction of users to authorized functions and data
- K5.7 Knowledge of the security controls related to hardware, system software (e.g., applications, operating systems) and database management systems.
- K5.8 Knowledge of risk and controls associated with virtualization of systems
- K5.9 Knowledge of risk and controls associated with the use of mobile and wireless devices, including personally owned devices (bring your own device [BYOD])
- K5.10 Knowledge of voice communications security (e.g., PBX, Voice-over Internet Protocol [VoIP])
- K5.11 Knowledge of network and Internet security devices, protocols and techniques
- K5.12 Knowledge of the configuration, implementation, operation and maintenance of network security controls
- K5.13 Knowledge of encryption-related techniques and their uses
- K5.14 Knowledge of public key infrastructure (PKI) components and digital signature techniques
- K5.15 Knowledge of risk and controls associated with peer-to-peer computing, instant messaging and web-based technologies (e.g., social networking, message boards, blogs, cloud computing)
- K5.16 Knowledge of data classification standards related to the protection of information assets
- K5.17 Knowledge of the processes and procedures used to store, retrieve, transport and dispose of confidential information assets
- K5.18 Knowledge of risk and controls associated with data leakage
- K5.19 Knowledge of security risk and controls related to end-user computing
- K5.20 Knowledge of methods for implementing a security awareness program
- K5.21 Knowledge of information system attack methods and techniques
- K5.22 Knowledge of prevention and detection tools and control techniques
- K5.23 Knowledge of security testing techniques (e.g., penetration testing, vulnerability scanning)

K5.24 Knowledge of processes related to monitoring and responding to security incidents (e.g., escalation procedures, emergency incident response team)

K5.25 Knowledge of the processes followed in forensics investigation and procedures in collection and preservation of the data and evidences (i.e., chain of custody).

K5.26 Knowledge of fraud risk factors related to the protection of information assets

Relationship of Task to Knowledge Statements

The task statements are what the CISA candidate is expected to know how to do. The knowledge statements delineate each of the areas in which the CISA candidate must have a good understanding in order to perform the tasks. The task and knowledge statements are mapped in **figure 5.1** insofar as it is possible to do so. Note that although there is often overlap, each task statement will generally map to several knowledge statements.

Figure 5.1—Task and Knowledge Statements Mapping

Task Statement	Knowledge Statements
T5.1 Evaluate the information security and privacy policies, standards and procedures for completeness, alignment with generally accepted practices and compliance with applicable external requirements.	K5.1 Knowledge of generally accepted practices and applicable external requirements (e.g., laws, regulations) related to the protection of information assets K5.2 Knowledge of privacy principles K5.3 Knowledge of the techniques for the design, implementation, maintenance, monitoring and reporting of security controls K5.4 Knowledge of physical and environmental controls and supporting practices related to the protection of information assets K5.5 Knowledge of physical access controls for the identification, authentication, and restriction of users to authorized facilities and hardware K5.6 Knowledge of logical access controls for the identification, authentication, and restriction of users to authorized functions and data K5.9 Knowledge of risk and controls associated with the use of mobile and wireless devices, including personally owned devices (bring your own device [BYOD]) K5.15 Knowledge of risk and controls associated with peer-to-peer computing, instant messaging, and web-based technologies (e.g., social networking, message boards, blogs, cloud computing) K5.16 Knowledge of data classification standards related to the protection of information assets K5.18 Knowledge of risk and controls associated with data leakage K5.19 Knowledge of security risk and controls related to end-user computing K5.20 Knowledge of methods for implementing a security awareness program K5.25 Knowledge of the processes followed in forensics investigation and procedures in collection and preservation of the data and evidences (i.e., chain of custody) K5.26 Knowledge of fraud risk factors related to the protection of information assets
T5.2 Evaluate the design, implementation, maintenance, monitoring and reporting of physical and environmental controls to determine whether information assets are adequately safeguarded.	K5.1 Knowledge of generally accepted practices and applicable external requirements (e.g., laws, regulations) related to the protection of information assets K5.2 Knowledge of privacy principles K5.3 Knowledge of the techniques for the design, implementation, maintenance, monitoring, and reporting of security controls K5.4 Knowledge of physical and environmental controls and supporting practices related to the protection of information assets K5.5 Knowledge of physical access controls for the identification, authentication, and restriction of users to authorized facilities and hardware K5.7 Knowledge of the security controls related to hardware, system software (e.g., applications, operating systems) and database management systems K5.10 Knowledge of voice communications security (e.g., PBX, Voice-over Internet Protocol [VoIP]) K5.17 Knowledge of the processes and procedures used to store, retrieve, transport and dispose of confidential information assets K5.18 Knowledge of risk and controls associated with data leakage K5.19 Knowledge of security risk and controls related to end-user computing K5.22 Knowledge of prevention and detection tools and control techniques K5.23 Knowledge of security testing techniques (e.g., penetration testing, vulnerability scanning) K5.26 Knowledge of fraud risk factors related to the protection of information assets

Figure 5.1—Task and Knowledge Statements Mapping *(cont.)*	
Task Statement	**Knowledge Statements**
T5.3 Evaluate the design, implementation, maintenance, monitoring and reporting of system and logical security controls to verify the confidentiality, integrity and availability of information.	K5.1 Knowledge of generally accepted practices and applicable external requirements (e.g., laws, regulations) related to the protection of information assets K5.2 Knowledge of privacy principles K5.3 Knowledge of the techniques for the design, implementation, maintenance, monitoring and reporting of security controls K5.6 Knowledge of logical access controls for the identification, authentication and restriction of users to authorized functions and data K5.7 Knowledge of the security controls related to hardware, system software (e.g., applications, operating systems) and database management systems. K5.8 Knowledge of risk and controls associated with virtualization of systems K5.9 Knowledge of risk and controls associated with the use of mobile and wireless devices, including personally owned devices (bring your own device [BYOD]) K5.10 Knowledge of voice communications security (e.g., PBX, Voice-over Internet Protocol [VoIP]) K5.11 Knowledge of network and Internet security devices, protocols, and techniques K5.12 Knowledge of the configuration, implementation, operation, and maintenance of network security controls K5.13 Knowledge of encryption-related techniques and their uses K5.14 Knowledge of public key infrastructure (PKI) components and digital signature techniques K5.15 Knowledge of risk and controls associated with peer-to-peer computing, instant messaging and web-based technologies (e.g., social networking, message boards, blogs, cloud computing) K5.18 Knowledge of risk and controls associated with data leakage K5.19 Knowledge of security risk and controls related to end-user computing K5.21 Knowledge of information system attack methods and techniques K5.22 Knowledge of prevention and detection tools and control techniques K5.23 Knowledge of security testing techniques (e.g., penetration testing, vulnerability scanning) K5.26 Knowledge of fraud risk factors related to the protection of information assets
T5.4 Evaluate the design, implementation and monitoring of the data classification processes and procedures for alignment with the organization's policies, standards, procedures and applicable external requirements.	K5.1 Knowledge of generally accepted practices and applicable external requirements (e.g., laws, regulations) related to the protection of information assets K5.2 Knowledge of privacy principles K5.3 Knowledge of the techniques for the design, implementation, maintenance, monitoring and reporting of security controls K5.9 Knowledge of risk and controls associated with the use of mobile & wireless devices, including personally owned devices (bring your own device [BYOD]) K5.13 Knowledge of encryption-related techniques and their uses K5.14 Knowledge of public key infrastructure (PKI) components and digital signature techniques K5.15 Knowledge of risk and controls associated with peer-to-peer computing, instant messaging and web-based technologies (e.g., social networking, message boards, blogs, cloud computing) K5.16 Knowledge of data classification standards related to the protection of information assets K5.18 Knowledge of risk and controls associated with data leakage K5.19 Knowledge of security risk and controls related to end-user computing K5.25 Knowledge of the processes followed in forensics investigation and procedures in collection and preservation of the data and evidences (i.e., chain of custody)

Figure 5.1—Task and Knowledge Statements Mapping *(cont.)*	
Task Statement	**Knowledge Statements**
T5.5 Evaluate the processes and procedures used to store, retrieve, transport and dispose of assets to determine whether information assets are adequately safeguarded.	K5.1 Knowledge of generally accepted practices and applicable external requirements (e.g., laws, regulations) related to the protection of information assets K5.2 Knowledge of privacy principles K5.3 Knowledge of the techniques for the design, implementation, maintenance, monitoring and reporting of security controls K5.4 Knowledge of physical and environmental controls and supporting practices related to the protection of information assets K5.5 Knowledge of physical access controls for the identification, authentication and restriction of users to authorized facilities and hardware K5.6 Knowledge of logical access controls for the identification, authentication, and restriction of users to authorized functions and data K5.7 Knowledge of the security controls related to hardware, system software (e.g., applications, operating systems) and database management systems K5.8 Knowledge of risk and controls associated with virtualization of systems K5.9 Knowledge of risk and controls associated with the use of mobile and wireless devices, including personally owned devices (bring your own device [BYOD]) K5.10 Knowledge of voice communications security (e.g., PBX, Voice-over Internet Protocol [VoIP]) K5.11 Knowledge of network and Internet security devices, protocols and techniques K5.12 Knowledge of the configuration, implementation, operation and maintenance of network security controls K5.13 Knowledge of encryption-related techniques and their uses K5.14 Knowledge of public key infrastructure (PKI) components and digital signature techniques K5.15 Knowledge of risk and controls associated with peer-to-peer computing, instant messaging, and web-based technologies (e.g., social networking, message boards, blogs, cloud computing) K5.17 Knowledge of the processes and procedures used to store, retrieve, transport and dispose of confidential information assets K5.18 Knowledge of risk and controls associated with data leakage K5.19 Knowledge of security risk and controls related to end-user computing K5.21 Knowledge of information system attack methods and techniques K5.22 Knowledge of prevention and detection tools and control techniques K5.23 Knowledge of security testing techniques (e.g., penetration testing, vulnerability scanning) K5.25 Knowledge of the processes followed in forensics investigation and procedures in collection and preservation of the data and evidences (i.e., chain of custody) K5.26 Knowledge of fraud risk factors related to the protection of information assets

Figure 5.1—Task and Knowledge Statements Mapping *(cont.)*

Task Statement	Knowledge Statements
T5.6 Evaluate the information security program to determine its effectiveness and alignment with the organization's strategies and objectives.	K5.1 Knowledge of generally accepted practices and applicable external requirements (e.g., laws, regulations) related to the protection of information assets K5.2 Knowledge of privacy principles K5.3 Knowledge of the techniques for the design, implementation, maintenance, monitoring and reporting of security controls K5.4 Knowledge of physical and environmental controls and supporting practices related to the protection of information assets K5.5 Knowledge of physical access controls for the identification, authentication and restriction of users to authorized facilities and hardware K5.6 Knowledge of logical access controls for the identification, authentication and restriction of users to authorized functions and data K5.7 Knowledge of the security controls related to hardware, system software (e.g., applications, operating systems) and database management systems K5.8 Knowledge of risk and controls associated with virtualization of systems K5.9 Knowledge of risk and controls associated with the use of mobile and wireless devices, including personally owned devices (bring your own device [BYOD]) K5.11 Knowledge of network and Internet security devices, protocols and techniques K5.12 Knowledge of the configuration, implementation, operation and maintenance of network security controls K5.13 Knowledge of encryption-related techniques and their uses K5.14 Knowledge of public key infrastructure (PKI) components and digital signature techniques K5.15 Knowledge of risk and controls associated with peer-to-peer computing, instant messaging and web-based technologies (e.g., social networking, message boards, blogs, cloud computing) K5.18 Knowledge of risk and controls associated with data leakage K5.19 Knowledge of security risk and controls related to end-user computing K5.20 Knowledge of methods for implementing a security awareness program K5.21 Knowledge of information system attack methods and techniques K5.22 Knowledge of prevention and detection tools and control techniques K5.23 Knowledge of security testing techniques (e.g., penetration testing, vulnerability scanning) K5.24 Knowledge of processes related to monitoring and responding to security incidents (e.g., escalation procedures, emergency incident response team) K5.25 Knowledge of the processes followed in forensics investigation and procedures in collection and preservation of the data and evidences (i.e., chain of custody) K5.26 Knowledge of fraud risk factors related to the protection of information assets

Knowledge Statement Reference Guide

Each knowledge statement is explained in terms of underlying concepts and relevance of the knowledge statement to the IS auditor. It is essential that the exam candidate understand the concepts. The knowledge statements are what the IS auditor must know in order to accomplish the tasks. Consequently, only the knowledge statements are detailed in this section.

The sections identified in K5.1 through K5.26 are described in greater detail in section two of this chapter.

K5.1 Knowledge of generally accepted practices and applicable external requirements (e.g., laws, regulations) related to the protection of information assets

Explanation	Key Concepts	Reference in Manual
There are a number of generally accepted practices related to the protection of information assets. The IS auditor should be aware of these practices. For example: • Security needs to be aligned with business objectives • Security should be led by senior management setting the "tone at the top" • Responsibilities for the protection of assets should be defined • Policies and procedures should be in place to: – Ensure the continued availability of information systems. – Ensure the integrity of information stored on its computer systems and while the information is in transit. – Preserve the confidentiality of sensitive data while stored and in transit. – Ensure compliance with applicable laws, regulations and standards. • Monitoring should be in place to ensure compliance with internal policies and any external requirements • A risk management process should be in place	Understanding the elements of information security management	5.2.1 Key Elements of Information Security Management 5.2.9 Critical Success Factors to Information Security Management
	Ability to assess the classification of information assets within an information security context	5.2.3 Classification of Information Assets

K5.2 Knowledge of privacy principles

Explanation	Key Concepts	Reference in Manual
IS auditors should be able to ensure adherence to trust and obligation requirements for any information relating to an identified or identifiable individual (i.e., data subject) in accordance with the applicable privacy policy, privacy laws and/or regulations.	Understanding of privacy principles	5.2.8 Privacy Principles and the Role of IS Auditors
	Knowledge of privacy laws and regulations	
	Understand how compliance is assured	

K5.3 Knowledge of the techniques for the design, implementation, maintenance, monitoring and reporting of security controls

Explanation	Key Concepts	Reference in Manual
The IS auditor should understand the different types of internal controls and their applicability. The design, implementation and monitoring of security should be aligned with business goals and objectives. The focus should be on those items that, if their security were compromised, would impact the organization in tangible (if not always quantifiable) ways. Controls generally incur a business cost, either directly or in their effect on business activities, and organizations should ensure that the cost of control does not materially exceed the business risk. The primary focus should be to ensure that risk which would have a material adverse impact on business is appropriately managed.	Understand the different types of controls (preventive, detective and corrective) and when to apply them	1.4.2 Internal Controls 5.2.5 Information Security Control Design 5.2.9 Critical Success Factors to Information Security Management
	Ability to assess the classification of information assets within an information security context	5.2.3 Classification of Information Assets
	Understanding information security as it applies to key network infrastructure components	5.4 Network Infrastructure Security

K5.4 Knowledge of physical and environmental controls and supporting practices related to the protection of information assets

Explanation	Key Concepts	Reference in Manual
Certain natural and manmade events have the ability to do great damage to an enterprise's information systems and business processes. Most data centers have mechanisms to prevent, detect or mitigate the impact of these threats. However, it is important that the readiness and sufficiency of these controls be periodically tested by management to ensure that they will function as intended. The IS auditor should understand the nature of these controls and how to ensure that they are functioning properly and are adequate to protect the enterprise. Environmental controls generally include fire and smoke detectors, fire suppression systems, water detectors, and temperature and humidity controls. The IS auditor should know the relative merits of different fire suppression systems and in what circumstances one type is more appropriate than another.	Understanding the common types of environmental controls and good practices for their deployment and periodic testing	5.7 Environmental Exposures and Controls

K5.5 Knowledge of physical access controls for the identification, authentication and restriction of users to authorized facilities and hardware

Explanation	Key Concepts	Reference in Manual
Physical security weaknesses can result in financial loss, legal repercussions or loss of credibility or competitive edge. Thus, information assets must be protected against physical attacks, such as vandalism and theft, through controls that restrict access to sensitive areas containing computer equipment or confidential data files. Such controls usually employ the use of access door locks that require the use of a password, key, token or biometric authentication of the person attempting entry. In high-security areas, access may require authentication through multiple means and the use of strong security measures such as the air-lock type or mantrap entrance. The IS auditor should understand the nature of physical controls and the ways in which they can be circumvented and the concept of "security boundary" to establish where such devices should be placed and how effective they must be.	Understanding physical access controls and their potential for circumvention	5.8 Physical Access Exposures and Controls

K5.6 Knowledge of logical access controls for the identification, authentication and restriction of users to authorized functions and data

Explanation	Key Concepts	Reference in Manual
Logical access controls are used to manage and protect information assets. Controls enact and substantiate policies and procedures designed by management to protect information assets, and controls are designed to reduce risk to a level acceptable to an enterprise. Controls exist at both the operating system level and the application level, so it is important to understand logical access controls as they apply to systems that may reside on multiple operating system platforms and involve more than one application system or authentication point. Logical security is often determined based on the job function of users. The success of logical access controls is tied to the strength of the authentication method (e.g., strong passwords). All user access to systems and data should be appropriately authorized and commensurate with the role of the individual. Authorization generally takes the form of signatures (physical or electronic) of relevant management. The strength of the authentication is proportional to the quality of the method used; strong authentication may include dual or multifactor authentication using user ID, password, tokens and biometrics.	Understanding the key elements of logical access controls	5.3 Logical Access

K5.7 Knowledge of the security controls related to hardware, system software (e.g., applications, operating systems) and database management systems

Explanation	Key Concepts	Reference in Manual
Access control software utilizes both identification and authentication (I&A): Once authenticated, the system or application then restricts access based on the specific role of the user. I&A is the process by which the system obtains the identity from a user and the credentials needed to authenticate the identity, and then I&A validates both pieces of information. I&A is a critical building block of computer security because it is needed for most types of access control and is necessary for establishing user accountability. For most systems, I&A is the first line of defense because it prevents unauthorized access (or unauthorized processes) to a computer system or an information asset. Logical access can be implemented in various ways. The IS auditor should be aware of the strengths and weaknesses of various architectures such as single sign-on (SSO), in which a single authentication will enable access to all authorized applications, identity management, multifactor authentication, etc. The IS auditor must understand the risk associated with the different architectures and how they may be addressed. For example, SSO may enable unauthorized access to applications and data if a single password is compromised. If this risk is considered manageable, it should drive the implementation of multifactor authentication.	Understanding good practices as they apply to identification and authentication	4.4 Information Systems Hardware 4.5.1 Operating Systems 4.5.5 Database Management Systems 5.3.5 Identification and Authentication

K5.8 Knowledge of risk and controls associated with virtualization of systems

Explanation	Key Concepts	Reference in Manual
Virtualization provides an enterprise with a significant opportunity to increase efficiency and to decrease costs in its IT operations. However, virtualization also introduces additional risk. IS auditors need to understand the advantages and disadvantages of virtualization and determine whether the enterprise has considered the applicable risk in its decision to adopt, implement and maintain this technology. At a high level, virtualization allows multiple operating systems (OSs), or guests, to coexist on the same physical server, or host, in isolation of one another. Virtualization creates a layer between the hardware and the guest OSs to manage shared processing and memory resources on the host. A management console often provides administrative access to manage the virtualized system. Although virtualization offers significant advantages, these advantages come with risk that an enterprise must manage effectively. Because the host in a virtualized environment represents a potential single point of failure within the system, a successful attack on the host could result in a compromise that is larger in both scope and impact.	Understanding the risk associated with virtualization	5.4.1 LAN Security

K5.9 Knowledge of risk and controls associated with the use of mobile and wireless devices, including personally owned devices (bring your own device [BYOD])

Explanation	Key Concepts	Reference in Manual
Portable and wireless devices present a ubiquitous threat to an enterprise's information assets and must be properly controlled. Policies and procedures and additional protection mechanisms must be put into place to ensure that data are protected to a greater extent on portable devices because such devices will most likely operate in environments in which physical controls are lacking or nonexistent. Most mobile devices, including tablets, smartphones, etc., are easily lost or stolen and, thus, require the use of encryption technologies and strong authentication. It also may be necessary to classify some data as inappropriate for storage on a mobile device. The IS auditor should understand that all such media and devices, including personal music (MP3) devices, can also be used by an individual to steal both data and programs for personal use or gain.	Understanding good practices for securing data on mobile computing devices	4.4.1 Computer Hardware Components and Architectures 4.6.6 Application of the OSI Model in Network Architectures 5.9 Mobile Computing

K5.10 Knowledge of voice communications security (e.g., PBX, Voice-over Internet Protocol [VoIP])

Explanation	Key Concepts	Reference in Manual
The increasing complexity and convergence of voice and data communications introduces additional risk that must be taken into account by the IS auditor. VoIP and PBX environments involve security risk (both within and outside the enterprise) that must be addressed to ensure the security and reliability of voice communications. The IS auditor should have enough understanding of these concepts to establish the business risk and identify appropriate controls.	Understanding the risk and associated controls related to voice communications and the impact of VoIP on overall network security	5.4.7 Voice-over IP 5.4.8 Private Branch Exchange

K5.11 Knowledge of network and Internet security devices, protocols and techniques

Explanation	Key Concepts	Reference in Manual
Application and evaluation of technologies to reduce risk and secure data are dependent on proper understanding of security devices, their functions and protocols used in delivering functionality. An enterprise implements specific applications of cryptographic systems to ensure confidentiality of important data. There are a number of cryptographic protocols that provide secure communications on the Internet. Additionally, the security landscape is filled with technologies and solutions to address a myriad of needs. Solutions include firewalls, intrusion detection and prevention devices, proxy devices, web filters, antivirus and antispam filters, data leak protection functionality, identity and access control mechanisms, secured remote access, and wireless security. Understanding the solution's function and its application to the underlying infrastructure requires knowledge of the infrastructure itself and the protocols in use. The IS auditor is not expected to possess a detailed, technical knowledge but rather a general understanding of the concepts, how they may be implemented and what business risk may be involved.	Understanding good practices for the implementation of encryption	5.4.5 Encryption
	Understanding the use and application of security devices and methods for securing data	4.6 IS Network Infrastructure 5.3.5 Identification and Authentication 5.4 Network Infrastructure Security

K5.12 Knowledge of the configuration, implementation, operation and maintenance of network security controls

Explanation	Key Concepts	Reference in Manual
Enterprises can effectively prevent and detect most attacks on their networks by employing perimeter security controls. Firewalls and intrusion detection systems (IDSs) provide protection and critical alert information at borders between trusted and untrusted networks. The proper implementation and maintenance of firewalls and IDSs is critical to a successful, in-depth security program. The IS auditor must understand the level of intruder detection provided by the different possible locations of the IDS and the importance of policies and procedures to determine the action required by security and technical staff when an intruder is reported.	Understanding network security threats and knowing the most appropriate controls to mitigate these threats	5.4.4 Internet Threats and Security

K5.13 Knowledge of encryption-related techniques and their uses

Explanation	Key Concepts	Reference in Manual
One of the best ways to protect the confidentiality of information is through the use of encryption. Effective encryption systems depend on: • Algorithm strength, secrecy and difficulty of compromising a key • The nonexistence of back doors by which an encrypted file can be decrypted without knowing the key • The inability to decrypt an entire ciphertext message if the way a portion of it decrypts is known (called a known-text attack) • Properties of the plaintext being known by a perpetrator Although the IS auditor is not expected to be an expert in how these algorithms are designed, the auditor should be able to understand how these techniques are used and the relative advantages and disadvantages of each.	Understanding the fundamentals of encryption techniques and the relative advantages and disadvantages of each	5.4.5 Encryption

K5.14 Knowledge of public key infrastructure (PKI) components and digital signature techniques

Explanation	Key Concepts	Reference in Manual
Encryption is the process of converting a plaintext message into a secure-coded form of text, called ciphertext, which cannot be understood without converting back via decryption (the reverse process) to plaintext. PKIs use encryption to facilitate the following: • Protect data in transit over networks from unauthorized interception and manipulation • Protect information stored on computers from unauthorized viewing and manipulation • Deter and detect accidental or intentional alterations of data • Verify authenticity of a transaction or document (e.g., when transmitted over a web-based connection in online banking, share dealing, etc.) • Protect data in such situations from unauthorized disclosure The IS auditor is not expected to have a detailed comprehension of cryptography but should understand the relationships between types of encryption (symmetric and asymmetric) and their respective algorithms (e.g., DES3, RSA) and the basic concepts and components of PKI in terms of business use. For example, if a message is encrypted with a private key, it provides authentication of the sender rather than privacy. Understanding the business use of digital signatures is also expected, especially its use in providing nonrepudiation of and replay protection to messages.	Understanding the key components of PKIs and how they are controlled	5.4.5 Encryption

K5.15 Knowledge of risk and controls associated with peer-to-peer computing, instant messaging and web-based technologies (e.g., social networking, message boards, blogs, cloud computing)

Explanation	Key Concepts	Reference in Manual
Peer-to-peer computing, instant messaging and web-based technologies (e.g., social networking, blogs) are technologies that introduce a unique type of risk to the enterprise. Information posted on social networking sites may inadvertently disclose confidential nonpublic information that may expose an organization (causing it to lose its competitive advantage) or may violate regulatory requirements such as financial securities or privacy laws. Peer-to-peer computing is inherently insecure, as it provides direct access to systems bypassing the network security controls, and may lead to the introduction of malicious code into an otherwise secure environment. The IS auditor should be familiar with the service models and deployment models available with cloud computing. The IS auditor should also be familiar with the key risk and controls associated with cloud computing including transborder issues, data disposal, exit strategy, etc.	Understanding risk and controls associated with peer-to-peer computing, instant messaging and web-based technologies (e.g., social networking, message boards, blogs)	5.10 Peer-to-peer Computing 5.11 Instant Messaging 5.12 Social Media
	Understanding risk and controls associated with cloud computing	2.9.2 Sourcing Practices 5.13 Cloud Computing

K5.16 Knowledge of data classification standards related to the protection of information assets

Explanation	Key Concepts	Reference in Manual
Information assets have varying degrees of sensitivity and criticality in meeting business objectives. Important first steps to data classification are discovery, inventory and risk assessment. The risk assessment should take into consideration that the value of the asset is directly proportional to its role in the strategy of the enterprise. Once this is accomplished, data classification can then be put into use. By assigning classes or levels of sensitivity and criticality to information resources and establishing specific security rules for each class, enterprises can define the level of access control and the retention time and destruction requirements that should be applied to each information asset. Data are, therefore, classified and protected in accordance with the degree of sensitivity and criticality assigned to them. The IS auditor should understand the process of classification and the interrelationship between data classification and the need for inventorying information assets and assigning responsibility to data owners. Data owner responsibilities should be clearly identified, documented and implemented.	Understanding data classification schemes and the need for assignment of data owners	4.3 IT Asset Management 5.2.3 Classification of Information Assets

K5.17 Knowledge of the processes and procedures used to store, retrieve, transport and dispose of confidential information assets

Explanation	Key Concepts	Reference in Manual
Confidential information assets are vulnerable during storage, retrieval and transport and must be disposed of properly. Management should define and implement procedures to prevent unauthorized access to or loss of sensitive information and software from computers, disks and other equipment or media when they are stored, transported or transmitted and during processing, retrieval and output. The IS auditor should also understand the need for correct disposal of information (and media) to ensure that no unauthorized person gains access to the information by restoration or recreation.	Understanding good practices for protecting information during storage, retrieval, transport and disposal	4.8.6 Backup and Restoration 5.3.7 Storing, Retrieving, Transporting and Disposing of Confidential Information

K5.18 Knowledge of risk and controls associated with data leakage

Explanation	Key Concepts	Reference in Manual
Data leakage is the risk that sensitive information may be inadvertently made public. It can occur in a variety of ways—from job postings that list the specific software and network devices with which applicants should have experience, to system administrators posting questions on technical web sites that include postings with specific details on the firewall or database version they are running and the IP addresses they are trying to connect. Other examples include posting organization charts and strategic plans on externally accessible web sites. At first glance, one would think that no enterprise would ever think of doing this, and yet there are governmental agencies and nonprofit organizations that have placed their organizations at risk in their zeal to be transparent. Data classification policies, security awareness training and periodic audits for data leakage are elements that the IS auditor will want to ensure are in place. The IS auditor should also be familiar with data leakage prevention (DLP) tools capabilities and risk.	Understanding how data leakage can occur and the methods for limiting data leakage	4.3 IT Asset Management 5.2.3 Classification of Information Assets 5.10 Peer-to-peer Computing 5.11 Instant Messaging 5.12 Social Media 5.14 Data Leakage

K5.19 Knowledge of security risk and controls related to end-user computing

Explanation	Key Concepts	Reference in Manual
It is necessary for the IS auditor to understand the security risk and controls associated with end-user computing (e.g. Microsoft® Excel, Access, etc.). This IS auditor should understand that these tools can be used to create key applications that are relied upon by the organization but not controlled by the IT department. This, in turn, means that they may not be secured, have logging enabled or sensitive data encrypted.	Understanding the security risk and controls associated with end-user computing	4.5.9 End-user Computing 5.15 End-user Computing Security Risk and Controls

K5.20 Knowledge of methods for implementing a security awareness program

Explanation	Key Concepts	Reference in Manual
The IS auditor should understand that risk in using IT systems is not only addressed through technical mechanisms. Security awareness programs can reduce risk through education. Security awareness programs should be aligned to the needs of the organization and focus on common user security concerns. Security awareness programs should also be tailored to specific groups. Security awareness programs can also be delivered through different media.	Understand the need for security awareness programs and the need to tailor them for organizational and user needs	5.2.1 Key Elements of Information Security Management
	Understand the different ways to implement security awareness programs	5.2.9 Critical Success Factors to Information Security Management

K5.21 Knowledge of information system attack methods and techniques

Explanation	Key Concepts	Reference in Manual
Risk arises from vulnerabilities (whether technical or human) within an environment. Attack techniques exploit those vulnerabilities and may originate either within or outside the enterprise. Computer attacks can result in proprietary or confidential data being stolen or modified, loss of customer confidence and market share, embarrassment to management, and legal actions against an enterprise. Understanding the methods, techniques and exploits used to compromise an environment provides the IS auditor with a more complete context for understanding the risk that an enterprise faces. Taking these techniques into consideration and understanding that they can be launched from any location allow for more thorough evaluations, ultimately providing a more secure environment. The IS auditor should understand enough of these attack types to recognize their risk to the business and how they should be addressed by appropriate controls. The IS auditor should understand the concept of social engineering as these attacks can circumvent the strongest technical security. The only effective control is regular user education.	Understanding general issues regarding attack methods and computer crime	5.2.12 Computer Crime Issues and Exposures
	Ability to identify controls that are most effective in preventing or detecting attacks involving social engineering, wireless access and threats originating from the Internet	5.4.3 Wireless Security Threats and Risk Mitigation 5.4.4 Internet Threats and Security

K5.22 Knowledge of prevention and detection tools and control techniques

Explanation	Key Concepts	Reference in Manual
Computer viruses and other malware continue to emerge at increasing rates and levels of sophistication and present significant threats to individuals and enterprises. Layered tools should be implemented and distributed throughout the environment to mitigate the ability of this malware to adversely impact the enterprise. Antivirus and antispam software are necessary and critical components of an enterprise's security program, providing a mechanism to detect, contain and notify whenever malicious code is detected. It is essential that the IS auditor understand not only the need for the implementation of anti-malware software, but that it should be constantly updated to ensure that it can detect and eradicate the latest attacks detected by the solutions providers.	Understanding the threats posed by malicious code and the good practices for mitigating these threats	5.4.4 Internet Threats and Security 5.4.6 Malware

K5.23 Knowledge of security testing techniques (e.g., penetration testing, vulnerability scanning)

Explanation	Key Concepts	Reference in Manual
Tools are available to assess the effectiveness of network infrastructure security. These tools permit identification of real-time risk to an information processing environment and corrective actions taken to mitigate the risk. Such risk often involves the failure to stay updated on patch management for operating systems or the misconfiguration of security settings. Assessment tools (whether open source or commercially produced) can quickly identify weaknesses that would have taken hundreds of hours to identify manually. The IS auditor should also be aware that security testing may be carried out by an approved third party (e.g., a company specializing in penetration testing).	Understanding how assessment tools can be used to identify vulnerabilities within the network infrastructure so that corrective actions can be taken to remediate risk	5.6 Auditing Network Infrastructure Security

K5.24 Knowledge of processes related to monitoring and responding to security incidents (e.g., escalation procedures, emergency incident response team)

Explanation	Key Concepts	Reference in Manual
A formal incident response capability should be established to minimize damage from security incidents, recover in a timely and controlled manner and learn from such incidents. The organization and management of an incident response capability should be coordinated or centralized with the establishment of key roles and responsibilities. While security management will typically be responsible for monitoring and investigating events and be the origination point for escalation procedures, other functions must be involved to ensure a proper response. Those functions must have well-defined and communicated processes in place that are tested periodically. These processes may include communications with executive management, forensic evidence collection, incident response and procedures to handle legal issues and public relations. The IS auditor should be aware of the need for enterprises to establish procedures to identify, report, record, respond, analyze, escalate and monitor security incidents.	Understanding the roles and responsibilities for incident response and the order and purpose of the key phases	5.2.13 Security Incident Handling and Response

K5.25 Knowledge of the processes followed in forensics investigation and procedures in collection and preservation of the data and evidences (i.e., chain of custody)

Explanation	Key Concepts	Reference in Manual
As electronic evidence is more dynamic than hard copy documents, security measures should be used to preserve the integrity of evidence collected and provide assurance that the evidence has not been altered in any way. In fraud investigations or legal proceedings, maintaining the integrity of evidence throughout the evidence life cycle may be referred to as the chain of custody when the evidence is classified as forensic. The IS auditor is expected to be aware of, rather than be a participant in, such specific evidence collection.	Factors to consider in collection, protection and chain of custody of evidence	5.5.4 Investigation Techniques

K5.26 Knowledge of fraud risk factors related to the protection of information assets

Explanation	Key Concepts	Reference in Manual
The IS auditor should be aware that the risk of fraud is increased where there is a perceived opportunity. An opportunity will be perceived where poor controls are in place. If an information asset is not properly protected it is more susceptible to fraud.	Relationship between controls and fraud risk	5.2.4 Fraud Risk Factors

SUGGESTED RESOURCES FOR FURTHER STUDY

Cendrowski, Harry; James P. Martin; Louis W. Petro; *The Handbook of Fraud Deterrence*, John Wiley & Sons Inc., USA, 2006

Davis, Chris; Mike Schiller; Kevin Wheeler; *IT Auditing: Using Controls to Protect Information Assets, 2nd Edition*, McGraw Hill, USA, 2011

Dubin, Joel; *The Little Black Book of Computer Security, 2nd Edition*, Penton Media Inc., USA, 2008

Harris, Shon; Allen Harper; Chris Eagle; Jonathan Ness; Gideon Lenkey; Terron Williams; *Gray Hat Hacking: The Ethical Hackers Handbook, 3rd Edition*, McGraw Hill, USA, 2011

ISACA, *The Business Model for Information Security*, USA, 2010

ISACA, *COBIT 5 for Information Security*, USA 2012, *www.isaca.org/cobit*

ISACA, *Security Considerations for Cloud Computing*, USA, 2013

International Organization for Standardization (ISO); *ISO/IEC 27002:2013: Information technology—Security techniques—Code of practice for information security controls*, Switzerland, 2013

Jaquith, Andrew; *Security Metrics: Replacing Fear, Uncertainty and Doubt*, Addison Wesley, USA, 2007

Killmeyer, Jan; *Information Security Architecture: An Integrated Approach to Security in the Organization, 2nd Edition*, Auerbach Publications, USA, 2006

Marcella Jr., Albert J.; Doug Menendez; *Cyber Forensics: A Field Manual for Collecting, Examining and Preserving Evidence of Computer Crime, 2nd Edition*, Auerbach Publications, USA, 2007

McClure, Stuart; Joel Scambray; George Kurtz; *Hacking Exposed 7: Network Security Secrets & Solutions*, McGraw Hill, USA, 2012

Natan, Ron Ben; *Implementing Database Security and Auditing*, Elsevier Digital Press, USA, 2005

Peltier, Thomas R.; *Information Security Risk Analysis, 3rd Edition*, Auerbach Publications, USA, 2010

Stamp, Mark; *Information Security: Principles and Practice, 2nd Edition*, John Wiley & Sons, USA, 2011

Stanley, Richard A.; *Managing Risk in the Wireless Environment: Security*, Audit and Control Issues, ISACA, USA, 2005

Vacca, John; *Biometric Technologies and Verification Systems*, Butterworth-Heinemann, USA, 2007

Wells, Joseph T.; *Fraud Casebook, Lessons From the Bad Side of Business*, John Wiley & Sons Inc., USA, 2007

Note: Publications in bold are stocked in the ISACA Bookstore.

SELF-ASSESSMENT QUESTIONS

CISA self-assessment questions support the content in this manual and provide an understanding of the type and structure of questions that have typically appeared on the exam. Questions are written in a multiple-choice format and designed for one best answer. Each question has a stem (question) and four options (answer choices). The stem may be written in the form of a question or an incomplete statement. In some instances, a scenario or a description problem may also be included. These questions normally include a description of a situation and require the candidate to answer two or more questions based on the information provided. Many times a question will require the candidate to choose the **MOST** likely or **BEST** answer among the options provided.

In each case, the candidate must read the question carefully, eliminate known incorrect answers and then make the best choice possible. Knowing the format in which questions are asked, and how to study and gain knowledge of what will be tested, will help the candidate correctly answer the questions.

5-1 An IS auditor reviewing the configuration of a signature-based intrusion detection system (IDS) would be **MOST** concerned if which of the following is discovered?

A. Auto-update is turned off.
B. Scanning for application vulnerabilities is disabled.
C. Analysis of encrypted data packets is disabled.
D. The IDS is placed between the demilitarized zone (DMZ) and the firewall.

5-2 Which of the following **BEST** provides access control to payroll data being processed on a local server?

A. Logging access to personal information
B. Using separate passwords for sensitive transactions
C. Using software that restricts access rules to authorized staff
D. Restricting system access to business hours

5-3 An IS auditor has just completed a review of an organization that has a mainframe computer and two database servers where all production data reside. Which of the following weaknesses would be considered the **MOST** serious?

A. The security officer also serves as the database administrator.
B. Password controls are not administered over the two database servers.
C. There is no business continuity plan for the mainframe system's noncritical applications.
D. Most local area networks (LANs) do not back up file-server-fixed disks regularly.

5-4 An organization is proposing to install a single sign-on facility giving access to all systems. The organization should be aware that:

A. maximum unauthorized access would be possible if a password is disclosed.
B. user access rights would be restricted by the additional security parameters.
C. the security administrator's workload would increase.
D. user access rights would be increased.

5-5 When reviewing an implementation of a Voice-over Internet Protocol (VoIP) system over a corporate wide area network (WAN), an IS auditor should expect to find:

A. an integrated services digital network (ISDN) data link.
B. traffic engineering.
C. wired equivalent privacy (WEP) encryption of data.
D. analog phone terminals.

5-6 An insurance company is using public cloud computing for one of its critical applications to reduce costs. Which of the following would be of **MOST** concern to the IS auditor?

A. The inability to recover the service in a major technical failure scenario
B. The data in the shared environment being accessed by other companies
C. The service provider not including investigative support for incidents
D. The long-term viability of the service if the provider goes out of business

5-7 Which of the following **BEST** determines whether complete encryption and authentication protocols for protecting information while being transmitted exist?

A. A digital signature with RSA has been implemented.
B. Work is being done in tunnel mode with the nested services of authentication header (AH) and encapsulating security payload (ESP).
C. Digital certificates with RSA are being used.
D. Work is being done in transport mode with the nested services of AH and ESP.

5-8 Which of the following concerns about the security of an electronic message would be addressed by digital signatures?

A. Unauthorized reading
B. Theft
C. Unauthorized copying
D. Alteration

5-9 Which of the following characterizes a distributed denial-of-service (DDoS) attack?

A. Central initiation of intermediary computers to direct simultaneous spurious message traffic at a specified target site
B. Local initiation of intermediary computers to direct simultaneous spurious message traffic at a specified target site
C. Central initiation of a primary computer to direct simultaneous spurious message traffic at multiple target sites
D. Local initiation of intermediary computers to direct staggered spurious message traffic at a specified target site

5-10 Which of the following is the **MOST** effective preventive antivirus control?

A. Scanning email attachments on the mail server
B. Restoring systems from clean copies
C. Disabling universal serial bus (USB) ports
D. An online antivirus scan with up-to-date virus definitions

ANSWERS TO SELF-ASSESSMENT QUESTIONS

5-1 **A. The most important aspect in a signature-based intrusion detection system (IDS) is its ability to protect against known (signature) intrusion patterns. Such signatures are provided by the vendor and are critical to protecting an enterprise from outside attacks.**
B. One of the key disadvantages of IDS is its inherent inability to scan for vulnerabilities at the application level.
C. An IDS cannot break encrypted data packets to identify the source of the incoming traffic.
D. A demilitarized zone (DMZ) is an internal network segment in which systems (e.g., a web server) accessible to the public are housed. In order to provide the greatest security and efficiency, an IDS should be placed behind the firewall so that it will detect only those attacks/intruders that enter the firewall.

5-2 A. Logging access to personal information is a good control in that it will allow access to be analyzed if there is concern of unauthorized access. However, it will not prevent access.
B. Restricting access to sensitive transactions will restrict access only to some of the data. It will not prevent access to other data.
C. The server and system security should be defined to allow only authorized staff members access to information about the staff whose records they handle on a day-to-day basis.
D. System access restricted to business hours only restricts when unauthorized access can occur and would not prevent such access at other times. It is important to consider that the data owner is responsible for determining who is allowed access via the written software access rules.

5-3 A. The security officer serving as the database administer, while a control weakness, does not carry the same disastrous impact as the absence of password controls.
B. The absence of password controls on the two database servers, where production data reside, is the most critical weakness.
C. Having no business continuity plan for the mainframe system's noncritical applications, while a control weakness, does not carry the same disastrous impact as the absence of password controls.
D. Most local area networks (LANs) not backing-up regularly, while a control weakness, does not carry the same disastrous impact as the absence of password controls.

5-4 **A. If a password is disclosed when single sign-on is enabled, there is a risk that unauthorized access to all systems will be possible.**
B. User access rights should remain unchanged by single sign-on, as additional security parameters are not implemented necessarily.
C. One of the intended benefits of single sign-on is the simplification of security administration.
D. One of the intended benefits of single sign-on is the unlikelihood of an increased workload.

5-5 A. The standard bandwidth of an integrated services digital network (ISDN) data link would not provide the quality of services required for corporate Voice-over Internet Protocol (VoIP) services.
B. To ensure that quality of service requirements are achieved, the VoIP service over the wide area network (WAN) should be protected from packet losses, latency or jitter. To reach this objective, the network performance can be managed to provide quality of service (QoS) and class of service (CoS) support using statistical techniques such as traffic engineering.
C. Wired equivalent privacy (WEP) is an encryption scheme related to wireless networking.
D. The VoIP phones are usually connected to a corporate local area network (LAN) and are not analog.

5-6 A. Benefits of cloud computing are redundancy and the ability to access systems and data in the event of a technical failure.

B. Considering that an insurance company must preserve the privacy/confidentiality of customer information, unauthorized access to information and data leakage are the major concerns.

C. The ability to investigate an incident is important, but most important is addressing the risk of an incident—the exposure of sensitive data.

D. If a cloud provider goes out of business, the data should still be available from backups.

5-7 A. A digital signature provides authentication and integrity.

B. Tunnel mode provides encryption and authentication of the complete IP package. To accomplish this, the authentication header (AH) and encapsulating security payload (ESP) services can be nested.

C. A digital certificate provides authentication and integrity.

D. The transport mode provides primary protection for the protocols' higher layers; that is, protection extends to the data field (payload) of an IP package.

5-8 A. Digital signatures will not identify, prevent or deter unauthorized reading.

B. Digital signatures will not identify, prevent or deter theft.

C. Digital signatures will not identify, prevent or deter unauthorized copying.

D. A digital signature includes an encrypted hash total of the size of the message as it was transmitted by its originator. This hash would no longer be accurate if the message was altered subsequently, indicating that the alteration had occurred.

5-9 **A. Choice A best describes a distribute denial-of-service (DDoS). Such attacks are centrally initiated and involve the use of multiple compromised computers. The attacks work by flooding the target site with spurious data, thereby overwhelming the network and other related resources. To achieve this objective the attacks need to be directed at a specific target and occur simultaneously.**

B. DDoS attacks are not locally initiated.

C. DDoS attacks are not initiated using a primary computer.

D. DDoS attacks are not staggered.

5-10 A. Scanning email attachments on the mail server is a preventive control. It will prevent infected email files from being opened by the recipients, which would cause their machines to become infected.

B. Restoring systems from clean copies is a preventive control. It will ensure that viruses are not introduced from infected copies or backups, which would reinfect machines.

C. Disabling universal serial bus (USB) ports is a preventive control. It prevents infected files from being copied from a USB drive onto a machine, which would cause the machine to become infected.

D. Antivirus software can be used to prevent virus attacks. By running regular scans, it can also be used to detect virus infections that have already occurred. Regular updates of the software are required to ensure it is able to update, detect and treat viruses as they emerge.

Section Two: Content

5.1 QUICK REFERENCE

Quick Reference Review

Chapter 5 addresses the need for the protection of information assets within an organization. Protection of information assets includes the key components that ensure confidentiality, integrity and availability (CIA) of information assets. The chapter evaluates design, implementation and monitoring of logical and physical access controls to ensure CIA. The chapter also evaluates network infrastructure security, environmental controls and processes and procedures used to store, retrieve, transport and dispose of confidential information assets. The chapter describes the various methods and procedures followed by organizations and focuses on the auditor's role in evaluating these procedures. Many of these topics may, on the surface, seem very familiar to candidates; however, it is important to note that the topics addressed in this chapter require a thorough knowledge of the technologies used and the potential control weaknesses that can be exploited by attackers. CISA candidates should be fully aware of and conversant with the components of network infrastructure security, logical access issues and the key elements of information security management.

CISA candidates should have a sound understanding of the following items, not only within the context of the present chapter, but also to correctly address questions in related subject areas. It is important to keep in mind that it is not enough to know these concepts from a definitional perspective. The CISA candidate must also be able to identify which elements may represent the greatest risk and which controls are most effective at mitigating this risk. Examples of key topics in this chapter are:

- Elements of information security management including senior management commitment and support, policies and procedures, organization, fraud risk factors, security control design, security awareness and education, monitoring and compliance, and incident handling and response
- General points of logical entry into a system including logical protection at the network, platform, database and application layers
- Identify how a failure at one layer could allow an unauthorized individual to bypass certain logical security mechanisms and gain access to confidential data
- Good practices for identification and authentication, including practices for handling default system accounts, normal user accounts and privileged user accounts, such as system administrators
- Various types of biometric technologies and the advantages and disadvantages of each
- Network infrastructure security including the various issues and risk associated with different technologies used in network infrastructures, and good practices for risk mitigation
  - Special attention should be focused on firewall implementation, the advantages and disadvantages of different types of intrusion detection/prevention systems, and encryption technologies.
- The importance of the proper maintenance of OS and other software, including using only known and acknowledged services and removing those that are not needed, patching the vulnerabilities and closing the ports that are not needed
- Environmental exposures and controls such as fire suppression systems, uninterruptible power supply (UPS), etc.
- Mobiles devices and the need for policies, procedures and encryption
- Social media and the risk in the enterprise
- The different models available with cloud computing including their risk and controls

5.2 INFORMATION SECURITY MANAGEMENT

Laying the foundation for effective information security management is the most critical factor in protecting information assets and privacy. Recent developments in the current environment—such as electronic trading through service providers and directly with customers, use of remote access facilities, and high-profile security exposures (e.g., viruses, denial-of-service [DoS] attacks, intrusions, identity theft, etc.)—have raised the profile of information and privacy risk and the need for effective information security management.

Security objectives to meet organization's business requirements include the following:

- Ensure the continued availability of their information systems and data.
- Ensure the integrity of the information stored on their computer systems and while in transit.
- Preserve the confidentiality of sensitive data while stored and in transit.
- Ensure conformity to applicable laws, regulations and standards.
- Ensure adherence to trust and obligation requirements in relation to any information relating to an identified or identifiable individual (i.e., data subject) in accordance with its privacy policy or applicable privacy laws and regulations.
- Ensure that sensitive data are adequately protected while stored and when in transit, based on organizational requirements.

COBIT 5 separates information goals into three subdimensions of quality:

- **Intrinsic quality**—The extent to which data values are in conformance with the actual or true values. It includes:
  - Accuracy—The extent to which information is correct and reliable
  - Objectivity—The extent to which information is unbiased, unprejudiced and impartial
  - Believability—The extent to which information is regarded as true and credible
  - Reputation—The extent to which information is highly regarded in terms of its source or content
- **Contextual and representational quality**—The extent to which information is applicable to the task of the information user and is presented in an intelligible and clear manner, recognizing that information quality depends on the context of use. It includes:
  - Relevancy—The extent to which information is applicable and helpful for the task at hand
  - Completeness—The extent to which information is not missing and is of sufficient depth and breadth for the task at hand
  - Currency—The extent to which information is sufficiently up to date for the task at hand
  - Appropriate amount of information—The extent to which the volume of information is appropriate for the task at hand
  - Concise representation—The extent to which information is compactly represented
  - Consistent representation—The extent to which information is presented in the same format

- Interpretability—The extent to which information is in appropriate languages, symbols and units, with clear definitions
- Understandability—The extent to which information is easily comprehended
- Ease of manipulation—The extent to which information is easy to manipulate and apply to different tasks

• **Security/accessibility quality**—The extent to which information is available or obtainable. It includes:
- Availability/timeliness—The extent to which information is available when required or is easily and quickly retrievable
- Restricted access—The extent to which access to information is restricted appropriately to authorized parties

It is important to recognize that these objectives are necessary, but not sufficient, because patterns often come into play regarding the objective of retaining competitive advantage. However, this section will not deal with ways and approaches to maintain such advantage, but rather how to protect the information systems from security pitfalls.

5.2.1 KEY ELEMENTS OF INFORMATION SECURITY MANAGEMENT

An IT system with state-of-the-art security features and devices will not be protected unless it is properly implemented and managed and carefully operated, monitored and reviewed. Security objectives cannot be met by only effecting technical and procedural protections. An educated security attitude and attention by all employees, management, and external service providers and external trusted IT users/partners are vital to the achievement of security objectives. Information security is more than just a mechanism. Information security also includes cultural aspects that must be embraced by all individuals within an organization for information security to be effective.

Information Security Management System

An information security management system (ISMS) is a framework of policies, procedures, guidelines and associated resources to establish, implement, operate, monitor, review, maintain and improve information security for all types of organizations. An ISMS is defined in the International Organization for Standardization (ISO)/International Electrotechnical Commission (IEC) 27000 series of standards and guidelines.

Introductory standard ISO/IEC 27000 defines the scope and vocabulary used throughout the ISMS standard and provides a directory of the publications that comprise the standard. This standard defines the requirements for an ISMS and establishes the basis for certification of an ISMS. ISO/IEC 27001 is the formal set of specifications against which organizations may seek independent certification of their information security management system. ISO/IEC 27002 contains a structured set of suggested controls that may be used by organizations as appropriate to address information security risk. Additional ISO/IEC 2700X publications offer guidance for managing information security in specific industries and situations.

The ISO/IEC 27000 series evolved from ISO/IEC 17799, which was based on the 1995 United Kingdom BSI standard BS7799 for the good practices of information security management. The ISO/IEC 27000 series may be purchased from ISO at www.iso.org or from the American National Standards Institute (ANSI) at *www.webstore.ansi.org*.

Note: For a detailed overview of Information Security Governance, please see chapter 2 Governance and Management of IT.

Figure 5.2 describes the related key elements of information security management.

Figure 5.2—Key Elements of Information Security Management

Senior management leadership, commitment and support	Commitment and support from senior management are important for successful establishment and continuance of an information security management program. This is commonly known as the "tone at the top."
Policies and procedures	The policy framework should be established with a concise top management declaration of direction, addressing the value of information assets, the need for security, and the importance of defining a hierarchy of classes of sensitive and critical assets. After approval by the governing body of the organization and by related roles and responsibilities, the information security program will be substantiated with the following: • Standards to develop minimum security baselines • Measurement criteria and methods • Specific guidelines, practices and procedures The policy should ensure resource conformity with laws and regulations. Security policies and procedures must be up to date and reflect business objectives, as well as generally accepted security standards and practices.
Organization	Responsibilities for the protection of individual assets should be clearly defined. The information security policy should provide general guidance on the allocation of security roles and responsibilities in the organization and, where necessary, detailed guidance for specific sites, assets, services and related security processes, such as IT recovery and business continuity planning.
Security awareness and education	All employees of an organization and, where relevant, third-party users should receive appropriate training and regular updates to foster security awareness and compliance with written security policies and procedures. For new employees, this training should occur before access to information or service is granted. A number of different mechanisms available for raising security awareness include: • Regular updates to written security policies and procedures • Formal information security training • Internal certification program for relevant personnel • Statements signed by employees and contractors agreeing to follow the written security policy and procedures, including nondisclosure obligations • Use of appropriate publication media for distribution of security-related material (e.g., company newsletter, web page, videos, etc.) • Visible enforcement of security rules and periodic audits • Security drills and simulated security incidents
Risk management	Processes should be in place to identify, assess, respond to and mitigate risk to information assets.

Figure 5.2—Key Elements of Information Security Management *(cont.)*	
Monitoring and compliance	IS auditors are usually charged to assess, on a regular basis, the effectiveness of an organization's security program(s). To fulfill this task, they must have an understanding of the protection schemes, the security framework and the related issues, including compliance with applicable laws and regulations. As an example, these issues may relate to organizational due diligence for security and privacy of sensitive information, particularly as it relates to specific industries (e.g., banking and financial institutions, health care).
Incident handling and response	A computer security incident is an event adversely affecting the processing of computer usage. This includes loss of confidentiality of information, compromise of integrity of information, denial of service, unauthorized access to systems, misuse of systems or information, theft and damage to systems. Other incidents include virus attacks and intrusion by humans within or outside the organization.

5.2.2 INFORMATION SECURITY MANAGEMENT ROLES AND RESPONSIBILITIES

All defined and documented responsibilities and accountabilities must be established and communicated to all relevant personnel and management. **Figure 5.3** presents roles and responsibilities of groups and individuals who may interact with information security management.

Figure 5.3—Roles and Responsibilities as Related to Information Security Management	
Information security steering committee	Security policies, guidelines and procedures affect the entire organization and, as such, should have the support and suggestions of end users, executive management, auditors, security administration, information systems personnel and legal counsel. Therefore, individuals representing various management levels should meet as a committee to discuss these issues and establish and approve security practices. The committee should be formally established with appropriate terms of reference. As an alternative, this role may be tasked to the IT strategy committee.
Executive management	Responsible for the overall protection of information assets, and for issuing and maintaining the policy framework.
Security advisory group	Responsible for defining the information security risk management process and acceptable level of risk and for reviewing the security plans of the organization. This group should include people involved in the business, provide comments on security issues to the chief security officer (CSO) and communicate to the business whether the security programs meet the business objectives.
Chief privacy officer (CPO)	A senior level corporate official responsible for articulating and enforcing the policies that companies use to protect their customers' and employees' privacy rights

Figure 5.3—Roles and Responsibilities as Related to Information Security Management *(cont.)*	
Chief information security officer (CISO)	The person in charge of information security within the enterprise
Chief security officer (CSO)	The person usually responsible for all security matters both physical and digital in an enterprise
Process owners	Ensure appropriate security measures are consistent with organizational policy and are maintained
Information asset owners and data owners	Ownership entails responsibility for the owned asset. This includes conducting a risk assessment, selecting appropriate controls to mitigate the risk to an acceptable level and accepting the residual risk.
Users	Follow procedures set out in the organization's security policy and adhere to privacy and security regulations, which are often specific to sensitive application fields (e.g., health care, finance, legal, etc.)
External parties	Follow procedures set out in the organization's security policy, and adhere to privacy and security regulations, which are often specific to sensitive application fields (e.g., health care, finance, legal, etc.)
Information security administrator	Staff level position responsible for providing adequate physical and logical security for IS programs, data and equipment. Normally, the information security policies will provide the basic guidelines under which the security administrator will operate.
Security specialists/ advisors	Assist with the design, implementation, management and review of the organization's security policy, standards and procedures
IT developers	Implement information security within their applications
IS auditors	Provide independent assurance to management on the appropriateness and effectiveness of information security objectives and the controls related to these objectives

5.2.3 CLASSIFICATION OF INFORMATION ASSETS

Effective control requires a detailed inventory of information assets. Creating this list is the first step in classifying assets and determining the level of protection needed for each asset.

Information assets have varying degrees of sensitivity and criticality in meeting business objectives. By assigning classes or levels of sensitivity and criticality to information resources and establishing specific security rules for each class, it is possible to define the level of access controls that should be applied to each information asset. Classification of information assets reduces the risk and cost of over- or under-protecting information resources in linking security to business objectives because it helps to build and maintain a consistent perspective of the security requirements for information assets throughout the organization.

The information owner is responsible for the information and should decide on the appropriate classification, based on the organization's data classification and handling policy. Classifications should be simple such as designations by differing

degrees for sensitivity and criticality. End-user managers and security administrators can then use these classifications in their risk assessment process to assist with determining who should be able to access what, and the most appropriate level of such access. Most organizations use a classification scheme with three to five levels of sensitivity. The number of classification categories should take into consideration the size and nature of the organization and the fact that complex schemes may become too impractical to use.

Data classification is a major part of managing data as an asset. Data classification as a control measure should define:
- The importance of the information asset
- The information asset owner
- The process for granting access
- The person responsible for approving the access rights and access levels
- The extent and depth of security controls

Data classification must take into account legal, regulatory, contractual and internal requirements for maintaining privacy, confidentiality, integrity and availability of information. Data classification is also useful to identify who should have access to the production data used to run the business versus those who are permitted to access test data and programs under development. For example, application programmers or system development programmers should not have access to production data or programs.

Adopting a classification scheme and assigning the information to one sensitivity level enables uniform treatment of data, through applying level-specific policies and procedures rather than addressing each type of information. It is highly difficult to follow information security policies if documents and media are not assigned to a sensitivity level and users are not instructed how to deal with each piece of information. If documents or media are not labeled according to a classification scheme, this is an indicator of a potential misuse of information. Users might reveal confidential information because they did not know that the requirements prohibited disclosure. Social engineering capitalizes on this kind of misunderstanding at the end user level. An example of classification of information is shown in **figure 5.4**.

Figure 5.4—Classification of Information	
Public Information	Company brochures
Private Information	Internal policies, procedures, normal business email messages, information controlled by legislation, etc.
Sensitive Information	Unpublished financials, company secrets, etc.

5.2.4 FRAUD RISK FACTORS

Fraud is the crime of using dishonest methods to take something valuable from a person or organization. There can be many reasons why a person commits fraud, but one of the more accepted models is the fraud triangle, which was developed by criminologist Donald R. Cressey in the 1950s. Cressey believed that the three key elements in the fraud triangle are opportunity, motivation and rationalization.

Motivation refers to a perceived financial (or other) need. The fraudster may be in debt, hold a personal grudge, have a problem with drugs or gambling, or want to enjoy status symbols, such as a bigger house or car.

Rationalization refers to the way the fraudster justifies the crime to himself/herself. Rationalization may include thoughts such as "I deserved the money," "I was only borrowing the money," "my family needs the money," "my employer has loads of money anyway," or "my employer treats me unfairly."

Opportunity refers to the method by which the crime is to be committed. Opportunity is created by abuse of position and authority, poor internal controls, poor management oversight, etc. Failure to establish procedures to detect fraud increases the likelihood of fraud occurring. Opportunity is the element over which organizations—and, by extension, IS auditors—have the most control. When considering information assets, the opportunities to commit fraud can be limited by security controls. These controls typically include logical access (including those for third parties), segregation of duties (SoD), human resources security, etc.

5.2.5 INFORMATION SECURITY CONTROL DESIGN

Information security is maintained through the use of controls. Controls may be **proactive**, meaning that they attempt to prevent an incident, or controls may be **reactive**, meaning that they allow the detection, containment and recovery from an incident. Proactive controls are often called safeguards, and reactive controls are known as countermeasures. For example, a sign that warns a person about a dangerous condition is a safeguard, whereas a fire extinguisher or sprinkler system is a countermeasure.

Every organization has some controls in place, and a risk assessment should document these controls and their effectiveness in mitigating risk. In some cases, the controls may be sufficient, whereas in others, the controls may need adjustment or replacement. An effective control is one that prevents, detects and/or contains an incident and enables recovery from an event.

It is common for an organization to have some situations where the controls currently in place are not sufficient to adequately protect the organization. In most cases, this requires the adjustment of the current controls or the implementation of new controls. However, it may not be feasible to reduce the risk to an acceptable level by either adjusting or implementing controls due to reasons such as cost, job requirements or availability of controls. An example of this could be found in a small organization when an individual is given administrator rights on a system and there is not adequate SoD. In this case, it may not be feasible to implement a new or enhanced control; some personnel need administrator rights to perform their jobs, and the risk cannot justify the cost of hiring new staff to address SoD. In such instances, compensating controls may be considered to reduce the risk. Compensating controls address the weaknesses in the existing controls through concepts such as layered defense, increased supervision, procedural controls, or increased audits and logging of system activity. These measures will work to compensate for the risk that could not be addressed in other ways.

Managerial, Technical and Physical Controls

Controls are often divided into three groups, as shown in **figure 5.5**.

Figure 5.5—Control Methods	
Category	**Description**
Managerial (administrative)	Controls related to the oversight, reporting, procedures and operations of a process. These include policy, procedures, balancing, employee development and compliance reporting.
Technical	Controls also known as logical controls and are provided through the use of technology, piece of equipment or device. Examples include firewalls, network or host-based intrusion detection systems (IDSs), passwords, and antivirus software. A technical control requires proper managerial (administrative) controls to operate correctly.
Physical	Controls that are locks, fences, closed-circuit TV (CCTV), and devices that are installed to physically restrict access to a facility or hardware. Physical controls require maintenance, monitoring and the ability to assess and react to an alert should a problem be indicated.

Further controls within these groups may be preventive, detective or corrective (see chapter 1 The Process of Auditing Information Systems, for more information). An example of a control matrix is shown in **figure 5.6**. Many controls may fit into more than one classification.

Figure 5.6—Control Matrix			
	Managerial	**Technical**	**Physical**
Preventive	User registration process	Login screen	Fence
Detective	Audit	Intrusion detection system (IDS)	Motion sensor
Corrective	Remove access	Network isolation	Close fire doors

Control Standards and Frameworks

The selection of controls requires the evaluation and implementation of the right control in the right way. Based on data collected through an analysis method (e.g., cost-benefit, return on investment [ROI], etc.), management will decide on the best available control, or group of controls, to mitigate a specific risk. However, a poorly implemented control may pose a significant risk to the organization by creating a false sense of security or leading to a denial of service if the control does not function correctly. The implementation of a technical control requires that the control is surrounded by proper procedures, the personnel that operate it are adequately trained, a person is assigned ownership of the control (often the person who owns the risk), and the control is monitored and tested to ensure its correct operation and effectiveness.

Many industries have standards that may be used as a benchmark for security across the industry sector. One example is the Payment Card Industry Data Security Standard (PCI DSS), which is used as a standard for all organizations that process payment cards (e.g., debit cards, credit cards, etc.). This is an example of an industry standard, but compliance is not required by law. Such standards, and the frameworks that implement those standards, are found in the health care, accounting, audit and telecommunications industries. In some regulated industries, regulations require compliance with a standard, such as the electrical power industry. To meet the requirements of the standard, a framework is often used to describe how an organization can achieve compliance.

A control framework is defined as a set of fundamental controls that facilitates the discharge of business process owner responsibilities to prevent financial or information loss in an enterprise. Therefore, it can be seen as the implementation of controls intended to support and protect business operations and preserve asset value.

Control Monitoring and Effectiveness

To support the ability to monitor and report on risk, the IS auditor should validate that processes, logs and audit hooks have been placed into the control framework. This allows for the monitoring and evaluation of controls. As controls are designed, implemented and operated, the IS auditor should ensure that logs are enabled, controls are able to be tested and regular reporting procedures are developed.

The IS auditor should also ensure that the capability to monitor a control and to support monitoring systems is addressed in control design. If the organization is using a managed security service provider (MSSP) or a security information and event management (SIEM) system, the ability to capture data, and the notification to the operations staff on the deployment of the system, are necessary.

5.2.6 SYSTEM ACCESS PERMISSION

System access permission is the prerogative to act on a computer resource. This usually refers to a technical privilege, such as the ability to read, create, modify or delete a file or data; execute a program; or open or use an external connection.

System access to computerized information resources is established, managed and controlled at the physical and/or logical level. Physical access controls restrict the entry and exit of personnel to an area such as an office building, suite, data center or room containing information processing equipment such as a local area network (LAN) server. There are many types of physical access controls including badges, memory cards, guard keys, true floor-to-ceiling wall construction fences, locks and biometrics. Logical system access controls restrict the logical resources of the system (transactions, data, programs, applications) and are applied when the subject resource is needed. On the basis of identification and authentication of the user that requires a given resource and by analyzing the security profiles of the user and the resource, it is possible to determine if the requested access is to be allowed (i.e., what information users can utilize, the programs or transactions they can run, and the modifications they can make). Such controls may be built into the operating system (OS), invoked through separate access control software and incorporated into application programs, database

systems, network control devices and utilities (e.g., real-time performance monitors).

Physical or logical system access to any computerized information should be on a documented need-to-know basis (often referred to as "role-based") where there is a legitimate business requirement based on least privilege. Other considerations for granting access are accountability (e.g., unique user ID) and traceability (e.g., logs). These principles should be used by IS auditors when they evaluate the appropriateness of criteria for defining permissions and granting security privileges. Organizations should establish such basic criteria for assigning technical access to specific data, programs, devices and resources, including who will have access and what level of access they will be allowed. For instance, it may be desirable for everyone in the organization to have access to specific information on the system such as the data displayed on an organization's daily calendar of meetings. The program that formats and displays the calendar might be modifiable by only a few system administrators, while the OS controlling that program might be directly accessible by still fewer.

The IT assets under logical security can be grouped in four layers—networks, platforms (OSs), databases and applications. This concept of layered security for system access provides greater scope and granularity of control to information resources. For example, network and platform layers provide pervasive general systems control over users authenticating into systems, system software and application configurations, data sets, load libraries, and any production data set libraries. Database and application controls generally provide a greater degree of control over user activity within a particular business process by controlling access to records, specific data fields and transactions.

The information owner or manager who is responsible for the accurate use and reporting of information should provide written authorization for users or defined roles to gain access to information resources under their control. The manager should hand over this documentation directly to the security administrator to ensure that mishandling or alteration of the authorization does not occur.

Logical access capabilities are implemented by security administration in a set of access rules that stipulate which users (or groups of users) are authorized to access a resource at a particular level (e.g., read-, update- or execute-only) and under which conditions (e.g., time of the day or a subset of computer terminals). The security administrator invokes the appropriate system access control mechanism upon receipt of a proper authorization request from the information owner or manager to grant a specified user the rights for access to, or use of, a protected resource. The IS auditor should be aware that access is granted to the organization's information systems utilizing the principles of need-to-know, least privilege and SoD.

Reviews of access authorization should be evaluated regularly to ensure that they are still valid. Personnel and departmental changes, malicious efforts, and just plain carelessness result in authorization creep and can impact the effectiveness of access controls. Many times, access is not removed when personnel leave an organization, thus increasing the risk of unauthorized access. For this reason, the information asset owner should review access controls periodically with a predetermined authorization matrix that defines the least-privileged access level and authority for an individual/role with reference to his/her job roles and responsibilities. Any access exceeding the access philosophy in authorized matrix or in actual access levels granted on a system should be updated and changed accordingly. One of the good practices is to integrate the review of access rights with human resource processes. When an employee transfers to a different function (i.e., promotions, lateral transfers or demotions), access rights are adjusted at the same time. Development of a security-conscious culture increases the effectiveness of access controls.

Nonemployees with access to corporate IS resources should also be held responsible for security compliance and be accountable for security breaches. Nonemployees include contract employees, vendor programmers/analysts, maintenance personnel, clients, auditors, visitors and consultants. It should be understood that nonemployees are also accountable to the organization's security requirements.

5.2.7 MANDATORY AND DISCRETIONARY ACCESS CONTROLS

Mandatory access controls (MACs) are logical access control filters used to validate access credentials that cannot be controlled or modified by normal users or data owners; they act by default. Controls that may be configured or modified by the users or data owners are called discretionary access controls (DACs).

MACs are a good choice to enforce a ground level of critical security without possible exception, if this is required by corporate security policies or other security rules. A MAC could be carried out by comparing the sensitivity of the information resources, such as files, data or storage devices, kept on a user-unmodifiable tag attached to the security object with the security clearance of the accessing entity such as a user or an application. With MACs, only administrators may make decisions that are derived from policy. Only an administrator may change the category of a resource, and no one may grant a right of access that is explicitly forbidden in the access control policy. MACs are prohibitive; anything that is not expressly permitted is forbidden.

DACs are a protection that may be activated or modified at the discretion of the data owner. This would be the case of data owner-defined sharing of information resources, where the data owner may select who will be enabled to access his/her resource and the security level of this access. DACs cannot override MACs; DACs act as an additional filter, prohibiting still more access with the same exclusionary principle.

When information systems enforce MAC policies, the systems must distinguish between MAC and the discretionary policies that offer more flexibility. This distinction must be ensured during object creation, classification downgrading and labeling.

5.2.8 PRIVACY PRINCIPLES AND THE ROLE OF IS AUDITORS

Privacy means freedom from unauthorized intrusion or disclosure of information about an individual (data subject). It is an organizationwide matter that, by its nature, requires a consistent

approach throughout the organization. Good practice to ensure this includes the following:
- Privacy should be considered from the outset and be built in by design. It should be systematically built into policies, standards and procedures from the beginning.
- Private data should be collected fairly in an open, transparent manner. Only the data required for the purpose should be collected in the first instance.
- Private data should be kept securely throughout their life cycle.
- Private data should only be used and/or disclosed for the purpose for which they were collected.
- Private data should be accurate, complete and up to date.
- Private data should be deleted when they are no longer required.

To best meet these challenges, management should perform a privacy impact analysis. IS auditors may be asked to support or perform this review. Such assessments should:
- Pinpoint the nature of personally identifiable information associated with business processes.
- Document the collection, use, disclosure and destruction of personally identifiable information.
- Ensure that accountability for privacy issues exists.
- Identify legislative, regulatory and contractual requirements for privacy.
- Be the foundation for informed policy, operations and system design decisions based on an understanding of privacy risk and the options available for mitigating that risk.

Based on the results, it should be possible to create a consistent format and structured process for analyzing technical and legal compliance with relevant regulations and internal policies. This structured process would provide a framework to ensure that privacy is considered in all IT projects, from the conceptual and requirements analysis stage to the final design approval, funding, implementation and communication stage, so that privacy compliance is built into projects rather than retrofitted.

The focus and extent of privacy impact analysis or assessment may vary depending on changes in technology, processes or people as shown in **figure 5.7**.

Figure 5.7—Changes That Impact Privacy

Technology	Processes	People
• New programs • Changes in existing programs • Additional system linkages • Data warehouse • New products	• Change management • Business process reengineering • Enhanced accessibility rules • New systems • New operations • Vendors	• Business partners • Service providers

The IS auditor may also be called on to give assurance on compliance with privacy policy, laws and other regulations. To fulfill this role, the IS auditor should:
- Identify and understand legal requirements regarding privacy from laws, regulations and contract agreements. Examples include the Organisation for Economic Co-operation and Development (OECD) Guidelines on the Protection of Privacy and Transborder Flows of Personal Data, European Union Data Protection Directives and the US-EU Safe Harbor Framework. Depending on the assignment, IS auditors may need to seek legal or expert opinion on these.
- Review management's privacy policy to ascertain whether it takes into consideration the requirement of these privacy laws and regulations.
- Check whether personal sensitive data are correctly managed in respect to these requirements.
- Verify that the correct security measures are adopted.

As laws and regulations vary from country to country, there may be a question as to how to approach privacy-related compliance requirements. *ISO/IEC 29100:2011: Information technology—Security techniques—Privacy framework* contains a description of the basic elements of a privacy framework and which privacy principles should be used. Furthermore, ISO/IEC 27018:2014 establishes commonly accepted control objectives, controls and guidelines for implementing measures to protect personally identifiable information (PII) in accordance with the privacy principles in ISO/IEC 29100 for the public cloud computing environment.

Note: The CISA exam does not test on specific privacy laws and standards because they vary by country.

5.2.9 CRITICAL SUCCESS FACTORS TO INFORMATION SECURITY MANAGEMENT

Managers and employees within an organization often tend to consider information security as a secondary priority if compared with their own efficiency or effectiveness matters because these have a direct and material impact on the outcome of their work.

For this reason, strong leadership, direction and commitment by senior management on security training is needed. This commitment should be supported with a comprehensive program of formal security awareness training.

Security Awareness, Training and Education

Risk that is inherent in using computing systems cannot be addressed through technical security mechanisms. An active security awareness program can greatly reduce risk by addressing the behavioral element of security through education and consistent application of awareness techniques. Security awareness programs should focus on common user security concerns—such as password selection, appropriate use of computing resources, email and web browsing safety, and social engineering—and the programs should be tailored to specific groups. In addition, users are the front line for the detection of threats that may not be detectable by automated means (e.g., fraudulent activity and social engineering). Employees should be educated on recognizing and escalating such events to enhance loss prevention.

An important aspect of ensuring compliance with the information security program is the education and awareness of the organization regarding the importance of the program. In addition to the need for information security, all personnel must be trained in their specific responsibilities related to information security. Particular attention must be paid to those job functions that

require virtually unlimited data access. People whose job is to transfer data may have access to data in most systems, and those doing performance tuning can change most OS configurations. People whose job is to schedule batch jobs have the authority to run most system jobs applications. Programmers have access to change application code. These functions are not typically managed by information security. Although it is possible to set up elaborate monitoring controls, it is not technically feasible or financially prudent for information security to provide oversight adequate to ensure that all data transfer jobs that transmit reports send them only to appropriately authorized recipients. Although information security can ensure that there is clear policy, develop applicable standards and assist in process coordination, management in all areas must assist in providing oversight.

Employee awareness should start from the point of joining the organization (e.g., through induction training) and continue regularly. Techniques for delivery need to vary to prevent them from becoming stale or boring and may also need to be incorporated into other organizational training programs.

Security awareness programs should consist of the following:
- Training (often administered online)
- Quizzes to gauge retention of training concepts
- Security awareness reminders such as posters, newsletters or screensavers
- A regular schedule of refresher training

In larger organizations, there may be a large enough population of middle and senior management to warrant special management-level training on information security awareness and operations issues.

All employees of an organization and, where relevant, third-party users must receive appropriate training and regular updates on the importance of security policies, standards and procedures in the organization. This includes security requirements, legal responsibilities and business controls, as well as training in the correct use of information processing facilities (e.g., login procedures, use of software packages). For new employees, this should occur before access to information or services is granted and be a part of new employee orientation.

A methodical approach should be taken to developing and implementing the education and awareness program with the following aspects being considered:
- Who is the intended audience (senior management, business managers, IT staff, end users)?
- What is the intended message (policies, procedures, recent events)?
- What is the intended result (improved policy compliance, behavioral change, better practices)?
- What communication method will be used (computer-based training [CBT], all-hands meeting, intranet, newsletters, etc.)?
- What is the organizational structure and culture?

A number of different mechanisms available for raising information security awareness include:
- Computer-based security awareness and training programs
- Email reminders and security tips
- Written security policies and procedures (and updates)
- Nondisclosure statements signed by the employee
- Use of different media in promulgating security (e.g., company newsletter, web page, videos, posters, login reminders)
- Visible enforcement of security rules
- Simulated security incidents for improving security
- Rewarding employees who report suspicious events
- Periodic reviews
- Job descriptions
- Performance reviews

A second critical success factor to information security management is that a professional risk-based approach must be used systematically to identify sensitive and critical information resources and to ensure that there is a clear understanding of threats and risk. Thereafter, appropriate risk assessment activities should be undertaken to mitigate unacceptable risk and ensure that residual risk is at an acceptable level. For more information, see sections 2.8 Risk Management, 4.3 IT Asset Management, and 5.2.3 Classification of Information Assets.

5.2.10 INFORMATION SECURITY AND EXTERNAL PARTIES

The security of the organization's information and information processing facilities that are accessed, processed, communicated to or managed by external parties should be maintained and should not be reduced by the introduction of external party products or services. Any access to the organization's information processing facilities and processing and communication of information by external parties should be controlled. Controls should be agreed to and defined in an agreement with the external party. Organizations shall gain the right to audit the implementation and operation of the resulting security controls.

These external party arrangements can include:
- Service providers such as Internet service providers (ISPs), network providers, telephone services, maintenance and support services
- Managed security services
- Customers
- Suppliers
- Outsourcing facilities and/or operations (e.g., IT systems, data collection services, call center operations)
- Management and business consultants and auditors
- Developers and suppliers (e.g., of software products and IT systems)
- Cleaning, catering and other outsourced support services
- Temporary personnel, student placement and other casual short-term appointments

Such agreements can help to reduce the risk associated with external parties.

Identification of Risk Related to External Parties

The risk to the organization's information and information processing facilities from business processes involving external parties should be identified and appropriate controls implemented before granting access. Where there is a need to allow an external party access to the information processing facilities or information of an organization, a risk assessment should be carried out to identify any requirements for specific controls. The identification of risk related to external party access should take into account the issues depicted in **figure 5.8**.

Figure 5.8—Risk Related to External Party Access

- The information processing facilities an external party is required to access
- The type of access the external party will have to the information and information processing facilities:
  - Physical access (e.g., to offices, computer rooms and filing cabinets)
  - Logical access (e.g., to an organization's databases and information systems)
  - Network connectivity between the organization's and the external party's network(s) (e.g., permanent connection and remote access)
  - Whether the access is taking place onsite or offsite
- The value and sensitivity of the information involved and its criticality for business operations
- The controls necessary to protect information that is not intended to be accessible by external parties
- The external party personnel involved in handling the organization's information
- How the organization or personnel authorized to have access can be identified, the authorization verified and how often this needs to be reconfirmed
- The different means and controls employed by the external party when storing, processing, communicating, sharing, exchanging and destroying information
- The impact of access not being available to the external party when required and the external party entering or receiving inaccurate or misleading information
- Practices and procedures to deal with information security incidents and potential damages and the terms and conditions for the continuation of external party access in the case of an information security incident
- Legal and regulatory requirements and other contractual obligations relevant to the external party that should be taken into account
- How the interests of any other stakeholders may be affected by the arrangements

Access by external parties to the organization's information should not be provided until the appropriate controls have been implemented and, where feasible, a contract has been signed defining the terms and conditions for the connection or access and the working arrangement. Generally, all security requirements resulting from work with external parties or internal controls should be reflected by the agreement with the external party. It should be ensured that the external party is aware of its obligations and accepts the responsibilities and liabilities involved in accessing, processing, communicating or managing the organization's information and information processing facilities.

External parties might put information at risk if their security management is inadequate. Controls should be identified and applied to administer external party access to information processing facilities. For example, if there is a special need for confidentiality of the information, nondisclosure agreements might be used. Organizations may face risk associated with interorganizational processes, management and communication, if a high degree of outsourcing is applied or where there are several external parties involved.

Addressing Security When Dealing With Customers

All identified security requirements should be addressed before giving customers access to the organization's information or assets.

The items presented in **figure 5.9** should be considered to address security prior to giving customers access to any of the organization's assets (depending on the type and extent of access given, not all of them may apply).

Figure 5.9—Customer Access Security Considerations

- Asset protection, including:
  - Procedures to protect the organization's assets, including information and software, and management of known vulnerabilities
  - Procedures to determine whether any compromise of the assets (e.g., loss or modification of data, has occurred)
  - Integrity
  - Restrictions on copying and disclosing information
- Description of the product or service to be provided
- The different reasons, requirements and benefits for customer access
- Access control policy, covering:
  - Permitted access methods and the control and use of unique identifiers such as user IDs and passwords
  - An authorization process for user access and privileges
  - A statement that all access that is not explicitly authorized is forbidden
  - A process for revoking access rights or interrupting the connection between systems
- Arrangements for reporting, notification and investigation of information inaccuracies (e.g., of personal details), information security incidents and security breaches
- The target level of service and unacceptable levels of service
- The right to monitor and revoke any activity related to the organization's assets
- The respective liabilities of the organization and the customer
- Responsibilities with respect to legal matters and ensuring that the legal requirements are met (e.g., data protection legislation), taking into account different national legal systems if the agreement involves cooperation with customers in other countries
- Intellectual property rights (IPRs), copyright assignment and protection of any collaborative work

The security requirements related to customers accessing organizational assets can vary considerably depending on the information processing facilities and information being accessed. These security requirements can be addressed using customer agreements that contain all identified risk and security requirements.

Agreements with external parties may also involve other parties. Agreements granting an external party access should include an allowance for designation of other eligible parties and conditions for their access and involvement.

Addressing Security in Third-party Agreements

Third-party agreements involving accessing, processing, communicating or managing the organization's information or information processing facilities or adding products or services to information processing facilities should cover all relevant security requirements. The agreement should ensure that there is no misunderstanding between the organization and the third party. The organization should ensure that the agreement includes

adequate indemnification provisions to protect against potential losses caused by the actions of the third party.

The contract terms listed in **figure 5.10** should be considered for inclusion in the agreement to satisfy the identified security requirements.

Figure 5.10—Recommended Contract Terms for Third-party Agreements

- Compliance with the organization's information security policy by the third party
- Controls to ensure asset protection, including:
  - Procedures to protect organizational assets, including information, software and hardware
  - Any required physical protection controls and mechanisms
  - Controls to ensure protection against malicious software
  - Procedures to determine whether any compromise of the assets (e.g., loss or modification of information, software and hardware) has occurred
  - Controls to ensure the return or destruction of information and assets at the end of or at an agreed point in time during the agreement
  - Confidentiality, integrity, availability and any other relevant property of the assets
  - Restrictions on copying and disclosing information, and using confidentiality agreements
- User and administrator training in methods, procedures and security
- A means to ensure user awareness of information security responsibilities and issues
- Provision for the transfer of personnel, where appropriate
- Responsibilities regarding hardware and software installation and maintenance
- A clear reporting structure and agreed reporting formats
- A clear and specified process for change management
- Access control policy, covering:
  - The different reasons, requirements and benefits that make the access by the third party necessary
  - Permitted access methods and the control and use of unique identifiers such as user IDs and passwords
  - An authorization process for user access and privileges
  - A requirement to maintain a list of individuals authorized to use the services being made available and what their rights and privileges are with respect to such use
  - A statement that all access that is not explicitly authorized is forbidden
  - A process for revoking access rights or interrupting the connection between systems
- Arrangements for reporting, notification and investigation of information security incidents and security breaches as well as violations of the requirements stated in the agreement
- A description of the product or service to be provided and a description of the information to be made available along with its security classification
- The target level of service and unacceptable levels of service
- The definition of verifiable performance criteria, their monitoring and reporting
- The right to monitor and revoke any activity related to the organization's assets
- The right to audit responsibilities defined in the agreement, to have those audits carried out by a third party and to enumerate the statutory rights of auditors (and, where appropriate, the provision of a service auditor's report)
- The establishment of an escalation process for problem resolution

Figure 5.10—Recommended Contract Terms for Third-party Agreements *(cont.)*

- Service continuity requirements, including measures for availability and reliability, in accordance with an organization's business priorities
- The respective liabilities of the parties to the agreement
- Responsibilities with respect to legal matters and ensuring that the legal requirements are met (e.g., data protection legislation), taking into account different national legal systems if the agreement involves cooperation with organizations in other countries
- IPRs and copyright assignment and protection of any collaborative work
- Involvement of the third party with subcontractors, and the security controls these subcontractors need to implement
- Conditions for renegotiation/termination of agreements such as:
  - A contingency plan in case either party wishes to terminate the relationship before the end of the agreements
  - A provision for renegotiation of agreements if the security requirements of the organization change
- Current documentation of asset lists, licenses, agreements or rights relating to them
- Non-assignability of the contract

In general, it is very difficult to ensure the return or destruction of confidential information disclosed to a third party at the end of the agreement. To prevent unauthorized copies or use, printed documents should be consulted on site. Using technical controls, such as digital rights management (DRM) where access control technologies are used by publishers, copyright holders and individuals to impose limitations on the usage of digital content and devices, should be considered to set up the desired constraints such as the printing of the document, copying, authorized readers or using it after a certain date.

The agreements can vary considerably for different organizations and among the different types of third parties. Therefore, care should be taken to include all identified risk and security requirements in the agreements. Where necessary, the required controls and procedures can be expanded in a security management plan.

If information security management is outsourced, the agreements should address how the third party will guarantee that adequate security, as defined by the risk assessment, will be maintained and how security will be adapted to identify and deal with changes to risk. Some of the differences between outsourcing and the other forms of third-party service provision include the question of liability, planning the transition period and potential disruption of operations during this period, contingency planning arrangements and due diligence reviews, and collection and management of information on security incidents. Therefore, it is important that the organization plans and manages the transition to an outsourced arrangement and has suitable processes in place to manage changes and the renegotiation/termination of agreements.

The procedures for continuing processing in the event that the third party becomes unable to supply its services need to be considered in the agreement to avoid any delay in arranging replacement services. Agreements with third parties may also involve other parties. Agreements granting third-party access should include allowance for designation of other eligible parties and conditions for their access and involvement.

A requirement for the third party to have certified compliance with recognized security standards (e.g., ISO 27001) may need to be considered.

Generally, agreements are primarily developed by the organization. There may be occasions in some circumstances where an agreement may be developed and imposed upon an organization by a third party. The organization needs to ensure that its own security is not unnecessarily impacted by third-party requirements stipulated in imposed agreements.

5.2.11 HUMAN RESOURCES SECURITY AND THIRD PARTIES

Proper information security practices should be in place to ensure that employees, contractors and third-party users understand their responsibilities and are suitable for their assigned roles. These practices can reduce the risk of theft, fraud or misuse of facilities. Specific security practices include:

- Security responsibilities should be addressed prior to employment in adequate job descriptions, and in terms and conditions of employment.
- All candidates for employment, contractors and third-party users should be adequately screened, especially for sensitive jobs.
- Employees, contractors and third-party users of information processing facilities should sign an agreement on their security roles and responsibilities, including the need to maintain confidentiality.

Security roles and responsibilities of employees, contractors and third-party users should be defined and documented in accordance with the organization's information security policy.

Screening

All candidates for employment, contractors or third-party users should be subject to background verification checks. These should be carried out and documented in accordance with relevant laws, regulations and ethics, and proportional to the business requirements, the classification of the information to be accessed and the perceived risk. When using an agency to provide contractors, the contract with the agency should clearly specify the agency's responsibilities for screening and the notification procedures they need to follow if screening has not been completed or if the results give cause for doubt or concern. In the same way, the agreement with the third party should clearly specify all responsibilities and notification procedures for screening.

Terms and Conditions of Employment

As part of their contractual obligation, employees, contractors and third-party users should agree and sign the terms and conditions of their employment, which should state their and the organization's responsibilities for information security. The terms and conditions of employment should reflect the organization's security policy in addition to clarifying and stating:

- That all employees, contractors and third-party users who are given access to sensitive information should sign a confidentiality or nondisclosure agreement prior to being given access to information processing facilities
- The employee, contractor and any other user's legal responsibilities and rights (e.g., regarding copyright laws or data protection legislation)
- Responsibilities for the classification of information and management of organizational assets associated with information systems and services handled by the employee, contractor or third-party user
- Responsibilities of the employee, contractor or third-party user for the handling of information received from other companies or external parties
- Responsibilities of the organization for the handling of personal information, including personal information created as a result of, or in the course of, employment with the organization
- Responsibilities that are extended outside the organization's premises and outside normal working hours (e.g., in the case of working at home)
- Actions to be taken if the employee, contractor or third-party user disregards the organization's security requirements

The organization should ensure that employees, contractors and third-party users agree to terms and conditions concerning information security appropriate to the nature and extent of access they will have to the organization's assets associated with information systems and services. Where appropriate, responsibilities contained within the terms and conditions of employment should continue for a defined period after the end of the employment.

During Employment

Management should require employees, contractors and third-party users to apply security in accordance with the established policies and procedures of the organization. Specific responsibilities should be documented in approved job descriptions. This will help ensure that employees, contractors and third-party users are aware of information security threats and concerns, their responsibilities and liabilities, and are equipped to support organizational security policy in the course of their normal work and to reduce the risk of human error. Management responsibilities should be defined to ensure that security is applied throughout an individual's employment within the organization. An adequate level of awareness, education and training in security procedures and the correct use of information processing facilities should be provided to all employees, contractors and third-party users to minimize possible security risk. A formal disciplinary process for handling security breaches should be established.

Termination or Change of Employment

When an employee, contractor or third-party user exits the organization, responsibilities should be in place to manage this process, including the return of all equipment and removal of all access rights. Communication of termination responsibilities should include ongoing security requirements and legal responsibilities. Where appropriate, responsibilities contained within any confidentiality agreement and the terms and conditions of employment continuing for a defined period after the end of the employee, contractor or third-party user's employment should also be communicated. Responsibilities and duties still valid after termination of employment should be contained in the employee, contractor or third-party user's contracts.

Removal of Access Rights

The access rights of all employees, contractors and third-party users to information and information processing facilities should be removed upon termination of their employment, contract or agreement, or adjusted upon change. The access rights that should be removed or adapted include physical and logical access, keys, identification cards, information processing facilities, subscriptions, and removal from any documentation that identifies them as a current member of the organization. This should include notifying partners and relevant third parties—if a departing employee has access to the third party premises. If a departing employee, contractor or third-party user has known passwords for accounts remaining active, these should be changed upon termination or change of employment, contract or agreement. Access rights for information assets and information processing facilities should be reduced or removed before the employment terminates or changes, depending on the evaluation of risk factors such as:

- Whether the termination or change is initiated by the employee, contractor or third-party user, or by management and the reason of termination
- The current responsibilities of the employee, contractor or any other user
- The value of the assets currently accessible

Procedures should be in place to ensure that information security management is promptly informed of all employee movements, including employees leaving the organization.

5.2.12 COMPUTER CRIME ISSUES AND EXPOSURES

Computer systems can be used to fraudulently obtain money, goods, software or corporate information. Crimes can also be committed when the computer application process or data are manipulated to accept false or unauthorized transactions. There is also the simple, nontechnical method of computer crime: stealing computer equipment.

Computer crime can be performed without anything physically being taken or stolen, and it can be done remotely. Simply viewing computerized data can provide an offender with enough intelligence to steal ideas or confidential information (intellectual property). In case of the systems connected to wide area networks (WANs) or the Internet, the crime scene could be anywhere in the world, making the investigation very difficult. Cyber-criminals take advantage of existing gaps in the legislation of different countries when planning cyberattacks in order to avoid prosecution.

Committing crimes that exploit the computer and the information it contains can be damaging to the reputation, morale and the continued existence of an organization. Loss of customers or market share, embarrassment to management and legal actions against the organization can result. Threats to business include the following:

- **Financial loss**—These losses can be direct, through loss of electronic funds, or indirect, through the costs of correcting the exposure.
- **Legal repercussions**—There are numerous privacy and human rights laws an organization should consider when developing security policies and procedures. These laws can protect the organization but also can protect the perpetrator from prosecution. In addition, not having proper security measures could expose the organization to lawsuits from investors and insurers if a significant loss occurs from a security violation. Most companies must also comply with industry-specific regulatory agencies' requirements. The IS auditor should obtain legal assistance when reviewing the legal issues associated with computer security.
- **Loss of credibility or competitive edge**—Many organizations, especially service firms such as banks, savings and loans and investment firms, need credibility and public trust to maintain a competitive edge. A security violation can damage this credibility severely, resulting in loss of business and prestige.
- **Blackmail/industrial espionage/organized crime**—By gaining access to confidential information or the means to adversely impact computer operations, a perpetrator can extort payments or services from an organization by threatening to exploit the security breach or publicly disclose the confidential information of the organization. Also, by gaining access, the perpetrator could obtain proprietary information and sell it to a competitor.
- **Disclosure of confidential, sensitive or embarrassing information**—As noted previously, such events can damage an organization's credibility and its means of conducting business. Legal or regulatory actions against the company may also be the result of disclosure.
- **Sabotage**—Some perpetrators are not looking for financial gain. They merely want to cause damage due to a dislike of the organization or for self-gratification. "Hacktivism" occurs when perpetrators make nonviolent use of illegal or legally ambiguous digital tools in pursuit of political ends.

It is important that the IS auditor knows and understands the differences between computer crime and computer abuse to support risk analysis methodologies and related control practices. What constitutes a crime depends on the jurisdiction and the court sentence. Certain breaches of security may be civil or criminal offenses. This brings into play requirements for what the organization needs to do should a crime be suspected (i.e., protection of evidence, reporting of a crime, etc.).

Perpetrators in computer crimes are often the same people who exploit physical exposures, although the skills needed to exploit logical exposures are more technical and complex. Possible perpetrators include:

- **Hackers (also referred to as crackers)**—Persons with the ability to explore the details of programmable systems and the knowledge to stretch or exploit their capabilities, whether ethical or not. Hackers are typically attempting to test the limits of access restrictions to prove their ability to overcome the obstacles. Some often do not access a computer with the intent of destruction, although this is often the result. Types of hackers include hacktivists and criminal hackers. Some hackers seek to commit a crime through their actions for some level of personal gain or satisfaction. The terms hack and crack are often used interchangeably.
- **Script kiddies**—Script kiddies refer to individuals who use scripts and programs written by others to perform their intrusions and are often incapable of writing similar scripts on their own.
- **Employees (authorized or unauthorized)**—Affiliated with the organization and given system access based on job responsibilities, these individuals can cause significant harm to an organization. Therefore, screening prospective employees through appropriate background checks is an important means of preventing computer crimes within the organization.
- **IT personnel**—These individuals have the easiest access to computerized information, as they are the custodians of this information. In addition to logical access controls, good SoD and supervision help in reducing logical access violations by these individuals.

- **End users**—Personnel who often have broad knowledge of the information within the organization and have easy access to internal resources
- **Former employees**—Former employees who have left on unfavorable terms may have access if it was not immediately removed at the time of the employee's termination or if the system has "back doors."
- **Nations**—As more critical infrastructure is controlled from the Internet (e.g., supervisory control and data acquisition [SCADA] systems) and more nation's key organizations and businesses rely on the Internet, it is not uncommon for nations to attack each other.
- **Interested or educated outsiders**—These may include:
  - Competitors
  - Terrorists
  - Organized criminals
  - Hackers looking for a challenge
  - Script kiddies for the purpose of curiosity, joyriding and testing their newly acquired tools/scripts and exploits
  - Crackers
  - Phreakers
- **Part-time and temporary personnel**—Remember that facility contractors such as office cleaners often have a great deal of physical access and could perpetrate a computer crime.
- **Third parties**—Vendors, visitors, consultants or other third parties who, through projects, gain access to the organization's resources and could perpetrate a crime
- **Opportunists**—Where information is inadvertently left unattended or left for destruction, a passerby can access same
- **Accidental unaware**—Someone who unknowingly perpetrates a violation

Other examples of criminals include small-time crooks, organized crime and state-sponsored criminal activities.

Although collaboration has been improved in solving cybercrimes committed from one country to another, political issues existing between some countries might hinder an investigation. Therefore, additional preventive measures should be taken to protect information systems vulnerable to international attacks.

Figures 5.11 and **5.12** describe common attack methods and techniques for computer crimes. Perpetrators may use one or more methods in tandem to commit a crime.

Figure 5.11—Computer Crimes

Source of the Attack	Target of the Attack	Examples
Computer is the target of the crime. Perpetrator uses another computer to launch an attack.	Specific identified computer	• Denial of service (DoS) • Hacking
Computer is the subject of the crime. Perpetrator uses computer to commit crime and the target is another computer.	Target may or may not be defined. Perpetrator launches the attack with no specific target in mind.	• Distributed DoS • Malware
Computer is the tool of the crime. Perpetrator uses computer to commit crime but the target is not the computer.	Target is data or information stored on the computer.	• Fraud • Unauthorized access • Phishing • Installing key loggers
Computer symbolizes the crime. Perpetrator lures the user of computers to get confidential information.	Target is user of the computers.	• Social engineering methods: – Phishing – Fake web sites – Scam mail – Spam mail – Fake resumes for employment

Figure 5.12—Common Attack Methods and Techniques	
Alteration Attack	Occurs when unauthorized modifications affect the integrity of the data or code Examples: Unauthorized alteration of binary code during the software development life cycle (SDLC) or addition of unauthorized libraries during recompilation of existing programs Cryptographic hash is a primary defense against alteration attacks.
Botnets	Comprise a collection of compromised computers (called zombie computers) running software, usually installed via worms, Trojan horses or back doors Examples: Denial-of-service (DoS) attacks, adware, spyware and spam
Brute Force Attack	Attack launched by an intruder, using many of the password-cracking tools available at little or no cost, on encrypted passwords and attempts to gain unauthorized access to an organization's network or host-based systems
Denial-of-service (DoS) Attack	Examples: ICMP flood attack: • Smurf attack—Occurs when misconfigured network devices allow packets to be sent to all hosts on a particular network via the broadcast address of the network • Ping flood—Occurs when the target system is overwhelmed with ping packets • SYN flood—Sends a flood of TCP/SYN packets with forged sender address, causing half-open connections and saturates available connection capacity of the target machine Teardrop attack—Involves sending mangled IP fragments with overlapping, oversized payloads to the target machine Peer-to-peer attack—Causes clients of large peer-to-peer file sharing hubs to disconnect from their peer-to-peer network and to connect to the victim's web site instead. As a result, several thousand computers may aggressively try to connect to a target web site, causing performance degradation.
Denial-of-service (DoS) Attack *(cont.)*	Permanent denial-of-service (PDoS) attack (also known as phlashing)—Damages a system hardware to the extent of replacement Application-level flood attack: • Buffer overflow consumes available memory or CPU time. • Brute force attack—Floods the target with an overwhelming flux of packets, oversaturating its connection bandwidth or depleting the target's system resources • Bandwidth-saturating flood attack—Relies on the attacker having higher bandwidth available than the victim • Banana attack—Redirects outgoing messages from the client back onto the client, preventing outside access, as well as flooding the client with the sent packets • Pulsing zombie—A DoS attack in which a network is subjected to hostile pinging by different attacker computers over an extended time period. This results in a degraded quality of service and increased workload for the network's resources. Nuke—A DoS attack against computer networks in which fragmented or invalid ICMP packets are sent to the target. Modified ping utility is used to repeatedly send corrupt data, thus slowing down the affected computer to a complete stop. Distributed denial-of-service attack (DDoS)—Occurs when multiple compromised systems flood the bandwidth or resources of the targeted system Reflected attack—Involves sending forged requests to a large number of computers that will reply to the requests. The source IP address is spoofed to that of the targeted victim, causing the replies to flood. Unintentional attack—Web site ends up denied, not due to a deliberate attack by a single individual or group of individuals, but simply due to a sudden enormous spike in popularity
Dial-in Penetration Attack/War Dialing	An intruder determines the dial-in phone number ranges from external sources, such as the Internet. The intruder may also employ social engineering tactics to get information from a company receptionist or a knowledgeable employee inside the company
Eavesdropping	An intruder gathers the information flowing through the network with the intent of acquiring and releasing the message contents for either personal analysis or for third parties who might have commissioned such eavesdropping. This is significant when considering that sensitive information, traversing a network, can be seen in real time by all other machines, including email, passwords and, in some cases, keystrokes. These activities can enable the intruder to gain unauthorized access, to fraudulently use information such as credit card accounts and to compromise the confidentiality of sensitive information that could jeopardize or harm an individual's or an organization's reputation.

Figure 5.12—Common Attack Methods and Techniques *(cont.)*	
Email Attacks and Techniques	Email Bombing—Characterized by abusers repeatedly sending an identical email message to a particular address Email spamming (also known as unsolicited commercial email (UCE) or junk email)—a variant of bombing and refers to sending email to hundreds or thousands of users (or to lists that expand to that many users). It may also occur innocently as a result of sending a message to mailing lists and not realizing that the list explodes to thousands of users or as a result of using a responder message, such as a vacation alert, that is not set up correctly. • Spam causes inconvenience and has severe impacts on productivity and thus is considered a business risk. • When spam is responded to, the email address of the recipient is validated and gives away information. • Spam may be combined with email spoofing (see below), making it more difficult to determine from whom the email is coming. • Spam is managed using the Sender Permitted Form (SPF) protocol and with the help of tools such as Bayesian filtering and greylisting. Email Spoofing—May occur in different forms, but all have a similar result: a user receives an email message that appears to have originated from one source but actually was sent from another source. Email spoofing is often an attempt to trick the user into making a damaging statement or releasing sensitive information such as passwords or account information. Examples of spoofed email that could affect the security of a site include: • Email claiming to be from a system administrator and requesting users to change their passwords to a specified string and threatening to suspend their account if they do not make the change • Email claiming to be from a person in authority and requesting users to send a copy of a password file or other sensitive information Phishing—The criminally fraudulent process of attempting to acquire sensitive information, such as usernames, passwords and credit card details, by masquerading as a trustworthy entity in an electronic communication. Phishing techniques include social engineering, link manipulation and web site forgery. Spear phishing—A pinpoint attack against a subset of people (users of a web site or product, employees of a company, members of an organization) to undermine that company or organization
Flooding	A DoS attack that brings down a network or service by flooding it with large amounts of traffic. The host's memory buffer is filled by flooding it with connections that cannot be completed.
Interrupt Attack	Occurs when a malicious action is performed by invoking the OS to execute a particular system call Example: A boot sector virus typically issues an interrupt to execute a write to the boot sector.
Malicious Codes	Trojan horses (often called Trojans)—Programs that are disguised as useful programs such as OS patches, software packages or games. Once executed, however, Trojans perform actions that the user did not intend, such as opening certain ports for subsequent access by the intruder. Logic bomb—A program or a section of a program that is triggered when a certain condition, time or event occurs. Logic bombs typically result in sabotage of computer systems and are commonly deployed by disgruntled insiders who have access to programs. For example, when terminated from an organization, a disgruntled software programmer could devise a logical bomb to delete critical files or databases. Logic bombs can also be used against attackers. Administrators sometimes intentionally install pseudo flaws, also called honey tokens, that look vulnerable to attack but really act as alarms or triggers of automatic actions when the intruder attempts to exploit the flaw. Trap doors—Commonly called back doors. Bits of code embedded in programs by programmers to quickly gain access during the testing or debugging phase. If an unscrupulous programmer purposely leaves in this code (or simply forgets to remove it), a potential security hole is introduced. Hackers often plant a back door on previously compromised systems to gain subsequent access. Threat vector analysis (a type of defense-in-depth architecture), SoD and code audits help to defend against logic bombs and trap/back doors.
Man-in-the-middle Attack	The following scenarios are possible: • The attacker actively establishes a connection to two devices. The attacker connects to both devices and pretends to each of them to be the other device. Should the attacker's device be required to authenticate itself to one of the devices, it passes the authentication request to the other device and then sends the response back to the first device. Having authenticated himself/herself in this way, the attacker can then interact with the device as he/she wishes. To successfully execute this attack, both devices have to be connectable. • The attacker interferes while the devices are establishing a connection. During this process, the devices have to synchronize the hop sequence that is to be used. The aggressor can prevent this synchronization so that both devices use the same sequence but a different offset within the sequence.

Figure 5.12—Common Attack Methods and Techniques *(cont.)*	
Masquerading	An active attack in which the intruder presents an identity other than the original identity. The purpose is to gain access to sensitive data or computing/network resources to which access is not allowed under the original identity. Masquerading also attacks the authentication attribute by letting a genuine session authentication take place and subsequently enters the information flow, masquerading as one of the authenticated users of the session. Impersonation both by people and machines falls under this category. Masquerading by machines (also known as IP spoofing)—A forged IP address is presented. This form of attack is often used as a means of breaking a firewall.
Message Modification	Involves the capturing of a message and making unauthorized changes or deletions (of full streams or parts of the message), changing the sequence or delaying transmission of captured messages. This attack can have disastrous effects if, for example, the message is an instruction to a bank to make a payment.
Network Analysis	An intruder applies a systematic and methodical approach known as footprinting to create a complete profile of an organization's network security infrastructure. During this initial reconnaissance phase, the intruder uses a combination of tools and techniques to build a repository of information about a particular company's internal network. This probably would include information about system aliases, functions, internal addresses, and potential gateways and firewalls. Next, the intruder focuses on systems within the targeted address space that responded to these network queries. Once a system has been targeted, the intruder scans the system's ports to determine what services and OS are running on the targeted system, possibly revealing vulnerable services that could be exploited.
Packet Replay	A combination of passive and active modes of attacks. The intruder passively captures a stream of data packets as the stream moves along an unprotected or vulnerable network. These packets are then actively inserted into the network as if the stream were another genuine message stream. This form of attack is effective particularly where the receiving end of the communication channel is automated and will act on receipt and interpretation of information packets without human intervention.
Pharming	An attack that aims to redirect the traffic of a web site to a bogus web site. Pharming can be conducted either by changing the host's file on a victim's computer or by exploiting a vulnerability in DNS server software. DNS servers are computers responsible for resolving Internet names into their real addresses—they are the "signposts" of the Internet. Compromised DNS servers are sometimes referred to as "poisoned." In recent years, both pharming and phishing have been used to steal identity information. Pharming has become a major concern to businesses hosting e-commerce and to online banking web sites. Sophisticated measures known as antipharming are required to protect against this serious threat. Antivirus software and spyware removal software cannot protect against pharming.
Piggybacking	The act of following an authorized person through a secured door or electronically attaching to an authorized telecommunications link to intercept and possibly alter transmissions. Piggybacking is considered a physical access exposure.
Race Conditions (also known as Time of Check [TOC]/Time of Use [TOU] attacks)	Exploit a small window of time between the time that the security control is applied and the time that the service is used. The exposure to a race condition increases in proportion to the time difference between TOC and TOU. Interference occurs when a device or system attempts to perform two or more operations at the same time, but the nature of the device or system requires the operations to happen in proper sequence. Race conditions occur due to interferences caused by the following conditions: • Sequence or nonatomic—These conditions are caused by untrusted processes, such as those invoked by an attacker, that may get in between the steps of the secure program. • Deadlock, livelock or locking failure—These conditions are caused by trusted processes running the same program. Since these different processes may have the same privileges, they may interfere with each other, if not properly controlled. Careful programming and good administration practices help to reduce race conditions.
Remote Maintenance Tools	If not securely configured and controlled, can be used as an attack method by malicious hackers to remotely gain elevated access and cause damage to the target system
Resource Enumeration and Browsing	When the attacker lists the various resources (names, directories, privileges, shares, policies) on targeted hosts and networks Browsing attack—A form of a resource enumeration attack and is performed by a manual search, frequently aided with commands and tools available in software, OSs or add-on utilities
Salami	Involves slicing small amounts of money from a computerized transaction or account. Similar to the rounding down technique. The difference between the rounding down technique and the salami technique is that, in rounding down, the program rounds off by the smallest money fraction. For example, in the rounding down technique, a US $1,235,954.39 transaction may be rounded to US $1,235,954.35. On the other hand, the salami technique truncates the last few digits from the transaction amount, so US $1,235,954.39 becomes US $1,235,954.30 or $1,235,954.00, depending on the algorithm/formula built into the program. In fact, other variations of the same technique are applied to rates and percentages.

Figure 5.12—Common Attack Methods and Techniques *(cont.)*	
Social Engineering	The human side of breaking into a computer system. Organizations with strong technical security countermeasures (such as authentication processes, firewalls and encryption) may still fail to protect their information systems. This situation may happen if an employee unknowingly gives away confidential information (e.g., passwords and IP addresses) by answering questions over the phone with someone they do not know or replying to an email message from an unknown person. Some examples of social engineering include impersonation through a telephone call, dumpster diving and shoulder surfing. The best means of defense for social engineering is an ongoing security awareness program, wherein all employees and third parties (who have access to the organization's facilities) are educated about the risk involved in falling prey to social engineering attacks.
Traffic Analysis	An inference attack technique that studies the communication patterns between entities in a system and deduces information. This typically is used when messages are encrypted and eavesdropping would not yield meaningful results. Traffic analysis can be performed in the context of military intelligence or counter-intelligence and is a concern in computer security.
Unauthorized Access Through the Internet or World Wide Web	Unauthorized access through the Internet or web-based services. Many Internet software packages contain vulnerabilities that render systems subject to attack. Additionally, many of these systems are large and difficult to configure, resulting in a large percentage of unauthorized access incidents. Examples include: • Email forgery (simple mail transfer protocol) • Telnet passwords transmitted in the clear (via path between client and server) • Altering the binding between IP addresses and domain names to impersonate any type of server. As long as the DNS is vulnerable and used to map universal resource locators (URLs) to sites, there can be no integrity on the Web. • Releasing common gateway interface (CGI) scripts as shareware. CGI scripts often run with privileges that give them complete control of a server. • Client-side execution of scripts (via Java™ in Java applets), which presents the danger of running code from an arbitrary location on a client machine
Viruses, Worms and Spyware/Malware	Viruses—Involve the insertion of malicious program code into other executable code that can self-replicate and spread from computer to computer, via sharing of removable computer media, USB removable devices, transfer of logic over telecommunication lines or direct link with an infected machine/code. A virus can harmlessly display cute messages on computer terminals, dangerously erase or alter computer files, or simply fill computer memory with junk to a point where the computer can no longer function. An added danger is that a virus may lie dormant for some time until triggered by a certain event or occurrence, such as a date or being copied a prespecified number of times, during which time the virus has silently been spreading. Worms—Destructive programs that may destroy data or use up tremendous computer and communication resources, but worms do not replicate like viruses. Such programs do not change other programs but can run independently and travel from machine to machine across network connections by exploiting vulnerability and application/system weaknesses. Worms also may have portions of themselves running on many different machines. Spyware/Malware—Similar to viruses. Examples are keystroke loggers and system analyzers that collect potentially sensitive information, such as credit card numbers, bank details, etc., from the host and then transmit the information to the originator when an online connection is detected.
War Chalking	The practice of marking a series of symbols (outward-facing crescents) on sidewalks and walls to indicate nearby wireless access points. These markings are used to identify hotspots, where other computer users can connect to the Internet wirelessly and at no cost. War chalking was inspired by the practice of unemployed migrant workers, during the Great Depression in the US, using chalk marks to indicate which homes were friendly.
War Driving	The practice of driving around businesses or residential neighborhoods while scanning with a laptop computer, hacking tool software and sometimes with a global positioning system (GPS) to search for wireless network names. While driving around the vicinity of a wireless network, an attacker might be able to see the wireless network name, but the use of wireless security will determine whether the attacker can do anything beyond viewing the wireless network name. With wireless security enabled and properly configured, war drivers cannot see the network name and are unable to send data, interpret data sent on the wireless network, access the shared resources of the wireless or wired network (shared files, private web sites), or use the Internet connection. Without wireless security enabled and properly configured, war drivers can send data, interpret data sent on the wireless network, access the shared resources of the wireless or wired network (shared files, private web sites), install viruses, modify or destroy confidential data, and use the Internet connection without the knowledge or consent of the owner. For example, a malicious user might use the Internet connection to send thousands of spam email messages or launch attacks against other computers. The malicious traffic could be traced back to the owner's home.
War Walking	Similar to war driving, but a vehicle is not used. The potential hacker walks around the vicinity with a handheld device. Currently, there are several free hacking tools that fit in these mini-devices.

5.2.13 SECURITY INCIDENT HANDLING AND RESPONSE

To minimize damage from security incidents and to recover and to learn from such incidents, a formal incident response capability should be established. Incident response should include the following phases:

- Planning and preparation
- Detection
- Initiation
- Recording
- Evaluation
- Containment
- Eradication
- Escalation
- Response
- Recovery
- Closure
- Reporting
- Postincident review
- Lessons learned

The organization and management of an incident response capability should be coordinated or centralized with the establishment of key roles and responsibilities. This should include:

- A coordinator who acts as the liaison to business process owners
- A director who oversees the incident response capability
- Managers who manage individual incidents
- Security specialists who detect, investigate, contain and recover from incidents
- Nonsecurity technical specialists who provide assistance based on subject matter expertise
- Business unit leader liaisons (legal, human resources, public relations, etc.)

In establishing this process, employees and contractors are made aware of procedures for reporting the different types of incidents (e.g., security breach, threat, weakness or malfunction) that might have an impact on the security of organizational assets. They should be required to report any observed or suspected incidents as quickly as possible to the designated point of contact. The organization should establish a formal disciplinary process for dealing with those who commit security breaches such as employees, third parties, etc. To address incidents properly, it is necessary to collect evidence as soon as possible after the occurrence. Legal advice may be needed in the process of evidence collection and protection.

Incidents occur because vulnerabilities are not addressed properly. Incident management processes should include vulnerabilities management practices. A postincident review phase should determine which vulnerabilities were not addressed and why, and input provided for improvement to the policies and procedures implemented to address vulnerabilities. Also, analyzing the cause of incidents may reveal errors in the risk analysis, indicating that the residual risk is higher than the calculated values and inappropriate countermeasures have been taken to reduce inherent risk.

Ideally, an organizational computer security incident response team (CSIRT) or computer emergency response team (CERT) should be formed with clear lines of reporting, and responsibilities for standby support should be established. Organizational CSIRT will act as an efficient detective and corrective control. Additionally, with its members' participation and involvement in security awareness programs, exercises and workshops, it can demonstrate a preventive control.

Organizational CSIRT should also disseminate security alerts such as recent threats, security guidelines and security updates to the users and assist them in understanding the security risk of errors and omissions. Organizational CSIRT should act as single point of contact for all incidents and issues related to information security, should also respond to abuse reports pertaining to the network of its constituency.

An IS auditor should ensure that the CSIRT is actively involved with users to assist them in the mitigation of risk arising from security failures and also to prevent security incidents. Auditors should also ensure that there is a formal, documented plan and that it contains vulnerabilities identification, reporting and incident response procedures to common, security-related threats/issues, such as:

- Virus outbreak
- Web defacement
- Abuse notification
- Unauthorized access alert from audit trails
- Security attack alerts from intrusion detection systems (IDSs)
- Hardware/software theft
- System root compromises
- Physical security breach
- Spyware/malware/Trojans detected on personal computers (PCs)
- Fake defamatory information in media, including on web sites
- Forensic investigations

Additionally, automated IDSs should be in place to notify administrators in a real-time manner of a potential incident and define a process for determining the severity of the incidents and the steps to take in high-risk situations. Please refer to section 2.12.5 Business Continuity Planning Incident Management, for more information.

5.3 LOGICAL ACCESS

Logical access controls are the primary means used to manage and protect information assets. They enact and substantiate management-designed policies and procedures intended to protect these assets and the controls are designed to reduce risk to a level acceptable to an organization. IS auditors need to understand this relationship. In doing so, IS auditors should be able to analyze and evaluate the effectiveness of a logical access control in accomplishing information security objectives and avoiding losses resulting from exposures. These exposures can result in minor inconveniences to a total shutdown of computer functions.

5.3.1 LOGICAL ACCESS EXPOSURES

Technical exposures are one type of exposure that exists due to accidental or intentional exploitation of logical access control weaknesses. Intentional exploitation of technical exposures might lead to computer crime. However, not all computer crimes exploit technical exposures.

Technical exposures are the unauthorized activities interfering with normal processing, such as implementation or modification of data and software, locking or misusing user services, destroying data, compromising system usability, distracting processing resources, or spying data flow or users activities at either the network, platform (OS), database or application level. Technical exposures include:

- **Data leakage**—Involves siphoning or leaking information out of the computer. This can involve dumping files to paper, or can be as simple as stealing computer reports and tapes. Unlike product leakage, data leakage leaves the original copy, so it may go undetected.
- **Wiretapping**—Involves eavesdropping on information being transmitted over telecommunications lines
- **Computer shutdown**—Initiated through terminals or personal computers connected directly (online) or remotely (via the Internet) to the computer. Only individuals who know a high-level logon ID usually can initiate the shutdown process, but this security measure is effective only if proper security access controls are in place for the high-level logon ID and the telecommunications connections into the computer. Some systems have proven to be vulnerable to shutting themselves down under certain conditions of overload.

Figure 5.12 presents common attack methods and techniques.

5.3.2 FAMILIARIZATION WITH THE ENTERPRISE'S IT ENVIRONMENT

For IS auditors to effectively assess logical access controls within their organization, they first need to gain a technical and organizational understanding of the organization's IT environment. The purpose of this is to determine which areas from a risk standpoint warrant IS auditing attention in planning current and future work. This includes reviewing the network, OS platform, database and application security layers associated with the organization's IT information systems architecture.

5.3.3 PATHS OF LOGICAL ACCESS

Access or points of entry to an organization's IS infrastructure can be gained through several avenues. Each avenue is subject to appropriate levels of access security.

These paths can be direct, as is the case for a PC terminal user tying directly into a mainframe. This happens when the IS environment is under direct control of the main system and when the users are locally known individuals, with well-defined access profiles. More complex is the case of a LAN, where many specific IS resources are tied to a common linking structure. The LAN resources may have different access paths/levels, normally mediated through LAN connectivity, and the network itself is considered an important IS resource at a higher access level. A combination of direct, local network and remote access paths is the most common configuration. Complexity is increased by a number of intermediate devices that act as "security doors" among the various environments. The need of crossing low-security or totally open IT spaces, such as the Internet, also necessitates increased complexity. An example of an access path through common nodes is a back-end or front-end interconnected network of systems for internally or externally based users.

Front-end systems are network-based systems connecting an organization to outside, untrusted networks, such as corporate web sites, where a customer can access the web site externally to initiate transactions that connect to a proxy server application which in turn connects to a back-end database system to update a customer database. Front-end systems can also be internally based to automate business, paperless processes that tie into back-end systems in a similar manner.

General Points of Entry

General points of entry to either front-end or back-end systems control the accesses from an organization's networking or telecommunications infrastructure into their information resources (e.g., applications, databases, facilities, networks). The approach followed is based on a client-server model. A large organization can have thousands of interconnected network servers. Connectivity in this environment needs to be controlled through a smaller set of primary domain controllers (servers), which enable a user to obtain access to specific secondary points of entry (e.g., application servers, databases, etc.).

General modes of access into this infrastructure occur through the following:

- **Network connectivity**—Access is gained by linking a PC to a segment of an organizations' network infrastructure, either through a physical or a wireless connection. At a minimum, such access requires user identification and authentication to a domain-controlling server. More specific access to a particular application or database may also require the users to identify and authenticate themselves to that particular server (secondary point of entry). Other modes of access into the infrastructure can also occur through network management devices, such as routers and firewalls, which should be strictly controlled.
- **Remote access**—A user connects remotely to an organization's server, which generally requires the user to identify and authenticate him/herself to the server for access to specific functions that can be performed remotely (e.g., email, File Transfer Protocol [FTP] or some application-specific function). Complete access to view all network resources usually requires a virtual private network (VPN), which allows a secure authentication and connection into those resources where privileges have been granted. Remote access points of entry can be extensive and should be centrally controlled where possible.

From a security standpoint, it is incumbent upon the organization to know all of the points of entry into its information resource infrastructure which, in many organizations, will not be a trivial task (e.g., thousands of remote access users). This is significant because any point of entry not appropriately controlled can potentially compromise the security of an organization's sensitive and critical information resources. When performing detailed network assessments and access control reviews, IS auditors should determine whether all points of entry are known and should support management's effort in obtaining the resources to identify and manage all access paths.

5.3.4 LOGICAL ACCESS CONTROL SOFTWARE

Information technology has made it possible for computer systems to store and contain large quantities of sensitive data, increase the capability of sharing resources from one system to another and

permit many users to access the system through Internet/intranet technologies. All of these factors have made organizations' IS resources more widely and promptly accessible and available.

To protect an organization's information resources, access control software has become even more critical in assuring the confidentiality, integrity and availability of information resources. The purpose of access control software is to prevent the unauthorized access and modification to an organization's sensitive data and the use of system critical functions.

To achieve this kind of control, it is necessary to apply access controls across all layers of an organization's IS architecture, including networks, platforms or OSs, databases, and application systems. Each of them usually features some form of identification and authentication, access authorization, checking of specific information resources, and logging and reporting of user activities.

The greatest degree of protection in applying access control software against internal and external users' unauthorized access is at the network and platform/OS levels. These systems are also referred to as general support systems, and they make up the primary infrastructure on which applications and database systems will reside.

OS access control software interfaces with network access control software and resides on network layer devices (e.g., routers, firewalls) that manage and control external access to organizations' networks. Additionally, OS access control software interfaces with database and/or application systems access controls to protect system libraries and user data sets.

General operating and/or application systems access control functions include the following:
- Create or change user profiles
- Assign user identification and authentication
- Apply user logon limitation rules
- Notification concerning proper use and access prior to initial login
- Create individual accountability and auditability by logging user activities
- Establish rules for access to specific information resources (e.g., system-level application resources and data)
- Log events
- Report capabilities

Database and/or application-level access control functions include the following:
- Create or change data files and database profiles
- Verify user authorization at the application and transaction level
- Verify user authorization within the application
- Verify user authorization at the field level for changes within a database
- Verify subsystem authorization for the user at the file level
- Log database/data communications access activities for monitoring access violations

In summary, access control software is provided at different levels within an IS architecture, where each level provides a certain degree of security. Properties of such relationships are that upper layers (applications, databases) are dependent on lower, infrastructure-type layers to protect general system resources. Upper layers provide the granularity needed at the application level in segregating duties by function.

5.3.5 IDENTIFICATION AND AUTHENTICATION

Identification and authentication (I&A) in logical access control software is the process of establishing and proving one's identity. It is the process by which the system obtains from a user his/her claimed identity and the credentials needed to authenticate this identity, and validates both pieces of information.

I&A is a critical building block of computer security because it is needed for most types of access control and is necessary for establishing user accountability. User accountability requires the linking of activities on a computer system to specific individuals and, therefore, requires the system to identify users. For most systems, I&A is the first line of defense because it prevents unauthorized people (or unauthorized processes) from entering a computer system or accessing an information asset. If users are not properly identified and authenticated, particularly in today's open-system–networked environments, organizations have a higher exposure to risk of unauthorized access.

Some of I&A's more common vulnerabilities that may be exploited to gain unauthorized system access include:
- Weak authentication methods (e.g., no enforcement of password minimum length, complexity and change frequency)
- Use of simple or easily guessed passwords
- The potential for users to bypass the authentication mechanism
- The lack of confidentiality and integrity for the stored authentication information
- The lack of encryption for authentication and protection of information transmitted over a network
- The user's lack of knowledge on the risk associated with sharing authentication elements (e.g., passwords, security tokens)

Authentication is typically categorized as "something you know" (e.g., password), "something you have" (e.g., token card) and "something you are (or do)" (a biometric feature). These techniques can be used independently or in combination to authenticate and identify a user. For example, a single-factor technique (something you know) involves the use of the traditional logon ID and password. Something you know, such as a personal identification number (PIN), combined and associated with something you have, such as a token card, is known as a two-factor authentication technique. Something you are is a biometric authentication technique, such as a palm or iris biometric scan. Each of these techniques is described in detail in the following sections.

A combination of more than one method, such as token and password (or PIN or token and biometric device), is referred to as "multifactor" authentication.

Identification and authentication are separate systems. They differ in respect to:
- Meaning
- Methods, peripherals and techniques supporting them
- Requirements in terms of secrecy and management
- Attributes—authentication does not have attributes in itself, while an identity may have a defined validity in time and other information attached to it.

- The fact that identity does not normally change, while authentication tokens bound to secrecy must be regularly replaced to preserve their reliability

Logon IDs and Passwords

Logon IDs and passwords are the components of a user identification and authentication process, where the authentication is based on something you know. The computer can maintain an internal list of valid logon IDs and a corresponding set of access rules for each logon ID. These access rules are related to the computer resources. As a minimum requirement, access rules are usually specified at the OS level (controlling access to files) or within individual application systems (controlling access to menu functions and types of data or transactions).

The logon ID should be restricted to provide individual, but not group identification. If a group of users is to be formed for interchangeability, the system usually offers the ability to attach a logon ID to a named group, with common rights. Each user gets a unique logon ID that can be identified by the system. The format of logon IDs is typically standardized.

FEATURES OF PASSWORDS

A password provides individual authentication. It should be easy for the user to remember, but difficult for an intruder to determine.

Initial passwords may be allocated by the security administrator or generated by the system itself. When the user logs on for the first time, the system should force a password change to improve confidentiality. Initial password assignments should be randomly generated. The ID and password should be communicated in a controlled manner to ensure that only the appropriate user receives this information. New accounts without an initial password assignment should be suspended.

If the wrong password is entered a predefined number of times, the logon ID should be automatically locked out. Locking-out may be made permanent (only the administrator may unlock the ID) or temporary (the system automatically unlocks the ID after a system-specified time period).

Users that have forgotten their password must notify a security administrator. This is the only person with sufficient privileges to reset the password and, in case this is necessary, to unlock the logon ID. The security administrator should reactivate the logon ID only after verifying the user's identification (challenge/response system), much like a bank verifies an account holder's ID before giving information over the phone (such as mother's maiden name). To verify, the security administrator should return the user's call after verifying his/her extension or calling his/her supervisor for verification.

Passwords should be hashed (a type of one-way encryption) and stored using a sufficiently strong algorithm. This allows checking passwords without the need of recording them explicitly. To reduce the risk of an intruder gaining access to other users' logon IDs, passwords should not be displayed in any form. Passwords are normally masked on a computer screen, and they are not shown on computer reports. Passwords should not be kept on index or card files or written on pieces of paper taped somewhere near the computer or inside a person's desk.

Passwords should be changed on a regular basis (e.g., every 30 days). The frequency should depend upon the criticality of the information access level, the nature of the organization, the IS architecture and technologies used, etc. Passwords should be changed by the user at his/her computer, rather than by the administrator or in any location where their new password might be observed. The best method is to force the change by notifying the user prior to the password expiration date. The risk of allowing voluntary password changes is that users will not change their passwords unless forced to do so. Password management is stronger if a history of previously used passwords is maintained by the system and their re-use prohibited for a period, such as no re-use of the last 12 passwords.

A password for a logon ID should only be known by the individual user; if a password is known to more than one person, the accountability of the user for all activity within the account cannot be enforced.

Special treatment should be applied to supervisor or administrator accounts. These accounts frequently allow full access to the system. Normally there are a limited number of such accounts per system/authentication level. For accountability, the administrator password should be known only by one individual. On the other hand, the organization should be able to access the system in an emergency situation when the administrator is not available. To enable this, practices such as keeping the administrator password in a sealed envelope, kept in a locked cabinet and available only to top managers should be implemented. This is sometimes referred to as a "firecall" ID.

All of the guidelines above should be formalized in a password policy and made as a mandatory requirement. An acceptable use policy should also include the requirement to follow the policy.

IDENTIFICATION AND AUTHENTICATION GOOD PRACTICES

Logon ID requirements include the following:

- Logon ID syntax should follow an internal naming rule, however this rule should be kept as confidential as the IDs themselves.
- Default system accounts—such as Guest, Administrator and Admin—should be renamed or disabled whenever technically possible.
- Logon IDs not used after a predetermined period of time should be deactivated to prevent possible misuse. This can be done automatically by the system or manually by the security administrator.
- The system should automatically disconnect or lock a logon session if no activity has occurred for a period of time. This reduces the risk of misuse of an active logon session left unattended, because the user went to lunch, left for home, went to a meeting or otherwise forgot to log off. This is often referred to as a session time out. Regaining access should require the reentry of the authentication method, password, token, etc.

Password syntax rules include:
- Ideally, passwords should be a minimum of eight characters in length. The length of the password will, at times, depend on the sensitivity of the systems and data to be protected and the capability of the system being used. A passphrase is generally accepted as a more secure password.
- Passwords should require a combination of at least three of the following characteristics: alphanumeric, upper and lower case letters and special characters.
- Passwords should not be particularly identifiable with the user (such as first name, last name, spouse name, pet's name, etc.). Some organizations prohibit the use of vowels, making word association/guessing of passwords more difficult.
- The system should enforce regular password changes every 30 days and not permit previous password(s) to be used for at least a year after being changed.

At a minimum, the above rules should be applied to individuals with privileged system account authority (system administrators, security administrators, etc.) versus general users. Users with privileged authority need such access in establishing and managing appropriate system configurations. However, such privileges enable the user to bypass any access control software restrictions that may exist on the system. The general rule to apply is that, the greater the degree of sensitivity of the access rights, the stricter the access controls should be.

Token Devices, One-time Passwords

In a common two-factor authentication technique, the user is assigned a microprocessor-controlled smart card, USB key or mobile device application synchronized with a specific authentication device on the system. This smart card/key/app is set to generate unique, time-dependent, pseudo-random strings that are called "session passwords" and are recognized by the authenticating device and program. They attest that the user is currently in possession of his/her own smart device. Each string is valid for only one logon session. Users must either physically read out and retype the string, or insert the smart card/USB key in a reader/USB slot along with typing in their own memorized password to gain access to the system. This technique involves something you have (a device subject to theft) and something you know (a personal identification number).

Biometrics

Biometric access controls are the best means of authenticating a user's identity based on a unique, measurable attribute or trait for verifying the identity of a human being. This control restricts computer access based on a physical (something you are) or behavioral (something you do) characteristic of the user. Due to advances in hardware efficiencies and storage, biometric systems are becoming a more viable option as an access control mechanism.

Using a biometric generally involves use of a reader device that interprets the individual's biometric features before permitting authorized access. However, this is not a flawless process because certain biometric features can change (e.g., scarred fingerprints, signature irregularities and change in voice). For this reason, biometric access control systems are not all equally effective and easy to use.

Entering a user's biometric into a system occurs through an enrollment process by storing a user's particular biometric feature. This occurs through an iterative averaging process of acquiring a physical or behavioral sample, extracting unique data from the sample (converted into a mathematical code), creating an initial template, comparing new sample(s) with what has been stored and developing a final template that can be used to authenticate the user. Subsequent samples will be used in determining whether a match or non-match condition exists for granting access.

Three percentage-based quantitative measures are used to determine the performance of biometric control devices. One measure, the false-rejection rate (FRR), or type-I error rate, is the number of times an individual granted authority to use the system is falsely rejected by the system. An aggregate measure of type-I error rates is the failure-to-enroll rate (FER), the proportion of people who fail to be enrolled successfully. The other, referred to as the false-acceptance rate (FAR), or type-II error rate, is the number of times an individual not granted authority to use a system is falsely accepted by the system. Each biometric system may be adjusted to lower FRR or FAR, but as a general rule when one decreases, the other increases (and vice versa), and there is an adjustment point where the two errors are equal. An overall metric related to the two error types is the equal error rate (EER), which is the percent showing when false rejection and acceptance are equal. The lower the overall measure the more effective the biometric.

PHYSICALLY ORIENTED BIOMETRICS

Generally, the ordering of biometric devices with the best response times and lowest EERs are palm, hand, iris, retina, fingerprint and voice, respectively.

Palm-based biometric devices analyze physical characteristics associated with the palm such as ridges and valleys. This biometric involves placing the hand on a scanner where physical characteristics are captured.

As one of the oldest biometric techniques, **hand geometry** is concerned with measuring the physical characteristics of the users' hands and fingers from a three-dimensional perspective. The user places his hand, palm-down, on a metal surface with five guidance pegs to ensure that fingers are placed properly and in the correct hand position. The template is built from measurements of physical geometric characteristics of a person's hand (usually 90 measurements)—for example, length, width, thickness and surface area.

Advantages of these systems are the social acceptance that they have received as well as the very little computer storage space that is required for the template, generally 10 to 20 bytes. The main disadvantage compared to other biometrics methods is the lack of uniqueness of hand geometry data. Moreover, an injury to the hand may cause the measurements to change, resulting in recognition problems.

An **iris**, which has patterns associated with the colored portions surrounding the pupils, is unique for every individual and, therefore, a viable method for user identification. To capture this information, the user is asked to center his/her eye onto a device by seeing the reflection of their iris in the device. Upon this alignment occurring, a camera takes a picture of the user's iris and compares it with a stored image. The iris is stable over time, having over 400 characteristics, although only approximately 260 of these are used to generate the template. As is the case with fingerprint scanning, the template carries less information than a high-quality image.

The key advantage to iris identification is that contact with the device is not needed, which contrasts with other forms of identification such as fingerprint and retinal scans. Disadvantages of iris recognition are the high cost of the system, as compared to other biometric technologies, and the high amount of storage requirements needed to uniquely identify a user.

Retina scan uses optical technology to map the capillary pattern of the eye's retina. The user has to put his eye within 0.4 to 0.8 inches (1 to 2 cm) of the reader while an image of the pupil is taken. The patterns of the retina are measured at over 400 points to generate a 96-byte template. Retinal scan is extremely reliable, and it has the lowest FAR among the current biometric methods. Disadvantages of retinal scanning include the need for fairly close physical contact with the scanning device, which impairs user acceptance, and the high cost.

Fingerprint access control is commonly used; the user places his/her finger on an optical device or silicon surface to get his/her fingerprint scanned. The template generated for the fingerprint, named "minutiae," measures bifurcations, divergences, enclosures, endings and valleys in the ridge pattern. It contains only specific data about the fingerprint (the minutiae), not the whole image of the fingerprint itself. Additionally, the full fingerprint cannot be reconstructed from the template. Depending on the provider, the fingerprint template may use between 250 bytes to more than 1,000 bytes. More storage space implies lower error rates. Fingerprint characteristics are described by a set of numeric values. While the user puts the finger in place for between two and three seconds, a typical image containing between 30 and 40 finger details is obtained and an automated comparison to the user's template takes place.

Advantages of fingerprint scanning are low cost, small size of the device, ability to physically interface into existing client-server–based systems, and ease of integration into existing access control methods. Disadvantages include the need for physical contact with the device and the possibility of poor-quality images due to residues, such as dirt and body oils, on the finger. Additionally, fingerprint biometrics are not as effective as other techniques.

In **face-recognition biometric** devices, the biometric reader processes an image captured by a video camera, which is usually within 24 inches (60 cm) of the human face, isolating it from the other objects captured within the image. The reader analyzes images captured for general facial characteristics. The template created is based on either generating two- or three-dimensional mapping arrays or by combining facial-metric measurements of the distance between specific facial features, such as the eyes, nose and mouth. Some vendors also include thermal imaging in the template.

The face is considered to be one of the most natural and most "friendly" biometrics, and it is acceptable to users because it is fast and easy to use. The main disadvantage of face recognition is the lack of uniqueness, which means that people who look alike may fool the device. Moreover, some systems cannot maintain high levels of performance as the database grows in size.

BEHAVIOR-ORIENTED BIOMETRICS

In **signature recognition**, also referred to as signature dynamics, the information from the reader is used to analyze two different areas of an individual's signature: the specific features of the signature and the specific features of the signing process. It includes speed, pen pressure, directions, stroke length and the points in time when the pen is lifted from the paper.

Advantages of this method are that it is fast, easy to use and has a low implementation cost. Other advantages are that even though a person might be able to duplicate the visual image of someone else's signature, it is difficult if not impossible to duplicate the dynamics (e.g., time duration in signing, pen-pressure, how often pen leaves signing block, etc.).

The main disadvantage is capturing the uniqueness of a signature particularly when a user does not sign his/her name in a consistent manner. For example this may occur due to illness/disease or use of initials versus a complete signature. Additionally, users' signing behavior may change when signing onto signature identification and authentication "tablets" versus writing the signature in ink onto a piece of paper.

Voice recognition involves taking the acoustic signal of a person's voice, saying a "passphrase," and converting it to a unique digital code that can then be stored in a template (approximately 1,500 to 3,000 bytes). Voice recognition incorporates several variables or parameters to recognize one's voice/speech pattern including pitch, dynamics and waveform.

The main attraction of this method is that it can be used for telephone applications, where it can be deployed with no additional user hardware costs. It also has a high rate of acceptance among users.

Disadvantages of this method include:

- The large volume of storage requirements
- Changes to people's voices
- The possibility of misspoken phrases
- A clandestine recording of the user's voice saying the passphrase could be made and played back to gain access.
- Background noises can interfere with the system.

MANAGEMENT OF BIOMETRICS

Management of biometrics should address effective security for the collection, distribution and processing of biometric data, encompassing:

- Data integrity, authenticity and nonrepudiation
- Management of biometric data across its life cycle—comprised of the enrollment, transmission, storage, verification, identification and termination processes

- Use of biometric technology, including one-to-one and one-to-many matching, for the identification and authentication of its users
- Application of biometric technology for internal and external, as well as logical and physical, access control
- Encapsulation of biometric data
- Techniques for the secure transmission and storage of biometric data
- Security of the physical hardware used throughout the biometric data life cycle
- Techniques for integrity and privacy protection of biometric data

Management should develop and approve a biometric information management and security (BIMS) policy. The auditor should use the BIMS policy to gain a better understanding of the biometric systems in use. With respect to testing, the auditor should make sure this policy has been developed and the biometric information is being secured appropriately.

As is the case with any critical information system, logical and physical controls including business continuity plans should address this area.

Life cycle controls for the development of biometric solutions should be in place to cover the enrollment request, the template creation and storage, and the verification and identification procedures. The identification and authentication procedures for individual enrollment and template creation should be specified in the BIMS policy. Management needs to have controls in place to ensure that these procedures are being followed in accordance with this policy. If the biometric device malfunctions or is inoperable, backup authentication methods should also be developed. Controls should also be in place to protect the sample data as well as the template from modification during transmission.

Single Sign-on

Users normally require access to a number of resources during the course of their daily routine. For example, users would first log into an OS and thereafter into various applications. For each OS application or other resource in use, users are required to provide a separate set of credentials to gain access. This can result in a situation where users' ability to remember passwords is significantly reduced. This also increases the chance that users will write them down on or near their workstation or area of work, thereby increasing the risk of a security breach within the organization. To address this situation, the concept of single sign-on (SSO) was developed. SSO is defined as the process for consolidating all organization platform-based administration, authentication and authorization functions into a single centralized administrative function. This function would provide the appropriate interfaces to the organization's information resources, which may include:
- Client-server and distributed systems
- Mainframe systems
- Network security including remote access mechanisms

The SSO process begins with the first instance where the user credentials are introduced into the organization's IT computing environment. The information resource or SSO server handling this function is referred to as the primary domain. Every other information resource, application or platform that uses those credentials is called a secondary domain.

The challenges in managing diverse platforms through SSO principally involve overcoming the heterogeneous nature of diverse networks, platforms, databases and applications often found in organizations when establishing a set of credentials acceptable to all of these information resources. To effectively integrate into the SSO process, SSO administrators need to obtain an understanding of how each system manages credentialing information, access control list (ACL) authorization rules, and audit logs and reports. Requirements developed in this regard should be based on security domain policies and procedures.

SSO advantages include:
- Multiple passwords are no longer required; therefore, a user may be more inclined and motivated to select a stronger password.
- It improves an administrator's ability to manage users' accounts and authorizations to all associated systems.
- It reduces administrative overhead in resetting forgotten passwords over multiple platforms and applications.
- It reduces the time taken by users to log into multiple applications and platforms.

SSO disadvantages include:
- Support for all major OS environments is difficult. SSO implementations will often require a number of solutions integrated into a total solution for an enterprise's IT architecture.
- The costs associated with SSO development can be significant when considering the nature and extent of interface development and maintenance that may be necessary.
- The centralized nature of SSO presents the possibility of a single point of failure and total compromise of an organization's information assets. For this reason, strong authentication in the form of complex password requirements and the use of biometrics is frequently implemented.

One example of SSO is Kerberos. Created by the Massachusetts Institute of Technology, it is an authentication service used to validate services and users in a distributed computing environment (DCE). The role of the authentication service is to allow principals to positively identify themselves and participate in a DCE. Both users and servers authenticate themselves in a DCE environment, unlike security in most other client-server systems where only users are authenticated. There are two distinct steps to authentication. At initial logon time, the Kerberos third-party protocol is used within DCE to verify the identity of a client requesting to participate in a DCE network. This process results in the client obtaining credentials initially registered with the trusted third party and cryptographically protected. These credentials form the basis for setting up secure sessions with DCE servers when the user tries to access resources.

SSO can also be addressed using the Security Assertion Markup Language (SAML). This is open standard data format using XML to exchange authentication and authorization information between services. The single most important requirement that SAML addresses is web browser SSO.

5.3.6 AUTHORIZATION ISSUES

The authorization process used for access control requires that the system be able to identify and differentiate among users.

Access rules (authorization) specify who can access what. For example, access control is often based on least privilege, which refers to the granting to users of only those accesses required to perform their duties. Access should be on a documented need-to-know and need-to-do basis by type of access.

Computer access can be set for various levels (i.e., files, tables, data items, etc.). When IS auditors review computer accessibility, they need to know what can be done with the access and what is restricted. For example, access restrictions at the file level generally include the following:

- Read, inquiry or copy only
- Write, create, update or delete only
- Execute only
- A combination of the above

The least dangerous type of access is read only, as long as the information being accessed is not sensitive or confidential. This is because the user cannot alter or use the computerized file beyond basic viewing or printing.

Access Control Lists

To provide security authorizations for the files and facilities listed previously, logical access control mechanisms utilize access authorization tables, also referred to as access control lists (ACLs) or access control tables. ACLs refer to a register of:

- Users (including groups, machines, processes) who have permission to use a particular system resource
- The types of access permitted

ACLs vary considerably in their capability and flexibility. Some only allow specifications for certain preset groups (e.g., owner, group and world), while more advanced ACLs allow much more flexibility such as user-defined groups. Also, more advanced ACLs can be used to explicitly deny access to a particular individual or group. With more advanced ACLs, access can be at the discretion of the policy maker (and implemented by the security administrator) or individual user, depending upon how the controls are technically implemented. When a user changes job roles within an organization, often their old access rights are not removed before adding their new required accesses. Without removing the old access rights, there could be a potential SoD issue.

Logical Access Security Administration

In today's client-server environment, the access identification and authentication, and the authorization process, can be administered either through a centralized or decentralized environment. The advantages of conducting security in a decentralized environment are:

- The security administration is onsite at the distributed location.
- Security issues are resolved in a timely manner.
- Security controls are monitored on a more frequent basis.

The risk associated with distributed responsibility for security administration includes:

- Local standards might be implemented rather than those required by the organization
- Levels of security management might be below what can be maintained by a central administration
- Unavailability of management checks and audits that are often provided by central administration to ensure that standards are maintained

There are many ways to control remote and distributed sites:

- Software controls over access to the computer, data files and remote access to the network should be implemented.
- The physical control environment should be as secure as possible, with additions such as lockable terminals and a locked computer room.
- Access from remote locations via modems and laptops to other microcomputers should be controlled appropriately.
- Opportunities for unauthorized people to gain knowledge of the system should be limited by implementing controls over access to system documentation and manuals.
- Controls should exist for data transmitted from remote locations such as sales in one location that update accounts receivable files at another location. The sending location should transmit control information, such as transaction control totals, to enable the receiving location to verify the update of its files. When practical, central monitoring should ensure that all remotely processed data have been received completely and updated accurately.
- When replicated files exist at multiple locations, controls should ensure that all files used are correct and current and, when data are used to produce financial information, that no duplication arises.

Remote Access Security

Remote access connectivity to their information resources is required for many organizations for different types of users, such as employees, vendors, consultants, business partners and customer representatives. In providing this capability, a variety of methods and procedures are available to satisfy an organization's business need for this level of access.

Remote access users can connect to their organization's networks with the same level of functionality that exists within their office. In doing so, the remote access design uses the same network standards and protocols applicable to the systems that they are accessing, Transmission Control Protocol/Internet Protocol (TCP/IP)-based systems and systems network architecture (SNA) systems, for the mainframe where the user uses terminal emulation software to connect to a mainframe-based legacy application. Support for these connections includes asynchronous point-to-point modem connectivity, integrated services digital network (ISDN) dial-on-demand connectivity, and dedicated lines (e.g., frame relay and digital subscriber lines [DSL]).

COMMON CONNECTIVITY METHODS

TCP/IP Internet-based remote access is a cost-effective approach that enables organizations to take advantage of the public network infrastructures and connectivity options available, under which ISPs manage modems and dial-in servers, and DSL and cable modems reduce costs further to an organization. To effectively use this option, organizations establish a virtual private network over the Internet to securely communicate data packets over

this public infrastructure. Available VPN technologies apply the Internet Engineering Task Force (IETF) IPSec standard (see section 5.4.5 Encryption for more details on IPSec). Advantages are their ubiquity, ease of use, inexpensive connectivity, and read, inquiry or copy only access. Disadvantages include that they are significantly less reliable than dedicated circuits, lack a central authority, and can be difficult to troubleshoot.

Organizations should be aware that using VPNs to allow remote access to their systems can create holes in their security infrastructure. The encrypted traffic can hide unauthorized actions or malicious software that can be transmitted through such channels. IDSs and virus scanners able to decrypt the traffic for analysis and then encrypt and forward it to the VPN endpoint should be considered as preventive controls. A good practice will terminate all VPNs to the same endpoint in a so called VPN concentrator, and will not accept VPNs directed at other parts of the network.

A less common method is to use dial-up lines (modem asynchronous point-to-point or ISDN) in accessing an organization's network access server (NAS) that works in concert with an organization's network firewall and router configuration. The NAS handles user authentication, access control and accounting, while maintaining connectivity. The most common protocol for doing this is the Remote Access Dial-in User Service (RADIUS) and Terminal Access Controller Access Control System (TACACS). In a typical NAS implementation, calls into the network are received, and as a good security practice, the call is terminated after recording the calling number and performing preliminary authentication procedures. The standard security practice has been for the NAS to initiate a call back to a predetermined number of the user. This control can be circumvented through effective implementation of call-forwarding mechanisms.

Dial-up connectivity, not based on centralized control and least preferred from a security and control standpoint, is an organization's server whose OS is set up to accept remote access, which is referred to as a remote access server (RAS). The latter approach is not recommended, as it is extremely difficult to control remote access from many servers using its own RAS capability.

Advantages of dial-up connectivity are its low end-user costs (local phone calls) and that it is intuitive and easy to use (familiarity). Disadvantages are related to performance; for example, reliability in establishing connections with the NAS (phone networks' electrical interference) and time-sensitive media-rich applications or a service's failure when data-rate throughput is low.

Another common connectivity method often used for remote access is dedicated network connections. Using private, often proprietary, network circuits is the approach generally considered the safest because the only network traffic carried belongs to the same organization. It is commonly used by branch/regional offices or with business partners.

Advantages of dedicated network connections include greater performance gains in data throughput and reliability, and data on a dedicated link belonging to the subscribing organization, where an intruder would have to compromise the telecommunications provider itself to access the data link. A disadvantage is that cost is typically two- to five-times higher than connections to the Internet.

Remote access risk includes:
- DoS where remote users may not be able to gain access to data or applications that are vital for them to carry out their day-to-day business
- Malicious third parties; these may gain access to critical applications or sensitive data by exploiting weaknesses in communications software and network protocols
- Misconfigured communications software, which may result in unauthorized access or modification of an organization's information resources
- Misconfigured devices on the corporate computing infrastructure
- Host systems not secured appropriately, which could be exploited by an intruder gaining access remotely
- Physical security issues over remote users' computers

Remote access controls include:
- Policy and standards
- Proper authorizations
- Identification and authentication mechanisms
- Encryption tools and techniques such as use of a VPN
- System and network management

Audit Logging in Monitoring System Access

Most access control software has security features that enable a security administrator to automatically log and report all levels of access attempts—successes and failures. For example, access control software can log computer activity initiated through a logon ID or computer terminal. This information provides management an audit trail to monitor activities of a suspicious nature, such as a hacker attempting brute force attacks on a privileged logon ID. Also, keystroke logging can be turned on for users that have sensitive access privileges. What is logged is determined by the action of the organization. Issues include what is logged, who/what has access to the logs and how long logs are retained (record-retention item).

ACCESS RIGHTS TO SYSTEM LOGS

Access rights to system logs for security administrators to perform the above activities should be strictly controlled.

Computer security managers and system administrators/managers should have access for review purposes; however, security and/or administration personnel who maintain logical access functions may not need to access audit logs.
It is particularly important to ensure the integrity of audit trail data against modification. This can be done using digital signatures, write-once devices or a security information and event management (SIEM) systems. The audit trail files need to be protected because intruders may try to cover their tracks by modifying audit trail records. Audit trail records should be protected by strong access controls to help prevent unauthorized

access. The integrity of audit trail information may be particularly important when legal issues arise, such as the use of audit trails as legal evidence. (This may, for example, require daily printing and signing of the logs.) Questions regarding such legal issues should be directed to the appropriate legal counsel.

The confidentiality of audit trail information may also be protected if the audit trail is recording information about users that may be disclosure-sensitive, such as transaction data containing personal information (e.g., before and after records of modification to income tax data). Strong access controls and encryption can be particularly effective in preserving confidentiality.

Media logging is used to support accountability. Logs can include control numbers (or other tracking data) such as the times and dates of transfers, names and signatures of individuals involved, and other relevant information. Periodic spot checks or audits may be conducted to determine that no controlled items have been lost and that all are in the custody of individuals named in control logs. Automated media tracking systems may be helpful for maintaining inventories of tape and disk libraries.

A periodic review of system-generated logs can detect security problems, including attempts to exceed access authority or gain system access during unusual hours. Certain reports are generated for security recorded in activity logs.

TOOLS FOR AUDIT TRAIL (LOGS) ANALYSIS

Many types of tools have been developed to help reduce the amount of information contained in audit records and to delineate useful information from the raw data.

On most systems, audit trail software can create large files, which can be extremely difficult to analyze manually. The use of automated tools is likely to be the difference between unused audit trail data and an effective review. Some of the types of tools include:

- **Audit reduction tools**—They are preprocessors designed to reduce the volume of audit records to facilitate manual review. Before a security review, these tools can remove many audit records known to have little security significance. (This alone may cut in half the number of records in the audit trail.) These tools generally remove records generated by specified classes of events—for example, records generated by nightly backups might be removed.
- **Trend/variance-detection tools**—They look for anomalies in user or system behavior. It is possible to construct more sophisticated processors that monitor usage trends and detect major variations. For example, if a user typically logs in at 09:00 but appears at 04:30 one morning, this may indicate a security problem that may need to be investigated.
- **Attack-signature-detection tools**—They look for an attack signature, which is a specific sequence of events indicative of an unauthorized access attempt. A simple example would be repeated failed logon attempts.
- **SIEM systems**—These tools capture audit trails or logs and perform real-time analysis on them. They can aggregate audit trails and logs from many different sources. These data can then be correlated and alerts provided if required. Some SIEM systems can also be configured to perform automated tasks based upon the alerts (e.g., launching a vulnerability scan or commanding the firewall to close a certain port).

COST CONSIDERATIONS

Audit trails involve many costs that factor into IT's determination as to how much logging is enough. First, some system overhead is incurred while recording the audit trail. Additional system overhead will be incurred to store and process the records. The more detailed the records, the more overhead is required. In some systems, logging every event could cause the system to lock up or slow to the point at which response time would be measured in minutes. Obviously, this is not acceptable if IT is to be properly aligned with the needs of the business. Another cost involves human and machine time required when performing the analysis. This can be minimized by using tools to perform most of the analysis. Many simple analyzers can be constructed quickly and inexpensively from system utilities, but they are limited to audit reduction and the identification of particularly sensitive events. More complex tools such as SIEM systems will be more expensive both to purchase and to implement.

The final cost of audit trails is the cost of investigating unexpected and anomalous events. If the system is identifying too many events as suspicious, administrators may spend undue time reconstructing events and questioning personnel.

The frequency of the security administrator's review of computer access reports should be commensurate with the sensitivity of the computerized information being protected. The IS auditor should ensure that the logs cannot be tampered with, or altered, without leaving an audit trail.

When reviewing or performing security access follow-up, the IS auditor should look for:

- Patterns or trends that indicate abuse of access privileges, such as concentration on a sensitive application
- Violations (such as attempting computer file access that is not authorized) and/or use of incorrect passwords

Once a violation has been identified:

- The person who identified the violator should refer the problem to the security administrator for investigation.
- The security administrator and responsible management should work together to investigate and determine the severity of the violation. Generally, most violations are accidental.
- If a violation attempt is serious, executive management should be notified, not law enforcement officials. Executive management normally is responsible for notifying law enforcement officials. Involvement of external agencies may result in adverse publicity that is ultimately more damaging than the original violation; therefore, the decision to involve external agencies should be left to executive management.
- Procedures should be in place to manage public relations and the press.
- To facilitate proper handling of access violations, written guidelines should exist that identify various types and levels of violations and how they will be addressed. This effectively provides direction for judging the seriousness of a violation.
- Disciplinary action should be a formal process that is applied consistently. This may involve a reprimand, probation or

immediate termination. The procedures should be legally and ethically sound to reduce the risk of legal action against the company.
- Corrective measures should include a review of the computer access rules, not only for the perpetrator but for interested parties. Excessive or inappropriate access rules should be eliminated.

Naming Conventions for Logical Access Controls

Access capabilities are implemented by security administration in a set of access rules that stipulates which users (or groups of users) are authorized to access a resource (such as a dataset or file) and at what level (such as read or update). The access control mechanism applies these rules whenever a user attempts to access or use a protected resource.

Access control naming conventions are structures used to govern user access to the system and user authority to access/use computer resources such as files, programs and terminals. These general naming conventions and associated files are required in a computer environment to establish and maintain personal accountability and SoD in the access of data. The owners of the data or application, with the help of the security officer, usually set up naming conventions. The need for sophisticated naming conventions over access controls depends on the importance and level of security that is needed to ensure that unauthorized access has not been granted. It is important to establish naming conventions that both promote the implementation of efficient access rules and simplify security administration.

Naming conventions for system resources (e.g., as datasets, volumes, programs and employees workstations) are an important prerequisite for efficient administration of security controls. Naming conventions can be structured so that resources beginning with the same high-level qualifier can be governed by one or more generic rule(s). This reduces the number of rules required to adequately protect resources which, in turn, facilitates security administration and maintenance efforts.

5.3.7 STORING, RETRIEVING, TRANSPORTING AND DISPOSING OF CONFIDENTIAL INFORMATION

Management should define and implement procedures to prevent access to, or loss of, sensitive information and software from computers, disks, and other equipment or media when they are stored, disposed of or transferred to another user.

This should be done for the following:
- **Backup files of databases**—Backup files on magnetic tapes are often unencrypted, so that even confidential information may be obtained by simply transferring backup databases to other systems for data analysis. Security problems of data media storage and transportation technologies involve ensuring that contractors used to transport and store backup tapes have adequate policies and procedures to protect the integrity and confidentiality of the information. It is good practice to fully encrypt all portable or backup media.
- **Data banks**—Research and commercial institutions collect the results of important survey or research projects on large tapes. These data have a high commercial value and may be subject to a requirement of availability and confidentiality that persist for many years, longer than the duration of the media containing them. Preserving the value requires precautions and possibly a planned media verification or duplication activity. In general, a solution is required to the problems of long-term computer storage of sensitive information. Optical disks are a possible medium, but their durability and standardization should be evaluated.
- **Disposal of media previously used to hold confidential information**—Procedures should be implemented to identify and erase the sensitive information and software inside computers, disks and other equipment or media that have been identified for disposal so that deleted data may not be retrieved by any internal or third party. Care must be taken not only to meet the requirements of data protection, but also when a machine is transferred to another user. The original user should remove any personal data that are confidential by nature. If previously held data were sensitive, the disk should be reformatted and then a secure wipe of the disk should be carried out to a defined standard.
  - In some cases, when information is highly confidential, it may prove insufficient to wipe the media. Random access memory (RAM) is included because favorable circumstances and appropriate technical analysis of these media could expose the data. This may require that such equipment or media should be disposed of in a secure manner (e.g., destruction). This may include "degaussing" (demagnetizing) the magnetic media, such as tapes or PC hard drives, and possibly their physical destruction.
- **Management of equipment sent for offsite maintenance**—Data files and proprietary software should be backed up, so that they can be erased from the equipment prior to sending it offsite for maintenance (e.g., computers, tablets, flash drives). Computers holding confidential data should not be sent out for repair, unless memories are withheld.
- **Public agencies and organizations concerned with sensitive, critical or confidential information**—These organizations may have particular obligations to develop a comprehensive records management program. Policies addressing these needs should reflect laws concerning availability, substance, degree of confidentiality and disposal compared to available technical solutions and organizational needs. For instance, public records may be destroyed only in accordance with precise record-retention schedules, and the record holder may not mutilate, destroy, sell, loan or otherwise dispose of any record, except under a record-retention schedule or with the written consent of the owner. Proper record retention requires the preparation of separate retention schedules depending on subject files (administrative vs. other legal requirements).
- **E-token electronic keys**—For such sensitive information, data transportation on removable media is not safe, and taking proper care of the media can significantly reduce the chances of data loss.
- **Storage records**—Many commercial organizations fulfill legal or institutional obligations to preserve specific types of records, which may be confidential in nature, for a given number of years. In some cases these obligations are fulfilled by preserving database images and the source of the documents, either online or on backup tapes. In these cases, the conditions of recreating the original document must be integrally retained as well.

Preserving Information During Shipment or Storage

Manufacturers publish recommended temperature and humidity levels in which to store media. These recommendations should be consulted and adhered to before storing or shipping important media. However, some general tips can be followed to help avoid potential damage to media during shipping and storage. The following environmental issues are applicable to all types of media:

- Keep out of direct sunlight.
- Keep free of dust.
- Keep free of liquids.
- Minimize exposure to magnetic fields, radio equipment or any sources of vibration.
- Do not air transport in areas and at times of exposure to a strong magnetic storm.

Media-specific Storage Precautions

Some precautions need to be considered regarding media-specific storage (see **figure 5.13**).

Figure 5.13—Media-specific Storage Precautions

Media Storage	Precautions
Hard drives	• Store hard drives in antistatic bags, and be sure that the person removing them from the bag is static-free. • If the original box and padding for the hard drive is available, use it for shipping. • Avoid Styrofoam packaging products or other materials that can cause static electricity. • Quick drops or spikes in temperature are a danger, because such changes can lead to hard drive crashes. • If the hard drive has been in a cold environment, bring it to room temperature prior to installing and using it. • Avoid sudden mechanical shocks or vibrations.
Tape cartridges	• Store cartridges vertically. • Store cartridges in protective containers for transport. • Write-protect cartridges immediately.
USB, flash and portable hard drives	• Avoid temperature and humidity extremes and strong magnetic fields.
CDs and DVDs	• Handle by the edges or by the hole in the middle. • Be careful not to bend the media. • Avoid long-term exposure to bright light. • Store in a hard jewel case, not in soft sleeves.

5.4 NETWORK INFRASTRUCTURE SECURITY

Communication networks (wide area or local area networks) generally include devices connected to the network as well as programs and files supporting the network operations. Control is accomplished through a network control terminal and specialized communications software.

The following are controls over the communication network:

- Network control functions should be performed by individuals possessing adequate training and experience.
- Network control functions should be separated, and the duties should be rotated on a regular basis, where possible.
- Network control software must restrict operator access from performing certain functions (e.g., the ability to amend/delete operator activity logs).
- Network control software should maintain an audit trail of all operator activities.
- Audit trails should be periodically reviewed by operations management to detect any unauthorized network operations activities.
- Network operation standards and protocols should be documented and made available to the operators and should be reviewed periodically to ensure compliance.
- Network access by the system engineers should be monitored and reviewed closely to detect unauthorized access to the network.
- Analysis should be performed to ensure workload balance, fast response time and system efficiency.
- A terminal identification file should be maintained by the communications software to check the authentication of a terminal when it tries to send or receive messages.
- Data encryption should be used, where appropriate, to protect messages from disclosure during transmission.
- Restrictions should be placed on remote printing facilities to ensure sensitive documents cannot be read by unauthorized personnel.

To improve the control and maintenance of the infrastructure and its use, besides the direct management of the network devices, consolidate the logs of these devices with the firewall's logs and the client-server OS's logs.

In recent years, the management of large capacity storage units is frequently based on fiber channel connections.

Systems security is improved when a dynamic inventory of the devices is possible. In the case of an incident, it is important to know which computer is used by whom.

Another important security improvement is the ability to identify users at every step of their activity. Some application packages use predefined names (e.g., SYSTEM). New monitoring tools have been developed to resolve this problem.

Adopting an IT governance practice enables an organization to comply with network security requirements effectively. The Information Technology Infrastructure Library (ITIL) is a framework of practice guidance in information technology service management that can be used in setting up service level agreements (SLAs), specifically for enterprise network operations, to maintain the uninterrupted operation of the network through controls, incident handling and auditing (see chapter 4 IS Operations, Maintenance and Service Management).

5.4.1 LAN SECURITY

LANs are computer networks that cover a limited area such as a home, office or campus. The security of a LAN is dependent on the security of its component parts.

As LANs facilitate the storage and retrieval of programs and data used by a group of people, the security of the LAN is also dependent on the security of the OS.

For more information on risk associated with networks and OSs, see chapter 4 IS Operations, Maintenance and Service Management.

Risk associated with use of LANs includes:
- Loss of data and program integrity through unauthorized changes
- Lack of current data protection through inability to maintain version control
- Exposure to external activity through poor user verification and potential public network access from remote connections
- Virus and worm infection
- Improper disclosure of data because of general access rather than need-to-know access provisions
- Illegal access by impersonating or masquerading as a legitimate LAN user
- Internal user's sniffing (obtaining seemingly unimportant information from the network that can be used to launch an attack such as network address information)
- Internal user's spoofing (reconfiguring a network address to pretend to be a different address)
- Lack of enabled detailed automated logs of activity (audit trails)
- Destruction of the logging and auditing data

The LAN security provisions available depend on the software product, product version and implementation. Commonly available network security administrative capabilities include:
- Declaring ownership of programs, files and storage
- Limiting access under the principle of least privilege (users can only access what they need to perform their role)
- Implementing record and file locking to prevent simultaneous update
- Enforcing user ID/password sign-on procedures, including the rules relating to password length, format and change frequency
- Using switches to implement port security policies rather than hubs or non-manageable routers. This will prevent unauthorized hosts, with unknown MAC addresses, to connect to the LAN.
- Encrypting local traffic using IPSec protocol

To gain a full understanding of the LAN, the IS auditor should identify and document the following:
- Users or groups with privileged access rights
- LAN topology and network design
- LAN administrator/LAN owner
- Functions performed by the LAN administrator/owner
- Distinct groups of LAN users
- Computer applications used on the LAN
- Procedures and standards relating to network design, support, naming conventions and data security

Virtualization

Virtualization provides an enterprise with a significant opportunity to increase efficiency and decrease costs in its IT operations. However, virtualization also introduces additional risk. IS auditors need to understand the advantages and disadvantages of virtualization to determine whether the enterprise has considered the applicable risk in its decision to adopt, implement and maintain this technology.

At a high level, virtualization allows multiple OSs (guests) to coexist on the same physical server (host) in isolation of one another. Virtualization creates a layer between the hardware and the guest OSs to manage shared processing and memory resources on the host. Often, a management console provides administrative access to manage the virtualized system. **Figure 5.14** summarizes several advantages and disadvantages of virtualization.

Figure 5.14—Advantages and Disadvantages of Virtualization

Advantages	Disadvantages
• Server hardware costs may decrease for both server builds and server maintenance. • Multiple OSs can share processing capacity and storage space that often goes to waste in traditional servers, thereby reducing operating costs. • The physical footprint of servers may decrease within the data center. • A single host can have multiple versions of the same OS, or even different OSs, to facilitate testing of applications for performance differences. • Creation of duplicate copies of guests in alternate locations can support business continuity efforts. • Application support personnel can have multiple versions of the same OS, or even different OSs, on a single host to more easily support users operating in different environments. • A single machine can house a multitier network in an educational lab environment without costly reconfigurations of physical equipment. • Smaller organizations that had performed tests in the production environment may be better able to set up logically separate, cost-effective development and test environments. • If set up correctly, a well-built, single access control on the host can provide tighter control for the host's multiple guests.	• Inadequate configuration of the host could create vulnerabilities that affect not only the host, but also the guests. • Exploits of vulnerabilities within the host's configuration, or a denial of service attack against the host, could affect all of the host's guests. • A compromise of the management console could grant unapproved administrative access to the host's guests. • Performance issues of the host's own OS could impact each of the host's guests. • Data could leak between guests if memory is not released and allocated by the host in a controlled manner. • Insecure protocols for remote access to the management console and guests could result in exposure of administrative credentials.

Although virtualization offers significant advantages, they come with risk that an enterprise must manage effectively. Because the host in a virtualized environment represents a potential single point of failure within the system, a successful attack on the host could result in a compromise that is larger in both scope and impact.

To address risk, an enterprise can often implement and adapt the same principles and good practices for a virtualized server environment that it would use for a server farm. These include the following:

- Strong physical and logical access controls, especially over the host and its management console
- Sound configuration management practices and system hardening for the host, including patching, antivirus, limited services, logging, appropriate permissions and other configuration settings
- Appropriate network segregation, including the avoidance of virtual machines in the demilitarized zone (DMZ) and the placement of management tools on a separate network segment
- Strong change management practices

5.4.2 CLIENT-SERVER SECURITY

A client-server is a network architecture in which each computer or process on the network is either a server (a source of services and data) or a client (a user of these services and data that relies on servers to obtain them). Client-server architectures can be two-tiered (includes the use of a thick client), three-tiered (includes the use of application servers and a thin client, probably a browser) or n-tiered (includes multiple applications servers, middleware, etc.).

The security of a client-server environment is dependent on the security of its component parts. This includes the security of the:
- LAN
- Client
- OS
- Database
- Middleware

In a client-server environment, several access routes exist, because application data may exist on the server or on the client. Therefore, each of these routes must be examined individually and in relation to each other to ensure that no exposures are left unchecked.

An additional risk to consider with the client-server model is the potential gaps between the components. In other words, how do the components connect to each other?

For example, in a two-tiered environment, the thick client must connect to the database. To achieve this, either (1) every user has a database account, in which case they may be able to bypass the client application (and hence the application controls) and connect directly to the database or (2) a proxy user (i.e., a single account that connects to the database on behalf of all others) is used, in which case the database password must be stored somewhere. This might be stored insecurely or unencrypted.

In a client-server environment the IS auditor should ensure that:
- Application controls cannot be bypassed.
- Passwords are always encrypted.
- Access to configuration or initialization files is kept to a minimum.
- Access to configuration or initialization files are audited.

Note: The IS auditor should be familiar with risk and exposures related to network infrastructure.

5.4.3 WIRELESS SECURITY THREATS AND RISK MITIGATION

The classification of security threats may be segmented into nine categories:
- Errors and omissions
- Fraud and theft committed by authorized or unauthorized users of the system
- Employee sabotage
- Loss of physical and infrastructure support
- Malicious hackers
- Industrial espionage
- Malicious code
- Foreign government espionage
- Threats to personal privacy

All of these represent potential threats in wired networks as well. However, the more immediate concerns for wireless communications are device theft, DoS, malicious hackers, malicious code, theft of service, and industrial and foreign government espionage.

Theft is likely to occur with wireless devices because of their portability. Authorized and unauthorized users of the system may commit fraud and theft; however, authorized users are more likely to carry out such acts. Because users of a system may know what resources a system has and the system's security flaws, it is easier for them to commit fraud and theft.

Malicious hackers, sometimes called crackers, are individuals who break into a system without authorization, usually for personal gain or to do harm. Such hackers may gain access to the wireless network access point by eavesdropping on wireless device communications.

Malicious code involves viruses, worms, Trojan horses, logic bombs or other unwanted software that is intended to damage files or bring down a system. Theft of service occurs when an unauthorized user gains access to the network and consumes network resources. In wireless networks, the unauthorized access threat stems from the relative ease with which eavesdropping can occur on radio transmissions.

Ensuring confidentiality, integrity, authenticity and availability are the prime objectives in wireless networks.

Security requirements include the following:
- **Authenticity**—A third party must be able to verify that the content of a message has not been changed in transit.
- **Nonrepudiation**—The origin or the receipt of a specific message must be verifiable by a third party.
- **Accountability**—The actions of an entity must be uniquely traceable to that entity.
- **Network availability**—The IT resource must be available on a timely basis to meet mission requirements or to avoid substantial losses. Availability also includes ensuring that resources are used only for intended purposes.

Risk in wireless networks is equal to the sum of the risk of operating a wired network plus the new risk introduced by weaknesses in wireless protocols. To mitigate the risk, an

organization must adopt security measures and practices that help bring risk to a manageable level.

To date, the list below includes some of the more salient threats and vulnerabilities of wireless systems:

- All the vulnerabilities that exist in a conventional wired network apply to wireless technologies.
- Weaknesses in wireless protocols increase the threat of disclosure of sensitive information. Many wireless networks are either not secure or use outdated encryption algorithms.
- Malicious entities may gain unauthorized access to an agency's computer or voice (IP telephony) network through wireless connections, potentially bypassing any firewall protections.
- Sensitive information that is not encrypted (or that is encrypted with poor cryptographic techniques) and is transmitted between two wireless devices may be intercepted and disclosed.
- DoS attacks may be directed at wireless connections or devices.
- Malicious entities may steal the identity of legitimate users and masquerade as them on internal or external corporate networks.
- Sensitive data may be corrupted during improper synchronization.
- Malicious entities may be able to violate the privacy of legitimate users and track their physical movements.
- Malicious entities may deploy unauthorized equipment (e.g., client devices and access points) to surreptitiously gain access to sensitive information.
- Mobile devices are easily stolen and can reveal sensitive information.
- Data may be extracted without detection from improperly configured devices.
- Viruses or other malicious code may corrupt data on a wireless device and be subsequently introduced to a wired network connection.
- Malicious entities may, through wireless connections, connect to other agencies for the purposes of launching attacks and concealing their activity.
- Interlopers, from inside or out, may be able to gain connectivity to network management controls and thereby disable or disrupt operations.
- Malicious entities may use a third-party, untrusted wireless network service to gain access to the network resources.

Currently, there are many ways that malicious entities may gain access to wireless devices. Those related to WLANs include, but are not limited to, war driving, war walking and war chalking as described in **figure 5.12**.

On the wireless personal area network (WPAN) side, one of the important types of risk is the man-in-the-middle attack as described in **figure 5.12**.

The other problem with WPANs is the uncontrolled propagation of radio waves; for example, the radio traffic on Bluetooth connections can be passively intercepted and recorded using Bluetooth protocol sniffers such as Red Fang, Bluesniff and others. If the device addresses are known, then even if the devices are currently in nondiscoverable mode, it is possible to synchronize to the frequency hopping sequence. All the layers of the Bluetooth protocol stack can be examined and analyzed offline. If encryption is not used, then it is possible to extract and monitor the transmitted user data. Use of an antenna with a strong directional characteristic and electronics capable of amplifying Bluetooth signals can make passive listening attacks possible from a greater distance than the functional range. Transmitting power control is optional and is not supported by every Bluetooth device.

The growing prevalence of people using Bluetooth-enabled equipment may follow the trend of Wi-Fi war driving, in which people try to identify inadequately secured networks by driving around with a laptop.

5.4.4 INTERNET THREATS AND SECURITY

The nature of the Internet makes it vulnerable to attack. The Internet is a global TCP/IP-based system that enables public and private heterogeneous networks to communicate with one another. Around 40 percent of the world's population is connected to the Internet (*www.internetlivestats.com*). Originally designed to allow for the freest possible exchange of information, it is widely used today for commercial purposes. This poses significant security problems for organizations when protecting their information assets. For example, hackers and virus writers try to attack the Internet and computers connected to the Internet. Some want to invade others' privacy and attempt to crack into databases of sensitive information or sniff information as it travels across Internet routes. Consequently, it becomes more important for IS auditors to understand the risk and security factors that are needed to ensure that proper controls are in place when a company connects to the Internet.

The IP is designed solely for the addressing and routing of data packets across a network. It does not guarantee or provide evidence on the delivery of messages; there is no verification of an address; the sender will not know if the message reaches its destination at the time it is required; the receiver does not know if the message came from the address specified as the return address in the packet. Other protocols correct some of these drawbacks.

Network Security Threats

One class of network attacks involves probing for network information. These passive attacks can lead to actual active attacks or intrusions/penetrations into an organization's network. By probing for network information, the intruder obtains network information that can be used to target a particular system or set of systems during an actual attack.

Passive Attacks

Examples of passive attacks that gather network information include network analysis, eavesdropping and traffic analysis as explained in **figure 5.12**.

Active Attacks

Once enough network information has been gathered, the intruder will launch an actual attack against a targeted system to either gain complete control over that system or enough control to cause certain threats to be realized. This may include obtaining unauthorized access to modify data or programs, causing a DoS, escalating privileges, accessing other systems, and obtaining sensitive information for personal gain. These types of penetrations or intrusions are known as active attacks. They affect the integrity, availability and authentication attributes of network security. Common forms of active attacks may include the following (explained in **figure 5.12**):

- Brute force attack
- Masquerading
- Packet replay
- Phishing
- Message modification
- Unauthorized access through the Internet or World Wide Web
- Denial of service (DoS)
- Dial-in penetration attacks
- Email bombing and spamming
- Email spoofing

Causal Factors for Internet Attacks

Generally, Internet attacks of both a passive and active nature occur for a number of reasons including:

- Availability of tools and techniques on the Internet or as commercially available software that an intruder can download easily. For example, to scan ports, an intruder can easily obtain network scanners such as strobe, netcat, jakal, nmap or Asmodeous (Windows). Additionally, password cracking programs such as John the Ripper and L0phtCrack are available free or at a minimal cost.
- Lack of security awareness and training among an organization's employees
- Exploitation of known security vulnerabilities in network- and host-based systems. Many organizations fail to properly configure their systems and to apply security patches or fixes when vulnerabilities are discovered. Most problems can be reduced significantly by keeping network- and host-based systems properly configured and up to date.
- Inadequate security over firewalls and host-based OSs allowing intruders to view internal addresses and use network services indiscriminately

With careful consideration when designing and developing network security controls and supporting processes, an organization can effectively prevent and detect most intrusive attacks on their networks. In this situation, it becomes important for IS auditors to understand the risk and security factors that are needed to ensure proper controls are in place when a company connects to the Internet. There are several areas of control risk that must be evaluated by the IS auditor to determine the adequacy of Internet security controls.

Internet Security Controls

To establish effective Internet security controls, an organization must develop controls within an information systems security framework from which Internet security controls can be implemented and supported. Generally, the process for establishing such a framework entails defining, through corporate policies and procedures, the rules the organization will follow to control Internet usage. For example, one set of rules should address appropriate use of Internet resources with rules that might reserve Internet privileges for those with a business need, define what information resources should be available for outside users, and define trusted and untrusted networks within and outside the organization.

Another set of rules should address the classification of the sensitivity or criticality of corporate information resources. This will help to determine what information will be available for use on the Internet and the level of security to be used for corporate resources of a sensitive or critical nature on the Internet.

From an evaluation of these issues, an organization will be able to develop guidelines specific to their situations for defining the level of security controls related to the confidentiality, integrity and availability of information resources (i.e., business applications) on the Internet. For example, OS security hardening guidelines can be developed which define how the OS should be configured, detail which Internet services should be blocked from use or exploitation by external untrusted users, and define how the system will be protected by firewalls. Additionally, supporting processes over these controls should be defined including:

- Risk assessments performed periodically over the development and redesign of Internet-based web applications
- Security awareness and training for employees, tailored to their levels of responsibilities
- Firewall standards and security to develop and implement firewall architectures
- Intrusion detection standards and security to develop and implement IDS architectures
- Remote access for coordinating and centrally controlling dial-up access on the Internet via corporate resources
- Incident handling and response for detection, response, containment and recovery
- Configuration management for controlling the security baseline when changes do occur
- Encryption techniques applied to protect information assets passing over the Internet
- A common desktop environment to control, in an automated fashion, what is displayed on a user's desktop
- Monitoring Internet activities for unauthorized usage and notification to end users of security incidents via CERT bulletins or alerts

In summary, Internet usage is drastically changing the way business is done and is creating opportunities for organizations to compete in what has become a global virtual market. To compete and survive in this new marketplace, organizations have to go through a paradigm shift in the way they regard security. Security, as it relates to the Internet, will have to be considered an enabler for success and treated as an essential business tool.

Firewall Security Systems

Every time a corporation connects its internal computer network to the Internet, it faces potential danger. Because of the Internet's openness, every corporate network connected to it is vulnerable to attack. Hackers on the Internet could theoretically break into the corporate network and do harm in a number of ways as described previously. Companies should build firewalls as one means of perimeter security for their networks. Likewise, this same principle holds true for sensitive or critical systems that need to be protected from untrusted users inside the corporate network (internal hackers). Firewalls are defined as a device installed at the point where network connections enter a site; they apply rules to control the type of networking traffic flowing in and out. Most commercial firewalls are built to handle the most commonly used Internet protocols.

To be effective, firewalls should allow individuals on the corporate network to access the Internet and, at the same time, stop hackers or others on the Internet from gaining access to the corporate network to cause damage. Generally, most organizations will follow a deny-all philosophy, which means that access to a given resource will be denied unless a user can provide a specific business reason or need for access to the information resource. The converse of this access philosophy, not widely accepted, is the accept-all philosophy under which everyone is allowed access unless someone can provide a reason for denying access.

Firewall General Features

Firewalls are hardware and software combinations that are built using routers, servers and a variety of software. They separate networks from each other and screen the traffic between them. Thus, along with other types of security, they control the most vulnerable point between a corporate network and the Internet, and they can be as simple or complex as the corporate information security policy demands. There are many different types of firewalls, but most enable organizations to:

- Block access to particular sites on the Internet
- Limit traffic on an organization's public services segment to relevant addresses and ports
- Prevent certain users from accessing certain servers or services
- Monitor communications and record communications between an internal and an external network
- Monitor and record all communications between an internal network and the outside world to investigate network penetrations or detect internal subversion
- Encrypt packets that are sent between different physical locations within an organization by creating a VPN over the Internet (i.e., IPSec, VPN tunnels)

The capabilities of some firewalls can be extended so they can also provide for protection against viruses and attacks directed to exploit known OS vulnerabilities.

Firewall Types

Generally, the types of firewalls available today fall into three categories:

- Packet filtering
- Application firewall systems
- Stateful inspection

PACKET FILTERING FIREWALLS

The simplest and earliest kinds of firewalls (i.e., first generation of firewalls) were packet filtering-based firewalls deployed between the private network and the Internet. In packet filtering, a screening router examines the header of every packet of data traveling between the Internet and the corporate network. Information contained in packet headers includes the IP address of the sender and receiver and the authorized port numbers (application or service) allowed to use the information transmitted. Based on that information, the router knows what kind of Internet service, such as web-based or FTP, is being used to send the data as well as the identities of the sender and receiver of the data. Using that information, the router can prevent certain packets from being sent between the Internet and the corporate network. For example, the router could block any traffic except for email or traffic to and from suspicious destinations.

The advantages of this type of firewall are its simplicity and generally stable performance as the filtering rules are performed at the network layer. Its simplicity is also a disadvantage, because it is vulnerable to attacks from improperly configured filters and attacks tunneled over permitted services. Because the direct exchange of packets is permitted between outside systems and inside systems, the potential for an attack is determined by the total number of hosts and services to which the packet filtering router permits traffic. Also, if a single packet filtering router is compromised, every system on the private network may be compromised and organizations with many routers may face difficulties in designing, coding and maintaining the rule base. This means that each host directly accessible from the Internet needs to support sophisticated user authentication and needs to be regularly examined by the network administrator for signs of attack.

Some of the more common attacks against packet filter firewalls are:

- **IP spoofing**—The attacker fakes the IP address of either an internal network host or a trusted network host so that the packet being sent will pass the rule base of the firewall. This allows for penetration of the system perimeter. If the spoofing uses an internal IP address, the firewall can be configured to drop the packet on the basis of packet flow direction analysis. However, if the attacker has access to a secure or trusted external IP address and spoofs on that address, the firewall architecture is defenseless.
- **Source routing specification**—It is possible to define the routing that an IP packet must take when it traverses from the source host to the destination host, across the Internet. In this process, it is possible to define the route so it bypasses the firewall. Only those that know of the IP address, subnet mask and default gateway settings at the firewall routing station can do this. A clear defense against this attack is to examine each packet and, if the source routing specification is enabled, drop that packet. However, if the topology permits a route, skipping the choke point, this countermeasure will not be effective.
- **Miniature fragment attack**—Using this method, an attacker fragments the IP packet into smaller ones and pushes it through the firewall in the hope that only the first of the sequence of fragmented packets would be examined and the others would pass without review. This is true if the default setting is to pass residual packets. This can be countered by configuring the firewall to drop all packets where IP fragmentation is enabled.

APPLICATION FIREWALL SYSTEMS

There are two types of application firewall systems. They are referred to as application- and circuit-level firewall systems and provide greater protection capabilities than packet filtering routers. Packet filtering routers allow the direct flow of packets between internal and external systems. Application and circuit gateway firewall systems allow information to flow between systems but do not allow the direct exchange of packets. The primary risk of allowing packet exchange between internal and external systems is that the host applications residing on the protected network's systems must be secure against any threat posed by the allowed packets.

Application firewall systems could be an appliance or sit atop hardened (tightly secured) OSs, such as Windows or UNIX. They work at the application level of the Open Systems Interconnection (OSI) model. The application-level gateway firewall is a system that analyzes packets through a set of proxies—one for each service (e.g., Hypertext Transmission Protocol [HTTP] proxy for web traffic, FTP proxy). An HTTP proxy is known as a web application firewall (WAF). This applies rules to HTTP conversations that cover known attacks such as cross-site scripting (XSS) and SQL injection. This kind of work could reduce network performance. Circuit-level firewalls are more efficient and also operate at the application level—where TCP and User Datagram Protocol (UDP) sessions are validated, typically through a single, general-purpose proxy before opening a connection. Commercially, circuit-level firewalls are quite rare.

Both application firewall systems employ the concept of bastion hosting in that they handle all incoming requests from the Internet to the corporate network, such as FTP or web requests. Bastion hosts are heavily fortified against attack. By having only a single host handling incoming requests, it is easier to maintain security and track attacks. Therefore, in the event of a break-in, only the firewall system has been compromised, not the entire network. In this way, none of the computers or hosts on the corporate network can be contacted directly for requests from the Internet, providing an effective level or layer of security.

Additionally, application-based firewall systems are set up as proxy servers to act on the behalf of someone inside an organization's private network. Rather than relying on a generic packet filtering tool to manage the flow of Internet services through the firewall, a special-purpose code called a proxy server is incorporated into the firewall system. For example, when someone inside the corporate network wants to access a server on the Internet, a request from the computer is sent to the proxy server, the proxy server contacts the server on the Internet, and the proxy server then sends the information from the Internet server to the computer inside the corporate network. By acting as a go-between, proxy servers can maintain security by examining a service's (e.g., FTP, Telnet) program code and modifying and securing it to eliminate known vulnerabilities. The proxy server can also log all traffic between the Internet and the network.

The application-level firewall implementation of proxy server functions is based on providing a separate proxy for each application service (e.g., FTP, Telnet, HTTP). This differs from circuit-level firewalls, which do not need a special proxy for each application-level service. In other words, one proxy server is used for all services.

Advantages of these types of firewalls are that they provide security for commonly used protocols and generally hide the internal network from outside untrusted networks. For example, a feature available on these types of firewall systems is the network address translation (NAT) capability. This capability takes private internal network addresses (unusable on the Internet) and maps them to a table of public IP addresses, assigned to the organization, which can be used across the Internet.

Disadvantages are poor performance and scalability as Internet usage grows. To offset this problem, the concept of load balancing is applicable in cases where a redundant fail-over firewall system may be used.

STATEFUL INSPECTION FIREWALLS

A stateful inspection firewall keeps track of the destination IP address of each packet that leaves the organization's internal network. Whenever the response to a packet is received, its record is referenced to ascertain and ensure that the incoming message is in response to the request that went out from the organization. This is done by mapping the source IP address of an incoming packet with the list of destination IP addresses that is maintained and updated. This approach prevents any attack initiated and originated by an outsider.

The advantages of this approach over application firewall systems is that stateful inspection firewalls control the flow of IP traffic by matching information contained in the headers of connection-oriented or connectionless IP packets at the transport layer, against a set of rules specified by the firewall administrator. This provides a greater degree of efficiency when compared to typical CPU-intensive, full-time application firewall systems' proxy servers, which may perform extensive processing on each data packet at the application level.

The disadvantages include that stateful inspection firewalls can be relatively complex to administer compared to the other two types of firewalls.

Examples of Firewall Implementations

Firewall implementations can take advantage of the functionality available in a variety of firewall designs to provide a robust layered approach in protecting an organization's information assets. Commonly used implementations available today include:

- **Screened-host firewall**—Utilizing a packet-filtering router and a bastion host, this approach implements basic network layer security (packet filtering) and application server security (proxy services). An intruder in this configuration has to penetrate two separate systems before the security of the private network can be compromised. This firewall system is configured with the bastion host connected to the private network with a packet filtering router between the Internet and the bastion host. Router filtering rules allow inbound traffic to access only the bastion host, which blocks access to internal systems. Because the inside hosts reside on the same network as the bastion host, the security policy of the organization determines whether inside systems are permitted direct access to the Internet or whether they are required to use the proxy services on the bastion host.
- **Dual-homed firewall**—This is a firewall system that has two or more network interfaces, each of which is connected to a different network. In a firewall configuration, a dual-homed firewall usually acts to block or filter some or all of the traffic trying to pass between the networks. A dual-homed firewall system is a more restrictive form of a screened-host firewall system, in which a dual-homed bastion host is configured with one interface established for information servers and another for private network host computers.

- **Demilitarized zone (DMZ) or screened-subnet firewall**—Utilizing two packet-filtering routers and a bastion host, this approach creates the most secure firewall system because it supports network- and application-level security while defining a separate DMZ network. The DMZ functions as a small, isolated network for an organization's public servers, bastion host information servers and modem pools. Typically, DMZs are configured to limit access from the Internet and the organization's private network. Incoming traffic access is restricted into the DMZ network by the outside router and protects the organization against certain attacks by limiting the services available for use. Consequently, external systems can access only the bastion host (and its proxying service capabilities to internal systems) and possibly information servers in the DMZ. The inside router provides a second line of defense, managing DMZ access to the private network, while accepting only traffic originating from the bastion host. For outbound traffic, the inside router manages private network access to the DMZ network. It permits internal systems to access only the bastion host and information servers in the DMZ. The filtering rules on the outside router require the use of proxy services by accepting only outbound traffic on the bastion host. The key benefits of this system are that an intruder must penetrate three separate devices, private network addresses are not disclosed to the Internet, and internal systems do not have direct access to the Internet.

Firewall Issues

Issues related to implementing firewalls include:

- A false sense of security may exist where management feels that no further security checks and controls are needed on the internal network (i.e., the majority of incidents are caused by insiders, who are not controlled by firewalls).
- The circumvention of firewalls through the use of modems may connect users directly to ISPs. Management should provide assurance that the use of modems when a firewall exists is strictly controlled or prohibited altogether.
- Misconfigured firewalls may allow unknown and dangerous services to pass through freely.
- What constitutes a firewall may be misunderstood (e.g., companies claiming to have a firewall merely have a screening router).
- Monitoring activities may not occur on a regular basis (i.e., log settings not appropriately applied and reviewed).
- Firewall policies may not be maintained regularly.
- Most firewalls operate at the network layer; therefore, they do not stop any application-based or input-based attacks. Examples of such attacks include structured query language (SQL) injection and buffer-overflow attacks. Newer-generation firewalls are able to inspect traffic at the application layer and stop some of these attacks.

Firewall Platforms

Firewalls may be implemented using hardware or software platforms. When implemented in hardware, it will provide good performance with minimal system overhead. Although hardware-based firewall platforms are faster, they are not as flexible or scalable as software-based firewalls. Software-based firewalls are generally slower with significant systems overhead; however, they are flexible with additional services. They may include content and virus checking, before traffic is passed to users.

It is generally better to use appliances, rather than normal servers, for the firewall. Appliances are normally installed with hardened OSs. When server-based firewalls are used, OSs in servers are often vulnerable to attacks. When the attacks on OSs succeed, the firewall would be compromised. Appliance-type firewalls are, generally, significantly faster to set up and recover.

Intrusion Detection Systems

Another element to securing networks complementing firewall implementations is an IDS. An IDS works in conjunction with routers and firewalls by monitoring network usage anomalies. It protects a company's IS resources from external as well as internal misuse.

An IDS operates continuously on the system, running in the background and notifying administrators when it detects a perceived threat. Broad categories of IDSs include:

- **Network-based IDSs**—They identify attacks within the monitored network and issue a warning to the operator. If a network-based IDS is placed between the Internet and the firewall, it will detect all the attack attempts, whether or not they enter the firewall. If the IDS is placed between a firewall and the corporate network, it will detect those attacks that enter the firewall (it will detect intruders). The IDS is not a substitute for a firewall, but it complements the function of a firewall.
- **Host-based IDSs**—They are configured for a specific environment and will monitor various internal resources of the OS to warn of a possible attack. They can detect the modification of executable programs, detect the deletion of files and issue a warning when an attempt is made to use a privileged command.

Components of an IDS are:

- Sensors that are responsible for collecting data, such as network packets, log files, system call traces, etc.
- Analyzers that receive input from sensors and determine intrusive activity
- An administration console
- A user interface

Types of IDSs include:

- **Signature-based**—These IDS systems protect against detected intrusion patterns. Identified intrusive patterns are stored as signatures.
- **Statistical-based**—These systems need a comprehensive definition of the known and expected behavior of systems.
- **Neural networks**—An IDS with this feature monitors the general patterns of activity and traffic on the network and creates a database. This is similar to the statistical model but with added self-learning functionality.

Signature-based IDSs will not be able to detect all types of intrusions due to the limitations of the detection rules. Statistical-based systems may report many events outside of the defined normal activity but which are normal activities on the network. A combination of signature- and statistical-based models provides better protection.

FEATURES

The features available in an IDS include:

- Intrusion detection

- Gathering evidence on intrusive activity
- Automated response (i.e., termination of connection, alarm messaging)
- Security policy
- Interface with system tools
- Security policy management

LIMITATIONS

An IDS cannot help with the following weaknesses:
- Weaknesses in the policy definition
- Application-level vulnerabilities
- Back doors into applications
- Weaknesses in identification and authentication schemes

In contrast to IDSs, which rely on signature files to identify an attack as (or after) it happens, an intrusion prevention system (IPS) predicts an attack before it can take effect. It does this by monitoring key areas of a computer system and looks for "bad behavior" such as worms, Trojans, spyware, malware and hackers. It complements firewall, antivirus and antispyware tools to provide complete protection from emerging threats. It is able to block new (zero-day) threats that bypass traditional security measures because it is not reliant on identifying and distributing threat signatures or patches.

POLICY

An IDS policy should establish the action to be taken by security personnel in the event that an intruder is detected.

Actions may include:
- **Terminate the access**—If there is a significant risk to the organization's data or systems, immediate termination is the usual procedure.
- **Trace the access**—If the risk to the data is low, the activity is not immediately threatening, or analysis of the entry point and attack method is desirable, the IDS can be used to trace the origin of the intrusion. This can be used to determine and correct any system weaknesses and to collect evidence of the attack which may be used in a subsequent court action.

In either case, the action required should be determined by management in advance and incorporated in a policy. This will save time when an intrusion is detected, which may impact the possible data loss.

Intrusion Prevention Systems

Intrusion prevention systems (IPSs) are closely related to IDSs and are designed to not only detect attacks, but also to prevent the intended victim hosts from being affected by the attacks. Whereas an IDS alerts or warns of an attack, requiring security personnel to take action, an IPS will make an attempt to stop the attack. For example, an IPS can disconnect an originating network or user session by blocking access to the target from the originating user account and/or IP address. Some IPSs can also reconfigure other security controls, such as a firewall or router, to block an attack. The intrusion prevention approach can be effective in limiting damage or disruption to systems that are attacked. However, as with an IDS, the IPS must be properly configured and tuned to be effective. Threshold settings that are too high or low will lead to limited effectiveness of the IPS. Some concerns have been raised that the IPS itself may constitute a threat because a clever attacker could send commands to a large number of hosts protected by an IPS in order to cause them to become dysfunctional. This could be catastrophic in environments where continuity of service is critical.

Honeypots and Honeynets

A honeypot is a software application that pretends to be a vulnerable server on the Internet and is not set up to actively protect against break-ins. It acts as a decoy system that lures hackers. The more a honeypot is targeted by an intruder, the more valuable it becomes. Although honeypots are technically related to IDSs and firewalls, they have no real production value as an active sentinel of networks.

There are two basic types of honeypots:
- **High-interaction**—Give hackers a real environment to attack
- **Low-interaction**—Emulate production environments and provide more limited information

A honeynet is a set of multiple, linked honeypots that simulate a larger network installation. Hackers infiltrate the honeynet, which allows investigators to observe their actions using a combination of surveillance technologies.

An IDS triggers a virtual alarm whenever an attacker breaches security of any networked computers. A stealthy keystroke logger watches everything the intruder types. A separate firewall cuts off the machines from the Internet anytime an intruder tries to attack another system from the honeynet.

All traffic on honeypots or honeynets are assumed to be suspicious because the systems are not meant for internal use and the information collected about these attacks are used proactively to update vulnerabilities on a company's live network.

If a honeypot is designed to be accessible from the Internet, there is a risk that external monitoring services that create lists of untrusted sites may report the organization's system as vulnerable, without knowing that the vulnerabilities belong to the honeypot and not to the system itself. Such independent reviews made public can affect the organization's reputation. Therefore, prior to implementing a honeypot in the network, careful judgment should be exercised.

5.4.5 ENCRYPTION

Encryption is the process of converting a plaintext message into a secure-coded form of text, called ciphertext, which cannot be understood without converting it back via decryption (the reverse process) to plaintext. This is done via a mathematical function and a special encryption/decryption password called the key.

Encryption generally is used to:
- Protect data in transit over networks from unauthorized interception and manipulation
- Protect information stored on computers from unauthorized viewing and manipulation
- Deter and detect accidental or intentional alterations of data
- Verify authenticity of a transaction or document

In many countries, encryption is subject to governmental laws and regulations.

Encryption is limited in that it cannot prevent the loss or modification of data. The protection of the keys is of paramount concern when using encryption systems. Therefore, even if encryption is regarded as an essential form of access control that should be incorporated into an organization's overall security landscape, it requires a thorough understanding of how schemes work as misuse or misconfiguration may significantly undermine the protection that an organization believes is in place.

Key Elements of Encryption Systems

Key elements of encryption systems include:

- **Encryption algorithm**—A mathematically based function that encrypts/decrypts data
- **Encryption keys**—A piece of information that is used by the encryption algorithm to make the encryption or decryption process unique. Similar to passwords, a user needs to provide the correct key to access or decrypt a message. The wrong key will decipher the message into an unreadable form.
- **Key Length**—A predetermined length for the key. The longer the key, the more difficult it is to compromise in a brute force attack.

Encryption schemes are susceptible to brute force attacks in which an attacker repeatedly tries to decrypt a piece of ciphertext using all the possible encryption keys until the correct one is found (i.e., brute forcing stops when the ciphertext does not decrypt to a non-sense message). As the amount of time required to search for the correct key depends exponentially on its length, it is fundamental to choose the key adequately in order to ensure the overall security of encryption scheme.

Attacks can also be mounted against the robustness of the underlying mathematical algorithms in order to speed-up the brute forcing process. Cryptanalysis is the science of finding such weaknesses. For example, an algorithm prone to a "known-plaintext attack" allows an attacker discard a large portion of the possible decryption keys if samples of ciphertexts and corresponding plaintexts are available. A variation of this attack consists of guessing parts of the plaintext leveraging on statistical properties of the encrypted data (e.g., spotting vowels or finding the word "the" in an English text).

The randomness of key generation is also a significant factor in the ability to compromise an encryption scheme. Common words or phrases significantly lessen the key space combinations required to search for the key, diminishing the strength of the encryption algorithm. Therefore, a 128-bit encryption algorithm's capabilities are diminished when encrypting keys are based on passwords, and the passwords lack randomness. This means that it is important that effective password syntax rules are applied, and easily guessed passwords are prohibited.

There are two types of encryption schemes: symmetric and asymmetric. Symmetric key systems use a unique key (usually referred to as the "secret key") for both encryption and decryption. The key is known as bidirectional because it encrypts and decrypts and it must be shared "out of band" (i.e., via a secure, alternative method to the encrypted message).

In asymmetric key systems the decryption key is different than the one used for encryption. The keys are unidirectional, they encrypt or decrypt, but complementary. In asymmetric key systems, the two parties (the sender and the recipient) are not expected to trust each other to keep the secret key. In fact, in asymmetric systems the encryption key is publicly disclosed while the decryption key is kept private (asymmetric systems are also known as public-key schemes).

Together with encryption algorithms, another important component of cryptographic protection schemes are hash functions. These functions transform a text of arbitrary length into one of fixed width called the "digest" or the "hash" of the input text (a basic example of a hash function is one that just truncates a text string after a fixed number of characters). To be used in cryptographic protection schemes, a hash function must be "one-way" (i.e., making it hard to find a piece of text that generates a given hash). Such functions can be used to augment encryption schemes with integrity and authenticity properties. Hashing algorithms are an accurate integrity check tool. The hash detects changes of even a single bit in a message. A hash algorithm will calculate a hash value from the entire input message. The output digest itself is a fixed length, so even though the input message can be of variable length, the output is always the same length. The length depends on the hash algorithm used. For example, MD5 generates a digest length of 128 bits; SHA-1, a digest of 160 bits; and SHA-512, a digest of 512 bits.

The most common message digest algorithms have been MD5, now moved to historic (datatracker.ietf.org/doc/rfc6331/) and SHA-1for which there are also security considerations (datatracker.ietf.org/doc/rfc6194/). For these reasons the industry is transitioning from SHA-1 to SHA-2. There are six hash functions available with SHA-2 with varying message digest lengths. SHA-3 has also been announced by the National Institute for Standards and Technology (NIST) in the event a successful attack is developed against SHA-2.

> **Note:** The IS auditor should be familiar with how a digital signature functions to protect data. The specific types of message digest algorithms are not tested on the CISA exam.

When a sender wants to send a message and ensure that it has not been altered, they can compute the digest of the message and send it along with the message to the receiver. When the receiver receives the message and its digest, he/she independently computes the digest of the received message and ensures that the digest computed is the same as the digest sent with the message (**figure 5.15**).

Symmetric Key Cryptographic Systems

Symmetric key cryptographic systems (**figure 5.16**) are based on a symmetric encryption algorithm, which uses a secret key to encrypt the plaintext to the ciphertext and the same key to decrypt the ciphertext to the corresponding plaintext. In this case, the key is said to be symmetric because the encryption key is the same as the decryption key.

Figure 5.15— Verifying Message Integrity Using a Hash Function

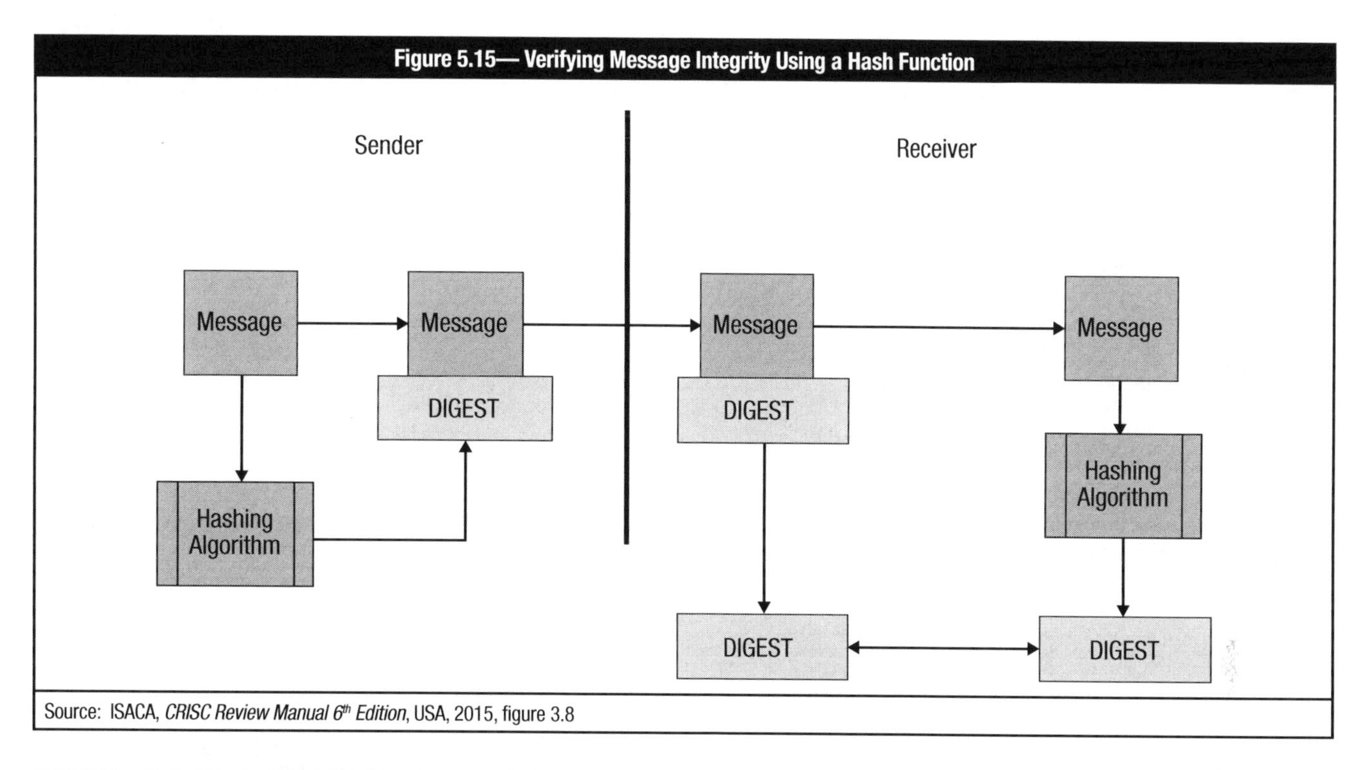

Source: ISACA, *CRISC Review Manual 6th Edition*, USA, 2015, figure 3.8

Figure 5.16—Symmetric Cryptography

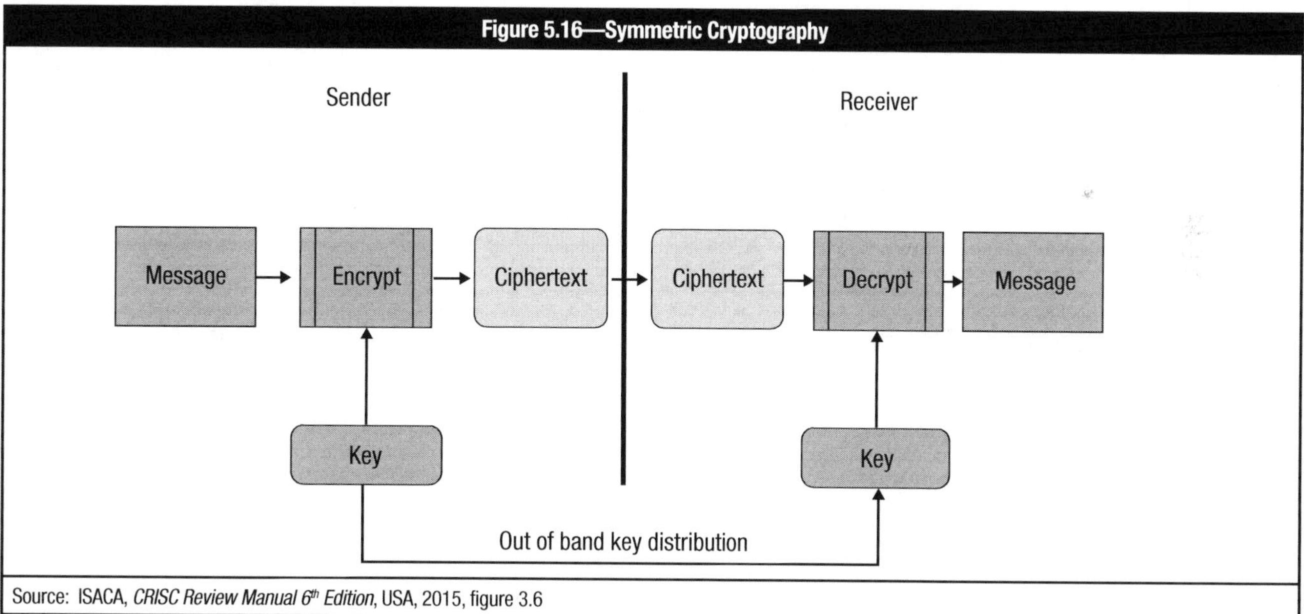

Source: ISACA, *CRISC Review Manual 6th Edition*, USA, 2015, figure 3.6

The most common symmetric key cryptographic system used to be the Data Encryption Standard (DES). DES is based on a public algorithm approved by NIST and employs keys of 56 bits (plus 8 bits used for parity checking). The bits in the plaintext are processed one 64-bit block at a time and, as such, DES belongs to the category of block-ciphers (as opposed to stream-ciphers, which encode one bit at a time).

DES has been withdrawn by NIST because it is no longer considered a strong cryptographic solution because its entire key space can be brute forced by a moderately large computer system within a relatively short period of time. Extensions of DES (Triple DES or 3DES) were proposed to extend the DES standard while retaining backward compatibility (it applies the DES cipher algorithm three times to each data block). In 2001, NIST replaced DES with the Advanced Encryption Standard (AES), a public algorithm that supports keys from 128 bits to 256 bits in size. Another commonly used symmetric key algorithm is RC4, a stream-cipher often used in SSL/TLS protocol sessions.

There are two main advantages of symmetric key systems such as 3DES or AES over asymmetric ones. The first is that keys are much shorter and can be easily remembered. The second is that symmetric key cryptosystems are generally less complicated and, therefore, use less processing power than asymmetric schemes. This makes symmetric key cryptosystems ideally suited for bulk data encryption. The major disadvantage of this approach is key

distribution, particularly in e-commerce environments where customers are unknown, untrusted entities. Also, a symmetric key cannot be used to sign electronic documents or messages due to the fact that the mechanism is based on a shared secret by at least two parties.

Public (Asymmetric) Key Cryptographic Systems

In public key cryptography (**figure 5.17**), two keys work together as a pair (they are inversely related to each other, based on mathematical integer factorization). One of the keys is kept private while the other one is publicly disclosed. Encryption works by feeding the public key to the underlying algorithm while the resulting ciphertext can be decoded using the private key. This scheme avoids requirement of the owner of the key pair to share a secret piece of information (the private key) with the other party of the communication. It is important to note that one key pair can be used in one-direction only (from the sender to the receiver). To implement a bidirectional communication between two parties, two key-pairs are required (one for each direction).

Public key systems were developed primarily to solve the problem of key distribution. In the first place only 2*N key-pairs are employed in a scenario in which communication happens between N parties: in the same scenario, a symmetric scheme would require roughly N^2 keys to be transmitted, one key for each pair of the involved parties. In addition, the exchanged keys are public, thus there is no confidentiality requirement to be fulfilled by the key distribution protocol.

The first practical implementation of a public key system was developed by Ron Rivest, Adi Shamir and Leonard Adleman (the RSA algorithm), which is a widespread asymmetric encryption scheme. The main drawback of this algorithm lies in the length of the keys (varying between 1024 and 4096 bits) and the complexity of the calculations involved for encoding and decoding. To address these issues, other encryption algorithms were developed. Promising alternatives like elliptic curve cryptography (ECC) are emerging because they have a much higher speed at encrypting/decrypting with significantly shorter keys (between 256 and 512 bits).

Quantum Cryptography

Quantum cryptography refers to the possibility of using properties of quantum computing (computer technology based on quantum theory) for cryptographic purposes, quantum key distribution (QKD) being the most important application. QKD schemes allow distribution of a shared encryption key between two parties who can detect when another unauthorized party is eavesdropping on the key exchange channel. Indeed, when this happens, the channel is inevitably disturbed and the exchanged key is tagged as compromised.

Quantum computing is also known to easily break the security of schemes like RSA. To overcome this drawback, post-quantum encryption algorithms have been developed which are resistant to a quantum attack.

Digital Signatures

An important property of public key systems is that the underlying algorithm works even if the private key is used for encryption and the public key for decryption. Even if this sounds counterintuitive, this way of using a public key system realizes a digital signature scheme able to authenticate the origin of an encoded message. Because the private key is known only by the owner of the key-pair, one can be sure that if a ciphertext is correctly decrypted using a public key, the owner of that public key cannot deny to have performed the encryption process. This important and peculiar property of public key cryptosystems is called non-repudiation.

Figure 5.17—Using Asymmetric Algorithms to Support Symmetric Cryptography

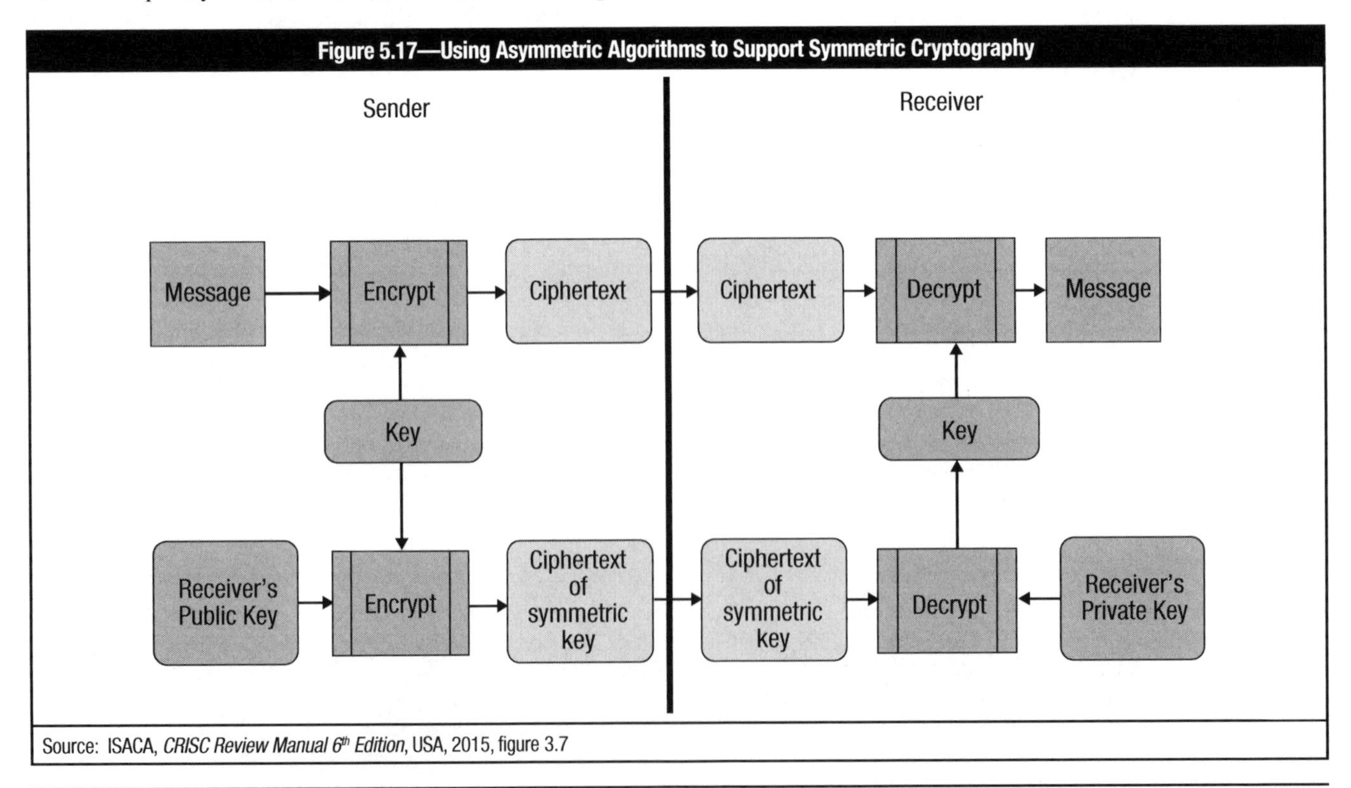

Source: ISACA, *CRISC Review Manual 6th Edition*, USA, 2015, figure 3.7

In most practical implementations of digital signature schemes (**figure 5.18**), the public key algorithm is never applied to the whole document as it would take a lot of processing power to calculate the signed data. Instead, a digest (or "pre hash") is first derived from the document to be signed; then the public key algorithm is applied to the digest in order to produce an encoded piece of data (the signature) that is sent alongside the document.

In order to authenticate the sender as the originator of the document, the same hashing function is applied by the recipient upon receiving and the resulting digest (or "post-hash") is compared with the decrypted pre-hash. In case of a match, the receiver can conclude that the document was actually signed by the owner of the public key.

Therefore, digital signature schemes ensure:
- **Data integrity**— Any change to the plaintext message would result in the recipient failing to compute the same document hash.
- **Authentication**—The recipient can ensure that the document has been sent by the claimed sender because only the claimed sender has the private key.
- **Nonrepudiation**— The claimed sender cannot later deny generating the document.

Notice that there is no guarantee that the owner of the public key actually sent the document. A malicious attacker could intercept the signed document and send it again to the recipient. To prevent this kind of attack (known as "replay attack"), a signed timestamping or a counter may be attached to the document.

Public Key Infrastructure

Public key encryption algorithms are a big step toward strengthening the trust of secure communications because private keys must not be shared by any of the parties involved in the system and no confidentiality requirements are imposed when distributing public keys.

However, public key systems are still vulnerable to man-in-the-middle (MITM) attacks in which the public keys are tampered with by an attacker (the man in the middle) controlling the communication channel. If this attacker replaces a genuine public key with his own key, any party sending a message to the owner of the tampered public key would instead be using the attacker's public key. This attacker is now able to intercept, read and modify any such message by decrypting and re-encrypting it using the genuine public key. The problem lies in the fact that the tampering of the public key cannot be detected by either the sender or the recipient. In other words, there is no guarantee of a binding between the public key and the identity of the owner.

To solve this problem, a trusted third party is introduced into the scheme from which any signed document is considered automatically authentic by the sender and the recipient. In the first place, this trusted party identifies the holder of a public key (the subject) and then signs this public key while appending details of the subject's identity. The resulting document is known as the public (or digital) certificate of the subject. The trusted third party is called a certification authority (CA). When a CA is introduced in a signature scheme it is known as public key infrastructure (PKI).

As well as issuing certificates, the CA maintains a list of compromised certificates (i.e., those whose private key has been leaked or lost) called the certificate revocation list (CRL). In some cases, certificates may also be marked as revoked in the CRL when the owner of the certificate voluntarily declares not to use the corresponding key pair any longer. This allows a party to

Figure 5.18—Verifying Message Integrity and Proof of Origin Using Digital Signatures

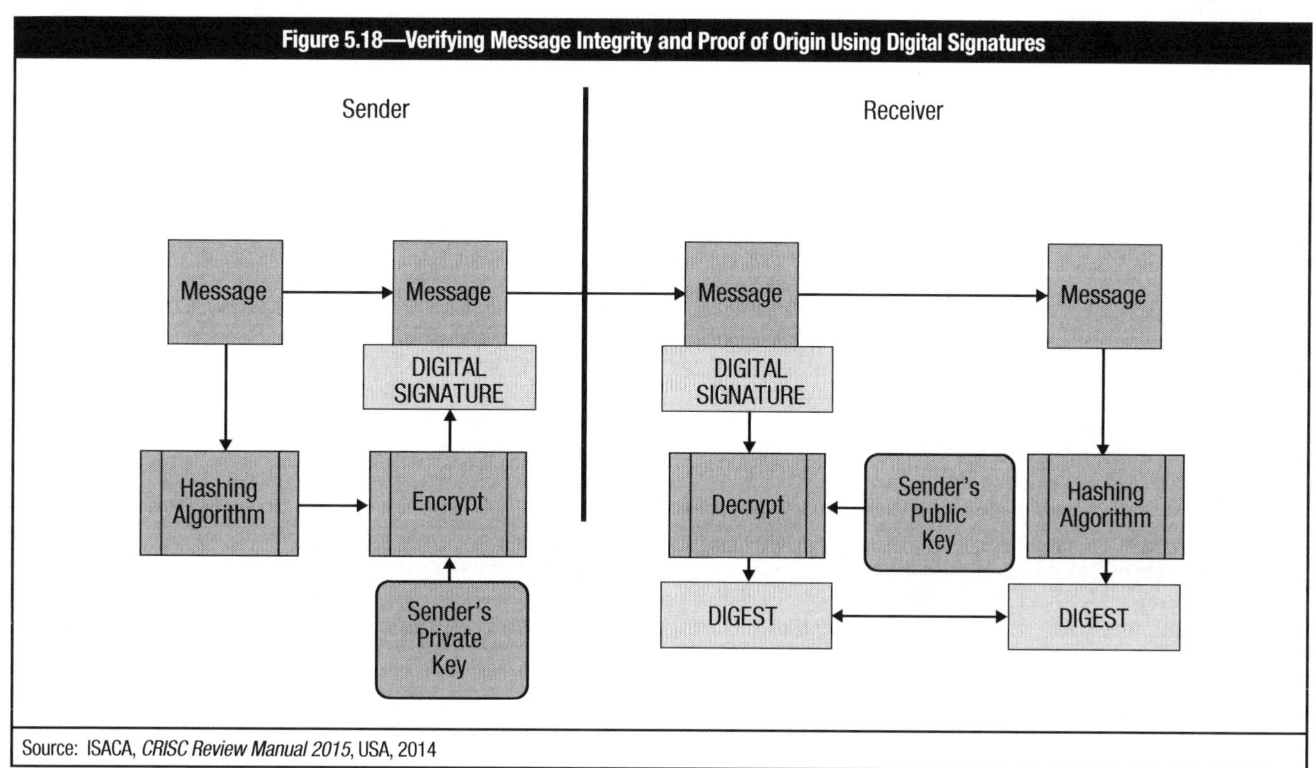

Source: ISACA, *CRISC Review Manual 2015*, USA, 2014

reject a signed document when the signature was generated after the private key has been compromised or revoked.

Certificates usually contain a certificate practice statement (CPS). This is a statement about the way a CA issues certificates. It may contain:
- The type of certificates issued
- Policies, procedures and processes for issuing, renewing and recovering certificates
- Cryptographic algorithms used
- The key length used for the certificate
- The lifetime of the certificate issued by the CA
- Policies for revoking certificates
- Policies for CRLs
- Policies for renewing certificates

Registration authorities (RA) are delegated some administrative functions for a specific community by the CA. For example, an international corporation may have a PKI setting in which national branches act as RAs for the employees in that nation.

The administrative functions that a particular RA implements will vary based on the needs of the CA but must support the principle of establishing or verifying the identity of the subscriber. These functions may include the following:
- Verifying information supplied by the subject (personal authentication functions)
- Verifying the right of the subject to requested certificate attributes
- Verifying that the subject actually possesses the private key being registered and that it matches the public key requested for a certificate (generally referred to as proof of possession [POP]).
- Reporting key compromise or termination cases where revocation is required
- Assigning names for identification purposes
- Generating shared secrets for use during the initialization and certificate pick-up phases of registration
- Initiating the registration process with the CA on behalf of the subject end entity
- Initiating the key recovery processing
- Distributing the physical tokens (such as smart cards) containing the private keys

Applications of Cryptographic Systems

Asymmetric and symmetric systems can be combined together to leverage on each system's peculiarities. A common scheme is to encrypt data using a symmetric algorithm with a secret key, which is randomly generated. The secret key is then encrypted using an asymmetric encryption algorithm to allow the secure distribution among those parties who need access to the encrypted data. Secure communication can thus enjoy both the speed of symmetric systems and the ease of key-distribution of asymmetric systems. In addition, because creating the secret key is an effortless operation, it can be employed just for a limited amount of data after which a new secret key can be chosen. This limits the possibilities of a malicious third-party to decrypt the whole set of data because he would be required to attack multiple secret keys. This combined scheme is used in protocols like SSL/TLS to protect web traffic and S/MIME for email encryption. In the latter case, the resulting document—the combination of the encrypted message and the encrypted secret key—is called a digital envelope.

A more comprehensive list of applications of such a method follows.

TRANSPORT LAYER SECURITY (TLS)

TLS is a cryptographic protocol that provides secure communications on the Internet. TLS is a session- or connection-layered protocol widely used for communication between browsers and web servers. Besides communication privacy, it also provides endpoint authentication. The protocols allow client-server applications to communicate in a way designed to prevent eavesdropping, tampering and message forgery.

TLS involves a number of basic phases:
- Peer negotiation for algorithm support
- Public-key, encryption-based key exchange and certificate-based authentication
- Symmetric cipher-based traffic encryption

During the first phase, the client and server negotiate which cryptographic algorithms will be used. Current implementations support the following choices:
- For public-key cryptography: RSA, Diffie-Hellman, DSA or Fortezza
- For symmetric ciphers: RC4, IDEA, Triple DES or AES
- For one-way hash functions: SHA-1 or SHA-2 (SHA-256)

TLS runs on layers above the TCP transport protocol and provides security to application protocols, even if it is most commonly used with HTTP to form Secure Hypertext Transmission Protocol (HTTPS). HTTPS serves to secure World Wide Web pages for applications. More, in electronic commerce, authentication may be used both in business-to-business (B-to-B) activities (for which both the client and the server are authenticated) and business-to-consumer (B-to-C) interaction (in which only the server is authenticated).

Besides TLS, Secure Socket Layer (SSL) protocol is also widely used in real-world applications, even though its use is now deprecated as a significant vulnerability was discovered in 2014. TLS and SSL are not interchangeable.

IP SECURITY (IPSEC)

IPSec is used for securing the communications at IP-level among two or more hosts, two or more subnets, or hosts and subnets.

This IP network layer packet security protocol establishes VPNs via transport and tunnel mode encryption methods. For the transport method, the data portion of each packet referred to as the encapsulation security payload (ESP) is encrypted, achieving confidentiality over the process. In the tunnel mode, the ESP payload and its header are encrypted. To achieve nonrepudiation, an additional authentication header (AH) is applied.

In establishing IPSec sessions in either mode, security associations (SAs) are established. SAs define which security parameters should be applied between the communicating parties as encryption algorithms, keys, initialization vectors, life span of keys, etc. Within either the ESP or AH header, respectively, an SA is established when a 32-bit security parameter index (SPI) field is defined within the sending host. The SPI is a unique identifier that enables the sending host to reference the security parameters to apply, as specified, on the receiving host.

IPSec can be made more secure by using asymmetric encryption through the use of Internet Security Association and Key Management Protocol/Oakley (ISAKMP/Oakley), which allows the key management, use of public keys, negotiation, establishment, modification and deletion of SAs and attributes. For authentication, the sender uses digital certificates. The connection is made secure by supporting the generation, authentication and distribution of the SAs and those of the cryptographic keys.

SECURE SHELL (SSH)

SSH is a client-server program that opens a secure, encrypted command-line shell session from the Internet for remote logon. Similar to a VPN, SSH uses strong cryptography to protect data, including passwords, binary files and administrative commands, transmitted between systems on a network. SSH is typically implemented between two parties by validating each other's credentials via digital certificates. SSH is useful in replacing Telnet and is implemented at the application layer, as opposed to operating at the network layer (IPSec implementation).

SECURE MULTIPURPOSE INTERNET MAIL EXTENSIONS (S/MIME)

S/MIME is a standard secure email protocol that authenticates the identity of the sender and receiver, verifies message integrity, and ensures the privacy of a message's contents, including attachments.

5.4.6 MALWARE

The term malware is generally applied to a variety of malicious computer programs that send out requests to the OS of the host system under attack to append the malware to other programs. In this way, malware are self-propagating to other programs. They can be relatively benign (e.g., web application defacement) or malicious (e.g., deleting files, corrupting programs or causing a DoS). Generally, malware attack four parts of the computer:
- Executable program files
- The file-directory system, which tracks the location of all the computer's files
- Boot and system areas, which are needed to start the computer
- Data files

Another variant of malware frequently encountered is a worm, which, unlike a virus, does not physically attach itself to another program. To propagate itself to the host systems, a worm typically exploits security weaknesses in OSs' configurations. These problems are particularly severe in today's highly decentralized client-server environments.

Currently, viruses or worms are transmitted easily from the Internet by downloading files to computers' web browsers. Malware are also transmitted as attachments to email, so that when the attachment opens, the system becomes infected if it is not using scanning software to review unopened attachments. Other methods of infection occur from files received through online services, social media, LANs and even shrink-wrapped software that the user may buy from a retail store.

Virus and Worm Controls

To effectively reduce the risk of computer viruses and worms infiltrating an organization, a comprehensive and dynamic anti-malware program needs to be established. There are two major ways to prevent and detect malware that infect computers and network systems. The first is by having sound policies and procedures in place (preventive controls) and the second is by technical means (detective controls), including anti-malware software. Neither is effective without the other.

Management Procedural Controls

Some of the policy and procedure controls that should be in place include the following:
- Build any system from original, clean master copies. Boot only from original media whose write protection has always been in place, if applicable.
- Allow no media (e.g., hard/flash drives) to be used until they have been scanned on a stand-alone machine that is used for no other purpose and is not connected to the network.
- Update malware software scanning definitions/signatures frequently.
- Protect removable media against theft and hazards.
- Have vendors run demonstrations on their machines.
- Enforce a rule of not using shareware without first scanning it thoroughly for malware.
- Scan before any new software is installed because commercial software occasionally includes a Trojan horse (viruses or worms).
- Insist that field technicians scan their disks on a test machine before they use any of their disks on the system.
- Ensure the network administrator uses workstation and server anti-malware software.
- Ensure all servers are equipped with an activated current release of the malware-detection software.
- Consider encrypting files and then decrypting them before execution.
- Ensure bridge, router and gateway updates are authentic.
- Because backups are a vital element of an anti-malware strategy, ensure a sound and effective backup plan is in place. This plan should account for scanning selected backup files for malware infection once malware has been detected.
- Educate users so they will heed these policies and procedures. For example, many malware today are propagated in the form of email attachments where the attachment, such as an executable Visual Basic script, infects the user's system upon opening the attachment. The hacker relies upon social engineering tactics in getting the user to open the attachment.
- Review anti-malware policies and procedures at least once a year.
- Prepare a malware eradication procedure and identify a contact person.

- Develop, rehearse and maintain clear incident management procedures in the event that anti-malware software reports an infection.

Technical Controls

Technical methods of preventing malware can be implemented through hardware and software means. The following are hardware tactics that can reduce the risk of infection:
- Use boot malware protection (i.e., built-in, firmware-based malware protection).
- Use remote booting (e.g., diskless workstations).
- Use a hardware-based password.
- Protect removable media against theft and hazards.
- Ensure that insecure protocols are blocked by the firewall from external segments and the Internet.

However, anti-malware software is, by far, the most common anti-malware tool and is considered the most effective means of protecting networks and host-based computer systems against malware. Anti-malware software is both a preventive and a detective control. Unless updated periodically, anti-malware software will not be an effective tool against malware.

Anti-malware software contains a number of components that address the detection of malware via scanning technologies from different angles. There are different types of anti-malware software.

Scanners look for sequences of bits called signatures that are typical of malware programs. The two primary types are:
- Malware masks or signatures—Anti-malware scanners check files, sectors and system memory for known and new (unknown to scanner) malware, on the basis of malware masks or signatures. Malware masks or signatures are specific code strings that are recognized as belonging to malware. For polymorphic viruses, the scanner sometimes has algorithms that check for all possible combinations of a signature that could exist in an infected file.
- Heuristic scanners—Analyzes the instructions in the code being scanned and decides on the basis of statistical probability whether it could contain malicious code. Heuristic scanning results could indicate that malware may be present (i.e., possibly infected). Heuristic scanners tend to generate a high level of false-positive errors (i.e., they indicate that malware may be present when, in fact, no malware is present).

Scanners examine memory, disk-boot sectors, executables, data files and command files for bit patterns that match a known malware. Scanners, therefore, need to be updated periodically to remain effective.

Active monitors interpret DOS and read-only memory (ROM) BIOS calls, looking for malware-like actions. Active monitors can be problematic because they cannot distinguish between a user request and a program or malware request. As a result, users are asked to confirm actions, including formatting a disk or deleting a file or set of files.

Integrity CRC checkers compute a binary number on a known malware-free program that is then stored in a database file. The number is called a cyclical redundancy check (CRC). On subsequent scans, when that program is called to execute, it checks for changes to the files as compared to the database and reports possible infection if changes have occurred. A match means no infection; a mismatch means a change in the program has occurred. A change in the program could mean malware within it. These scanners are effective in detecting infection; however, they can do so only after infection has occurred (i.e., it is often too late to save files). Also, CRC checkers can only detect subsequent changes to files, because they assume files are malware free in the first place. Therefore, they are ineffective against new files that are malware-infected and that are not recorded in the database. Integrity checkers take advantage of the fact that executable programs and boot sectors do not change often, if at all.

Behavior blockers focus on detecting potentially abnormal behavior, such as writing to the boot sector or the master boot record or making changes to executable files. Blockers can potentially detect malware at an early stage. Most hardware-based anti-malware mechanisms are based on this concept.

Immunizers defend against malware by appending sections of themselves to files—somewhat in the same way that file malware append themselves. Immunizers continuously check the file for changes and report changes as possible malware behavior. Other types of immunizers are focused to a specific virus and work by giving the malware the impression that the malware has already infected the computer. This method is not always practical because it is not possible to immunize files against all known malware.

Once malware has been detected by anti-malware software, an eradication program can be used to wipe the malware from the hard disk. Sometimes eradication programs can kill the malware without having to delete the infected program or data file, while other times those infected files must be deleted. Still, other programs, sometimes called inoculators, do not allow a program to be run if it contains malware.

Anti-malware Software Implementation Strategies

Organizations have to develop malware implementation strategies to effectively control and prevent the spread of malware throughout their IS infrastructure. An important means of controlling the spread of malware is to detect the malware at its point of entry—before it has the opportunity to cause damage. This includes everything from networks, server platforms and end-user workstations.

The user server or workstation level could include screening of software and data as they enter the machine, where anti-malware programs can be set to perform:
- Scheduled malware scans (e.g., daily, weekly, etc.)
- Manual/on-demand scans, where the malware scan is requested by the user
- Continuous/on-the-fly scanning, where files are scanned as they are processed

At the corporate network level, in cases of interconnected networks, malware scanning software is used as an integral part of firewall technologies, referred to as malware walls. Malware walls scan incoming traffic with the intent of detecting and removing malware before they enter the protected network. Malware walls normally work at the following levels:
- SMTP protection, to scan inbound and outbound SMTP traffic for malware in coordination with the mail server
- HTTP protection, to prevent malware-infected files from being downloaded and to offer protection against malicious Java and ActiveX programs
- FTP protection, to prevent infected files from being downloaded

Malware walls most often are updated automatically with new malware signatures by their vendors on a scheduled basis or on an as-needed basis when dangerous, new malware emerge. Malware walls also provide facilities to log malware incidents and deal with the incident in accordance with preset rules. The presence of malware walls does not preclude the necessity for malware-detection software to be installed on computers within a network because the malware wall only addresses one channel through which malware enter the network. Malware-detection software should be loaded on all computers within the network. Malware signature files should be kept updated. The facility of automatic "live update" has become fairly popular and allows organizations to update the malware scanner signature files as soon as updates are available.

For malware scanners to be acceptable and viable, they should have the following features:
- Reliability and quality in the detection of malware
- Memory resident, which is a continuous checking facility
- Efficiency, such as a reasonable working speed and usage of resources

5.4.7 VOICE-OVER IP

IP telephony, also known as Internet telephony, is the technology that makes it possible to have a voice conversation over the Internet or over any dedicated IP network instead of dedicated voice transmission lines. The protocols used to carry the signal over the IP network are commonly referred to as Voice-over IP (VoIP). VoIP is a technology where voice traffic is carried on top of existing data infrastructure. Sounds are digitized into IP packets and transferred through the network layer before being decoded back into the original voice. VoIP has significantly reduced long-distance costs in a number of large organizations.

VoIP allows the elimination of circuit switching and the associated waste of bandwidth. Instead, packet switching is used, where IP packets with voice data are sent over the network only when data needs to be sent.

It has advantages over traditional telephony:
- Unlike traditional telephony, VoIP innovation progresses at market rates rather than at the rates of the multilateral committee process of the International Telecommunications Union (ITU)
- Lower costs per call or even free calls, especially for long-distance calls
- Lower infrastructure costs. Once IP infrastructure is installed, no or little additional telephony infrastructure is needed.

VoIP introduces security risk and opportunities. VoIP has a different architecture than traditional circuit-based telephony, and these differences result in significant security issues.

VoIP systems take a wide variety of forms, including traditional telephone handsets, conferencing units and mobile units. In addition to end-user equipment, VoIP systems include a variety of other components, including call processors/call managers, gateways, routers, firewalls and protocols. Most of these components have counterparts used in data networks, but the performance demands of VoIP mean that ordinary network software and hardware must be supplemented with special VoIP components.

When designing a VoIP system, the backup has to be considered. While telecom companies usually operate under the requirement to have 99.9999 percent uptime, data traffic normally has less reliability. For this reason, the backup has to be designed to ensure that communication will not be interrupted should undesirable events occur on the data backbone. Bandwidth capacity should be baselined to determine the current levels of data traffic and adjust the necessary additional bandwidth for voice traffic. Quality of service will need to be defined so that voice traffic will be given priority over data traffic. Other considerations are laws and regulations. Certain countries may ban the use of VoIP.

VoIP Security Issues

With the introduction of VoIP, the need for security is more important because it is needed to protect two assets—the data and the voice.

Protecting the security of conversations in VoIP is vital now. In a conventional office telephone system, security is a more valid assumption. Intercepting conversations requires physical access to telephone lines or compromise of the office private branch exchange (PBX). Only particularly security-sensitive organizations bother to encrypt voice traffic over traditional telephone lines. It cannot be said for Internet-based connections. In VoIP, packets are sent over the network from a user's computer or VoIP phone to similar equipment on the other end. Packets may pass through several intermediate systems that are not under the control of the user's ISP. The current Internet architecture does not provide the same physical wire security as the phone lines. The key to securing VoIP is to use the security mechanisms such as those deployed in data networks (e.g., firewalls, encryption) to emulate the security level currently used by public switched telephone network (PSTN) network users.

The main concern with VoIP solutions is that while, in the case of traditional telephones, if the data system is disrupted, then different sites of the organization could still be reached via telephone. With VoIP, a computer system disruption also terminates the telephone because both are supported by the same devices. In this case, only mobile phones will function. Thus, a backup communications facility should be planned for if the availability of communications is vital to the organization. This would be the case with branches of financial institutions.

Another issue might arise with the fact that IP telephones and their supporting equipment require the same care and maintenance as computer systems do.

Thus, OS patches and virus signature updates must be promptly applied to prevent a potential system outage. To enhance the protection of the telephone system and data traffic, the VoIP infrastructure should be segregated using virtual local area networks (VLANs). Any connections between these two infrastructures should be protected using firewalls that can interpret VoIP protocols.

In many cases, session border controllers (SBCs) are utilized to provide security features for VoIP traffic similar to that provided by firewalls. SBCs can be configured to filter specific VoIP protocols, monitor for DoS attacks, and provide network address and protocol translation features.

5.4.8 PRIVATE BRANCH EXCHANGE

A PBX is a sophisticated computer-based switch that can be thought of as essentially a small, in-house phone company for the organization that operates it. Protection of the PBX is, thus, a high priority. Failure to secure a PBX can result in exposing the organization to toll fraud, theft of proprietary or confidential information, loss of revenue, or legal entanglements.

PBXs have been a part of organizations' communication infrastructures since the early 1920s, originally using analog technology. PBXs of today use digital technology; digital signals are converted to analog for outside calls on the local loop using Plain Old Telephone Service (POTS), which refers to the standard telephone service that most homes use.

Digital PBXs are widespread throughout industry and public organizations, having replaced their analog predecessors. The advent of software-based PBXs has provided a wealth of communications capabilities within these switches. Today, even the most basic PBX systems have a wide range of capabilities that were previously available only in large-scale switches. These new features have opened up many new opportunities for an intruder to attempt to exploit the PBX, particularly by usage of these features for a purpose that was never intended.

Attributes of today's PBXs include:
- More than two telephone trunk (multiple phone) lines that terminate at the PBX
- The use of digital phones that permit integrated voice/data workstations
- Scalable computer-based PBX systems with memory that manages the switching of the calls within the PBX
- Distributed architecture with multiple switches in a hierarchical or meshed configuration with distributed intelligence providing enhanced reliability
- Nonblocking configurations where all attached devices can be engaged in calls simultaneously
- The network of lines within the PBX
- An operator console or switchboard for a human operator

One of the principal purposes of a PBX is to save the cost of requiring a line for each user to the telephone company's central office. Also, it is easier to call someone within a PBX because only three or four digits need to be dialed.

PBX Risk

PBX environments involve many security risk, presented by people both internal and external to the organization. If a PBX is not correctly configured, back doors can be easily established for unauthorized purposes. The threats to PBX telephone systems are many, depending on the goals of these attackers, and include:
- **Theft of service**—Toll fraud, probably the most common of motives for attackers
- **Disclosure of information**—Data disclosed without authorization, either by deliberate action or by accident. Examples include eavesdropping on conversations and unauthorized access to routing and address data.
- **Data modification**—Data altered in some meaningful way by reordering, deleting or modifying it. For example, an intruder may change billing information or modify system tables to gain additional services.
- **Unauthorized access**—Actions that permit an unauthorized user to gain access to system resources or privileges
- **Denial of service**—Actions that prevent the system from functioning in accordance with its intended purpose. A piece of equipment or entity may be rendered inoperable or forced to operate in a degraded state; operations that depend on timeliness may be delayed.
- **Traffic analysis**—A form of passive attack in which an intruder observes information about calls (although not necessarily the contents of the messages) and makes inferences (e.g., from the source and destination numbers or frequency and length of the messages). For example, an intruder observes a high volume of calls between a company's legal department and patent office and concludes that a patent is being filed.

PBXs are sophisticated computer systems, and many of the threats and vulnerabilities associated with OSs are shared by PBXs. But there are two important ways in which PBX security is different from conventional OS security:
- **External access/control**—As with larger telephone switches, PBXs typically require remote maintenance by the vendor. Instead of relying on local administrators to make OS updates and patches, organizations normally have updates installed remotely by the switch manufacturer. This, of course, requires remote maintenance ports and access to the switch by a potentially large pool of outside parties.
- **Feature richness**—The wide variety of features available on PBXs, particularly administrative features and conference functions, provide the possibility of unexpected attacks. A feature may be used by an attacker in a manner that was not intended by its designers. Features may also interact in unpredictable ways, leading to system compromise, even if each component of the system conforms to its security requirements and the system is operated and administered correctly.

Some additional control weaknesses include:
- Uncontrolled definition of direct inward dial (DID) lines, which would allow an external party to request a dial tone locally, enabling that person to make unauthorized long-distance phone calls

- Lack of system access controls over long-distance phone calls (e.g., default system vendor passwords unchanged, 24/7 availability of PBX lines)
- Lack of blocking controls for long-distance phone calls to particular numbers (e.g., hot numbers, cellular numbers, etc.)
- Lack of control over the numbers destined for fax machines and modems
- Not activating the option to register calls, which enables the use of call-tracking logs

Although most features are common from PBX to PBX, the implementation of these features may vary and the degree of vulnerability, if any, will depend on how each feature is implemented. For example, many PBX vendors have proprietary designs for the Digital Signaling Protocol between the PBX and the user instruments.

Knowing the design implementation will aid in determining if an intruder or an insider have an easy way to exploit weaknesses or normal functions.

PBX Audit

When planning a PBX audit, the type of skills, the number of auditors and the length of time required to perform the audit cannot be determined without a preliminary assessment of the PBX system, because these depend on the size and complexity of the chosen PBX. The type of perceived threat and the seriousness of any discovered vulnerabilities must be decided by the auditor. Consequently, any corrective actions must also be determined based on the cost of the loss compared with the cost of the corrective action.

A list of critical items of PBX structure, usage and setup will be given, together with specific risk and applicable controls.

PBX System Features

Many features sometimes available to the system may be used by phreakers (security crackers) or intruders for illegal purposes, including:
- Eavesdropping on conversations, without the other parties being aware of it
- Eavesdropping on conference calls
- Illegally forwarding calls from specific instruments to remote numbers
- Forwarding a user's instrument to an unused or disabled number, thereby making it unreachable by external calls

PBX System Attacks

PBX system features and capabilities may present significant vulnerabilities. This occurs because, with such a large number of features available, it becomes difficult for the manufacturer to consider all individual risk and the potential problems caused by the manner in which different features may interact. This may result in vulnerabilities that allow an intruder unwanted access to the PBX and its instruments. **Figure 5.19** shows PBX system features and corresponding risk.

Figure 5.19—PBX System Features and Risk

System Feature	Description	Risk
Automatic call distribution	Allows a PBX to be configured so that incoming calls are distributed to the next available agent or placed on hold until one becomes available	Tapping and control of traffic
Call forwarding	Allows specifying an alternate number to which calls will be forwarded based on certain conditions	User tracking
Account codes	Used to: • Track calls made by certain people or for certain projects for appropriate billing • Dial-in system access (user dials from outside and gains access to the normal features of the PBX) • Changing the user class of service so a user can access a different set of features (i.e., the override feature)	Fraud, user tracking, nonauthorized features
Access codes	Key for access to specific features from the part of users with simple instruments (i.e., traditional analog phones)	Nonauthorized features
Silent monitoring	Silently monitors other calls	Eavesdropping
Conferencing	Allows for conversation among several users	Eavesdropping, by adding unwanted/unknown parties to a conference
Override (Intrude)	Provides for the possibility to break into a busy line to inform another user of an important message	Eavesdropping
Autoanswer	Allows an instrument to automatically go when called—usually gives an audible or visible warning which can easily be turned off.	Gaining information not normally available, for various purposes (i.e., eavesdropping through the automatic answering of an instrument in a conference room)

Figure 5.19—PBX System Features and Risk *(cont.)*

System Feature	Description	Risk
Tenanting	Limits system user access to only those users who belong to the same tenant group—useful when one company leases out part of its buildings to other companies and tenants share an attendant, trunk lines, etc.	Illegal usage, fraud, eavesdropping
Voice mail	Stores messages centrally and—by using a password—allows for retrieval from inside or outside lines	Disclosure or destruction of all messages of a user when that user's password is known or discovered by an intruder, disabling of the voice mail system and even the entire switch by lengthy messages or embedded codes, illegal access to external lines
Privacy release	Supports shared extensions among several devices, ensuring that only one device at a time can use an extension. Privacy release disables the security by allowing devices to connect to an extension already in use.	Eavesdropping
Nonbusy extensions	Allows calls to an in-use extension to be added to a conference when that extension is on conference and already off-hook	Eavesdropping a conference in progress
Diagnostics	Allows for bypassing normal call restriction procedures. This kind of diagnostic is sometimes available from any connected device. It is a separate feature, in addition to the normal maintenance terminal or attendant diagnostics	Fraud and illegal usage
Camp-on or call waiting	When activated, sends a visual or audible warning to an off-hook instrument that is receiving another call. Another option of this feature is to conference with the camped-on or call-waiting party.	Making the called individual a party to a conference without knowing it
Dedicated connections	Connections made through the PBX without using the normal dialing sequence. It can be used to create hot-lines between devices (i.e., one rings when the other goes off-hook). It is also used for data connections between devices and the central processing facility.	Eavesdropping on a line

Protecting against all of the risk is not easy. Knowing that a given vulnerability in fact exists is already a vital indication. A conservative approach of enabling only the needed features is advisable. Following are some controls to minimize PBX system attacks:

- Where possible, configure and secure separate and dedicated administrative ports.
- Control the definition of DID lines to avoid an external party requesting a dial tone locally, disabling that person's ability to make unauthorized long-distance phone calls.
- Establish system access controls over long-distance phone calls (e.g., change default system vendor passwords, limit the 24/7 availability of PBX lines).
- Block controls for long-distance phone calls to particular numbers (e.g., hot numbers, cellular numbers).
- Establish control over the numbers destined for fax machines and modems.
- Activate the option to register calls, enabling the use of call-tracking logs.

Hardware Wiretapping

A PBX's susceptibility to tapping depends on the methods used for communication between the PBX and its attached devices. This communication may include voice, data and signaling information. The signaling information is typically composed of commands to the devices (e.g., turn on indicators, microphones and speakers) and status from the devices (e.g., hook status and keys pressed). Communications methods use analog voice with or without separate control signals, analog voice with inclusive control signals, and digital voice with inclusive control signals.

Tapping or intrusion in the control sequences is possible using various appropriate hardware technologies, which often include device modification.

The following measures provide controls to minimize risk:

- Physical security of the PBX facilities
- Usage of appropriate anti-tamper devices on critical hardware components

Hardware Conferencing

When implemented in hardware, the conferencing feature may employ a circuit card known as a conference bridge or a signal processor chip. This allows multiple lines to be bridged to create a conference where all parties can both speak and listen. Some PBXs have a mute feature where all parties can hear, but only certain parties can speak. An intruder could try to obtain a connection to the bridge where the conference could be overheard. A hardware modification to the bridge itself may make it possible to cause the output of the bridge available to a specific port. As in device modifications, some additional steps must be taken to receive this information. This may include modifying the database to make the intruder a permanent member of the bridge so any conference on that bridge could be overheard.

The following measures provide controls to minimize risk:
- Establish a strong physical security to prevent unauthorized access to telephone closets and to the PBX facilities. Whenever possible, the PBX should be kept in a locked room with restricted access.
- Lock critical hardware with anti-tamper devices.

Remote Access

Remote access is frequently an unavoidable necessity of maintenance, but it can represent a serious vulnerability. The maintenance features may be accessible via a remote terminal with a modem, a system console or other device or over an outside dial-in line. This allows for systems to be located over a large area (perhaps around the world) and have one central location from which maintenance can be performed. Often it is necessary for the switch manufacturer to have remote access to the switch, to install software upgrades or to restart a switch that has experienced service degradation. Dial-back modem usage is a basic precaution but does not offer full protection. When possible, remote access should be left closed and should be opened only upon need or upon a verified request from the organization that performs the maintenance. It is easy for attackers to contact the switch manufacturer on the pretext of needing help with a particular type of switch, obtain the names of the manufacturer's remote maintenance personnel, and then masquerade as these personnel to obtain access to the victim's switch.

The following measures provide controls to minimize risk:
- A dial-back scheme
- Careful scrutiny and proper authentication of requests to open the remote control

Maintenance

A common maintenance feature is maintenance out of service (MOS). This feature allows maintenance personnel to place a line out of service for maintenance. It is typically used when a problem is detected with a line or when it is desired to disable a line. However, if a line is placed into MOS while it is in operation, the PBX may terminate its signaling communication with the instrument and leave the instrument's voice channel connection active, even after the telephone device is placed on-hook. If the MOS feature were to function in this manner, the potential exists for someone to use the MOS feature to establish a live microphone connection to a user's location without the user's knowledge and, thereby, eavesdrop on the area surrounding the user's telephone.

Another common maintenance feature is the ability to connect two lines together to transmit data from one line to the other and verify whether or not the second line receives the data properly. This feature would allow someone with maintenance access to connect a user's instrument to an instrument at another location to eavesdrop on the area surrounding the user's telephone without the user's knowledge.

Also, the PBX may support some maintenance features that are not normally accessible to the owner/operator of the PBX for several reasons. These types of utilities vary greatly from one PBX to another, so a general approach to finding them cannot be detailed. Some suggested courses of action to verify the existence of such features are listed below:
- Ask the manufacturer or maintenance company if any such features exist.
- Attempt to learn about undocumented usernames/passwords.
- Attempt to search the system's programmable read-only memory (PROM) or disks for evidence of such features.
- View the system load files with a binary editor to determine whether this reveals the names of undocumented commands among a list of known maintenance commands, which can be recognized in the binaries.
- Verify the existence of alarm features.
- Enable and review usage and intervention logs.

Special Manufacturer's Features

These types of features would most likely be accessible via undocumented username/password access to the maintenance and/or administrative tools. Some possible undocumented features and their associated risk are listed below:
- **Database upload/download utility**—This utility allows the manufacturer to download the database from a system that is malfunctioning and examine it at their location to try to determine the cause of the malfunction. It also allows the manufacturer to upload a new database to a PBX in the event that the database became so corrupted that the system became inoperable. Compromise of such a utility could allow an adversary to download a system's database, insert a Trojan horse or otherwise modify it to allow special features to be made available to the adversary, and upload the modified database back into the system.
- **Database examine/modify utility**—This utility allows the manufacturer to remotely examine and modify a system's database to repair damage caused by incorrect configuration, design bugs or tampering. This utility can also provide an intruder with the ability to modify the database to gain access to special features.
- **Software debugger/update utility**—This utility gives the manufacturer the ability to remotely debug a malfunctioning system. It also allows the manufacturer to remotely update systems with bug fixes and software upgrades. It could also grant an adversary the same abilities. This is perhaps the most dangerous vulnerability because access to the software would give an adversary virtually unlimited access to the PBX and its associated instruments.

Manufacturer's Development and Test Features

There may be features that were added to the system during its development phase that were forgotten and not removed when production versions were released. There also may be hidden features that were added by a person on the development team with the intent of creating a back door into the customer's systems. The test features are probably easy to access for ease of development and have few restrictions to reduce development time.

Potential forms of attack include:
- Undocumented username/passwords
- Entering out-of-range values in database fields
- Dialing undocumented access codes on instruments
- Pressing certain key sequences on instruments

Measures that provide controls to minimize risk include:
- Establish strong authentication of external technicians.
- Keep maintenance terminals in a locked, restricted area.
- Turn off maintenance features when not needed, if possible.

Software Loading and Update Tampering

When software is initially loaded onto a PBX and when any software updates/patches are loaded, the PBX is particularly vulnerable to software tampering. A software update sent to a PBX administrator could be intercepted by an adversary. The update could be modified to allow special access or special features to the adversary. The modified update would then be sent to the PBX administrator who would install the update and unknowingly give the adversary unwanted access to the PBX.

A control for software loading and updates would be strong modification—tamper detection based on cryptography used in software packages. Conventional error detection codes, such as checksums or CRCs, are not sufficient to ensure tamper detection.

Crash-restart Attacks

System crashes may indicate a DoS vulnerability. The means by which a system may be crashed vary significantly from one system to another. The following list suggests a few features and conditions that can sometimes trigger a system crash:
- Call forwarding
- Voicemail
- Physical removal of hardware or media from the PBX
- Use of administrative/maintenance terminal functions
- Direct modification of the system or the database. This may be possible if the media can be read by utilities typically found on a PC or workstation.
- Normal system shutdown procedures

These approaches should be tested as possible ways of exposing the weaknesses discussed in the remainder of this section. One possible additional weakness is that a crash may interrupt the control flow and leave microphones open, so an intruder could overhear what is said after the crash. A further danger is that, in some cases, embedded logons and passwords are restored upon rebooting the system, making it possible for a remote operator to complete the remote restart. However, this also makes it possible for an attacker to gain administrator privileges on a system by crashing the system, then applying a known embedded logon ID/password combination.

Controls for these types of attacks include:
- Crash-restart vulnerability tests and preventing or forbidding, if possible, the triggering of events
- Restart procedures that eliminate the vulnerability from loss of control. This may require doing a cold start (i.e., complete shutdown, power-off, and restart) in the event of a system crash.
- If embedded passwords are found, patching the load module to replace them. Authorized manufacturer personnel can be given the new password, if needed.
- A PBX firewall to enhance the protection of the PBX. In recent years, specialized firewalls have been developed specifically for the protection of PBX systems.

Passwords

Most PBXs grant administrative access to the system database through a system console or a generic dumb terminal. Username/password combinations are often used to protect the system from unwanted changes to the database. If remote access to the maintenance features is available, it is usually restricted by some form of password protection. There may be a single fixed maintenance account, multiple fixed maintenance accounts or general user-defined maintenance accounts. The documentation provided with the PBX should state what type of maintenance access is available. The documentation should also indicate how passwords function. Dangers from improper definition and usage of passwords are the loss of control and confidentiality, illegal usage and tampering of the database.

Controls for passwords include:
- Passwords resistant to cracking by automated tools. A password generator that creates random passwords can assist in defeating password crackers.
- Monitoring of multilevel password rights
- Setting an appropriate time-out period for logins

5.5 AUDITING INFORMATION SECURITY MANAGEMENT FRAMEWORK

Auditing the information security framework of an organization involves the audit of logical access, the use of techniques for testing security and the use of investigation techniques.

5.5.1 AUDITING INFORMATION SECURITY MANAGEMENT FRAMEWORK

The information security management framework should be reviewed per the basic elements in an information security framework.

Reviewing Written Policies, Procedures and Standards

Policies and procedures provide the framework and guidelines for maintaining proper operation and control. The IS auditor should review the policies and procedures to determine if they set the tone for proper security and provide a means for assigning responsibility for maintaining a secure computer processing environment. This policy review should also include reviewing the date of the last update to ensure that documents remain current and meet organizational information security needs.

Logical Access Security Policies

These policies should encourage limiting logical access on a need-to-know basis. They should reasonably assess the exposure to the identified concerns.

Formal Security Awareness and Training

Effective security will always be dependent on people. As a result, security can only be effective if employees know what is expected of them and what their responsibilities are. They should know why various security measures, such as locked doors and use of logon IDs, are in place and the repercussions of violating security.

Promoting security awareness is a preventive control. Through this process, employees become aware of their responsibilities for maintaining good physical and logical security. This can also be a detective measure, because it encourages people to identify and report possible security violations.

Training should start with the new employee orientation or induction process. Ongoing awareness can be provided in company newsletters through visible and consistent security enforcement and short reminders during staff meetings. The security administrator should direct the program. To determine the effectiveness of the program, the IS auditor should interview a sample of employees to determine their overall awareness.

Data Ownership

Data ownership refers to the classification of data elements and the allocation of responsibility for ensuring that they are kept confidential, complete and accurate. A key point of ownership is that, by assigning responsibility for protecting computer data to particular employees, accountability is established. The IS auditor can use this information to determine if proper ownership has been assigned and whether the data owner is aware of the assignment. The IS auditor should also review a sample of job descriptions to ensure that responsibilities and duties are consistent with the information security policy. The auditor should review the classification of data and evaluate their appropriateness, as they relate to the area under review.

Data Owners

Data owners are generally managers and directors responsible for using information for running and controlling the business. Their security responsibilities include authorizing access, ensuring that access rules are updated when personnel changes occur, and regularly review access rules for the data for which they are responsible.

Data Custodians

Data custodians are responsible for storing and safeguarding the data, and include IS personnel such as systems analysts and computer operators.

Security Administrator

Security administrators are responsible for providing adequate physical and logical security for IS programs, data and equipment. (The physical security may be handled by someone else, not always by the security administrator.) Normally, the information security policy will provide the basic guidelines under which the security administrator will operate.

New IT Users

New IT users (employees or third parties) and, in general, all new users assigned PCs or other IT resources should sign a document containing the main IT security obligations that they are thereby engaged to know and observe. These are:

- Reading and agreeing to follow security policies
- Keeping logon IDs and passwords secret
- Creating quality passwords according to policy
- Locking their terminal screens when not in use
- Reporting suspected violations of security
- Maintaining good physical security by keeping doors locked, safeguarding access keys, not disclosing access door lock combinations and questioning unfamiliar people
- Conforming to applicable laws and regulations
- Use of IT resources only for authorized business purposes

Data Users

Data users, including the internal and the external user communities, are the actual users of the computerized data. Their levels of access into the computer should be authorized by the data owners and restricted and monitored by the security administrator. Their responsibilities regarding security are to be vigilant regarding the monitoring of unauthorized people in the work areas and comply with general security guidelines and policies.

Documented Authorizations

Data access should be identified and authorized in writing. The IS auditor can review a sample of these authorizations to determine if the proper level of written authority was provided. If the facility practices data ownership, only the data owners provide written authority.

Terminated Employee Access

Termination of employment can occur in the following circumstances:

- On the request of the employee (voluntary resignation from service)
- Scheduled (on retirement or completion of contract)
- Involuntary (forced by management in special circumstances)

In case of involuntary termination of employment, the logical and physical access rights of employees to the IT infrastructure should either be withdrawn completely or highly restricted as early as possible, before the employee becomes aware of the termination or its likelihood. This ensures that terminated employees cannot continue to access potentially confidential or damaging information from the IT resources or perform any action that would result in damage of any kind to the IT infrastructure, applications and data. Similar procedures should be in place to terminate access for third parties upon terminating their activities with the organization.

When it is necessary for employees to continue to have access, such access must be monitored carefully and continuously and should take place with senior management's knowledge and authorization.

In case of voluntary or scheduled termination of employment, it is management's prerogative to decide whether access is restricted or withdrawn. This depends on:

- The specific circumstances associated with each case
- The sensitivity of the employee's access to the IT infrastructure and resources
- The requirements of the organization's information security policies, standards and procedures

Security Baselines

A baseline security plan is meant to be used as a first step to IT security. The baseline plan should be followed with a full security evaluation and plan. **Figure 5.20** illustrates baseline security topics and their associated recommendations.

Figure 5.21 depicts a checklist for a baseline security evaluation.

Access Standards

Access standards should be reviewed by the IS auditor to ensure they meet organizational objectives for separating duties, prevent fraud or error, and meet policy requirements for minimizing the risk of unauthorized access.

Standards for security may be defined:

- At a generic level (e.g., all passwords must be at least eight characters long)
- For specific machines (e.g., all UNIX machines can be configured to enforce password changes)
- For specific application systems (e.g., sales ledger clerks can access menus that allow entry of sales invoices, but may not access menus that allow check authorization)

5.5.2 AUDITING LOGICAL ACCESS

When evaluating logical access controls the IS auditor should:

- Obtain a general understanding of the security risk facing information processing, through a review of relevant documentation, inquiry, observation, risk assessment and evaluation techniques
- Document and evaluate controls over potential access paths into the system to assess their adequacy, efficiency and effectiveness by reviewing appropriate hardware and software security features and identifying any deficiencies or redundancies
- Test controls over access paths to determine whether they are functioning and effective by applying appropriate audit techniques
- Evaluate the access control environment to determine if the control objectives are achieved by analyzing test results and other audit evidence
- Evaluate the security environment to assess its adequacy by reviewing written policies, observing practices and procedures, and comparing them with appropriate security standards or practices and procedures used by other organizations

Figure 5.20—IT Security Baseline Recommendations

Topics	Objective	Recommendations
Inventory	Establish and maintain an inventory	Users are expected to follow standards for managing computers connected to the network and have registered network addresses. The OS and owner should be included along with the data provided.
Malware	Install antivirus software with automatic updating	Antivirus software with an automatic DAT file should be updated at regular intervals—no less than weekly.
Passwords	Recognize the importance of passwords	Users must use only strong passwords. The IT department should provide password guidance. Departmental accounts are created for workgroups to prevent/avoid password sharing.
Patching	Make it automatic—less work necessary, less chance for compromise	Each machine should be configured to patch automatically for OS and basic software patching. A process should be set up that works for the department and minimizes disruptions at inconvenient times. Workstations should be more automated to enable system administrators the time to give servers the attention required to minimize the impact on services offered.
Minimizing services offered by systems	Eliminate unnecessary services— reducing security risk and saving time in the long run	To improve basic security and minimize effort to maintain systems, workstations should offer only needed services. Many OSs are installed with services turned on. By removing services, a workstation's chances of being compromised are reduced and security risk is minimized.
Addressing vulnerabilities	Eliminate many vulnerabilities with good system administration	System compromises can be time-consuming and damage credibility and the business's integrity. Information from enterprisewide scans helps to identify vulnerabilities on each system and provide a baseline for comparison when system integrity is in question.
Backups	Allow easy recovery from user mistakes and hardware failure with backups	Backups should be made offsite for increased security.

Figure 5.21—Baseline Security Evaluation Checklist

Topics	Evaluation Questions
Environment/inventory	• What types of data are maintained by the enterprise (e.g., financial, statistical, graphical)? • In what form are they maintained (e.g., spreadsheets, databases, etc.)? • Is there any critical or confidential information maintained or handled? If so, how is it protected? • Are there any specific requirements for handling data? (legal or regulatory requirements) • Have you identified machines that store or require access to confidential information? • What type of operating systems exist? • How many subnets exist? • How many workstations/servers exist? • In how many locations is there IT infrastructure? • Has the wireless infrastructure been deployed? How is it secured? • Is staff instructed on how to lock workstations when they step away? • Are users aware that unexpected email attachments should not be opened? • Is staff aware that many compromises are due to social engineering and the sharing of information? • Does the enterprise have a network diagram that includes IP addresses, room numbers and responsible parties? • Has the enterprise limited and secured physical and remote access to network services? • Is corporate hardware upgraded at regular intervals? • Does the enterprise have a current documented inventory of hardware and software? • Is all corporate software licensed? • Is license documentation available (licenses, purchase orders) if a software audit is required?
Antivirus	• Does the enterprise have an antivirus policy? • Are all workstations running the latest version of antivirus software, the scanning engine and the virus signature file? • Are DAT files downloaded automatically or manually? If manually, how often and why? • Does staff know whom to contact when a virus is found? • Does the antivirus system have a way to defend against zero-day attacks?
Passwords	• Is there a corporate policy requiring strong passwords? • Is the enterprise using software that enforces strong passwords? • Is password caching disabled on all workstations? • Are passwords changed? If so, how often? • Are employees aware that passwords and accounts are not to be shared? • Does the system administrator have written authorization to check for weak passwords?
Patching	• Are software patches applied to all operating systems automatically when possible? If done manually, how often? • Are patches applied to web browsers and applications? If yes, how frequently? • Do you back up each machine before applying a patch? • Do you test patches prior to applying? • Does the department have a documented process for patching? • Do you subscribe to sufficient newsletters and groups to be aware of patches to all relevant hardware and software?
Minimizing services offered by systems	• Have you identified services that each user needs to accomplish job assignments? • Have you removed unnecessary services that were installed by default? • Does the technical staff review security settings and policies? • Have you identified what services your systems are offering? • Have you taken security measures for remote access? • Have you transitioned to secure services?
Addressing vulnerabilities/auditing	• Have you resolved vulnerabilities discovered by enterprisewide scans? • Who is the contact for vulnerability scans? • Does the IT staff complete an independent vulnerability scan for the enterprise? • Has the enterprise deployed any form of firewalls or IDS (host or network-based)? Are any under consideration?
Backup and recovery/business continuity	• Are files regularly backed up? • Are files kept onsite in a secure location? • Are backup files sent offsite to a physically secure location? • Are backup files periodically restored as a test to verify whether they are a viable alternative? • Can you ensure that any forms of media containing confidential and sensitive information are sanitized before disposal? • Is there redundant hardware to allow work to continue in the event of a single hardware failure? • Does the enterprise have the ability to continue to function if central services is not available? • Does the enterprise have the ability to continue to function in the event of a wide area network failure? • Have you responded to and recovered from any abuse issues/incidents?

Figure 5.21—Baseline Security Evaluation Checklist *(cont.)*	
Topics	**Evaluation Questions**
IT staff	• How many IT staff are employed full-time/part-time? • Does each IT staff member have a current job description? • Do job descriptions and evaluations include IT security duties? • Does the department have sufficient documentation to ease the transition of incoming/outgoing staff? • Does the enterprise have a privacy policy? • Are all staff aware of privacy considerations? • Are management/department users aware of the types of (private/nonpublic) information available to systems administrators? • Does the enterprise have a privacy policy to address this privileged information (confidentiality agreement/nondisclosure agreement)? • Does the enterprise have a firewall or IDS, or other software for network diagnosis? Does the enterprise have tools requiring privileges and access to confidential information acquired via routers, switches, IDS, firewalls, etc.?

Familiarization With the IT Environment

This is the first step of the audit and involves obtaining a clear understanding of the technical, managerial and security environment of the IS processing facility. This typically includes interviews, physical walk-throughs, review of documents and risk assessments.

Assessing and Documenting the Access Paths

The access path is the logical route an end user takes to access computerized information. This starts with a terminal/workstation and typically ends with the data being accessed. Along the way, numerous hardware and software components are encountered. The IS auditor should evaluate each component for proper implementation and physical and logical access security.

Special consideration should be given to the:
- Origination and authorization of the data
- Validity and correctness of the input data
- Maintenance of the affected OSs (patching, hardening and closing the unnecessarily open ports)

The typical sequence of the components is as follows:
- A PC, which is part of the LAN, is used by an end user to sign on. The PC should be physically secure and the logon ID/password used for sign-on should be subject to the restrictions identified previously.
  - The OS running on the PC should be patched according to the suggestions of the supplier of the OS and the malware defense must also be updated. Out-of-date OS versions and out-of-date virus defenses can be exploited by attackers. The PC OS must be hardened—deleting the unnecessary services (e.g., those connected with remote procedure calls, sending mail, or network management) and library routines. The parameter settings and configuration of the OS must also be investigated. The ports that are not used should be closed.
- One or more servers from which the applications to be used are invoked. The OS running on the servers should be patched according to the recommendations of the supplier of the OS and the virus defense must also be updated. Out-of-date OS versions and out-of-date virus defenses can be exploited by attackers. The server OS must be hardened—deleting unnecessary services (e.g., those connected with remote procedure calls, sending mail, or network management) and library routines. The parameter settings and configuration of the OS must also be investigated. The ports that are not used should be closed.
- The telecommunications software (LAN server or terminal emulator if connecting to a mainframe) intercepts the logon to direct it to the appropriate telecommunication link. The telecommunication software can restrict PCs to specific data or application software. A key audit issue with telecommunication software is to ensure that all applications have been defined within the software and that the various optional telecommunication control and processing features used are appropriate and approved by management. This analysis typically requires the assistance of a system software analyst.
- The transaction processing software may be the next component in the access path. This software routes transactions to the appropriate application software. Key audit issues include ensuring proper identification/authentication of the user (logon ID and password) and authorization of the user to gain access to the application. This analysis is performed by reviewing internal tables that reside in the transaction processing software or in separate security software. Access to these should be restricted to the security administrator.
- The application software then is encountered and should process transactions in accordance with program logic. Audit issues include restricting access to the production software library to only the implementation coordinator.
- The database management system (DBMS) directs access to the computerized information. Audit issues include ensuring that all data elements are identified in the data dictionary, that access to the data dictionary is restricted to the database administrator (DBA) and that all data elements are subject to logical access control. The application data now can be accessed.
- The access control software can wrap logical access security around all of the above components. This is done via internal security tables. Audit issues include ensuring all of the above components are defined to the access control software, providing access rules that define who can access what on a need-to-know basis and restricting security table access to the security administrator.

> **Note:** The development of the application systems must be disciplined. The IS auditor should evaluate the control objectives, referring to the origination and authorization of the applications data, and should evaluate the control measures used in data input and processing. Omitting these control objectives and measures makes the applications vulnerable to attacks either from within or from the outside, especially from the Internet. Firewalls do not protect applications against the types of attacks that come with the HTTP communication that is usually permitted on the applications.

Interviewing Systems Personnel

To control and maintain the various components of the access path, as well as the OS and computer mainframe, technical experts often are required. These people can be a valuable source of information to the IS auditor when gaining an understanding of security. To determine who these people are, the IS auditor should meet with the IS manager and review organizational charts and job descriptions. Key people include the security administrator, network control manager and systems software manager.

The security administrator should be asked to identify the responsibilities and functions of the position. If the answers provided to this question do not support sound control practices or do not adhere to the written job description, the IS auditor should compensate by expanding the scope of the testing of access controls. Also, the IS auditor should determine whether the security administrator is aware of the logical accesses that must be protected, has the motivation and means to actively monitor logons to account for employee changes, and is knowledgeable in how to maintain and monitor access.

A sample of end users should be interviewed to assess their awareness of management policies regarding logical security and confidentiality.

Reviewing Reports From Access Control Software

The reporting features of access control software provide the security administrator with the opportunity to monitor adherence to security policies. By reviewing a sample of security reports, the IS auditor can determine whether enough information is provided to support an investigation and if the security administrator is performing an effective review of the report.

Unsuccessful access attempts should be reported and should identify the time, terminal, logon and file or data element for which access was attempted.

Reviewing Application Systems Operations Manual

An application systems manual should contain documentation on the programs that generally are used throughout a data processing installation to support the development, implementation, operations and use of application systems. This manual should include information about the platform the application can run on, DBMSs, compilers, interpreters, telecommunication monitors and other applications that can run with the application.

5.5.3 TECHNIQUES FOR TESTING SECURITY

Auditors can use different techniques for testing security. Some methods are described in the following subsections.

Terminal Cards and Keys

The IS auditor can take a sample of these cards or keys and attempt to gain access beyond that which is authorized. Also, the IS auditor will want to know if the security administrator followed up on any unsuccessful attempted violations.

Terminal Identification

The IS auditor can work with the network manager to get a listing of terminal addresses and locations. This list can then be used to inventory the terminals, looking for incorrectly logged, missing or additional terminals. The IS auditor should also select a sample of terminals to ensure that they are identified in the network diagram.

Logon IDs and Passwords

To test confidentiality, the IS auditor could attempt to guess the password of a sample of employees' logon IDs (although this is not necessarily a test). This should be done discreetly to avoid upsetting employees. The IS auditor should tour end-user and programmer work areas looking for passwords taped to the side of terminals, the inside of desk drawers or located in card files. Another source of confidential information is the wastebasket. The IS auditor might consider going through the office wastebasket looking for confidential information and passwords. Users could be asked to give their password to the IS auditor. However, unless specifically authorized for a particular situation and supported by the security policy, no user should ever disclose his/her password. Another way to test password strength is to analyze global configuration settings for password strength in the system application and compare this with the organization's security policy.

To test encryption, the IS auditor should work with the security administrator to attempt to view the internal password table. If viewing is possible, the contents should be unreadable. Being able to view encrypted passwords can still be dangerous. Although passwords on some systems are impossible to decipher, if an individual can obtain the encryption program, they can encrypt common passwords and look for matches. This was a method used to break into UNIX computers prior to the development of shadow password files. Application logs should also be reviewed to ensure that passwords and logon IDs are not recorded in a clear form.

To test access authorization, the IS auditor should review a sample of access authorization documents to determine if proper authority has been provided and if the authorization was granted on a need-to-know basis. Conversely, the IS auditor should get a computer-generated report of computer access rules, take a sample to determine if the access is on a need-to-know basis, and attempt to match the sample of these rules to authorizing documents. If no written authorization is found, this indicates a breakdown in control and may warrant further review to determine the exposures and implications.

Account settings for minimizing unauthorized access should be available from most access control software or from the OS. To verify that these settings actually are working, the IS auditor can perform the following manual tests:

- To test periodic change requirements, the IS auditor can draw on his/her experiences using the system and interview a sample of users to determine if they are forced to change their password after the prescribed time interval.
- To test for disabling or deleting of inactive logon IDs and passwords, the IS auditor should obtain a computer-generated list of active logon IDs. On a sample basis, the IS auditor should match this list to current employees, looking for logon IDs assigned to employees or consultants who are no longer with the company.
- To test for password syntax, the IS auditor should attempt to create passwords in a format that is invalid, such as too short, too long, repeated from the previous password, incorrect mix of alpha or numeric characters, or the use of inappropriate characters.
- To test for automatic logoff of unattended terminals, the IS auditor should log on to a number of terminals. The IS auditor then simply waits for the terminals to disconnect after the established time interval. Before beginning this test, the IS auditor should verify with the security administrator that this automatic logoff feature applies to all terminals.
- To test for automatic deactivation of terminals after unsuccessful access attempts, the IS auditor should attempt to log on, purposefully entering the wrong password a number of times. The logon ID should deactivate after the established number of invalid passwords has been entered. The IS auditor will be interested in how the security administrator reactivates the logon ID. If a simple telephone call to the security administrator with no verification of identification results in reactivation, then this function is not controlled properly.
- To test for masking of passwords on terminals, the IS auditor should log on to a terminal and observe if the password is displayed when entered.

Controls Over Production Resources

Computer access controls should extend beyond application data and transactions. There are numerous high-level utilities, macro or job control libraries, control libraries, and system software parameters for which access control should be particularly strong. Access to these libraries would provide the ability to bypass other access controls.

The IS auditor should work with the system software analyst and operations manager to determine if access is on a need-to-know basis for all sensitive production resources. Working with the security administrator, the IS auditor should determine who can access these resources and what can be done with this access.

Logging and Reporting of Computer Access Violations

To test the reporting of access violations, the IS auditor should attempt to access computer transactions or data for which access is not authorized. The attempts should be unsuccessful and identified on security reports. This test should be coordinated with the data owner and security administrator to avoid violation of security regulations.

Follow-up Access Violations

To test the effectiveness and timeliness of the security administrator and data owner's responses to reported violation attempts, the IS auditor should select a sample of security reports and look for evidence of follow-up and investigation of access violations. If such evidence cannot be found, the IS auditor should conduct further interviews to determine why this situation exists.

Bypassing Security and Compensating Controls

This is a technical area of review. As a result, the IS auditor should work with the system software analyst, network manager, operations manager and security administrator to determine ways to bypass security. This typically includes bypass label processing, special system maintenance logon IDs, OS exits, installation utilities and input/output (I/O) devices. Working with the security administrator, the IS auditor should determine who can access these resources and what can be done with this access. The IS auditor should determine if access is on a need-to-know/have basis or if compensating detective controls exist.

There should be restrictions and procedures of monitoring access to computer features that bypass security. Generally, only system software programmers should have access to these features:

- **Bypass label processing (BLP)**—BLP bypasses the computer reading of the file label. Because most access control rules are based on file names (labels), this can bypass access control programs.
- **System exits**—This system software feature permits the user to perform complex system maintenance, which may be tailored to a specific environment or company. They often exist outside of the computer security system and, thus, are not restricted or reported in their use.
- **Special system logon IDs**—These logon IDs often are provided by vendors. The names can be determined easily because they are the same for all similar computer systems (i.e., "system"). Passwords should be changed immediately upon installation to secure the systems.

Because many of these bypassing security features can be exploited by technically sophisticated intruders, the IS auditor should also ensure that:

- All uses of these features are logged, reported and investigated by the security administrator or system software manager
- Unnecessary bypass security features are deactivated
- If possible, the bypass security features are subject to additional logical access controls

Review Access Controls and Password Administration

Access controls and password administration are reviewed to determine that:

- Procedures exist for adding individuals to the list of those authorized to have access to computer resources, changing their access capabilities and deleting them from the list.
- Procedures exist to ensure that individual passwords are not inadvertently disclosed.
- Passwords issued are of an adequate length, cannot be easily guessed and do not contain repeating characters.
- Passwords are periodically changed.
- User organizations periodically validate the access capabilities currently provided to individuals in their department.

- Procedures provide for the suspension of user identification codes (logon IDs or accounts) or the disabling of terminal, microcomputer or data entry device activity—after a particular number of security procedure violations.

5.5.4 INVESTIGATION TECHNIQUES

Investigation techniques include the investigation of computer crime and the protection of evidence and chain of custody, among others.

Investigation of Computer Crime

Computer crimes are not reported in most cases because they are not detected. In many cases where computer crimes are detected, companies hesitate to report them because they generate a large amount of negative publicity that can affect their business. In such cases, the management of the affected company seeks to fix the vulnerabilities used for the crime and resume operations. In addition, in many countries current laws are directed toward protecting physical property. It is very difficult to use such laws against computer crime. Even in jurisdictions where the laws have been updated, the investigation procedures are not always widely known and the necessary hardware and software tools are not always available to collect the digital evidence.

In the aftermath of a computer crime, it is very important that proper procedures are used to collect evidence from the crime scene. If data and evidence is not collected in the proper manner, it could be damaged and, even if the perpetrator is eventually identified, prosecution will not be successful in the absence of undamaged evidence. Therefore, after a computer crime, the environment and evidence must be left unaltered and specialist law enforcement officials must be called in. If the incident is to be handled in-house, the company must have a suitably qualified and experienced incident response team.

Computer Forensics

Computer forensics is defined as the "process of identifying, preserving, analyzing and presenting digital evidence in a manner that is legally acceptable in any legal proceedings (i.e., a court of law)," according to D. Rodney McKemmish in Computer and Intrusion Forensics. An IS auditor may be required or asked to be involved in a forensic analysis in progress to provide expert opinion or to ensure the correct interpretation of information gathered.

Computer forensics includes activities that involve the exploration and application of methods to gather, process, interpret and use digital evidence that help to substantiate whether an incident happened such as:
- Providing validation that an attack actually occurred
- Gathering digital evidence that can later be used in judicial proceedings

Any electronic document or data can be used as digital evidence, provided there is sufficient manual or electronic proof that the contents of digital evidence are in their original state and have not been tampered with or modified during the process of collection and analysis.

It is very important to preserve evidence in any situation. Most organizations are not well equipped to deal with intrusions and electronic crimes from an operational and procedural perspective, and they respond to it only when the intrusion has occurred and the risk is realized. The evidence loses its integrity and value in legal proceedings if it has not been preserved and subject to a documented chain of custody. This happens when the incident is inappropriately managed and responded to in an ad hoc manner.

For evidence to be admissible in a court of law, the chain of custody needs to be maintained professionally. The chain of evidence essentially contains information regarding:
- Who had access to the evidence (chronological manner)
- The procedures followed in working with the evidence (e.g., disk duplication, virtual memory dump)
- Proving that the analysis is based on copies that are identical to the original evidence (e.g., documentation, checksums or timestamps)

It is important to use industry-specified good practices, proven tools and due diligence to provide reasonable assurance of the quality of evidence.

It is also important to demonstrate integrity and reliability of evidence for it to be acceptable to law enforcement authorities. For example, if the IS auditor "boots" a computer suspected of containing stored information that might represent evidence in a court case, the auditor cannot later deny that they wrote data to the hard drive because the boot sequence writes a record to the drive. This is the reason specialist tools are used to take a true copy of the drive, which is then used in the investigation.

There are four major considerations in the chain of events in regards to evidence in computer forensics:
- **Identify**—Refers to the identification of information that is available and might form the evidence of an incident.
- **Preserve**—Refers to the practice of retrieving identified information and preserving it as evidence. The practice generally includes the imaging of original media in presence of an independent third party. The process also requires being able to document chain-of-custody so that it can be established in a court of law.
- **Analyze**—Involves extracting, processing and interpreting the evidence. Extracted data could be unintelligible binary data after it has been processed and converted into human readable format. Interpreting the data requires an in-depth knowledge of how different pieces of evidence may fit together. The analysis should be performed using an image of media and not the original.
- **Present**—Involves a presentation to the various audiences such as management, attorneys, court, etc. Acceptance of the evidence depends upon the manner of presentation (as it should be convincing), qualifications of the presenter, and credibility of the process used to preserve and analyze the evidence.

The IS auditor should give consideration to key elements of computer forensics during audit planning. These key elements are described in the following subsections.

DATA PROTECTION
To prevent sought-after information from being altered, all measures must be in place. It is important to establish specific protocols to inform appropriate parties that electronic evidence will be sought and to not destroy it by any means.

Infrastructure and processes for incident response and handling should be in place to permit an effective response and forensic investigation if an event or incident occurs.

DATA ACQUISITION
All information and data required should be transferred into a controlled location; this includes all types of electronic media such as fixed disk drives and removable media. Each device must be checked to ensure that it is write-protected. This may be achieved by using a device known as a write-blocker.

It is also possible to get data and information from witnesses or related parties by recorded statements.

By volatile data, investigators can determine what is currently happening on a system. This kind of data includes open ports, open files, active processes, user logons and other data present in RAM. This information is lost when the computer is shut down.

IMAGING
Imaging is a process that allows one to obtain a bit-for-bit copy of data to avoid damage of original data or information when multiple analyses may be performed. The imaging process is made to obtain residual data, such as deleted files, fragments of deleted files and other information present, from the disk for analysis. This is possible because imaging duplicates the disk surface, sector by sector.

With appropriate tools, it is sometimes possible to recover destroyed information (erased even by reformatting) from the disk's surface.

EXTRACTION
This process consists of identification and selection of data from the imaged data set. This process should include standards of quality, integrity and reliability. The extraction process includes software used and media where an image was made.

The extraction process could include different sources such as system logs, firewall logs, IDS logs, audit trails and network management information.

INTERROGATION
Interrogation is used to obtain prior indicators or relationships, including telephone numbers, IP addresses and names of individuals, from extracted data.

INGESTION/NORMALIZATION
This process converts the information extracted to a format that can be understood by investigators. It includes conversion of hexadecimal or binary data into readable characters or a format suitable for data analysis tools.

It is possible to create relationships from data by extrapolation, using techniques such as fusion, correlation, graphing, mapping or time lining, which could be used in the construction of the investigation's hypothesis.

REPORTING
The information obtained from computer forensics has limited value when it is not collected and reported in the proper way.

When an IS auditor writes the report, he/she must include why the system was reviewed, how the computer data were reviewed and what conclusions were made from this analysis.

The report should achieve the following goals (from Mandia, Kevin; Matt Pepe; Chris Prosise; *Incident Response & Computer Forensics, 2nd Edition*, McGraw Hill/Osborne, USA, 2003):
- Accurately describe the details of an incident
- Be understandable to decision-makers
- Be able to withstand a barrage of legal scrutiny
- Be unambiguous and not open to misinterpretation
- Be easily referenced
- Contain all information required to explain conclusions reached
- Offer valid conclusions, opinions or recommendations when needed
- Be created in a timely manner

The report should also identify the organization, sample reports and restrictions on circulation (if any) and include any reservations or qualifications that the IS auditor has with respect to the assignment.

Protection of Evidence and Chain of Custody
The evidence of a computer crime exists in the form of log files, file time stamps, contents of memory, etc. Rebooting the system or accessing files could result in such evidence being lost, corrupted or overwritten. Therefore, one of the first steps taken should be copying one or more images of the attacked system. Memory content should also be dumped to a file before rebooting the system. Any further analysis must be performed on an image of the system and on copies of the memory dumped—not on the original system in question.

In addition to protecting the evidence, it is also important to preserve the chain of custody. Chain of custody refers to documenting, in detail, how evidence is handled and maintained, including its ownership, transfer and modification. This is necessary to satisfy legal requirements that mandate a high level of confidence regarding the integrity of evidence.

5.6 AUDITING NETWORK INFRASTRUCTURE SECURITY

When performing an audit of the network infrastructure, the IS auditor should:
- Review network diagrams (campus LAN networks, WANS, metropolitan area networks [MANs]) that identify the organizations internetworking infrastructure, which would include gateways, firewalls, routers, switches, hubs, access servers, modems, etc. This information is important because the

IS auditor will want to evaluate these links to determine whether proper physical and logical access controls are in effect and to inventory the various terminal connections to ensure that the diagram is accurate.
- Identify the network design implemented, including the IP strategy used, segmentation of routers and switches for campus environments, and WAN configurations and protocols.
- Determine that applicable security policies, standards, procedures and guidance on network management and usage exist and have been distributed to staff, network management and administration. He/she should also ensure that staff members have been trained in their duties and responsibilities.
- Identify who is responsible for security and operation of Internet connections, and evaluate whether they have sufficient knowledge and experience to undertake this role.
- Determine whether consideration has been given to the legal problems arising from use of the Internet. Considerations should include liability arising from inaccurate web pages; legislation regarding sale or advertising of regulated products, such as financial services; implications of selling/buying in different countries; and the state of application of standard contract terms to electronic trading.
- Determine whether a vulnerability scanning process is in place. Vulnerability scanning refers to an automated process to proactively identify security weaknesses in a network or individual system. It can detect known vulnerabilities and recommend patches, upgrades, fixes or workarounds.
- If the service is outsourced, review SLAs to ensure that they include provisions for security in addition to availability and quality of service.
- Review network administrator procedures to ensure that hardware and software components are upgraded in response to new vulnerabilities.
- Review the transmission medium used for the LAN and its physical security protection to establish vulnerability to wiretapping.
- Review the network topological design to ensure it is sufficiently resilient to maintain business continuity in the event of disruption (e.g., a ring network is more resilient than a star).
- Review the network design to identify single points of failure, such as all WAN connections entering a building at the same place.

5.6.1 AUDITING REMOTE ACCESS

Remote use of information resources dramatically improves business productivity but generates control issues and security concerns. IS auditors should determine that all remote access capabilities used by an organization provide for effective security of the organization's information resources. Remote access security controls should be documented and implemented for authorized users operating outside of the trusted network environment.

In reviewing existing remote access architectures, IS auditors should assess remote access points of entry in addressing how many (known/unknown) exist and whether greater centralized control of remote access points is needed. IS auditors should also review access points for appropriate controls, such as in the use of VPNs, authentication mechanisms, encryption, firewalls and IDSs.

As part of this review, the IS auditor should also test dial-up access controls. To test for dial-up access authorization, the IS auditor should dial the computer from a number of authorized and unauthorized telephone lines. If controls are adequate, successful connection will occur with the authorized numbers only. The IS auditor should test the logical controls invoked after authorized connections to the computer are achieved by using the successful dial-up connections to attempt to gain unauthorized file access. Performance of this test should be coordinated through the security administrator to avoid violating security regulations.

In reviewing future remote access initiatives, IS auditors should first determine whether design and development of remote access approaches are based on a cost-effective, risk-based solution taking into account business requirements. This includes assessing the types of remote environments applicable, the integrity and availability of telecommunication services, and required measures to take in protecting the corporate infrastructure.

Auditing Internet Points of Presence

When auditing an organization's presence on the Internet, the IS auditor should review the use of the Internet to ensure that a business case has been demonstrated for the following possible uses:
- Email (i.e., communications to/from business partners, customers and the general public)
- Marketing (e.g., mechanism for communicating customer values such as online home shopping catalogue)
- Sales channel/electronic commerce (e.g., electronic ordering of goods/services, purchasing of goods from home shopping catalogue using credit cards or electronic transmission of standard electronic data interchange (EDI)-formatted order messages by business partners)
- Channel of delivery for goods/services (such as online bookstores and Internet banking)
- Information gathering (e.g., staff browsing the web for information)

Network Penetration Tests

Combinations of procedures, whereby an IS auditor uses the same techniques as a hacker, are called penetration tests, intrusion tests or ethical hacking. These are effective methods of identifying the real-time risk to an information processing environment. During penetration testing, an auditor attempts to circumvent the security features of a system and exploits the vulnerabilities to gain access that would otherwise be unauthorized.

Scope can vary based on the terms and conditions of the client and requirements. However, from an audit risk perspective, the following should be mentioned clearly in the audit scope:
- Precise IP addresses/ranges to be tested
- Host restricted (i.e., hosts not to be tested)
- Acceptable testing techniques (i.e., social engineering, DoS/distributed denial of service [DDoS], SQL injections, etc.)
- Acceptance of proposed methodology from management
- Timing of attack simulation (i.e., business hours, off hours, etc.)
- IP addresses of the source of attack simulation (to identify between approved simulated attack and actual attack)
- Point of contact for both the penetration tester/auditor and the targeted system owner/administrator

- Handling of information collected by the penetration tester/auditor (i.e., nondisclosure agreement [NDA] or reference to standard rules of engagement)
- Warning notification from penetration tester/auditor, before the simulation begins to avoid false alarms to law enforcement bodies

The different phases of penetration testing appear in **figure 5.22** and the corresponding procedures in **figure 5.23**.

Penetration testing is intended to mimic an experienced hacker attacking a live site. It should only be performed by experienced and qualified professionals who are aware of the risk of undertaking such work and can limit the damage resulting from a successful break-in to a live site (e.g., avoidance of DoS attacks). It is a simulation of a real attack and maybe restricted by the law, an organization's policy and federal regulations; therefore, it is imperative to obtain management's consent in writing before finalization of the test/engagement scope.

Figure 5.22—Penetration Testing Phases

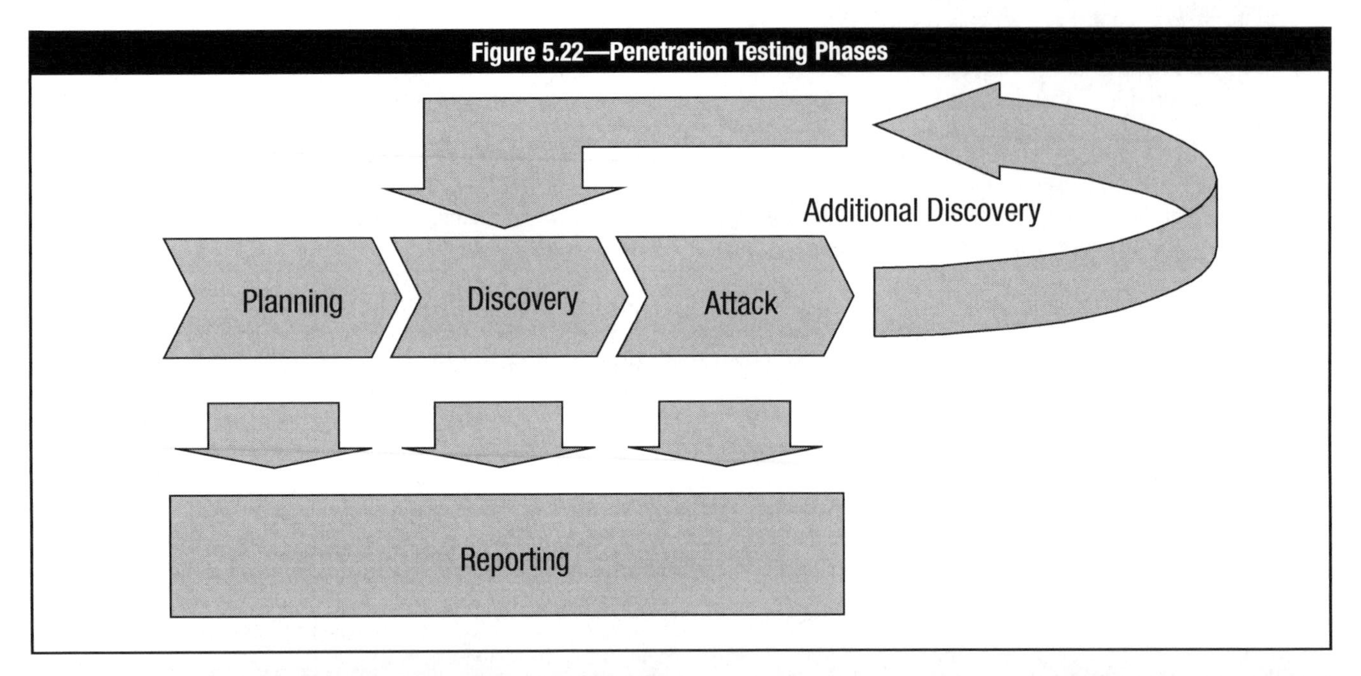

Figure 5.23—Penetration Testing Phases and Procedures	
Phase	**Procedures**
Planning	• Rules of engagement • Management approval/finalization • Adopted testing methodology • Intrusive or nonintrusive testing • Goals/objectives identified and agreed upon • Timelines/deadlines agreed upon • Milestones identified • Assignment time tracking technique understood and communicated • Deliverables agreed upon • Tools collected/installed/tested in a test environment
Reconnaissance/discovery	• Network mapping • DNS interrogation • WHOIS queries • Searching target's web site for information • Searching target's related data on search engines • Searching target's related data and employees on social media (reveals system related details) • Searching resume/curriculum vitae of target's current and formal employees (reveals system related details) • Packet capture/sniffing (during internal testing only) • Host detection (Internet control message protocol [ICMP], DNS, WHOIS, PingSweep, TCP/UDP Sweep, etc.) • Service detection (port scanning, stealth scanning, error/banner detection, etc.) • Network topology detection (ICMP, etc.) • OS detection (TCP stack analysis, etc.) • Web site mapping • Web page analysis

Figure 5.23—Penetration Testing Phases and Procedures *(cont.)*	
Phase	**Procedures**
Reconnaissance/discovery *(cont.)*	• Unused pages/scripts • Broken links • Hidden links/files accessible • Application logic/use • Points of input error page banner grabbing • Vulnerability classification (based on information collected in previous steps, vulnerabilities are searched on available search engines or custom-built repositories) Some of the attack techniques are as follows: • Directory browsing • Show code • Error injection • Type and bound checks on input
Attacks	• Special character injection (meta-characters, escape characters, etc.) • Cookie/session IDs analysis • Authentication circumvention • Long input • System functions (shell escapes, etc.) • Logic alteration (SQL injection, etc.) • Cookie/session IDs manipulation • Internet service exploits (bind, mdac, unicode, apache-http, statd, sadmind, etc.) • OS exploits • Network exploits (SYN flooding, ICMP redirects, DNS poisoning, etc.) Furthermore, once an attack is successful, it typically follows these subprocedures of an attack phase: • Privilege escalation—If only a user-level access was gained previously, then the tester will attempt to obtain super-level access (i.e., root on UNIX/Linux and administrator on Windows) • Information gathering from inside—The attacker will probe further systems on the network efficiently utilizing the compromised system as a launch pad and thereby attempt to gain access to trusted/high-risk systems. • Installation of further attack tools inside the system—Attacker may require installation of additional tools and penetration testing software to gain further access to the resources, trusted or high-risk systems.
Reporting	This phase simultaneously occurs with the rest of the three phases. In the planning phase, rules of engagement, written consent and test plans are developed, discussed and reported. In the discovery phase, written logs are kept and periodic reports on the status of assignment are reported to management, as appropriate. Following the attack phase, the vulnerabilities and weaknesses discovered are reported with risk rating based on probability derived from ease of exploitation and impact derived from attack results or official advisories and resources from the vendor. In addition, the recommendations contain steps to mitigate the risk and to effectively rectify the weaknesses.

There are several types of penetration tests depending upon the scope, objective and nature of the test. Generally accepted and common types of penetration tests are:
- **External testing**—Refers to attacks and control circumvention attempts on the target's network perimeter from outside the target's system (i.e., usually the Internet)
- **Internal testing**—Refers to attacks and control circumvention attempts on the target from within the perimeter. The objective is to identify what would occur if the external network perimeter was successfully compromised and/or an authorized user from within the network wanted to compromise security of a specific resource on the network.
- **Blind testing**—Refers to the condition of testing when the penetration tester is provided with limited or no knowledge of the target's information systems. Such testing is expensive, because penetration testers have to research the target and profile it based on publicly available information only.
- **Double blind testing**—Refers to an extension of blind testing, because the administrator and security staff at the target are also not aware of the test. Such testing can effectively evaluate the incident handling and response capability of the target.
- **Targeted testing**—Refers to attacks and control circumvention attempts on the target, while both the target's IT team and penetration testers are aware of the testing activities. Penetration testers are provided with information related to target and network design. Additionally, they may also be provided with a limited-privilege user account to be used as a starting point to identify privilege-escalation possibilities in the system.

Although management may sponsor the activities of penetration testing, some of the associated risk includes the following:
- Penetration testing does not provide assurance that all vulnerabilities are discovered and may fail to discover significant vulnerabilities.

- Miscommunication may result in the test objectives not being achieved.
- Testing activities may inadvertently trigger escalation procedures that may not have been appropriately planned.
- Sensitive information may be disclosed, heightening the target's exposure level.
- Without proper background and qualification checks of penetration testers, the penetration tester may damage the information assets or misuse the information gained for personal benefits.

Additionally, these techniques are becoming more popular for testing the reliability of firewall access controls. The IS auditor should be extremely careful if attempting to break into a live production system because, if successful, the IS auditor may cause the system to fail. Permission for the use of such techniques should always be obtained from top-level senior management. Permission from top-level senior management is also required to determine what other tests can be performed without informing the staff who are responsible for the monitoring and reporting of security violations (if any are aware that the attack will take places, they are likely to be more vigilant than normal).

Full Network Assessment Reviews

Upon completion of penetration testing, comprehensive review of all network system vulnerabilities should occur to determine whether threats to confidentiality, integrity and availability have been identified. The following reviews should occur:

- Security policy and procedures should be reviewed to determine good practices are in place.
- The network and firewall configuration should be evaluated to ensure that they have been designed to support the security of the services being provided (e.g., screening routers, dual/multihomed host, screened subnet, demilitarized zone proxy servers).
- The logical access controls should be evaluated to ensure that they support segregation of duties (e.g., development vs. operation, security administration vs. audit).
- The following should be determined:
  - Intrusion detection software is in place.
  - Filtering is being performed.
  - Encryption is being used (consider VPNs/tunneling, digital signatures for email, etc.).
  - Strong forms of authentication are being used (consider use of smart cards, biometrics, etc., for authentication to firewalls, to internal software/hardware within the network, and to external hardware/software).
  - The firewalls have been configured properly (consider removal of all unnecessary software, addition of security and auditing software, removal of unnecessary logon IDs, disabling of unused services).
  - The application- or circuit-level gateways in use are running proxy servers for all legitimate services (e.g., teletype network [Telnet], HTTP, FTP).
  - Virus scanning is being used.
  - Periodic penetration testing is being completed.
  - Audit logging is undertaken for all key systems (e.g., firewalls, application gateways, routers, etc.) and audit logs are copied to secure file systems (consider the use of SIEM software).
  - The security administrators are keeping up to date with the latest known vulnerabilities via the organizations' vendors, their local and international CERT, and vulnerability databases (e.g., the National Vulnerability Database operated by the NIST).

Development and Authorization of Network Changes

Network configuration changes to update telecommunications lines, terminals, modems and other network devices should be authorized in writing by management and implemented in a timely manner. The IS auditor can test this change control by:

- Sampling recent change requests, looking for appropriate authorization and matching the request to the actual network device
- Matching recent network changes, such as new telecommunication lines, to added terminals and authorized change requests

As an added control, the IS auditor should determine who can access the network change software. This access should be restricted to senior network administrators.

Specific development and change control procedures should be in place for network components' hardware and software. Procedures should cover:

- Firewalls
- Routers
- Switches
- Application gateways
- DNS/network topology
- Client software
- Network management software
- Web server hardware and configuration
- Application software
- Web pages

Unauthorized Changes

One of the most important objectives of change control procedures is to prevent or detect unauthorized changes to software, configurations or parameters, and data. Unauthorized changes include any changes to software or configurations/parameters that occur without conforming to change control procedures. They include situations where changes are made to software code without authorization, in addition to legitimate changes made in accordance with change control procedures.

Controls to prevent unauthorized changes to software and software configurations include:

- SoD between software development, software administration and computer operations
- Restricting the software development team's access to the development environment only
- Restricting access to the software source codes

Controls to detect unauthorized changes to software include software code comparison utilities. Unauthorized changes to configurations/parameters can be detected through logging and monitoring system administrator activities.

Changes to data normally are controlled through the applications. Application access control mechanisms and built in application controls normally prevent unauthorized access to data. These controls can be circumvented by direct access to data. For this reason, direct access to data (specifically "write" or "change" access) should be restricted and monitored.

5.7 ENVIRONMENTAL EXPOSURES AND CONTROLS

As with any other manmade objects, IT infrastructure and, hence, information assets are exposed to the environment. The IS auditor should be aware of these exposures and the controls used to mitigate them.

5.7.1 ENVIRONMENTAL ISSUES AND EXPOSURES

Environmental exposures are due primarily to naturally occurring events such as lightning storms, earthquakes, volcanic eruptions, hurricanes, tornados and other types of extreme weather conditions. The result of such conditions can lead to many types of problems. One particular area of concern is power failures of computer and supporting environmental systems. Generally, power failures can be grouped into four distinct categories, based on the duration and relative severity of the failure:

- **Total failure (blackout)**—A complete loss of electrical power, which may span from a single building to an entire geographical area and is often caused by weather conditions (e.g., storm, earthquake) or the inability of an electrical utility company to meet user demands (e.g., during summer months)
- **Severely reduced voltage (brownout)**—The failure of an electrical utility company to supply power within an acceptable range (i.e., 108-125 volts AC in the US). Such failure places a strain on electronic equipment and may limit their operational life or even cause permanent damage.
- **Sags, spikes and surges**—Temporary and rapid decreases (sags) or increases (spikes and surges) in voltage levels. These anomalies can cause loss of data, data corruption, network transmission errors or physical damage to hardware devices (e.g., hard disks or memory chips).
- **Electromagnetic interference (EMI)**—Caused by electrical storms or noisy electrical equipment (e.g., motors, fluorescent lighting, radio transmitters). This interference may cause computer systems to hang or crash as well as damages similar to those caused by sags, spikes and surges.

Short-term interruptions, such as sags, spikes and surges, which last from a few millionths to a few thousandths of a second, can be prevented by using properly placed surge protectors. Intermediate-term interruptions, which last from a few seconds to 30 minutes, can be controlled by UPS devices. Finally, long-term interruptions, which last from a few hours to several days, require the use of alternate power generators. These generators may be portable devices or part of the building's infrastructure and are powered by alternative sources of energy such as diesel, gasoline or propane.

Another area of concern deals with water damage/flooding. This is a concern even with facilities located on upper floors of high-rise buildings because water damage typically occurs from broken water pipes.

Manmade concerns include terrorist threats/attacks, vandalism, electrical shock and equipment failure.

Some questions that organizations must address related to environmental issues and exposures include the following:

- Is the power supply to the computer equipment properly controlled to ensure that power remains within the manufacturer's specifications?
- Are the air conditioning, humidity and ventilation control systems for the computer equipment adequate to maintain temperatures within manufacturers' specifications?
- Is the computer equipment protected from the effects of static electricity, using an antistatic rug or antistatic spray?
- Is the computer equipment kept free of dust, smoke and other particulate matter such as food?
- Do policies exist that prohibit the consumption of food, beverage and tobacco products near computer equipment?
- Are backup media protected from damage due to temperature extremes, the effects of magnetic fields and water damage?

5.7.2 CONTROLS FOR ENVIRONMENTAL EXPOSURES

Environmental exposures should be afforded the same level of protection as physical and logical exposures.

Alarm Control Panels

An alarm control panel should ideally be:

- Separated from burglar or security systems located on the premises
- Accessible to fire department personnel at all times
- Located in a weatherproof box
- In accordance with temperature requirements set by the manufacturer
- Situated in a controlled room to prevent access by unauthorized personnel
- Allocated power from a dedicated and separate circuit
- Able to control or disable separate zones within the facilities
- In adherence with local and national regulations and approved by local authorities

Water Detectors

In the computer room, water detectors should be placed under raised floors and near drain holes, even if the computer room is on a high floor (because of possible water leaks). Any unattended equipment storage facilities should also have water detectors. When activated, the detectors should produce an audible alarm that can be heard by security and control personnel. The location of the water detectors should be marked on the raised computer room floor for easy identification and access. On hearing the alarm, specific individuals should be responsible for investigating the cause and initiating remedial action; other staff should be made aware by security and control personnel that there is a risk of electric shock.

Handheld Fire Extinguishers

Fire extinguishers should be in strategic locations throughout the facility. They should be tagged for inspection and inspected at least annually.

Manual Fire Alarms

Hand-pull fire alarms should be placed strategically throughout the facility. These are normally located near exit doors to ensure personnel safety. The resulting audible alarm should be linked to a monitored guard station.

Smoke Detectors

Smoke detectors should be installed above and below the ceiling tiles throughout the facilities and below the raised computer room floor. The detectors should produce an audible alarm when activated and be linked to a monitored station (preferably by the

fire department). The location of the smoke detectors above the ceiling tiles and below the raised floor should be marked on the tiling for easy identification and access. Smoke detectors should supplement, not replace, fire suppression systems.

Fire Suppression Systems

These systems are designed to automatically activate immediately after detection of high heat, typically generated by fire. Like smoke detectors, the system should produce an audible alarm when activated and be linked to a central guard station that is regularly monitored. The system should also be inspected and tested annually. Testing intervals should comply with industry and insurance standards and guidelines. Ideally, the system should automatically trigger other mechanisms to localize the fire. This includes closing fire doors, notifying the fire department, closing off ventilation ducts and shutting down nonessential electrical equipment. In addition, the system should be segmented so a fire in one part of a large facility does not activate the entire system.

Broadly speaking, there are two methods for applying an extinguishing agent: total flooding and local application.

Systems working under *a total flooding* principle apply an extinguishing agent to a three-dimensional enclosed space in order to achieve a concentration of the agent (volume percent of the agent in air) adequate to extinguish the fire. These types of systems may be operated automatically by detection and related controls or manually by the operation of a system actuator.

Systems working under a *local application* principle apply an extinguishing agent directly onto a fire (usually a two-dimensional area), or into the three-dimensional region immediately surrounding the substance or object on fire. The main difference between local application and total flooding designs is the absence of physical barriers enclosing the fire space in the local application design.

In the context of automatic extinguishing systems, local application does normally not refer to the use of manually operated wheeled or portable fire extinguishers, although the nature of the agent delivery is similar.

The medium for fire suppression varies, but is usually one of the following:

- **Water-based systems** are typically referred to as sprinkler systems. These systems are effective but are also unpopular because they damage equipment and property. The system can be dry-pipe or charged (water is always in the system piping). A charged system is more reliable but has the disadvantage of exposing the facility to expensive water damage if the pipes leak or break.
- **Dry-pipe sprinkling systems** do not have water in the pipes until an electronic fire alarm activates the water pumps to send water into the system. This is opposed to a fully charged water pipe system. Dry-pipe systems have the advantage that a failure in the pipe will not result in water leaking into sensitive equipment from above. Because water and electricity do not mix, these systems must be combined with an automatic switch to shut down the electricity supply to the area protected.
- **Halon systems** release pressurized Halon gases that remove oxygen from the air, thus starving the fire. Halon was popular because it is an inert gas and does not damage equipment like water does. There should be an audible alarm and brief delay before discharge to permit personnel time to evacuate the area or to override and disconnect the system. Halon systems were prevalent for many years, but because Halon adversely affects the ozone layer, it was banned by the Montreal (Canada) Protocol of 1987. As a banned gas, all Halon installations are required by international agreements to be removed. Popular replacements are FM-200® and Argonite®.
- **FM-200®**—also called heptafluoropropane, HFC-227 or HFC-227ea (ISO name)—is a colorless odorless gaseous halocarbon, which is safe to be used where people are present. It is commonly used as a gaseous fire suppression agent. The HFC-227 fire suppression agent was the first nonozone-depleting replacement for Halon 1301. In addition, HFC-227 leaves no residue on valuable equipment after discharge. Trade names include FE-227™ (DuPont™), FM-200® (DuPont) and Solkaflam® 227 (Solvay Chemicals). This agent suppresses fire by discharging as a gas onto the surface of combusting materials. Large amounts of heat energy are absorbed from the surface of the burning material, lowering its temperature below the ignition point. FM-200 fire suppression systems have low atmospheric lifetimes, global warming and ozone depletion potentials.
- **Argonite®** is the brand name (a registered trademark owned by Ginge-Kerr) for a mixture of 50 percent argon (Ar) and 50 percent nitrogen (N2). It is an inert gas used in gaseous fire suppression systems for extinguishing fires where damage to equipment is to be avoided. Although argon is nontoxic, it does not satisfy the body's need for oxygen and is a simple asphyxiant. People have suffocated by breathing argon by mistake.

  Unlike carbon dioxide (CO_2) fire suppression systems, which are unable to sustain human life, FM-200 and Argonite systems are environmentally friendly. They provide an effective, safe method of fire suppression, and unlike charged sprinkler systems, they do not suffer from the disadvantage of exposing the expensive information processing facility (IPF) to water damage during the firefighting or if the pipe leaks or breaks.
- CO_2 systems release pressurized CO_2 gas into the area protected to replace the oxygen required for combustion. Unlike Halon and its later replacements, CO_2 is unable to sustain human life. Therefore, in most countries, it is illegal for such systems to be set to automatic release if any human may be in the area. Because of this, these systems are usually discharged manually, introducing an additional delay in combating the fire. CO_2 installations are permitted where no humans are regularly present, such as unmanned data centers (or "dark sites"), but there should be an automated facility for shutting down the system when anyone enters the area.

Strategically Locating the Computer Room

To reduce the risk of flooding, the computer room should not be located in the basement or top floor. If located in a multistory building, studies show that the best location for the computer room—the location which reduces the risk of fire, smoke and water damage—is on the middle floors (e.g., third, fourth, fifth or sixth floor). Adjacent water or gas pipes should be avoided except in the case of fire suppression systems. Care should be taken to avoid locating computer rooms adjacent to areas where functions carrying a high risk are carried out, such as paper storage. The activity of neighboring organizations should be considered when establishing a computer facility. Locations adjacent to, or on the final path to, an airport or a chemical works where explosive gases may be present, for example, should be avoided.

Where a data center is already located in an area vulnerable to flooding, such as a basement, an alternative to costly removal is the provision of a plastic sheet, or "umbrella," covering the area, which diverts any water flow away from the sensitive equipment.

Regular Inspection by Fire Department

To ensure that all fire detection systems comply with building codes, the fire department should inspect the system and facilities annually. Also, the fire department should be notified of the location of the computer room, so it can be prepared with equipment appropriate for electrical fires.

Fireproof Walls, Floors and Ceilings of the Computer Room

Walls surrounding the information processing facility should contain or block fire from spreading. The surrounding walls should be from true floor to the true ceiling and should have at least a two-hour fire resistance rating.

Electrical Surge Protectors

These electrical devices reduce the risk of damage to equipment due to power spikes. Voltage regulators measure the incoming electrical current and either increase or decrease the charge to ensure a consistent current. Such protectors are typically built into the UPS system.

Uninterruptible Power Supply/Generator

A UPS system consists of a battery or gasoline-powered generator that interfaces with the electrical power entering the facility and the electrical power entering the computer. The system typically cleanses the power to ensure that voltage into the computer is consistent. The UPS continues providing electrical power from the generator to the computer for a defined length of time should a power failure occur. Depending on the sophistication of the UPS, electrical power could continue to flow for days or for just a few minutes to permit an orderly computer shutdown. A UPS system can be built into a computer or can be an external piece of equipment.

Emergency Power-off Switch

There may be a need to immediately shut off power to the computer and peripheral devices such as during a computer room fire or emergency evacuation. Two emergency power-off switches should serve this purpose—one in the computer room, the other near, but outside, the computer room.

Switches should be clearly labeled and easily accessible, for this purpose, and yet they should still be secure from unauthorized people. The switches should be shielded to prevent accidental activation.

Power Leads From Two Substations

Electrical power lines that feed into the facility are exposed to many environmental hazards—water, fire, lightning, cutting due to careless digging, etc. To reduce the risk of a power failure due to these events that, for the most part, are beyond the control of the organization, redundant power lines should feed into the facility. In this way, interruption of one power line does not adversely affect electrical supply.

Fully Documented and Tested Business Continuity Plan

See section 2.12.10 Plan Testing for a description of testing BCPs.

Wiring Placed in Electrical Panels and Conduit

To reduce the risk of an electrical fire occurring and spreading, wiring should be placed in fire-resistant panels and conduit. This conduit generally lies under the fire-resistant raised computer room floor.

Inhibited Activities Within the Information Processing Facility

Food, drink and tobacco use can cause fires, the buildup of contaminants or damage to sensitive equipment (especially in the case of liquids). They should be prohibited from the information processing facility (IPF). This prohibition should be overt, such as with a sign on the entryway.

Fire-resistant Office Materials

Wastebaskets, curtains, desks, cabinets and other general office materials in the IPF should be fire-resistant. Cleaning fluids for desktops, console screens and other office furniture/fixtures should not be flammable.

Documented and Tested Emergency Evacuation Plans

Evacuation plans should emphasize human safety but should not leave IPFs physically unsecured. Procedures should be in place for a controlled shutdown of the computer in an emergency situation, if time permits.

5.7.3 AUDITING ENVIRONMENTAL CONTROLS

The following testing procedures should also be applied to any offsite storage and processing facilities. When this facility is outsourced to a third party, a contractual right of audit may be required. Assurance may also be provided through other methods, such as through Service Organization Control (SOC) reports.

Note that an IS auditor's first concern should be to establish the environmental risk by assessing the location of the data center. Higher risk items have a greater need for specific environmental controls.

Water and Smoke Detectors

Visual verification of the presence of water and smoke detectors in the computer room is needed. Whether the power supply to these detectors is sufficient should be determined, especially in instances of battery-operated devices. Also, the locations of the devices should be placed to give early warning of a fire, such as immediately above the computer equipment they are protecting, and should be clearly marked and visible.

Handheld Fire Extinguishers

Handheld fire extinguishers should be in strategic highly visible locations throughout the facility and should be inspected annually.

Fire Suppression Systems
Fire suppression systems are expensive to test and, therefore, the IS auditor's ability to determine operability is limited. IS auditors may need to limit their tests to reviewing documentation to ensure that the system has been inspected and tested within the last year. The exact testing interval should comply with industry and insurance standards and guidelines.

Regular Inspection by Fire Department
The person responsible for fire equipment maintenance should be contacted and asked if a local fire department inspector or insurance evaluator has been recently invited to tour and inspect the facilities. If so, a copy of the report should be obtained, and how to address the noted deficiencies should be determined.

Fireproof Walls, Floors and Ceilings of the Computer Room
With the assistance of building management, the documentation that identifies the fire rating of the walls surrounding the IPF should be located. These walls should have at least a two-hour fire resistance rating.

Electrical Surge Protectors
The presence of electrical surge protectors on sensitive and expensive computer equipment should be visually observed.

Power Leads From Two Substations
With the assistance of building management, documentation concerning the use and placement of redundant power lines into the IPF should be located.

Fully Documented and Tested Business Continuity Plan
See section 2.12.10 Plan Testing for a description of testing BCPs.

Wiring Placed in Electrical Panels and Conduit
Wiring in the IPF should be placed in fire-resistant panels and conduit.

UPS/Generator
The most recent test date should be determined and the test reports should be reviewed.

Documented and Tested Emergency Evacuation Plans
A copy of the emergency evacuation plan should be obtained. It should be examined to determine whether it describes how to leave the IPFs in an organized manner that does not leave the facilities physically insecure. A sample of IS employees should be interviewed to determine if they are familiar with the documented plan. The emergency evacuation plans should be posted throughout the facilities.

Humidity/Temperature Control
The IPF should be visited on regular intervals to determine whether temperature and humidity are adequate.

5.8 PHYSICAL ACCESS EXPOSURES AND CONTROLS

Physical exposures could result in financial loss, legal repercussions, loss of credibility or loss of competitive edge. They primarily originate from natural and man-made hazards, and can expose the business to unauthorized access and unavailability of the business information.

5.8.1 PHYSICAL ACCESS ISSUES AND EXPOSURES
Physical access issues are a major concern in security. Exposures and possible perpetrators are described in the following subsections.

Physical Access Exposures
Exposures that exist from accidental or intentional violation of these access paths include:
- Unauthorized entry
- Damage, vandalism or theft to equipment or documents
- Copying or viewing of sensitive or copyrighted information
- Alteration of sensitive equipment and information
- Public disclosure of sensitive information
- Abuse of data processing resources
- Blackmail
- Embezzlement

Possible Perpetrators
Possible perpetrators include employees with authorized or unauthorized access who are:
- Disgruntled (upset by or concerned about some action by the organization or its management)
- On strike
- Threatened by disciplinary action or dismissal
- Addicted to a substance or gambling
- Experiencing financial or emotional problems
- Notified of their termination

Other possible perpetrators could include:
- Former employees
- Interested or informed outsiders such as competitors, thieves, organized crime and hackers
- An accidental ignorant (e.g., someone who unknowingly perpetrates a violation)

The most likely source of exposure is from the uninformed, accidental or unknowing person, although the greatest impact may be from those with malicious or fraudulent intent.

Other questions and concerns to consider include the following:
- Are hardware facilities reasonably protected against forced entry?
- Are keys to the computer facilities adequately controlled to reduce the risk of unauthorized access?
- Are computer terminals locked or otherwise secured to prevent removal of boards, chips and the computer itself?
- Are authorized equipment passes required before computer equipment can be removed from its normal secure surroundings?

From an IS perspective, facilities to be protected include:
- Programming area
- Computer room
- Operator consoles and terminals
- Tape library, tapes, disks and all magnetic media
- Storage rooms and supplies
- Offsite backup file storage facility
- Input/output control room
- Communications closets
- Telecommunications equipment (including radios, satellites, wiring, modems and external network connections)
- Microcomputers and PCs
- Power sources
- Disposal sites
- Minicomputer establishments
- Dedicated telephones/telephone lines
- Control units and front-end processors
- Portable equipment (handheld scanners and coding devices, bar code readers, laptop computers, printers, pocket LAN adapters and others)
- Onsite and remote printers
- Local area networks

Additionally, system, infrastructure or software application documentation should be protected against unauthorized access.

For these safeguards to be effective, they must extend beyond the computer facility to include any vulnerable access points within the entire organization and at organizational boundaries/interfaces with external organizations. This may include remote locations and rented, leased or shared facilities. Additionally, the IS auditor may require assurances that similar controls exist within service providers or other third parties, if they are potentially vulnerable access points to sensitive information within the organization.

5.8.2 PHYSICAL ACCESS CONTROLS

Physical access controls are designed to protect the organization from unauthorized access. These controls should limit access to only those individuals authorized by management. This authorization may be explicit, as in a door lock for which management has authorized you to have a key, or implicit, as in a job description that implies a need to access sensitive reports and documents.

Bolting door locks require the traditional metal key to gain entry. The key should be stamped "do not duplicate" and should be stored and issued under strict management control.

Combination door locks (cipher locks) use a numeric key pad or dial to gain entry and are often seen at airport gate entry doors and smaller server rooms. The combination should be changed at regular intervals or whenever an employee with access is transferred, fired or subject to disciplinary action. This reduces the risk of the combination being known by unauthorized people.

Electronic door locks use a magnetic or embedded chip-based plastic card key or token entered into a sensor reader to gain access. A special code internally stored in the card or token is read by the sensor device that then activates the door locking mechanism. Electronic door locks have the following advantages over bolting and combination locks:
- Through the special internal code, cards can be assigned to an identifiable individual.
- Through the special internal code and sensor devices, access can be restricted based on the individual's unique access needs. Restrictions can be assigned to particular doors or to particular hours of the day.
- They are difficult to duplicate.
- Card entry can be easily deactivated in the event an employee is terminated or a card is lost or stolen. Silent or audible alarms can be automatically activated if unauthorized entry is attempted. Issuing, accounting for and retrieving the card keys is an administrative process that should be carefully controlled. The card key is an important item to retrieve when an employee leaves the firm. An example of a common technique used for card entry is the swipe card. A swipe card is a physical control technique that uses a plastic card with a magnetic strip containing encoded data to provide access to restricted or secure locations. The encoded data can be read by a slotted electronic device. After a card has been swiped, the application attached to the slotted electronic device prevents unauthorized physical access to those sensitive locations, as well as logs all card users that try to gain access to the secure location.

Biometric door locks are activated by an individual's unique body features, such as voice, retina, fingerprint, hand geometry or signature. This system is used in instances when extremely sensitive facilities must be protected such as in the military.

Manual logging means all visitors are required to sign a visitor's log indicating their name, company representative, reason for visiting, person to see and date and time of entry and departure. Logging is typically done at the front reception desk and entrance to the computer room. Before gaining access, visitors should also be required to provide verification of identification such as a driver's license, business card or vendor identification tag.

Electronic logging is a feature of electronic and biometric security systems. All access can be logged, with unsuccessful attempts being highlighted.

Identification badges (photo IDs) should be worn and displayed by all personnel. Visitor badges should be a different color from employee badges for easy identification. Sophisticated photo IDs can also be used as electronic card keys. Issuing, accounting for and retrieving the badges is an administrative process that must be carefully controlled.

Video (CCTV) cameras should be located at strategic points and monitored by security guards. Sophisticated video cameras can be activated by motion. The video surveillance recording should be retained for possible future playback, and it should be recorded in sufficient resolution to permit enlarging the image to identify an intruder.

Security guards are very useful if supplemented by video cameras and locked doors. Guards supplied by an external agency should be bonded to protect the organization from loss.

Controlled visitor access means all visitors should be escorted by a responsible employee. Visitors include friends, maintenance personnel, computer vendors, consultants (unless long-term, in which case special guest access may be provided) and external auditors.

All service contract personnel, such as cleaning people and offsite storage services, should be **bonded personnel**. This does not improve physical security but limits the financial exposure of the organization.

Deadman doors, also referred to as a mantrap or airlock entrance, uses two doors and is typically found in entries to facilities, such as computer rooms and high-security areas. For the second door to operate, the first entry door must close and lock, with only one person permitted in the holding area. This reduces the risk of piggybacking, when an unauthorized person follows an authorized person through a secured entry. In some installations, this same effect is accomplished through the use of a full height turnstile. Deadman doors may also be used for delivery and dispatch areas where outer doors open to admit a truck and the inner doors cannot be opened to load or unload until the outer doors are closed and locked.

Computer workstation locks secure the device to the desk, prevent the computer from being turned on or disengage keyboard recognition, thus preventing use. Another available feature is locks that prevent turning on a PC workstation until a key lock is unlocked by a turnkey or card key. This is sometimes seen in the case of high-security workstations, such as those that process payroll.

A **controlled single entry point**, monitored by a receptionist, should be used by all incoming personnel. Multiple entry points increase the risk of unauthorized entry. Unnecessary or unused entry points, such as doors to outside smoking or break areas, should be eliminated. Emergency exits can be wired to an alarmed panic bar for quick evacuation.

An **alarm system** should be linked to inactive entry points, motion detectors, and the reverse flow of enter- or exit-only doors. Security personnel should be able to hear the alarm when activated.

Secured report/document distribution carts such as mail carts, should be covered and locked and should not be left unattended.

Facilities such as computer rooms should not be visible or identifiable from the outside; there should be no windows or directional signs. The building or department directory should discreetly identify only the general location of the information processing facility. If windows are present, they should be constructed of reinforced glass and, if on the ground floor of the building, further protected for example, by bars.

5.8.3 AUDITING PHYSICAL ACCESS

Touring the computer site is useful for the auditor to gain an overall understanding and perception of the installation being reviewed. As with environmental controls where the site is owned by a third party, a contractual right of audit may be required. This tour provides the opportunity to begin reviewing physical access restrictions (e.g., control over employees, visitors, intruders and vendors).

The computer site (i.e., computer room, developers' area, media storage, printer stations and management offices) and any offsite storage facilities should be included in this tour.

Much of the testing of physical safeguards can be achieved by visually observing the previously noted safeguards. Documents to assist with this effort include emergency evacuation procedures, inspection tags (recent inspection?), fire suppression system test results (successful? recently tested?) and key lock logs (all keys accounted for and not outstanding to former employees or consultants?).

Testing should extend beyond the computer room to include the following related facilities:
- Location of all operator consoles
- Printer rooms
- Computer storage rooms (this includes equipment, paper and supply rooms)
- UPS/generator
- Location of all communications equipment identified on the network diagram
- Media storage
- Offsite backup storage facility

To complete a thorough test, the IS auditor should look above the ceiling panels and below the raised floor in the computer operations center, observing smoke and water detectors, general cleanliness, and walls that extend all the way to the real ceiling (not just the fake/suspended ceiling). For a ground-floor computer room, the auditor may also consider walking around the outside of the room, viewing the location of any windows, examining emergency exit doors for evidence that they are routinely used (such as the presence of cigarette stubs or litter) and examining the air conditioning units. The auditor should also consider whether any additional threats exist close to the room, such as storage of dangerous or flammable material.

The following paths of physical entry should be evaluated for proper security:
- All entry doors
- Emergency exit doors
- Glass windows and walls
- Movable walls and modular cubicles
- Above suspended ceilings and beneath raised floors
- Ventilation systems
- Over a curtain, fake wall

5.9 MOBILE COMPUTING

Mobile computing refers to devices that are transported or moved during normal usage. Common mobiles devices include tablets, smartphones, laptops, USB storage devices, digital cameras and other similar technologies. Their mobility makes it more difficult to implement logical and physical access controls.

Figure 5.24 presents some known vulnerabilities and associated threats that need to be understood when dealing with mobile devices.

In addition, the advent of bring your own device (BYOD), where enterprises encourage staff to use their own mobile devices for company business, adds another layer of complexity when protecting these devices.
The following controls will reduce the risk of disclosure of sensitive data stored on mobile devices. Many of these can be enforced by mobile device management (MDM) systems and/or secure containers (a separately authenticated, encrypted area of the mobile device that is used to keep sensitive enterprise data segregated from the personal data) for both corporate and personal devices:

- **Device registration**—All mobile devices authorized for business use should be registered in a database. Devices that are personally owned should be flagged. Organizations can push updates or manage authorized devices and exclude personally owned mobile devices.
- **Tagging**—Physically tagging the device with an asset ID may result in its return should it be lost; however, there is risk in identifying the organization that owns the device.
- **Physical security**—If the device is stationary and permits it, use a cable locking system or a locking system with a motion detector that sounds an audible alarm.
- **Data storage**—Only store what is absolutely needed on the device. With the ability to remotely access central servers the requirement to store any data locally should be questioned. If it is not stored locally, it will not be an issue if the device is lost or stolen. The data that are stored should be backed up on a regular basis, preferably to shared folders on the company's file server.
- **Virus detection and control**—The threat associated with viruses applies to all mobile devices. The enterprise should update the mobile device antivirus software to prevent perpetuation of malware.
- **Encryption**—Mobile devices used to store sensitive or confidential information should be encrypted in accordance with the organization's information security policies, mandating use of a strong encryption mechanism.
- **Compliance**—Mobile devices should comply with the security requirements as defined in corporate standards. All mobile devices should require a password. Two-factor authentication could be used to further enhance security.
- **Approval**—Mobile device use should be appropriately authorized and approved in accordance with the organization's policies and procedures.
- **Acceptable use policy**—A security policy should exist for mobile devices. The enterprise should have a policy addressing mobile device use and specifying the type of information and kind of devices and information services that may be accessible through the devices.

Figure 5.24—Mobile Device Vulnerabilities, Threats and Risk

Vulnerability	Threat	Risk
Information travels across wireless networks that are often less secure than wired networks.	Malicious outsiders can do harm to the enterprise.	Information interception resulting in a breach of sensitive data, enterprise reputation, adherence to regulation or legal action
Mobility provides users with the opportunity to leave enterprise boundaries and thereby eliminates many security controls.	Mobile devices cross boundaries and network perimeters, carrying malware, and can bring this malware into the enterprise network.	Malware propagation, which may result in data leakage, data corruption and unavailability of necessary data
Bluetooth technology is very convenient for many users to have hands-free conversations; however, it is often left on and then is discoverable.	Hackers can discover the device and launch an attack.	Device corruption, lost data, call interception, possible exposure of sensitive information
Unencrypted information is stored on the device.	In the event that a malicious outsider intercepts data in transit or steals a device or if the employee loses the device, the data are readable and usable.	Exposure of sensitive data, resulting in damage to the enterprise, customers or employees
Lost data may affect employee productivity.	Mobile devices may be lost or stolen due to their portability. Data on these devices are not always backed up.	Workers dependent on mobile devices unable to work in the event of broken, lost or stolen devices and data that are not backed up
The device has no authentication requirements applied.	In the event that the device is lost or stolen, outsiders can access the device and all of its data.	Data exposure, resulting in damage to the enterprise and liability and regulation issues
The enterprise is not managing the device.	If no mobile device strategy exists, employees may choose to bring in their own unsecured devices.	Data leakage, malware propagation or unknown data loss in the case of device loss or theft.
The device allows for installation of unsigned third-party applications.	Applications may carry malware that propagates Trojans or viruses; the applications may also transform the device into a gateway for malicious outsiders to enter the enterprise network.	Malware propagation, data leakage or intrusion on enterprise network

Source: ISACA, *Securing Mobile Devices*, USA, 2012

- **Due care**—Employees should exercise due care within office environments and especially during travel. Any loss or theft of a mobile device must be treated as a security breach and reported immediately in accordance to security management policies and procedures.
- **Awareness training**—Employee orientation and security awareness training should include coverage of mobile device policy and guidelines. The training will allow propagation of awareness that mobile devices are important business tools when used properly and have risk associated with them, if not managed accordingly.
- **Network authentication, authorization and accounting**—IT organizations should adopt a solution that allows them to tie devices connecting to the network with each user's identity and role and then apply role-based policies to grant proper access privileges. This enables IT to differentiate access for different levels of employees or guests or even by device type. It also lets IT take a proactive stance on tracking and monitoring how mobile devices are being used within their network.
- **Secure transmission**—Mobile devices should connect to the enterprise network via a secure connection, such as over a VPN.
- **Standard mobile device applications**—Configuration and use of the mobile device should be baselined and controlled. Only applications that either meet with the corporate security architecture or are delivered as standard on the mobile device should be authorized for use, and all software applications must be appropriately licensed and installed by the organization's IS support team. MDM solutions support this.
- **Geolocation tracking**—There are many debates about the privacy concerns of GPS tracking, but location capabilities inherent in mobile devices can be invaluable in the case of loss or theft.
- **Remote wipe and lock**—Due to the nature of mobile devices, many device management solutions are focused on securing the device if it is lost or stolen. Some MDM solutions allow IT to send an alarm to the device to help identify the location for a user, and if truly lost, IT can then remotely wipe and lock the device and/or container.
- **BYOD**—An employee BYOD agreement or acceptable use agreement (AUA) should require the employee to agree with the items in the policy before the device can be used for business purposes. It may also state that devices can be seized if necessary for a legal matter. An AUA ensures that maintaining security when using personal devices is a responsibility that is shared between both the user and IT. In addition, BYOD should be approved by executive management and be subject to oversight and monitoring.
- **Secure remote support**—Employees relying on personal devices to conduct work will often be out of the office. Having a secure way to support and fix these devices from a remote location is imperative to maintain employee satisfaction. Depending on device type, remote support solutions allow help desks to configure devices, chat, transfer files, and even remotely see and control the device. It is important to select a solution that supports a wide variety of devices and keeps all access and activity logs behind the company's firewall to ensure security.

5.10 PEER-TO-PEER COMPUTING

Peer-to-peer (P2P) computing is a distributed architecture where tasks or workloads are shared between peers. In P2P computing, there is no specific server to which one connects. For the most part, the connection is established between two peers—a connection between any two or more systems for a common interest. P2P networks are used almost exclusively for file sharing. Enterprises should strongly consider the risk against any perceived advantages before allowing access to P2P networks (**figure 5.25**).

Figure 5.25—Risk of Peer-to-peer Computing

Threats and Vulnerabilities	Risk	Controls
Introduction of viruses and malware to the organizational network	• Data leakage/theft • "Owned" systems (zombies) • System downtime • Resources required to clean systems	• Ensure that antivirus and anti-malware controls are installed on all systems and updated daily. • Block P2P traffic. • Prevent installation of P2P clients. • Establish or update policies and standards. • Develop and conduct awareness training and campaigns to inform employees of the risk involved with P2P computing.
Copyrighted content held on the enterprises network	• Regulatory sanctions and fines • Adverse legal actions • Licensing issues • Reputational damage	• Block P2P traffic. • Prevent installation of P2P clients. • Establish or update policies and standards. • Develop and conduct awareness training and campaigns to inform employees of the risk involved with P2P computing.
Excessive use of P2P in the workplace	• Network utilization issues • Productivity loss	• Restrict P2P usage. • Establish or update policies and standards. • Develop and conduct awareness training and campaigns to inform employees of the risk involved with P2P computing.
IP address exposure	• IP spoofing • Traffic sniffing • Other IP-based attacks	• Block P2P traffic. • Prevent installation of P2P clients. • Establish or update policies and standards. • Network address translation.

5.11 INSTANT MESSAGING

Instant messaging (IM) is a communications service that enables a user to chat in real time over a network on the Internet. It is a popular mechanism for collaborating and keeping in touch. One can connect with another user and chat with prompt acknowledgment and response, rather than sending numerous email messages. However, there is risk associated with IM (**figure 5.26**).

5.12 SOCIAL MEDIA

Social media technology involves the creation and dissemination of content through social networks using the Internet. The differences between traditional and social media are defined by the level of interaction and interactivity available to the consumer. For example, a viewer can watch the news on television with no interactive feedback mechanisms, while social media tools allow consumers to comment, discuss and even distribute the news. Use of social media has created highly effective communication platforms where any user, virtually anywhere in the world, can freely create content and disseminate this information in real time to a global audience ranging in size from a handful to literally millions.

There are many types of social media tools: blogs (e.g., WordPress), image and video sharing sites (e.g., Flickr and YouTube), social networking (e.g., Facebook), and professional networking (e.g., LinkedIn). The common link among all forms of social media is that the content is supplied and managed by individual users who leverage the tools and platforms provided by social media sites.

Enterprises are using social media to increase brand recognition, sales, revenue and customer satisfaction; however, there is risk associated with its usage. These are divided into those enterprises with a corporate social media presence (**figure 5.27**) and those whose employees engage in social media (**figure 5.28**).

Figure 5.26—Risk of Instant Messaging

Threats and Vulnerabilities	Risk	Controls
Introduction of viruses and malware to the organizational network (especially through phishing)	• Data leakage/theft • "Owned" systems (zombies) • System downtime • Resources required to clean systems	• Ensure that antivirus and anti-malware controls are installed on all systems and updated daily. • Block IM traffic. • Only allow an enclosed corporate IM solution. • Establish or update policies and standards. • Develop and conduct awareness training and campaigns to inform employees of the risk involved with IM.
Eavesdropping	• Data leakage/theft	• Encrypt IM traffic. • Only allow an enclosed corporate IM solution. • Establish or update policies and standards. • Develop and conduct awareness training and campaigns to inform employees of the risk involved with IM.
Excessive use of IM in the workplace	• Network utilization issues • Productivity loss	• Restrict IM usage. • Establish or update policies and standards. • Develop and conduct awareness training and campaigns to inform employees of the risk involved with IM.

Figure 5.27—Risk of a Corporate Social Media Presence

Threats and Vulnerabilities	Risk	Controls
Introduction of viruses and malware to the organizational network	• Data leakage/theft • "Owned" systems (zombies) • System downtime • Resources required to clean systems	• Ensure that antivirus and anti-malware controls are installed on all systems and updated daily. • Use content filtering technology to restrict or limit access to social media sites. • Ensure that appropriate controls are also installed on mobile devices such as smartphones. • Establish or update policies and standards. • Develop and conduct awareness training and campaigns to inform employees of the risk involved with using social media sites.
Exposure to customers and the enterprise through a fraudulent or hijacked corporate presence	• Customer backlash/adverse legal actions • Exposure of customer information • Reputational damage • Targeted phishing attacks on customers or employees	• Engage a brand protection firm that can scan the Internet and search out misuse of the enterprise brand. • Give periodic informational updates to customers to maintain awareness of potential fraud and to establish clear guidelines regarding what information should be posted as part of the enterprise social media presence.

Figure 5.27—Risk of a Corporate Social Media Presence *(cont.)*		
Threats and Vulnerabilities	**Risk**	**Controls**
Unclear or undefined content rights to information posted to social media sites	• Enterprise's loss of control/legal rights of information posted to the social media sites	• Ensure that legal and communications teams carefully review user agreements for social media sites that are being considered. • Establish clear policies that dictate to employees and customers what information should be posted as part of the enterprise social media presence. • If feasible and appropriate, ensure that there is a capability to capture and log all communications.
A move to a digital business model may increase customer service expectations.	• Customer dissatisfaction with the responsiveness received in this arena, leading to potential reputational damage for the enterprise and customer retention issues	• Ensure that staffing is adequate to handle the amount of traffic that could be created from a social media presence. • Create notices that provide clear windows for customer response.
Mismanagement of electronic communications that may be impacted by retention regulations or e-discovery	• Regulatory sanctions and fines • Adverse legal actions	• Establish appropriate policies, processes and technologies to ensure that communications via social media that may be impacted by litigation or regulations are tracked and archived appropriately. • Note that, depending on the social media site, maintaining an archive may not be a recommended approach.
Source: ISACA, *Social Media: Business Benefits and Security, Governance and Assurance Perspectives*, USA, 2010, figure 1		

Figure 5.28—Risk of Employee Personal Use of Social Media		
Threats and Vulnerabilities	**Risk**	**Controls**
Use of personal accounts to communicate work-related information	• Privacy violations • Reputational damage • Loss of competitive advantage	• Work with the human resources (HR) department to establish new policies or ensure that existing policies address employee posting of work-related information. • Work with the HR department to develop awareness training and campaigns that reinforce these policies.
Employee posting of pictures or information that link them to the enterprise	• Brand damage • Reputational damage	• Work with the HR department to develop a policy that specifies how employees may use enterprise related images, assets and intellectual property (IP) in their online presence.
Excessive employee use of social media in the workplace	• Network utilization issues • Productivity loss • Increased risk of exposure to viruses and malware due to longer duration of sessions	• Manage accessibility to social media sites through content filtering or by limiting network throughput to social media sites.
Employee access to social media via enterprise-supplied mobile devices (smartphones, tablets)	• Infection of mobile devices • Data theft from mobile devices • Circumvention of enterprise controls • Data leakage	• If possible, route enterprise smartphones through corporate network filtering technology to restrict or limit access to social media sites. • Ensure that appropriate controls are also installed and continuously updated on mobile devices such as smartphones. • Establish or update policies and standards regarding the use of smartphones to access social media. • Develop and conduct awareness training and campaigns to inform employees of the risk involved with using social media sites.
Source: ISACA, *Social Media: Business Benefits and Security, Governance and Assurance Perspectives*, USA, 2010, figure 2		

5.13 CLOUD COMPUTING

As discussed in section 2.9.2 Sourcing Practices, there are different service models (IaaS, PaaS, and SaaS) and deployment models (private cloud, community cloud, public cloud and hybrid cloud) available when considering cloud computing.

Regardless of the models deployed, the security objectives required to meet organization's business requirements must still be met. These include the following:
- Ensure the continued availability of their information systems and data.
- Ensure the integrity of the information stored on their computer systems and while in transit.
- Preserve the confidentiality of sensitive data while stored and in transit.
- Ensure conformity to applicable laws, regulations and standards.
- Ensure adherence to trust and obligation requirements in relation to any information relating to an identified or identifiable individual (i.e., data subject) in accordance with its privacy policy or applicable privacy laws and regulations.
- Ensure that sensitive data are adequately protected while stored and when in transit, based on organizational requirements.

Risk associated with cloud computing and associated controls is described in **figure 5.29**.

Figure 5.29—Risk Associated with Cloud Computing

Risk	Description	Control
Legal transborder requirements	Cloud service providers (CSPs) are often transborder, and different countries have different legal requirements, especially concerning personal private information. The enterprise might be committing a violation of regulations in other countries when storing, processing or transmitting data within the CSP's infrastructure without the necessary compliance controls. Furthermore, government entities in the hosting country may require access to the enterprise's information with or without proper notification.	• Request the CSP's list of infrastructure locations and verify that regulation in those locations is aligned with the enterprise's requirements. • Include terms in the contract to restrict the moving of enterprise assets to only those areas known to be compliant with the enterprise's own regulation. • Prevent disclosure, encrypt any asset prior to migration to the CSP and ensure proper key management is in place.
Physical security	Physical security is required in any infrastructure. When the enterprise moves assets to a cloud infrastructure, those assets are still subject to the corporate security policy, but they can also be physically accessed by the CSP's staff, which is subject to the CSP's security policy. There could be a gap between the security measures provided by the CSP and the requirements of the enterprise.	• Request the CSP's physical security policy and ensure that it is aligned with the enterprise's security policy. • Request that the CSP provide proof of independent security reviews or certification reports that meet the enterprise's compliance requirements (e.g., SOC reports, SOX, PCI DSS, HIPAA, ISO certification). • Include in the contract language that requires the CSP to be aligned with the enterprise's security policy and to implement necessary controls to ensure it. • Request the CSP's disaster recovery plans and ensure that they contain the necessary countermeasures to protect physical assets during and after a disaster.
Data disposal	Proper disposal of data is imperative to prevent unauthorized disclosure. If appropriate measures are not taken by the CSP, information assets could be sent (without approval) to countries where the data can be legally disclosed due to different regulations concerning sensitive data. Disks could be replaced, recycled or upgraded without proper cleaning so that the information still remains within storage and can later be retrieved. When a contract expires, CSPs should ensure the safe disposal or destruction of any previous backups. Any of the data fed into the CSP's application must be erased immediately using the necessary tools to avoid disclosures and confidentiality breaches (forensic cleaning may be required for sensitive data).	• Request CSP's technical specifications and controls that ensure that data are properly wiped and backup media are destroyed when requested. • Include terms in the contract that require, upon contract expiration or any event ending the contract, a mandatory data wipe carried out under the enterprise's supervision.

Figure 5.29—Risk Associated with Cloud Computing *(cont.)*		
Risk	**Description**	**Control**
Multi-tenancy and isolation failure	One of the primary benefits of the cloud is the ability to perform dynamic allocation of physical resources when required. The most common approach is a multi-tenant environment (public cloud), where different entities share a pool of resources, including storage, hardware and network components. For example, when allocated storage is no longer needed by a client it can be freely reallocated to another enterprise. In that case, sensitive data could be disclosed if the storage has not been scrubbed thoroughly (e.g., using forensic software). Furthermore, malicious entities in the cloud could take advantage of shared information—for example, by utilizing shared routing tables to map the internal network topology of an enterprise, preparing the way for an internal attack.	• Request the CSP's technical details for approval and require additional controls to ensure data privacy, when necessary. • A contractual agreement is necessary to officially clarify who is allowed to access the enterprise's information, naming specific roles for CSP employees and external partners. All controls protecting the enterprise's information assets must be clearly documented in the contract agreement or service level agreement (SLA). • Use a private cloud deployment model (no multi-tenancy).
Application disposal	When applications are developed in a PaaS environment, originals and backups should always be available. In the event of a contract termination, the details of the application could be disclosed and used to create more selective attacks on applications or could be copied violating the enterprise's IP.	• Include terms in the contract that require the proper disposal of applications including objects, source and backups. • Include non-compete clauses in the contract.
Lack of visibility into software systems development life cycle (SDLC)	Enterprises that use cloud applications have little visibility into the software SDLC. Customers do not know in detail how the applications were developed and what security considerations were taken into account during the SDLC. This could lead to an imbalance between the security provided by the application and the security required by customers/users.	• If possible include a right of audit in the contract. • Include in the contract language that requires the CSP to be aligned with the enterprise's security policy and to implement necessary controls to ensure it. • Require SLAs that include a schedule of software changes.
Lack of control of the release management process	CSPs are able to introduce patches in their applications quickly. These deployments are often done without the approval (or even the knowledge) of the application users for practical reasons: if an application is used by hundreds of different enterprises, it would take an extremely long time for a CSP to look for the formal approval of every customer. In this case, the enterprise could have no control (or no view) of the release management process and could be subject to unexpected side effects.	• If possible, include a right of audit in the contract. • Include in the contract language that requires the CSP to be aligned with the enterprise's security policy and to implement necessary controls to ensure it. • Require SLAs that include a schedule of patches and software releases.
Identity and access management (IAM)	Information assets could be accessed by unauthorized entities due to faulty or vulnerable access management measures or processes. This could result from a forgery/theft of legitimate credentials or a common technical practice (e.g., administrator permissions override).	• If possible, include a right of audit in the contract. • Include in the contract language that requires the CSP to be aligned with the enterprise's security policy and to implement necessary controls to ensure it. • Request that the CSP provide proof of independent security reviews or certification reports that meet the enterprise's compliance requirements (e.g., SOC reports, SOX, PCI DSS, HIPAA, ISO certification).
Service Orientated Architecture (SOA)–related vulnerabilities	Security for SOA presents new challenges because vulnerabilities arise not only from the individual elements, but also from their mutual interaction. Because the SOA libraries are under the responsibility of the CSP and are not completely visible to the enterprise, there may be unnoticed application vulnerabilities.	• If possible include a right of audit in the contract. • Include in the contract language that requires the CSP to be aligned with the enterprise's security policy and to implement necessary controls to ensure it. • Request that the CSP provide proof of independent security reviews or certification reports that meet the enterprise's compliance requirements (e.g., SOC reports, SOX, PCI DSS, HIPAA, ISO certification).

Figure 5.29—Risk Associated with Cloud Computing *(cont.)*		
Risk	**Description**	**Control**
Exit strategy	CSPs tools facilitate bring data to the cloud or CSP but rarely the other way around. This can make it very difficult for the enterprise to migrate from one CSP to another or to bring services back in-house. It can also result in serious business disruption or failure should the CSP go bankrupt, face legal action or be the potential target for an acquisition (with the likelihood of sudden changes in CSP policies and any agreements in place). If the organization decides to bring the data back in-house, the question of how to securely render the data becomes critical because the in-house applications may have been decommissioned or "sunsetted" and there is no application available to render the data. Another possibility is the "run on the banks" scenario, in which there is a crisis of confidence in the CSP's financial position resulting in a mass exit and withdrawal on first-come, first-served basis. If there are limits to the amount of content that can be withdrawn in a given time frame, then the enterprise might not be able to retrieve all its data in the time specified.	• Ensure by contract or SLA with the CSP an exit strategy that specifies the terms that should trigger the retrieval of the enterprise's assets in the time frame required by the enterprise. • Implement a disaster recovery plan, taking into account the possibility of complete CSP disruption.
Ease to contract SaaS	Business organizations may contract cloud applications without proper procurement and approval oversight, thus bypassing compliance with internal enterprise policies.	• Require that the purchase of cloud services follow the established procedures. Ensure executive management support for this.
Collateral damage	If one tenant of a public cloud is attacked, there could be an impact to the other tenants of the same CSP, even if they are not the intended target (e.g., DDoS). Another possible scenario of collateral damage could be a public cloud IaaS that is affected by an attack exploiting vulnerabilities of software installed by one of the tenants.	• Ask the CSP to include the enterprise in its incident management process that deals with notification of collateral events. • Include contract clauses and controls to ensure that the enterprise's contracted capacity is always available and cannot be directed to other tenants without approval. • Use a private cloud deployment model (no multi-tenancy).
Hypervisor attacks	Hypervisors are vital for server virtualization. They provide the link between virtual machines and the underlying physical resources required to run the machines by using hypercalls (similar to system calls, but for virtualized systems). An attacker using a virtual machine in the same cloud could fake hypercalls to inject malicious code or trigger bugs in the hypervisor. This could potentially be used to violate confidentiality or integrity of other virtual machines or crash the hypervisor (similar to a DDoS attack).	• If possible include a right of audit in the contract. • Include in the contract language that requires the CSP to be aligned with the enterprise's security policy and to implement necessary controls to ensure it.
Support for audit and forensic investigations	Security audits and forensic investigations are vital to the enterprise to evaluate the security measures of the CSP (preventive and corrective), and in some cases the CSP itself (for example, to authenticate the CSP). This raises several issues because performing these actions requires extensive access to the CSP's infrastructure and monitoring capabilities, which are often shared with other CSP's customers.	• Request the CSP the right to audit as part of the contract or SLA. If this is not possible, request security audit reports by trusted third parties. • Request that the CSP provide appropriate and timely support (logs, traces, hard disk images, etc.) for forensic analysis as part of the contract or SLA. If this is not possible, request to authorize trusted third parties to perform forensic analysis when necessary.
Source: Data from ISACA; *Security Considerations for Cloud Computing*, USA, 2012.		

5.14 DATA LEAKAGE

As previously discussed, data leakage involves siphoning or leaking information out of the computer. This includes dumping files to paper or stealing computer reports and tapes. Unlike product leakage, data leakage leaves the original copy, so it may go undetected.

Fundamentally, data leakage involves the unauthorized transfer of sensitive or proprietary information from an internal network to the outside world.

Ways that this information can leave the enterprise include P2P networks, IM, social media, email, cloud storage and file sharing solutions.

Common controls to prevent data leakage have also been covered including identifying assets, classifying them and an ISMS, including policies and procedures.

Despite these controls, many enterprises still leak sensitive information. These leaks create risk to enterprises, their customers and business partners negatively impact an enterprise's reputation, compliance, competitive advantage and finances.

Concerns over the need to better control and protect sensitive information have given rise to a new set of solutions. These solutions vary in their capabilities and methodologies, but collectively they have been placed in a category known as data leak prevention (or protection).

5.14.1 DATA LEAK PREVENTION

Data leak prevention (DLP) is suite of technologies and associated processes that locate, monitor and protect sensitive information from unauthorized disclosure. Most DLP solutions include a suite of technologies that facilitates three key objectives:

- Locate and catalog sensitive information stored throughout the enterprise.
- Monitor and control the movement of sensitive information across enterprise networks.
- Monitor and control the movement of sensitive information on end-user systems.

These objectives are associated with three primary states of information: data at rest, data in motion and data in use. Each of these three states of data is addressed by a specific set of technologies provided by DLP solutions.

Data at Rest

A basic function of DLP solutions is the ability to identify and log where specific types of information (e.g., credit card or social security numbers) are stored throughout the enterprise. To accomplish this, most DLP systems use crawlers, which are applications that are deployed remotely to log onto each end system and "crawl" through data stores, searching for and logging the location of specific information sets based on a set of rules that have been entered into the DLP management console.

Data in Motion (Network)

To monitor data movement on enterprise networks, DLP solutions use specific network appliances or embedded technology to selectively capture and analyze network traffic. When files are sent across a network they are typically broken into packets. To inspect the information being sent across the network the DLP solution must be able to passively monitor the network traffic, recognize the correct data streams to capture, assemble the collected packets, reconstruct the files carried in the data stream, and perform the same analysis that is done on the data at rest to determine whether any portion of the file contents is restricted by its rule set. At the core of this ability is a process known as deep packet inspection (DPI). DPI goes beyond the basic header information of a packet to read the contents within the packet's payload (akin to a letter within a postal envelope). If sensitive data are detected flowing to an unauthorized destination, the DLP solution has the capability to alert and optionally block the data flows in real or near real time, again based on the rule set defined within its central management component. Based on the rule set, the solution may also quarantine or encrypt the data in question.

Data in Use (Endpoint)

Data in use primarily refers to monitoring data movement stemming from actions taken by end users on their workstations, whether that would entail copying data to a flash drive, sending information to a printer or even cutting and pasting between applications. DLP solutions typically accomplish this through the use of a software program known as an agent, which is ideally controlled by the same central management capabilities of the overall DLP solution.

To be considered a full DLP solution, the capability to address the three states of information must exist and be integrated by a centralized management function. The range of services available in the management console varies between products but many have functions in common, such as those outlined in the following sections.

Policy Creation and Management

Policies (rule sets) dictate the actions taken by the various DLP components. Most DLP solutions come with preconfigured policies (rules) that map to common regulations. It is just as important to be able to customize these policies or build completely custom policies. These should be built upon the asset management and data classifications exercises performed by the enterprise.

Directory Services Integration

Integration with directory services allows the DLP console to map a network address to a named end user.

Workflow Management

Most full DLP solutions provide the capacity to configure incident handling, allowing the central management system to route specific incidents to the appropriate parties based on violation type, severity, user and other such criteria.

Backup and Restore

Backup and restore features allow for preservation of policies and other configuration settings.

Reporting

A reporting function may be internal or may leverage external reporting tools.

5.14.2 DLP RISK, LIMITATIONS AND CONSIDERATIONS

- **Improperly tuned network DLP modules**—Proper tuning and testing of the DLP system should occur before enabling actual blocking of content. Enabling the system in monitor-only mode will allow for tuning and provide the opportunity to alert users to out-of-compliance processes and activities, so they may make adjustments accordingly. Involving the appropriate business and IT stakeholders in the planning and monitoring stages will help ensure that disruptions to processes will be anticipated and mitigated. Finally, establish some means of accessibility in the event there is critical content being blocked during off-hours when the team managing the DLP solution is not available.
- **Excessive reporting and false positives**—Similar to an improperly configured IDS, DLP solutions may register significant amounts of false positives, which overwhelm staff and can obscure valid hits. Avoid excessive use of template patterns or "black box" solutions that allow for little customization. The greatest feature of a DLP solution is the ability to customize rules or templates to specific organizational data patterns. It is also important that the system be rolled out in phases, focusing on the highest risk areas first. Trying to monitor too many data patterns or enabling too many detection points early on can quickly overwhelm resources.
- **Encryption**—DLP solutions can only inspect encrypted information that they can first decrypt. To do this, DLP agents, network appliances and crawlers must have access to, and be able to utilize, the appropriate decryption keys. If users have the ability to use personal encryption packages where keys are not managed by the enterprise and provided to the DLP solution, the files cannot be analyzed. To mitigate this risk, policies should forbid the installation and use of encryption solutions that are not centrally managed, and users should be educated that anything that cannot be decrypted for inspection (meaning that the DLP solution has the encryption key) will ultimately be blocked.
- **Graphics**—DLP solutions cannot intelligently interpret graphics files. Short of blocking or manually inspecting all such information, a significant gap will exist in an enterprise's control of its information. Sensitive information scanned into a graphics file or intellectual property that exists in a graphics format, such as design documents, would fall into this category. Enterprises that have significant intellectual property in a graphics format should develop strong policies that govern the use and dissemination of this information. While DLP solutions cannot intelligently read the contents of a graphics file, they can identify specific file types, their source and destination. This capability, combined with well-defined traffic analysis, can flag uncharacteristic movement of this type of information and provide some level of control.

5.15 END-USER COMPUTING SECURITY RISK AND CONTROLS

As noted in chapter 4, end-user computing (EUC) refers to the ability of end users to design and implement their own information system utilizing computer software products. Notwithstanding the aforementioned benefits the lack of IT department oversight can lead to security risk. Examples include:

- **Authorization**—There may be no secure mechanism to authorize access to the system
- **Authentication**—There may be no secure mechanism to authenticate users to the system
- **Audit logging**—This is not available on standard EUC solutions (e.g. Microsoft Excel, Access, etc.)
- **Encryption**—The application may contain sensitive data which has not been encrypted or otherwise protected

The IS auditor should ensure that policies for the use of EUC exist. According to chapter 4 IT Asset Management, an inventory of all such applications should exist. In most instances EUC applications will not pose a significant risk to the enterprise. Nonetheless, management should define risk criteria to determine the criticality of the application. These should also be subject to data classification with those deemed critical enough subject to the same controls as any other application.

5.16 CASE STUDIES

The following case studies are included as a learning tool to reinforce the concepts introduced in this chapter.

5.16.1 CASE STUDY A

Management is currently considering ways in which to enhance the physical security and protection of its data center. The IS auditor has been asked to assist in this process by evaluating the current environment and making recommendations for improvement. The data center consists of 15,000 square feet (1,395 square meters) of raised flooring on the ground floor of the corporate headquarters building. A total of 22 operations personnel require regular access. Currently, access to the data center is obtained using a proximity card, which is assigned to each authorized individual. There are three entrances to the data center, each of which utilizes a card reader and has a camera monitoring the entrance. These cameras feed their signals to a monitor at the building reception desk, which cycles through these images along with views from other cameras inside and outside the building. Two of the doors to the data center also have key locks that bypass the electronic system so that a proximity card is not required for entry. Use of proximity cards is written to an electronic log. This log is retained for 45 days. During the review, the IS auditor noted that 64 proximity cards are currently active and issued to various personnel. The data center has no exterior windows, although one wall is glass and overlooks the entry foyer and reception area for the building.

Case Study A Questions	
A1.	Which of the following risk would be mitigated by supplementing the proximity card system with a biometric scanner to provide two-factor authentication? A. Piggybacking or tailgating B. Sharing access cards C. Failure to log access D. Copying of keys
A2.	Which of the following access mechanisms would present the greatest difficulty in terms of user acceptance? A. Hand geometry recognition B. Fingerprints C. Retina scanning D. Voice recognition
See answers and explanations to the case study questions at the end of the chapter (page 414).	

5.16.2 CASE STUDY B

A company needed to enable remote access to one of its servers for remote maintenance purposes. Firewall policy did not allow any external access to the internal systems. Therefore, it was decided to install a modem on that server and to activate the remote access service to permit dial-up access. As a control, a policy has been implemented to manually power on the modem only when the third party was requesting access to the server and powered off by the company's system administrator when the access is no longer needed. As more and more systems are being maintained remotely, the company is asking an IS auditor to evaluate the current risk of the existing solution and to propose the best strategy for addressing future connectivity requirements.

Case Study B Questions	
B1.	What test is **MOST** important for the IS auditor to perform as part of the review of dial-up access controls? A. Dial the server from authorized and unauthorized telephone lines B. Determine bandwidth requirements of remote maintenance and the maximum line capacity C. Check if the availability of the line is guaranteed to allow remote access any time D. Check if call back is not used and the cost of calls is charged to the third party
B2.	What is the **MOST** significant risk that the IS auditor should evaluate regarding the existing remote access practice? A. Modem is not powered on/off whenever is needed B. A nondisclosure agreement was not signed by the third party C. Data exchanged over the line is not encrypted D. Firewall controls are bypassed
B3.	Which of the following recommendations is **MOST** likely to reduce the current level of remote access risk? A. Maintain an access log with the date and time when the modem was powered on/off B. Encrypt the traffic over the telephone line C. Migrate the dial-up access to an Internet VPN solution D. Update firewall policies and implement an IDS system
B4.	What control should be implemented to prevent an attack on the internal network being initiated though an Internet VPN connection? A. Firewall rules are periodically reviewed B. All VPNs terminate at a single concentrator C. An IDS capable to analyze encrypted traffic is implemented D. Antivirus software is installed on all production servers
See answers and explanations to the case study questions at the end of the chapter (page 414).	

5.16.3 CASE STUDY C

"My Music" is a company dedicated to the production and distribution of video clips specializing in jazz music. Born in the Internet era, the company has actively supported the use of laptops computers by its staff so they can use them when traveling and when working from home. Through the Internet they can access the company databases and provide online information to customers. This decision has resulted in an increase in productivity and high morale among employees who are allowed to work up to two days a week from home. Based on written procedures and a training course, employees learn security procedures to avoid the risk of unauthorized access to company data. Employees' access to the company data includes using logon IDs and passwords to the application server through a VPN. Initial passwords are assigned by the security administrator. When the employee logs on for the first time, the system forces a password change to improve confidentiality.

Management is currently considering ways to improve security protection for remote access by employees. The IS auditor has been asked to assist in this process by evaluating the current environment and making recommendations for improvement.

Case Study C Questions	
C1.	Which of the following levels provides a higher degree of protection in applying access control software to avoid unauthorized access risk? A. Network and OS level B. Application level C. Database level D. Log file level
C2.	When an employee notifies the company that he/she has forgotten his/her password, what should be done **FIRST** by the security administrator? A. Allow the system to randomly generate a new password B. Verify the user's identification through a challenge/ response system C. Provide the employee with the default password and explain that it should be changed as soon as possible D. Ask the employee to move to the administrator terminal to generate a new password in order to assure confidentiality
See answers and explanations to the case study questions at the end of the chapter (page 415).	

5.16.4 CASE STUDY D

A major financial institution has just implemented a centralized banking solution (CBS) in one of its branches. It has a secondary concern to look after marketing of the bank. Employees of a separate legal entity work on the bank premises, but they have no access to the bank's solution software. Employees of other branches get training on this solution from this branch and for training purposes temporary access credentials are also given to such employees. IS auditors observed that employees of the separate legal entity also access the CBS software through the branch employees access credentials. IS auditors also observed that there are numerous active IDs of employees who got training from the branch and have since been transferred to their original branch.

Case Study D Questions	
D1.	Which of the following should IS auditors recommend to effectively eliminate such password sharing? A. Assimilation of security need to keep password secret B. Stringent rules prohibiting sharing of password C. Use of smart card along with strong password D. Use of smart card along with employee's terminal ID
D2.	Which of the following **BEST** addresses user ID management of trainee employees? A. Unused user ID shall be automatically deleted periodically B. To integrate access rights with human resource process C. Password of unused but active user ID shall be suspended D. Active user ID register shall checked frequently
See answers and explanations to the case study questions at the end of the chapter (page 415).	

5.17 ANSWERS TO CASE STUDY QUESTIONS

ANSWERS TO CASE STUDY A QUESTIONS

A1. **B** Two-factor authentication involving the use of biometrics would effectively prevent the sharing of access cards because these cards would be ineffective without the corresponding biometric. Piggybacking or tailgating would not be mitigated because traffic flow would remain unchanged. Because two entrances utilize key locks that override the electronic entry system, individuals entering using keys would not be logged by the electronic system, and keys could still potentially be copied.

A2. **C** Although the highest in terms of accuracy, many individuals feel uncomfortable with the idea of having a device scan the inside of their eye. So, even though retina scanning may be highest in terms of effectiveness from a control perspective, its lack of user acceptance may make it inappropriate for applications where customer acceptance is of prime importance. Fingerprints, hand geometry and voice recognition are all less invasive and, therefore, not as subject to adverse negative reaction by users. The objective of this area is to ensure that the CISA candidate understands and can provide assurance that the security architecture (policies, standards, procedures and controls) ensures the confidentiality, integrity and availability of information assets.

ANSWERS TO CASE STUDY B QUESTIONS

B1. **A** Dial-up access should be possible only from authorized telephone lines as a preventive control for unauthorized access when logon credentials are compromised or misused by third-party personnel. Initiating the connection by the server to an authorized phone number using the call back feature would be one implementation of this requirement. Options B, C and D address performance issues and not access control issues.

B2. **D** The company's security infrastructure relies on controls implemented on the firewall. The fact that someone from the outside can connect directly to an internal system, bypassing firewall rules, could expose the internal network to the third party, thereby facilitating unauthorized access. Choices A, B and C are types of risk to be considered by the IS auditor, but concern only the server being maintained remotely, and not the entire internal system.

B3. **C** Using an Internet VPN solution will eliminate the vulnerabilities of the dial-up access such as lack of encryption and bypassing firewall controls. Option A and B will address punctual issues and Option D will have no effect since security infrastructure controls are bypassed by the direct dial-up access.

B4. **C** An IDS should be able to analyze the encrypted traffic of the VPN connection to determine potential attacks. A firewall rules review and ending all VPNs in a single concentrator will prevent unauthorized connections to the internal network, but this will not prevent an attack occurring through an authorized VPN connection. Antivirus software will prevent contamination by computer viruses, but the internal system is still vulnerable to many other threats.

ANSWERS TO CASE STUDY C QUESTIONS

C1. **A** The greatest degree of protection in applying access control software against internal and external users' unauthorized access is at the network and platform/OS levels. These systems are also referred to as general support systems, and they make up the primary infrastructure on which applications and database systems will reside.

C2. **B** When an employee notifies that he/she has forgotten his/her password, the security administrator should start a password process generation procedure only after verifying the user's identification using a challenge/response system or similar procedure. To verify, it is advised that the security administrator should return the user's call after verifying his/her extension or calling his/her supervisor for verification.

ANSWERS TO CASE STUDY D QUESTIONS

D1. **A** Assimilation of security need to keep password secret can only effectively refrain such password sharing and such assimilation is possible only through continuous and conscientious security awareness and education programs. Without assimilation of security need stringent rules prohibiting sharing of password cannot effectively stop password sharing. Use of smart card along with strong password and use of smart card along with employee's terminal ID do not deter password sharing.

D2. **B** Integration of access rights with human resource process is the best way to address user ID management. Automatic periodic deletion of unused user ID, suspension of password of unused but active user ID and frequently checking active user ID register are not the best way since vulnerability persists during the period in which user IDs remain active.

Page intentionally left blank

APPENDIX A: IS AUDIT AND ASSURANCE STANDARDS, GUIDELINES AND TOOLS AND TECHNIQUES

The specialized nature of IS audit and assurance and the skills necessary to perform such audits require standards that apply specifically to IS audit and assurance. One of the goals of ISACA is to advance globally applicable standards to meet its vision. The development and dissemination of the IS Audit and Assurance Standards is a cornerstone of the ISACA professional contribution to the audit and assurance community. The framework for the IS Audit and Assurance Standards provides multiple levels of guidance.

Standards define mandatory requirements for IS audit and assurance. They inform:
- IS audit and assurance professionals of the minimum level of acceptable performance required to meet the professional responsibilities set out in the ISACA Code of Professional Ethics.
- Management and other interested parties of the profession's expectations concerning the work of practitioners.
- Holders of the CISA designation of requirements. Failure to comply with these standards may result in an investigation into the CISA holder's conduct by the ISACA Board of Directors or appropriate ISACA group and, ultimately, in disciplinary action.

The IS Audit and Assurance Guidelines provide guidance in applying the IS Audit and Assurance Standards. The IS audit and assurance professional should consider the guidelines in determining how to: achieve implementation of the standards, use professional judgment in their application, and be prepared to justify any departure. The objective of the IS Audit and Assurance Guidelines is to provide further information on how to comply with the IS Audit and Assurance Standards.

The IS Audit and Assurance Tools and Techniques provide examples of procedures that an IS audit and assurance professional may follow in an audit engagement. They provide information on how to meet the IS Audit and Assurance Standards when performing IS auditing work, but do not set requirements. The objective of the IS Audit and Assurance Tools and Techniques is to provide further information on how to comply with the IS Audit and Assurance Standards.

COBIT is a framework covering the governance of enterprise IT and is a supporting tool set that allows managers to bridge the gaps among control requirements, technical issues and business risks. COBIT enables clear policy development and good practice for IS control throughout enterprises. It emphasizes regulatory compliance, helps enterprises increase the value attained from IT, enables alignment and simplifies implementation of the COBIT framework concepts. COBIT is intended for use by business and IT management as well as by IS auditors. Its use enables the understanding of business objectives, the communication of good practices, and the ability to make recommendations around a commonly understood and well-respected framework.
COBIT is available for download on the ISACA web site, *www.isaca.org/COBIT.*

RELATIONSHIP OF STANDARDS TO GUIDELINES AND TOOLS AND TECHNIQUES

The IS Audit and Assurance Standards are brief, mandatory reports on requirements regarding the audit and its findings. The IS Audit and Assurance Guidelines and Tools and Techniques are detailed guidance on how to follow those standards. The tools and techniques examples show the steps performed by an IS audit and assurance professional and are more detailed than the IS Audit and Assurance Guidelines. The examples are constructed to follow the IS Audit and Assurance Standards and the IS Audit and Assurance Guidelines and to provide information on following the IS Audit and Assurance Standards. To some extent, they also establish best practices for the tools and techniques to be followed.

Codification:
- Standards are divided into three categories:
  - **General standards (1000 series)**—Are the guiding principles under which the IS assurance profession operates. They apply to the conduct of all assignments and deal with the IS audit and assurance professional's ethics, independence, objectivity and due care as well as knowledge, competency and skill.
  - **Performance standards (1200 series)**—Deal with the conduct of the assignment, such as planning and supervision, scoping, risk and materiality, resource mobilization, supervision and assignment management, audit and assurance evidence, and the exercising of professional judgment and due care.
  - **Reporting standards (1400 series)**—Address the types of reports, means of communication and the information communicated.
- Guidelines are also divided into three categories:
  - **General guidelines (2000 series).**
  - **Performance guidelines (2200 series)**
  - **Reporting guidelines (2400 series).**
- Tools and Techniques take a variety of forms, such as discussion documents, technical direction, white papers, audit programs or books—e.g., the ISACA publication on SAP, which provides guidance on enterprise resource planning (ERP) systems.

Please refer to *www.isaca.org/standards* for a complete listing of the IS Audit and Assurance Standards, Guidelines, and Tools and Techniques.

USE

It is suggested that the IS audit and assurance professional review the standards to ensure compliance with them during the annual audit program, as well as during individual reviews throughout the year. The IS audit and assurance professional is encouraged to refer to the ISACA standards in the report, stating that the review was conducted in compliance with the laws of the country, applicable audit regulations and ISACA standards.

Page intentionally left blank

APPENDIX B: CISA EXAM GENERAL INFORMATION

ISACA is a professional membership association composed of individuals interested in IS audit, assurance, control, security and governance. The CISA Certification Working Group is responsible for establishing policies for the CISA certification program and developing the exam.

> **Note:** Because information regarding the CISA examination, requirements and locations and dates may change, please refer to *www.isaca.org/certification* for the most up-to-date information.

REQUIREMENTS FOR CERTIFICATION

The CISA designation is awarded to those individuals who have met the following requirements: (1) a passing score on the CISA exam, (2) submitting verified evidence of IS auditing, control, assurance or security experience, (3) abiding by the *Code of Professional Ethics*, (4) abiding by the continuing professional education policy, and (5) abiding by the IS Auditing Standards as adopted by ISACA.

SUCCESSFUL COMPLETION OF THE CISA EXAM

The exam is open to all individuals who wish to take it. Successful exam candidates are not certified until they apply for certification (and demonstrate that they have met all requirements) and receive approval from ISACA.

EXPERIENCE IN IS AUDITING, CONTROL AND SECURITY

A minimum of five years professional IS auditing, control, assurance and security work experience is required for certification. Substitutions and waivers of such experience may be obtained as follows:

- A maximum of one year of IS audit, control, assurance or security experience may be substituted for:
  - One full year of non-IS audit or information systems experience, or
  - A bachelor's or master's degree from a university that enforces the ISACA-sponsored Model Curriculum.
  - An Associate's degree (60 semester college credits or its equivalent).
- Two years IS audit, control, assurance or security experience may be substituted for a bachelor's degree (120 semester college credits or its equivalent) or master's degree.
- A master's degree in information security or information technology from an accredited university can be substituted for one year of experience.
- Two years as a full-time university instructor in a related field (e.g., computer science, accounting or information systems auditing) can be substituted for every one year of IS audit, control, assurance or security experience.
- Two year waiver for CIMA (Chartered Institute of Management Accountants) full certification.
- Two year waiver for ACCA member status from the Association of Chartered Certified Accountants.

Experience must have been gained within the 10-year period preceding the application date for certification or within five years from the date of initially passing the exam. A completed application for certification must be submitted within five years from the passing date of the CISA exam. All experience must be independently verified with employers. Please note that certification application decisions are not final as there is an appeal process for certification application denials. Appeals undertaken by a certification exam taker, certification applicant or by a certified individual are undertaken at the discretion and cost of the exam taker, applicant or individual. Inquiries of denials of certification can be sent to *certification@isaca.org*.

DESCRIPTION OF THE EXAM

The CISA Certification Working Group oversees the development of the exam and ensures the currency of its content. Questions for the CISA exam are developed through a multitiered process designed to enhance the ultimate quality of the exam. Once the CISA Certification Working Group approves the questions, they go into the item pool from which all CISA exam questions are selected.

The purpose of the exam is to evaluate a candidate's knowledge and experience in conducting IS audits and reviews. The exam consists of 150 multiple-choice questions, administered during a four-hour session. The CISA exam is offered in 11 languages. A proctor speaking the primary language used at each test site is available.

REGISTRATION FOR THE CISA EXAM

The CISA exam is administered three times annually. Please refer to the ISACA Exam Candidate Information Guide at www.isaca.org/examguide for specific exam registration dates, language offerings and deadlines as well as important key information for exam day. Exam registrations can be placed online at *www.isaca.org/examreg.*

CISA PROGRAM ACCREDITATION RENEWED UNDER ISO/IEC 17024:2012

The American National Standards Institute (ANSI) has voted to continue the accreditation for the CISA, CISM, CGEIT and CRISC certifications, under ISO/IEC 17024:2012, General Requirements for Bodies Operating Certification Systems of Persons. ANSI, a private, nonprofit organization, accredits other organizations to serve as third-party product, system and personnel certifiers.

ISO/IEC 17024 specifies the requirements to be followed by organizations certifying individual against specific requirements. ANSI describes ISO/IEC 17024 as "expected to play a prominent role in facilitating global standardization of the certification community, increasing mobility among countries, enhancing public safety, and protecting consumers."

ANSI's accreditation:
- Promotes the unique qualifications and expertise ISACA's certifications provide
- Protects the integrity of the certifications and provides legal defensibility
- Enhances consumer and public confidence in the certifications and the people who hold them
- Facilitates mobility across borders or industries

Accreditation by ANSI signifies that ISACA's procedures meet ANSI's essential requirements for openness, balance, consensus and due process. With this accreditation, ISACA anticipates that significant opportunities for CISAs, CISMs, CGEITs and CRISCs will continue to open in the United States and around the world.

ANSI Accredited Program
PERSONNEL
CERTIFICATION

PREPARING FOR THE CISA EXAM

The CISA exam evaluates a candidate's practical knowledge of the job practice domains listed in this manual and online at *www.isaca.org/cisajobpractice*. That is, the exam is designed to test a candidate's knowledge, experience and judgment of the proper or preferred application of IS audit, security and control principles, methods and practices. Since the exam covers a broad spectrum of information systems audit, control and security issues, candidates are cautioned not to assume that reading CISA study guides and reference publications will fully prepare them for the exam. CISA candidates are encouraged to refer to their own experiences when studying for the exam and refer to CISA study guides and reference publications for further explanation of concepts or practices with which the candidate is not familiar.

No representation or warranties are made by ISACA in regard to CISA exam study guides, other ISACA publications, references or courses assuring candidates' passage of the exam.

TYPES OF EXAM QUESTIONS

CISA exam questions are developed with the intent of measuring and testing practical knowledge and the application of general concepts and standards. All questions are multiple choice and are designed for one best answer.

Every question has a stem (question) and four options (answer choices). The candidate is asked to choose the correct or best answer from the options. The stem may be in the form of a question or incomplete statement. In some instances, a scenario may also be included. These questions normally include a description of a situation and require the candidate to answer two or more questions based on the information provided. The candidate is cautioned to read each question carefully, eliminate known incorrect answers and then make the best choice possible. To gain a better understanding of the types of question that might appear on the exam and how these questions are developed, refer to the Item Writing Guide available at *www.isaca.org/itemwriting*. Representations of CISA exam questions are available at *www.isaca.org/cisaassessment*.

ADMINISTRATION OF THE EXAM

ISACA has contracted with an internationally recognized testing agency. This not-for-profit corporation engages in the development and administration of credentialing exams for certification and licensing purposes. It assists ISACA in the construction, administration and scoring of the CISA exam.

SITTING FOR THE EXAM

Candidates are to report to the testing site at the report time indicated on their admission ticket. NO CANDIDATE WILL BE ADMITTED TO THE TEST CENTER ONCE THE CHIEF EXAMINER BEGINS READING THE ORAL INSTRUCTIONS. Candidates who do not attend or arrive after the oral instructions have begun will not be allowed to sit for the exam and will forfeit their registration fee. To ensure that candidates arrive in time for the exam, it is recommended that candidates become familiar with the exact location of, and the best travel route to, the exam site prior to the date of the exam. Candidates can use their admission tickets only at the designated test center on the admission ticket.

To be admitted into the test site, candidates must bring the email printout OR a printout of the downloaded admission ticket and an acceptable form of photo identification such as a driver's license, passport or government-issued ID. This ID must be a current and original government-issued identification that is not handwritten and that contains both the candidate's name as it appears on the admission ticket and the candidate's photograph. Candidates who do not provide an acceptable form of identification will not be allowed to sit for the exam and will forfeit their registration fee.

The following conventions should be observed when completing the exam:
- Do not bring study materials (including notes, paper, books or study guides) or scratch paper or notepads into the exam site. For further details regarding what personal belongings can (and cannot) be brought into the test site, please visit *www.isaca.org/cisabelongings*.Items brought to the test site may not be accessed during the exam administration.
- Candidates are not allowed to bring any type of communication device (e.g., cell phones, tablets, smart watches or glasses, mobile devices, etc.) into the test center. If candidates are viewed with any such device during the exam administration, their exams will be voided and they will be asked to immediately leave the exam site.
- Candidates who leave the testing area without authorization or accompaniment by a test proctor will not be allowed to return to the testing room and will be subject to disqualification.
- Candidates should bring several no. 2 pencils since pencils will not be provided at the exam site.
- Include your exam identification number as it appears on your admission ticket and any other requested information. Failure to do so may result in a delay or errors.

- Read the provided instructions carefully before attempting to answer questions. Skipping over these directions or reading them too quickly could result in missing important information and possibly losing credit points.
- Mark the appropriate area when indicating responses on the answer sheet. When correcting a previously answered question, fully erase a wrong answer before writing in the new one.
- Remember to answer all questions since there is no penalty for wrong answers. Grading is based solely on the number of questions answered correctly. Do not leave any question blank. The exam will be scored based on the answer sheet recordings only.
- Identify key words or phrases in the question (**MOST**, **BEST**, **FIRST**, etc.) before selecting and recording the answer.
- The chief examiner or designate at each test center will read aloud the instructions for entering information on the answer sheet. It is imperative that candidates include their exam identification number as it appears on their admission ticket and any other requested information on their exam answer sheet. Failure to do so may result in a delay or errors.

BUDGETING YOUR TIME

The following are time-management tips for the exam:

- It is recommended that candidates become familiar with the exact location of, and the best travel route to, the exam site prior to the date of the exam.
- Candidates must arrive at the exam testing site by the time indicated on their admission ticket. This will give candidates time to be seated and get acclimated.
- The exam is administered over a four-hour period. This allows for a little over 1.5 minutes per question. Therefore, it is advisable that candidates pace themselves to complete the entire exam. In order to do so, candidates should complete an average of 38 questions per hour.
- Candidates are urged to record their answers on their answer sheet. No additional time will be allowed after the exam time has elapsed to transfer or record answers should candidates mark their answers in the question booklet.

RULES AND PROCEDURES

- Candidates will be asked to sign the answer sheet to protect the security of the exam and maintain the validity of the scores.
- Upon the discretion of the CISA Certification Working Group, any candidate can be disqualified who is discovered engaging in any kind of misconduct, such as giving or receiving help; using notes, papers, or other aids; attempting to take the examination for someone else; or removing test materials or notes from the testing room. The testing agency will provide the CISA Certification Working Group with records regarding such irregularities. The Working Group will review reported incidents, and all Working Group decisions are final.
- Additional information on exam day rules is available in the ISACA Exam Candidate Information Guide (*www.isaca.org/examguide*).
- Candidates may **not** take the exam question booklet after completion of the exam.

GRADING THE EXAM

The CISA exam consists of 150 items. Candidate scores are reported as a scaled scored. A scaled score is a conversion of a candidate's raw score on an exam to a common scale. ISACA uses and reports scores on a common scale from 200 to 800.

A candidate must receive a score of 450 or higher to pass the exam. A score of 450 represents a minimum consistent standard of knowledge as established by ISACA's CISA Certification Working Group. A candidate receiving a passing score may then apply for certification if all other requirements are met.

Passing the exam does not grant the CISA designation. To become a CISA, each candidate must complete all requirements, including submitting an application and receiving approval for certification.

The CISA examination contains some questions which are included for research and analysis purposes only. These questions are not separately identified and the candidate's final score will be based only on the common scored questions. There are various versions of each exam but only the common questions are scored for your results.

A candidate receiving a score less than 450 is not successful and can retake the exam by registering and paying the appropriate exam fee for any future exam administration. To assist with future study, the result letter each candidate receives will include a score analysis by content area. There are no limits to the number of times a candidate can take the exam.

Approximately five weeks after the test date, the official exam results will be mailed to candidates. Additionally, with the candidate's consent during the registration process, an email containing the candidates pass/fail status and score will be sent to paid candidates. This email notification will only be sent to the address listed in the candidate's profile at the time of the initial release of the results. To ensure the confidentiality of scores, exam results will not be reported by telephone or fax. To prevent the email notification from being sent to the candidate's spam folder, the candidate should add *exam@isaca.org* to his/her address book, whitelist or safe senders list.

In order to become CISA-certified, candidates must pass the CISA exam and must complete and submit an application for certification (and must receive confirmation from ISACA that the application is approved). The application is available on the ISACA web site at *www.isaca.org/cisaapp*. Once the application is approved, the applicant will be sent confirmation of the approval. The candidate is not CISA-certified, and cannot use the CISA designation, until the candidate's application is approved. A processing fee must accompany CISA applications for certification.

The score report contains a subscore for each job domain. The subscores can be useful in identifying those areas in which the candidate may need further study before retaking the exam. Unsuccessful candidates should note that taking either a simple or weighted average of the subscores does not derive the total scaled score. Candidates receiving a failing score on the exam may request a rescoring of their answer sheet. This procedure ensures that no stray marks, multiple responses or other conditions interfered with computer scoring. Candidates should understand, however, that all scores are subjected to several quality control checks before they are reported; therefore, rescores most likely will not result in a score change. Requests for hand scoring must be made in writing to the certification department within 90 days following the release of the exam results. Requests for a hand score after the deadline date will not be processed. All requests must include a candidate's name, exam identification number and mailing address. A fee of US $75 must accompany this request.

Page intentionally left blank

GLOSSARY

Note: Glossary terms are provided for reference within the CISA Review Manual. As definitions of terms may evolve due to the changing technological environment, please see *www.isaca.org/glossary* for the most up-to-date terms and definitions.

A

Abend—An abnormal end to a computer job; termination of a task prior to its completion because of an error condition that cannot be resolved by recovery facilities while the task is executing

Acceptable use policy—A policy that establishes an agreement between users and the enterprise and defines for all parties' the ranges of use that are approved before gaining access to a network or the Internet

Access control—The processes, rules and deployment mechanisms that control access to information systems, resources and physical access to premises

Access control list (ACL)—An internal computerized table of access rules regarding the levels of computer access permitted to logon IDs and computer terminals. Also referred to as access control tables.

Access control table—An internal computerized table of access rules regarding the levels of computer access permitted to logon IDs and computer terminals

Access method—The technique used for selecting records in a file, one at a time, for processing, retrieval or storage. The access method is related to, but distinct from, the file organization, which determines how the records are stored.

Access path—The logical route an end user takes to access computerized information. Typically, it includes a route through the operating system, telecommunications software, selected application software and the access control system.

Access rights—The permission or privileges granted to users, programs or workstations to create, change, delete or view data and files within a system, as defined by rules established by data owners and the information security policy

Access servers—Provides centralized access control for managing remote access dial-up services

Address—Within computer storage, the code used to designate the location of a specific piece of data

Address space—The number of distinct locations that may be referred to with the machine address. For most binary machines, it is equal to 2n, where n is the number of bits in the machine address.

Addressing—The method used to identify the location of a participant in a network. Ideally, addressing specifies where the participant is located rather than who they are (name) or how to get there (routing).

Administrative controls—The rules, procedures and practices dealing with operational effectiveness, efficiency and adherence to regulations and management policies

Adware—A software package that automatically plays, displays or downloads advertising material to a computer after the software is installed on it or while the application is being used. In most cases, this is done without any notification to the user or without the user's consent. The term adware may also refer to software that displays advertisements, whether or not it does so with the user's consent; such programs display advertisements as an alternative to shareware registration fees. These are classified as adware in the sense of advertising supported software, but not as spyware. Adware in this form does not operate surreptitiously or mislead the user, and provides the user with a specific service.

Alpha—The use of alphabetic characters or an alphabetic character string

Alternative routing—A service that allows the option of having an alternate route to complete a call when the marked destination is not available. In signaling, alternate routing is the process of allocating substitute routes for a given signaling traffic stream in case of failure(s) affecting the normal signaling links or routes of that traffic stream.

American Standard Code for Information Interchange—See ASCII.

Analog—A transmission signal that varies continuously in amplitude and time, and is generated in wave formation. Analog signals are used in telecommunications.

Anonymous File Transfer Protocol (FTP)—A method for downloading public files using the File Transfer Protocol. Anonymous FTP is called anonymous because users do not need to identify themselves before accessing files from a particular server. In general, users enter the word "anonymous" when the host prompts for a username; anything can be entered for the password such as the user's email address or simply the word "guest." In many cases, an anonymous FTP site will not even prompt users for a name and password.

Antivirus software—An application software deployed at multiple points in an IT architecture. It is designed to detect and potentially eliminate virus code before damage is done and repair or quarantine files that have already been infected.

Applet—A program written in a portable, platform independent computer language such as Java, JavaScript or Visual Basic. An applet is usually embedded in a Hypertext Markup Language (HTML) page downloaded from web servers and then executed by a browser on client machines to run any web-based application (e.g., generate web page input forms, run audio/video programs, etc.). Applets can only perform a restricted set of operations, thus preventing, or at least minimizing, the possible security compromise of the host computers. However, applets expose the user's machine to risk if not properly controlled by the browser, which should not allow an applet to access a machine's information without prior authorization of the user.

Application—A computer program or set of programs that perform the processing of records for a specific function. Contrasts with systems programs, such as an operating system or network control program, and with utility programs, such as copy or sort.

Application controls—The policies, procedures and activities designed to provide reasonable assurance that objectives relevant to a given automated solution (application) are achieved

Application layer—In the Open Systems Interconnection (OSI) communications model, the application layer provides services for an application program to ensure that effective communication with another application program in a network is possible. The application layer is not the application that is doing the communication; a service layer that provides these services.

Application program—A program that processes business data through activities such as data entry, update or query. Contrasts with systems programs, such as an operating system or network control program, and with utility programs such as copy or sort.

Application programming—The act or function of developing and maintaining applications programs in production

Application programming interface (API)—A set of routines, protocols and tools referred to as "building blocks" used in business application software development. A good API makes it easier to develop a program by providing all the building blocks related to functional characteristics of an operating system that applications need to specify, for example, when interfacing with the operating system (e.g., provided by Microsoft Windows, different versions of UNIX). A programmer utilizes these APIs in developing applications that can operate effectively and efficiently on the platform chosen.

Application software tracing and mapping—Specialized tools that can be used to analyze the flow of data through the processing logic of the application software and document the logic, paths, control conditions and processing sequences. Both the command language or job control statements and programming language can be analyzed. This technique includes program/system: mapping, tracing, snapshots, parallel simulations and code comparisons.

Arithmetic logic unit (ALU)—The area of the central processing unit that performs mathematical and analytical operations

Artificial intelligence—Advanced computer systems that can simulate human capabilities, such as analysis, based on a predetermined set of rules

ASCII—Representing 128 characters, the American Standard Code for Information Interchange (ASCII) code normally uses 7 bits. However, some variations of the ASCII code set allow 8 bits. This 8-bit ASCII code allows 256 characters to be represented.

Assembler—A program that takes as input a program written in assembly language and translates it into machine code or machine language

Asymmetric key (public key)—A cipher technique in which different cryptographic keys are used to encrypt and decrypt a message (See public key encryption)

Asynchronous Transfer Mode (ATM)—A high-bandwidth low-delay switching and multiplexing technology that allows integration of real-time voice and video as well as data. It is a data link layer protocol. ATM is a protocol-independent transport mechanism. It allows high-speed data transfer rates at up to 155 Mbit/s. The acronym ATM should not be confused with the alternate usage for ATM, which refers to an automated teller machine.

Asynchronous transmission—Character-at-a-time transmission

Attribute sampling—An audit technique used to select items from a population for audit testing purposes based on selecting all those items that have certain attributes or characteristics (such as all items over a certain size)

Audit evidence—The information used to support the audit opinion

Audit objective—The specific goal(s) of an audit. These often center on substantiating the existence of internal controls to minimize business risk.

Audit plan—
1. A plan containing the nature, timing and extent of audit procedures to be performed by engagement team members in order to obtain sufficient appropriate audit evidence to form an opinion. Includes the areas to be audited, the type of work planned, the high-level objectives and scope of the work, and topics such as budget, resource allocation, schedule dates, type of report and its intended audience and other general aspects of the work
2. A high-level description of the audit work to be performed in a certain period of time

Audit program—A step-by-step set of audit procedures and instructions that should be performed to complete an audit

Audit risk—The probability that information or financial reports may contain material errors and that the auditor may not detect an error that has occurred

Audit trail—A visible trail of evidence enabling one to trace information contained in statements or reports back to the original input source

Authentication—The act of verifying the identity of a user and the user's eligibility to access computerized information. Authentication is designed to protect against fraudulent logon activity. It can also refer to the verification of the correctness of a piece of data.

B

Backbone—The main communications channel of a digital network. The part of the network that handles the major traffic. Employs the highest-speed transmission paths in the network and may also run the longest distances. Smaller networks are attached to the backbone, and networks that connect directly to the end user or customer are called "access networks." A backbone can span a geographic area of any size from a single building to an office complex to an entire country. Or, it can be as small as a backplane in a single cabinet.

Backup—Files, equipment, data and procedures available for use in the event of a failure or loss, if the originals are destroyed or out of service

Badge—A card or other device that is presented or displayed to obtain access to an otherwise restricted facility, as a symbol of authority (e.g., police) or as a simple means of identification. Also used in advertising and publicity.

Balanced scorecard (BSC)—Developed by Robert S. Kaplan and David P. Norton as a coherent set of performance measures organized into four categories that includes traditional financial measures, but adds customer, internal business process, and learning and growth perspectives

Bandwidth—The range between the highest and lowest transmittable frequencies. It equates to the transmission capacity of an electronic line and is expressed in bytes per second or Hertz (cycles per second).

Bar code—A printed machine-readable code that consists of parallel bars of varied width and spacing

Base case—A standardized body of data created for testing purposes. Users normally establish the data. Base cases validate production application systems and test the ongoing accurate operation of the system.

Baseband—A form of modulation in which data signals are pulsed directly on the transmission medium without frequency division and usually utilize a transceiver. The entire bandwidth of the transmission medium (e.g., coaxial cable) is utilized for a single channel.

Batch control—Correctness checks built into data processing systems and applied to batches of input data, particularly in the data preparation stage. There are two main forms of batch controls: sequence control, which involves consecutively numbering the records in a batch so that the presence of each record can be confirmed, and control total, which is a total of the values in selected fields within the transactions.

Batch processing—The processing of a group of transactions at the same time. Transactions are collected and processed against the master files at a specified time.

Baud rate—The rate of transmission for telecommunications data, expressed in bits per second (bps)

Bayesian filter—A method often employed by antispam software to filter spam based on probabilities. The message header and every word or number are each considered a token and given a probability score. Then the entire message is given a spam probability score. A message with a high score will be flagged as spam and discarded, returned to its sender or put in a spam directory for further review by the intended recipient.

Benchmarking—A systematic approach to comparing organization performance against peers and competitors in an effort to learn the best ways of conducting business. Examples include benchmarking of quality, logistic efficiency and various other metrics.

Binary code—A code whose representation is limited to 0 and 1

Biometrics—A security technique that verifies an individual's identity by analyzing a unique physical attribute such as a handprint

Black box testing—A testing approach that focuses on the functionality of the application or product and does not require knowledge of the code intervals

Bridge—A device that connects two similar networks together

Broadband—Multiple channels are formed by dividing the transmission medium into discrete frequency segments. Broadband generally requires the use of a modem.

Brouters—Devices that perform the functions of both a bridge and a router. A brouter operates at both the data link and the network layers. It connects same data link type local area network (LAN) segments as well as different data link ones, which is a significant advantage. Like a bridge, it forwards packets based on the data link layer address to a different network of the same type. Also, whenever required, it processes and forwards messages to a different data link type network based on the network protocol address. When connecting same data link type networks, it is as fast as a bridge and is able to connect different data link type networks.

Buffer—Memory reserved to temporarily hold data to offset differences between the operating speeds of different devices, such as a printer and a computer. In a program, buffers are reserved areas of random access memory (RAM) that hold data while they are being processed.

Bus—Common path or channel between hardware devices. Can be located between components internal to a computer or between external computers in a communications network.

Bus configuration—All devices (nodes) are linked along one communication line where transmissions are received by all attached nodes. This architecture is reliable in very small networks, as well as easy to use and understand. This configuration requires the least amount of cable to connect the computers together and, therefore, is less expensive than other cabling arrangements. It is also easy to extend, and two cables can be easily joined with a connector to make a longer cable for more computers to join the network. A repeater can also be used to extend a bus configuration.

Business case—Documentation of the rationale for making a business investment, used both to support a business decision on whether to proceed with the investment and as an operational tool to support management of the investment through its full economic life cycle

Business continuity plan (BCP)—A plan used by an organization to respond to disruption of critical business processes. Depends on the contingency plan for restoration of critical systems.

Business impact analysis (BIA)—A process to determine the impact of losing the support of any resource. The BIA assessment study will establish the escalation of that loss over time. It is predicated on the fact that senior management, when provided reliable data to document the potential impact of a lost resource, can make the appropriate decision.

Business process reengineering (BPR)—The thorough analysis and significant redesign of business processes and management systems to establish a better performing structure, more responsive to the customer base and market conditions, while yielding material cost savings

Business risk—A probable situation with uncertain frequency and magnitude of loss (or gain)

Bypass label processing (BLP)—A technique of reading a computer file while bypassing the internal file/data set label. This process could result in bypassing of the security access control system.

C

Capability Maturity Model (CMM)—CMM for software, from the Software Engineering Institute (SEI), is a model used by many organizations to identify best practices useful in helping them assess and increase the maturity of their software development processes.

Capacity stress testing—Testing an application with large quantities of data to evaluate its performance during peak periods. Also called volume testing.

Card swipe—A physical control technique that uses a secured card or ID to gain access to a highly sensitive location. If built correctly, card swipes act as a preventive control over physical access to those sensitive locations. After a card has been swiped, the application attached to the physical card swipe device logs all card users who try to access the secured location. The card swipe device prevents unauthorized access and logs all attempts to enter the secured location.

Central processing unit (CPU)—Computer hardware that houses the electronic circuits that control/direct all operations of the computer system

Certificate (certification) authority (CA)—A trusted third party that serves authentication infrastructures or organizations and registers entities and issues them certificates

Certificate revocation list (CRL)—An instrument for checking the continued validity of the certificates for which the certification authority (CA) has responsibility. The CRL details digital certificates that are no longer valid. The time gap between two updates is very critical and is also a risk in digital certificates verification.

Certification practice statement (CPS)—A detailed set of rules governing the certificate authority's operations. It provides an understanding of the value and trustworthiness of certificates issued by a given CA. In terms of the controls that an organization observes, the method it uses to validate the authenticity of certificate applicants and the CA's expectations of how its certificates may be used.

Chain of custody—A legal principle regarding the validity and integrity of evidence. It requires accountability for anything that will be used as evidence in a legal proceeding to ensure that it can be accounted for from the time it was collected until the time it is presented in a court of law. Includes documentation as to who had access to the evidence and when, as well as the ability to identify evidence as being the exact item that was recovered or tested. Lack of control over evidence can lead to it being discredited. Chain of custody depends on the ability to verify that evidence could not have been tampered with. This is accomplished by sealing off the evidence, so it cannot be changed, and providing a documentary record of custody to prove that the evidence was at all times under strict control and not subject to tampering.

Challenge/response token—A method of user authentication that is carried out through use of the Challenge Handshake Authentication Protocol (CHAP). When a user tries to log into the server using CHAP, the server sends the user a "challenge," which is a random value. The user enters a password, which is used as an encryption key to encrypt the "challenge" and return it to the server. The server is aware of the password. It, therefore, encrypts the "challenge" value and compares it with the value received from the user. If the values match, the user is authenticated. The challenge/response activity continues throughout the session and this protects the session from password sniffing attacks. In addition, addition, CHAP is not vulnerable to "man-in-the-middle" attacks because the challenge value is a random value that changes on each access attempt.

Change management—A holistic and proactive approach to managing the transition from a current to a desired organizational state, focusing specifically on the critical human or "soft" elements of change. Includes activities such as culture change (values, beliefs and attitudes), development of reward systems (measures and appropriate incentives) incentives), organizational design, stakeholder management, human resources (HR) policies and procedures, executive coaching, change leadership training, team building and communication planning and execution

Channel Service Unit/Digital Service Unit (CSU/DSU)—Interfaces at the physical layer of the open systems interconnection (OSI) reference model, data terminal equipment (DTE) to data circuit terminating equipment (DCE), for switched carrier networks

Check digit—A numeric value, which has been calculated mathematically, that is added to data to ensure that original data have not been altered or that an incorrect, but valid match has occurred. Check digit control is effective in detecting transposition and transcription errors.

Checklist—A list of items that is used to verify the completeness of a task or goal. Used in quality assurance (and, in general, in information systems audit) to check process compliance, code standardization and error prevention, and other items for which consistency processes or standards have been defined.

Checkpoint restart procedures—A point in a routine at which sufficient information can be stored to permit restarting the computation from that point.

Checksum—A mathematical value that is assigned to a file and used to "test" the file at a later date to verify that the data contained in the file have not been maliciously changed. A cryptographic checksum is created by performing a complicated series of mathematical operations (known as a cryptographic algorithm) that translates the data in the file into a fixed string of digits called a hash value, which is then used as the checksum. Without knowing which cryptographic algorithm was used to create the hash value, it is highly unlikely that an unauthorized person would be able to change data without inadvertently changing the corresponding checksum. Cryptographic checksums are used in data transmission and data storage. Cryptographic checksums are also known as message authentication codes, integrity check-values, modification detection codes or message integrity codes.

Ciphertext—Information generated by an encryption algorithm to protect the plaintext and that is unintelligible to the unauthorized reader.

Circuit-switched network—A data transmission service requiring the establishment of a circuit-switched connection before data can be transferred from source data terminal equipment (DTE) to a sink DTE. A circuit-switched data transmission service uses a connection network.

Circular routing—In open systems architecture, circular routing is the logical path of a message in a communication network based on a series of gates at the physical network layer in the open systems interconnection (OSI) model.

Client-server—A group of computers connected by a communications network, in which the client is the requesting machine and the server is the supplying machine. Software is specialized at both ends. Processing may take place on either the client or the server, but it is transparent to the user.

Cloud computing—A model for enabling convenient, on-demand network access to a shared pool of configurable computing resources (e.g., networks, servers, storage, applications and services) that can be rapidly provisioned and released with minimal management effort or service provider interaction

Coaxial cable—Composed of an insulated wire that runs through the middle of each cable, a second wire that surrounds the insulation of the inner wire like a sheath, and the outer insulation which wraps the second wire. Has a greater transmission capacity than standard twisted-pair cables, but has a limited range of effective distance.

Cohesion—The extent to which a system unit—subroutine, program, module, component, subsystem—performs a single dedicated function. Generally, the more cohesive are units, the easier it is to maintain and enhance a system because it is easier to determine where and how to apply a change.

Cold site—An IS backup facility that has the necessary electrical and physical components of a computer facility, but does not have the computer equipment in place. The site is ready to receive the necessary replacement computer equipment in the event the users have to move from their main computing location to the alternative computer facility.

Communication processor—A computer embedded in a communications system that generally performs basic tasks of classifying network traffic and enforcing network policy functions. An example is the message data processor of a digital divide network (DDN) switching center. More advanced communications processors may perform additional functions.

Comparison program—A program for the examination of data, using logical or conditional tests to determine or to identify similarities or differences

Compensating control—An internal control that reduces the risk of an existing or potential control weakness resulting in errors and omissions

Compiler—A program that translates programming language (source code) into machine executable instructions (object code)

Completely connected (mesh) configuration—A network topology in which devices are connected with many redundant interconnections between network nodes (primarily used for backbone networks)

Completeness check—A procedure designed to ensure that no fields are missing from a record

Compliance testing—Tests of control designed to obtain audit evidence on both the effectiveness of the controls and their operation during the audit period

Components (as in component-based development)—Cooperating packages of executable software that make their services available through defined interfaces. Components used in developing systems may be commercial off-the-shelf software (COTS) or may be purposely built. However, the goal of component-based development is to ultimately use as many predeveloped, pretested components as possible.

Comprehensive audit—An audit designed to determine the accuracy of financial records as well as evaluate the internal controls of a function or department

Computer emergency response team (CERT)—A group of people integrated at the organization with clear lines of reporting and responsibilities for standby support in case of an information systems emergency. This group will act as an efficient corrective control, and should also act as a single point of contact for all incidents and issues related to information systems.

Computer forensics—The application of the scientific method to digital media to establish factual information for judicial review. This process often involves investigating computer systems to determine whether they are or have been used for illegal or unauthorized activities. As a discipline, it combines elements of law and computer science to collect and analyze data from information systems (e.g., personal computers, networks, wireless communications and digital storage devices) in a way that is admissible as evidence in a court of law.

Computer sequence checking—Verifies that the control number follows sequentially and that any control numbers out of sequence are rejected or noted on an exception report for further research

Computer-aided software engineering (CASE)—The use of software packages that aid in the development of all phases of an information system. System analysis, design programming and documentation are provided. Changes introduced in one CASE chart will update all other related charts automatically. CASE can be installed on a microcomputer for easy access.

Computer-assisted audit technique (CAAT)—Any automated audit technique, such as generalized audit software (GAS), test data generators, computerized audit programs and specialized audit utilities

Concurrency control—Refers to a class of controls used in database management systems (DBMS) to ensure that transactions are processed in an atomic, consistent, isolated and durable manner (ACID). This implies that only serial and recoverable schedules are permitted, and that committed transactions are not discarded when undoing aborted transactions.

Configuration management—The control of changes to a set of configuration items over a system life cycle

Console log—An automated detail report of computer system activity

Contingency planning—Process of developing advance arrangements and procedures that enable an enterprise to respond to an event that could occur by chance or unforeseen circumstances

Continuity—Preventing, mitigating and recovering from disruption. The terms "business resumption planning," "disaster recovery planning" and "contingency planning" also may be used in this context; they all concentrate on the recovery aspects of continuity.

Continuous auditing approach—This approach allows IS auditors to monitor system reliability on a continuous basis and to gather selective audit evidence through the computer.

Continuous improvement—The goals of continuous improvement (Kaizen) include the elimination of waste, defined as "activities that add cost, but do not add value;" just-in-time (JIT) delivery; production load leveling of amounts and types; standardized work; paced moving lines; right-sized equipment. A closer definition of the Japanese usage of Kaizen is "to take it apart and put back together in a better way." What is taken apart is usually a process, system, product or service. Kaizen is a daily activity whose purpose goes beyond improvement. It is also a process that, when done correctly, humanizes the workplace, eliminates hard work (both mental and physical), and teaches people how to do rapid experiments using the scientific method and how to learn to see and eliminate waste in business processes.

Control group—Members of the operations area that are responsible for the collection, logging and submission of input for the various user groups

Control objective—A statement of the desired result or purpose to be achieved by implementing control procedures in a particular process

Control practice—Key control mechanism that supports the achievement of control objectives through responsible use of resources, appropriate management of risk and alignment of IT with business

Control risk—The risk that a material error exists that would not be prevented or detected on a timely basis by the system of internal controls

Control section—The area of the central processing unit (CPU) that executes software, allocates internal memory and transfers operations between the arithmetic-logic, internal storage and output sections of the computer

Cookie—A message kept in the web browser for the purpose of identifying users and possibly preparing customized web pages for them. The first time a cookie is set, a user may be required to go through a registration process. Subsequent to this, whenever the cookie's message is sent to the server, a customized view based on that user's preferences can be produced. The browser's implementation of cookies has, however, brought several security concerns, allowing breaches of security and the theft of personal information (e.g., user passwords that validate the user's identity and enable restricted web services).

Corporate governance—The system by which organizations are directed and controlled. The board of directors are responsible for the governance of their organizations. It consists of the leadership and organizational structures and processes that ensure the organization sustains and extends strategies and objectives.

Corrective control—Designed to correct errors, omissions and unauthorized uses and intrusions once they are detected

Countermeasure—Any process that directly reduces a threat or vulnerability

Coupling—Measure of interconnectivity among structure of software programs. Coupling depends on the interface complexity between modules. This can be defined as the point at which entry or reference is made to a module, and what data pass across the interface. In application software design, it is preferable to strive for the lowest possible coupling between modules. Simple connectivity among modules results in software that is easier to understand and maintain, and less prone to a ripple or domino effect caused when errors occur at one location and propagate through a system.

Critical infrastructure—Systems whose incapacity or destruction would have a debilitating effect on the economic security of an enterprise, community or nation

Critical success factor (CSF)—The most important issue or action for management to achieve control over and within its IT processes

Customer relationship management (CRM)—A way to identify, acquire and retain customers. CRM is also an industry term for software solutions that help an organization manage customer relationships in an organized manner.

D

Data communications—The transfer of data between separate computer processing sites/devices using telephone lines, microwave and/or satellite links

Data custodian—Individual(s) and department(s) responsible for the storage and safeguarding of computerized information. This typically is within the IS organization.

Data dictionary—A database that contains the name, type, range of values, source, and authorization for access for each data element in a database. It also indicates which application programs use those data so that when a data structure is contemplated, a list of the affected programs can be generated. May be a stand-alone information system used for management or documentation purposes, or it may control the operation of a database.

Data diddling—Changing data with malicious intent before or during input into the system

Data Encryption Standard (DES)—An algorithm for encoding binary data. It is a secret key cryptosystem published by the National Bureau of Standards (NBS), the predecessor of the US National Institute of Standards and Technology (NIST). DES was defined as a Federal Information Processing Standard (FIPS) in 1976 and has been used commonly for data encryption in the forms of software and hardware implementation. (See private key cryptosystem.)

Data leakage—Siphoning out or leaking information by dumping computer files or stealing computer reports and tapes

Data owner—Individual(s), normally a manager or director, who have responsibility for the integrity, accurate reporting and use of computerized data

Data security—Those controls that seek to maintain confidentiality, integrity and availability of information

Data structure—The relationships among files in a database and among data items within each file

Database—A stored collection of related data needed by organizations and individuals to meet their information processing and retrieval requirements

Database administrator (DBA)—An individual or department responsible for the security and information classification of the shared data stored on a database system. This responsibility includes the design, definition and maintenance of the database.

Database management system (DBMS)—A software system that controls the organization, storage and retrieval of data in a database

Database replication—The process of creating and managing duplicate versions of a database. Replication not only copies a database but also synchronizes a set of replicas so that changes made to one replica are reflected in all of the others. The beauty of replication is that it enables many users to work with their own local copy of a database, but have the database updated as if they were working on a single centralized database. For database applications in which, geographically users are distributed widely, replication is often the most efficient method of database access.

Database specifications—These are the requirements for establishing a database application. They include field definitions, field requirements, and reporting requirements for the individual information in the database.

Data-oriented systems development—Focuses on providing ad hoc reporting for users by developing a suitable accessible database of information and to provide useable data rather than a function

Decentralization—The process of distributing computer processing to different locations within an organization

Decision support system (DSS)—An interactive system that provides the user with easy access to decision models and data, to support semistructured decision-making tasks

Decryption—A technique used to recover the original plaintext from the ciphertext such that it is intelligible to the reader. The decryption is a reverse process of the encryption.

Decryption key—A piece of information used to recover the plaintext from the corresponding ciphertext by decryption

Degauss—The application of variable levels of alternating current for the purpose of demagnetizing magnetic recording media. The process involves increasing the alternating current field gradually from zero to some maximum value and back to zero, leaving a very low residue of magnetic induction on the media. Degauss loosely means to erase.

Demodulation—The process of converting an analog telecommunications signal into a digital computer signal.

Detection risk—The risk that material errors or misstatements that have occurred will not be detected by the IS auditor

Detective control—Exists to detect and report when errors, omissions and unauthorized uses or entries occur.

Dial-back—Used as a control over dial-up telecommunications lines. The telecommunications link established through dial-up into the computer from a remote location is interrupted so the computer can dial back to the caller. The link is permitted only if the caller is from a valid phone number or telecommunications channel.

Dial-in access control—Prevents unauthorized access from remote users who attempt to access a secured environment. Ranges from a dial-back control to remote user authentication.

Digital certificate—A piece of information, a digitized form of signature, that provides sender authenticity, message integrity and nonrepudiation. A digital signature is generated using the sender's private key or applying a one-way hash function.

Digital signature—A piece of information, a digitized form of a signature, that provides sender authenticity, message integrity and nonrepudiation. A digital signature is generated using the sender's private key or applying a one-way hash function.

Disaster recovery plan (DRP)—A set of human, physical, technical and procedural resources to recover, within a defined time and cost, an activity interrupted by an emergency or disaster

Disaster tolerance—The time gap during which the business can accept the non-availability of IT facilities.

Discovery sampling—A form of attribute sampling that is used to determine a specified probability of finding at least one example of an occurrence (attribute) in a population

Discretionary access control (DAC)—A means of restricting access to objects based on the identity of subjects and/or groups to which they belong. The controls are discretionary in the sense that a subject with a certain access permission is capable of passing that permission (perhaps indirectly) on to any other subject.

Diskless workstations—A workstation or PC on a network that does not have its own disk, but instead stores files on a network file server

Distributed data processing network—A system of computers connected together by a communications network. Each computer processes its data and the network supports the system as a whole. Such a network enhances communication among the linked computers and allows access to shared files.

Diverse routing—The method of routing traffic through split cable facilities or duplicate cable facilities. This can be accomplished with different and/or duplicate cable sheaths. If different cable sheaths are used, the cable may be in the same conduit and, therefore, subject to the same interruptions as the cable it is backing up. The communication service subscriber can duplicate the facilities by having alternate routes, although the entrance to and from the customer premises may be in the same conduit. The subscriber can obtain diverse routing and alternate routing from the local carrier, including dual entrance facilities. However, acquiring this type of access is time-consuming and costly. Most carriers provide facilities for alternate and diverse routing, although the majority of services are transmitted over terrestrial media. These cable facilities are usually located in the ground or basement. Ground-based facilities are at great risk due to the aging infrastructures of cities. In addition, cable-based facilities usually share room with mechanical and electrical systems that can impose great risks due to human error and disastrous events.

Domain name system (DNS)—A hierarchical database that is distributed across the Internet that allows names to be resolved into IP addresses (and vice versa) to locate services such as web and email servers

Domain name system (DNS) poisoning—Corrupts the table of an Internet server's DNS, replacing an Internet address with the address of another vagrant or scoundrel address. If a web user looks for the page with that address, the request is redirected by the scoundrel entry in the table to a different address. Cache poisoning differs from another form of DNS poisoning in which the attacker spoofs valid email accounts and floods the "n" boxes of administrative and technical contacts. Cache poisoning is related to URL poisoning or location poisoning, in which an Internet user behavior is tracked by adding an identification number to the location line of the browser that can be recorded as the user visits successive pages on the site. It is also called DNS cache poisoning or cache poisoning.

Downloading—The act of transferring computerized information from one computer to another computer

Downtime report—A report that identifies the elapsed time when a computer is not operating correctly because of machine failure

Dry-pipe fire extinguisher system—Refers to a sprinkler system that does not have water in the pipes during idle usage, unlike a fully charged fire extinguisher system that has water in the pipes at all times. The dry-pipe system is activated at the time of the fire alarm and water is emitted to the pipes from a water reservoir for discharge to the location of the fire.

Dumb terminal—A display terminal without processing capability. Dumb terminals are dependent on the main computer for processing. All entered data are accepted without further editing or validation.

Dynamic Host Configuration Protocol (DHCP)—A protocol used by networked computers (clients) to obtain IP addresses and other parameters such as the default gateway, subnet mask and IP addresses of domain name system (DNS) servers from a DHCP server. The DHCP server ensures that all IP addresses are unique (e.g., no IP address is assigned to a second client while the first client's assignment is valid [its lease has not expired]). Thus, IP address pool management is done by the server and not by a human network administrator.

E

Echo checks—Detects line errors by retransmitting data back to the sending device for comparison with the original transmission

E-commerce—The processes by which enterprises conduct business electronically with their customers, suppliers and other external business partners, using the Internet as an enabling technology. E-commerce encompasses both business-to-business (B2B) and business-to-consumer (B2C) e-commerce models, but does not include existing non-Internet Internet e-commerce methods based on private networks such as electronic data interchange (EDI) and Society for Worldwide Interbank Financial Telecommunication (SWIFT).

Edit control—Detects errors in the input portion of information that is sent to the computer for processing. May be manual or automated, and allow the user to edit data errors before processing.

Editing—Ensures that data conform to predetermined criteria and enable early identification of potential errors

Electronic data interchange (EDI)—The electronic transmission of transactions (information) between two organizations. EDI promotes a more efficient paperless environment. EDI transmissions can replace the use of standard documents, including invoices or purchase orders.

Electronic funds transfer (EFT)—The exchange of money via telecommunications. EFT refers to any financial transaction that originates at a terminal and transfers a sum of money from one account to another.

Email/interpersonal messaging—An individual using a terminal, PC or an application can access a network to send an unstructured message to another individual or group of people

Embedded audit module (EAM)—Integral part of an application system that is designed to identify and report specific transactions or other information based on predetermined criteria. Identification of reportable items occurs as part of real-time processing. Reporting may be real-time online or may use store and forward methods. Also known as integrated test facility or continuous auditing module.

Encapsulation (objects)—The technique used by layered protocols in which a lower-layer protocol accepts a message from a higher-layer protocol and places it in the data portion of a frame in the lower layer.

Encryption—The process of taking an unencrypted message (plaintext), applying a mathematical function to it (encryption algorithm with a key) and producing an encrypted message (ciphertext)

Encryption key—A piece of information, in a digitized form, used by an encryption algorithm to convert the plaintext to the ciphertext

End-user computing—The ability of end users to design and implement their own information system utilizing computer software products

Enterprise resource planning (ERP)—A packaged business software system that allows an organization to automate and integrate the majority of its business processes, share common data and practices across the entire organization, and produce and access information in a real-time environment. Examples of ERP include SAP, Oracle Financials and J.D. Edwards.

Escrow agent—A person, agency or organization that is authorized to act on behalf of another to create a legal relationship with a third party in regards to an escrow agreement; the custodian of an asset according to an escrow agreement. As it relates to a cryptographic key, an escrow agent is the agency or organization charged with the responsibility for safeguarding the key components of the unique key.

Escrow agreement—A legal arrangement whereby an asset (often money, but sometimes other property such as art, a deed of title, web site, software source code or a cryptographic key) is delivered to a third party (called an escrow agent) to be held in trust or otherwise pending a contingency or the fulfillment of a condition or conditions in a contract. Upon the occurrence of the escrow agreement, the escrow agent will deliver the asset to the proper recipient; otherwise the escrow agent is bound by his/her fiduciary duty to maintain the escrow account. Source code escrow means deposit of the source code for the software into an account held by an escrow agent. Escrow is typically requested by a party licensing software (e.g., licensee or buyer) to ensure maintenance of the software. The software source code is released by the escrow agent to the licensee if the licensor (e.g., seller or contractor) files for bankruptcy or otherwise fails to maintain and update the software as promised in the software license agreement.

Ethernet—A popular network protocol and cabling scheme that uses a bus topology and carrier sense multiple access/collision detection (CSMA/CD) to prevent network failures or collisions when two devices try to access the network at the same time

Evidence—The information an auditor gathers in the course of performing an IS audit; relevant if it pertains to the audit objectives and has a logical relationship to the findings and conclusions it is used to support

Exception reports—An exception report is generated by a program that identifies transactions or data that appear to be incorrect. Exception reports may be outside a predetermined range or may not conform to specified criteria.

Exclusive-OR (XOR)—The exclusive-OR operator returns a value of TRUE only if just one of its operands is TRUE. The XOR operation is a Boolean operation that produces a 0 if its two Boolean inputs are the same (0 and 0 or 1 and 1) and it produces a 1 if its two inputs are different (1 and 0). In contrast, an inclusive-OR operator returns a value of TRUE if either or both of its operands are TRUE.

Executable code—The machine language code that is generally referred to as the object or load module

Expert system—The most prevalent type of computer system that arises from the research of artificial intelligence. An expert system has a built in hierarchy of rules, which are acquired from human experts in the appropriate field. Once input is provided, the system should be able to define the nature of the problem and provide recommendations to solve the problem.

Exposure—The potential loss to an area due to the occurrence of an adverse event

Extended Binary-coded Decimal Interchange Code (EBCDIC)—An 8-bit code representing 256 characters; used in most large computer systems

Extensible Markup Language (XML)—Promulgated through the World Wide Web Consortium, XML is a web-based application development technique that allows designers to create their own customized tags, thus enabling the definition, transmission, validation and interpretation of data between applications and organizations

Extranet—A private network that resides on the Internet and allows a company to securely share business information with customers, suppliers, or other businesses as well as to execute electronic transactions. different from an intranet in that it is located beyond the company's firewall. Therefore, an extranet relies on the use of securely issued digital certificates (or alternative methods of user authentication) and encryption of messages. A virtual private network (VPN) and tunneling are often used to implement extranets, to ensure security and privacy.

F

Fallback procedures—A plan of action or set of procedures to be performed if a system implementation, upgrade or modification does not work as intended. May involve restoring the system to its state prior to the implementation or change. Fallback procedures are needed to ensure that normal business processes continue in the event of failure and should always be considered in system migration or implementation.

False authorization—Also called false acceptance; occurs when an unauthorized person is identified as an authorized person by the biometric system.

False enrollment—Occurs when an unauthorized person manages to enroll into the biometric system. Enrollment is the initial process of acquiring a biometric feature and saving it as a personal reference on a smart card, a PC or in a central database.

Fault tolerance—A system's level of resilience to seamlessly react to hardware and/or software failure

Feasibility study—A phase of a system development life cycle (SDLC) methodology that researches the feasibility and adequacy of resources for the development or acquisition of a system solution to a user need

Fiber-optic cable—Glass fibers that transmit binary signals over a telecommunications network. Fiber-optic systems have low transmission losses as compared to twisted-pair cables. They do not radiate energy or conduct electricity. They are free from corruption and lightning-induced interference, and they reduce the risk of wiretaps.

Field—An individual data element in a computer record. Examples include employee name, customer address, account number, product unit price and product quantity in stock.

File—A named collection of related records

File allocation table (FAT)—A table used by the operating system to keep track of where every file is located on the disk. Since a file is often fragmented, and thus subdivided into many sectors within the disk, the information stored in the FAT is used when loading or updating the contents of the file.

File layout—Specifies the length of the file's record, and the sequence and size of its fields. Also will specify the type of data contained within each field: for example, alphanumeric, zoned decimal, packed and binary.

File server—A high-capacity disk storage device or a computer that stores data centrally for network users and manages access to that data. File servers can be dedicated so that no process other than network management can be executed while the network is available; file servers can be nondedicated so that standard user applications can run while the network is available.

File Transfer Protocol (FTP)—A protocol used to transfer files over a Transmission Control Protocol/Internet Protocol (TCP/IP) network (Internet, UNIX, etc.)

Financial audit—An audit designed to determine the accuracy of financial records and information

Firewall—A system or combination of systems that enforces a boundary between two or more networks typically forming a barrier between a secure and an open environment such as the Internet

Firmware—Memory chips with embedded program code that hold their content when power is turned off

Foreign key—A value that represents a reference to a tuple (a row in a table) containing the matching candidate key value. The problem of ensuring that the database does not include any invalid foreign key values is known as the referential integrity problem. The constraint that values of a given foreign key must match values of the corresponding candidate key is known as a referential constraint. The relation (table) that contains the foreign key is referred to as the referencing relation and the relation that contains the corresponding candidate key as the referenced relation or target relation. (In the relational theory it would be a candidate key, but in real database management systems (DBMSs) implementations it is always the primary key.)

Format checking—The application of an edit, using a predefined field definition to a submitted information stream; a test to ensure that data conform to a predefined format

Fourth-generation language (4GL)—High-level, user-friendly, nonprocedural computer languages used to program and/or read and process computer files

Frame relay—A packet-switched wide-area network (WAN) technology that provides faster performance than older packet-switched WAN technologies. Best suited for data and image transfers. Because of its variable-length packet architecture, it is not the most efficient technology for real-time voice and video. In a frame-relay network, end nodes establish a connection via a permanent virtual circuit (PVC).

Function point analysis—A technique used to determine the size of a development task, based on the number of function points. Function points are factors such as inputs, outputs, inquiries and logical internal sites.

G

Gateway—A device (router, firewall) on a network that serves as an entrance to another network

General computer control—A control, other than an application control, that relates to the environment within which computer-based application systems are developed, maintained and operated, and that is therefore applicable to all applications. The objectives of general controls are to ensure the proper development and implementation of applications and the integrity of program and data files and of computer operations. Like application controls, general controls may be either manual or programmed. Examples of general controls include the development and implementation of an IS strategy and an IS security policy, the organization of IS staff to separate conflicting duties and planning for disaster prevention and recovery.

Generalized audit software (GAS)—Multipurpose audit software that can be used for general processes such as record selection, matching, recalculation and reporting

Geographical information system (GIS)—A tool used to integrate, convert, handle, analyze and produce information regarding the surface of the earth. GIS data exist as maps, tridimensional virtual models, lists and tables.

Governance—Ensures that stakeholder needs, conditions and options are evaluated to determine balanced, agreed-on enterprise objectives to be achieved; setting direction through prioritization and decision making; and monitoring performance and compliance against agreed-on direction and objectives. Conditions can include the cost of capital, foreign exchange rates, etc. Options can include shifting manufacturing to other locations, sub-contracting portions of the enterprise to third parties, selecting a product mix from many available choices, etc.

H

Hacker—An individual who attempts to gain unauthorized access to a computer system

Handprint scanner—A biometric device that is used to authenticate a user through palm scans

Hardware—The physical components of a computer system

Hash total—The total of any numeric data field in a document or computer file. This total is checked against a control total of the same field to facilitate accuracy of processing.

Hardware—The physical components of a computer system

Help desk—A service offered via phone/Internet by an organization to its clients or employees that provides information, assistance, and troubleshooting advice regarding software, hardware, or networks. A help desk is staffed by people that can either resolve the problem on their own or escalate the problem to specialized personnel. A help desk is often equipped with dedicated customer relationship management (CRM) software that logs the problems and tracks them until they are solved.

Heuristic filter—A method often employed by antispam software to filter spam using criteria established in a centralized rule database. Every email message is given a rank, based upon its header and contents, which is then matched against preset thresholds. A message that surpasses the threshold will be flagged as spam and discarded, returned to its sender or put in a spam directory for further review by the intended recipient.

Hexadecimal—A numbering system that uses a base of 16 and uses 16 digits: 0, 1, 2, 3, 4, 5, 6, 7, 8, 9, A, B, C, D, E and F. Programmers use hexadecimal numbers as a convenient way of representing binary numbers.

Hierarchical database—A database structured in a tree/root or parent/child relationship. Each parent can have many children, but each child may have only one parent.

Honeypot—A specially configured server, also known as a decoy server, designed to attract and monitor intruders in a manner such that their actions do not affect production systems.

Hot site—A fully operational offsite data processing facility equipped with both hardware and system software to be used in the event of a disaster

Hypertext Markup Language (HTML)—A language designed for the creation of web pages with hypertext and other information to be displayed in a web browser. HTML is used to structure information—denoting certain text as headings, paragraphs, lists and so on—and can be used to describe, to some degree, the appearance and semantics of a document.

I

Image processing—The process of electronically inputting source documents by taking an image of the document, thereby eliminating the need for key entry

Impact assessment—A review of the possible consequences of a risk

Impersonation—A security concept related to Windows NT that allows a server application to temporarily "be" the client in terms of access to secure objects. Impersonation has three possible levels: identification, letting the server inspect the client's identity; impersonation, letting the server act on behalf of the client; and delegation, the same as impersonation but extended to remote systems to which the server connects (through the preservation of credentials). Impersonation by imitating or copying the identification, behavior or actions of another may also be used in social engineering to obtain otherwise unauthorized physical access.

Incident—Any event that is not part of the standard operation of a service and that causes, or may cause, an interruption to, or a reduction in, the quality of that service.

Incident response—The response of an enterprise to a disaster or other significant event that may significantly affect the enterprise, its people, or its ability to function productively. An incident response may include evacuation of a facility, initiating a disaster recovery plan (DRP), performing damage assessment, and any other measures necessary to bring an enterprise to a more stable status.

Incremental testing—Deliberately testing only the value added functionality of a software component

Independence—An IS auditor's self-governance and freedom from conflict of interest and undue influence. The IS auditor should be free to make his/her own decisions, not influenced by the organization being audited and its people (managers and employees).

Indexed sequential access method (ISAM)—A disk access method that stores data sequentially while also maintaining an index of key fields to all the records in the file for direct access capability

Information processing facility (IPF)—The computer room and support areas

Information security—Ensures that within the enterprise, information is protected against disclosure to unauthorized users (confidentiality), improper modification (integrity), and non access when required (availability)

Information security governance—The set of responsibilities and practices exercised by the board and executive management with the goal of providing strategic direction, ensuring that objectives are achieved, ascertaining that risk is managed appropriately and verifying that the enterprise's resources are used responsibly

Information systems (IS)—The combination of strategic, managerial and operational activities involved in gathering, processing, storing, distributing and using information and its related technologies. Information systems are distinct from information technology (IT) in that an information system has an IT component that interacts with the process components.

Inherent risk—The risk level or exposure without taking into account the actions that management has taken or might take (e.g., implementing controls)

Inheritance (objects)—Database structures that have a strict hierarchy (no multiple inheritance). Inheritance can initiate other objects irrespective of the class hierarchy, thus there is no strict hierarchy of objects

Initial program load (IPL)—The initialization procedure that causes an operating system to be loaded into storage at the beginning of a workday or after a system malfunction

Input control—Techniques and procedures used to verify, validate and edit data, to ensure that only correct data are entered into the computer

Instant messaging (IM)—An online mechanism or a form of real-time communication among two or more people based on typed text and multimedia data. The text is conveyed via computers or another electronic device (e.g., cell phone or handheld device) connected over a network, such as the Internet.

Integrated services digital network (ISDN)—A public end-to-end, digital telecommunications network with signaling, switching, and transport capabilities supporting a wide range of service accessed by standardized interfaces with integrated customer control. The standard allows transmission of digital voice, video and data over 64 kbps lines.

Integrated test facilities (ITF)—A testing methodology where test data are processed in production systems. The data usually represent a set of fictitious entities such as departments, customers and products. Output reports are verified to confirm the correctness of the processing.

Integrity—The guarding against improper information modification or destruction, and includes ensuring information nonrepudiation and authenticity

Interface testing—A testing technique that is used to evaluate output from one application while the information is sent as input to another application

Internal controls—The policies, procedures, practices and organizational structures designed to provide reasonable assurance that business objectives will be achieved and undesired events will be prevented or detected and corrected

Internet—1) Two or more networks connected by a router; 2) the world's largest network using Transmission Control Protocol/Internet Protocol (TCP/IP) to link government, university and commercial institutions.

Internet Engineering Task Force (IETF)—An organization with international affiliates as network industry representatives that sets Internet standards. This includes all network industry developers and researchers concerned with the evolution and planned growth of the Internet.

Internet packet (IP) spoofing—An attack using packets with the spoofed source Internet packet (IP) addresses. This technique exploits applications that use authentication based on IP addresses. This technique also may enable an unauthorized user to gain root access on the target system.

Internet Protocol Security (IPSec)—A set of protocols developed by the Internet Engineering Task Force (IETF) to support the secure exchange of packets

Irregularity—Intentional violations of established management policy or regulatory requirement. It may consist of deliberate misstatements or omissions of information concerning the area under audit or the organization as a whole; gross negligence or unintentional illegal acts.

Internet Security Association and Key Management Protocol (ISAKMP)—A protocol for sharing a public key

IT governance framework—A model that integrates a set of guidelines, policies and methods that represent the organizational approach to the IT governance. Per COBIT, IT governance is the responsibility of the board of directors and executive management. It is an integral part of institutional governance, and consists of the leadership and organizational structures and processes that ensure that the organization's IT sustains and extends the organization's strategy and objectives.

IT incident—Any event that is not part of the ordinary operation of a service that causes, or may cause, an interruption to, or a reduction in, the quality of that service

IT infrastructure—The set of hardware, software and facilities that integrates an organization's IT assets. Specifically, the equipment (including servers, routers, switches, and cabling), software, services and products used in storing, processing, transmitting and displaying all forms of information for the organization's users.

IT steering committee—An executive-management-level committee that assists in the delivery of the IT strategy, oversees day-to-day management of IT service delivery and IT projects, and focuses on implementation aspects

IT strategic plan—A long-term plan (i.e., three- to five-year horizon) in which business and IT management cooperatively describe how IT resources will contribute to the enterprise's strategic objectives (goals)

IT strategy committee—A committee at the level of the board of directors to ensure that the board is involved in major IT matters and decisions. The committee is primarily accountable for managing the portfolios of IT-enabled investments, IT services and other IT resources. The committee is the owner of the portfolio.

J

Judgment sampling—Any sample that is selected subjectively or in such a manner that the sample selection process is not random or the sampling results are not evaluated mathematically

K

Key goal indicator (KGI)—A measure that tells management, after the fact, whether an IT process has achieved its business requirements; usually expressed in terms of information criteria

Key management practice—Management practices that are required to successfully execute business processes

Key performance indicator (KPI)—A measure that determines how well the process is performing in enabling the goal to be reached. A lead indicator of whether a goal will likely be reached or not, and a good indicator of capabilities, practices and skills. It measures the activity goal, which is an action that the process owner must take to achieve effective process performance.

L

Leased line—A communication line permanently assigned to connect two points, as opposed to a dial up line that is only available and open when a connection is made by dialing the target machine or network. Also known as a dedicated line.

Librarian—The individual responsible for the safeguard and maintenance of all program and data files

Licensing agreement—A contract that establishes the terms and conditions under which a piece of software is being licensed (i.e., made legally available for use) from the software developer (owner) to the user

Life cycle—A series of stages that characterize the course of existence of an organizational investment (e.g, product, project, program)

Limit check—Tests specified amount fields against stipulated high or low limits of acceptability. When both high and low values are used, the test may be called a range check.

Literals—Any notation for representing a value within programming language source code (e.g., a string literal); a chunk of input data that is represented "as is" in compressed data

Local area network (LAN)—Communication network that serves several users within a specified geographical area. A personal computer LAN functions as distributed processing system in which each computer in the network does its own processing and manages some of its data. Shared data are stored in a file server that acts as a remote disk drive for all users in the network.

Log—To record details of the information or events in an organized record-keeping system, usually sequenced in the order in which they occurred

Logical access controls—The policies, procedures, organizational structure and electronic access controls designed to restrict access to computer software and data files

Logon—The act of connecting to the computer, which typically requires entry of a user ID and password into a computer terminal

M

Magnetic card reader—Reads cards with a magnetic surface on which data can be stored and retrieved

Malware—Short for malicious software. Designed to infiltrate, damage or obtain information from a computer system without the owner's consent. Malware is commonly taken to include computer viruses, worms, Trojan horses, spyware and adware. Spyware is generally used for marketing purposes and, as such, is not really malicious although it is generally unwanted. Spyware can, however, be used to gather information for identity theft or other clearly illicit purposes.

Management information system (MIS)—An organized assembly of resources and procedures required to collect, process and distribute data for use in decision making

Mandatory access controls (MAC)—A means of restricting access to data based on varying degrees of security requirements for information contained in the objects and the corresponding security clearance of users or programs acting on their behalf

Mapping—Diagramming data that is to be exchanged electronically, including how they are to be used and what business management systems need them. Mapping is a preliminary step for developing an applications link. (See application tracing and mapping.)

Masking—A computerized technique of blocking out the display of sensitive information, such as passwords, on a computer terminal or report

Master file—A file of semi-permanent information that is used frequently for processing data or for more than one purpose

Materiality—An auditing concept regarding the importance of an item of information with regard to its impact or effect on the functioning of the entity being audited. An expression of the relative significance or importance of a particular matter in the context of the organization as a whole.

Maturity—In business, indicates the degree of reliability or dependency that the business can place on a process achieving the desired goals or objectives

Maturity model—See capability maturity model (CMM).

Media Access Control (MAC)—Applied to the hardware at the factory and cannot be modified, MAC is a unique, 48-bit, hard-coded address of a physical layer device, such as an Ethernet local area network (LAN) or a wireless network card.

Media oxidation—The deterioration of the media on which data are digitally stored due to exposure to oxygen and moisture. Tapes deteriorating in a warm, humid environment are an example of media oxidation. Proper environmental controls should prevent, or significantly slow, this process.

Memory dump—The act of copying raw data from one place to another with little or no formatting for readability. Usually, dump refers to copying data from the main memory to a display screen or a printer. Dumps are useful for diagnosing bugs. After a program fails, one can study the dump and analyze the contents of memory at the time of the failure. A memory dump will not help unless each person knows what to look for because dumps are usually output in a difficult-to-read form (binary, octal or hexadecimal).

Message switching—A telecommunications methodology that controls traffic in which a complete message is sent to a concentration point and stored until the communications path is established

Microwave transmission—A high-capacity line-of-sight transmission of data signals through the atmosphere which often requires relay stations

Middleware—Another term for an application programmer interface (API). It refers to the interfaces that allow programmers to access lower- or higher-level services by providing an intermediary layer that includes function calls to the services.

Milestone—A terminal element that marks the completion of a work package or phase. Typically marked by a high-level event such as project completion, receipt, endorsement or signing of a previously-defined deliverable or a high-level review meeting at which the appropriate level of project completion is determined and agreed to. A milestone is associated with some sort of decision that outlines the future of a project and, for an outsourced project, may have a payment to the contractor associated with it.

Mission-critical application—An application that is vital to the operation of the organization. The term is very popular for describing the applications required to run the day-to-day business.

Mobile site—The use of a mobile/temporary facility to serve as a business resumption location. The facility can usually be delivered to any site and can house information technology and staff.

Modulation—The process of converting a digital computer signal into an analog telecommunications signal

Monetary unit sampling—A sampling technique that estimates the amount of overstatement in an account balance

N

Network—A system of interconnected computers and the communications equipment used to connect them

Network administrator—Responsible for planning, implementing and maintaining the telecommunications infrastructure; also may be responsible for voice networks. For smaller organizations, the network administrator may also maintain a local area network (LAN) and assist end users.

Network attached storage (NAS)—Utilize dedicated storage devices that centralize storage of data. NAS devices generally do not provide traditional file/print or application services.

Network interface card (NIC)—A communication card that when inserted into a computer, allows it to communicate with other computers on a network. Most NICs are designed for a particular type of network or protocol.

Noise—Disturbances in data transmissions, such as static, that cause messages to be misinterpreted by the receiver

Nondisclosure agreement (NDA)—A legal contract between at least two parties that outlines confidential materials the parties wish to share with one another for certain purposes, but wish to restrict from generalized use; a contract through which the parties agree not to disclose information covered by the agreement. Also called a confidential disclosure agreement (CDA), confidentiality agreement or secrecy agreement. An NDA creates a confidential relationship between the parties to protect any type of trade secret. As such, an NDA can protect non-public business information. In the case of certain governmental entities, the confidentiality of information other than trade secrets may be subject to applicable statutory requirements and, in some cases, may be required to be revealed to an outside party requesting the information. Generally, the governmental entity will include a provision in the contract to allow the seller to review a request for information the seller identifies as confidential and the seller may appeal such a decision requiring disclosure. NDAs are commonly signed when two companies or individuals are considering doing business together and need to understand the processes used in one another's businesses solely for the purpose of evaluating the potential business relationship. NDAs can be "mutual," meaning both parties are restricted in their use of the materials provided, or they can only restrict a single party. It is also possible for an employee to sign an NDA or NDA-like agreement with a company at the time of hiring; in fact, some employment agreements will include a clause restricting "confidential information" in general.

Normalization—The elimination of redundant data

Numeric check—An edit check designed to ensure that the data element in a particular field is numeric

O

Object code—Machine-readable instructions produced from a compiler or assembler program that has accepted and translated the source code

Object orientation—An approach to system development in which the basic unit of attention is an object, which represents an encapsulation of both data (an object's attributes) and functionality (an object's methods). Objects usually are created using a general template called a class. A class is the basis for most design work in objects. A class and its objects communicate in defined ways. Aggregate classes interact through messages, which are directed requests for services from one class (the client) to another class (the server). A class may share the structure or methods defined in one or more other classes—a relationship known as inheritance.

Objectivity—The ability of the IS auditor to exercise judgment, express opinions and present recommendations with impartiality

Offsite storage—A facility located away from the building housing the primary information processing facility (IPF), used for storage of computer media such as offline backup data and storage files

Online data processing—Achieved by entering information into the computer via a video display terminal. With online data processing, the computer immediately accepts or rejects the information as it is entered.

Open system—System for which detailed specifications of the composition of its component are published in a nonproprietary environment, thereby enabling competing enterprises to use these standard components to build competitive systems. The advantages of using open systems include portability, interoperability and integration.

Operating system (OS)—A master control program that runs the computer and acts as a scheduler and traffic controller. The operating system is the first program copied into the computer's memory after the computer is turned on; it must reside in memory at all times. It is the software that interfaces between the computer hardware (disk, keyboard, mouse, network, modem, printer) and the application software (word processor, spreadsheet, email), which also controls access to the devices and is partially responsible for security components and sets the standards for the application programs that run in it.

Operational audit—An audit designed to evaluate the various internal controls, economy and efficiency of a function or department

Operational control—Deals with the everyday operation of a company or organization to ensure that all objectives are achieved.

Operator console—A special terminal used by computer operations personnel to control computer and systems operations functions. Operator console terminals typically provide a high level of computer access and should be properly secured.

Optical scanner—An input device that reads characters and images that are printed or painted on a paper form into the computer

Outsourcing—A formal agreement with a third party to perform IS or other business functions for an enterprise

P

Packet—Data unit that is routed from source to destination in a packet-switched network. A packet contains both routing information and data. Transmission Control Protocol/Internet Protocol (TCP/IP) is such a packet-switched network.

Packet switching—The process of transmitting messages in convenient pieces that can be reassembled at the destination

Paper test—A walk-through of the steps of a regular test, but without actually performing the steps. Usually used in disaster recovery and contingency testing; team members review and become familiar with the plans and their specific roles and responsibilities.

Parallel testing—The process of feeding test data into two systems, the modified system and an alternative system (possibly the original system), and comparing results to demonstrate the consistency and inconsistency between two versions of the application

Parity check—A general hardware control that helps to detect data errors when data are read from memory or communicated from one computer to another. A 1-bit digit (either 0 or 1) is added to a data item to indicate whether the sum of that data item's bit is odd or even. When the parity bit disagrees with the sum of the other bits, the computer reports an error. The probability of a parity check detecting an error is 50 percent.

Partitioned file—A file format in which the file is divided into multiple sub files and a directory is established to locate each sub file

Passive assault—Intruders attempt to learn some characteristic of the data being transmitted. With a passive assault, intruders may be able to read the contents of the data so the privacy of the data is violated. Alternatively, although the content of the data itself may remain secure, intruders may read and analyze the plaintext source and destination identifiers attached to a message for routing purposes, or they may examine the lengths and frequency of messages being transmitted.

Password—A protected, generally computer-encrypted string of characters that authenticate a computer user to the computer system

Patch management—An area of systems management that involves acquiring, testing, and installing multiple patches (code changes) to an administered computer system in order to maintain up-to-date software and often to address security risk. Patch management tasks include the following: maintaining current knowledge of available patches; deciding what patches are appropriate for particular systems; ensuring that patches are installed properly; testing systems after installation; and documenting all associated procedures, such as specific configurations required. A number of products are available to automate patch management tasks. Patches are sometimes ineffective and can sometimes cause more problems than they fix. Patch management experts suggest that system administrators take simple steps to avoid problems such as performing backups and testing patches on non-critical systems prior to installations. Patch management can be viewed as part of change management.

Payroll system—An electronic system for processing payroll information and the related electronic (e.g., electronic timekeeping and/or human resources system), human (e.g., payroll clerk), and external party (e.g., bank) interfaces. In a more limited sense, it is the electronic system that performs the processing for generating payroll checks and/or bank direct deposits to employees.

Penetration testing—A live test of the effectiveness of security defenses through mimicking the actions of real life attackers

Performance driver—A measure that is considered the "driver" of a lag indicator. It can be measured before the outcome is clear and, therefore, is called a "lead indicator." There is an assumed relationship between the two that suggests that improved performance in a leading indicator will drive better performance in the lagging indicator. They are also referred to as key performance indicators (KPIs) and are used to indicate whether goals are likely to be met.

Performance testing—Comparing the system's performance to other equivalent systems, using well-defined benchmarks

Peripherals—Auxiliary computer hardware equipment used for input, output and data storage. Examples of peripherals include disk drives and printers.

Personal identification number (PIN)—A type of password (i.e., a secret number assigned to an individual) that, in conjunction with some means of identifying the individual, serves to verify the authenticity of the individual. PINs have been adopted by financial institutions as the primary means of verifying customers in an electronic funds transfer (EFT) system.

Phishing—This is a type of electronic mail (email) attack that attempts to convince a user that the originator is genuine, but with the intention of obtaining information for use in social engineering. Phishing attacks may take the form of masquerading as a lottery organization advising the recipient or the user's bank of a large win; in either case, the intent is to obtain account and personal identification number (PIN) details. Alternative attacks may seek to obtain apparently innocuous business information which may be used in another form of active attack.

Phreakers—Those who crack security, most frequently phone and other communication networks

Plaintext—Digital information, such as cleartext, that is intelligible to the reader

Point-of-sale (POS) systems—Enable the capture of data at the time and place of transaction. POS terminals may include use of optical scanners for use with bar codes or magnetic card readers for use with credit cards. POS systems may be online to a central computer or may use stand-alone terminals or microcomputers that hold the transactions until the end of a specified period when they are sent to the main computer for batch processing.

Point-to-point protocol (PPP)—A protocol used for transmitting data between two ends of a connection

Policy—1. Generally, a document that records a high-level principle or course of action that has been decided on. The intended purpose is to influence and guide both present and future decision making to be in line with the philosophy, objectives and strategic plans established by the enterprise's management teams. In addition to policy content, policies need to describe the consequences of failing to comply with the policy, the means for handling exceptions, and the manner in which compliance with the policy will be checked and measured.
2. Overall intention and direction as formally expressed by management (COBIT 5 perspective)

Portfolio—A grouping of "objects of interest" (investment programs, IT services, IT projects, other IT assets or resources) managed and monitored to optimize business value (The investment portfolio is of primary interest to Val IT. IT service, project, asset and other resource portfolios are of primary interest to COBIT.)

Preventive control—An internal control that is used to avoid undesirable events, errors and other occurrences that an enterprise has determined could have a negative material effect on a process or end product

Privacy—The rights of an individual to trust that others will appropriately and respectfully use, store, share and dispose of his/her associated personal and sensitive information within the context, and according to the purposes, for which it was collected or derived. What is appropriate depends on the associated circumstances, laws and the individual's reasonable expectations. An individual also has the right to reasonably control and be aware of the collection, use and disclosure of his\her associated personal and sensitive information.

Private branch exchange (PBX)—A telephone exchange that is owned by a private business, as opposed to one owned by a common carrier or by a telephone company

Private key cryptosystem—Used in data encryption, it utilizes a secret key to encrypt the plaintext to the ciphertext. Private key cryptosystems also use the same key to decrypt the ciphertext to the corresponding plaintext. In this case, the key is symmetric such that the encryption key is equivalent to the decryption key.

Problem escalation procedure—The process of escalating a problem up from junior to senior support staff, and ultimately to higher levels of management. Problem escalation procedure is often used in help desk management, when an unresolved problem is escalated up the chain of command, until it is solved.

Procedure—A document containing a detailed description of the steps necessary to perform specific operations in conformance with applicable standards. Procedures are defined as part of processes.

Process—Generally, a collection of activities influenced by the enterprise's policies and procedures that takes inputs from a number of sources (including other processes), manipulates the inputs and produces outputs. Processes have clear business reasons for existing, accountable owners, clear roles and responsibilities around the execution of the process, and the means to measure performance.

Production program—Program used to process live or actual data that were received as input into the production environment

Production software—Software that is being used and executed to support normal and authorized organizational operations. Production software is to be distinguished from test software, which is being developed or modified, but has not yet been authorized for use by management.

Professional competence—Proven level of ability, ability, often linked to qualifications issued by relevant professional bodies and compliance with their codes of practice and standards

Program Evaluation and Review Technique (PERT)—A project management technique used in the planning and control of system projects

Program flowchart—Shows the sequence of instructions in a single program or subroutine. The symbols used in program flowcharts should be the internationally accepted standard. Program flowcharts should be updated when necessary.

Program narrative—Provides a detailed explanation of program flowcharts, including control points and any external input

Project—A structured set of activities concerned with delivering a defined capability (that is necessary but not sufficient, to achieve a required business outcome) to the enterprise based on an agreed-on schedule and budget

Project portfolio—The set of projects owned by a company. It usually includes the main guidelines relative to each project, including objectives, costs, time lines and other information specific to the project.

Protocol—The rules by which a network operates and controls the flow and priority of transmissions

Protocol converter—Hardware devices, such as asynchronous and synchronous transmissions, that convert between two different types of transmission

Prototyping—The process of quickly putting together a working model (a prototype) in order to test various aspects of a design, illustrate ideas or features and gather early user feedback. Prototyping uses programmed simulation techniques to represent a model of the final system to the user for advisement and critique. The emphasis is on end-user screens and reports. Internal controls are not a priority item since this is only a model.

Proxy server—A server that acts on behalf of a user. Typical proxies accept a connection from a user, make a decision as to whether the user or client IP address is permitted to use the proxy, perhaps perform additional authentication, authentication and complete a connection to a remote destination on behalf of the user.

Public key cryptosystem—Used in data encryption, it uses an encryption key, as a public key, to encrypt the plaintext to the ciphertext. It uses a different decryption key, as a secret key, to decrypt the ciphertext to the corresponding plaintext. In contrast to a private key cryptosystem, the decryption key should be secret; however, the encryption key can be known to everyone. In a public key cryptosystem, the two keys are asymmetric, such that the encryption key is not equivalent to the decryption key.

Public key encryption—A cryptographic system that uses two keys: one is a public key, which is known to everyone, and the second is a private or secret key, which is only known to the recipient of the message

Public key infrastructure (PKI)—A series of processes and technologies for the association of cryptographic keys with the entity to whom those keys were issued

Q

Quality assurance (QA)—A planned and systematic pattern of all actions necessary to provide adequate confidence that an item or product conforms to established technical requirements (ISO/IEC 24765).

Queue—A group of items that are waiting to be serviced or processed

R

Radio wave interference—The superposition of two or more radio waves resulting in a different radio wave pattern that is more difficult to intercept and decode properly

Random access memory (RAM)— The computer's primary working memory. Each byte of RAM can be accessed randomly regardless of adjacent bytes.

Range check—Range checks ensure that data fall within a predetermined range

Rapid application development—A methodology that enables enterprises to develop strategically important systems faster, while reducing development costs and maintaining quality by using a series of proven application development techniques, within a well-defined methodology

Real-time processing—An interactive online system capability that immediately updates computer files when transactions are initiated through a terminal

Reasonable assurance—A level of comfort short of a guarantee, but considered adequate given the costs of the control and the likely benefits achieved

Reasonableness check—Compares data to predefined reasonability limits or occurrence rates established for the data

Reciprocal agreement—Emergency processing agreement between two or more enterprises with similar equipment or applications. Typically, participants of a reciprocal agreement promise to provide processing time to each other when an emergency arises.

Record—A collection of related information treated as a unit. Separate fields within the record are used for processing the information.

Recovery point objective (RPO)—Determined based on the acceptable data loss in case of a disruption of operations. It indicates the earliest point in time to which it is acceptable to recover the data. The RPO effectively quantifies the permissible amount of data loss in case of interruption.

Recovery strategy—An approach by an enterprise that will ensure its recovery and continuity in the face of a disaster or other major outage. Plans and methodologies are determined by the enterprise's strategy. There may be more than one methodology or solution for an enterprise's strategy. Examples of methodologies and solutions include: contracting for hot site or cold site, building an internal hot site or cold site, identifying an alternate work area, a consortium or reciprocal agreement, contracting for mobile recovery or crate and ship, and many others.

Recovery testing—A test to check the system's ability to recover after a software or hardware failure

Recovery time objective (RTO)—The amount of time allowed for the recovery of a business function or resource after a disaster occurs

Redundancy check—Detects transmission errors by appending calculated bits onto the end of each segment of data

Redundant Array of Inexpensive Disks (RAID)— Provides performance improvements and fault-tolerant capabilities via hardware or software solutions, by writing to a series of multiple disks to improve performance and/or save large files simultaneously

Reengineering—A process involving the extraction of components from existing systems and restructuring these components to develop new systems or to enhance the efficiency of existing systems. Existing software systems thus can be modernized to prolong their functionality. An example of this is a software code translator that can take an existing hierarchical database system and transpose it to a relational database system. Computer-aided software engineering (CASE) includes a source code reengineering feature.

Registration authority (RA)—The individual institution that validates an entity's proof of identity and ownership of a key pair

Regression testing—A testing technique used to retest earlier program abends or logical errors that occurred during the initial testing phase

Remote access service (RAS)—Refers to any combination of hardware and software to enable the remote access to tools or information that typically reside on a network of IT devices. Originally coined by Microsoft when referring to their built-in NT remote access tools, RAS was a service provided by Windows NT that allowed most of the services that would be available on a network to be accessed over a modem link. Over the years, many vendors have provided both hardware and software solutions to gain remote access to various types of networked information. In fact, most modern routers include a basic RAS capability that can be enabled for any dial-up interface.

Remote Procedure Call (RPC)—The traditional Internet service protocol widely used for many years on UNIX-based operating systems and supported by the Internet Engineering Task Force (IETF) that allows a program on one computer to execute a program on another (e.g., server). The primary benefit derived from its use is that a system developer need not develop specific procedures for the targeted computer system. For example, in a client-server arrangement, the client program sends a message to the server with appropriate arguments, and the server returns a message containing the results of the program executed. Common Object Request Broker Architecture (CORBA) and Distributed Component Object Model (DCOM) are two newer object-oriented methods for related RPC functionality.

Repeaters—A physical layer device that regenerates and propagates electrical signals between two network segments. Repeaters receive signals from one network segment and amplify (regenerate) the signal to compensate for signals (analog or digital) distorted by transmission loss due to reduction of signal strength during transmission (i.e., attenuation).

Replication—In its broad computing sense, involves the use of redundant software or hardware elements to provide availability and fault-tolerant capabilities. In a database context, replication involves the sharing of data between databases to reduce workload among database servers, thereby improving client performance while maintaining consistency among all systems.

Repository—The enterprise database that stores and organizes data

Request for proposal (RFP)—A document distributed to software vendors, requesting them to submit a proposal to develop or provide a software product

Requirements definition—A technique used in which the affected user groups define the requirements of the system for meeting the defined needs. Some of these are business-, regulatory- and security-related requirements as well as development-related requirements.

Resilience—The ability of a system or network to resist failure or to recover quickly from any disruption, usually with minimal recognizable effect

Return on investment (ROI)—A measure of operating performance and efficiency, computed in its simplest form by dividing net income by the total investment over the period being considered

Reverse engineering—A software engineering technique whereby existing application system code can be redesigned and coded using computer-aided software engineering (CASE) technology

Ring configuration—Used in either token ring or fiber-distributed database interface (FDDI) networks, all stations (nodes) are connected to a multistation access unit (MSAU), which physically resembles a star-type topology. A ring configuration is created when these MSAUs are linked together in forming a network. Messages in this network are sent in a deterministic fashion from sender and receiver via a small frame, referred to as a token ring. To send a message, a sender obtains the token with the right priority as the token travels around the ring, with receiving nodes reading those messages addressed to it.

Ring topology—A type of local area network (LAN) architecture in which the cable forms a loop, with stations attached at intervals around the loop. In ring topology, signals transmitted around the ring take the form of messages. Each station receives the messages and each station determines, on the basis of an address, whether to accept or process a given message. However, after receiving a message, each station acts as a repeater, retransmitting the message at its original signal strength.

Risk—The combination of the probability of an event and its consequence (ISO/IEC 73)

Risk analysis—The initial steps of risk management: analyzing the value of assets to the business, identifying threats to those assets and evaluating how vulnerable each asset is to those threats. It often involves an evaluation of the probable frequency of a particular event, as well as the probable impact of that event.

Risk appetite—The amount of risk, on a broad level, that an entity is willing to accept in pursuit of its mission.

Risk assessment—A process used to identify and evaluate risks and their potential effects. Includes assessing the critical functions necessary for an organization to continue business operations, defining the controls in place to reduce organization exposure and evaluating the cost for such controls. Risk analysis often involves an evaluation of the probabilities of a particular event.

Risk evaluation—The process of comparing the estimated risk against given risk criteria to determine the significance of the risk. [ISO/IEC Guide 73:2002]

Risk management—The coordinated activities to direct and control an organization with regard to risk (in this International Standard the term 'control' is used as a synonym for 'measure'). [ISO/IEC Guide 73:2002]

Risk mitigation—The management of risk through the use of countermeasures and controls

Risk tolerance—The acceptable level of variation that management is willing to allow for any particular risk as the enterprise pursues its objectives

Risk transfer—The process of assigning risk to another enterprise, usually through the purchase of an insurance policy or by outsourcing the service. Also known as risk sharing.

Risk treatment—The process of selection and implementation of measures to modify risk [ISO/IEC Guide 73:2002]

Rounding down—A method of computer fraud involving a computer code that instructs the computer to remove small amounts of money from an authorized computer transaction by rounding down to the nearest whole value denomination and rerouting the rounded off amount to the perpetrator's account

Router—A networking device that can send (route) data packets from one local area network (LAN) or wide area network (WAN) to another, based on addressing at the network layer (Layer 3) in the open systems interconnection (OSI) model. Networks connected by routers can use different or similar networking protocols. Routers usually are capable of filtering packets based on parameters, such as source address, destination address, protocol and network application (ports).

RSA—A public key cryptosystem developed by R. Rivest, A. Shamir and L. Adleman used for both encryption and digital signatures. The RSA has two different keys, the public encryption key and the secret decryption key. The strength of RSA depends on the difficulty of the prime number factorization. For applications with high-level security, the number of the decryption key bits should be greater than 512 bits.

Run-to-run totals—Provide evidence that a program processes all input data and that it processed the data correctly

S

Salami technique—A method of computer fraud involving a computer code that instructs the computer to slice off small amounts of money from an authorized computer transaction and reroute this amount to the perpetrator's account

Scheduling—A method used in the information processing facility (IPF) to determine and establish the sequence of computer job processing

Scope creep—Also called requirement creep; this refers to uncontrolled changes in a project's scope. Scope creep can occur when the scope of a project is not properly defined, documented and controlled. Typically, the scope increase consists of either new products or new features of already approved products. Hence, the project team drifts away from its original purpose. Because of one's tendency to focus on only one dimension of a project, scope creep can also result in a project team overrunning its original budget and schedule. For example, scope creep can be a result of poor change control, lack of proper identification of what products and features are required to bring about the achievement of project objectives in the first place, or a weak project manager or executive sponsor.

Screening routers—A router configured fi to permit or deny traffic based on a set of permission rules installed by the administrator

Secure Sockets Layer (SSL)—A protocol that is used to transmit private documents through the Internet. The SSL protocol uses a private key to encrypt the data that is to be transferred through the SSL connection.

Security administrator—The person responsible for implementing, monitoring and enforcing security rules established and authorized by management

Security awareness—The extent to which every member of an enterprise and every other individual who potentially has access to the enterprise's information understand:
- Security and the levels of security appropriate to the enterprise
- The importance of security and consequences of a lack of security
- Their individual responsibilities regarding security (and act accordingly)

This definition is based on the definition for IT security awareness as defined in *Implementation Guide: How to Make Your Organization Aware of IT Security*, European Security Forum (ESF), London, 1993

Security incident—A series of unexpected events that involves an attack or series of attacks (compromise and/or breach of security) at one or more sites. A security incident normally includes an estimation of its level of impact. A limited number of impact levels are defined and, for each, the specific actions required and the people who need to be notified are identified.

Security policy—A high-level document representing an enterprise's information security philosophy and commitment

Security procedures—The formal documentation of operational steps and processes that specify how security goals and objectives set forward in the security policy and standards are to be achieved

Security testing—Ensuring that the modified or new system includes appropriate controls and does not introduce any security holes that might compromise other systems or misuses of the system or its information

Segregation/separation of duties (SoD)—A basic internal control that prevents or detects errors and irregularities by assigning to separate individuals the responsibility for initiating and recording transactions and for the custody of assets. Segregation/separation of duties is commonly used in large IT organizations so that no single person is in a position to introduce fraudulent or malicious code without detection.

Sequence check—Verification that the control number follows sequentially and any control numbers out of sequence are rejected or noted on an exception report for further research. Can be alpha or numeric and usually utilizes a key field.

Sequential file—A computer file storage format in which one record follows another. Records can be accessed sequentially only. It is required with magnetic tape.

Service bureau—A computer facility that provides data processing services to clients on a continual basis

Service level agreement—An agreement, preferably documented, between a service provider and the customer(s)/user(s) that defines (SLA) minimum performance targets for a service and how they will be measured

Service set identifier (SSID)—A 32-character unique identifier attached to the header of packets sent over a wireless local area network (WLAN) that acts as a password when a mobile device tries to connect to the basic service set (BSS). The SSID differentiates one WLAN from another so all access points and all devices attempting to connect to a specific WLAN must use the same SSID. A device will not be permitted to join the BSS unless it can provide the unique SSID. Because an SSID can be sniffed in plaintext from a packet, it does not supply any security to the network. An SSID is also referred to as a network name, because it is a name that identifies a wireless network.

Servlet—A Java applet or a small program that runs within a web server environment. A Java servlet is similar to a common gateway interface (CGI) program, but unlike a CGI program, once started, it stays in memory and can fulfill multiple requests, thereby saving server execution time and speeding up the services.

Session border controller (SBC)—Provide security features for Voice-over Internet Protocol (VoIP) traffic similar to that provided by firewalls. SBCs can be configured to filter specific VoIP protocols, monitor for denial-of-service (DOS) attacks, and provide network address and protocol translation features.

Sign-on procedure—The procedure performed by a user to gain access to an application or operating system. If the user is properly identified and authenticated by the system's security, the user will be able to access the software.

Simple Object Access Protocol (SOAP)—A platform-independent formatted protocol based on extensible markup language (XML) enabling applications to communicate with each other over the Internet. Use of SOAP may provide a significant security risk to web application operations since use of SOAP piggybacks onto a web-based document object model and is transmitted via Hypertext Transfer Protocol (HTTP) (port 80) to penetrate server firewalls, which are usually configured to accept port 80 and port 21 File Transfer Protocol (FTP) requests. Web-based document models define how objects on a web page are associated with each other and how they can be manipulated while being sent from a server to a client browser. SOAP typically relies on XML for presentation formatting and also adds appropriate HTTP-based headers to send it. SOAP forms the foundation layer of the web services stack, providing a basic messaging framework on which more abstract layers can build. There are several different types of messaging patterns in SOAP but, by far the most common is the Remote Procedure Call (RPC) pattern, in which one network node (the client) sends a request message to another node (the server), and the server immediately sends a response message to the client.

Slack time (float)—Time in the project schedule, the use of which does not affect the project's critical path; the minimum time to complete the project based upon the estimated time for each project segment and their relationships. Slack time is commonly referred to as "float" and generally is not "owned" by either party to the transaction.

SMART—Stands for specific, measurable, attainable, realistic and timely, generally used to describe appropriately set goals

Smart card—A small electronic device that contains electronic memory, and possibly an embedded integrated circuit. Smart cards can be used for a number of purposes including the storage of digital certificates or digital cash, or they can be used as a token to authenticate users.

Software—Programs and supporting documentation that enable and facilitate use of the computer. Software controls the operation of the hardware and the processing of data.

Source code—The language in which a program is written. Source code is translated into object code by assemblers and compilers. In some cases, source code may be converted automatically into another language by a conversion program. Source code is not executable by the computer directly. It must first be converted into machine language.

Source documents—The forms used to record data that have been captured. A source document may be a piece of paper, a turnaround document or an image displayed for online data input.

Source lines of code (SLOC)—Often used in deriving single-point software size estimations.

SPOOL (simultaneous peripheral operations online)—An automated function that can be based on operating system or application in which electronic data being transmitted between storage areas are spooled or stored until the receiving device or storage area is prepared and able to receive the information. SPOOL allows more efficient electronic data transfers from one device to another by permitting higher speed sending functions, such as internal memory, to continue on with other operations instead of waiting on the slower speed receiving device such as a printer.

Spyware—Software whose purpose is to monitor a computer user's actions (e.g., web sites they visit) and report these actions to a third party, without the informed consent of that machine's owner or legitimate user. A particularly malicious form of spyware is software that monitors keystrokes to obtain passwords or otherwise gathers sensitive information such as credit card numbers, which it then transmits to a malicious third party. The term has also come to refer more broadly to software that subverts the computer's operation for the benefit of a third party.

Standard—A mandatory requirement, code of practice or specification approved by a recognized external standards organization, such as International Organization for Standardization (ISO)

Standing data—Permanent reference data used in transaction processing. These data are changed infrequently such as a product price file or a name and address file.

Star topology—A type of local area network (LAN) architecture that utilizes a central controller to which all nodes are directly connected. With star topology, all transmissions from one station to another pass through the central controller which is responsible for managing and controlling all communication. The central controller often acts as a switching device.

Statistical sampling—A method of selecting a portion of a population, by means of mathematical calculations and probabilities, for the purpose of making scientifically and mathematically sound inferences regarding the characteristics of the entire population

Storage area networks (SANs)—A variation of a local area network (LAN) that is dedicated for the purpose of connecting storage devices to servers and other computing devices. SANs centralize the process for the storage and administration of data.

Structured Query Language (SQL)—The primary language used by both application programmers and end users in accessing relational databases

Substantive testing—Obtaining audit evidence on the completeness, accuracy or existence of activities or transactions during the audit period

Supply chain management (SCM)—A concept that allows an organization to more effectively and efficiently manage the activities of design, manufacturing, distribution, service and recycling of products and services to its customers

Surge suppressor—Filters out electrical surges and spikes

Suspense file—A computer file used to maintain information (transactions, payments, or other events) until the proper disposition of that information can be determined. Once the proper disposition of the item is determined, it should be removed from the suspense file and processed in accordance with the proper procedures for that particular transaction. Two examples of items that may be included in a suspense file are receipt of a payment from a source that is not readily identified or data that do not yet have an identified match during migration to a new application.

Switches—Typically associated as a data link layer device, switches enable local area network (LAN) segments to be created and interconnected, which also has the added benefit of reducing collision domains in Ethernet-based networks.

Synchronous transmission—Block-at-a-time data transmission

System software—A collection of computer programs used in the design, processing and control of all applications. The programs and processing routines that control the computer hardware, including the operating system and utility programs.

System testing—Testing conducted on a complete, integrated system to evaluate the system's compliance with its specified requirements. System test procedures typically are performed by the system maintenance staff in their development library.

System development life cycle (SDLC)—The phases deployed in the development or acquisition of a software system. SDLC is an approach used to plan, design, develop, test and implement an application system or a major modification to an application system. Typical phases of the SDLC include the feasibility study, requirements study, requirements definition, detailed design, programming, testing, installation and postimplementation review.

System exit—Special system software features and utilities that allow the user to perform complex system maintenance. Use of system exits often permits the user to operate outside of the security access control system.

System flowchart—Graphic representations of the sequence of operations in an information system or program. Information system flowcharts show how data from source documents flow through the computer to final distribution to users. Symbols used should be the internationally accepted standard. System flowcharts should be updated when necessary

T

Table look-up—Used to ensure that input data agree with predetermined criteria stored in a table

Tape management system (TMS)—A system software tool that logs, monitors and directs computer tape usage.

Telecommunications—Electronic communication by special devices over distances or around devices that preclude direct interpersonal exchange

Test data—Simulated transactions that can be used to test processing logic, computations and controls actually programmed in computer applications. Individual programs or an entire system can be tested. This technique includes Integrated Test Facilities (ITFs) and Base Case System Evaluations (BCSEs).

Test programs—Programs that are tested and evaluated before approval into the production environment. Test programs, through a series of change control moves, migrate from the test environment to the production environment and become production programs.

Third-party review—An independent audit of the control structure of a service organization, such as a service bureau, with the objective of providing assurance to the users of the service organization that the internal control structure is adequate, effective and sound

Threat—Anything (e.g., object, substance, human) that is capable of acting against an asset in a manner that can result in harm. A potential cause of an unwanted incident (ISO/IEC 13335).

Throughput—The quantity of useful work made by the system per unit of time. Throughput can be measured in instructions per second or some other unit of performance. When referring to a data transfer operation, throughput measures the useful data transfer rate and is expressed in kbps, Mbps and Gbps.

Token—A device that is used to authenticate a user, typically in addition to a username and password. A token is usually a device the size of a credit card that displays a pseudo random number that changes every few minutes.

Token ring topology—A type of local area network (LAN) ring topology in which a frame containing a specific format, called the token, is passed from one station to the next around the ring. When a station receives the token, it is allowed to transmit. The station can send as many frames as desired until a predefined time limit is reached. When a station either has no more frames to send or reaches the time limit, it transmits the token. Token passing prevents data collisions that can occur when two computers begin transmitting at the same time.

Topology—The physical layout of how computers are linked together. Examples of topology include ring, star and bus.

Transaction—Business events or information grouped together because they have a single or similar purpose. Typically, a transaction is applied to a calculation or event that then results in the updating of a holding or master file.

Transaction log—A manual or automated log of all updates to data files and databases

Transmission Control Protocol/Internet Protocol (TCP/IP)—Provides the basis for the Internet; a set of communication protocols that encompass media access, packet transport, session communications, file transfer, electronic mail (email), terminal emulation, remote file access and network management.

Trap door—Unauthorized electronic exit, or doorway, out of an authorized computer program into a set of malicious instructions or programs

Trojan horse—Purposefully hidden malicious or damaging code within an authorized computer program. Unlike viruses, they do not replicate themselves, but they can be just as destructive to a single computer.

Tunneling—Commonly used to bridge between incompatible hosts/routers or to provide encryption, a method by which one network protocol encapsulates another protocol within itself. When protocol A encapsulates protocol B, a protocol A header and optional tunneling headers are appended to the original protocol B packet. Protocol A then becomes the data link layer of protocol B. Examples of tunneling protocols include IPSec, Point-to-point Protocol Over Ethernet (PPPoE), and Layer 2 Tunneling Protocol (L2TP).

Tuple—A row or record consisting of a set of attribute value pairs (column or field) in a relational data structure

Twisted pairs—A low-capacity transmission medium; a pair of small, insulated wires that are twisted around each other to minimize interference from other wires in the cable.

U

Unicode—A standard for representing characters as integers. Unicode uses 16 bits, which means that it can represent more than 65,000 unique characters; this is necessary for languages such as Chinese and Japanese.

Uninterruptible power supply (UPS)—Provides short-term backup power from batteries for a computer system when the electrical power fails or drops to an unacceptable voltage level

Unit testing—A testing technique that is used to test program logic within a particular program or module. The purpose of the test is to ensure that the internal operation of the program performs according to specification. It uses a set of test cases that focus on the control structure of the procedural design.

Universal Serial Bus (USB)—An external bus standard that provides capabilities to transfer data at a rate of 12 Mbps. A USB port can connect up to 127 peripheral devices.

User awareness—The training process in security-specific issues to reduce security problems; users are often the weakest link in the security chain.

Utility programs—Specialized system software used to perform particular computerized functions and routines that are frequently required during normal processing. Examples include sorting, backing up and erasing data.

Utility script—A sequence of commands input into a single file to automate a repetitive and specific task. The utility script is then executed, either automatically or manually, to perform the task. In UNIX, these are known as shell scripts.

V

Vaccine—A program designed to detect computer viruses

Validity check—Programmed checking of data validity in accordance with predetermined criteria

Value-added network (VAN)—A data communication network that adds processing services such as error correction, data translation and/or storage to the basic function of transporting data

Variable sampling—A sampling technique used to estimate the average or total value of a population based on a sample; a statistical model used to project a quantitative characteristic such as a monetary amount

Verification—Checks that data are entered correctly

Virus—A program with the ability to reproduce by modifying other programs to include a copy of itself. A virus may contain destructive code that can move into multiple programs, data files or devices on a system and spread through multiple systems in a network.

Voice mail—A system of storing messages in a private recording medium which allows the called party to later retrieve the messages

Voice-over Internet Protocol (VoIP)—Also called IP Telephony, Internet Telephony and Broadband Phone, a technology that makes it possible to have a voice conversation over the Internet or over any dedicated Internet Protocol (IP) network instead of dedicated voice transmission lines

Vulnerability—A weakness in the design, implementation, operation or internal control of a process that could expose the system to adverse threats from threat events

Vulnerability analysis—A process of identifying and classifying vulnerabilities

W

WAN switch—A data link layer device used for implementing various WAN technologies such as asynchronous transfer mode, point-to-point frame relay solutions, and integrated services digital network (ISDN). These devices are typically associated with carrier networks providing dedicated WAN switching and router services to organizations via T-1 or T-3 connections.

Warm site—Similar to a hot site but not fully equipped with all of the necessary hardware needed for recovery

Waterfall development—Also known as traditional development, a procedure-focused development cycle with formal sign-off at the completion of each level

Web Services Description Language (WSDL)—A language formatted with extensible markup language (XML). Used to describe the capabilities of a web service as collections of communication endpoints capable of exchanging messages; WSDL is the language used by Universal Description, Discovery and Integration (UDDI). See also Universal Description, Discovery and Integration (UDDI).

White box testing—A testing approach that uses knowledge of a program/module's underlying implementation and code intervals to verify its expected behavior

Wide area network (WAN)—A computer network connecting different remote locations that may range from short distances, such as a floor or building, to extremely long transmissions that encompass a large region or several countries

Wide area network (WAN) switch—A data link layer device used for implementing various WAN technologies such as asynchronous transfer mode, point-to-point frame relay solutions, and integrated services digital network (ISDN). WAN switches are typically associated with carrier networks providing dedicated WAN switching and router services to enterprises via T-1 or T-3 connections.

Wi-Fi Protected Access (WPA)—A class of systems used to secure wireless (Wi-Fi) computer networks. WPA was created in response to several serious weaknesses researchers found in the previous system, Wired Equivalent Privacy (WEP). WPA implements the majority of the IEEE 802.11i standard, and was intended as an intermediate measure to take the place of WEP while 802.11i was prepared. WPA is designed to work with all wireless network interface cards, but not necessarily with first generation wireless access points. WPA2 implements the full standard, but will not work with some older network cards. Both provide good security with two significant issues. First, either WPA or WPA2 must be enabled and chosen in preference to WEP; WEP is usually presented as the first security choice in most installation instructions. Second, in the "personal" mode, the most likely choice for homes and small offices, a pass phrase is required that, for full security, must be longer than the typical six to eight character passwords users are taught to employ.

Wired Equivalent Privacy (WEP)—A scheme that is part of the IEEE 802.11 wireless networking standard to secure IEEE 802.11 wireless networks (also known as Wi-Fi networks). Because a wireless network broadcasts messages using radio, it is particularly susceptible to eavesdropping. WEP was intended to provide comparable confidentiality to a traditional wired network (in particular, it does not protect users of the network from each other), hence the name. Several serious weaknesses were identified by cryptanalysts, and WEP was superseded by Wi-Fi Protected Access (WPA) in 2003, and then by the full IEEE 802.11i standard (also known as WPA2) in 2004. Despite the weaknesses, WEP provides a level of security that can deter casual snooping.

Wiretapping—The practice of eavesdropping on information being transmitted over telecommunications links

X

X.25 Interface—An interface between data terminal equipment (DTE) and data circuit-terminating equipment (DCE) for terminals operating in the packet mode on some public data networks

> **Note:** The CISA candidate may also want to be familiar with ISACA's Glossary, which can be viewed at *www.isaca.org/glossary*. Also available is a list of CISA exam terminology in different languages, which can be viewed at *www.isaca.org/examterm.*

ACRONYMS

The following is a list of common acronyms used throughout the *CISA Review Manual*. These may be defined in the text for clarity.

4GL	Fourth-generation language
ACID	Atomicity, consistency, isolation and durability
ACL	Access control list
AES	Advanced Encryption Standard
AH	Authentication header
AI	Artificial intelligence
AICPA	American Institute of Certified Public Accountants
ALE	Annual loss expectancy
ALU	Arithmetic-logic unit
ANSI	American National Standards Institute
API	Application programming interface
ARP	Address Resolution Protocol
ASCII	American Standard Code for Information Interchange
ASIC	Application-specific integrated circuit
ATDM	Asynchronous time division multiplexing
ATM	Asynchronous Transfer Mode
ATM	Automated teller machine
B-to-B	Business-to-business
B-to-C	Business-to-consumer
B-to-E	Business-to-employee
B-to-G	Business-to-government
BCI	Business Continuity Institute
BCM	Business continuity management
BCP	Business continuity plan
BCP	Business continuity planning
BDA	Business dependency assessment
BI	Business intelligence
BIA	Business impact analysis
BIMS	Biometric Information Management and Security
BIOS	Basic Input/Output System
BIS	Bank for International Settlements
Bit	Binary digit
BLP	Bypass label process
BNS	Backbone network services
BOM	Bill of materials
BOMP	Bill of materials processor
BPR	Business process reengineering
BRP	Business recovery (or resumption) plan
BSC	Balanced scorecard
C-to-G	Consumer-to-government
CA	Certificate authority
CAAT	Computer-assisted audit technique
CAD	Computer-assisted design
CAE	Computer-assisted engineering
CAM	Computer-aided manufacturing
CASE	Computer-aided software engineering
CCK	Complementary Code Keying
CCM	Constructive Cost Model
CCTV	Closed-circuit television
CD	Compact disk
CD-R	Compact disk-recordable
CD-RW	Compact disk-rewritable
CDDF	Call Data Distribution Function
CDPD	Cellular Digital Packet Data
CEO	Chief executive officer
CERT	Computer emergency response team
CGI	Common gateway interface
CIA	Confidentiality, integrity and availability
CIAC	Computer Incident Advisory Capability
CICA	Canadian Institute of Chartered Accountants
CIM	Computer-integrated manufacturing
CIO	Chief information officer
CIS	Continuous and intermittent simulation
CISO	Chief information security officer
CMDB	Configuration management database
CMM	Capability Maturity Model
CMMI	Capability Maturity Model Integration
CNC	Computerized Numeric Control
COCOMO2	Constructive Cost Model
CODASYL	Conference on Data Systems Language
COM	Component Object Model
COM/DCOM	Component Object Model/Distributed Component Object Model
COOP	Continuity of operations plan
CORBA	Common Object Request Broker Architecture
CoS	Class of service
COSO	Committee of Sponsoring Organizations of the Treadway Commission
CPM	Critical Path Methodology
CPO	Chief privacy officer
CPS	Certification practice statement
CPU	Central processing unit
CRC	Cyclic redundancy check
CRL	Certificate revocation list
CRM	Customer relationship management
CSA	Control self-assessment
CSF	Critical success factor
CSIRT	Computer security incident response team
CSMA/CA	Carrier sense Multiple Access/Collision Avoidance

CSMA/CD	Carrier sense Multiple Access/Collision Detection
CSO	Chief security officer
CSU-DSU	Channel service unit/digital service unit
DAC	Discretionary access control
DASD	Direct access storage device
DAT	Digital audio tape
DBA	Database administrator
DBMS	Database management system
DCE	Data communications equipment
DCE	Distributed computing environment
DCOM	Distributed Component Object Model (Microsoft)
DCT	Discrete Cosine Transform
DD/DS	Data dictionary/directory system
DDL	Data Definition Language
DDN	Digital Divide Network
DDoS	Distributed denial of service
DECT	Digital Enhanced Cordless Telecommunications
DES	Data Encryption Standard
DFD	Data flow diagram
DHCP	Dynamic Host Configuration Protocol
DID	Direct inward dial
DIP	Document image processing
DLL	Dynamic link library
DMS	Disk management system
DMZ	Demilitarized zone
DNS	Domain name system
DoS	Denial of service
DRII	Disaster Recovery Institute International
DRM	Digital rights management
DRP	Disaster recovery plan
DRP	Disaster recovery planning
DSL	Digital subscriber lines
DSS	Decision support systems
DSSS	Direct Sequence Spread Spectrum
DTE	Data terminal equipment
DTR	Data terminal ready
DVD	Digital video disc
DVD-HD	Digital video disc-high definition/high density
DW	Data warehouse
EA	Enterprise architecture
EAC	Estimates at completion
EAI	Enterprise application integration
EAM	Embedded audit module
EAP	Extensible Authentication Protocol
EBCDIC	Extended Binary-coded for Decimal Interchange Code
EC	Electronic commerce
ECC	Elliptical Curve Cryptography
EDFA	Enterprise data flow architecture
EDI	Electronic data interchange
EER	Equal-error rate
EFT	Electronic funds transfer
EIGRP	Enhanced Interior Gateway Routing Protocol
EJB	Enterprise java beans
EMI	Electromagnetic interference
EMRT	Emergency response time
ERD	Entity relationship diagram
ERP	Enterprise resource planning
ESP	Encapsulating security payload
EVA	Earned value analysis
FAR	False-acceptance rate
FAT	File allocation table
FC	Fibre channels
FDDI	Fiber Distributed Data Interface
FDM	Frequency division multiplexing
FEA	Federal enterprise architecture
FEMA (USA)	Federal Emergency Management Association
FER	Failure-to-enroll rate
FERC	Federal Energy Regulatory Commission (USA)
FFIEC	Federal Financial Institutions Examination Council (USA)
FFT	Fast Fourier Transform
FHSS	Frequency-hopping spread spectrum
FIPS	Federal Information Processing Standards
FP	Function point
FPA	Function point analysis
FRAD	Frame relay assembler/disassembler
FRB	Federal Reserve Board (USA)
FRR	False-rejection rate
FTP	File Transfer Protocol
GAS	Generalized audit software
Gb	Gigabit
GB	Gigabyte
GID	Group ID
GIS	Geographic information systems
GPS	Global positioning system
GSM	Global system for mobile communications
GUI	Graphical user interface
HA	High availability
HD-DVD	High definition/high density-digital video disc
HDLC	High-level data link control
HIPAA	Health Insurance Portability and Accountability Act (USA)
HIPO	Hierarchy input-process-output
HMI	Human machine interfacing
HTML	Hypertext Markup Language
HTTP	Hypertext Transmission Protocol
HTTPS	Secured Hypertext Transmission Protocol
HW/SW	Hardware/software
I/O	Input/output
I&A	Identification and authentication

ICMP	Internet Control Message Protocol
ICT	Information and communication technologies
ID	Identification
IDE	Integrated development environment
IDEF1X	Integration Definition for Information Modeling
IDS	Intrusion detection system
IETF	Internet Engineering Task Force
IMS	Integrated manufacturing systems
IP	Internet protocol
IPF	Information processing facility
IPL	Initial program load
IPMA	International Project Management Association
IPRs	Intellectual property rights
IPS	Intrusion prevention system
IPSec	IP Security
IPX	Internetwork Packet Exchange
IR	Infrared
IRC	Internet relay chat
IrDA	Infrared Data Association
IRM	Incident response management
IS/DRP	IS disaster recovery planning
ISAKMP/ Oakley	Internet Security Association and Key Management Protocol/Oakley
IS	Information systems
ISAM	Indexed Sequential Access Method
ISDN	Integrated services digital network
ISO	International Organization for Standardization
ISP	Internet service provider
IT	Information technology
ITF	Integrated test facility
ITGI	IT Governance Institute
ITIL	Information Technology Infrastructure Library
ITSM	IT service management
ITT	Invitation to tender
ITU	International Telecommunications Union
IVR	Interactive voice response
JIT	Just in time
Kb	Kilobit
KB	Kilobyte
KB	Knowledge base
KGI	Key goal indicator
KPI	Key performance indicator
KRI	Key risk indicator
L2TP	Layer 2 Tunneling Protocol
LAN	Local area network
LCP	Link Control Protocol
M&A	Mergers and acquisition
MAC	Mandatory access control
MAC	Message Authentication Code
MAC Address	Media Access Control Address
MAN	Metropolitan area network
MAP	Manufacturing accounting and production
MIS	Management information system
MODEM	Modulator/demodulator
MOS	Maintenance out of service
MPLS	Multiprotocol label switching
MRP	Manufacturing resources planning
MSAUs	Multistation access units
MTBF	Mean time between failures
MTS	Microsoft's Transaction Server
MTTR	Mean time to repair
NAP	Network access point
NAS	Network access server
NAS	Network attached storage
NAT	Network address translation
NCP	Network Control Protocol
NDA	Nondisclosure agreement
NFPA	National Fire Protection Agency (USA)
NFS	Network File System
NIC	Network interface card
NIST	National Institute of Standards and Technology (USA)
NNTP	Network News Transfer Protocol
NSP	Name Server Protocol
NSP	Network service provider
NT	New technology
NTFS	NT file system
NTP	Network Time Protocol
OBS	Object breakdown structure
OCSP	Online Certificate Status Protocol
ODC	On-demand computing
OECD	Organization for Economic Cooperation and Development
OEP	Occupant emergency plan
OLAP	Online analytical processing
OOSD	Object-oriented system development
ORB	Object request broker
OS	Operating system
OSI	Open Systems Interconnection
OSPF	Open Shortest Path First
PAD	Packet assembler/disassembler
PAN	Personal area network
PBX	Private branch exchange
PC	Personal computer/microcomputer
PCR	Program change request
PDCA	Plan-do-check-act
PDN	Public data network
PER	Package-enabled reengineering
PERT	Program Evaluation Review Technique
PICS	Platform for Internet Content Selection
PID	Process ID
PID	Project initiation document

PIN	Personal identification number
PKI	Public key infrastructure
PLC	Programmable logic controllers
PMBOK	Project Management Body of Knowledge
PMI	Project Management Institute
POC	Proof of concept
POP	Proof of possession
POS	Point of sale (or Point-of-sale systems)
POTS	Plain old telephone service
PPP	Point-to-point Protocol
PPPoE	Point-to-point Protocol Over Ethernet
PPTP	Point-to-Point Tunneling Protocol
PR	Public relations
PRD	Project request document
PRINCE2	Projects in Controlled Environments 2
PROM	Programmable Read-only Memory
PSTN	Public switched telephone network
PVC	Permanent virtual circuit
QA	Quality assurance
QAT	Quality assurance testing
RA	Registration authority
RAD	Rapid application development
RADIUS	Remote Access Dial-in User Service
RAID	Redundant Array of Inexpensive Disks
RAM	Random access memory
RAS	Remote access service
RBAC	Role-based access control
RDBMS	Relational database management system
RF	Radio frequencies
RFI	Request for information
RFID	Radio frequency identification
RFP	Request for proposal
RIP	Routing Information Protocol
RMI	Remote method invocation
ROI	Return on investment
ROLAP	Relational online analytical processing
ROM	Read-only memory
RPC	Remote procedure call
RPO	Recovery point objective
RSN	Robust secure network
RST	Reset
RTO	Recovery time objective
RTU	Remote terminal unit
RW	Rewritable
S/HTTP	Secure Hypertext Transfer Protocol
S/MIME	Secure Multipurpose Internet Mail Extensions
SA	Security Association
SAN	Storage area network
SANS	SysAdmin, Audit, Network, Security
SAS	Statement on Auditing Standards
SBC	Session border controller
SCADA	Supervisory Control and Data Acquisition
SCARF	Systems Control Audit Review File
SCARF/EAM	Systems Control Audit Review File and Embedded Audit Modules
SCM	Supply chain management
SCOR	Supply chain operations reference
SDLC	System development life cycle
SD/MMC	Secure digital multimedia card
SDO	Service delivery objective
SEC	Securities and Exchange Commission (USA)
SET	Secure electronic transactions
SIP	Service improvement plan
SLA	Service level agreement
SLIP	Serial Line Internet Protocol
SLM	Service level management
SLOC	Source lines of code
SMART	Specific, measurable, attainable, realistic, timely
SME	Subject matter expert
SMF	System management facility
SMTP	Simple Mail Transport Protocol
SNA	Systems network architecture
SNMP	Simple Network Management Protocol
SO	Security officer
SOA	Service-oriented architectures
SOAP	Simple Object Access Protocol
SOHO	Small office-home office
SOW	Statement of work
SPI	Security parameter index
SPICE	Software Process Improvement and Capability dEetermination
SPOC	Single point of contact
SPOOL	Simultaneous peripheral operations online
SQL	Structured Query Language
SSH	Secure Shell
SSID	Service set identifier
SSL	Secure Sockets Layer
SSO	Single sign-on
SVC	Switched virtual circuits
SYSGEN	System generation
TACACS	Terminal Access Controller Access Control System
TCO	Total cost of ownership
TCP	Transmission Control Protocol
TCP/IP	Transmission Control Protocol/Internet Protocol
TCP/UDP	Transmission Control Protocol/User Datagram Protocol
TDM	Time-division multiplexing
TELNET	Teletype network
TES	Terminal emulation software
TFTP	Trivial File Transport Protocol
TKIP	Temporal Key Integrity Protocol

TLS	Transport layer security
TMS	Tape management system
TP monitors	Transaction processing monitors
TQM	Total quality management
TR	Technical report
UAT	User acceptance testing
UBE	Unsolicited bulk email
UDDI	Universal description, discovery and integration
UDP	User Datagram Protocol
UID	User ID
UML	Unified Modeling Language
UPS	Uninterruptible power supply
URI	Uniform resource identifier
URL	Uniform resource locator
URN	Uniform resource name
USB	Universal Serial Bus
VLAN	Virtual local area network
VoIP	Voice-over IP
VPN	Virtual private network
WAN	Wide area network
WAP	Wireless Application Protocol
WBS	Work breakdown structure
WEP	Wired Equivalent Privacy
WLAN	Wireless local area network
WML	Wireless Markup Language
WORM	Write Once and Read Many
WP	Work package
WPA	Wi-Fi Protected Access
WPAN	Wireless personal area network
WSDL	Web Services Description Language
WWAN	Wireless wide area network
WWW	World Wide Web
X-to-X	Exchange-to-exchange
XBRL	Extensible Business Reporting Language
XML	Extensible Markup Language
XOR	Exclusive-OR
Xquery	XML query
XSL	Extensible Stylesheet Language

Page intentionally left blank

INDEX

A

B

C

CISA
Certified Information Systems Auditor
An ISACA Certification

D

E

F

CISA
Certified Information Systems Auditor
An ISACA Certification

J

K

L

M

N

O

CISA
Certified Information Systems Auditor
An ISACA Certification

P

Q

R

S

CISA
Certified Information Systems Auditor
An ISACA Certification

T

U

V

W

X

Printed and bound by PG in the USA